ECONOMIC
MINERAL DEPOSITS

BOOKS BY A. M. BATEMAN

The Formation of Mineral Deposits
Economic Mineral Deposits. Second Edition

The Argonauts at Colchis by the Euxine examining the Golden Fleece on which the gold is being collected. (*From Agricola: De Re Metallica.*)

ECONOMIC
MINERAL DEPOSITS

ALAN M. BATEMAN

Silliman Professor of Geology, Yale University
Editor, Economic Geology

Second Edition

JOHN WILEY & SONS, Inc.
New York · London · Sydney

PRINTED IN THE UNITED STATES OF AMERICA

PREFACE

The first edition of this book appeared at the end of 1942. Since that time war and post-war readjustment has forcibly demonstrated the extent to which the materials of the mineral kingdom constitute the backbone of industrial life and peace-time economic development of nations. Former bountiful supplies now show serious depletion, and nations of the world are looking farther and farther afield for those mineral supplies necessary to their subsistence. This book deals with such mineral deposits, how they are formed, what they are, how and where they occur, and what they are used for. Its chief purpose is as a textbook, designed both for elementary and more advanced courses. The first edition was in demand also as a source of information to all those interested in mineral deposits and in the mineral industry.

The organization of the first edition has been retained. The book is divided into three parts: (I) Principles and Processes, (II) Ore Deposits, and (III) Nonmetallic Mineral Deposits. Each part can be used separately or conjointly. The heart of the book is devoted to the principles and processes of formation of mineral deposits (Part I), and the results of these processes are exemplified in the occurrences described in Parts II and III. For advanced courses Part I can be expanded with contemporaneous collateral assignments chosen from Parts II and III along with other selected readings. The use of the book presupposes some knowledge of general geology and mineralogy.

The treatment of mineral deposits according to processes of formation instead of by a classification of mineral deposits is again followed in this edition. In the author's experience it is more satisfactory from the standpoint of the student and of the field worker as well as for practical considerations of ore finding. Increasing population and increasing mechanization, both in the home and in industry, are making greater and greater demands upon mineral resources and are requiring more scientific methods of exploration for new deposits to replace those being depleted. It is hoped that this book may provide fundamental knowledge for such purposes. For this the author has drawn largely upon his own field studies in many countries of the world, and a large number of the descriptions of the mineral deposits are based upon personal knowledge.

In this edition, as in the former one, magmatic deposits are treated in new detail, as are oxidation and supergene enrichment. Considerable space is devoted to mineral deposits that arise from evaporation and sedimentation; and metamorphism is assigned a place in mineral formation. The chapter on ground water is retained. Statistics are eliminated, in general.

In this revision, the greatest changes have been in Part I. The subchapter on contact metasomatism has had numerous changes, and the term " contact metasomatism " has replaced " contact metamorphism." The former subchapters on replacement and cavity filling have been consolidated under hydrothermal processes, with a new preliminary discussion of hydrothermal processes fundamental to both cavity filling and replacement. The included sections have been much revised and consolidated. The subchapter on sedimentation has been completely rewritten, and much of the material formerly included has been transferred to Parts II and III. The subchapters on residual concentration and mechanical concentration have been consolidated into one subchapter with a preliminary treatment of principles applicable to both. Each of the other chapters has undergone considerable revision for purposes of betterment and to bring the subject matter up to date. The war years brought many new mineral developments and changes, and these have been incorporated. In response to many requests, the selected references at the end of each chapter have been expanded to a total of 41 pages, covering up to the end of 1949. Many new illustrations are included.

For the material presented in this book the author has drawn upon his own field, teaching, government, and editorial experience, and he gratefully acknowledges the heritage of learning of those who have preceded him and of his contemporaries. Since footnotes are omitted, acknowledgments lacking in the text are here recorded with pleasure. His thanks are also due for permission for the use of illustrations, and to those who kindly sent in suggestions for changes.

YALE UNIVERSITY
April, 1950

CONTENTS

ix

CHAPTER 1

INTRODUCTION

Economic geology deals with the materials of the mineral kingdom that man wrests from the earth for his necessities of life and comfort. The search for them has given rise to voyages of discovery and settlement of new lands; their ownership has resulted in commercial or political supremacy or has caused strife and war. In the quest for these mineral substances knowledge of their distribution, character, occurrence, and uses has gradually accumulated, and this knowledge has led to theories regarding their origin. Thus, the subject of mineral deposits developed and as such was taught as one phase of mining in the early mining schools. As greater attention was paid to the rocks that enclosed the ore deposits, to deciphering their character, structure, and origin, and to the land forms developed upon the rocks, the broader science of *geology* gradually arose. Today *economic geology* is a separate branch of geology, as are *mineralogy, petrology, paleontology* and *stratigraphy, structural geology,* and *physiography* or *geomorphology.*

The future of our mineral industry, which is basic to the national economy, rests largely upon the functioning of economic geology for a continued supply of materials. As C. K. Leith expresses it,

With the advent of the industrial revolution in England, a century ago, began the real exploitation of earth materials in a way to influence essentially our material civilization. In this short time, at an ever accelerating rate, minerals have become the fundamental basis of industrialism . . . In these hundred years the production of pig iron has increased 100-fold, of mineral fuels 75-fold, and of copper, 63-fold.

Those countries abundantly supplied with mineral resources became the great industrial nations, and the insatiable demand for minerals to sustain industrialized life has caused the world to dig and consume more minerals within the period embracing the two world wars than in all previous history. Former adequate sources of supply are beginning to look small, and large sources are becoming fewer.

This alarming consumption of our mineral resources and the exhaustion of known reserves means that new supplies must be discovered to take their place if the industry is to persist unimpaired. With waning

1

discovery of obvious mineral outcrops, search must be directed to the less obvious deposits, of which vast numbers must be hidden by the ubiquitous overburden. Every art of geology must be employed to this end, and it promises to become the important work of the economic geologist. In this connection the petroleum geologist has already made an enviable record in the adaptation of geophysical methods and instruments to the discovery of petroleum.

The scope of economic geology includes not only metallic *ore deposits* but the broader field of *nonmetallics,* whose value today is three times that of the metallic ores. In addition, it includes the general application of geology to the uses of man. Thus, it deals with practical problems of the industries and arts, the occurrence of subsurface waters and soils, and the application of geologic principles to important engineering projects. The construction of any large dam, for example, involves questions of the suitability of the foundation rock, of leakage, of subsurface water flow, and of the character and resources of materials that enter into its construction. The subject of economic geology is related also to *geography* and *economics,* since it furnishes information regarding the geographic distribution and resources of the earth materials that are the foundation of the extractive industries.

The early kinship with *mining,* and metal mining in particular, with which economic geology grew up, has persisted, and there is now a specialized subdivision of economic geology known as *mining geology* which deals especially with the problems of ore deposits and their relation to metal mining and, to some extent, with *metallurgy.* This relation may be understood better by considering that the desired metals are locked up in ore minerals, which are admixed with undesired minerals or rock to form ores, and their separation involves the art of metallurgy; the extraction of ores from the ground falls within the realm of mining; and the study of the occurrence, localization, and origin of these ores and their relation to the enclosing rock is the domain of mining geology. The mining geologist functions early in mining operations when he is called upon to determine the probable shape, size, and value of mineral deposits, and particularly their extensions with depth. In addition, he cooperates with the mining engineer during mining operations in the exploration and development of mineral deposits, in finding faulted bodies and in other ways helping to maintain ore reserves, and in the proper location of mine workings to avoid caving ground. In the future he will be called upon more and more to apply geology to mineral finding in districts of waning mines. His knowledge is sought also by the metallurgist to help solve problems of ore and metal extraction and to obtain suitable ore mixtures for economical smelting.

Another important subdivision of economic geology is *petroleum geology*. It deals specifically with the many problems of the location, occurrence, migration, and origin of petroleum and gas. The petroleum geologist is called upon to determine probable oil-containing formations, to unravel their structure by geological or geophysical methods, and to locate prospecting wells. For this purpose he invokes a knowledge of structural geology, stratigraphy, paleontology, and the occurrence of petroleum.

These examples indicate the broad scope of economic geology. Since it deals with the basic materials underlying the extractive industries, its problems are intertwined with those of diverse industries. It enters into phases of transportation, international trade, and engineering. It also embraces many interesting scientific problems in its own field in which intellectual curiosity plays a greater part than utilitarian problems. The problems of the genesis of different mineral deposits hold opportunity for long-continued research.

Only certain phases of the broad field of economic geology are covered in this book, which confines itself largely to mineral deposits and the principles underlying their occurrence and formation. The technology of extraction (mining) and treatment (ore dressing and metallurgy) of the mineral substances are not considered in detail, and little space is given to statistics. Also the geological features of other than mineral deposits is beyond the scope of this book. The mineral substances are not followed far into industry save to indicate their uses.

Of the great variety of mineral substances won from the earth for the uses of man, coal is the most valuable, followed by metallic minerals, petroleum and natural gas, and other nonmetallic substances such as clay and gypsum.

For ease in study and ready reference, these materials are divided in this book into two parts: *metalliferous deposits*, such as gold, copper, iron, or nickel (Part II); and *nonmetallic substances*, such as coal, clay, petroleum, or gemstones (Part III). The metalliferous deposits, or *ore deposits*, are sought for the metals they contain, which are extracted generally in the metallic state. These deposits are subdivided according to the individual metals. Typical examples are described in order that their content, occurrence, and origin may be studied. The nonmetallic, or earthy substances, on the other hand, are not generally desired for their content of metal but are utilized principally, after suitable processing, in the form in which they are extracted. For example, clay is not mined for its aluminum content or asbestos for its magnesium; but clay is used as a compound in making porcelain or pottery, and asbestos is used as the mineral asbestos. Their physical properties, more than their chemical, for the most part determine their utilization. Both graphite and diamond, for example, consist of carbon, but neither is desired for its carbon content. It is their physical properties that make one a coveted gem and the other a heat- and chemical-resisting substance desired for metallurgical purposes.

There are so many utilized nonmetallic substances of such diverse character and origin that they defy simple classification. For the purpose of this book, however, they are grouped according to their important uses, as, for example, under mineral fuels, ceramic materials, or metallurgical materials. Such an arrangement offers the advantage, for an introductory book, of assembling many diversified materials that have common use under well-known utilitarian groups susceptible of ready reference for both the student of geology and the interested reader. The mode of occurrence and what constitutes workable deposits of these materials will be described under each group.

Economic mineral deposits are geologic bodies that may be worked for one or more minerals or metals. They are exceptional features, sparsely scattered in the rocks or on the earth's surface; they constitute only an infinitesimal part of the earth's crust, but they assume an importance far in excess of their relative volume because of the highly valuable materials they supply to national wealth and industry. They have been concentrated in the rocks under peculiar and exceptional conditions, which it will be our purpose to study. No two mineral deposits are alike in all respects; nevertheless, certain broad principles control their formation. To understand properly how a gold vein or clay deposit has been formed it is necessary to understand first the constitution of mineral deposits and the processes that operate within and upon the earth to form them. Consequently, Part I of this volume is devoted primarily to a general consideration of the principles and processes of the formation of mineral deposits.

General references are found at the end of the book. Selected references are found at the end of each chapter.

CHAPTER 2

BRIEF HISTORY OF THE USE OF MINERALS AND OF THE DEVELOPMENT OF ECONOMIC GEOLOGY

Ancient Times

Economic geology probably had its inception with the ancient utilization of mineral products. Long ages must have passed, however, before the early crude knowledge became a craft, later to develop into a science. The early incentive for the acquisition of such knowledge was undoubtedly utilitarian, but later it was raised to an intellectual plane by the Greek philosophers.

The first earth materials used by primitive man were nonmetallic substances — flint, chert, quartz, and certain hard and soft stones such as quartzite, soapstone, or limestone — sought for their use in weapons, implements, utensils, and for carving. Clay was widely and extensively used, first for pottery and later for bricks. Unquestionably clay represents the first large-scale mineral industry, an industry that has persisted continuously through the ages. Burned clay figures believed to be Aurignacian (30,000–20,000 B.C.) have been discovered in Moravia, and excellent Paleolithic pottery of the Solutrean period (+ 10,000 B.C.) has been found in Egypt. Brick, tile, and clay tablets were extensively used by the Chaldeans, Babylonians, and early Egyptians for building their cities, for irrigation, and for writing materials. The early Asiatic and African dwellings were built with bricks made of clay. Later, building stones were extensively used. During the building of the pyramids (2980–2925 B.C.) this extractive industry must have been on a grand scale, as the Pyramid of Gizeh contains 2,300,000 blocks of stone averaging 2½ tons apiece.

Paleolithic man between 100,000 and 7000 B.C., according to S. H. Ball, used 13 varieties of minerals — chalcedony, quartz, rock crystal, serpentine, obsidian, pyrite, jasper, steatite, amber, jadeite, calcite, amethyst, and fluorspar. He also utilized ochers or mineral paints. At about the time Neolithic man became acquainted with gold and copper, he also used nephrite, sillimanite, and turquois. These nonmetallic materials are mostly common substances that probably were found by accident and whose quest neither greatly stimulated human

5

curiosity nor created specialized knowledge regarding their occurrence. They were accepted as found, and utilized. Economic geology had not yet arisen; it was the pre-dawn stage.

EGYPTIAN, GREEK, AND RELATED CULTURES

As the desire for gemstones and metals became more urgent, however, economic geology probably had its inception. Facts of occurrence were noted and recorded; crude theories of origin were evolved; expeditions were organized for the discovery and exploitation of deposits; and ownership and barter of these substances became an important part of the life of the people, even more important relatively than it is today. The use of gemstones and the mining of them reached a high art among the early Egyptians, Babylonians, Assyrians, and Indians. Gemstones were greatly prized, and, living or dead, the Egyptian was bedecked with jewels, which attained important significance among a people obsessed with mysticism. In pre-Dynastic times (+ 3400 B.C.), it was the color rather than the substance that the Egyptian prized most. The Theban craftsmen created pleasing color schemes, utilizing the azure of the lapis lazuli, the red of the carnelian, the purple of the amethyst, the green of the malachite, the yellow of the jasper, and the blue of the turquois. He also used agate, beryl, chalcedony, and garnet and shaped and polished hard stones, producing not only ovoid but also faceted beads. All these stones except lapis have come from Egypt itself. Even in those remote times there must have been international barter, since the lapis was probably obtained from Afghanistan, some 2,400 miles away.

According to Ball, other stones are known to have appeared, such as onyx in the 2nd Dynasty; azurite and jade in the 3rd Dynasty; and amber in the 6th Dynasty (2625–2475 B.C.). The stele of Nebona (18th Dynasty) reads: "I have consecrated numerous gifts in the temple of my father Osiris in silver in gold in lapis lazuli in copper and in precious stones." (Ball.) Later, under Greek influence, in the time of the Ptolemies, several other stones were introduced, including some Indian gems, such as sapphire, zircon, and topaz.

The oldest form of mining was for gems and decorative stones, and for over 2,000 years the Pharaohs dispatched expeditions including engineers and prospectors to the Sinai Peninsula for turquois, and into the Sudan. Ball identifies as the first economic geologist the Egyptian, Captain Haroeris, who about 2000 B.C. led an expedition to Sinai and after 3 months' prospecting discovered and extracted large quantities of turquois. The ancient Egyptians (from 1925 B.C.) sank hundreds of shafts for emeralds on the Egyptian coast of the Red Sea;

certain workings are said to have been 800 feet deep and sufficiently large to permit 400 men to work at a time therein.   (Ball.)

The first metals used were probably gathered as native metals from streams by primitive man.   Gold is presumed to have been used before copper, and copper is considered by some to have been discovered 18,000 years before Christ; certainly copper was known to the Egyptians in 12,000 B.C. and was widely used in Europe about 4000 B.C. Strabo tells us that " in the country of Saones, where is Colchis, the winter torrents bring down gold which the barbarians collect in troughs pierced with holes and lined with fleeces." Hence the legend of the Golden Fleece.   Such fleeces, hung on the trees to dry so that the fine gold could be beaten out of them, spurred Jason and the Argonauts in the ship *Argo* to seek the Golden Fleece near the shore of the Euxine. This is the earliest record of placer gold mining and a poetic expression of an early mining adventure.   Even today somewhat similar methods are utilized to extract fine placer gold in South America.

At the ancient mines of Cassandra, Greece, which Sagui estimates to have been mined from about 2500 to 356 B.C., the skillful extraction of the gold-silver ores was based upon a knowledge of their localization at the intersections of fissures, toward which tunnels were run below the oxidized zones.   Also, the complications of faulting were sufficiently understood to trace the displaced end of a lode beyond an important fault.   A beginning had been made in understanding the occurrence of ores.

A knowledge of the occurrence of ores and the beginning of curiosity regarding their genesis is shown in the writings of the Greek and Roman philosophers.   Herodotus (484?–425 B.C.) told of the occurrence of gold in quartz veins in the Krissites district, Greece, later described by Diodorus.   Theophrastus (372–287 B.C.), a pupil of Aristotle, in his *Book of Stones*, the first mineralogy textbook, described 16 minerals, grouped as *metals, stones,* and *earths*.   Strabo, writing in A.D. 19, says in reference to alluvial mining in Spain: " Gold is not only dug from mines, but likewise collected; sand containing gold being washed down by the rivers and torrents . . . at the present day more gold is produced by washing than by digging it from the mines." (H. C. Hamilton and W. Falconer.)   Many descriptions of ore occurrences in Spain are given in Pliny's elaborate technical descriptions. He also tells us that Hannibal had a silver mine, named the Baebulo, in southern Spain, in a mountain that had been penetrated 1,500 paces. Pliny said it yielded 300 pounds of silver daily.   The production of silver-lead ores was an important industry in Attica at a remote period, the famous mines of Laurium having been worked long before the days of Xenophon, who wrote a report upon them in 365 B.C. The ancients sank here more than 2,000 shafts, one of which is 386 feet deep, and their locations disclose an accurate knowledge of the occurrence of the ores. Throughout the Dark Ages little appears to have

Quid Medici poſſent manibus ? quas iungere plagas
Vlceribus ſordes . ſigna mouere loco ?
Extitit hic ſolus qui pondera . viſcera Terræ
Rimatus , nobis bella metalla fodit .

Georgius Agricola

FIG. 2–1. Portrait and signature of Agricola. (*From Hofmann; Adams.*)

been added to the knowledge of the early philosophers, except by Avicenna (980–1037), the Arabian translator of Aristotle, who grouped minerals as *Stones, Sulphur minerals, Metals,* and *Salts* (Crook), thus definitely recognizing the sulphide group of minerals.

Commencement of the Scientific Eras

The first reasonable theory of ore genesis was formulated by Georgius Agricola (Bauer) (1494–1555). Born in Saxony amidst the mines of the Erzgebirge, he became a keen observer of minerals and a careful recorder (Fig. 2–1). Although some of his views were fantastic, he showed in his *De Re Metallica* that lodes originated by deposition of minerals in " canales " (fissures) from circulating underground waters, largely of surface origin, that had become heated within the earth and had dissolved the minerals from the rocks. He made a clear distinction between *homogeneous minerals* (minerals) and *heterogeneous minerals* (rocks) ; the former he divided into *Earths, Salts, Gemstones, Metals,* and *Other minerals.* He also classified ore deposits genetically into veins (vena profunda), beds, stocks, and stringers. Prior to Agricola, most writers thought that lodes were formed at the same time as the earth, but he recognized clearly that they were of different age from the enclosing rocks, as he states: " to say that lodes are of the same age as the earth itself is the opinion of the vulgar " (*De Re Metallica,* 1556). This knowledge of earth materials led him to be the first to refute vigorously the efficacy of the forked hazel stick commonly used at that time in attempting to find metals and water. He made accurate observations on the weathering of rocks and the surface decomposition of metallic sulphide ores. Many of his careful observations are quaintly portrayed by interesting woodcuts (Fig. 2–2). His description and cuts of veins were drawn freely from von Kalbe's *Bergbüchlein* of 1518.

Agricola's writings are among the most original contributions to the study of ore genesis. They were a marked advance in scientific thought and influenced greatly the thought of later writers.

Contributions of the Seventeenth and Eighteenth Centuries

Although knowledge of minerals and rocks must have continued to accumulate in the extraordinary mining atmosphere of the Erzgebirge and the Harz Mountains, little information was recorded from the time of Agricola to that of Descartes, whose *Principia Philosophae* was published in 1644. His conception of the earth as a cooled star with a hot interior led him to suggest that the ore minerals were driven upward from a deep metalliferous shell by interior heat in the form of exhalations and resurgent surface waters, to be deposited as lodes in the fissures of the outer stony crust. This conception is clearly the forerunner of some of the ideas held today.

In the eighteenth century, the accumulated factual knowledge further incited human curiosity about the genesis of ore materials. Under the stimulus of inspiring leaders, hypotheses of origin burst forth, which at the end of the century led to vigorous controversies. Most of these theories emanated from the German mining districts, but the Swedes also made early contributions. Becher (1703) and Henkel (1725) attributed the origin of ore veins to the action on

stony materials of vapors arising from " fermentation " in the bowels
of the earth.   Henkel's idea of " transmutation " had in it the germ
of modern metasomatism.   In 1749 Zimmermann also anticipated the
idea of metasomatic replacement when he ascribed the origin of lodes
to the transformation of rocks into metallic minerals and veinstones
by the action of solutions that entered through innumerable small

FIG. 2-2.   Medieval miners employing the divining rod.
(*Agricola, De Re Metallica.*)

rents and other openings in the rocks.   This idea had in it also a
suggestion of the subsequent lateral-secretion hypothesis.   To von
Oppel (1749) belongs the credit for having shown that veins were
mainly the filling of fault fissures whose formation preceded the circulation of the ore-depositing solutions.   His ideas, however, escaped
attention for a long time.   Lehman, in 1753, explained that the upward
branching of veins indicated deposition from exhalations and vapors
that emanated from the earth's interior and rose through the crust, like
sap rising from the roots into the branches of a tree.   Such were the
theories of the origin of mineral deposits before 1756.

In that year an event occurred which profoundly affected the subsequent development of economic geology. The famous Mining Academy of Freiberg, Germany, was founded in the midst of the varied mineral deposits of the Erzgebirge. Its famed teachers conducted excursions to study the nearby ores and enclosing rocks; extensive mineral collections were made and studied in the Academy. Scholars flocked to study under its masters. Here the newer geology flourished, and for over a century and a quarter its teachings in ore deposits influenced world thought.

Throughout most of the eighteenth century the prevalent view was that mineral deposits were formed by exhalations from the earth's interior that brought up the metals from depth and deposited them in fissures, or by substitution of rock matter. Lassius, however, in 1789, following Delius (1770) and Gerhard (1781), explained ore solutions as diffused ascending water that dissolved scattered grains of metals from the rocks through which it passed. Thus, most of the germs of theories of mineral deposits had been considered, even if unscientifically, before the controversial time of Hutton and Werner.

When Abraham Gottlob Werner (Fig. 2–3) (1749–1807) became the professor of mineralogy and geology at the Freiberg Mining Academy in 1775 he discarded the theories of an interior source for the metals and became an insistent advocate of the theory that mineral veins were formed by descending percolating waters derived from the primeval universal ocean, from which, according to his views, not only sediments but also all the igneous and metamorphic rocks were precipitated. These waters, he thought, descended from above into fissures and there deposited the vein materials by chemical precipitation. His stimulating personality and fiery lectures caused students from all Europe to flock to him and to return as zealous disciples to defend his Neptunist views. His thought then dominated in all matters relating to ore genesis, particularly after the publication, in 1791, of his classic treatise on the origin of veins. Werner's enthusiastic lectures perhaps carried conviction to his hearers more by his personality and oratory than by the soundness of his facts. In one sense his leadership retarded the advancement of thought regarding mineral genesis, but at the same time his dogmatic statements aroused vigorous opposition and thus stimulated wider consideration of other ideas, of which the most noteworthy was the Plutonist or Vulcanist school headed by Hutton.

Hutton (born in 1726), a quiet Scotchman, averse to publishing, was a careful observer and investigator. In his *Theory of the Earth* (1788) he first defined the true origin of plutonic and metamorphic rocks, and his proponents waged bitter controversy with the Neptunist school.   Hutton also applied his theory of the magmatic origin of igneous rocks to all mineral deposits.   He claimed that ore minerals were not soluble in water but were igneous injections.   In the words of his advocate, Playfair (1802), " The materials which fill the mineral veins were melted by heat and forcibly injected into the clefts and

Fig. 2–3.   Portrait of Werner.   (*Beck; Adams.*)

fissures of the strata."   Hutton's observations were confined largely to rocks rather than veins, and with Werner it was the reverse.   Hutton's correct conclusions of the origin of igneous rocks made him go too far in attributing all veins to melted injections and discarding water as a possible agent.   Werner's correct conclusions regarding the role of water in the formation of veins made him go too far in ascribing granites and basalts to water deposition.   The Plutonists won out with respect to the rocks; the Neptunists prevailed with respect to the dominance of water in the formation of mineral veins, although, owing to the disrepute of Werner's explanation of rocks, his ideas regarding ores were overlooked for some time.

NINETEENTH AND TWENTIETH CENTURIES

The Plutonist and Neptunist controversy quickened observations regarding the occurrence of minerals and rocks, and many data, particularly regarding ore veins, became available. Hutton's igneous injection theory of mineral veins was quickly forgotten. The early nineteenth century writers reverted to the pre-Werner ideas of mineral formation by exhalations from the interior of the earth, recognizing more, however, the significance of water in the formation of veins. Gradually, water of igneous derivation was considered to play the important role. This view received confirmation by the work of Necker (1832), who, believing that the intrusions generated the vein materials, demonstrated the close relationship, in various regions, between igneous rocks and mineral deposits.

The connection of the mineral-forming solutions with magmas was given further emphasis by the French geologists Daubrée, Scheerer, and Elie de Beaumont. The brilliant Daubrée introduced the first experimental methods into the study of mineral deposits. In 1841 he produced artificial cassiterite (tinstone) from stannous chloride and inferred that vapors or mineralizers containing water, fluorine, and boron generated at depth had deposited the tin ores and associated minerals.

Scheerer in 1847, following Scrope, who concluded that magmatic water played a part in the formation of igneous rocks, stated clearly that water was an important constituent of granite magmas and that mineral veins were formed by exudations of aqueous solutions from granite intrusions.

A few months later an important paper by Elie de Beaumont appeared, which Thomas Crook stated was perhaps the " most important and influential paper ever published on the theory of ore deposits." Elie de Beaumont might well be called the father of our modern thought on the formation of mineral deposits. He was the first to show that most mineral deposits must be regarded as just one phase of igneous activity. He recognized that water vapor was an essential feature in volcanic activity and that metalliferous veins were formed as incrustations upon the walls of fissures from hot waters of igneous origin. He distinguished these veins from dikes injected in a molten condition. He cited occurrences of segregations of magnetite and chromite in basic igneous rocks, which he considered had crystallized out during the cooling of the intrusive. Elie de Beaumont thus recognized clearly the igneous affiliation of many types of mineral deposits — that some were formed as segregations during crystallization of the magma and that others were formed from hot aqueous emanations that escaped upward from the igneous intrusion. Essentially similar views are held today, but for many years the clear statements of Elie de Beaumont were overlooked.

The conflicting opinions of these times were carefully weighed by Von Cotta, whose excellent treatise on ore deposits appeared from Freiberg in 1859 (in English in 1870) and remained a standard textbook on mineral deposits for two decades. He presented concise information regarding the content, character, and structural and textural features of mineral deposits and careful descriptions of the outstanding mineral districts. He examined judiciously the various theories of mineral genesis and correctly concluded that no one theory was applicable to all deposits. In his concluding observations he remarks (Prime's translation, 1870) " Thus the formation of lodes shows itself to be . . . very manifold; and appears to have always stood in some connection with neighboring . . . eruptions of igneous rocks. The local reaction of the igneous-fluid interior of the earth created fissures, forced igneous-fluid masses into many of the same, caused gaseous emanations and sublimations in others; and in addition, during long periods of time impelled the circulation of heated water, which acted, dissolving at one point and again depositing the dissolved substances at another, dissolving new ones in their stead. The whole process is not confined to any particular geological period, or any particular locality." Such statements might well be a part of a modern textbook. He shows clearly that most mineral deposits were the products of deep-seated igneous action. He recognized not only a definite zonal arrangement of minerals dependent upon temperature and pressure conditions of deposition but also certain superficial changes imposed upon mineral deposits by weathering. Von Cotta's balanced treatise exerted a profound influence upon the subject of economic geology.

A trend back toward the Hutton views of a straight igneous origin for mineral deposits is to be noted in the writings of Fournet (1844, 1856) and Belt (1861), who considered that veins and many mineral deposits were the result of an igneous injection into fissures in a molten state, and were thus *ore magmas,* to use the term later proposed by Spurr (1923). Belt, however, believed that water played a part in helping to lower the temperature of the " liquefaction " (fusions) of such minerals as quartz.

In the meantime these views of hydrothermal and igneous action in the formation of mineral deposits were partly obscured during the advocacy of another startling hypothesis of the origin of ores.

The earlier assumptions of Delius, Gerhard, and Lassius, that water percolating through the rocks had dissolved out certain ingredients and afterward precipitated them in fissures, was later taken up by Bischoff (1847) and put forward as a reasoned theory of lateral secretion by waters of meteoric origin. This theory was supported by data regarding the dissemination of lode minerals in superficial rocks and their significance in ore genesis. Somewhat similar ideas were advanced in America by T. Sterry Hunt in 1861 and in England by J. A. Phillips in 1875, but they were advocated in an extreme form in Germany in 1882 by Sandberger, who sought to establish two facts: (1) that the gangue of ore veins corresponded with the wall rocks; and (2) that traces of the heavy metals occurred in the wall rocks. Meteoric waters were supposed to search out these ingredients from the surrounding wall rocks and deposit them in the fissures. Daubrée in his later comprehensive study of underground waters (1887) concluded that hot waters were the most important agent in the formation of mineral deposits and that these waters, for the most part, were not magmatic but were meteoric waters that had become heated in depth and had risen again. These were the views of Hunt and Phillips, and likewise later of S. F. Emmons, who in 1886 explained the origin of the ore deposits of Leadville, Colo., by surface waters that had leached metallic ingredients from the neighboring rocks. Similar views were elaborated in 1901 by C. R. Van Hise, who concluded that magmatic waters played a minor role and that most mineral deposits resulted from surface waters that had descended to depths where they became heated, dissolved metals from the rocks, and rose again to deposit their metallic content in fissures or other openings — a circulation resembling that of a hot-water heating system.

There then followed animated discussions on the respective merits of the *descensionist-*, *ascensionist-*, and *lateral-secretionist* theories.

The supporters of lateral secretion, however, lost ground rapidly under the attacks of Stelzner (1879), Patera (1888), Pošepný (1894), De Launay (1893), and others who contended that the mineral substances were not dissolved from the surrounding rocks by meteoric waters but rather were deposited there by ascending hot waters that had carried the materials up from the deep-seated sources. Both Pošepný and De Launay, following the earlier ideas of Elie de Beaumont, sought a source for the metals and waters in deep-seated eruptive rocks or still deeper sources in the barysphere. They were hot-water ascensionists, although Pošepný clearly distinguished certain types of deposits formed by the circulation of meteoric waters in the vadose zone. The vigorous discussions that followed the presentation of Pošepný's classical paper in Chicago in 1893 directed attention once more to the close association between mineral deposits and igneous rocks, and an igneous origin for mineralizing solutions was urged by J. F. Kemp (1901), Waldemar Lindgren (1901), W. H. Weed (1903), and others whose ideas became rather generally accepted.

In the meantime J. H. L. Vogt of Norway had been laying the foundations of our present-day conceptions of the origin of mineral deposits. Starting from the ideas of Elie de Beaumont, he delved further, utilizing physico-chemical principles, into the source of ascending hot mineralizing solutions. On the basis of careful field studies of magnetite, chromite, nickel, pyrrhotite, and pyrite he concluded that such substances were igneous injections of material derived from their igneous source by processes of magmatic differentiation (1893). He also concluded that the hot mineralizing waters were likewise derived by magmatic differentiation. Thus he linked, more clearly than his predecessors, the processes of petrology and ore formation and recognized magmatic differentiation as a process of ore formation that gave rise to (1) ore segregations or injections, (2) mineralizing gases and vapors, and (3) hot mineralizing waters. In recent years these processes have been elaborated, clarified, and experimentally verified by the work of such investigators as J. F. Kemp, Waldemar Lindgren, V. Goldschmidt, and particularly by the able members of the Geophysical Laboratory of the Carnegie Institute of Washington. Such conceptions are the ones generally current today to explain the origin of most primary mineral deposits. It is recognized, however, as was pointed out by Von Cotta long ago, that there are many types of mineral deposits, which have been formed by different processes.

A recent contribution to the theories of the genesis of mineral deposits is the extreme magmatic view of J. E. Spurr (1923) who, following the earlier conceptions of Thomas Belt (1861), postulates that many or most ore deposits have resulted from the injection and rapid freezing of highly concentrated magmatic residues, for which he proposed the terms " ore magmas " and " vein dikes." Spurr's ideas are not generally accepted, and the idea of magmatic waters holds almost undisputed sway today.

The advance in the study of ore deposits has in large part centered around the development of theories of origin, since each new theory quickened field observations directed to prove or disprove it. Consequently, a great mass of material regarding the character, distribution, and localization of mineral deposits has accumulated.

A review of the development of ideas concerning the genesis of mineral deposits is enlightening to the student, because it presents the background of the current prevailing ideas and discloses that most of them have been advanced in previous generations and that at present we are only elaborating upon or amplifying earlier conceptions.

Selected References on Development of Economic Geology

History of the Theory of Ore Deposits. THOMAS CROOK. Thos. Murby and Co., London, 1935. *An interesting, brief, historical treatise.*

Study of Ore Deposits. F. H. HATCH. Allen and Unwin, London, 1929. Chap. 1, pp. 13–20. *Brief summary of some early writings.*

Man and Metals. T. A. RICKARD. McGraw-Hill, New York, 1932. *Scattered parts of general historical interest.*

Historical Notes on Gem Mining. S. H. BALL. Econ. Geol. 26:681–738, 1931. *The most thorough treatment of this subject.*

General Reference 9. Chap. I. Historical Review of Geology as Related to Western Mining, by F. L. RANSOME. *Development of geological thought as related to Western United States mineral deposits.*

The Birth and Development of the Geologic Sciences. F. D. ADAMS. Williams and Wilkins, Baltimore, 1938. *An exhaustive treatment dealing with pre-nineteenth century development of ideas; particularly Chap. 9, The Origin of Metals and Their Ores.*

Fiftieth Anniversary Volume, 1888–1938. Geol. Soc. Amer., 1941. *Development of progress of thought in the various branches of geology during the last 50 years, by 21 authors.*

History of the Study of Ore Minerals. ELLIS THOMSON. Am. Mineral. 24:137–154, 1939.

The Romance of Mining. T. A. RICKARD. Macmillan Co., Toronto, 1944. *Many references to historical development.*

General Reference 13. Chap. 3, Minerals in History before the Industrial Era. *Early use, distribution, and trade in minerals.*

A Source Book in Greek Science. M. R. COHEN and I. E. DRABKIN. McGraw-Hill Book Co., New York, 1948. *Quotations from writings of Greek scholars relating to geology.*

Bergwerk- und Probierbüchlein. Translation of U. R. von Kalbe's German edition of 1518 by A. G. SISCO and C. S. SMITH. A.I.M.E., New York, 1949. *Translation of first known treatises on technology of minerals, metals, and veins.*

CHAPTER 3

MATERIALS OF MINERAL DEPOSITS AND THEIR FORMATION

Mineral deposits, whether metalliferous or nonmetallic, are accumulations or concentrations of one or more useful substances that are for the most part sparsely distributed in the earth's outer crust. They are mostly extremely rare in igneous rocks and may be absent from sedi-

TABLE 1

COMMON MINERALS OF EARTH'S CRUST IN PERCENT

MINERAL	LITHOSPHERE	IGNEOUS ROCKS	SEDIMENTARY ROCKS
Feldspar	49	50	16
Quartz	21	21	35
Pyroxene, amphibole, olivine	15	17	..
Mica	8	8	15
Magnetite	3	3	..
Titanite and ilmenite	1	1	..
Others	3	..	3
Kaolin (clay)	..	..	9
Dolomite	..	..	9
Chlorite	..	..	5
Calcite	..	..	4
Limonite	..	..	4
	100	100	100

mentary rocks. A few deposits, however, consist of concentrations of common rock-making minerals such as feldspar or mica.

The elements that enter into the materials of mineral deposits have been derived either from the rocks of the earth's outer crust or from molten bodies (magmas) that have cooled to form igneous rocks. Originally, all the elements except those that may have persisted from the primitive atmosphere were derived from magmas or igneous rocks of the outer rocky shell of the earth. Of the **98** known elements, only **8**, according to Clarke and Washington, are present in the earth's crust in amounts exceeding 1 percent, and **99.5** percent of the earth's outer crust (10 miles deep) is made up of the following 13 elements:

oxygen, silicon, aluminum, iron, calcium, sodium, potassium, magnesium, titanium, phosphorus, hydrogen, carbon, and manganese. The remaining elements, constituting only 0.5 percent, include all the precious and useful substances such as platinum, gold, silver, copper, lead, zinc, tin, nickel, and others. It is thus clear that geologic processes of concentration are necessary to collect these diffuse elements into workable mineral deposits, and it will be our purpose to study some of these exceptional and interesting processes.

The minerals that constitute the bulk of the earth's crust are also few in number. Over 1,600 mineral species are known; about 50 of these are rock-making minerals, of which only 29 are common ones. About 200 are classed as economic minerals. The remainder come mostly from mineral deposits. The abundance of the common minerals that constitute the bulk of the earth's crust may be seen from Table 1.

Since there is considerable difference in the constitution of metalliferous and nonmetallic deposits, and the terminology used is different for each, the materials of the two will be considered separately.

Materials of Metalliferous Deposits

Metalliferous deposits represent, in general, extreme concentrations of formerly diffuse metals. The desired metals are generally chemically bound with other elements to form *ore minerals*. These in turn are commonly interspersed with nonmetallic minerals or rock matter, called *gangue*. The mixture of ore minerals and gangue constitutes *ore*, which generally is enclosed in *country rock*. Some metalliferous deposits, however, may lie upon the surface and thus are not enclosed within country rock.

Ore Minerals. An ore mineral is one that may be used to obtain one or more metals. Most of them are metallic minerals, such as galena, which is mined for its lead. A few are nonmetallic minerals, such as malachite, bauxite, or cerussite, ore minerals of copper, aluminum, and lead. The ore minerals occur as native metal, of which gold and platinum are examples, or as combinations of the metals with sulphur, arsenic, oxygen, silicon, or other elements. Combinations are the most common.

A single metal may be extracted from several different ore minerals. Thus, there are several ore minerals of copper, such as chalcocite, bornite, chalcopyrite, cuprite, native copper, and malachite; one or more of these may occur in an individual deposit. Also, more than one metal may be obtained from a single ore mineral; stannite, for example, yields both copper and tin. An individual ore deposit, therefore, may yield several metals from different ore minerals.

The metals of commerce are derived from many metallic combinations. Most of the world's gold has come from native gold; consequently its removal from admixed minerals is a relatively simple process, and offered no serious problems of extraction, even to the ancients. Silver, on the other hand, is derived not only from the native metal but also from combinations with sulphur and other elements. This is also true of copper, lead, zinc, and most of the other metals. The vast quantity of iron used in industry is obtained almost entirely from combinations with oxygen. It is from such simple combinations that the human race has been supplied with desired metals for over 2,000 years. In addition, there are more complex metallic combinations that yield considerable quantities of the common metals as well as many of the minor metals.

Ore minerals are also classed as *primary* or *hypogene*, and *secondary* or *supergene*. The former were deposited during the original period or periods of metallization; the latter are alteration products of the former as a result of weathering or other surficial processes resulting from descending surface waters. The term *primary* has also been used to designate the earliest of a sequence of ore minerals as contrasted with later minerals of the same sequence, which some writers have called secondary. This gave rise to some confusion, and to avoid this, Ransome proposed the terms *hypogene* and *supergene*. Primary and hypogene are generally considered synonymous, but hypogene, as the word implies, indicates formation by ascending solutions. All hypogene minerals are necessarily primary, but not all primary ore minerals are hypogene; sedimentary hematite, for example, is of primary deposition but has not been formed from ascending solutions. Similarly, confusion has arisen with the use of the word *secondary*, which is eliminated by the better term *supergene*.

Some of the important ore minerals of several of the metals are listed in Table 2.

Gangue Minerals. Gangue minerals are the associated nonmetallic materials of a deposit. They may be introduced minerals, or the enclosing rock, and are usually discarded in the treatment of the ore. The *gangue*, in customary usage, includes only nonmetallic minerals, but in technical usage it includes also some metallic minerals, such as pyrite, which are usually discarded as worthless. Certain gangue materials, however, may at times be collected as by-products and utilized. For example, rock gangue may be utilized for " road metal "; fluorspar for flux; quartz for abrasive or concrete; pyrite

TABLE 2

LIST OF THE COMMON ORE MINERALS

Metal	Ore Mineral	Composition	Percent Metal	Primary	Super-gene
Gold	Native gold	Au	100	X	X
	Calaverite	$AuTe_2$	39	X	
	Sylvanite	$(Au,Ag)Te_2$	..	X	
Silver	Native silver	Ag	100	X	X
	Argentite	Ag_2S	87	X	X
	Cerargyrite	AgCl	75		X
Iron	Magnetite	$FeO \cdot Fe_2O_3$	72	X	
	Hematite	Fe_2O_3	70	X	X
	" Limonite "	$Fe_2O_3 \cdot H_2O$	60		X
	Siderite	$FeCO_3$	48	X	X
Copper	Native copper	Cu	100	X	X
	Bornite	Cu_5FeS_4	63	X	X
	Brochantite	$CuSO_4 \cdot 3Cu(OH)_2$	62		X
	Chalcocite	Cu_2S	80	X	X
	Chalcopyrite	$CuFeS_2$	34	X	X
	Covellite	CuS	66	X	X
	Cuprite	Cu_2O	89		X
	Enargite	$3Cu_2S \cdot As_2S_5$	48	X	
	Malachite	$CuCO_3 \cdot Cu(OH)_2$	57		X
	Azurite	$2CuCO_3 \cdot Cu(OH)_2$	55		X
	Chrysocolla	$CuSiO_3 \cdot 2H_2O$	36		X
Lead	Galena	PbS	86	X	
	Cerussite	$PbCO_3$	77		X
	Anglesite	$PbSO_4$	68		X
Zinc	Sphalerite	ZnS	67	X	
	Smithsonite	$ZnCO_3$	52		X
	Hemimorphite	H_2ZnSiO_5	54		X
	Zincite	ZnO	80	X	
Tin	Cassiterite	SnO_2	78	X	?
	Stannite	$Cu_2S \cdot FeS \cdot SnS_2$	27	X	?
Nickel	Pentlandite	$(Fe,Ni)S$	22	X	
	Garnierite	$H_2(Ni,Mg)SiO_3 \cdot H_2O$	..		X
Chromium	Chromite	$FeO \cdot Cr_2O_3$	68	X	
Manganese	Pyrolusite	MnO_2	63		X
	Psilomelane	$Mn_2O_3 \cdot xH_2O$	45		X
	Braunite	$3Mn_2O_3 \cdot MnSiO_3$	69	?	X
	Manganite	$Mn_2O_3 \cdot H_2O$	62		X
Aluminum	Bauxite	$Al_2O_3 \cdot 2H_2O$	39		X
Antimony	Stibnite	Sb_2S_3	71	X	
Bismuth	Bismuthinite	Bi_2S_3	81	X	X
Cobalt	Smaltite	$CoAs_2$	28	X	
	Cobaltite	CoAsS	35	X	
Mercury	Cinnabar	HgS	86	X	
Molybdenum	Molybdenite	MoS_2	60	X	
	Wulfenite	$PbMoO_4$	39		X
Tungsten	Wolframite	$(Fe,Mn)WO_4$	76	X	
	Huebnerite	$MnWO_4$	76	X	
	Scheelite	$CaWO_4$	80	X	

for sulphur; and limestone for fertilizer or flux. Some gangue minerals considered worthless today may prove of value tomorrow. The gangue, even though worthless, may so influence the method or cost of treatment that it determines the value of the ore.

Some of the common gangue minerals are listed in Table 3.

TABLE 3

List of Common Gangue Minerals

Class	Name	Composition	Primary	Supergene
Oxides	Quartz	SiO_2	X	X
	Other silica	SiO_2	X	X
	Bauxite, etc.	$Al_2O_3 \cdot 2H_2O$		X
	Limonite	$Fe_2O_3 \cdot H_2O$		X
Carbonates	Calcite	$CaCO_3$	X	X
	Dolomite	$(Ca,Mg)CO_3$	X	X
	Siderite	$FeCO_3$	X	X
	Rhodochrosite	$MnCO_3$	X	
Sulphates	Barite	$BaSO_4$	X	
	Gypsum	$CaSO_4 + 2H_2O$		X
Silicates	Feldspar	. . . .	X	
	Garnet	. . . .	X	
	Rhodonite	$MnSiO_3$	X	
	Chlorite	. . . .	X	
	Clay minerals	. . . .	X	X
Miscellaneous	Rock matter		X	
	Fluorite	CaF_2	X	
	Apatite	$(CaF)Ca_4(PO_4)_3$	X	
	Pyrite	FeS_2	X	X
	Marcasite	FeS_2	X	X
	Pyrrhotite	$Fe_{1-x}S$	X	
	Arsenopyrite	$FeAsS$	X	

Ore. The term " ore " is often loosely used to designate anything that is mined. Technically, it is an aggregation of ore minerals and gangue from which one or more metals may be extracted at a profit. To be ore, material must, therefore, be payable, and this involves economic considerations as well as geologic. Obviously a body of valueless pyrrhotite devoid of gold would not be ore, even though pyrrhotite is a metallic mineral, and a body of pyrite containing gold may or may not be ore, depending upon the amount of gold present and whether the value of the recoverable gold is greater than the cost of extraction.

The question of profit depends upon the amount and price of the metal and upon the cost of mining, treating, transporting, and marketing the product. This in turn depends in part upon the geographic location of the deposit. A high-grade hematite body located in Arctic America, for example, would not be iron ore because the cost of extraction and transportation would be greater than the value of the iron. However, it might become ore in the future. Likewise, increased efficiency and lower cost of metallurgical technique or mining practice may enable present worthless material to be classed as ore in the future. Also, new uses may transform worthless materials into valuable ones. The discovery that beryllium makes a fatigue-resisting alloy with copper has changed beryl from a mineralogical curiosity to a much-sought, valuable mineral.

What constitutes ore may also depend upon the gangue or upon minor constituents. Certain materials may be profitably worked for their metallic content only if some part of the gangue can be utilized. The presence of small quantities of bismuth, cadmium, or arsenic may make otherwise valuable deposits of lead, zinc, or copper worthless.

The relative proportions of ore minerals and gangue vary enormously. In average ores, gangue greatly predominates. It is costly to smelt valueless gangue in order to obtain the enclosed metal; so it is customary to subject the ores to ore-dressing processes (milling) whereby the ore minerals are concentrated and the waste gangue discarded. Thus, from 5 to 30 tons of crude ore will yield 1 ton of concentrates containing most of the metallic content of the original lot. This is then smelted for its metallic content, thereby saving the cost of treating the 4 to 29 tons of discarded gangue. For example, if a gold-copper ore contains $6.00 worth of metal per ton and freight and smelting charges are $6.00 per ton, there would be no profit in smelting the crude ore, but if it were concentrated 10 to 1 (the ratio of concentration), then there would be freight and smelting charges for only 1 ton of concentrates instead of for 10 tons of ore, equivalent to about 60 cents per ton of original ore. Consequently, the ratio of concentration is of vital importance in determining whether a material is ore; the gangue may thus play fully as important a part as the ore minerals.

The amount of metalliferous minerals present varies greatly in ores of different metals and also in ores of the same metal. High-grade iron ores may consist of 100 percent hematite; copper ores range from less than 1 percent to 75 percent of metalliferous minerals, those with low percentages being concentrated, and those with higher percentages being smelted directly. In contrast, gold ores may contain only an infinitesimal amount of gold. For example, the Alaska Juneau gold mine mined with profit ores that contained only 0.00016 percent of gold. The amount of metal that must be present to constitute ore obviously depends upon the price of the metal. During periods of low metal prices, only ore with a metallic content higher than normal can be classed as ore. Conversely, when the price of gold was increased much material that previously was valueless later became good gold ore. Because of the fixed price of gold and rising mining costs, some former ore is today not commercial.

Associated Metals in Ores. Ores may yield a single metal (simple ores) or several metals (complex ores). Ores that are generally worked for only a single metal are those of iron, aluminum, chromium, tin, mercury, manganese, tungsten, and some ores of copper. Gold ores may yield only gold, but silver is a common associate. Much gold, however, is extracted as a by-product from other ores. Ores that commonly yield either two or three metals are those of gold, silver, copper, lead, zinc, nickel, cobalt, antimony, and manganese. Some complex ores may yield four or five metals, such as copper-gold-silver-lead, silver-lead-zinc-copper-gold, tin-silver-lead-zinc, or nickel-copper-gold-platinum. Many minor metals are not won directly from their ores but are obtained as by-products from ores of other metals during smelting or refining operations. This is true of arsenic, bismuth, cadmium, selenium, and others. Where precious metals accompany base metals, their presence may make good ore of otherwise uneconomic material.

Some of the common associations of metals in ores are: gold and silver; silver and lead; lead and zinc; lead, zinc, and copper; copper and gold; iron and manganese; iron and titanium; nickel and copper; nickel and cobalt; chromium and platinum; tin and tungsten; molybdenum and copper; and zinc and cadmium.

Tenor of Ores. The metal content of an ore is called the *tenor*, which is generally expressed in percentage or, in the case of precious metals, in ounces per ton. The tenor varies with the price of a metal, with the cost of production, with ores of different metals, and also with ores of the same metal. The higher the price of a metal the lower the metal content necessary to make it profitable. Most iron ore to be profitable must have a tenor of 35 to 50 percent of iron, whereas copper ore need contain only 0.8 percent copper or gold ore only 1/1000 of 1 percent of gold. The tenor of ore need have no upper limit; the richer, the better. The lower limit, however, is fixed by economic considerations and varies according to the nature and size of a deposit, its location, metal price, and cost of extraction. Identical material may be good ore in one locality and worthless in another.

The tenor of ores of the commoner metals, also their prices and units, are given in Table 4.

TABLE 4

DATA ON METALS AND THEIR ORES

Metal	Unit of Measure	Tenor		Common Associates	Commercial Unit	Price Ranges,	
		Low	Average			1925–40	1941–49
Gold	oz/ton	0.15	0.2–0.3	Ag	oz Troy	20.67–35.00	35.00
Silver	oz/ton	10	12–30	Au, Pb	oz Troy	0.25–0.70	0.35–0.90
Platinum	oz/ton	0.1	0.3	Pt group	oz Troy	31.00–67.65	36.00–93.00
Iron	% Fe	30	40–60	Mn	ton iron	15.00–24.00	20.00–34.00
Copper	% Cu	0.7	1–5	Au, Ag	lb Cu	0.05–0.21	0.12–0.235
Lead	% Pb	3	5–10	Zn, Ag	lb Pb	0.03–0.09	0.06–0.215
Zinc	% Zn	3	10–30	Pb	lb Zn	0.03–0.08	0.075–0.175
Tin	% Sn	0.5	1–5	W	lb Sn	0.22–0.65	0.52–1.03
Nickel	% Ni	1.5	1.5–3	Cu, Pb	lb Ni	0.35–0.39	0.31–0.35
Aluminum	% Al$_2$O$_3$	30	55–65	....	lb Al	0.17–0.24	0.15–0.17
Antimony	% Sb	20	40–60	Ag	lb Sb	0.05–0.17	0.14–0.417
Bismuth	% Bi	BP*	40–60	W	lb Bi	0.85–1.30	1.25–2.00
Beryllium	% BeO	8	10–12	....	unit BeO	30.00–35.00	26.00–47.00
Arsenic	% As$_2$O$_3$	BP	.....	....	lb As$_2$O$_3$	0.01–0.03	0.04–0.06
Cobalt	% Co	5	8–10	Ag, Cu	lb Co	1.10–3.00	1.50–1.80
Chromium	% Cr$_2$O$_3$	32	35–50	....	ton Cr$_2$O$_3$	17.00–47.00	34.00–48.00
Cadmium	% Cd	BP	.....	Zn	lb Cd	0.55–1.42	0.75–2.00
Manganese	% Mn	35	45–55	Fe	units/ton	0.18–0.55	0.70–0.73
Mercury	% Hg	0.5	1–3	....	flask–76 lb	58.00–202.00	76.00–196.00
Molybdenum	% MoS$_2$	0.4	1–3	....	lb MoS$_2$	0.34–0.45	0.45–0.54
Titanium	% TiO$_2$	3	4–40	Fe	lb TiO$_2$	....	5.00–6.00
Tungsten	% Wo$_3$	....	60–70	....	unit WO$_3$	9.20–20.61	24.00–28.50
Vanadium	% V$_2$O$_5$	2	3–8	....	lb V$_2$O$_5$	0.20–1.05	0.27–0.275

*BP = by-product.

Materials of Nonmetallic Deposits

The materials of nonmetallic deposits consist of solids, liquids, and gases. The term " ore " is not generally applied to such substances; they are referred to by the name of the substance itself, as for example, mica, asbestos, or petroleum. Neither is the term " ore mineral " applied to the desired materials, but " gangue " is often used to denote the undesired material, although it is more generally called *waste.*

Since nonmetallic materials are, in general, common substances, their price is correspondingly low as compared with metals. Except for a few substances, such as gemstones, the deposits consist predominantly or entirely of the useful material; there is little or no waste. Coal and gypsum deposits consist entirely of the desired materials;

feldspar, barite, or fluorspar deposits include considerable waste that must be removed by " processing." Gemstones, asbestos, or graphite generally constitute only a small part of the deposit.

The nonmetallic materials consist of a vast array of substances utilized in modern civilization. These include fuels, rocks, earthy materials, sand, salts, minerals of pegmatite dikes, and many other nonmetallics such as asbestos, gypsum, fluorspar, mica, barite, graphite, and sulphur. The associated gangue or waste consists mostly of enclosing rock, or parts of the nonmetallic products themselves that are discarded as unfit for use because of physical or chemical defects. The separation of the two, called " processing," consists of hand-sorting, simple mechanical concentration, flotation, or washing. Since smelting, or involved metallurgical treatment, is unnecessary, the determination of what is economic material is not so dependent upon the associated gangue as in the case of ores. Rather, it is dependent upon the price and the physical and chemical properties of the products themselves. A metal is just a metal, but clay, for example, to be usable must meet definite specifications regarding plasticity, specific gravity, fusibility, shrinkage, and tensile strength. Its value depends more upon its physical than its chemical properties, and not all clay fulfills the requirements. Different requirements exist for each nonmetallic product. Consequently, the properties that determine the commercial use of nonmetallic products are multitudinous as compared with the few that determine ore.

The nonmetallic products are not generally associated in groups as are the metals in ores. Some common associations are: petroleum and gas; potash, salt, and gypsum; feldspar and mica; and soapstone and talc.

Determination of Materials

The materials that make up mineral deposits can, for the most part, be determined visually. However, for more exact determination, precise methods are necessary, such as assaying, chemical analyses, microscopic examination, X-ray examination, spectroscopic examination, thermal analyses, or physical tests.

In ores the metallic content of the desired constituents is determined usually by fire or by wet (chemical) assaying, and the results are expressed in the units given in Table 4. As this method expresses only the metallic content, without regard to the mineralogical constituents, a supplemental microscopic examination is usually desirable to reveal the identity of the minerals and their relation to each other. Important minute metalliferous constituents can thus be detected. Partial or complete chemical analyses are sometimes made in order to determine the quantities of other ingredients, particularly those that affect the metallurgical treatment. Thus, for smelting, it is necessary to know the proportions of the oxides of silicon, calcium, magnesium, and iron in order that correct proportions of fluxes may be added to the ore or concentrates. Also, the quantities of undesired ingredients, such as arsenic, are determined by chemical analyses. In some cases minute quantities of rare elements are determined spectroscopically.

In nonmetallic products, the materials are determined by chemical analyses, physical tests, visual inspection, or microscopic examination. Occasionally, for such fine materials as clays, X-ray examinations and thermal analyses are employed. Different methods are employed for different products. Thus, physical characters, such as strength, grain size, hardness, specific gravity, plasticity, fusibility, and electrical conductivity, are determined for such materials as building stones, sands, clays, abrasives, mica, or asbestos. On the other hand, partial analyses are necessary for such materials as fuels, fertilizers, bauxite, cements, magnesite, or lime. But for such substances as gemstones, roofing slates, quartz, or barite, generally visual inspection alone is necessary. Microscopic examinations supplement the other tests.

The Formation of Minerals and Mineral Products

An understanding of mineral deposits necessitates a knowledge of the manner in which the constituents have been formed. Much information has accumulated in recent years regarding the conditions of formation, particularly temperature and pressure. In consequence, the study of minerals has taken on a new and wider geological significance in that the presence of certain minerals may supply definite information as to temperature, pressure, or chemical character of the mineralizing agencies, which thus aids in deciphering the origin of the deposits that contain them. For example, the presence of digenite in an ore deposit is of interest not because of its physical properties but because it indicates formation at moderate temperatures and is, therefore, of hypogene rather than of supergene origin.

As it is the intent to consider here the formation only of those minerals that are of economic importance, the following discussion deals with the formation of the *materials* of mineral deposits and not with the deposits themselves.

TEMPERATURE AND PRESSURE

The formation of a mineral generally indicates a change from a mobile to a solid state. As most minerals have been precipitated from solutions, either liquid or gaseous, temperature and pressure play important roles.

Changes in temperature affect the solubility of materials in solution, and therefore their precipitation. In general, decrease of temperature promotes precipitation from aqueous solutions or magmas. The more soluble salts will tend to stay in solution longer and be precipitated later than the less soluble, thereby supplying one explanation of the sequences of minerals in mineral deposits and mineral zoning. Minerals once precipitated may be redissolved and reprecipitated. Van't Hoff's laws show that, for precipitation from solutions in which several reactions may take place, that one occurs which is attended by the greatest evolution of heat. Solution is generally *endothermic* (uses up heat), and precipitation is *exothermic* (gives off heat).

Changes in pressure are also important, although less effective than temperature changes in promoting precipitation. In general, increase of pressure promotes solubility, and decrease of pressure, such as that which occurs when solutions ascend in the earth, promotes precipitation. Gases in solution are very sensitive to a change of pressure. For example, carbon dioxide, held in water by pressure, promotes the solubility of calcium carbonate, and its release under lessened pressure causes precipitation of calcium carbonate. Precipitation of material from a gas can be effected by decrease in pressure alone. Likewise, the escape of gases from magmatic fluids under decreasing pressure promotes precipitation.

MODES OF FORMATION

The constituents of mineral deposits are formed in the different ways discussed below. Temperature, pressure, and water play an important part in the deposition of the vast majority of minerals.

Crystallization from Magmas. Since a magma is a molten or fluid silicate solution, crystallization from it follows the same laws as with an aqueous solution. When a magma cools and the saturation point of the solution is exceeded for any given mineral, that mineral will crystallize, provided the temperature at the existing pressure is below the fusion point of the mineral. Thus, from certain magmas, economic minerals such as apatite, magnetite, or chromite have formed by crystallization. The complicated processes involved are discussed in Chapter 5·1.

Sublimation. The heat of igneous activity may cause the volatilization of certain substances that later are deposited as sublimates around volcanic vents, fumaroles, or shallow intrusions. Reactions between gases may also be involved. Native sulphur is not an un-common sublimate.

Distillation. Some geologists consider that petroleum and natural gas have been formed by slow distillation of organic material deposited with marine sediments.

Evaporation and Supersaturation. Salts in solution are precipitated when evaporation of the solvent brings about supersaturation, as, for example, in the formation of salt deposits by evaporation of brines. Other familiar examples are the formation in mines, through evaporation, of efflorescences of sulphates of copper, iron, zinc, magnesium, calcium, and other salts. Indeed, extensive deposits of copper ores have been formed by evaporation in Chile, and evaporation also was effective in forming the many minerals that constitute the valuable nitrate deposits of Chile.

Reaction of Gases with Other Gases, Liquids, or Solids. Igneous activity is accompanied by the release of vast quantities of gaseous emanations that contain many elements and compounds found in mineral deposits. Zies found among the fumarolic incrustations of the Valley of Ten Thousand Smokes large quantities of magnetite, also base-metal sulphides, fluorides, borates, sulphur, molybdenite, and other minerals. The metals clearly had been brought up in the vapor phase since no liquid solution had at any time been active, and their deposition resulted from reactions between different gases and vapors. For example, native sulphur and hematite may be formed at high temperatures according to the following reactions:

$$2H_2S + SO_2 \rightleftarrows 3S + 2H_2O$$

$$Fe_2Cl_6 + 3H_2O \rightleftarrows Fe_2O_3 + 6HCl$$

This demonstration that gaseous emanations are effective transporters of minerals is enlightening in connection with the origin of mineral deposits.

Gases also react with liquids to form minerals, both at high and at normal temperatures. The familiar precipitation of copper sulphide from cupric sulphate mine waters by hydrogen sulphide is an example.

More important still are reactions between gaseous emanations and solids that yield a variety of high-temperature minerals, such, for example, as the assemblage of unusual contact-metasomatic silicates, oxides, and sulphides formed by the action of magmatic emanations upon intruded rocks, particularly carbonate rocks.

Reaction of Liquids with Liquids and Solids. The large volume of magmatic fluids given off by consolidating intrusives carries vast quantities of mineral matter in solution. These solutions are, or become, liquid and are responsible for the formation of most of the materials of mineral deposits. In their ascent they may meet surface waters of different composition and also wall rock of varying reactibility, or they may intermingle with other magmatic solutions. When two such intermingling liquids contain common ions, precipitation of minerals of difficult solubility ensues according to *Nernst's* law, which states that the solubility of a salt is decreased by the presence in solution of another salt that has a common ion. For example, the solubility of smithsonite or siderite in a saturated solution is decreased by the presence of calcite in solution, and their precipitation results. R. C. Wells showed that if an alkaline sulphide is added in excess to a solution containing iron and copper, both metals are completely precipitated. Many supergene minerals of the oxidized zone, and likewise presumably many hypogene minerals, are formed in this manner.

Intermingling of different solutions may also cause precipitation of minerals by reduction. Bischoff's early experiment showed that a sealed mixture of sodium sulphate and ferrous carbonate yielded pyrite. The reducing action of organic matter in precipitating gold, silver, copper, and other minerals is well known.

Ransome suggests that the ore constituents of the Goldfield deposits were brought up in hot solutions containing hydrogen sulphide that oxidized near the surface to sulphuric acid, which then percolated downward to mingle with the uprising currents; and that " the precipitation of the richest ores took place in the zone where the two solutions mingled, and as a consequence of such mingling." Acidification of an alkaline sulphide solution by sulphuric acid precipitates sulphides of heavy metals, as was demonstrated by Grout.

If a concentrated mineral solution is diluted by ground water, metals carried as complexes are released and precipitation ensues.

Many other reactions between solutions to produce precipitation of minerals are too well known in chemistry to need further discussion.

Reactions between solutions and solids are probably the most important natural processes in the formation of both hypogene and supergene minerals. Hypogene and supergene waters are continuously in contact with rocks and minerals; chemical action takes place, and ore and gangue minerals are precipitated. Several processes are involved, namely, metasomatism or replacement, relative solubility, reduction or oxidation, direct deposition, catalytic action, adsorption, base exchange, chemical complexes, and others.

Metasomatism or *metasomatic replacement* or simply *replacement,* as it is generally called, is a process of essentially simultaneous capillary solution and deposition by which new minerals are substituted for earlier minerals or rocks. A single mineral may replace and retain the exact form and size of a replaced mineral (pseudomorph), or silica may replace wood, retaining the woody structure (petrification), or a large body of ore minerals may take the place of an equal volume of rock. The replacing mineral (metasome) need not have a common ion with the replaced substance. The metasomes are carried in solution, and the replaced substances are carried away in solution: it is an open circuit, not a closed one. If, in a brick wall, each brick were removed one by one and a silver brick of similar size substituted for each, the end result would be a wall of the same size and form, even to the minutiae of brick pattern, save that it would be composed of silver instead of clay. This is how replacement proceeds, except that the parts are infinitesimally small — of molecular or ionic size. Consequently, the shape, size, structure, and texture may be faithfully preserved in the replacing substance.

Replacement is the most important process in the formation of epigenetic minerals and mineral deposits, or those formed later than the rocks enclosing them. It is the dominating process of mineral formation in the hydrothermal mineral deposits. The ore minerals of contact-metasomatic deposits (Chap. 5·3) have been emplaced by this process. Replacement is the controlling or dominating process in the formation of most supergene mineral deposits (Chap. 5·8). It plays a major role in the extensive rock alteration that accompanies most epigenetic metallization. Replacement may occur through the action of hot vapors and gases, or by either hot or cold water solutions. Cold copper sulphate solution may replace limestone to yield copper carbonate, or pyrite to yield chalcocite.

The *relative solubility* of solid and solute determines the precipitation of many minerals from solution. For example, if a copper sulphate solution meets sphalerite, which is more soluble, copper sulphide will be deposited at the expense of the sphalerite, which will go into solution; should the copper sulphate meet cinnabar, which is less soluble, no deposition will occur.

Reduction and oxidation may cause precipitation when a solution contacts a solid. Organic matter or pyrite reduces gold from auriferous solutions, and the organic matter reduces ferrous carbonate (siderite) from iron solutions. Cupriferous solutions may be oxidized by ferric iron, causing deposition of native copper; and the familiar oxidation of pyrite yields limonite.

Direct deposition in open space, without involving replacement, may result by change of temperature and pressure, and by mineralizing solutions coming in contact with solids. The country rock traversed may change the solutions from acid to alkaline; certain wall rocks of fissures favor deposition of vein filling opposite them; and materials dissolved from such solids may bring about deposition.

Catalytic action, in which certain substances cause precipitation from solution without themselves entering into such solution, is another cause of mineral deposition. Many catalysts, such as platinum, are used chemically, and on a large scale pyrolusite has been effectively employed by Zapffe as a catalytic agent to precipitate manganese and iron from the municipal water supply of Brainerd, Minn.

Adsorption is the taking up of one substance at the surface of another. For instance, kaolin adsorbs copper from solution to form much of the so-called chrysocolla, and silica gel adsorbs ferric oxide. In part, the change involves chemical reaction between the substances.

Base exchange occurs between solids and liquids whereby cations are exchanged, producing changed characteristics of both.

Precipitation by Bacteria. This is exemplified by the well-known precipitation of iron by bacteria. Harder recognizes three types of such bacteria, i.e., those that precipitate (1) ferric hydroxide from ferrous bicarbonate solutions by absorbing carbon dioxide, (2) ferric hydroxide without requiring ferrous carbonate, (3) ferric hydroxide or basic ferric salts from iron salts of organic acids. They are mostly thread and soil bacteria, of which *Crenothrix* is a common form. Such bacteria are considered by many geologists to have caused the precipitation of extensive deposits of iron ore (Chap. 5·5). Zapffe found that soil bacteria are effective precipitants of manganese from the subsurface waters of Brainerd, Minn. Anaerobic bacteria, or those that live without oxygen, precipitate sulphides and sulphur. Algae, another form of life, precipitate travertine and in some places silica.

Unmixing of Solid Solutions. Natural solutions of one solid in another are well known. Much gold contains silver in solid solution, and the ease with which gold unites with mercury to form amalgam is commercially utilized in the extraction of free gold from its ores. Solid solutions of magnetite and ilmenite, chalcocite and covellite, and other mineral pairs, are common. Some solid solutions form, and remain stable, at low temperatures. Others form only at high temperatures and become unstable at lower temperatures. When these cool slowly, one mineral may separate out of the other at a certain point in the cooling-temperature curve. This is known as unmixing, or *exsolution.* Thus, ilmenite plates separate out from solid solution in magnetite; covellite laths from chalcocite; argentite from galena; chalcopyrite from sphalerite and stannite; and cubanite from chalcopyrite. The minerals that form by exsolution remain as inclusions in the host and generally are visible only with a microscope.

Colloidal Deposition. Although the chemistry of colloidal solutions is as yet imperfectly known, there are many geologic examples of the deposition of *colloids* from colloidal solutions. Colloids are matter in a particular state rather than a particular class of substances. The solutions are two-phase systems: one is called the continuous phase, or dispersion medium, and is generally a liquid; the other, known as the dispersed phase, may be either solid, dispersed liquid, or gas, and is disseminated in the other in minute particles but not in true molecular solution. Such systems are generally called *sols;* if the dispersed phase is a solid they are called *suspensoids;* if a liquid, *emulsoids.* In both, the particles are of submicroscopic size and all those in the same sol carry similar electrical charges and, therefore, repel one another. Their viscosity may be little greater than that of pure liquid, and they differ from ordinary electrolytic solutions in that they do not pass through membranes. Many substances of difficult solubility can readily be dispersed to form highly concentrated sols, and this is an important feature in mineral deposition. The suspensoid sols have low viscosity and do not gelatinize. The emulsoid sols have higher viscosity and by coagulation, cooling, or evaporation yield a gelatinous mass called *gel.*

Deposition from sols is readily brought about by the addition of small quantities of electrolytes, which neutralize the charges on the contained particles. In suspensoids this quickly causes the solid to flocculate, and it is not readily returned to a colloidal state; in emulsoids, coagulation to a gel takes place, and this can readily be returned to the colloidal state. Colloids are readily precipitated from natural solutions as flocculent or gelatinous masses because such waters contain dissolved salts, which are electrolytes. These may harden to rounded or colloform masses, and such forms as botryoidal, reniform, mammillary, nodular, or pisolitic are often thought to have resulted from colloidal deposition. The solidified colloids may persist in an amorphous state, such as opal, or, more commonly, acquire crystallinity and become *metacolloids,* such as marcasite, malachite, or psilomelane. Commonly, such recrystallized colloids exhibit a radial structure. Lindgren considers it proof of colloidal origin if single radial crystals cross colloform banding.

At present there is a tendency among geologists to attach much importance to colloidal phenomena in mineral formation. Many hypogene and supergene minerals are considered to have had a colloidal origin, such, for example, as bauxite, chrysocolla, malachite, native arsenic, wurtzite, and others. Lindgren also believes that colloid minerals may replace older minerals. However, there is still much to be learned about action of colloids in mineral deposits.

Weathering Processes. Weathering is much more important in the formation of economic minerals than is generally realized. It is a complex operation that involves several distinct processes, such as disintegration, oxidation, hydration, reactions of solutions and gases with other solutions, gases, and solids, and evaporation — processes that may operate singly or jointly. Weathering is generally subdivided into mechanical and chemical action; usually both operate together. Mechanical action, although important in yielding valuable placer deposits, does not create the useful minerals; it merely frees and concentrates minerals already formed. However, it facilitates chemical weathering by reducing the size of the materials, thus creating more specific surface available for attack. Chemical weathering, however, actually creates useful minerals by acting upon (1) pre-existing economic mineral deposits; (2) submarginal mineral bodies; (3) gangue minerals; (4) rocks (see also Chap. 5·7).

(1) *Pre-existing mineral deposits* subjected to weathering yield new minerals that are stable under surface or near-surface conditions. These changes are brought about by the action of surface waters and atmospheric water, oxygen, and carbon dioxide. Some minerals are altered *in situ;* others are taken into solution, carried below, and there reprecipitated as new minerals. Common sulphides are particularly susceptible to attack, and in the zone of oxidation alter to limonite, native metals, and the familiar oxides, carbonates, silicates, sulphates, and chlorides of the common metals. Below the zone of weathering, supergene sulphides, such as chalcocite, covellite, argentite, and others, are precipitated, producing an entirely new group of valuable ore minerals. The details of these changes are given in Chapter 5·8.

(2) *Sub-marginal bodies* of low-grade disseminated minerals, such as pyrite and chalcopyrite, are similarly converted into commercial deposits of chalcocite and covellite. Many of the large copper deposits of the world have been formed in this manner.

(3) *Gangue minerals,* such as rhodochrosite and siderite, the carbonates of manganese and iron, and feldspar, weather to usable oxides of manganese and iron and to china clay.

(4) *Rocks* weather into newly formed minerals, of which some form valuable mineral deposits. Feldspathic igneous rocks and shales yield clay deposits. Aluminous rocks in warm, moist climates yield deposits of bauxite, the ore of aluminum. Serpentine in Cuba yields iron laterite, which forms extensive iron deposits at Mayari. Man-

ganese laterites have been formed in India. The new minerals have been created by chemical weathering. The details of the changes are discussed in Chapter 5·7.

Metamorphism. The agencies of metamorphism, mainly pressure, heat, and water, act upon rocks and minerals, causing recombination and recrystallization of the ingredients into new minerals that are stable under the imposed conditions. Many of them are of economic value. For example, garnet, graphite, and sillimanite minerals are created by metamorphism.

Stability of Minerals

Minerals, like life, are responsive to their environment. They are for the most part stable under the conditions under which they were created but become unstable with different environment. Most of those formed under conditions of high temperature and pressure perish at the surface, and, vice versa, many of those formed near the surface change to more stable forms under high temperature and pressure. Some few less sensitive minerals such as the diamond or cassiterite persist under changed conditions. Others called *persistent minerals,* such as pyrite or gold, form under many different conditions. Such sensitivity to conditions of formation or environment is utilized in interpreting the history of the mineral deposits that contain them. It is the earmark of progressive or retrograde metamorphism, of high or low temperature of formation, of hydrothermal alteration, of surface weathering, or other earth processes. Subjected to temperature changes or solvent action, minerals change to other substances more stable under the new conditions. A mixture of water, charcoal, and sulphur (gunpowder) is stable under low temperature, but touch a match to it and it changes instantly to a gaseous form with explosive force. And so minerals, less spectacularly, change their form. A bit of iron, cradled in a fiery furnace, if thrown out, succumbs quickly to mere moisture and oxygen; it rusts, or changes to the more stable form of limonite. So does pyrite.

The most noticeable changes of stability occur during weathering. Under the relentless attack of atmospheric water, oxygen, and carbon dioxide, few hypogene minerals survive. They are altered to native metals, oxides, carbonates, sulphates, chlorides, silicates, and other forms. Those minerals that already are oxides suffer least; sulphides suffer most. Some of our most beautiful mineral specimens are thus formed from more somber predecessors, and they are generally stable under their new environment. Their presence denotes the action of surface agencies. They presage a change in the nature of the mineral deposit below the depth of supergene alteration.

Common substances, such as clays, reared on the earth's surface and later buried deeply beneath a pressing load of sediments will. under the new environment of increased pressure and temperature, change to more stable forms of muscovite or other silicates. Mica schists or garnets may thus be formed, and most of the metamorphic minerals belong to this group.

A monzonite, freshly consolidated, fissured, and coursed by metallizing solutions arising from below, will have its shining feldspar and scintillating biotite changed to dull sericite — a mineral more stable under these conditions. This tells a story of hydrothermal alteration as convincingly as though the change had actually been observed. The feldspar is not stable in such environment, but the newly formed sericite is.

Some minerals respond to a change in environment by undergoing a molecular change. Chalcocite, for example, is orthorhombic if formed below 105° C and digenite ("isometric chalcocite") is formed above 80° C; upon cooling, the isometric form unmixes to chalcocite and covellite. Pyrite and marcasite also have the same composition but different crystallographic structure representing stabilities for different conditions. Also high- and low-quartz crystallize above or below their inversion point of 573° C. Similarly, sillimanite, andalusite, and kyanite are triplets of the same composition, and with increased temperature change to mullite. The fields of stability of many minerals are now known, thus broad generalizations can be made concerning weathering, hydrothermal alteration, metamorphism, and temperatures of formation of mineral deposits.

Geologic Thermometers

Minerals that yield information as to the temperatures of their formation, and of the enclosing deposits, are termed *geologic thermometers*. They are of scientific and practical importance for a proper understanding of the origin of mineral deposits and their classification. This information has been obtained by direct observation, by laboratory experiments, and from the repeated observations of association of certain minerals with other previously determined diagnostic minerals. Some of the methods by which geologic thermometry has been determined are given below.

Direct Measurements. The measurement of the temperatures of lavas, fumaroles, and hot springs yields maximum temperatures of formation for the minerals contained therein. Temperatures as high as 1185° C have been recorded for basic lava at Kilauea by Perret, Day, and Shepherd and up to 1140° C at Vesuvius by Perret. Washington estimated the more acidic lava of Santorini between 800° and 900° C. In general, the earliest minerals of the more basic rocks, according to Bowen, form in part above 870° C but principally between 870° and 600° C, decreasing with increasing silica content. Pyrogenic ore minerals, such as chromite, for example, form within the range of magma consolidation. Also, contact-metamorphic minerals would not ordinarily form at temperatures higher than those of the magmatic emanations that produce them. The gas temperatures of fumaroles likewise indicate maximum temperatures for fumarole minerals. A temperature of 645° C has been measured in the fumaroles of Katmai, and magnetite and other minerals have been deposited about their conduits. With waning fumarolic activity, lower temperatures occur.

The temperatures of hot springs extend downward from the boiling point of water, and maximum temperatures of formation can be assigned to opal, gypsum, cinnabar, stibnite, and many others that have been observed in spring deposits.

Melting Points. The melting points of minerals indicate maximum temperatures at which they can crystallize, or upper limits of their range of formation-temperature. The presence of other substances generally lowers the melting point. Examples of melting points are orthoclase at 1150° C, stibnite at 546° C, and bismuth at 271° C. Because bismuth occurs in the ores of Cobalt, Ontario, it means that those minerals contemporaneous with or later than the bismuth must have formed below 271° C.

Dissociation. Minerals that lose volatile constituents at certain temperatures may serve as geologic thermometers. However, the temperature of dissociation is increased by pressure. Most zeolites indicate low temperatures of formation because when heated they lose their water content, provided the pressure is not too high. Calcite dissociates under atmospheric pressure at 900° C, but according to Smyth and Adams, only 40 meters of rock pressures is required to prevent dissociation at 1100° C. However, silica available for combination with the calcium oxide lowers the dissociation temperature. Thus, the presence of intermixed calcite and quartz indicates a temperature below their combining point, giving consideration to pressure.

Similarly, pyrite dissociates into pyrrhotite and sulphur vapor, and, if the partial pressure of sulphur is 20 mm, pyrite cannot form above 615° C, and pyrrhotite is deposited.

Inversion Points. The most useful temperature indicators are inversion points since they are little affected by pressure and the changes for the most part are readily recognizable. Many inversion points are known within the temperature range prevalent in the formation of most mineral deposits. Silica is the most readily utilized. It is of widespread occurrence and has four stable crystalline modifications whose ranges of stability are known. The utilization of tridymite and cristobalite is attended with complications. Quartz is the most reliable; it forms only below 870° C and therefore designates a definite upper temperature limit to the deposits in which it occurs. At about 573° C, high quartz changes over or inverts to low quartz (and vice versa), with recognizably different symmetry. Thus, low quartz may have been formed originally below 573° C or it may originally have been high quartz that has inverted to the low form. The two forms can be distinguished. Wright and Larsen found that geode quartz and much vein and pegmatite quartz were formed below 573° and that the quartz of igneous rocks is high quartz. Much pegmatite quartz, formerly thought to be high quartz, has been found by Ingerson in many cases to be formed below 250° C.

A high-temperature inversion point is that between wollastonite and pseudo-wollastonite at 1125° C ±.[1] Pseudo-wollastonite, although known in slags, is not known in nature. Therefore, it is inferred that the presence of wollastonite indicates temperature conditions below 1125° C (or 1300° C). Similarly, isometric sphalerite inverts to hexagonal wurtzite (ZnS) at 1020° C, but the presence of 17 percent iron in the sphalerite lowers this temperature to 880° C. Wurtzite, however, under certain conditions forms and persists below the inversion points (e.g., under acid conditions) and does not necessarily indicate high-temperature formation.

N. W. Buerger has shown that orthorhombic chalcocite inverts at 105° C to a hexagonal form but reverts on cooling. He also showed that what was formerly called "isometric chalcocite" is digenite (Cu_9S_5), which above 78° C dissolves covellite. If digenite is found containing demonstrably unmixed covellite it may be assumed to have been formed above 78° C and to be hypogene. Much of the digenite of Kennecott, Alaska, shows such unmixing and is therefore of hypogene origin; the chalcocite of Bristol, Conn., was originally deposited as orthorhombic crystals and must have been formed as a low-temperature mineral below 105° C, but it has been shown, nevertheless, to be of hypogene origin.

[1] When $CaMg(SiO_3)_2$ is absent; if present in sufficient quantity, the inversion point is raised above 1300° C.

Argentite (isometric) and acanthite (orthorhombic) represent respectively the high- and low-temperature forms of Ag_2S, with an inversion point of 179° C. The external form of argentite crystals is isometric. Therefore, it follows that they were formed above 179° C, and the anomalous anisotropism commonly ascribed to argentite indicates that such argentite was originally isometric and later inverted to the orthorhombic acanthite. Unquestionably, much Ag_2S that is called argentite is really acanthite. Ramdohr has shown that cubanite I inverts to cubanite II at 235° C. (See Table 5 for other examples.)

Exsolution. Minerals that form natural solid solutions in each other, and at determined lower temperatures unmix to yield distinguishable mineral intergrowths, serve as geologic thermometers, indicating a temperature of formation above that at which exsolution takes place. For example, Schwartz has shown that chalcopyrite and bornite unmix at 475° C, cubanite and chalcopyrite at 450° C, cubanite and pentlandite at 450° C, and bornite and chalcocite at 175° to 225° C. Borchert has shown that chalcopyrrhotite exsolves below 255° C into chalcopyrite, cubanite, and pyrrhotite. Similarly, allemontite exsolves into arsenic and antimony at 200° to 250° C, and chalcocite and covellite at 75° C. Unmixing of magnetite and ilmenite, and hematite and ilmenite, have been demonstrated by Ramdohr, although his temperatures of exsolution, 600° C to 700° C and 675° C respectively, have been shown by Greig to be too high. Other examples are given in Table 5.

Recrystallization. This change is somewhat similar to inversion and exsolution but applies more specifically to native metals. Carpenter and Fisher have found that native copper undergoes a recognizable recrystallization at about 450° C. Microscopic examination reveals that most native copper has been formed below this temperature. Similarly, they found that native silver recrystallizes at about 200° C and that the silver ores of Cobalt, Ontario, were formed above this temperature, thus adding further evidence that these ores were of hypogene and not supergene origin.

Liquid Inclusions. Sorby long ago showed that liquid inclusions in cavities of crystals indicate the approximate temperature of formation of the crystals by the amount of contraction of the liquid, assuming that the liquid originally filled the cavity. W. H. Newhouse has applied this method to determine the temperature of formation of various sphalerites, which were heated on a microscope stage until the liquid filled the cavities, when the temperature was read. He found that the sphalerite of the Tri-State District, for example, had been formed at temperatures of 115° to 135° C, indicating thereby that such sphalerite had a hypogene origin. W. S. Twenhofel similarly determined that a fluorite crystal from New Mexico started growth at 202° C and continued to 150° C. Determinations by Earl Ingerson on pegmatite quartz disclosed surprisingly low temperatures of formation, none more than 250° C, even with pressure corrections. H. S. Scott developed a procedure for opaque minerals by which mineral powders are heated until the inclusions burst, thereby giving the upper temperature of formation.

The deposition of crystals of salts from such included solutions has also been utilized to indicate minimum temperatures of enclosure. Figures of around 500° C for sodium chloride and 300° C for potassium chloride have been obtained.

Changes in Physical Properties. Some minerals at certain temperatures undergo recognizable changes in certain of their physical properties. The pleochroic haloes in mica are destroyed at 480° C; smoky quartz and amethyst lose color between 240° C and 260° C; and the color disappears from fluorite at around 175° C.

Associated Minerals. The repeated association of certain minerals in deposits that contain one or more geologic thermometers enables them to be ranked roughly as high-, intermediate-, or low-temperature minerals. There are many such inferred rough geologic thermometers. One such mineral by itself may not be diagnostic, but an association of two or more such minerals may be as diagnostic as an established geologic thermometer. Among such semi-diagnostic minerals are the following common examples, arranged roughly according to temperature:

HIGH	INTERMEDIATE	LOW	
Magnetite	Chalcopyrite	Stibnite	Ruby silver
Specularite	Arsenopyrite	Realgar	Marcasite
Pyrrhotite	Galena	Cinnabar	Adularia
Tourmaline	Sphalerite	Tellurides	Chalcedony
Cassiterite	Tetrahedrite	Selenides	Rhodochrosite
Garnet		Argentite	Siderite
Pyroxene			
Amphibole			
Topaz			

General Considerations. The previous discussion shows that there are several classes of geologic thermometers, a few of which record, fairly accurately, specific temperature conditions of formation; some provide an upper or a lower temperature above or below which they do not form; others provide a range of temperature within which

TABLE 5

List of Geologic Thermometers

Temperature (°C)	Mineral	Nature	Remarks	Authority
1890	Olivine (forsterite)	Melting point		Bowen
1713	Cristobalite	Melting point		Wright-Larsen
1550	Anorthite	Melting point		Bowen
1470	Tridymite to cristobalite	Inversion point		Wright-Larsen
1391	Diopside	Melting point		Bowen
1248	Nepheline to carnegite	Inversion point		Bowen
1185	Basalt lava at Kilauea	Measured		Day-Shepherd
1157–1187	Pyrrhotite	Melting point		Bowen
1150±20	Orthoclase	Melts incongruently	Yields leucite	Morey-Bowen
1125±	Wollastonite to pseudowollastonite	Inversion point	$CaMg(SiO_3)_2$ in solution raises to 1300°	Osborn-Schairer
1120	Galena melts	Melting point		Jaeger-Van Klooster
1120	Albite melts	Melting point		Bowen
1045	PbS-ZnS eutectic melts	Melting point	PbS = 94%	Ramdohr
1020	Sphalerite to wurtzite	Inversion point	If 17% Fe T is 880°	Allen-Crenshaw-Merwin
1000	Sillimanite, kyanite, andalusite		Yields mullite	Posnjak-Bowen
955–1140	Orthorhombic pyroxene to monoclinic pyroxene	Inversion point	Upper limit of orth. pyr.	Bowen-Schairer
990	Aegirine	Melts	Incongruently	Bowen
900	Tremolite	Dissociation	Yields diopside	Posnjak-Bowen
900?	Calcite dissociates at 1 atm	Dissociation point	40 atm = 1100°	Smith-Adams
870	High-quartz to tridymite	Inversion point	Sluggish	Sosman
842	Argentite	Melting point		Edwards
830–900	Cobaltite inverts	Inversion point		Ramdohr
800	Garnet loses birefringence			Lindgren
800	Magnetite-spinel	Exsolution		Ramdohr
600–700	Carbon driven out of limestone		800° at 40 atm	Lindgren
700	Magnetite-ilmenite unmixing	Exsolution	Doubtful — too high	Ramdohr
685	Pyrite to pyrrhotite-sulphur (1 atm)	Dissociation		Bowen
675	Hematite-ilmenite unmixing	Exsolution	Doubtful — too high	Ramdohr
630	Galena-argentite eutectic melts	Melting point		Bowen
630	Antimony	Melting point		Edwards
609	Jamesonite melts	Melting point	Incongruently	Jaeger, Van Klooster
605	Pigment of limestone expelled			Erdmannsdörffer
603	α-Leucite to β-leucite	Inversion		Schairer
600	Chalcopyrite-pyrrhotite	Exsolution		Hewitt-Schwartz
580	Cinnabar	Sublimes		Edwards
573	Low quartz to high quartz	Inversion point	Enantiotropic	Wright-Larsen
550?	Sphalerite and chalcopyrite unmix	Exsolution?		Borchert
550	Maghemite-hematite	Recrystallization		Ramdohr
546	Stibnite melts	Melting point		Jaeger, Van Klooster

TABLE 5 — *Continued*

Temperature (°C)	Mineral	Nature	Remarks	Authority
530	Brucite	Stable up to		Gillingham
500	Stannite-chalcopyrite	Exsolution		Ahlfeld
500	Chalcopyrite-tetrahedrite	Exsolution		Edwards
500	Sphalerite in chalcopyrite	Exsolution		Borchert
500	Chalcopyrite	?		Ramdohr
485	Plagionite-stibnite eutectic	Melting point		Bowen
483	Pyrargyrite	Melting point		Edwards
481	Mica — pleochroic haloes destroyed			Science, 1935
475	Bornite-chalcopyrite unmixes	Exsolution	Formed above 475°	Schwartz
473	Pyrargyrite-proustite solid solution	Melting point	Minimum	Bastin
472	Calaverite melts	Melting point		Pelabon
210–465	Wollastonite	Recombination	Indefinite	Morey-Ingerson, Gillingham
450	Cubanite-pentlandite unmixes	Exsolution		Schwartz
450–425	Pyrrhotite-pentlandite unmixes	Exsolution		Newhouse
450	Chalcopyrite-cubanite unmixes	Exsolution	?	Schwartz
450	Marcasite to pyrite	Inversion	Monotropic	Allen et al.
450–300	Pyrr. cp. soln. to chalcopyrrhotite	Inversion		Borchert
400	Native copper recrystallizes above —	Recrystallization		Carpenter-Fisher
400	Microcline	Exsolution		E. Spencer
400	Adularia	Maximum temperature of formation		E. Spencer
400	Metacinnabar to cinnabar	Inversion		Ramdohr
400–500	Lime silicates, formation of			Lindgren
400–500	Sodium chloride, solubility curve	Fluid inclusions		Lindgren
360	Gold	Recrystallization above		Edwards
350–550	Pyrrhotite-chalcopyrite	Exsolution		Borchert
350–400	Siderite to hematite	Dissociation	Not proved	Schneiderhöhn
350–400	Chalcopyrite in sphalerite	Exsolution		Buerger
310–320	Orpiment and realgar melt	Melting point		Borgström
300	Smoky quartz — color disappears			Lindgren
300	Chalcocite-stromeyerite	Exsolution		Schwartz
275–350	Silver-dyscrasite	Exsolution		Carpenter-Fisher
275	Bornite-tetrahedrite	Exsolution		Edwards
271	Bismuth melts	Melting point	(289°-Ramdohr)	Johnson, Adams
268	Carnallite — inversion	Inversion point		Ramdohr
265	Boracite, orthorhombic to isometric	Inversion point	Prompt	Mugge
262	Ag-Bi eutectic	Melting point		Bowen
255–235	Chalcopyrrhotite to chalcopyrite and pyrrhotite	Unmixing?		Borchert
250	Ag_3Sb eutectic	Melting point		Ramdohr
240–260	Smoky quartz and amethyst lose color			Holden

TABLE 5 — *Continued*

Tempera-ture (°C)	Mineral	Nature	Remarks	Authority
235	Cubanite I to cubanite II	Inversion		Ramdohr
235	Chalcopyrrhotite to cubanite and pyrrhotite	Unmixing?		Borchert
215	Ilmenite inversion	Unmixing		Koenigsberger
210–330	Galena-matildite	Exsolution		Ramdohr
210	Matildite, low ortho. to high cubic	Inversion	Prompt	Ramdohr
200–250	Allemontite unmixes to As, Sb	Exsolution		Ramdohr
200	Silver, recrystallizes above —	Recrystallization		Carpenter
184	Calaverite	Inversion		Borchert
179	Acanthite-argentite inversion	Inversion point		Schneiderhöhn
175	Fluorite — color disappears			Lindgren
175–225	Bornite-chalcocite unmix	Exsolution	Slow	Schwartz
168	Carnallite	Melts incongruently		Van't Hoff-Meyerhoffer
150	Ag₂Te			Ramdohr
149	Hessite	Inversion point		
144–139	Pyrrhotite	Inversion point	Low hex. to high ortho.	Roberts
135	Sphalerite of Tri-State	Vacuoles		Newhouse
133	Ag₂S₂			Ramdohr
130	Goethite — unstable above —			Posnjak-Merwin
119	Sulphur melts	Melting point		Wigand
100	Stromeyerite inversion	Inversion point		Ramdohr
100±	Zeolites — max. limit of formation		Low pressure	Bowen
100+	Adularia		Lower limit	
93–105	Chalcocite, ortho. to hexagonal	Inversion point	Prompt	Zies et al.
75	Bismuth inversion	Inversion point		Ramdohr
70–75	Chalcocite-covellite unmix	Exsolution		Bateman
70	Bismuth, α to β	Inversion point		Edwards
−43	Aragonite to calcite	Inversion point	Monotropic	Bäckström

they may form; still others serve only as rough indicators. It should be understood that the temperature figures are subject to slight variation due to pressure or other factors. The presence of two or more of the less precise geologic thermometers in a mineral deposit may establish a fairly narrow range of temperature of formation for the deposit as a whole; one may fix a minimum temperature of formation, another a maximum.

Table 5 gives a list of some of the best-established geologic thermometers. A number of the higher-temperature ones, based particularly upon melting points, have been omitted, because that range of temperature is seldom evidenced in mineral deposits.

Selected References

General Reference 8. Chaps. 1, 2. *Relates to distribution and formation of minerals.*

Deposits of the Useful Minerals and Rocks. F. BEYSCHLAG, J. H. L. VOGT, and P. KRUSCH. Macmillan, London, 1914. Pp. 126–146. *Some features of mineral formation from magmas and solutions.*

Geologic Thermometry. N. L. BOWEN. In **Laboratory Investigation of Ores.** McGraw-Hill, New York, 1928. Chap. 10. *Data regarding several geologic thermometers.*

Dana's Textbook of Mineralogy. W. E. FORD, 4th Ed. John Wiley & Sons, New York, 1932. *A general textbook for references on mineralogy.* Also, **Dana's Manual of Mineralogy.** C. S. HURLBUT, 15th Ed. John Wiley & Sons, 1941. *A more modern treatment of selected minerals.*

Pneumatolytic and Hydrothermal Alteration and Synthesis of Silicates. G. W. MOREY and EARL INGERSON. Econ. Geol. 32: Supp. to No. 5, 1937. *Compilation and digests of research relating to synthesis and alteration of silicate minerals.*

Textures of the Ore Minerals, and Their Significance. A. B. EDWARDS, Aust. Inst. Min. and Met. Melbourne, Australia, 1947. Chaps. 1, 4, 6. *Deals with several geologic thermometers based upon recrystallization, solid solution and exsolution, inversion points, and melting points.*

Liquid Inclusions in Geologic Thermometry. EARL INGERSON. Am. Mineral. 32:375–388, 1947. *Discussion of determining temperatures of crystallization by liquid inclusions and pressure corrections; quartz determinations.*

Progress in the Study of Exsolution in Ore Minerals. G. M. SCHWARTZ. Econ. Geol. 37:345–364, 1942. *Criteria and examples of exsolution.*

The Pyrite Geo-thermometer. F. G. SMITH. Econ. Geol. 42:515–523, 1947. *Determination of temperature of formation of pyrite by measuring thermoelectric potential.*

Decrepitation Method Applied to Minerals with Fluid Inclusions. H. S. SCOTT. Econ. Geol. 43:637–654, 1948. *Opaque minerals with fluid inclusions are heated until they burst thus giving audible record of temperature of formation.*

CHAPTER 4

MAGMAS, ROCKS, AND MINERAL DEPOSITS

It is now generally considered that most metalliferous deposits and also many nonmetallic deposits result from igneous activity. Magmas are the parents, mineral deposits the offspring. Many are immediate descendants that originated soon after the consolidation of the parent magma; others, immature at birth, have been built up during one or more periods of growth by diverse processes to yield subsequent fully developed deposits. A placer gold deposit, for example, may have been originally a lean magmatic offspring that first suffered disintegration and was then later concentrated by moving waters into a workable deposit. Another intermediate stage may have intervened; the earliest surficial concentration may have collected the gold into a sedimentary formation, still too lean to be worked, and then later reworking may have further concentrated it into eagerly sought placer deposits. The progenitor of the gold, however, was some parental magma. The wily prospector, aware of this, has traced the errant placer gold upstream to its source in order to search for the " mother lode," as he terms it. Many profitable gold mines attest his success. Still other lean magmatic offspring have not matured until they have been exposed by erosion to the beneficent activity of the atmosphere, whose leaching action has removed the metals from undesired companions and concentrated them into cleaner, fatter, richer deposits of supergene origin, but of magmatic ancestry nevertheless.

There have also been widely noted associations of certain metals and minerals with certain kinds of igneous rocks, for example, chromite with peridotite or tin with granite. Here the kinship is even closer. The offspring have remained within the body of the parent or close to its edge. The repeated association cannot be coincidence; it must be genetic. Slightly more wayward offspring escape from the parent body but nestle close to its fringe, and their high-temperature birth clearly indicates a source from adjacent igneous masses. The most wayward offspring wandered far, perhaps many thousands of feet, but also contain minerals that point to moderate temperatures of formation and igneous affiliations. Thus, the valuable materials of most mineral deposits, whether hypogene or supergene, have directly or

indirectly come from magmas. It is, therefore, of the utmost importance for a proper understanding of the origin and nature of mineral
deposits to consider briefly some features of magmas and their crystallization.

Magmas

Magmas are masses of molten matter within the earth's crust from
which igneous rocks crystallize. Their composition, however, is not
the same as that of the rocks to which they give rise, because magmas
contain water and other small but important quantities of volatile
substances that escape before complete consolidation occurs. Strictly
speaking, they are high-temperature mutual solutions of silicates,
silica, metallic oxides, and always some dissolved volatile substances.
They obey the laws of ordinary chemical solutions. Their temperature, according to Larsen, ranges from 600° C for rhyolite magmas up
to 1,250° C for basaltic magmas.

Their composition is as variable as the vast array of rocks that
they yield; it ranges from extreme silicic to extreme basic. The
volatiles consist chiefly of predominating water, carbon dioxide, sulphur, chlorine, fluorine, and boron. Although minor in amount, the
volatiles play an important role in decreasing viscosity, in lowering
the melting point, in collecting and transporting metals, and in the
formation of mineral deposits. These constituents are largely expelled
when the magma consolidates.

Magmas are temporary features within the earth's crust. They
form in magma pockets or reservoirs, are forced upward, and then
consolidate. The melting is local; no continuous molten layer underlies the solid crust. The prevalent view is that at depth the rock
temperature, due to internal heat of the earth, is above its melting
point but the enormous pressure induced by the superincumbent rock
load prevents melting. However, should the pressure be relieved by
upward buckling, faulting, or removal of the overlying load by erosion, then melting would follow and magma would result. It should
be mentioned, however, that other views have been advanced to
account for the melting, such as accumulated heat of radioactivity,
or heat generated by crustal movements. Nevertheless, the general
occurrence of igneous intrusions and lava outpourings along great zones
of weakness in the earth's crust where crustal disturbances have occurred indicates that such places where relief of pressure occurs favor
magma formation and its movement.

Liquid magma, like any liquid under pressure, tends to move to the
place of least pressure. The direction of movement is thus dominantly

upward. Movement may be induced by the relief of pressure caused by forces already mentioned, by earth movements that tend to squeeze the magma out; or the magma may melt and assimilate the overlying cover rock and " eat " its way upward, aided by gas pressure within itself. In its upward movement, the magma may pry off blocks of roof rock that sink into the liquid, giving rise to *magmatic stoping;* it may wedge apart weak rocks, or it may be squirted into fractures or along bedding planes, forming laccoliths, dikes, or sills. It may be expelled to the surface, giving rise to volcanic extrusives; or it might solidify at depth, forming such large intrusives as stocks or batholiths.

Crystallization

Inasmuch as magmas are solutions and obey the laws of aqueous solutions, crystallization of the constituent minerals, which is dependent upon their solubility in the rest of the magma, will commence when the magma temperature falls below their individual saturation points. Crystallization of the minerals is not determined by their temperatures of fusion, although obviously no mineral can crystallize above its fusion point. Also, the solution of one constituent in another may give a lower melting point than that of either constituent. Consequently, a magma may remain fluid at a temperature below the melting point of all its constituents. Crystallization commences below this point and as the temperature drops complete crystallization eventually ensues, giving rise to individual minerals or isomorphous mixtures.

Order of Crystallization. The most insoluble substances crystallize first, and these in general are the *accessory minerals,* such as apatite, zircon, titanite, rutile, ilmenite, magnetite, and chromite. In general, the order of crystallization is one of decreasing basicity. Olivine and orthorhombic pyroxene are among the earliest *essential minerals* to crystallize, followed by clino-pyroxenes, basic plagioclase, hornblende, medium plagioclase, acid plagioclase, orthoclase, mica, and quartz. This is the normal succession, but there may be exceptions. Naturally, not all of these minerals occur in the one rock, but represent sequences from ultrabasic to silicic rocks.

With the subtraction of the more basic minerals from the magma, the residual magma in general becomes progressively more silicic. Granitic residual magmas are solutions rich in silica, alkalies, and water; some of this may be squeezed out into fissures to form pegmatites. With basic magmas, however, the residual magma may be rich in iron. Volatile substances or *mineralizers,* such as fluorine, boron, chlorine, along with tin, concentrate in the *mother liquors* of silicic rest-magmas and may become tapped off to form pegmatite dikes rich in rare minerals.

With progressing crystallization the final aqueous extracts gather the metals that originally were sparsely contained in the magma, along with the rare elements, the rare earths, and chlorine, boron, fluorine, hydrogen, sulphur, arsenic, and other substances.

These mother liquors become expelled upon final crystallization and constitute the magmatic solutions that give rise to most economic mineral deposits. Consequently, they are a part of the magma of particular interest to economic geologists.

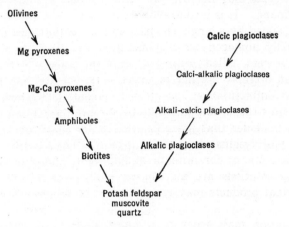

FIG. 4–1. Bowen's reaction series in subalkaline rocks.

Reaction Series. Crystallization is not in all cases a simple formation of minerals that persist as such. Bowen has shown, on the contrary, that certain minerals, once formed, may continue to react with the enclosing liquid magma, with the result that their composition is continually being modified and new minerals or solid solutions result. A definite sequence (Fig. 4–1) of reactions has been determined, which Bowen has called the *reaction series*. For example, olivine may crystallize early, but by reaction with the remaining liquid it changes to pyroxene. Similarly, any part or all of the sequence shown in Fig. 4–1 may result. The minerals on the left in Fig. 4–1 form a *discontinuous* reaction series, those on the right a *continuous* series.

The reaction principle indicates that olivine, if present, should precede pyroxene, which should precede hornblende, and that basic plagioclase should precede intermediate or acid plagioclase, thus explaining the " normal " order of crystallization. Also, according to the reaction principle, a greatly extended temperature range of consolidation will result. The reaction principle operates where fractional crystallization occurs and not in a eutectic liquid.

Differentiation

Observations have disclosed that neighboring volcanoes erupt unlike lavas and that even the same volcano yields several successive effusions of unlike lavas. Lassen Peak, for example, has erupted four lavas of quite distinct composition. Also, different rocks make up a single intrusive body; the border may differ from the center, and gradational types may lie between. Many other, similar relationships have been found. It is inconceivable that originally different magmas have been tapped by a single conduit, and the conclusion is irresistible that originally homogeneous magmas have split up into unlike fractions — a process called *magmatic differentiation*. Some geologists believe that original magma is universally basaltic in composition and that its differentiation has yielded magmas of different composition and these, in turn, may undergo still further differentiation. Many petrologists consider that primary granite, for example, is the silicic differentiate of original basaltic magma. Rocks of different kinds, bodies of iron ore, of chromite, or of sulphides, vapors and gases, and mineralizing solutions are all *magmatic products* of differentiation. The individual products may crystallize or be tapped off at different times.

Differentiation may occur in a single stage or in several stages. There may be deep-seated differentiation in the original magma chamber and consolidation of the differentiates *in situ*. Or some of the differentiates may be squeezed out before consolidation to form intrusive or extrusive bodies and then crystallize directly as already differentiated bodies, such as a stock of uniform monzonite. Or, such separate fractions may, before consolidation, undergo further differentiation in their last resting place and, upon consolidation, give rise to a series of distinct but related rocks such as are commonly observed in many small intrusives. Before complete consolidation, however, some of these later differentiates may again be tapped off to form satellitic dikes, or other minor intrusions. Last of all, the aqueous solutions that have accumulated with the residual fractions and which have gradually become more and more concentrated with the metals, are tapped off and supply the constituents of veins and other mineral bodies.

During the progress of differentiation certain metallic substances, such as oxides of iron, may be collected into fractions in which are concentrated these substances and consolidate either as part of the intrusion or as separately injected bodies to form *magmatic ore deposits*. Thus differentiation yields not only different rock types from a common magma but also magmatic mineral deposits of great economic importance and in part the mineralizing solutions that form the vast majority of metallic mineral deposits.

Although the fact of differentiation is definitely recognized, the process by which it has been achieved is but imperfectly understood. Several hypotheses have been advanced. Most geologists regard crystallization as the prime cause of magmatic differentiation, but it seems probable that other processes may also in part enter in, particularly in the case of local or unusual bodies of rock or minerals. Several of the theories advanced to explain differentiation involve crystallization; the most favored theories will be considered briefly.

Crystallization Differentiation. It has been shown that after a magma begins to crystallize certain crystals tend to form first. The liquid portion of the magma thus becomes depleted in the constituents that enter such early formed crystals. Normally, the early formed crystals are chiefly heavy minerals such as magnetite or olivine, and these will sink in the lighter liquid. Gradually, as crystallization proceeds, the early formed minerals will settle out and a liquid of different composition, or a residual magma, will remain above. Should the early formed minerals be lighter than the remaining liquid, as in the case of certain basic magmas, they would rise toward the top and a similar separation would occur. This is the theory that has been so ardently upheld by Bowen, who, by careful experiments, examination of field evidence, and able reasoning, has compelled assent of most geologists to the dominant part played by gravitational crystallization differentiation.

With further crystallization, still other minerals will similarly separate and collect, again depriving the residual liquid of those constituents and leaving a residual magma of different composition. Finally, the residual magma, enriched in the minerals latest to crystallize, and generally more silicic, will consolidate to form silicic rocks of quite different composition from the earlier crystallizations. In some basic magmas, however, iron becomes concentrated in the residual magma.

Where early formed crystals consist of magnetite or chromite, important deposits of these minerals may accumulate (see Chap. 5·1). Some investigators hold that the early formed crystals of magnetite that settle may be redissolved in the deeper, hotter portions of the magma chamber. Bowen and others, however, maintain that remelting of sunken magnetite and chromite crystals could not occur because (1) the temperature required would be too high, and (2) the walls of the magma chamber would melt before the crystals, and would become miscible with the oxides. Other features of crystallization differentiation related to the formation of deposits of magnetite and chromite are discussed in Chapter 5·1.

Filter Pressing. A variation of crystallization differentiation occurs where a magma has partly crystallized and consists of a mush of crystals with residual fluid in the interstices. If subjected to pressure by earth movements or other means, the crystals may be mashed together and the residual liquid squeezed out, thus forming fractions of distinct chemical composition. This process was recognized by Barrow, and Bowen has aptly termed it filtration or filter pressing. It has been advanced by Osborne to explain the formation of Adirondack magnetite deposits in anorthosite, a magnetite-rich fraction being the residual liquid filtered out after crystallization of calcic plagioclase.

An iron-rich residual magma might also drain out from among early formed silicates and gravitate to the bottom of the magma chamber, thus giving rise to a *liquid gravitative accumulation* (see Chap. 5·1).

Zonal Growth Differentiation. Another variation of crystallization differentiation is by the formation of zoned crystals, by which process the character of the magma undergoes continuous change, by the abstraction from it of the constituents of the different zonal growths. This process, however, operates only where the crystals remain in contact with the remaining magma.

Flowage Differentiation. A variant of the crystallization-differentiation process has been advanced by Balk to account for the anorthosite and magnetite deposits of the Adirondacks. Balk, following Bowen, assumes the formation of a mush of crystals of plagioclase, or of magnetite, in a liquid, but considers that the separation of crystals and liquid is brought about by flowage friction. As the magma with its enclosed crystals moves under pressure along relatively stationary walls, or particularly through restricted openings, the suspended crystals because of their greater surface friction are relatively retarded. This causes them to group into clusters and coalesce into larger masses that become compacted in the magma chamber, from which the liquid in the interstices is drained or crystallizes. Thus, crystals and residual liquid are separated, resulting in differentiation.

Diffusion and Convection. An earlier concept to explain differentiation involved the "Soret principle" and fractional crystallization.

In a magma chamber crystallization will start first near the coolest part, which is the border. Here, the magma becomes impoverished in the constituents of the first crystals to form and a composition difference ensues. This lowering in concentration causes a diffusion of particles of the early formed constituents from the hot interior toward the cooler border. The early border crystals will continue to grow from the material constantly fed to them, and a border phase will be formed whose composition differs from that of the remaining magma. It has been demonstrated, however, that diffusion is too slow to permit significant differentiation by this means. But this process may account for small, insignificant basic borders of intrusives, and, particularly, it may explain certain border deposits of magnetite or chromite, which although of economic importance, represent only a trivial part of the magma as a whole and require only a small amount of differentiation to produce them.

Convection has been called upon as a more rapid process than diffusion for feeding new supplies of magma to a cool border. This involves a slow movement of magma itself rather than of molecules or ions as in diffusion; it would be upward in the hotter, more gaseous, central portion and downward along the cooler border. New material would thus continually be supplied to the growing border crystals and larger marginal deposits might be built up than by diffusion alone.

Possibly also, mechanical flow of magma under movement toward its site of consolidation might continuously supply new materials to a locus of border crystallization and thus build up marginal segregations.

Diffusion, convection, and the suggested mechanical flow are merely variations of fractional crystallization; they are other methods by which the separation of crystals and liquid may be accomplished.

Liquid Immiscibility.[1] Vogt accounts for the origin of certain sulphide deposits by considering that dissolved sulphides with lowering temperature separate in part as immiscible droplets that settle out as a molten fraction, in the same manner that molten copper matte settles to the bottom of a copper furnace while the silicate melt (slag) floats at the top. Such a sulphide melt may consolidate where it settles or be forced into the chilled border of adjacent rocks and there consolidate. The process does not apply to the formation of different rock types, however, because molten silicates are miscible in each other. Limited immiscibility would be a simple explanation of the origin of magmatic magnetite deposits, but unfortunately magnetite likewise is held to be miscible in all proportions with molten silicates. This question, however, may not be a closed one.

[1] By immiscibility is meant the inability to mix, as oil and water.

Gaseous Transfer. Some geologists think differentiation is effected by the streaming of volatiles and gases through a magma reservoir from the lower toward the upper levels, where escape is possible. The gaseous bubbles are considered to carry away material when they escape, and to make selective transfer of materials from lower to higher levels.

Such bubbles, according to Fenner, rising through a magma " serve as collection chambers for other gases of very small vapor pressures. The latter are distilled into the bubble essentially as if it were a vacuum, no matter what pressure it may be under."

This idea applied to large-scale rock differentiation is not entirely clear and is as strongly disputed as it is advocated. However, as claimed by Fenner, gaseous transfer must be an effective means of collecting and transporting metals from the magma to the top of the magma chamber or into the overlying rocks, and thus plays an important role in the formation of mineral deposits.

Assimilation of Foreign Material. The fusion and assimilation of overlying country rock or of engulfed blocks would produce liquids that would alter the composition of the original magma. These secondary melts, along with admixed magma, would tend to rise toward the top of the magma chamber and there collect before diffusion would render them miscible. This hypothesis is not generally entertained to account for large-scale differentiation but is considered a probable explanation by many geologists for small intrusions of alkaline rocks. It has also been advanced, without much acceptance, to account for some types of magnetite deposits.

The Igneous Rocks

As a result of crystallization and differentiation, associations of minerals are formed that yield many kinds of igneous rocks; the varieties far exceed the number of the relatively few rock-making minerals that enter into their composition. The rock texture is determined chiefly by the rate of cooling and also by the amount of " mineralizers " present during consolidation. Slower cooling gives opportunity for large crystals to grow and results in coarse or *granular* texture. If cooling is rapid the crystals are small and the texture is *aphanitic;* if very rapid no crystallization occurs and *glass* forms, as in the case of some lavas. Interrupted crystallization may give a *porphyritic* texture consisting of large crystals (*phenocrysts*) in a finer-grained matrix. This texture was probably caused by initial crystallization of early formed minerals, followed by movement of the magma into another place where the remaining liquid underwent complete crystallization. Thus, different varieties of igneous rocks result from both compositional and textural differences. A granitic source, depending upon the rate of cooling, may yield granite, granite porphyry, rhyolite porphyry, rhyolite, obsidian, or tuff.

Many of the igneous rocks are in themselves economic mineral deposits, since they are used for building stones, road and building construction, and other purposes. Also, certain ore deposits of magnetite, chromite, and other minerals are merely unusual varieties of igneous rocks. A grouping of the common igneous rocks is shown in Table 6.

Pegmatites. With progressing crystallization the late residual liquid of a granite, for example, is made up principally of low-melting silicates and considerable water, along with other low-melting compounds and volatiles, and a relative concentration of many of the substances that enter into mineral deposits of igneous origin. In addition to water, the volatile substances consist of compounds of boron, fluorine, chlorine, sulphur, phosphorus, and other rarer elements. They aid crystallization by decreasing the viscosity of the magma and by lowering the freezing point of minerals. This is an aqueo-igneous stage — a transition between a strictly igneous stage and a hydrothermal stage, leaning more to the igneous — and is referred to as the *pegmatitic stage.* Withdrawals of the early liquid yield simple *pegmatite dikes* that are varieties of igneous rocks; later withdrawals of a more aqueous stage yield pegmatites, commonly characterized by druses, and may contain compounds of tungsten, tin, uranium, titanium, beryllium, phosphorus, chlorine, fluorine, and other elements, and some of the minerals of ore deposits; sulphides are uncommon. These pegmatites have often been referred to as pegmatite veins rather than dikes. Some, indeed, are valuable mineral deposits and are discussed later. There is probably a still later withdrawal that is the chief ore-forming fluid. Under deep-seated conditions there would be, according to Morey, a continuous passage from magmatic melt to hydrothermal solution.

As the pegmatite liquid cools and crystallizes, the earliest formed minerals will be those of latest formation in granite, such as potash, feldspar, quartz, and mica. This change enriches the residual pegmatite liquid in water, soda (its compounds being less volatile than potash compounds), lithium, and other substances; the potash feldspar becomes unstable in it and is replaced by albite. Similarly, wholesale replacements of the original pegmatite minerals take place by later hot aqueous solutions of the pegmatite and the next liquid stage. In consequence, many of the minerals of many pegmatite dikes are of

TABLE 6

COMMON IGNEOUS ROCKS

Texture	Mineral Composition					
	Chiefly Light-Colored Minerals				Chiefly Dark Minerals	Entirely Dark Minerals
	Orthoclase, Chief Feldspar		Plagioclase, Chief Feldspar			
	Quartz	No Quartz	Quartz	No Quartz	No Quartz	
Granular	Granite	Syenite	Granodiorite, Quartz monzonite	Diorite	Gabbro, Dolerite	Peridotite, Dunite
Granular, porphyritic	Granite porphyry	Syenite porphyry	Granodiorite, Quartz monzonite porphyry	Diorite porphyry	Gabbro porphyry	Pyroxenite, Hornblendite
Aphanitic, porphyritic	Rhyolite	Trachyte	Quartz latite	Andesite	Basalt porphyry	
Aphanitic	Felsite				Basalt	
Glassy	Obsidian, pitchstone, perlite, pumice				Basalt glass	
Fragmental	Volcanic tuff, breccia, agglomerate					

secondary replacement origin. Transitions have been noted from aqueo-igneous pegmatite dikes to pegmatitic quartz and then to hydro-thermal quartz veins carrying ore minerals. The pegmatites are important containers of many nonmetallic industrial minerals and some metallic ones such as tantalite, columbite, and cassiterite.

Gaseous Emanations and Liquids

Fenner has shown that a magma tends to separate into (1) immiscible sulphide liquids that settle and form magmatic sulphide deposits; (2) crystals of silicates and oxides that form igneous rocks or ore deposits; (3) gaseous emanations that escape; and (4) residual liquids. The first two have already been considered. The last two are of particular interest to economic geologists, as they are the collectors and transporters of most of the constituents of mineral deposits. By themselves, or mingled with meteoric waters, they constitute the *mineralizing solutions* or *hydrothermal solutions* to which most metallic mineral deposits owe their formation.

The gases and liquids were originally in solution in the magma, and as long as the magma remained liquid and the pressure unchanged, they remained in solution. But rise of the magma to levels of lesser pressures, or crystallization, brings about a separation of the gaseous phase, and bubbles rise to the highest part of the magma chamber. With progressing crystallization, the liquids also tend to collect with the residual magma. In traversing the magma, both gases and liquids collect metals and other minor constituents present, which become concentrated in the mother liquids. These in part may escape before consolidation of the magma is complete. With final consolidation the remaining fluids are expelled under great pressure (except as they may enter into the composition of the rock or become imprisoned in the rock minerals) and seek an exit toward the surface. A part actually reaches the surface in the form of fumaroles and hot springs, after undergoing change in the passage through the rocks. Studies of the constituents of fumaroles and hot springs have contributed much toward an understanding of the genesis of mineral deposits.

Gases and Vapors. Enormous quantities of gas and vapor escape from the magma into the atmosphere during volcanism; fumaroles likewise represent discharges from beneath the surface. Deep-seated intrusives must similarly yield large emissions, of which only the effects become visible. Of the observed emissions, water constitutes 80 to 90 percent of the volume; CO_2, H_2S, and S are abundant; and CO, H_2, N, Cl, F, B, and others occur. In addition, there are many volatile and other compounds, such as the chlorides of H, Na, K, Si, Ca,

NH_4, and the metals, fluorides, tellurides, arsenides, and sulphur dioxide.

The fumaroles of the Valley of Ten Thousand Smokes have given much information not only of the gases and compounds emitted but also of the vast quantities of metallic minerals dispersed in the porous tuffs. Such minerals as magnetite, specularite, molybdenite, pyrite, galena, sphalerite, covellite, cotunnite, and others were found around the fumarole conduits, where no liquid solution had at any time been active. Zies found that the magnetite contained also lead, copper, zinc, tin, molybdenum, nickel, cobalt, and manganese. Also there were deposited NaCl, KCl, S, Se, Te, and evidences of As, Bi, Th, and B_2O_3. Zies estimated that these fumaroles yield annually 1,250,000 tons of hydrochloric acid and 200,000 tons of hydrofluoric acid. These fumaroles demonstrate the ability of gaseous emanations to collect, transport, and deposit the metals.

High-temperature gaseous action is called *pneumatolytic*, which has been redefined by Fenner as " pertaining to processes or results effected by gases evolved from igneous magmas or entrained with gases of that origin." Fenner considers that the gaseous emanations are " the agents best adapted to effect primary separation of materials from the magma and transport it outward into the surrounding rocks;" and is inclined to assign to them the major role in such work. He also considers that " pneumatolytic processes are capable of effecting important metamorphic and metasomatic results in portions of the igneous mass and in the contact zone by adding or subtracting ingredients and bringing about recrystallization." Certain types of ore deposits, such as contact-metasomatic deposits, have long been considered to be due to gaseous action, and Fenner's careful studies suggest that its role in the genesis of mineral deposits is probably much greater than has hitherto been suspected. T. E. Gillingham's experiments prove that many nonvolatile substances can be dissolved and transported by high-temperature steam. When gases cool and condense, the metallic load is transferred from the gaseous to the more common aqueous environment.

It should be remembered, however, that not all magmas may contain sufficient volatiles to effect large-scale metallization and further that not all magmas may contain the metals in sufficient quantities to yield large deposits.

Residual Liquids. It is universally agreed that most mineral deposits of igneous affiliations result from hot waters of magmatic derivation. Furthermore, the hydrothermal solutions are considered to spring, directly or indirectly, from the magma in consequence of crystallization and differentiation. Some geologists believe that they left the magma chamber as liquids; others, that they are the condensation of gaseous emanations; still others believe that they have both derivations. That gaseous emissions take place is well known; and that they can and do collect, transport, and deposit minerals has been demonstrated; that they must condense into hydrothermal solutions is obvious. Gases and vapors are clearly one source of metallizing hydrothermal solutions. But, as is shown below, opinions are held that hydrothermal liquids need not have passed through a gaseous phase.

The nature of the residual liquids can be determined only by inference because they cannot be observed directly, as can gaseous emanations. Hot springs give little indication of the original liquid because their waters may have condensed from volatiles to start with; and undoubtedly they must have undergone great change in passing through the rocks. Also, magmatic spring waters have probably been changed by intermingling with meteoric waters.

There is no settled opinion as to the immediate source of the metallizing hydrothermal solutions. Bowen and other geologists believe that they are the residue of pegmatite injections left after the pegmatite constituents have crystallized. Most economic geologists, however, believe that they represent a stage beyond the pegmatitic liquids — the last stage of differentiation in the magma chamber — and that the residual liquors have been greatly modified by the crystallization of the last rock minerals. The withdrawal and crystallization of pegmatitic liquid would unquestionably leave a tenuous liquid residue outside of the magma chamber, from which quartz and other minerals may be precipitated. It is not accepted, however, that this residual liquid is the ore-forming fluid. The pegmatitic liquid must have been expelled from the magma chamber while some residual magma was still present and when quartz and alkalic feldspar were still forming in the parent magma. After withdrawal of the pegmatitic liquid there would still remain some residual liquid undergoing differentiation and enrichment in mineralizers and metals until the cessation of crystallization; this liquid most probably constitutes the chief ore-forming fluid. H. Neumann has shown that during crystallization, after the residual magma has become saturated with water, there will exist a solid rock phase, a molten magma phase saturated with water, and a fluid water phase which would constitute the hydrothermal solutions.

As to the mode of exit of the residual magmatic fluids, Day and Allen and Fenner consider that hydrothermal liquids would not escape from the magma as such but only as a gas phase. On the other hand, it has long been held that the mineralizing solutions escape from the magmas as liquids carrying with them the ingredients of mineral deposits, a view strongly supported by Lindgren and later upheld by C. S. Ross, Graton, and others.

C. S. Ross argues that in deposits of the Ducktown type certain minerals are precipitated in the order characteristic of a liquid but not of a gas phase. He cites as evidence that the carbonate sequence of calcium, magnesium, and iron, with calcium carried farthest, indicates magmatic segregation, transportation, and deposition by a liquid hydrothermal phase, whereas this order would be reversed in a gas phase in which calcium would have the lowest vapor pressure. This argument, however, does not carry conviction since materials carried in a gas phase would, after condensation of the gas phase, be transferred to a liquid phase. Further, Ingerson and Morey point out that Ross adduces evidence only against one assumed group of volatile compounds and this evidence does not apply to a compressed gaseous water-rich phase. Lindgren points out that the transportation of large amounts of difficultly volatile compounds of potassium, sodium, calcium, and magnesium indicates liquids, not gases, as the transporters. Graton argues at length that the mineralizing fluids left the magma as liquids at a late stage and that the solutions initially and continuously were alkaline. He considers a gas phase as improbable and impotent to transport the huge volumes of materials that make up a single deposit, such as the huge Cerro de Pasco sulphide body. The same volumetric argument can be applied to dilute liquid solutions.

Ingerson and Morey consider that there is a late orthomagmatic liquid stage devoid of critical phenomena and a coexisting vapor phase of different composition that would exhibit critical phenomena. Morey believes that the magma (or orthomagmatic-liquid phase) remains liquid until it reaches the hydrothermal or hot-spring stage, but he points out that a volatile stage would probably be present if the external pressure were less than the vapor pressure and the enclosing rock were somewhat porous.

Field evidence indicates that in many types of deposits there are two distinct phases of mineralization that might correspond to the ideas of Ingerson and Morey. The earlier phase is barren of metals; the later phase is the ore carrier. In some contact-metasomatic deposits, for example, the barren aureole of alteration that has universally preceded the metallization may have been caused by an escaping and penetrating vapor phase, and the introduction of late metals of lower temperature of formation may, in some cases, represent an invasion of the coexisting orthomagmatic residual liquid. At Butte,

Mont., in the central zone of intense mineralization (see Zoning, Chap. 6) the widespread and pervasive barren rock alteration that preceded the vein filling may possibly have been caused by a penetrating low-temperature vapor phase as an advancing surge of hydrothermal liquids from which the metals were deposited. It is not impossible also that the similar widespread pervasive, barren, rock alteration of "porphyry copper" deposits that has universally preceded the copper introduction might have been caused by an advancing surge of a vapor phase at relatively low temperatures.

It seems clear that magmas yield (1) volatiles that later condense to hydrothermal solutions, (2) probably hydrothermal liquids, and (3) earlier pegmatitic liquids that in turn yield pegmatites, volatiles, and hydrothermal solutions low in metals. The first two are the important ore carriers.

Magmas and Mineral Deposits

So far, the relationship between magmas and mineral deposits has been assumed. The evidence consists of the occurrence of igneous rocks which are themselves ores; of the relations of specific igneous rocks and metals; of the depositions from volcanoes, fumaroles, and hot springs; of mineral zoning about igneous centers; and of the character of mineralizing solutions as evidenced by the minerals deposited.

Igneous Rocks as Ores. Some igneous rocks are themselves bodies of ore, such as some magnetite deposits and bodies of chromite, ilmenite, corundum, or diamonds. Since these are merely igneous rocks, although of unusual composition, they indicate a direct relationship between magmas and mineral deposits.

Relationship Between Certain Metals and Specific Rocks. Field observations disclose an association of certain ore minerals with specific rocks that is so general and widespread as to preclude coincidence. This association establishes a definite relationship between the ore minerals and rocks, indicating thereby a magmatic source for both. For example, primary platinum deposits occur only in ultrabasic rocks, such as dunite or peridotite; diamonds in kimberlite; chromite in peridotite or serpentine; ilmenite in gabbro or anorthosite; titaniferous magnetite in gabbro or anorthosite; corundum in quartz-free rocks, such as nepheline syenite; nickeliferous sulphides in norite or gabbro; tin in silicic granites; and beryl in granite pegmatite. Other associations are given in Chap. 6 and Fig. 4–2. These associations are so universal as to render inescapable the conclusion that the rock and associated ore issued from the same magma.

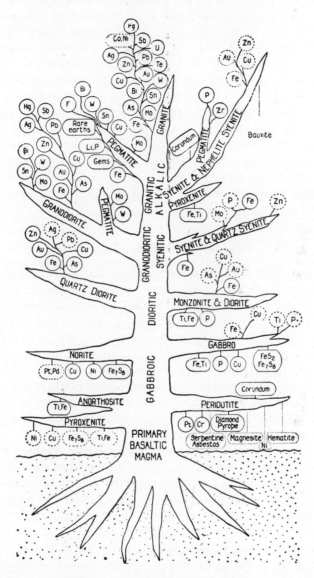

Fig. 4-2. Relation of igneous rocks to mineral deposits. (*Grout, modified by Buddington.*) The elements that branch upward occur in deposits formed from magmatic emanations; those hanging down are magmatic concentrations; fallen ones are weathered products. Full circles indicate major associations.

Relation to Volcanoes. Volcanism offers opportunity for direct observation of relationships between mineral deposits and magmas, although the deposits formed are relatively unimportant economically. Scientifically, however, they demonstrate a magmatic derivation for many minerals and metals. Deposits of native sulphur found in and about volcanic craters are well known. Among the sublimates from Vulcano have been noted sulphur, realgar, glauber salt, tellurium, cobalt, tin, zinc, lead, copper, bismuth, and phosphorus. From Vesuvius have been recorded specular hematite, tenorite, sodium, iron, copper, boracic acid, and other substances. Similarly, other compounds noted elsewhere clearly establish a magmatic source for these materials of mineral deposits.

Fumaroles. Even better than volcanoes, whose violence is not congenial for deposition of minerals, is the evidence of the sublimates and other depositions from fumaroles. Those from Katmai have already been mentioned. They demonstrate not only a magmatic source for such substances but also that the substances are carried in a gas phase. Fumaroles about Etna, Vesuvius, and Stromboli have yielded such chlorides as those of sodium, potassium, iron, and copper; sodium carbonate and sulphate; fluorides; sulphides; and compounds of cobalt, manganese, lead, zinc, tin, copper, bismuth, tellurium, arsenic, potassium, rubidium, caesium, thallium, and other compounds. Most of the elements that make up mineral deposits have been observed in such volcanic emanations.

Hot Springs. Hot springs associated with volcanism have long enticed the geologist and chemist because they contain many mineral substances and demonstrate that such hot waters dissolve, transport, and deposit them.

Hot-spring waters may be meteoric, magmatic, or both. The distinction is often difficult. Temperature alone is no criterion because hot springs may originate without volcanic heat, and even those in areas of volcanism may be meteoric waters heated by hot volcanic rocks or by small additions of volcanic gases. Probably few, if any, surface hot springs consist wholly of magmatic waters because of inescapable dilution by near-surface meteoric waters. The best indication of magmatic derivation is the presence of magmatic constituents, as shown by E. T. Allen for the Yellowstone Park waters (see below).

Lindgren divides "juvenile" springs into sodium chloride-silicate waters and sodium carbonate waters. Sodium carbonate springs are widely distributed, with many representatives in western North America, France, and Germany. The Carlsbad Springs of Germany contain As, Sb, B, Br, Co, Cu, Cr, F, Au, I, Li, Ni, P, Se, Sr, Sn, Ti, and

Zn. The chloride-silicate waters are characteristic of the three great geyser regions of the world. They all deposit abundant silica or geyserite. It was formerly thought, after the work of Arnold Hague, that the Yellowstone Park waters were entirely meteoric, having received their heat and mineral content from contact with the underlying hot rocks. The careful restudy by Allen and Fenner, however, based upon the waters, contained gases, and rock alteration (from bore holes), showed: (1) the presence in the waters of silica, soda, potash, calcium, cerium, borates, fluorides, arsenates, FeS_2, and Fe_2O_3; (2) that soda predominates over potash; (3) that silica and potash have been added to the underlying rock and soda abstracted; (4) that hydrogen sulphide and carbon dioxide are abundant; (5) that the dissolved substances include six elements of which there are only traces in the underlying rhyolite, namely, C, S, B, As, Cl, and F, and these are typical magmatic substances found in the gaseous state in fumaroles; (6) that superheated steam is present. This led to the conclusion that hot springs are caused by the heating of meteoric waters by condensation in them of magmatic vapors and gases that contributed some water and the magmatic constituents. About 10 to 15 percent of the water is considered magmatic. This new interpretation of the hot springs throws light on the origin of some types of mineral deposits.

The New Zealand geyser waters also carry gold, silver, and mercury. The Steamboat Springs of Nevada have deposited chalcedony, quartz, calcium carbonate, precious metals and sulphides of the base metals, antimony, mercury, arsenic, and traces of cobalt, nickel, and manganese. There are many other examples of this type of hot spring. In both types of springs the rare metals are prominent, and borates, arsenates, and fluorides are relatively abundant. Both types are clustered about recent volcanic centers, and the contents alone strongly indicate expiring magmatic activity. These waters may well be the type that produce shallow-seated veins. Indeed, the issuance of such thermal waters from mineral veins at Plombières, France, Comstock Lode, and Silver Cliff suggest a direct connection between the present thermal waters and the underlying ore deposits.

Most so-called juvenile spring waters are probably in large part meteoric, with minor magmatic water and contributed magmatic constituents. It is a mistake, however, to assume that these hot-spring waters are representative of most hydrothermal mineralizing solutions. They may be characteristic of solutions that form shallow-seated deposits within the zone of ground water, but most hypogene ore deposits are formed at depths far below the reach of ground water, where magmatic waters dominate.

Mineral Zoning. In many mining districts the ores are arranged zonally outward from an igneous center, with the higher-temperature and least soluble minerals nearest the source and low-temperature and most soluble minerals farther out. Temperature has been an important factor but not the only one in the arrangement. The zonal distribution points definitely to a hot center of mineralization, and minerals of igneous affiliation constitute a clear-cut indication of kinship with a magma. Details and examples of zonal distribution are given in Chapter 6.

Character of Hypogene Mineralizing Solutions. Hypogene mineralizing solutions, as has been stated previously, may be either gaseous or liquid, or both, and the liquids may be either aqueous or pegmatitic. The thermal springs that transport and deposit metals at the surface are alkaline, and some of the metals have been carried as alkaline sulphides. Consequently, it is inferred that the ore-forming fluids also were alkaline.

The residual pegmatitic liquid of a differentiating magma must be alkaline, according to Bowen, owing to the interaction between water and silicates. However, the gas phase that escapes from such liquid must be acid, since it will contain an excess of HCl, HF, H_2S, CO_2, H_3BO_4, H_2SO_4, and other volatile acids. Bowen states that there should be present in it H, O, Cl, S, F, B, K, Na, Fe, Ti, and Al, along with Sn, Pb, Cu, Zn, Ag, and other metals. Direct observations of the hot gases escaping from the underlying magma at Katmai corroborate this. They are acid and actually yield vast quantities of hydrochloric acid.

The hot acid solutions formed from the condensation of the gaseous emanations would readily attack most rock minerals except quartz and alkaline feldspars. Carbonate, and basic igneous and metamorphic rocks, would be particularly susceptible to attack and in turn sulphides and certain oxides of the heavy metals, which are most readily transported in acid solutions, would be deposited as the solutions gradually become alkaline by reaction with the rock minerals. As stated by Bowen, " At greater distances the [acid] solutions will be alkaline, for this is the ultimate fate of hot waters in contact with silicate and other rock minerals."

Therefore, the waters that eventually emerge as juvenile thermal springs in most cases must become alkaline whether or not they were so originally. The attack of acid solutions upon silicates would release an ample supply of free silica and this may be a source of much gangue quartz. Many geologists now hold that silica has been carried in colloidal solution and precipitated as a gel, which crystallizes to quartz. Lindgren, especially, has strongly advocated this and has advanced the idea that the metals may have been largely transported as colloidal solutions, in which colloidal silica may have helped keep the metals in a dispersed state. Gold and other metals can readily be brought into colloidal solution. The simplicity of transportation and particularly of deposition renders the idea intriguing. Certainly there is much to suggest it in the many colloform minerals present in ore deposits, particularly in those of low-temperature formation. Colloidal processes afford an alternative explanation for the origin of many gold quartz veins.

Summary

The magma is the direct source of most of the materials of hypogene mineral deposits. Through crystallization and differentiation some constituents collect as crystal aggregates or molten liquids, before final consolidation, to form magmatic oxide and sulphide deposits. Progressive differentiation of silicic magmas results in a residual alkaline liquid that becomes more and more enriched in the volatiles and other constituents, including metals, that formerly were dispersed throughout the magma. Before final consolidation some or all of this liquid may be tapped off to form pegmatites, whose crystallization leaves an aqueous residue that may form one class of mineralizing solutions. Continued crystallization of the magma produces an aqueous residue charged with volatiles, metals, and other constituents. When consolidation is complete, or nearly complete, this mobile, nonviscous, aqueous residue may be expelled as hydrothermal mineralizing fluids that later deposit their load to form various types of mineral deposits.

If the consolidating magma is rather close to the surface, under slight pressure, the volatiles will boil off from the residual aqueous liquid and may reach the surface directly as acid gases, forming fumaroles, such as those of Katmai. They transport and dissipate the metals that they formerly collected; no appreciable metallization results.

At less shallow depths, boiling may occur, giving rise to an acid distillate that rises through the fractured rocks. Reactions with the wall rocks will cause deposition of some ore minerals. These vapors condense to acid hydrothermal solutions that also react with the minerals of the wall rocks and deposit more of their load. The attack upon the silicates, particularly the basic ones, will render the solutions neutral and finally alkaline, in which condition, after mingling with meteoric waters, they may emerge as juvenile hot springs still carrying the most soluble part of their mineral load.

If the external pressure is greater than the vapor pressure of the residual liquid, no vapor phase will result, and the mother liquors will be alkaline liquids, which may be expelled as such to form rising hydrothermal metallizing solutions.

Under deep-seated conditions and high pressure a vapor phase may be absent, and the last residual alkaline liquids, which may or may not contain sufficient concentrations of metals to form economic metallizers, may be tapped off as hydrothermal solutions, or they may be excluded by final crystallization of the residual magma, or they may be absorbed in the formation of the last minerals to crystallize.

It is thus evident that there are two distinct schools of thought regarding the state of hydrothermal solutions. The first school contends that they leave the magma as gaseous emanations that later condense to hydrothermal liquids from which ores are deposited. Fumaroles attest that such gases do escape and do transport and deposit metals; contact-metasomatic deposits attest that they transport and deposit ores. This explanation offers a simple and likely method for the transportation of valuable constituents from the magma. Moreover, it accounts satisfactorily for the observed effects of solutions upon wall rocks and constitutes a ready means of supplying colloidal silica to hydrothermal solutions.

Members of the other school are proponents of hydrothermal solutions leaving the magma as attenuated alkaline liquids charged with metallic constituents, although the methods of ejection and uprising are left somewhat vague. The arguments mostly advanced are that hypogene deposits give evidence of having been deposited from hot waters, that juvenile thermal springs are alkaline, and that metals are soluble in alkaline solutions. But, as has been pointed out, some of these arguments lose force since most solutions would eventually become alkaline if in contact with alkaline rocks, and deposition from hot waters does not imply that such waters necessarily left the magma as liquids; they may have passed through a vapor phase.

It seems probable that hydrothermal solutions originated in both ways. If the rock pressure is less than the vapor pressure of the residual magmatic liquors, and the confining rock is permeable to the high penetrating power of a vapor under pressure, a vapor phase results, later to condense to a liquid phase. If the rock pressure is greater than the vapor pressure the residual aqueous concentrations will be in the state of, and emerge as, liquid hydrothermal mineralizing solutions.

Selected References

The Evolution of the Igneous Rocks. N. L. BOWEN. Princeton University Press, 1928. *A thoughtful discussion of differentiation and evolution of igneous rocks.*

Igneous Rocks and the Depths of the Earth. R. A. DALY. McGraw-Hill, New York, 1933. *Origin and differentiation of igneous rocks.*

General Reference 9. Chap. III, Part I. Pneumatolytic Processes in the Formation of Minerals and Ores, by C. N. FENNER. *A scholarly discussion of magma end phases and of the dominant role played by gaseous emanations in transporting minerals from a magma.* Part II. Magmatic Differentiation Briefly Told, by N. L. BOWEN. *A concise summary of differentiation and pegmatite evolution.* Part III. Differentiation as a Source of Vein and Ore-Forming Materials, by C. S. Ross. *A thesis that hydrothermal mineralizers leave the magma as liquids.* Part IV. Pegmatites, by W. T. SCHALLER. *Compositional characteristics and origin.*

General Reference 8. Chaps. 7, 8, and 10. *Mineral springs, underground water, and magmas.*

Nature of the Ore-Forming Fluid. L. C. GRATON. Econ. Geol. 35: Supp. to No. 2, 1940. *Lengthy discussion of origin, state, composition, and migration of fluids; ore fluids leave magma as alkaline liquids.*

Discussions of above article by E. INGERSON and G. W. MOREY, Econ. Geol. 35:772–785; C. N. FENNER, pp. 883–904, 1940; J. BICHAN, 36:212–217, 1941.

Geyser Basins and Igneous Emanations. E. T. ALLEN. Econ. Geol. 30:1–13, 1935. *Derivation of geyser waters from meteoric and magmatic contributions.*

Hot Springs of the Yellowstone National Park. E. T. ALLEN and A. L. DAY. Carnegie Institution, Washington, 1935. *A comprehensive discussion of spring waters and their origin.*

Rocks and Rock Minerals. L. V. PIRSSON and A. KNOPF. John Wiley & Sons, New York, 1947. Pp. 138–148. *Brief description of magmatic differentiation and of the origin of igneous rocks and of pegmatites.*

Eruptive Rocks. S. J. SHAND. John Wiley & Sons, New York, 1947. Chaps. 1, 3–7, 10–12. *Magmas — temperature and pressure conditions, fugitive constituents, crystallization; late-magmatic and post-magmatic reactions; the genesis of pegmatites; and the relation between eruptive rocks and ore deposits. Good bibliographies.*

The Solubility and Transfer of Silica and Other Nonvolatiles in Steam. T. E. GILLINGHAM. Econ. Geol. 43:241–272, 1948. *Experiments demonstrating the solution, transportation, and deposition of silica and other substances in high-temperature steam, and mineral synthesis.*

On Hydrothermal Differentiation. HEINRICH NEUMANN. Econ. Geol. 43:77–83, 1948. *Hydrothermal solutions are formed as a separate phase later than the pegmatitic stage.*

Transport and Deposition of Nonsulphide Vein Minerals. III. Phase Relations at the Pegmatitic Stage. F. GORDON SMITH. Econ. Geol. 43:535–546, 1948. *Experiments show that a granitic rest-magma during crystallization would separate into two liquid phases between 290° C and 550° C forming two immiscible residual solutions — one constituting the hydrothermal solutions.*

CHAPTER 5

PROCESSES OF FORMATION OF MINERAL DEPOSITS

General Remarks

The formation of mineral deposits is complex. There are many types of deposits, generally containing several ore and gangue minerals. No two are alike; they differ in mineralogy, texture, content, shape, size, and other features. They are formed by diverse processes, and more than one process may enter into the formation of an individual deposit. The modes of formation of minerals discussed in Chapter 3, although many, are but part of the larger processes that build up mineral deposits. Certain features of some processes of formation of mineral deposits were discussed in Chapter 4, but they relate only to deposits of direct igneous derivation; there are also many others.

Among the agencies that enter into the formation of mineral deposits, water plays a dominant role. It may be in the form of water vapor, hot magmatic water, cold meteoric water, or ocean, lake, or river water. Temperature, likewise, plays an important part, but many processes operate at surface temperature and pressure. Other agencies are magmas, gases, vapors, solids in solution, the atmosphere, organisms, and country rock.

The various processes that yield mineral deposits are considered in detail because a knowledge of them is fundamental to an understanding of mineral deposits. The deposits are susceptible of field classification according to the processes that operate to form them. Broad economic generalizations can be applied to each group; the mode of occurrence of one representative may indicate favorable places to search for similar unknown deposits; the characteristics of one well-explored representative can be applied to slightly explored members of the same group; conclusions as to probable size, continuity, and value may be drawn, based upon experience gained from other deposits formed by the same processes. For example, if a given deposit were formed by contact metasomatism of a favorable bed of limestone by a monzonite intrusive, then search for similar deposits is warranted at other places where that same bed is in contact with the monzonite.

Should ore be found, then it may be assumed to exhibit mineralogy, distribution, and continuity similar in general to the initial deposit. It was such realization of the controls of mineralization that led, for example, to the remarkably rapid development of great tonnages of copper ore in the Northern Rhodesian Copper Belt, where the geographic situation would ordinarily have been a deterrent. Past experience with deposits formed by one process can be applied quickly to similar new deposits under consideration.

Different processes may operate to produce distinct types of deposits of the same metal. Deposits of iron ore, for example, may be formed by magmatic, contact-metasomatic, replacement, sedimentary, and supergene processes. The distinction between them is vital, particularly from an economic standpoint. An iron deposit formed by sedimentary processes may be expected to exhibit the characteristics of a sedimentary rock — lateral continuity, thinness, general uniformity of composition — whereas one formed by hydrothermal replacement would be laterally restricted, of smaller tonnage, and probably of irregular shape. Early lack of such distinction greatly retarded the development of the Clinton sedimentary iron ores in the eastern United States.

Various Processes

The various processes that have given rise to mineral deposits are:

1. Magmatic concentration.
2. Sublimation.
3. Contact metasomatism.
4. Hydrothermal processes.
    Cavity filling.
    Replacement.
5. Sedimentation (exclusive of evaporation).
6. Evaporation.
7. Residual and mechanical concentration.
8. Oxidation and supergene enrichment.
9. Metamorphism.

Two or more of these processes may have combined, either simultaneously or at different times, to produce many of the mineral deposits. Replacement and cavity filling commonly operate together; fissures become filled and the walls are replaced. Sedimentation yields a bed of low-grade iron ore; weathering enriches it, and metamorphism alters it.

Deposits formed at the same time as the rocks that enclose them, such as magmatic or sedimentary iron deposits, are often referred to as *syngenetic;* those formed later than the rocks that enclose them are referred to as *epigenetic.* These words are not used here for classifying mineral deposits but merely as useful descriptive terms.

The various processes of formation of mineral deposits are considered in detail, starting with original magmatic materials, through those of lower-temperature conditions of formation, to secondary processes.

5·1   MAGMATIC CONCENTRATION

Certain accessory or uncommon constituents of magmas may become concentrated into bodies of sufficient size and richness to constitute valuable mineral deposits. Some of these are large and rich; some, such as chromite and platinum, are the sole source of these minerals; many, however, are of greater scientific than commercial interest. Representatives of magmatic concentration are many and widespread, but there are relatively few types; their mineralogy is simple, and the products yielded are not numerous. Although individual deposits are of great value, as a whole they are overshadowed in importance by those resulting from other processes.

Magmatic ore deposits are characterized by their close relationship with intermediate or deep-seated intrusive igneous rocks. Actually, they themselves are igneous rocks whose composition happens to be of particular value to man. They constitute either the whole igneous mass, a part of it, or form offset bodies. They are magmatic products that crystallize from magmas. They are also termed magmatic segregations, magmatic injections, or igneous syngenetic deposits.

Mode of Formation

Magmatic deposits result from simple crystallization, or from concentration by differentiation, of intrusive igneous masses. They were firmly established by Vogt as one of the major types of mineral deposits, at which time they were considered to be simply an ore facies of igneous rock, formed by earliest crystallization of the ore minerals. Subsequently, it was learned that in many of the deposits the ore minerals actually crystallized later than the rock minerals, and for such deposits the former simple concept had to be modified. It is now realized that there are several modes of formation of magmatic deposits and that they originate during different periods of magma crystallization. In some, the ore minerals crystallized early, in others late; in still others they remained as immiscible liquids until after crystallization of the host rock. Graton and McLaughlin proposed the term *pneumotectic* for deposits in which mineralizers played a part and *orthotectic* for the early straight magmatic ones, which is the same as Niggli's term *orthomagmatic*. A revised terminology is given in Table 7.

In the formation of magmatic concentrations the processes of differentiation discussed in Chapter 4 apply.

Much confusion exists in the literature on magmatic deposits, in part because of fuzzy terminology. "Magmatic segregation" has often been assumed to be the only kind of magmatic deposit; early and

TABLE 7

CLASSIFICATION OF MAGMATIC MINERAL DEPOSITS AND PROCESSES

TYPE	PROCESS	EXAMPLES
I. Early Magmatic:		
A. Dissemination.	Disseminated crystallization without concentration.	Diamond pipes; some corundum deposits.
B. Segregation.	Crystallization differentiation and accumulation.	Bushveld chromite.
C. Injection.	Differentiation and injection.	Kiruna, Sweden(?)
II. Late Magmatic:		
A. Residual liquid segregation.	Crystallization differentiation and residual-magma accumulation.	Bushveld titanomagnetite; Taberg(?); Iron Mt., Wyo.(?); Bushveld platinum.
B. Residual liquid injection.	Same, with filter pressing, and, or, injection.	Adirondack magnetite; Kiruna, Sweden(?); pegmatites.
C. Immiscible liquid segregation.	Immiscible liquid separation and accumulation.	Insizwa, S. Africa.
D. Immiscible liquid injection.	Same, with injection.	Vlackfontein, Bushveld, S. Africa.

late magmatic too often have not been distinguished. As a result, some authors have considered that certain deposits could not be magmatic simply because they were not segregations or because ore minerals later than rock silicates precluded separation by crystallization. Unnecessary conflict of interpretation arose and still exists.

Some of this confusion need not occur if it is realized that there can be more than one period of magmatic ore formation within the magmatic period and that there may be more than one process of differentiation by which magmatic deposits may result.

In an attempt to clarify the matter the author proposes the classification shown in Table 7 for magmatic deposits and for the processes that give rise to them, retaining familiar terminology.

EARLY MAGMATIC DEPOSITS

The early magmatic deposits resulted from straight magmatic processes — those that have been termed orthotectic and orthomagmatic. These deposits have been formed by (1) simple crystallization without concentration, (2) segregation of early formed crystals, and (3) injections of materials concentrated elsewhere by differentiation. The ore minerals have crystallized earlier than the rock silicates and in part presumably have separated by crystallization differentiation.

Dissemination. Simple crystallization of a deep-seated magma *in situ* will yield a granular igneous rock in which early formed crystals may be disseminated throughout it. If such crystals are valuable and abundant, the result is a magmatic mineral deposit. The whole rock mass, or a part of it, may constitute the deposit, and the individual crystals may or may not be phenocrysts.

The diamond pipes of South Africa are examples. The diamonds are sparsely disseminated throughout kimberlite rock and the whole kimberlite pipe is the mineral deposit (Chap. 24). The diamonds are phenocrysts, and in this example, as far as known, they crystallized in a former magma chamber and were transported with the enclosing magma and perhaps even continued to grow before final consolidation occurred in the present pipes. There was, however, no concentration of diamonds before the final magma consolidation. The disseminated corundum in nepheline syenite in Ontario is another example.

The resulting deposits of this class have the shape of the intrusive, which may be a dike, pipe, or small stocklike mass. Their size is large in comparison with most mineral deposits.

The same process may also yield noncommercial bodies of valuable minerals which, however, may undergo later concentration by mechanical or residual processes to yield economic placers or residual concentrations. Placer deposits of ilmenite, monazite, and some gemstones are examples.

Segregation. The term " segregation " has often been loosely used to designate magmatic deposits as contrasted with those formed by solutions or other means. In the sense here used, however, following the original meaning, it is restricted to concentrations of early crystallizing minerals in place and is to be distinguished from an *injection,* where the differentiate has undergone a change in position before consolidation. Early magmatic segregations are early concentrations of valuable constituents of the magma that have taken place as a result of gravitative crystallization differentiation. Consequently, they have often been referred to as " liquid magmatic." Such constituents as chromite may crystallize early and become segregated in bodies of sufficient size and richness to constitute economic deposits. The segregation may take place by the sinking of heavy early formed crystals to the lower part of the magma chamber, by marginal accumulation, or by constrictional flowage, as explained by Balk. Details of the methods of accumulation are discussed in Chapter 4.

The early demonstration of magmatic ore deposits at Taberg by Sjögren, Törnebohm, and Igelström, and their later establishment by J. H. L. Vogt in 1891–1893 led to the conclusion that all magmatic ore deposits were segregations formed as described above. Then it was found that in many deposits, particularly those of iron, the ore minerals were later than the silicates and, therefore, could not have originated by the separation of early formed crystals. This led the writer to arrange a restudy of certain Adirondack titaniferous magnetite deposits which, since the work of Kemp in 1897, had been considered type examples of segregations. The work was undertaken by F. F. Osborne, who found that those deposits are injections and not segregations and are either concordant or discordant with the primary structure of the host rock. He found that the ore minerals are definitely later than the silicates and, therefore, could not be segregations of early crystals.

The restudy of other magnetite deposits has led to similar conclusions, and today few representatives of early magmatic segregations, as defined above, remain. Perhaps some of these will prove to be injections instead of segregations; others may prove to be late magmatic deposits.

Chromite deposits have long been considered unimpeachable illustrations of early magmatic segregation. Field and microscopic evidence testify that in many places this is the case. However, Diller, J. T. Singewald, Jr., Sampson, C. S. Ross, and others cite evidence of some late magmatic or even hydrothermal chromite. Similarly, several magmatic ilmenite deposits, long considered typical early segregations, have been shown to be late.

Some nickeliferous sulphide deposits were also thought to be early magmatic segregations, but the sulphides are now known to be later than the silicates and, therefore, cannot be considered as early magmatic. They are either late magmatic or hydrothermal.

Few new deposits have been added to the class of early magmatic segregations. Many have been removed and few remain. It is probable that with further detailed work even more examples will be moved

FIG. 5·1–1. Chromite bands (black) in stratified anorthosite. Dwars River, Transvaal. Note convergence of some chromite bands and inclusion of anorthosite surrounded by chromite. (*Photo by author.*)

into other classes and that this group will hold only a small number of chromite deposits.

The mineral deposits formed by early magmatic segregation are generally lenticular and of relatively small size. Commonly, they are disconnected pod-shaped lenses, stringers, and bunches. Less commonly they form layers in the host rock. This is notably the case in the Bushveld Igneous Complex of South Africa, where stratiform bands of chromite (Fig. 5·1–1) of remarkably uniform thickness lie parallel to the pseudo-stratification of the enclosing basic igneous rocks and can be traced for miles (Fig. 5·1–2). Even more remarkable are other thin bands of chromite in the "Merensky Reef" horizon that contain economic quantities of platinum. The South African ge-

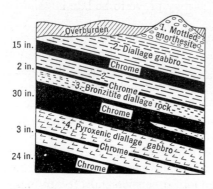

FIG. 5·1–2. Upper chromite horizon in Bushveld Complex, Pritchard's workings, Kroonendal, Transvaal. (*Kupferbürger et al., South Africa Geol. Survey.*)

ologists have considered these Bushveld chromite deposits to be early magmatic segregations formed by the settling of early crystals, but there is much evidence, some of which is shown in Fig. 5·1–1, to indicate that they are late magmatic.

Injections. Many magmatic deposits were formerly thought to belong to this group. The ore minerals have been concentrated presumably by crystallization differentiation (see Chap. 4). They are

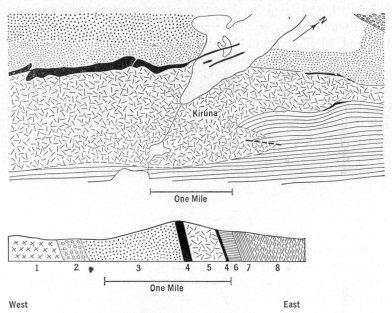

One Mile

West   East

Fig. 5·1–3. Plan and cross section of magnetite deposit, Kiruna, Sweden. 1, Greenstone; 2, conglomerate; 3, syenite porphyry; 4, magnetite; 5, quartz porphyry; 6, 7, 8, sediments. (*Geijer, Econ. Geol.*)

earlier than, or contemporaneous with, associated pyrogenic (igneous) minerals. They have not remained at the place of original accumulation, however, but have been injected into the host rock or into surrounding rocks. The structural relations of the deposits to the enclosing rocks show clearly that they have been injected; they transect enclosing rock structures, include rock fragments, or occur as dikes or other intrusive bodies in foreign rocks; they even metamorphose the wall rocks. Examples that have been tentatively assigned to this class are the titaniferous magnetite dike of Cumberland, R. I., and the largest deposit of magnetite in the world, at Kiruna, Sweden (Fig. 5·1–3). Some of the Bushveld chromite looks suspiciously like an injection.

As these large concentrations of magnetite and ilmenite are injections, the differentiation must have occurred elsewhere. Generally, magnetite and ilmenite have been considered to have crystallized before the silicates and settled in the magma to accumulate as solid crystals. But solid crystals cannot be injected. Consequently, J. H. L. Vogt and Lindgren have maintained that the crystals were remelted, and then injected. Bowen, however, has pointed out that there is insufficient heat to remelt magnetite, and further, to obtain remelting, the high temperature required would also remelt the surrounding silicates, and miscibility would ensue. Others have suggested that the iron oxides and silicates form immiscible liquids, but petrologic investigations do not support this. Some geologists have suggested that ultrabasic intrusives have been injected as a mush of early formed crystals lubricated by minor residual liquid. Possibly some of the magmatic ore deposits may have been similarly emplaced. The author has suggested elsewhere that in differentiation of basic igneous magmas rich in iron, titanium, volatiles, and fluxes, the ore deposits may form as a residual magma, and thus be available for injection. If so, there may be no representatives of the class of early magmatic injections.

LATE MAGMATIC DEPOSITS

Late magmatic deposits are bodies of pyrogenic minerals that have crystallized toward the close of the magmatic period. They are the consolidated parts of the igneous fractions that remained after the crystallization of the early formed rock silicates and in this respect differ from concentrations of early formed ore minerals. Consequently, the ore minerals of late magmatic deposits are later than the rock silicates and cut across them, embay them, and react with them, yielding *reaction rims*. These changes, termed *deuteric alterations,* occurred before the final consolidation of the igneous body and are to be distinguished from somewhat similar effects produced by later pneumatolytic or hydrothermal solutions. Some of these effects, however, may have been caused by solutions excluded from the consolidating end products.

The late magmatic deposits are dominantly associates of basic igneous rocks and have resulted from variations of crystallization differentiation, gravitative accumulation of a heavy residual liquid, liquid immiscibility, or other modes of differentiation. This group now includes most magmatic deposits. Primary pegmatites, which occur chiefly as injections but also as segregations, also should fall within this group, since they result from the crystallization of residual liquids.

Residual Liquid Segregation. It was shown in Chapter 4 that in a magma undergoing differentiation the residual magma generally becomes progressively richer in silica, alkalies, and water but that, in certain types of basic magmas, the residual magma may become enriched in iron and titanium. This is particularly true of magmas that yield anorthosite; the plagioclase crystallizes first, and iron oxides, with or without pyroxene, crystallize last. This residual liquid may drain out, or otherwise segregate from the crystal interstices, within the magma chamber and there crystallize, without further movement, as the last pyrogenic minerals. Under quiescent conditions, this liquid forms late magmatic segregations in the central portion of the magma chamber, or in bottom layers. Such bodies may be sufficiently large and rich to form valuable ore deposits. The deposits commonly parallel any primary igneous structure that may be present in the host rock. The host rocks are commonly anorthosite, norite, gabbro, or related rocks. Representative examples of late magmatic segregations would be the extensive titaniferous magnetite bands of the Bushveld Complex in South Africa, some platinum deposits, and perhaps Iron Mountain, Wyo., and Taberg, Sweden, with its 450 million tons of titaniferous iron ore.

Residual Liquid Injection. In this process the iron-rich, residual liquid accumulates as described above, and under conditions of concomitant disturbance such as often accompanies igneous intrusion may (1) be squirted out toward places of less pressure in the overlying consolidated portions of the mother rock or into the enclosing rocks, or (2) if liquid accumulation had not occurred, interstitial iron-rich residual liquid might be filter pressed out, to form late magmatic injections. The injected nature of such deposits distinguishes them from segregations. The resulting ore bodies may be irregularly shaped masses, sills, or dikes, and generally transect the primary structure of the host rocks or cross cut the invaded rocks. They exhibit the intrusive relationships of normal igneous intrusives, and the ore minerals surround, cut across, corrode, and react with the earlier formed magmatic silicates. These reactions are strictly magmatic, however; they take place before final consolidation. If the injected iron-rich fluids are rich in volatiles some pneumatolytic action may occur.

Where filter pressing has occurred the early formed silicates that constituted the crystal mush from which the residual liquid was ejected show evidence of the squeezing that they have undergone during the filter-pressing movements. The silicate crystals are mashed, their corners are broken off, and they are fractured, producing what is termed a *sideronitic* texture.

Many deposits of magnetite and ilmenite, which were formerly called segregations, are now known to be injections of this class. Examples are the titaniferous magnetite deposits associated with basic igneous rocks in the Adirondack region of New York, and many other iron oxide deposits in which the ore minerals have crystallized later than the silicates. Some geologists consider that the titaniferous magnetite deposits of Iron Mountain, Wyo., are late magmatic injections. The huge magnetite deposit of Kiruna, Sweden, is definitely an injection, and it may prove to have accumulated as a late differentiate at depth and to be a late magmatic injection.

S. J. Shand points out in connection with magmatic ore injections that, in most deep-seated magmas, ferrous oxide is more abundant than ferric oxide and that crystallization of a gabbroic magma leaves a residual solution enriched in ferrous iron. This, however, is not in the form of magnetite but as *ferrous hydroxide hydrosol*, which upon dehydration undergoes self-oxidation, yielding magnetite and hydrogen. This intriguing suggestion is advanced to account for injected bodies of magnetite and titanium oxides. There seems no question as to the validity of the chemical reaction that yields magnetite from ferrous hydroxide hydrosol, but there is a question whether ferrous hydroxide hydrosol would or could exist in a deep-seated magma, as has been pointed out by C. H. White.

Immiscible Liquid Segregation. Although metallic oxides apparently cannot form immiscible solutions in silicate magmas, Vogt has shown that iron-nickel-copper sulphides are soluble up to 6 or 7 percent in basic magmas and that upon cooling they may in part separate out as immiscible drops, which accumulate at the bottom of the magma chamber to form liquid sulphide segregations. The sulphides remain liquid until after the silicates crystallize, hence they penetrate and corrode the silicates and crystallize around them. Such sulphides are the latest pyrogenic minerals to crystallize, and because they penetrate and corrode the earlier silicates they give rise to relationships that have often been interpreted as hydrothermal.

The accumulated sulphide need not necessarily be a pure sulphide melt. Rather it must be thought of as an enrichment in sulphide of the lowest part of the magma, which upon consolidation gives rise to a basic igneous rock with sulphide content up to 10 or 20 percent. This, however, may be quite sufficient to constitute a valuable ore deposit, and most sulphide segregations are of this type. Rarely, an almost pure sulphide melt may occur, giving rise to bodies of massive sulphide.

The deposits formed in this manner consist of pyrrhotite-chalcopyrite-pentlandite nickel-copper ores, with accompanying platinum, gold, silver, and other elements. They are confined to basic igneous rocks of the gabbro family that generally display pronounced differentiation. The deposits occur commonly as disconnected bodies along the lower margins of the differentiated intrusives, notably where there are depressions in the floor. Their size is proportionate to that of the mother intrusive. The mineralogy is simple and monotonously uniform. Pyrrhotite, chalcopyrite, and pentlandite are the chief metallic minerals; sperrylite and lead and zinc sulphides are rare; the platinum metals and gold, tellurium, and selenium are generally present. Vogt noted a relatively uniform ratio of nickel to copper.

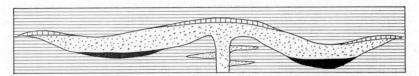

Fig. 5·1–4. Generalized diagram of Insizwa type intrusion. Black, differentiated basal zone carrying nickel-copper sulphides. (*After D. L. Scholtz, Geol. Soc. South Africa, 1936.*)

Examples of this class of segregation are: the nickel-copper sulphide deposits of the Insizwa type, South Africa (Fig. 5·1–4), the nickeliferous sulphide deposits of the Bushveld, South Africa, and of Norway, and possibly some of the early mined, small, marginal deposits of Sudbury, Ontario.

Immiscible Liquid Injection. If the sulphide-rich fraction accumulated in the manner described above should be subjected to disturbance before consolidation, it might be squirted out toward places of less pressure, such as sheared or brecciated areas along the margins of the consolidated mother rock or in the enclosing rocks. There it will consolidate to form immiscible liquid injections. Such deposits give unmistakable evidence of late magmatic age. They intrude older rocks and enclose brecciated fragments of host and foreign rock. They exhibit the intrusive relations of dikes. The ore minerals penetrate, corrode, alter, and even replace the silicates. The exudations of consolidation produce some hydrothermal alteration of the surrounding silicates. The deposits are irregular or dikelike in form. If the residual fraction is rich in volatiles the resulting deposits might display transitions into hydrothermal types.

Examples of this class of deposit are the Vlackfontein mine of South Africa and probably some of the Norway nickel deposits. Perhaps also the Frood deposit of Sudbury may come to be considered as belonging to this class.   If there are such deposits as pyritic sulphide injections, of which the author is dubious, they also would belong in this class.

ECONOMIC PRODUCTS

The products of magmatic processes may be divided into native metals, oxides, sulphides, and gemstones. The various deposits and important minerals are listed below.

DEPOSIT	MINERALS	DEPOSIT	MINERALS
NATIVE METALS		SULPHIDES	
Platinum	Platinum, and with chromite or Ni-Cu-Co sulphides	Nickel-copper	Chalcopyrite, pentlandite, polydymite, sperrylite, with pyrrhotite, and precious metals
Platinum metals	Osmium, iridium, palladium, and others	Nickel	Pentlandite and polydymite, with pyrrhotite
Gold, silver	By-product metals		
Iron-nickel	Native metals (rare)		
OXIDES		Copper	Bornite and chalcopyrite, with pyrite (rare)
Iron	Magnetite, some hematite	Molybdenum	Molybdenite (rare)
Iron-titanium	Titaniferous magnetite, hematite	GEMSTONES	
Titanium	Ilmenite	Diamond	Diamond
Chromium	Chromite	Garnet	Pyrope, almandite
Tungsten	Wolframite	Peridot	Peridot
Corundum	Corundum		

One or more of the products listed may be associated in the same deposit. For example, titanium is associated with iron oxide, and nickel and copper are intimate associates. Platinum and the platinum metals occur together, and these in turn may be associated with chromite, as in South Africa, or with nickel-copper, as in the Merensky Reef of the Bushveld in South Africa. The nickel-copper deposits of Sudbury, Ontario, yield most of the platinum of the world, a feature of possible significance in considering their origin. All commercial chromium is of magmatic origin, as are also much magnetic iron ore and most titanium ores. The amount of copper of magmatic origin other than that associated with some nickel is small; this is true also of molybdenum and tungsten; magmatic corundum is largely restricted to that associated with certain nepheline syenites.

Occurrences and examples of deposits are given under individual metals and nonmetallic minerals in Parts II and III.

Associations of Rocks and Mineral Products

Definite associations exist between specific magmatic ores and certain kinds of rocks. Platinum occurs only with basic to ultrabasic rock, such as the varieties of norite, peridotite, or their alteration products. Chromite, with rare exceptions, is found only in peridotite, anorthosite, and similar basic rocks. Titaniferous magnetite and ilmenite are mothered by gabbro and anorthosite, and magmatic magnetite deposits occur with syenite. Nickel-copper deposits are associated universally with norite, and magmatic corundum with nepheline syenite. The diamond occurs in commercial quantities only in kimberlite, a variety of peridotite. It is thus seen that the deep-seated basic rocks are predominantly the associates of nearly all important magmatic mineral deposits, which may indicate a genetic relationship during the early history of the basic rocks.

Selected Readings on Magmatic Concentration

Magmatic Segregations. J. T. Singewald, Jr. In **General Reference 9.** Chap. 11, Part I. *General discussion of United States deposits.*

General Reference 8. Chap. 31. *Theoretical discussion and description of some deposits.*

Magmas and Igneous Ore Deposits. J. H. L. Vogt. Econ. Geol. 21:207–233, 309–332, 469–497, 1926. *Fundamental discussion of magmatic processes.*

A Study of Magmatic Sulphide Ores. C. F. Tolman and A. F. Rogers. Stanford Univ. Press, 1916. *For development of ideas.*

Certain Magmatic Titaniferous Ores and Their Origin. F. F. Osborne. Econ. Geol. 23:724–761, 895–922, 1928. *Filter pressing applied to titaniferous magnetites.*

Anorthosite Area of the Laramie Mountains, Wyoming. K. S. Fowler. Amer. Jour. Sci. 5–19:305–315, 373–403, 1930. *Segregation of titaniferous magnetites.*

Studies of Domestic Chromite Deposits. J. S. Diller. A. I. M. E. Trans. 63: 105–149, 1920. *General occurrences.*

May Chromite Crystallize Late? E. Sampson. Econ. Geol. 24:632–641, 1929; 26: 662–669, 833–839, 1931; 27:113–144, 1932, and accompanying discussions. *Occurrence and thesis of late origin for magmatic chromite.*

The Iron Ores of the Kiruna Type. Per Geijer. Swedish Geol. Surv. No. 367, Stockholm, 1931. *Occurrence and origin.*

Magmatic Nickel Deposits of the Bushveld Igneous Complex. P. A. Wagner. South Africa Geol. Surv. Mem. 21, 1924.

The Bushveld Igneous Complex, Transvaal. A. L. Hall. South Africa Geol. Surv. Mem. 28, 1932. *Detailed occurrence and origin of this great magmatic body and its included magnetite, chromite, nickel, and platinum.*

Ore Deposits of Magmatic Origin. P. Niggli. Trans. by H. C. Boydell. Thos. Murby and Co., London, 1921. *Fundamental theoretical discussion.*

Magmas and Ores. Alan M. Bateman. Econ. Geol. 37:1–15, 1942. *Theoretical discussion of various origins of magmatic ore deposits.*

The Genesis of Intrusive Magnetite and Related Ores. S. J. SHAND. Econ. Geol. 42:634–636, 1947. *Residual solutions from gabbroic magmas do not contain magnetite but ferrous hydroxide hydrosol, which by self-oxidation changes to magnetite.* Discussion of above title by C. H. WHITE. Econ. Geol. 43:232, 1948.

Critical Zone of the Bushveld Igneous Complex. G. S. J. KUSCHKE. Geol. Soc. S. Africa Trans. 42:57–82, 1939. *Study of petrographic variations.*

The Chromite Deposits of the Bushveld Igneous Complex. W. KUPFERBÜRGER and B. V. LOMBAARD. South Africa Dept. Mines, Geol. Ser. B No. 10, 48 pp., 1937. *Detailed geology of all chrome deposits of South Africa.*

See also references with Chapter 4, and under Iron, Nickel, Chromium, Platinum in Part II.

5·2 SUBLIMATION

Sublimation, a very minor process in the formation of mineral deposits, is included here for the sake of completeness of processes. Sublimation applies only to compounds that are volatilized and subsequently redeposited from vapor at lower temperature or pressure. It involves direct transition from the solid to the gaseous state, or vice versa, without passing through the liquid state which usually intervenes between the two. It does not include minerals formed by reactions of gases and vapors. Many compounds cannot be sublimed in the presence of oxygen. The process is associated with volcanism and fumaroles.

There are many sublimates deposited around volcanoes and fumaroles but they are seldom in sufficient abundance to make workable mineral deposits. Sulphur supposedly of this origin has been mined in Italy, Japan, and elsewhere, and sodium chloride or common salt has been extracted locally around volcanoes. Common, but noncommercial sublimates are the chlorides of iron, copper, zinc, oxides of iron and copper, boracic acid, and various salts of the alkali metals and ammonium. Most of these, however, are quickly blown away or washed away.

5·3 CONTACT METASOMATISM

Following magmatic stage processes, next to be considered are the contact effects of escaping high-temperature gaseous emanations during or shortly after the consolidation of intrusive magmas. These effects have been divided by Barrell into two types: (1) the effects of heat, without appreciable accessions, giving rise to *contact metamorphism;* and (2) the effects of heat combined with accessions from the magma chamber, giving rise to *contact metasomatism.* The two are to be sharply distinguished since contact metamorphism does not give rise to mineral deposits except a few rare cases of nonmetallic deposits such as sillimanite, and contact metasomatism may give rise to valuable and distinctive mineral deposits.

Contact metamorphism manifests itself by (1) *endogene* or internal effects upon the margins of the intrusive body itself, and by (2) *exogene* or external effects upon the rocks invaded by the igneous mass. The endogene effects consist chiefly of textural and mineral changes in the border zone; pegmatite minerals such as tourmaline, beryl, or garnets may be present.

The exogene effects of large intrusive bodies are generally pronounced. They consist of a baking or hardening of the surrounding rocks and generally their thorough transformation. Old minerals are broken up, and their ions recombine to form new minerals that are stable under the changed conditions. For example, original minerals AB and CD may recombine to AC and BD. In an impure limestone consisting of carbonates of calcium, magnesium, and iron and quartz and clay, the calcium oxide and quartz may combine to form wollastonite; dolomite, quartz, and water form tremolite, or, by addition of iron, actinolite; or calcite, clay, and quartz form grossularite garnet. These, and other changes, yield simple recrystallized rocks such as quartzite from sandstone or marbles from limestone or dolomite, and more highly metamorphosed rocks, such as hornfels, from shale or slate and complex silicate rocks from impure limestones. In such alteration the chemical composition of the rock undergoes little change, except for the loss of some volatiles and the addition of magmatic volatiles such as boron or fluorine. The alteration is most intense nearest the intrusive and less intense farther away. The result of these changes is to form around the intrusive a contact-metamorphic *aureole* that varies in shape and size according to the size and shape of the intrusive and the character and structure of the invaded rocks.

Contact metasomatism differs from contact metamorphism in that important accessions from the magma are involved, which by metasomatic reaction with the contact rocks form new minerals under conditions of high temperature and pressure. To the effects of the heat changes of contact metamorphism are added those of metasomatism, by which the replacing minerals may be composed in part or in whole of constituents added from the magma. The resulting mineralogy is thus more varied and complex than that by heat metamorphism alone. If the magmatic emanations are highly charged with the constituents of mineral deposits, *contact-metasomatic deposits* result, particularly in the favorable environment of calcareous rocks. Such deposits have been frequently designated as "contact-metamorphic deposits"; but, as pointed out by Lindgren, they are not metamorphic — they are metasomatic, and their substance has largely been derived from the magma and not the invaded rock. Consequently, the use of the term metasomatic instead of metamorphic will eliminate some of the confusion that has existed. Lindgren proposed the term *pyrometasomatic deposits* and defined them as " formed by metasomatic changes in rocks, principally in limestone, at or near intrusive contacts under influence of magmatic emanations." Therefore, contact-metasomatic deposits and pyrometasomatic deposits are essentially the same, except that Lindgren includes under pyrometasomatic numerous deposits remote from intrusive contacts, many of which (e.g., Ducktown, Tenn.) are considered by other investigators to be hypothermal replacement deposits. The term contact-metasomatic deposits is favored, therefore, because it connotes the much-described relationship with intrusive contacts and a class of highly characteristic deposits recognized by their distinctive assemblage of high-temperature minerals so widely referred to in the older literature as contact-metamorphic deposits.

The process was first realized by Von Cotta in 1865, from a study of the deposits at Banat, Hungary, and was definitely recognized as a distinct ore-forming process by Von Groddeck in 1879. By 1894, Vogt established it firmly as the important ore-forming process in the formation of the Kristiania ores. The main concept was familiar to Lacroix by 1900. Lindgren, the first to recognize the type in the United States, in 1899, at Seven Devils, Idaho, redefined it. Since the nineties the work of Kemp, Blake, Weed, Barrell, Lindgren, Goldschmidt, and others led to the realization of the importance of this type of deposit and helped unravel the intricacies of the mode of formation. Barrell, in 1907, distinguished contact metamorphism from contact metasomatism at Marysville, Mont., and in 1911 Goldschmidt, in his study of the Kristiania region, Norway, separated contact metamorphism into (1) normal heat metamorphism, involving only recrystallization and recombination of original rock constituents; and (2) pneumatolytic metamorphism, involving in addition gaseous transfer of materials from the magma.

Process and Effects

General. The high-temperature effects of deep-seated magma intrusions upon the invaded rocks result from the heat transferred by the magmatic emanations and to a minor extent by conduction, which is very slow. Entire beds of carbonate rocks may be changed to complex silicate rocks called *tactite,* or to *skarn* with addition of iron oxides, or to garnet rock. Huge additions and subtractions of materials are involved. The emanations may carry the constituents of mineral deposits that replace the invaded rock to form metallic and nonmetallic mineral deposits distributed spasmodically within the contact aureole. Not all magma intrusions yield mineral deposits. For their formation, certain types of magma are essential; the magma must contain the ingredients of mineral deposits; it must be intruded at depths not too shallow; and it must contact reactive rocks. The changes in the intrusive itself are minor.

Temperatures. The temperature at the immediate contact must have been that of the intruding magma, which in siliceous magmas ranges from 500° to 1,100° C. Outward from the magma contact the temperature would gradually decline, owing to the cooling effect of the invaded rocks. Also, there would be a gradual lowering of the temperature during the slow cooling of the intrusive. Consequently, there may have been a range of several hundred degrees in the temperature of formation of the contact minerals. The presence of wollastonite indicates that the temperature did not exceed 1,125° C (Chap. 3). Wollastonite begins to form between 660° and 800° C. Similarly, the presence of andalusite indicates a temperature of less than 1,000° C, since above this temperature it changes to mullite. Merwin showed that contact-metamorphic andradite commonly exhibits anomalous birefringence and that heating to 800° C makes it isotropic, hence the temperature must have been below 800° C. High quartz has been identified in a contact-metamorphic zone, proving a temperature of formation above 573° C. Also, low quartz has been identified, but this may have been formed among the last minerals after the temperature had dropped below 573° C. Lindgren and Whitehead concluded that the contact-metasomatic deposit at Zimapan, Mexico, had been formed between 400° and 500° C. Thus, contact metasomatism takes place at temperatures from 400° to 800° C, perhaps even higher.

Recrystallization, Recombination and Accessions. Recrystallization and recombination of rock minerals take place in the alteration halo. Limestone and dolomite are recrystallized to marble; carbonaceous impurities may form graphite; sandstone may be converted to quartzite, and shale to hornfels. Recrystallization alone generally indicates mild contact action or feebler action in the outer zone of alteration. Usually, recombination of rock ions occurs also. With additions of materials from the magma, the original minerals AB and CD that recombined to AC and BD may become ACX and BDY, where X and Y represent accessions. N. L. Bowen has shown that heating dolomite and quartz gives rise first to tremolite and, in order of rising temperature, forsterite, diopside, periclase, wollastonite, monticellite, okemanite, spurrite, merwinite, and larnite. Thus many complex calcio-silicates are formed such as amphiboles, pyroxenes, vesuvianite, tourmaline, or garnets. The ore minerals, of course, result almost entirely from accessions from the magmas.

The ore minerals and those with constituents foreign to the invaded rocks give abundant testimony of noteworthy additions to the contact zone from the magma.

The magmatic accessions consist chiefly of the metals, silica, sulphur, boron, chlorine, fluorine, potassium, magnesium, and some sodium.

Volume Changes. Lindgren, in his classical work on the contact-metasomatic deposits at Morenci, Ariz., has shown that vast quantities of material have been added to, and subtracted from, the invaded rocks. He shows that if all the CaO in 1 cc of $CaCO_3$ were converted to andradite garnet the volume would become 1.40 cc, or a volume increase of nearly one-half, but the field relations show no volume change. He concludes that for every cubic meter of the altered limestones 460 kg of CaO and 1,190 kg of CO_2 have been carried away, and 1,330 kg of SiO_2 and 1,180 kg of Fe_2O_3 have been added. These astonishing figures of 1.8 tons of material removed and 2.8 tons added per cubic meter attest to the vigorous wholesale transfer of material and the quantity of accessions during contact metasomatism. Additional quantities of metallic minerals were also introduced.

Stages of Formation. Contact metasomatism apparently commences shortly after intrusion and continues until well after consolidation of the outer part of the intrusive. In general, the first stage, which is one of heat, consists of recrystallization and recombination with or without accessions from the magma. This gives rise to many of the silicates. Magnetite and hematite form with the silicates and later than them, but they generally precede the formation of the sulphides, although Butler found magnetite later than chalcopyrite in Utah. The sulphides mostly form later than the silicates and oxides. A common order is pyrite and arsenopyrite followed by pyrrhotite, molybdenite, sphalerite, chalcopyrite, galena, and sulpho-salts last. In some places, however, the sulphides form contemporaneously with the silicates, but the older idea that the silicates, oxides, and sulphides were formed simultaneously no longer seems to hold, and successive stages of development have generally been noted. The stages of development, however, seem to vary considerably with different magmas.

Mode and Timing of Transfer. The recrystallization and part of the recombination may have been achieved by heat alone immediately after the intrusion. The main transfer of materials by magmatic fluids, however, must have occurred at a later period, after the freezing of the chill zone of the intrusive and during the accumulation of the rest-magma in which the mineralizers were being concentrated.

Opinion is not yet unanimous in regard to the mode of transfer of the magmatic materials. It is shown in Chapter 4 that gaseous emanations can and do transfer materials outward from magmas. Fenner has concluded that they are the agents best adapted to separating ore materials from magmas and to transporting them outward and further that pneumatolytic processes are capable of producing important contact metasomatism. Ingerson and Morey, and T. E. Gillingham have demonstrated the effectiveness of gaseous water in transporting silica. If the pressure is such that a vapor phase is formed, high-temperature gaseous emanations are to be expected and pneumatolytic processes will prevail. If no vapor phase can be formed, magmatic fluids may emanate as high-temperature liquids and the contact metasomatism and accompanying mineral deposition may be produced by them. Hickok concluded that the magnetite ore of the Cornwall deposits was formed from liquids that emanated from the adjacent intrusive, and Graton has advocated that the ore fluids leave the magma as liquids. Unfortunately, there are no criteria to establish conclusively whether mineral deposits have been actually deposited from gaseous emanations. In the past it has generally been considered that contact-metasomatic ore deposits have been formed from gaseous emanations. Field evidence, however, indicates that if deposits have been so formed liquid emanations have operated in the later phases of their formation.

Relation to Intrusives

Contact metasomatism that gives rise to mineral deposits does not occur indiscriminately with all magmas. It is restricted exclusively to intrusive magmas, it appears to depend upon the composition of the magma, and it is related to the size and depth of formation of the intrusive body. Extrusive bodies may produce a little baking, hardening, or other effect at the contact but never mineral deposits.

Composition of Intrusive. Although widespread contact-metamorphic effects may be produced by magmas of wide variation in composition, those that yield mineral deposits are mostly silicic ones of intermediate composition, such as quartz monzonite, monzonite, granodiorite, or quartz diorite. Highly silicic rocks, such as normal granite, rarely yield mineral deposits. Neither are contact-metasomatic deposits found with ultrabasic rocks, and only rarely with basic rocks. Two noted examples of such deposits in basic rocks are

deposits with quartz diabase at Cornwall, Pa., and with a white gabbro
at Hedley, British Columbia. The lack of contact-metasomatic
deposits around the great basic Bushveld Igneous Complex of South
Africa is noteworthy, especially since so many ore deposits are included
within the intrusive.

The reason why contact-metasomatic deposits are produced from
silicic rather than from basic intrusives is probably because the silicic
material has a high content of water, whereas the basic is relatively
dry. Schwartz has shown that greenstone in Minnesota develops
hydrous minerals where intruded by granite and nonhydrous minerals where intruded by gabbro, indicating that the gabbro was deficient in water. Since water in magma is the chief collector and
transporter of metals, its low content in basic rocks probably accounts
for the paucity of contact-metasomatic deposits with basic intrusives.

Size and Form of Intrusive. Most contact-metasomatic deposits are
associated with stocks, batholiths, and intrusive bodies of similar
size; they are rarely associated with laccoliths and large sills and are
absent from small sills and dikes. Intrusive bodies whose flanks dip
gently produce wider zones of contact metasomatism than those with
steeply dipping flanks. Also, those whose roofs are irregular in form
and are marked by cupolas and roof pendants give rise to extensive
zones of pronounced contact metasomatism.

Depth of Intrusion. The depth of intrusion appears to be an important factor in the formation of contact-metasomatic deposits, because deposits are found only with rocks of granular groundmass,
which indicates relatively slow cooling at considerable depth. The
absence of deposits with rocks of glassy and aphanitic texture, indicative of rapid cooling at shallow depths, shows that near-surface
conditions are not favorable to the formation of contact-metasomatic
deposits, probably because of rapid loss of the magmatic emanations
at shallow depth.

The actual depth of consolidation of granular intrusives is known
only imperfectly. Most of them have probably crystallized at depths
greater than 5,000 feet. At Philipsburg, Mont., the depth was 6,500
feet. Brögger showed that near Oslo, Norway, some intrusives had
crystallized at depths of less than 3,300 feet; Barrell concluded that
at Marysville, Mont., the batholith had reached within 4,000 feet
of the surface, and similar depths were ascertained for the intrusives
associated with contact-metasomatic deposits in Utah and New
Mexico. The oft-quoted shallow depth of 1,000 feet for the contact-
metasomatic magnetite deposits of Cornwall, Pa., has been shown to
be incorrect.

Alteration of Intrusive. In general, the intrusive itself has been little affected during contact metamorphism. Rarely its margins may be so altered as to obscure the exact boundary between intrusive and altered intruded rock. Epidote, the chief mineral formed in the intrusive, presumably results from the absorption of CaO and CO_2 released from the invaded rock. Less commonly garnet, vesuvianite, chlorite, diopside, and other minerals occur. Sericitization of the intrusive is common; this, however, is probably an after effect caused by later emanations of hot waters through the frozen margins of the intrusive.

Relation to Invaded Rocks

The character and extent of the alteration of the invaded rocks depend upon their composition and in part upon their structure. Certain rocks are more susceptible than others and exhibit extreme effects; some are only slightly changed.

Relation to Composition. Carbonate rocks are the ones most affected by intrusion of magma. Pure limestone and dolomite readily recrystallize and recombine with introduced elements. Impure carbonate rocks are affected even more, since such impurities as silica, alumina, and iron are ingredients available to enter into new combinations with calcium oxide. The entire rock adjacent to the intrusive may be converted to a mass of garnet rock, silicates, and ore.

Sandstones are but little affected. They recrystallize to quartzite and may contain sparse contact-metasomatic minerals.

Shales and slates are baked and hardened, or altered to hornfels, generally with andalusite, sillimanite, and staurolite. The degree of alteration varies with their purity, being greatest with carbonate varieties. The argillaceous rocks seldom contain important mineral deposits.

Invaded igneous rocks do not contain contact-metasomatic deposits. Their slight alteration is most pronounced if their composition is markedly unlike that of the intrusive. For example, granodiorite may intrude granite, with little effect, but a basic rock such as gabbro might suffer considerable change. Likewise, the metamorphic rocks are uncongenial for ore and undergo little change. This is especially true of the crystalline schists, whose constituents have already been metamorphosed to minerals that are stable under conditions of fairly high pressure and temperature.

In general, then, the rocks most susceptible to contact metasomatism and most favorable for the formation of ore deposits are the sediments, particularly the impure calcareous ones.

Relation to Structure. The structure of the invaded rocks, and faults, affect the extent and position of the contact-metasomatic zone. Where sedimentary beds dip into the intrusive, as in Fig. 5·3–1, good up-dip channelways are provided for escaping emanations in contrast to the slower migration across the bedding; ore deposits are likely to be larger and more widely distributed up the dip than across it.

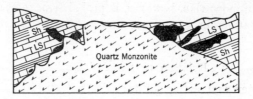

FIG. 5·3–1.   Relation of contact-metasomatic deposits (black) to monzonite intrusive and structure of limestone.

Faults that extend outward and upward from the intrusive serve as through channelways that concentrate and conduct the emanations far from the intrusive and localize the metamorphism and mineralization in tabular form.

Resulting Mineral Deposits

The mineral deposits that result from contact metasomatism constitute a distinctive class characterized by an unusual assemblage of ore and gangue minerals. They contribute to the world's mineral production and supply many of the uncommon mineral products. The deposits generally consist of several disconnected bodies. They are mostly small as compared with " porphyry coppers," or sedimentary deposits. They are vexatious deposits to exploit because of their relatively small size, their capricious distribution within the contact aureole, and their abrupt terminations. Like the scattered plums in a plum pudding, they are difficult to find; they exhibit few " signboards " pointing to their presence, and costly exploration and development are necessary to discover and outline them. Their development must be undertaken with caution, and the optimism attendant upon mining such concentrated and often rich bodies frequently gives way quickly to disappointment upon the sudden termination of the ore body.

Position. The ore bodies occur within the contact aureole and generally relatively close to the intrusive contact. J. B. Umpleby, however, has pointed out that in several sulphide deposits the ore bodies lie on the outer or limestone side of the contact aureole, perhaps because of the neutralizing effect of the limestone upon acid generated by reaction between metal chlorides and hydrogen sulphide.

The deposits are generally scattered irregularly around the contact but tend to be concentrated upon the side of the intrusive that has the gentler dip. If the dip of the intrusive is low, the deposits may lie a considerable distance horizontally from the contact and still be relatively close to it. Roof pendants (Fig. 5·3–2) and large inclusions are especially favorable loci, as at Mackay, Idaho. Where prominent

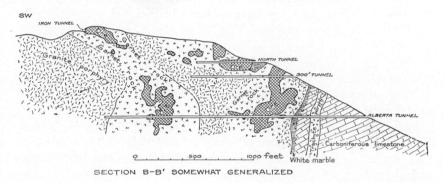

FIG. 5·3–2. Cross section of contact-metasomatic deposits at Mackay, Idaho, showing ore (cross-hatched) in garnet rock developed from huge limestone inclusions. (*After J. B. Umpleby, U. S. Geol. Survey.*)

faults extend outward from the contact, deposits may be aligned along them and extend 2,000 feet or more from the contact. This is the case at Bisbee, Ariz., and Knopf describes contact-metasomatic deposits at Rochester, Nev., along a fault zone at a distance of 2,000 feet from the intrusive.

Form and Size. Contact-metasomatic deposits are notably irregular in outline. They may have almost any shape. Generally, they are more or less equidimensional, but ramifying tongues may project outward along bedding planes, joints, or fissures enhancing the irregularity of outline. Those more irregular in shape occur chiefly in thick beds of limestone. Roughly tabular bodies are formed where deposits are aligned along the contact for a distance, or along fault zones, or where they have been formed by selective replacement of individual susceptible limestone beds. Just why certain beds prove more congenial to ore formation than others is uncertain.

In general, contact-metasomatic deposits are of comparatively small size, with dimensions of 100 to 400 feet, and contain from a few tens of thousands to a few hundreds of thousands of tons of ore; a few deposits contain several million tons.

Texture. Commonly, the ores are coarse in texture, containing large crystals or clusters of crystals. Few of the minerals, except garnet, show crystal outlines, and even garnet appears often in sugarlike masses of irregular outline. Columnar and radiating minerals may show crystal faces several inches in length; calcite is generally in coarse grains. Magnetite may occur in bunches composed of large grains of shapeless outline. The individual minerals appear to be closely interlocked; open space is rare. Crustification and banding are absent, but orbicular structures have been noted. The metallic minerals, except pyrite and arsenopyrite, generally lack crystal outlines.

In some cases, however, as in parts of the deposits of Hanover, N. Mex., the minerals are extremely fine grained and the ore has a flinty appearance, resembling dense hornfels; the individual grains of the intimately admixed gangue and metallic minerals cannot be distinguished by the naked eye.

Mineralogy. The outstanding feature of the mineralogy is the distinctive assemblage of gangue minerals characteristic of high-temperature formation. These include grossularite and andradite garnet, hedenbergite, hastingsite, tremolite, actinolite, wollastonite, epidote, zoisite, vesuvianite, ilvaite, diopside, forsterite, anorthite, albite, fluorite, chlorite, and micas. Quartz and carbonates are generally present. In addition, silicates containing the mineralizers, such as tourmaline, axinite, scapolite, ludwigite, chondrodite, and topaz, may be present.

The ore minerals consist of oxides, native metals, and sulphides, arsenides, and sulpho-salts. The oxides are represented by magnetite, ilmenite, hematite (specularite), corundum, and spinels. Magnetite is particularly abundant. Graphite, gold, and platinum represent the native minerals, but the last two are rare. The sulphides consist chiefly of base-metal sulphides. Sulpho-arsenides and antimonides are rare, as are the tellurides. In addition, scheelite and wolframite occur.

Mineral deposits formed by contact metasomatism include the minerals listed on page 93.

Selected References on Contact Metasomatism

General Reference 9. Chap. 11, Part III, pp. 537–556. *Brief review of western occurrences illustrating Lindgren's classification.*

General Reference 8. Chap. 28.

Some Observations on Contact Metamorphic Deposits. BASIL PRESCOTT. Econ. Geol. 10:55–69, 1915. *Pointed general remarks and Mexican occurrences.*

TYPES OF MINERAL DEPOSITS FORMED BY CONTACT METASOMATISM WITH THE CHIEF CONSTITUENT MINERALS AND EXAMPLES

Deposit	Chief Minerals	Examples of Deposit
Iron	Magnetite and hematite	Cornwall, Pa.; Iron Springs, Utah; Fierro, N. Mex.; Banat, Hungary; Gora, Urals; Elba, Italy
Copper	Chalcopyrite and bornite with pyrite, pyrrhotite, sphalerite, molybdenite, and iron oxides	Some deposits of Morenci and Bisbee, Ariz. and Bingham, Utah; Cananea, Matehuala, Mexico; Suan, Korea
Zinc	Sphalerite with magnetite, sulphides of iron and lead	Hanover, N. Mex.; Long Lake, Ontario; Kamioka, Japan
Lead	Galena, magnetite, and sulphides of iron, copper, and zinc	Magdalena, N. Mex.; Inyo County, Calif.
Tin	Cassiterite, wolframite, magnetite, scheelite, pyrrhotite	Pitkäranta, Finland; Saxony; Dartmoor, England; Yak, Alaska
Tungsten	Scheelite and minor sulphides, or wolframite with molybdenite and minor sulphides	Mill City, Nev.; Inyo County, Bishop, Calif.; King Id; Australia
Molybdenum	Molybdenite, pyrite, garnet	Yetholm, Australia; Azegour, Morocco; Buckingham, Quebec
Graphite	Graphite and contact silicates	Adirondacks, N. Y.; Buckingham, Quebec; Ceylon; South Australia
Gold	Gold with arsenopyrite, magnetite, and sulphides of iron and copper	Cable, Mont.; Hedley, British Columbia; Suan, Korea
Manganese	Manganese and iron oxides and silicates	Längban, Sweden
Emery	Magnetite and corundum, with ilmenite and spinel	Virginia; Peekskill, N. Y.; Turkey; Greece
Garnet	Garnet and silicates	
Corundum	Corundum with magnetite, garnet, and other silicates	Peekskill, N. Y.; Chester, Mass.

Clifton-Morenci District, Arizona. WALDEMAR LINDGREN. U. S. Geol. Surv. Prof. Paper 43, 1905. *Outstanding work on contact metasomatism, particularly the transfer of materials from magma to invaded rocks.*

Occurrence of Ore on Limestone Side of Garnet Zones. J. B. UMPLEBY. Univ. of Calif. (Berkeley) Pub. 10:25–37, 1916. *Review of contact-metasomatic deposits showing that ore bodies lie on the outside of the garnet zone.*

Contakt Metamorphose im Kristianiagebiet. V. M. GOLDSCHMIDT. 1911. *Classical treatise of contact metamorphism.*

Ore Deposits at the Contacts of Intrusive Rocks and Limestones. J. F. KEMP. Econ. Geol. 2:1–13, 1907.

Geology and Ore Deposits of the Mackay Region, Idaho. J. B. UMPLEBY. U. S. Geol. Surv. Prof. Pap. 97, 1917. *Detailed account of contact-metasomatic deposits and estimates of magmatic accessions.*

Contact Metamorphism at Bingham, Utah. WALDEMAR LINDGREN. Geol. Soc. Amer. Bull. 35:507–534, 1924. *Metasomatic changes and estimates of materials added and subtracted.*

Contact Metamorphic Tungsten Deposits of the United States. F. L. HESS and E. S. LARSEN. U. S. Geol. Surv. Bull. 725-D, 1922.

Iron Ore Deposits at Cornwall, Pa. W. O. HICKOK. Econ. Geol. 28:193–255, 1933. *Contact-metasomatic magnetite deposits.*

Rocks and Rock Minerals. L. V. PIRSSON and ADOLPH KNOPF. 3rd Ed. 349 pp. John Wiley & Sons. New York, 1947. Chap. 13.

Contact Pyrometasomatic Aureoles. H. SCHMITT. Am. Inst. Mining Met. Engrs. Tech. Pub. 2357, 1948. *Localization of tactite zones by structure.*

See also references under the ores listed above in Parts II and III.

5·4 HYDROTHERMAL PROCESSES

It is shown in Chapter 4 that magmatic differentiation gives rise to an end product of magmatic fluids in which there may be concentrated the metals originally present in the magma. These hydrothermal solutions carry out the metals from the consolidating intrusive to the site of metal deposition and are considered to be the major factor in the formation of epigenetic mineral deposits. They are, or become, liquids and gradually lose heat with increased distance from the intrusive. Thus, they give rise to high-temperature hydrothermal deposits nearest the intrusive, intermediate-temperature deposits at some distance outward, and low-temperature deposits farthest outward. Lindgren has designated these three groups as *hypothermal, mesothermal,* and *epithermal* deposits, according to the temperatures, pressures, and geologic relations under which they were formed, as indicated by the contained minerals (see Chap. 8).

The source of the hydrothermal solutions has been from early times a subject of inquiry. Prior to Werner (see Chap. 2) the ore solutions were considered to be diffuse ascending waters of uncertain source. Werner advocated that they were descending waters derived from the primeval ocean. The concept of mineralizing solutions being of magmatic origin gained ground rapidly in the middle of the last century and has persisted to the present despite intervening controversies regarding lateral secretion, descending meteoric waters, and ascending heated meteoric waters. It is realized, however, that rising magmatic hydrothermal solutions, as they near the surface, may mingle with meteoric waters, and also that rarely shallow-seated intrusives might discharge gases and vapors into groundwater to create mineralizing solutions. The evidence from deep mines that are dry indicates that, at the great depths at which most mineral deposits are thought to have been formed, the hydrothermal ore solutions had little or no contact with meteoric waters.

The concept of the formation of mineral deposits from water solutions has evolved slowly as supporting evidence has accumulated. It was only natural that early investigators turned to familiar chemical reactions in solutions to account for the natural deposition of ore minerals in the rocks.

It was demonstrated in the laboratory that the common ore minerals could be precipitated from solutions, and most of the minerals of hydrothermal deposits have been produced artificially. Deposition of gold, silver, copper, lead, zinc, takes place almost before our eyes from the hot waters of Steamboat Springs, Nevada. These, and many other hot springs, demonstrate beyond question that the constituents of the hydrothermal type of mineral deposits are taken into solution, transported, and deposited from hot waters.

Other evidence consists of liquid inclusions so often found in the minerals of ore deposits, which disclose not only that the minerals were deposited from solutions but also the kind of solution and even the temperature of deposition. Moreover, the direction of flow of the mineralizing solutions can be deduced from the shape of crystals and the arrangement of smaller crystals perched on the up-current side of larger crystals.

These features, and others not enumerated here, constitute the evidence for deposition of minerals from hydrothermal solutions.

In their journey through the rocks, the hydrothermal solutions may lose their mineral content by deposition in the various kinds of openings in the rocks, to form *cavity-filling deposits,* or by metasomatic replacement of the rocks to form *replacement deposits.* The filling of openings by precipitation may at the same time be accompanied by some replacement of the walls of the openings. Thus, there may be a gradation between these two types of mineral deposits. In general, replacement dominates under the conditions of high temperatures and pressures nearer the intrusive where hypothermal deposits are formed, and cavity filling dominates under the conditions of low temperatures and pressures where the epithermal deposits are formed; both are characteristic of the mesothermal zone.

PRINCIPLES OF HYDROTHERMAL PROCESSES

Geologists attribute to hydrothermal processes that vast array of metallic mineral deposits that supply the major part of our useful metals and minerals. From such deposits are won most of the gold and silver, copper, lead and zinc, mercury, antimony and molybdenum, most of the minor metals, and many nonmetallic minerals. Consequently, it is these deposits that have been mined, investigated, and written about far more than any other group. They have given rise to many of the great mining districts of the world; the lore of mining sprang from them.

Essentials for the formation of hydrothermal deposits are: (1) available mineralizing solutions capable of dissolving and transporting mineral matter, (2) available openings in rocks through which the solutions may be channeled, (3) available sites for the deposition of the mineral content, (4) chemical reaction that results in deposition, and (5) sufficient concentrations of deposited mineral matter to constitute workable deposits.

CHARACTER OF SOLUTIONS

The nature of hydrothermal solutions must be interpreted by inference and by analogy with certain types of hot springs. Their action is visible only in the form of mineral deposits or wall rock alteration. As the term hydrothermal implies, they are hot waters that probably range in temperature from 500° C down to 50° C. Those of higher temperature also would be under higher pressure. The mineral substances are presumed to be carried in chemical solution, although colloidal solutions are thought to have been operative. Their chemical character is discussed in Chapter 4.

OPENINGS IN ROCKS

The movement of hydrothermal solutions from source to site of deposition is dependent largely upon available openings in the rocks. The deposition of large bodies of extraneous minerals involves the necessity of continuing supplies of new material, and this means that through channelways must be available. The openings must be interconnected. Furthermore cavity-filling deposits obviously cannot form unless there are available cavities to be filled. Equally obviously, replacement deposits cannot form unless the solutions can reach the rock undergoing replacement. Consequently, openings in rocks are fundamental to the formation of epigenetic deposits. Likewise, they are essential for the existence of bodies of groundwater, oil, or gas.

The various types of openings in rocks that may serve as receptacles for ore or permit the movement of solutions or their constituents through the rocks may be classified as shown below.

Original Cavities

1. Pore spaces
2. Crystal lattices
3. Vesicles or "blow holes"
4. Lava drain channels

5. Cooling cracks
6. Igneous breccia cavities
7. Bedding planes

Induced Cavities

1. Fissures, with or without faulting
2. Shear-zone cavities
3. Cavities due to folding and warping
    a. Saddle reefs
    b. Pitches and flats
    c. Anticlinal and synclinal cracking
       and slumping

4. Volcanic pipes
5. Tectonic breccias
6. Collapse breccias
7. Solution caves
8. Rock alteration openings

Most of the kinds of cavities listed above may in rare cases under special conditions become filled to form various types of mineral deposits. Some serve only as conduits for mineralizing solutions; some permit ingress of solution for replacement; others serve as receptacles or conduits for water, oil, and gas; still others such as pore spaces, shear zones, and rock alteration openings play a more important role in replacement than in cavity filling. The formation of epigenetic mineral deposits is dependent upon them.

Pore Spaces. Rock pores are interstitial openings between grains, capable of absorbing water. They make rocks permeable and serve as containers for ores, petroleum, gas, and water. The oil and gas pools of the world are contained for the most part in rock pores, as are also subsurface water supplies.

Porosity. Porosity of a rock is the volume of pore space measured in percentage of the rock volume. It ranges from practically nothing in hard rocks to a maximum of 52.94 per cent in loose-textured soils. King and Schlichter have shown that with spheres of uniform size the minimum porosity is 25.95 percent and the maximum 47.64 percent. The maximum occurs when the arrangement of the spheres is in the unstable form, with the center of one directly above the center of another. The minimum occurs with the stable form of packing, similar to the arrangement obtained when cannon balls are stacked in a pyramid. The same maximum and minimum porosities are obtained regardless of the diameters of the spheres. The spheres in a container arranged in unstable packing take on the stable arrangement when jarred. Tests made on natural and artificial materials gave minimum and maximum porosities of 25.55 and 52.94 percent. Angular materials have greater porosity than spherical ones, and *finer* angular materials have considerably greater porosity than *coarser* angular materials, i.e., fine clayey sand contains 52.94 percent porosity and coarser material contains only about 33 percent. In other words, a coal bin filled with fine coal will contain less coal than one filled with coarse coal, or fine sediments have greater porosity than coarse sediments.

Permeability. Permeability of a rock exists by virtue of its porosity, but a rock may be porous and not permeable. Permeability does not increase in direct proportion to the porosity but is dependent upon the size of the pores, the total amount of pore space, and particularly interconnection of pore spaces. A light bulb has much porosity but no permeability. The smaller the pores, the greater the surface exposed, the greater the friction, and the smaller the flow. In spheres, if the diameter is halved, the number of pores is increased 4 times, according to King; and Schlichter states that, if the same sand is packed to give 26 percent and 47 percent porosity, the flow through the latter will be 7 times greater than through the former. Thus, a coarse sandstone with only 15 percent porosity may be much more permeable than a shale with 30 percent porosity. Wet clay, because of its very fine pores, exerts a very tight hold upon the contained water. Consequently, water-wet clays and shales are essentially impermeable. Coarse-pored rocks, on the other hand, even if of low porosity, are quite permeable, if the pores are connected.

Porosity of Rocks. Sediments composed of fine, angular grains, as a rule, have greater porosity and less permeability than those composed of coarse grains. In general, therefore, clays and shales are less permeable than sandstones, which are by far the more important mineral receptacles.

The percentage of porosity of some common rock samples are:

	AVERAGE	MAXIMUM	MINIMUM
Granites (25)	0.369	0.62	0.19
Building limestones (25)	4.88	13.36	0.53
Building sandstones (35)	15.9	28.28	4.81
Oil sands (29)	19.4		
Clays (14)	28.43		

Crystal Lattices. The openings between the atoms of a crystal may permit the diffusion through such openings of ions of smaller ionic radii. This diffusion may permit replacement or additions to occur within crystals.

Bedding Planes. These are well-known features of all sedimentary formations. They permit ingress of hydrothermal solutions and replacement by ore of adjacent walls.

Vesicles or " blow holes " are openings produced by expanding vapors typical of the upper part of many basaltic lava flows. They are tubular in shape, roughly circular in cross section, and may be spaced from ½ to 2 inches apart. If the vesicles are filled, the rock is called an amygdaloid. If they are closely crowded together, they form a cellular rock like a sponge, called *scoria*.

Volcanic flow drains form in lava flows when the outside of the lava has solidified and liquid lava in the center drains out leaving a pipe, or tunnel.

Cooling cracks are formed as a result of contraction in cooling igneous rocks. They may be regularly spaced joints that divide the rock into blocks, or parallel platy partings, or irregular cracklings.

Igneous Breccias Cavities. Igneous breccias are of two types: volcanic breccias forming agglomerates, and intrusion breccias. Both consist of coarsely and angularly fragmented igneous rocks with finer interstitial material. They may be quite permeable.

Fissures. Fissures are continuous tabular openings in rocks, generally of considerable length and depth. They are formed by compressive, tensile, or torsional forces operating on rocks and may or may not be accompanied by faulting. Thus, faults are fissures, but all fissures need not be faults. They may constitute long and continuous channelways for solutions. When occupied by metals or minerals they form *fissure veins*.

Shear-Zone Cavities. Shear zones result where fractures, instead of being concentrated in one or two single breaks, are expressed in innumerable closely spaced and more or less parallel, discontinuous surfaces of deep-seated rupture and crushing. Faulting is present generally. The thin, sheetlike openings, mostly of infinitesimal size, make excellent channelways for solutions, as is evidenced by the copious water flows where they are cut in tunnels and mines. Because of the minute openings, only minor open-space deposition can occur, but the large specific surface available makes shear zones favorable localizers of replacement deposits.

Folding and Warping. Flexing and folding of sedimentary strata give rise to: (1) *saddle reef* openings (p. 126) at the crests of closely folded, narrow anticlines; (2) *pitches* (p. 128), which are highly inclined, and *flats* which are openings formed by the parting of beds under gentle slumping; (3) longitudinal cracks along the crests of anticlines and synclines.

Volcanic Pipes. When explosive volcanic activity bores pipelike openings, the material blown out may fall back or be washed back into the opening, forming an angular breccia with spaces between fragments. These form excellent confined conduits for mineralizing solutions from which cavity filling or replacement deposits may form.

Breccias may be formed by the crushing of any brittle rock caused by folding, faulting, intrusion, or other tectonic forces, forming *tectonic breccias*, or by collapse of rock overlying an opening, giving rise to *collapse breccias*. As with other breccias, the openings between the angular fragments provide space for circulation of solutions, cavity filling, or replacement.

Solution openings, such as caves and enlarged joints or fissures in soluble rocks, supply channelways and open spaces for cavity filling.

Rock Alteration Openings. Wall rocks that have been altered by solutions are found by tests to be generally more porous than unaltered rocks and permit ingress of mineralizing solutions.

The part played by the various openings mentioned above is discussed in a later part of this subchapter.

Movement of Solutions through Rocks

The movement of hydrothermal solutions through rocks appears to take place most readily where long continuous openings, such as fissures, are available, or where smaller openings are interconnected as in shear zones, vesicular lava beds, or permeable porous sediments. Since many deposits contain millions or even hundreds of millions of tons of ore, vast quantities of solutions must have been necessary to transport this substance. Hence, fairly large confined channelways must have been available. Diffusion through crystal lattices is known to be too slow and too incapable of acting in large volume over long distances. However, diffusion is an important factor in permitting ingress of replacing ions into crystals at actual foci of replacement where voluminous moving solutions could not gain entry. Moreover, the flow of solutions, particularly prior to reaching the site of deposition, must have been relatively confined or their mineral matter would have become too dispersed. A widespread volcanic breccia, for example, might be quite permeable throughout, but, just because of this widespread permeability, mineralizing solutions would be so spread out that the minerals deposited from them would be too diffuse to constitute ore.

To supply a large amount of new mineral matter at the point of deposition, fissures, channelways, and permeable beds serve as the main freight lines, and small fractures, cleavage planes, and pore spaces further distribute it to the front lines, where diffusion acting over short distances may deliver it to the point of deposition. As is shown under "Replacement," diffusion is essential for the entry of ions through an already formed wall of replacement ore, but it is not essential for cavity filling. Thus, rock openings are essential to transmit large volumes of mineral matter to the site of deposition.

ROCK-PARTICLE SIZE AND SPECIFIC SURFACE

The size of rock particles is important not only in the flow of solutions through rocks but particularly in chemical reactions between rock matter and solutions. The surface area of a unit mass of a finely divided substance is termed the *specific surface*. With large surface area and small particle size, pores are small and permeability is low. But these features are opportune for mineral deposition since large specific surface permits greater contact between rock and solution and, therefore, greater opportunity for reaction between them. An angular open breccia, for example, offers to mineralizing solutions simultaneously many times greater rock surface than confining walls of a fissure vein and, hence, greater opportunity for deposition or replacement. A permeable bed, such as an arkose with connected but small pores, presents large specific surface for reaction between rock grains and solutions, and in addition the small pores retard the rate of flow and afford more time for reaction.

EFFECT OF HOST ROCKS

Field evidence indicates that, in many cases the world over, reactive wall rocks not in equilibrium with solutions exert a profound effect upon hydrothermal mineral deposition, more particularly in replacement than in cavity filling. Certain carbonate beds, for reasons not entirely known, are congenial to ore deposition whereas others are not. Metallizing solutions seem almost to have been empowered with peculiar discrimination in passing by certain carbonate beds and replacing others at Leadville, Colo., Bisbee, Ariz., Bingham, Park City, or Tintic in Utah, or Kennecott, Alaska. The preference of ore for dolomitic limestone over pure limestone in lead-zinc-silver ores of Mexico has been pointed out by Hayward and Triplett. However, at Santa Eulalia, Mexico, the reverse is true, and massive sulphide ore in limestone terminates abruptly at a dolomite contact. Likewise, the selective mineralization of "greenstone" in contrast to adjacent granite and "porphyry" has been demonstrated repeatedly in the gold camps of northern Canada. Wandke and Hoffman point out that at Guanajuato, Mexico, bonanza silver ores occur in fissures opposite andesite or porphyry but not opposite schist. Weston-Dunn shows that the Giblin tin lode at Mt. Bischoff, Australia, widens in slate but pinches in quartzite. Similarly, other rocks are known definitely to favor ore deposition, and innumerable examples can be drawn from all over the world indicating that ore deposition, particularly replacement, is influenced or localized by the character of the host rock.

Factors Affecting Deposition

The modes of formation of the minerals of the various types of mineral deposits are considered on pages 28 to 35. Those that apply to deposition from hydrothermal solutions are dominantly chemical changes in the solutions, reactions between the solutions and wall rocks or vein matter, and changes in temperature and pressure.

Chemical Changes and Reactions. Mineralizing solutions in their long journey upward inevitably must undergo some chemical change by reaction with the rocks through which they pass. Silicate rocks would make them alkaline or more alkaline. Their hydrogen-ion concentration (pH) may determine when reaction with rocks or deposition may occur.

In replacement, the substitution of new minerals for older ones can, of course, take place only by reaction between solution and solid. Strongly reactive wall rocks, such as limestone, out of equilibrium with the solutions bring about a rapid chemical change accompanied by deposition.

Temperature and Pressure. The most important factors in promoting hydrothermal deposition from solutions are *changes* in temperature and pressure. In general, a drop in temperature decreases solubility and causes precipitation.

Hydrothermal solutions start their journey with heat supplied by the magma, and this is slowly lost in passage through the rocks. The temperature drop depends upon the rate of loss of heat to the wall rocks, which in turn depends upon the amount of solution moving past, exothermic reactions, and particularly the capacity of the wall rock to conduct heat away. The higher the thermal diffusivity of a rock the more rapidly will heat be conducted away and the greater will be the temperature drop in the solutions. In the initial stages of circulation with cool wall rocks, the temperature drop will be relatively rapid, but continued flow of solutions will heat the wall rocks to the temperature of the solutions, at which time the loss of heat will slow down.

The nature of the rock openings also affects the heat loss. Rapid flow through a straight-walled open fissure would entail less loss of heat than flow through the intricate openings of a breccia with large specific surface where the initial drop in temperature would be rapid. Once heated up, however, the breccia would not remove much heat from the solution. The greater the volume of new solution passing a given point, the greater would be the supply of fresh heat and the slower the drop in the temperature of the solutions. Thus, in a fissure with characteristic constrictions and open spaces, the temperature of the solutions would drop less in the constricted than in the wider areas. Such features are also important in causing and localizing deposition of mineral content.

The solutions are initiated under the high pressures that prevail at the great depths where they originate. Their upward journey, through zones of lower pressure, normally is accompanied by a drop in pressure, which promotes precipitation. But other factors may also result in changes in pressure. Constrictions in channelways, or partial filling by mineral deposition, or barriers may build up excess pressure below them. Escape of the solutions to more open areas above the constrictions would lower the pressure and promote deposition. Thus may changes in the physical character of the openings through which solutions pass play an important role in causing and localizing deposition of minerals from hydrothermal solutions.

Van Nieuwenburg and van Zon have shown, and Gillingham has proved, for nonvolatile substances in supercritical steam, that at constant temperature there is an increase in solubility with increase in pressure, and at constant pressure there is a decrease in solubility with increase of temperature. Thus, up to a certain point, the solubility of a nonvolatile substance in a gas is more dependent on pressure than on temperature. These startling conclusions lead to the interesting corollary that, in contrast to the precipitation of minerals from a liquid by decrease in temperature, precipitation *from a gas can be effected by decrease in pressure alone.* This means that, at the depths of vein formation where temperature is relatively constant, if a mineral-laden gas enters a porous rock where pressure release takes place, a sudden dumping of mineral content can occur. It also follows that a gas at constant pressure could dissolve substances from the rock walls while a drop in temperature occurs through loss of heat to the wall rocks.

WALL-ROCK ALTERATION

It has long been observed that hydrothermal mineral deposits are generally accompanied by a band of alteration of the wall rocks

readily visible to the eye. For example, in a fissure vein the alteration zone lies parallel to the walls of the fissure, is of relatively uniform
width, and varies in width according to the size of the vein. The
nature and intensity of the alteration depend also upon the nature
of the wall rock and the chemical character and temperature and
pressure of the mineralizing solutions. If the veins are closely spaced
the alteration halo of one vein may merge with that of another, and
the intervein space is entirely altered, as in the central zone of Butte,
Mont. This is particularly striking in the case of " porphyry coppers "
where the host rock between the numerous intersecting veinlets has
been altered intensely to dimensions measurable in thousands of feet,
as at Bingham, Utah, Ray, Ariz., or Santa Rita, N. Mex.

In an epithermal vein the visible altered zone is generally narrow
and the change may be hardly discerned, but in a mesothermal vein
the zone is likely to be wide and intense, gradually merging into fresh
rock outward from the vein. Under hypothermal conditions such
high-temperature minerals as tourmaline, topaz, pyroxenes, and amphiboles may be developed in the wall rocks.

The nature of the alteration also varies somewhat with the kind of
rock, but it is surprising that in the case of a mesothermal copper
deposit the alteration product of a quartz monzonite resembles that
derived from diorite or crystalline schist, as at Miami, Ariz. With
most rocks, except limestone and quartzite, the end product of alteration is a rock composed mostly of sericite and quartz. The original
feldspars, ferromagnesian minerals, and micas are all changed to sericite, and silica is generally added. This is referred to as *sericitization.*
Recent work has disclosed that in many of the " porphyry coppers "
and at Butte, Mont., the outer zone of alteration is characterized by
what Lovering has termed *argillic alteration,* namely the development
of such clay minerals as dickite and montmorillonite.

In many places, two stages of rock alteration may be superimposed.
To illustrate, the hypogene copper mineralization of Ray, Ariz., was
accompanied by widespread intense sericitization of the host rocks.
The subsequent supergene sulphide enrichment was accompanied by
an equally widespread superimposed kaolinization of the sericite. In
fact, the location and intensity of each alteration are usable indicators
of the location and intensity of both the hypogene and supergene
metallization. Such supergene kaolinization is common to all supergene-enriched copper deposits, and also to many other types of
deposits.

Some of the effects of hydrothermal alteration of different kinds of
rocks under different temperature conditions are tabulated below:

Conditions	Wall Rock	Some Alteration Products
Epithermal	Limestone	Silicification
	Lavas	Alunite, chlorite, pyrite, some sericite, clay minerals
	Igneous intrusives	Chlorite, epidote, calcite, quartz, some sericite, clay minerals
Mesothermal	Limestone	Silicified to jasperoid; dolomites, siderite
	Shales, lavas	Silicification, clay minerals
	Silicic igneous rocks	Largely sericite and quartz; some clay minerals
	Basic igneous rocks	Serpentinized, epidote, chlorite
Hypothermal	Granitic rocks; schists; lavas	Greisen; topaz, white mica, tourmaline; pyroxenes, amphibole

The presence of a halo of alteration has been used as a guide to ore-finding. Such indicators of hidden ore are particularly valuable in the case of ore deposits where weathering has leached out the metallic minerals near the surface, but the alteration halo has persisted to indicate their former presence. It has been used also by the author to distinguish unmetallized faults that contain " drag ore " from veins.

Localization of Hydrothermal Mineralization

The cause of the localization of hydrothermal deposits is of both scientific interest and practical importance. It varies, of course, with each district and may be due to one or more factors acting together. For the most part, it is controlled by the chemical and physical character of the host rock, by structural features, by intrusives, by depth of formation, by changes in size of rock openings, or by a combination of them. In some cases, the cause of localization is clearly defined; in others, it is an enigma.

Intrusives. Since most hydrothermal solutions are of magmatic derivation, the position of the parent intrusive may determine the localization of the ore. Cupolas on the intrusive, or apophyses, may localize ore in their vicinity; volcanoes may do the same (see also p. 304).

Character of Host Rock. Hydrothermal deposits may be formed in any kind of host rock, but certain ones influence ore deposition more than others, as shown on pages 101 and 308. In the case of cavity-filling deposits it is the location of the opening, more than the character of the containing rock, that localizes the ore, although both the physical and chemical nature of the host rock may determine the location and shape of the cavity. For example, brittle rocks shatter more readily than nonbrittle rocks and, thereby, localize fractures or breccias; carbonate rocks permit solution openings. Regardless of how chemically favorable a host rock may be, ore deposition cannot occur unless rock openings are available to afford sites for cavity filling or to permit entry of solutions to the front for replacement. Permeability is necessary, and this may be supplied by original pore space, fissility, cleavage planes of minerals, brecciation, joints, minor fractures, or other features. The influence of the host rock in causing ore localization may thus be chemical or physical or both. Specific surface is also an important factor.

Structural Features. Structural features are important localizers of hydrothermal deposits. *Fissures* serve as loci in themselves and also as channelways to conduct ore fluids to rocks susceptible of replacement. The intersection of fissures with favorable rocks is utilized in searching for replacement deposits. Both are necessary. A limestone favorable for replacement may exist, but no replacement can occur in it if the solutions cannot gain access to it, and, conversely, an excellent fissure conduit may be available, but if its walls are unfavorable no replacement will occur. A coincidence of the two is necessary. Thus, in many mining districts, even minute fissures are followed by drifts to their intersection with known favorable beds in the expectation of discovering workable deposits at such sites.

Multiple fissures and *shear zones* localize mineral deposits in a manner similar to fissures.

Fissure intersections are particularly favorable sites for ore deposition. Close, pitching *folds* and *drag folds* have been important localizers for replacement deposits as at the Homestake Mine, S. Dak. (Fig. 5·4–35), and Chambishi Mine, N. Rhodesia. *Breccias* are very favorable loci for both cavity-filling and replacement deposits. Features due to *sedimentation,* such as bedding planes, lamination or continuous permeable beds, unconformities, or overlying impervious beds may control the location of deposits by affording channelways for mineralizing solutions.

MINERAL SEQUENCE — PARAGENESIS

In the formation of all mineral deposits of magmatic affiliations the individual minerals are formed in orderly sequence, and this sequential arrangement is termed *paragenesis*. It is quite simple in magmatic and contact-metasomatic deposits where the gangue minerals are early, the oxides later, and sulphides latest. The relationship is best studied, however, in the hydrothermal deposits where sequential deposition, and often repetition and overlapping, is so characteristic of minerals deposited in cavities or formed by replacement. In cavity-filled deposits, the ore is built up in successive layers, or crust by crust, called *crustification,* a younger crust being deposited upon an older one; in some deposits eight or ten minerals may occur, or recur, in one mineral sequence. The same order of deposition is repeated in numerous localities the world over. Among the common minerals of ore deposits, generally the sequence starts with quartz, followed by iron sulphides or arsenides, sphalerite, enargite, chalcopyrite, bornite, galena, gold, and complex silver minerals. Later generations of gangue and ore minerals may appear. The latest minerals are found perched on older crystals in unfilled central portions of the deposit called *vugs.* In replacement deposits the sequence occurs by the successive replacement of early minerals by later ones. Thus, at Butte, Mont., each of the following sulphides may replace any preceding one: quartz, pyrite, sphalerite, enargite, tennantite, chalcopyrite, bornite, and chalcocite.

The cause of such mineral sequences in cavity fillings is generally considered to be the decreasing solubility of the minerals in solution under decreasing temperature and pressure; i.e., the most soluble minerals stay in solution longest, and the least soluble precipitate first. In sulphide replacement deposits, it is a question of relative solubility in which the mineral being replaced is more soluble than the ore being deposited; otherwise replacement would not occur. The pH of the solutions is also a factor since different minerals may be deposited under alkaline or neutral conditions. In complex ores, other factors must enter in, such as deposition in decreasing order of the potentials of the elements, or the action of mineral complexes.

A study of the paragenesis of a deposit clarifies the nature, continuity in time, and changing character of the mineralizing solutions. It also gives an indication of possible changes in character of ore in depth inasmuch as the earlier minerals of a sequence might be expected in greater quantity in hotter, deeper portions.

CAVITY FILLING

The filling of open spaces or cavities in the rocks was early thought to be the only mode of formation of mineral deposits and was long considered the most important. Particularly, a filled fissure, called fissure vein (see p. 111) was almost synonymous with ore deposit and is still considered the representative type of mineral deposit. More types of mineral deposits are formed by cavity filling than by any other process, and of these the fissure vein is by far the most common and important.

However, as stated previously, fissure filling is generally accompanied by replacement.

THE PROCESS AND CHARACTERISTIC FEATURES

Cavity filling consists in deposition from solutions of minerals in rock openings. The solutions may be dilute or concentrated, hot or cold, and of magmatic or meteoric derivation. Mostly they are hot, dilute, and magmatic. Precipitation of the minerals is brought about by the processes considered in Chapter 3, of which changes in chemical character and temperature and pressure of the mineralizing solutions are the chief ones.

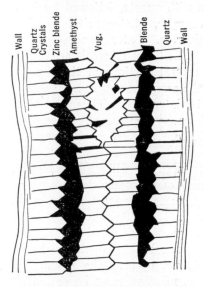

FIG. 5·4–1. Growth of quartz crystals from fissure walls toward center, forming comb structure, and a vug in center containing latest crystals.

The first mineral to be deposited lines the walls of the cavity and grows inward generally with the development of crystal faces pointed toward the supplying solution (see quartz crystals of Fig. 5·4–1). In some cases the same mineral or minerals may be deposited continuously on both walls until the cavity becomes filled or nearly so. Such a type of filling gives rise to homogeneous or *massive* ore. Generally, however, successive crusts of different minerals are deposited upon the first one, perhaps with repetition of earlier minerals, until the filling is complete, and this gives rise to *crustification;* if the cavity is a fissure, a *crustified vein* results (Figs. 5·4–1, 5·4–3, 5·4–4). If the crusts surround breccia fragments, *cockade* ore may result (Fig. 5·4–2). If prominent crystals project from the walls, as in Fig. 5·4–1, it forms *comb structure.* Commonly, the filling may not be complete and open *vugs* (Fig. 5·4–1) remain in the center; some are large enough to admit a man. The vugs may contain one or more sequences of crystals perched on the walls and are eagerly sought by mineral collectors because they often prove to be treasure houses of the beautiful and rare crystals that adorn mineral museums. Such crustified deposits and vugs permit the paragenesis of the minerals to be worked out.

Vein crustification may be *symmetrical* with similar crusts on both

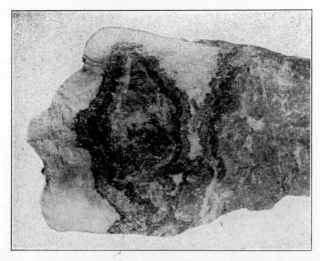

FIG. 5·4–2. Cockade ore formed by deposition of successive crusts of minerals around earlier ore fragments. Length, 8 in. Ikuno mine, Japan. Center, quartz, cassiterite, wolframite; white crust, quartz; black crust, sphalerite; gray crust, pyrite; outside white, calcite. Courtesy Mitsubishi Min. Co.

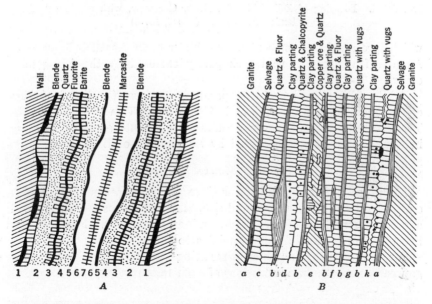

FIG. 5·4–3. Crustified veins. A, Symmetrical crustification, Freiberg, Germany (*after Von Wissenbach*); B, asymmetrical crustification in tin vein, Redruth, Cornwall (*after Phillips*).

sides (Fig. 5·4–3*A*) or *asymmetrical* (Figs. 5·4–3*B*, 5·4–4) with unlike layers on each side. The latter is generally caused by reopening of the fissure, permitting further deposition, as between layers *a* and *b* of Fig. 5·4–3*B*. Several reopenings may occur. Many quartz veins are characterized by *ribbon structure*, consisting of narrow layers of quartz separated by thin dark seams of smeared altered wall rock,

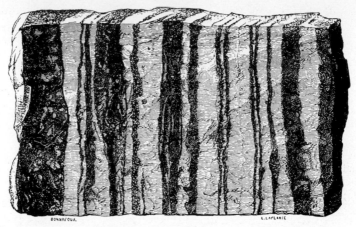

FIG. 5·4–4.   Drawing of Zellerfeld vein, Harz, Germany, with crusts of galena, blende and calcite.   (*Burat, 1870.*)

which presumably represent successive reopenings and movement. Angular fragments of wall rock partly enclosed by the filling may occur.

Cavity filling involves two separate processes: (1) the formation of the opening to start with and (2) the deposition of the minerals. The two may operate almost simultaneously, but generally they are independent processes separated by an interval of time.

Resulting Mineral Deposits

The process of cavity filling has given rise to a vast number of mineral deposits of diverse form and size, and such deposits have yielded a great assemblage of metals and mineral products. Much of the literature on economic geology relates to them.

For convenience, the deposits resulting from cavity filling may be grouped as below, and discussions of each follow in sequence:

1. Fissure veins.
2. Shear-zone deposits.
3. Stockworks.
4. Saddle reefs.

5. Ladder veins.
6. Pitches, and flats; fold cracks.
7. Breccia-filling deposits: volcanic, collapse, tectonic.
8. Solution cavity fillings: cave, channel, gash veins.
9. Pore-space fillings.
10. Vesicular fillings.

1. Fissure Veins

A fissure vein is a tabular ore body that occupies one or more fissures; two of its dimensions are much greater than the third. Fissure veins are the most widespread and most important of the cavity fillings and yield a great variety of minerals and metals. There are several varieties that differ from each other chiefly in form. They are the earliest described type of bedrock deposit, around which the lore of mining has grown.

Formation. The formation of a fissure vein involves (1) the formation of the fissure itself and (2) the ore-forming processes — a distinction often overlooked. The two may have been separated by a long interval of time. Neither can result in the formation of a fissure vein by itself — a coincidence of the two is necessary.

Fissures may be formed by stresses operating within the earth's crust and may or may not be accompanied by faulting. Also, they may be formed or enlarged at the time of mineralization by the intrusive force of the mineralizing solutions which acts as a wedge from below and spreads the rocks apart along some crack or line of weakness. It has been contended that the force of growing crystals may wedge apart the walls of a crack to make a wider fissure. Growing crystals do exert a definite pressure, within the limit of their strength, upon confining substances, but it seems extremely doubtful that this force is sufficient to form fissures, except in yielding rocks under light load near the surface.

Varieties. The varieties of fissure veins are: simple, composite, linked, sheeted, dilated, and chambered; each of these may be massive or crustified.

The simple fissure vein occupies a single fissure whose walls are relatively straight and parallel. Where the walls are irregular and brecciated, particularly the hanging wall, owing to formation under light load near the surface, it is often called a *chambered* vein (Fig. 5·4–5A).

Dilation or *lenticular veins* (Fig. 5·4–5B) are fat lenses in schists. Generally several occur together, like a string of sausages, or they may be disconnected en echelon lenses (Fig. 5·4–5D). They are thought to be caused by the bulging or dilation of schistose rocks due to pressure transmitted by the mineralizing solutions. Some are due to the pulling apart of a pre-existing vein during later metamorphism of the enclosing rock.   In width they range from a few inches to several tens of feet.

A group of closely spaced, distinct, parallel fractures is a *sheeted vein* (Figs. 5·4–5C, 12–2, 12–9).   Each fracture is filled with mineral matter and is separated by layers of barren rock, and the whole is mined as a single deposit lode whose width may be several tens of feet. If individual fractures are linked by diagonal veinlets a *linked vein* is formed (Figs. 5·4–5E, 12–8).

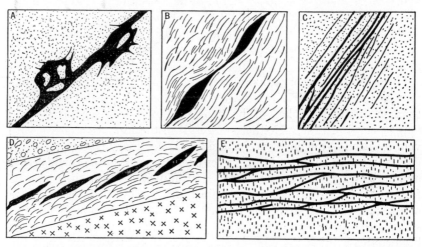

Fig. 5·4–5.   Varieties of fissure veins.   *A*, Chambered vein (*after Becker*); *B*, dilation veins in schist; *C*, sheeted vein, Cripple Creek, Colo.; *D*, en echelon veins in schist; *E*, linked vein.

A *composite vein* or *lode* is a large fracture zone, up to many tens of feet in width, consisting of several approximately parallel ore-filled fissures and connecting diagonals, whose walls and intervening country rock have undergone some replacement.   The Comstock Lode of Nevada is an example.

Physical Features.   Most fissure veins are narrow and range in length from a few hundreds of feet to a few miles.   Few are vertical; most are highly inclined and apex at the surface.   Consequently, their outcrop is the trace of an approximate plane upon the surface.   If the surface is flat, the outcrop will be relatively straight; if irregular, the outcrop will be irregularly curved, depending upon the relief and the relation of the dip to the surface slope, as in Fig. 5·4–6.   Two parallel veins may thus form seemingly divergent outcrops.   Greater disparity results if two veins have parallel strikes but different dips.

To determine the true relation of the veins to each other it is necessary to plot their strike and dip, or project them to a horizontal plane. Misconceptions of the true strike and dip in regions of pronounced topographic relief have given rise to many erroneous deductions regarding the extensions of veins in strike and dip. Many a tunnel has been projected to intersect a vein at depth but has missed it because the outcrop has been confused with the strike.

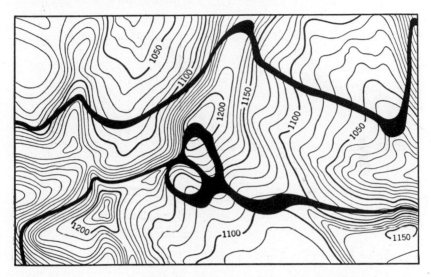

Fig. 5·4–6. Relation of outcrop of 10-foot southerly dipping inclined veins to topography, accurately plotted. The two veins are parallel in strike but of different dip. One would not suspect that the peculiar outcrop of the lower vein could result from the intersection of an inclined plane with an irregular surface.

Veins are seldom planes, notwithstanding the customary textbook illustration. Most of them curve, along both strike and dip, but the dip is apt to be straighter than the strike.

Most veins exhibit irregularity in width, or *pinches* and *swells*, owing to movement of one wall past the other (Figs. 5·4–7, 12–22). A swell followed by a pinch is apt to be followed by another swell. Ordinarily, it is difficult to conceive of wide, straight-walled, inclined veins having existed as open yawning cavities before being filled, but the support of juxtaposed protuberances, as in Fig. 5·4–7, may allow wide pre-mineral spaces to persist.

Displacement along fissure veins is mostly small, although generally present. It is noteworthy that large faults are seldom mineralized, although there are exceptions, such as the Comstock Lode. Ransome has pointed out that in the Coeur d'Alene district the veins exhibit small displacement and that the large faults of the region are barren. Most fault fissure veins have displacements of only a few feet to a few tens of feet, rarely a few hundred feet, and only exceptionally over a thousand feet.

Fissure veins branch, divide, and join again, enclosing *horses* of country rock, split into stringers, and form *brecciated* zones.

The *walls* of fissure veins are commonly marked by a band of *selvage* or *gouge*, which is a claylike or gummy substance formed by movement of one wall upon the other and subsequently altered. The vein filling

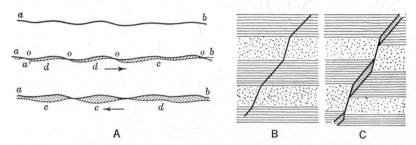

FIG. 5·4–7.   Pinches and swells in fissures, produced by movement along irregular fissures, and with walls supported at juxtaposed protuberances.   (*A, from De la Beche.*)

often scales off cleanly from a gouge wall constituting what is called a *free wall.* In contrast, the walls may be *frozen;* i.e., the vein matter, particularly quartz, adheres to the wall so tenaciously that in mining it cannot readily be separated, and the ore becomes diluted with country rock. The vein matter may be sharply delimited against the country rock or it may merge into it, forming commercial walls whose limits are determined by the economic workable ore.

The *vein matter* may consist of several minerals. Generally both gangue and ore minerals are present. Cavity fillings, unlike most other classes of deposits, generally contain more than one gangue mineral, such as quartz, calcite, and rhodochrosite. Several metallic minerals are also commonly present.

Relation of Fissure Veins to Each Other. Fissures seldom occur alone but tend to occur in groups (Fig. 5·4–8), and if the fissures of a group are of the same age and have approximately parallel strike and dip they constitute a *fissure system* (Fig. 5·4–8A). Fissures formed at the same time are *cognate fissures* as in Fig. 5·4–8D; those with parallel strike but intersecting dip are said to be *conjugated* (Fig. 5·4–8F). Different systems intersect each other; at Butte, Mont.

(Figs. 13–2, 13–3), there are seven systems each of which displaces the earlier ones. Other notable examples of intersecting systems are

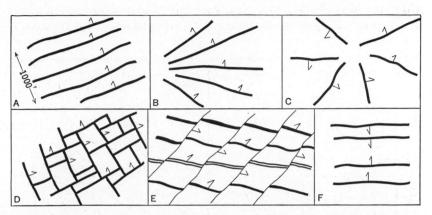

FIG. 5·4–8. Classes of fissure systems. *A*, Parallel system; *B*, fan-shaped; *C*, radial; *D*, intersecting cognate; *E*, intersecting systems; *F*, conjugated.

those of Freiberg, Germany, and the Silverton-Telluride district, Colo. (Figs. 5·4–9, 5·4–10).

Two intersecting systems may differ in age (Fig. 5·4–8*E*), or they

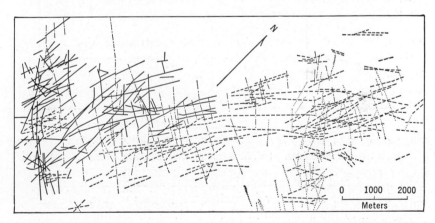

FIG. 5·4–9. Part of mineral district of Freiberg, Germany, showing five systems of veins. 1 and 2 are rich silver-lead veins trending N–S and NW (solid black); 3 (dash lines), quartz-lead veins; 4 (dotted), barite-lead veins; and 5 (dots and dashes), barren veins.

may be cognate and of the same age (5·4–8*D*), like the cleavage of augite. Those differing in age generally displace each other, carry different ores, and have independent legal rights (see p. 117); those of the same age carry similar ores, and do not fault each other. To distinguish between them is important. Either they are (1) cognate; (2) different systems that displace each other but both are pre-mineral; or (3) different systems, and older vein fillings faulted by younger.

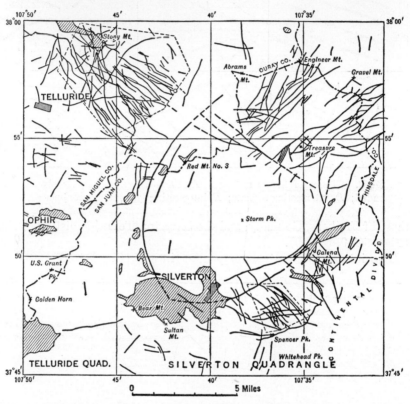

Fig. 5·4–10. Vein systems of the Silverton-Telluride district, Colo. Shaded areas are intrusives. (*W. S. Burbank, Colo. Sci. Soc. Proc.*)

In cases 1 and 2 the intersecting systems would be filled simultaneously by similar ores, the vein matter would pass without break from one system to the other, as in Fig. 5·4–11, and would not be faulted; and the junctions would legally be branches and not faulted ends. In A no displaced part would be found beyond the junction, but in B the pre-mineral displaced end is present.

In case 3 (above) the older veins (Fig. 5·4–12) may carry different ores than the younger, as in the classical example of Freiberg, Germany; and since the veins themselves are faulted they carry different legal ownership rights. The filling of the different systems may have been widely separated, or the mineralization may have been a long-drawn-out, continuous event, within which different periods of fissuring and faulting occurred, as at Butte, Mont., where similar filling occurs in three fissure systems that displace each other. The relative ages of the systems may be determined from the intersections, the fillings, and their relation to rocks of known age. The older veins may be slickensided, bent, and crushed; and the younger veins may cut

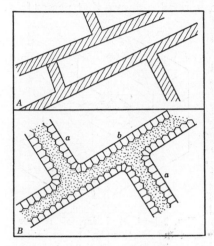

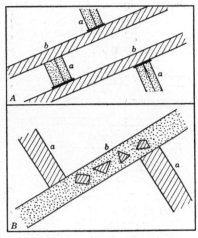

Fig. 5·4–11. Simultaneous filling of pre-mineral fissures. *A*, Member of intersecting cognate system the parts of which do not displace each other; *B*, fissure of system *a* displaced by *b*, both of which are pre-mineral. Members of both *A* and *B* form junctions.

Fig. 5·4–12. Veins of earlier system *a* displaced by those of later system *b* and containing different ores (also drag ore from *a*). These do not constitute junctions.

sharply across the older and may contain *drag ore* from the older veins (Fig. 5·4–12). The older veins are commonly displaced by the younger in the same direction and for similar distances, as, for example, at Butte, where all the older veins of the east-west system are thrown to the left by members of the younger northwest system.

The legal features just mentioned relate to the apex rule, in American mining law, by which the prior locator of the top or apex of a vein is entitled to follow it down dip, within the end lines of the mining claim, beneath the land of another. If, down the dip, the vein branches, the holder of the apex is entitled to the ore beneath the junction; if, however, the junction is a faulted end, the holder of the apex may not be entitled to the ore beneath the junction. Much mining litigation has resulted in the interpretation of junctions.

Effects of Change of Formation on Fissure Veins. Fissure veins tend to change in both fissuring and filling when they pass from one rock formation into another. This is due chiefly to the different physical behavior of different rocks toward stresses, but the chemical composition of the rocks also influences mineral deposition.

If alternate layers of rubber and glass, as in Fig. 5·4–13, are subject to stress, each layer of brittle glass will fracture, but the intervening rubber will bend. Similarly a strong, regular fissure that passes from

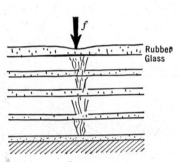

Fig. 5·4–13. Alternate layers of glass and rubber. Under an applied force (*f*) the layers of incompetent rubber bend but the layers of brittle glass become fissured.

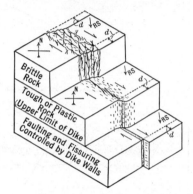

Fig. 5·4–14. Relation of fissures to each other and to enclosing rocks, Arrastre Basin, Colo. (*W. S. Burbank, Colo. Sci. Soc. Proc.*)

a tough rock, such as greenstone, into a brittle rock like sandstone is likely to break up into a series of irregular, interlacing cracks as in Fig. 5·4–15*A* or in Fig. 5·4–15*F*. This feature is remarkably well shown (Fig. 5·4–14) by Burbank at Arrastre Basin, Colo. Conversely, a strong fissure passing from a brittle rock to a yielding rock, such as shale, is likely to be distorted or disappear (Fig. 5·4–15*B*). Or, such a fissure may undergo constriction where it passes through a narrow band of different rock, such as a dike (Fig. 5·4–15*C*). Again, it may divide into a group of stringers, as in Fig. 5·4–15*D*. If a fissure enters or passes through a strongly schistose or sheared rock it may break up into an en echelon arrangement of disconnected lenses.

Where a fissure enters another formation at a low angle of incidence it may be reflected or refracted as in Fig. 5·4–15*E*, but if the angle of incidence is high, little or no change may be expected. Rarely a fissure may undergo no change in passing from one formation to another.

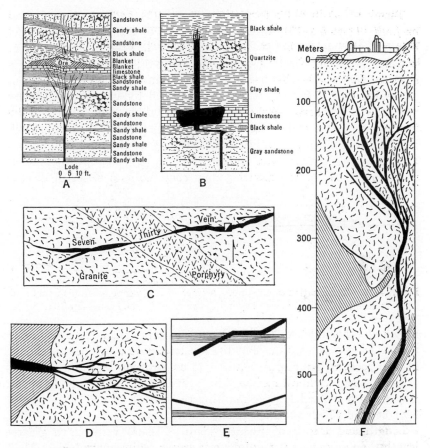

Fig. 5·4–15.   Relation of fissures to physical features of rocks.   *A*, Strong vein in competent rocks breaking up into stringers and disappearing upon entering in-competent rocks.   (*Ransome, U. S. Geol. Survey.*)   *B*, Strong fissure at Ouray deflected in shale and disappearing in black shale above.   (*Irving, U. S. Geol. Survey.*)   *C*, Vein at Georgetown, Colo., constricted in passage through porphyry.   (*Spurr and Garrey, U. S. Geol. Survey.*)   *D*, Division of strong vein into stringers upon entering schist.   Gottlob Morgengang vein, Freiberg, Germany.   (*Beck.*)   *E*, Refraction and diversion (upper) and deflection (lower) of a vein upon encoun-tering an incompetent rock.   *F*, Strong fissure in schist diverging upward into minor fissures in brittle andesite, Mazarron, Spain.   (*After Pilz, Z. prakt. Geol.*)

Terminations of Fissure Veins. Since fissures cannot enter rock formations later in age than themselves, unless there is later movement along them, but may continue into older rock formations, obviously it is essential to know their relative age in the geologic history of the region.

Terminations in older formations occur: (1) *against other fissures or faults* (Fig. 5·4–12), where the presence or absence of fault phenomena will determine whether the vein has been displaced, or whether it simply abuts a pre-mineral fissure; (2) *upon entering a different formation*, a fissure may fray out into stringers and die out gradually (Fig. 5·4–15*A, B*), or be absorbed by incompetent rocks; (3) *within the same formation*, termination may be indicated by fraying out at the ends (Fig. 5·4–15*D*), or by a gradual pinching out, in which case three possibilities exist: it may be the final termination (Fig. 5·4–16*a*); it may be a temporary pinch beyond which is another swell (5·4–16*b*); or the fissuring may continue as a series of discon-

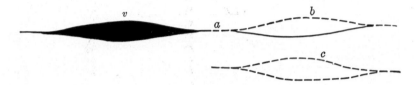

FIG. 5·4–16.   Termination of a fissure.   A pinch of vein *v* may mean (*a*) final termination; (*b*) another swell beyond the pinch; (*c*) an en echelon offset.

nected en echelon lenses (5·4–16*c*) offset to either side.   In many cases the fissuring may continue but the ore terminates, owing to a change in wall rock, to structural features, to constriction of width, or other causes.   The termination of ore where the fissuring continues, however, may be only a local feature marking the end of an ore shoot (see p. 123) beyond which another shoot may occur.

Length and Depth of Fissure Veins.   Fissure veins have greater length than any other type of deposit except sedimentary beds.   Most of them, however, do not exceed a few hundred feet in length; many are a few thousand feet in length; and a few are several miles in length. For example, the Halsbrücker Spat vein at Freiberg, Germany, has been followed for 5 miles and the Rosenbofer vein in the Harz Mountains, Germany, for 10 miles; the Veta Madre at Guanajuato, Mexico, is 7 miles long, the Comstock Lode, 4 miles; and veins many miles long occur in the Mother Lode of California and in the Linares district, Spain.   The great barren quartz veins near Great Bear Lake, Canada, are reported to be 50 miles long.   Similarly, the barren quartz Pfal vein in Bavaria is stated by Lindgren to be 140 kilometers long.

The *depth* of fissure veins is generally less than their length.   Most of them disappear within a few hundred feet of the surface, but those that continue to depths of several thousand feet are numerous.   Some of the deep mines of the world are on fissure veins.   The Kolar mine in India, for example, is over a mile deep, as are also mines on quartz veins in California and northern Ontario.

The relation between length and depth of veins is of vital economic importance to mining men. In general, long, strong fault fissure veins are likely to extend to great depths, and, conversely, short, weak veins are likely to be shallow. However, there is no mathematical ratio between length and depth. Generally the length is much better known

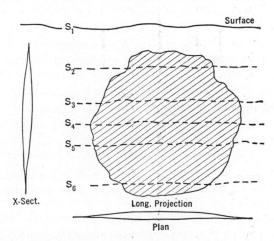

FIG. 5·4–17. Relation of length and depth of fissure vein to outcrop length. Shaded area is a longitudinal projection of an ideal nonfaulted simple fissure vein that pinches out in all directions. With surface at S_1, the vein is " blind " or has a subsurface apex; at S_2 the depth is greater than outcrop length; at S_3, S_4, or S_5 the depth is approximately one-half of the outcrop length (a diameter as compared with a radius); at S_6, the outcrop length is greater than the depth and the vein roots are exposed.

than the depth since it can be tested with greater ease and less cost, whereas the actual depth is seldom determined because, in general, the ore has given out before the bottom of the fissure has been reached, and further tests of depth would involve costly exploration that is seldom justified. An old rule is often quoted, namely, that the depth will be equal to the length. This does not appear to hold. However, if the rule were to be restated, that " the depth may equal about *half* the length," it would probably prove nearer correct for most veins, since the length of outcrop is, in general, a function of the depth of erosion, as is shown below.

Ordinarily, an ideal simple fissure vein is as shown in Fig. 5·4–17. If the surface of erosion is at S_1, it has no outcrop and is a *blind vein*. If the erosion surface has reached S_2, there is a short outcrop of vein whose length is much less than the depth; if it is in positions S_3, S_4, or S_5, the outcrop length is approximately twice the depth. But, if the surface is at S_6, only the roots outcrop, and the depth is less than half of the outcrop length. On the average, the chances are that erosion will intersect the vein somewhere in its wide medial portion rather than at its extreme top or bottom, and, therefore, it follows that in general the depth will be about one-half the outcrop length. This would apply, however, only to nonfault veins within a single homogeneous formation.

Predictions as to Depth. If it can be ascertained whether the outcrop of a vein represents the top, medial portion, or roots, then approximate predictions may be made as to the expected depth on the basis of the rule given above. Lacking this information, the rule itself can be applied in want of more precise data.

In general it is difficult to determine which part of a vein outcrops, but certain features may suggest it. (1) The top may be indicated if it is known that the vein is confined to a formation that has not undergone much erosion. For example, some Tertiary veins near Chinapas, Mexico, occur in an andesite flow and die out in overlying tuff. The tuff has been eroded only slightly below the andesite floor. Therefore, the present vein outcrop is near its top. Blind veins, of course, give a definite top. (2) In rare cases of young fault veins, if fault scarps along the veins persist, they indicate scant erosion. (3) Associated placer deposits may give a clue as to the amount of erosion if it can be shown that the placers were derived from the vein or veins under consideration. Such a check in Alaska confirmed a conclusion that the roots of certain veins were exposed and that only shallow depth might be expected. The placer gold ceased upstream above the vein outcrops. The total placer gold, extracted and in the ground, was estimated. Many samples gave an average value of the veins; their area was computed. To yield the amount of placer gold, thousands of feet of the veins must have been eroded even under the assumption that higher up their value and their area were three times as great as at the present surface. (4) The degree of supergene enrichment may give a clue. Where rich supergene sulphide ores overlie lean protore, it indicates much erosion (Chap. 5·8), since a high degree of enrichment is possible only by the oxidation of much of the overlying vein. For example, a copper vein with hypogene ore averaging 1 percent, overlain by 500 feet of 5 percent supergene ore, must have undergone a minimum of 2,000 feet of oxidation, provided the tenor and width persisted unchanged above; if the oxidized zone is now 100 feet deep, then at least 1,900 feet has been eroded.

From such data, approximations as to the depth of erosion and rough predictions as to the expected depth may be made. It should be remembered that fissure veins are essentially shallow features and that all of them will terminate at depth.

Distribution of Values. The valuable hypogene minerals are never equally distributed throughout a fissure vein. They tend to be concentrated toward hanging or foot wall, or in the center. Horizontally,

SOME IMPORTANT FISSURE VEIN DEPOSITS

ORES	EXAMPLES
Gold	Cripple Creek, Camp Bird, San Juan, Colo.; Mother Lode, Grass Valley, Calif.; El Oro, Veta Madre, Mexico; Porcupine, Kirkland Lake, Ontario; Kalgoorlie, Australia
Silver	Sunshine, Id.; Tonopah, Nev.; Fresnillo, Pachuca, Guanajuato, Mexico; Potosi, Huanchaca, Bolivia; Cobalt, Ontario; Tintic, Utah
Silver-lead	San Juan, Colo.; Clausthal, Freiberg, Germany; Przibram, Austria
Copper	Parts of Butte, Mont. and Cerro de Pasco, Peru; Walker Mine, Calif.
Lead	Clausthal, Freiberg, Germany; Przibram, Austria; Linares, Spain
Zinc	Butte, Mont.; Sardinia
Tin	Llallagua, Huanani, Oruro, Bolivia; Cornwall, England; Erzgebirge, Germany
Antimony	Hunan, China; Mayenne, France
Cobalt	Cobalt, Ontario; Annaberg, Germany
Mercury	New Idria, Calif.
Molybdenum	Temiskaming County, Quebec
Radium, uranium	Great Bear Lake, Canada; Joachimsthal, Czechoslovakia; Katanga, Belgian Congo
Tungsten	Boulder County, Colo.; Kiangsi, China; Chicote, Bolivia
Fluorspar	Illinois-Kentucky region
Barite	Missouri; Harz, Germany
Gems	Colombia

some parts may be rich or lean or barren, and, vertically, they decrease or increase with depth, or some intermediate levels may be rich or lean or barren. Where large-scale concentrations occur they are termed *ore shoots*, which are described on pages 158 to 163. These vagaries of distribution of values constitute one of the hazards of mining fissure veins.

Examples of Important Fissure Vein Deposits. Some of the world's most famed deposits, both ancient and recent, are fissure veins. Included among them are some of the deepest and richest mines of the world, as well as some of the oldest worked deposits. Although they do not contain the enormous tonnages of some magmatic, sedimentary, or replacement deposits, vast treasures of gold and silver have been won from them. They have also contributed largely to the world's production of copper, lead, zinc, tin, tungsten, mercury, and fluorspar. They have been the chief contributors of antimony, cobalt, germanium, uranium, and radium. Two areas that are claimed to be the " richest hills on earth," namely Butte, Mont., and Potosi, Bolivia, are mostly fissure vein deposits.

A brief list of important fissure vein deposits is shown on page 123.

2. SHEAR-ZONE DEPOSITS

The thin, sheetlike, connected openings of a shear zone serve as excellent channelways for mineralizing solutions, and some deposition takes place within the seams and crevices in the form of fine grains or thin plates of minerals. The open space is insufficient to accommodate enough of the nonferrous metals to constitute ore, but gold with pyrite forms workable deposits, as at Otago, New Zealand. Many lodes and " fahlbands " formerly thought to have been formed by impregnation of the shear openings are now known to have originated by replacement of the sheared rock. Shear zones, because of their large specific surface, are particularly favorable for replacement, and this process has given rise to many large and valuable ore deposits.

3. STOCKWORKS

A stockwork is an interlacing network of small ore-bearing veinlets traversing a mass of rock. The individual veinlets rarely exceed an inch or so in width or a few feet in length, and they are spaced a few inches to a few feet apart. The interveinlet areas may in part be impregnated by ore minerals. The entire rock mass is mined. In general, the veinlets consist of open-space fillings that exhibit comb structure, crustification, and druses.

Stockworks may occur as separate bodies, or in association with other types of deposits. As separate bodies they may attain a large size, as at Altenberg, Germany (Fig. 5·4–18), where the tin stockworks have a diameter of over 3,000 feet. Such deposits, in general, yield low-grade ore because the entire rock mass is mined, but the large tonnages compensate for this. Outlines are irregular, and the walls merge into unprofitable country rock. Stockworks occur as large bulges on the hanging wall of the great Veta Madre gold vein in Mexico (Fig. 12–23) ; in association with replacement deposits at Leadville, Colo.; and associated with veins at Victoria, Australia; Slocan, British Columbia; and tin-tungsten veins in Bolivia.

Stockworks yield ores of tin, gold, silver, copper, molybdenum, cobalt, lead, zinc, mercury, and asbestos. Much of the world's supply of lode tinstone in the past has come from stockworks at Altenberg and Zinnwald, Germany; Mount Bischoff, Tasmania; Cornwall, England; New South Wales; and South Africa; disintegrated stockworks have yielded much alluvial tin in the Netherlands Indies and Malaya. Gold is won from stockworks in Victoria, Mexico, and Juneau, Alaska. Bastin describes a gold-silver-bearing stockwork at Quartz Hill, Gilpin County, Colo., that is 500 to 800 feet across and has been traced downward for 1,600 feet. Large low-grade, gold stockworks in granite

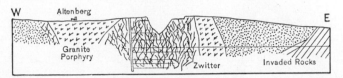

FIG. 5·4–18. The tin stockworks of Altenberg. (*After Dalmer, Z. prakt. Geol.*)

porphyry have been explored at Lake Athabaska, Canada. These consist of quartz veinlets carrying pyrite, a little chalcopyrite, and free gold. Most of the gold of the large low-grade Alaska Juneau mine is contained in irregular quartz stringers. Silver-cobalt ores are won from a large stockwork in Schneeberg, Germany, according to Müller, and a silver stockwork is mined at Fresnillo, Mexico.

Stockwork veinlets are formed by: (1) Crackling upon cooling of the upper and marginal parts of intrusive igneous rocks. (2) Irregular fissures produced by tensional or torsional forces. For example, fault movement downward along a curved fissure produces a crackling where the hanging wall moves over humps on the footwall, as shown by Wandke and Martinez for the great Veta Madre vein in Guanajuato, Mexico.

4. SADDLE REEFS

If a thick stack of writing paper is sharply arched, openings form between the sheets at the crest of the arch. Similar ore receptacles are formed when alternating beds of competent and incompetent rocks, such as quartzite and slate, are closely folded. When filled by ore they resemble the cross section of a saddle, hence their name (Fig. 5·4–19). The saddle reefs of Bendigo, Australia, are the type examples and have yielded over 300 million dollars in gold.

The saddles are repeated in bed after bed down the axial plane or "center-country," and this may be vertical or inclined (Fig. 12–14).

Fifteen " lines of reef " are known in Bendigo, of which five lines of superimposed saddles have been highly productive. The saddles are 20 to 50 feet across and have irregular crests, and the leg depth is mostly less than 100 feet. The cross sections are small, but one saddle has been followed horizontally for 9,000 feet in Bendigo, and mining has reached 4,600 feet vertically. Superimposed saddles are

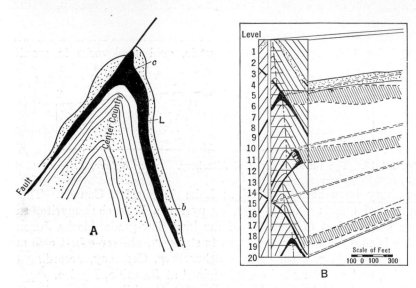

Fig. 5·4–19. *A*, Typical saddle reefs of Bendigo, Australia, *c*, cap; *L*, leg; *b*, back; black is gold quartz ore; *B*, three types of Bendigo reefs and methods of mining them. See also Fig. 12–14, page 450. (*After Report by Bendigo Mines, Ltd.*)

generally connected by faults (Fig. 5·4–19*B*), which according to Stone conducted the ore solutions to the saddles; ore seldom is found where faults are lacking.

The ore emplacement has been chiefly by cavity filling, but Stillwell and Stone believe that considerable replacement has also taken place.

Similar reefs occur in other parts of Victoria and in New Zealand, and smaller pitching saddle reefs occur in Nova Scotia.

5. LADDER VEINS

" Ladder vein " is the name applied to more or less regularly spaced, short, transverse fractures in dikes (Fig. 5·4–20). These generally extend roughly parallel to each other, from wall to wall of the dike. Their width is thus restricted, but they may extend for considerable distance along the dikes. Such openings may become filled with mineral matter to form commercial deposits, as the Morning Star gold-bearing dike in Victoria, Australia. The individual fissures may form separate veins, or if they are closely spaced the dike as a whole may be mined. Such transverse veins in a vertical dike resemble the rungs of a ladder, hence their name.

The fractures that constitute ladder veins have generally been considered to be contraction joints. However, some fissures extend be-

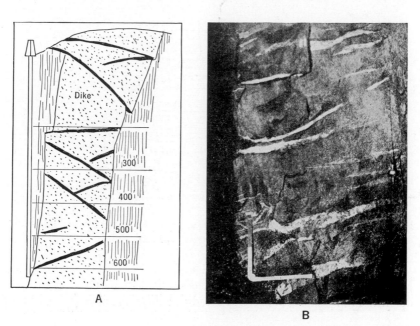

FIG. 5·4–20. *A*, Morning Star dike, Woods Point, Victoria. (*After Whitelaw-Junner, Econ. Geol.*) *B*, Ladder vein; ladder quartz in dike. All Nations mine, Matlock, Victoria. (*Whitelaw-Junner, Econ. Geol.*)

yond the dikes into the walls, which indicates that cooling contraction may not be their cause. Also, Grout found that the total width of quartz filling in ladder veins in Minnesota greatly exceeded the possible contraction through cooling. He suggested that tangential movement of weak wall rocks opened up transverse tension cracks in an enclosed or brittle dike. Tangential movement is probably more important than contraction.

Ladder veins are not numerous or important, but examples are the gold quartz ladder veins of the Morning Star, Waverly, and All Nations mines in Victoria, Australia; molybdenite veins in New South Wales; and the copper ladder veins of Telemarken, Norway.

6. Pitches and Flats — Fold Cracks

Under light load, slumping or gentle synclinal folding of brittle sedimentary beds gives rise to a series of connected tension cracks or openings collectively known as *pitches and flats*. Gentle open folding also forms *anticlinal tension cracks* at the crests of anticlines (or troughs of synclines).

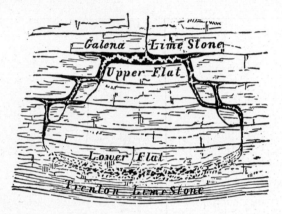

Fig. 5·4–21. Pitches and flats, containing lead-zinc ores, Mississippi Valley. (*After Chamberlin.*)

Pitches and Flats. Figures 5·4–21 and 5·4–22 show typical openings of the upper Mississippi Valley. These openings occur in the Galena dolomitic limestone and are encrusted or filled with zinc and lead ores.

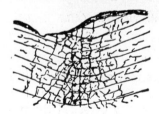

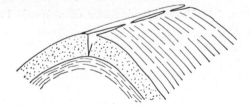

Fig. 5·4–22. Brecciation at bottom of syncline, Preston Point, Wis. (*After Chamberlin.*)

Fig. 5·4–23. Tension cracks along an anticline.

Large museum crystals of galena and sphalerite project from the walls into the open spaces, and unfilled central cavities or vugs are the rule. The mineralized area may be 75 feet wide, 50 feet high, and up to 1,000 feet long. These are considered by H. F. Bain to have been formed by settling or slumping of the Galena limestone due to decrease in bulk of the underlying oil rock.

Fold Cracks. Anticlinal tension cracks, and synclinal ones, produced by folding under light load are shown in Figs. 5·4–22 and 5·4–23. The anticlinal cracks generally have small vertical extent but may

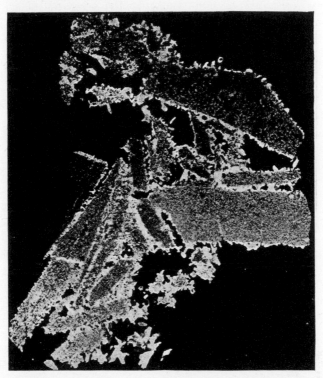

Fig. 5·4–24. Breccia-filling deposit. Breccia cavities are lined with quartz crystals and filled with chalcopyrite. Latite ore, Pilares mine, Sonora, Mexico. (*Locke, Econ. Geol.*)

extend along the axis of the fold, as a series of disconnected fractures, for considerable distances. They are generally wedge shaped and commonly terminate abruptly, either up or down, against another bed. Such anticlinal tension cracks contain gold ores in Victoria, Australia, zinc ores in Mexico, and asphalt in Utah. Some of the copper bodies at Kennecott, Alaska, are in synclinal tension cracks.

7. BRECCIA-FILLING DEPOSITS

The haphazard arrangement of the angular rock fragments in breccias gives rise to numerous openings that permit the entry of solutions and mineral deposition, forming breccia-filling deposits (Fig. 5·4–24). The breccias may result from volcanism, collapse, or shattering.

Volcanic Breccia Deposits. Explosive volcanic activity gives rise to bedded breccia deposits and breccia pipes or craters. The unconfined bedded deposits are unimportant ore loci, but the confined pipes offer an opportunity for concentrated mineralization.

The pipes are vertical or highly inclined, roughly oval-shaped, and are filled by breccias. They are ideal channelways for mineralizing solutions and for ore deposition. Beautiful crustification of gangue

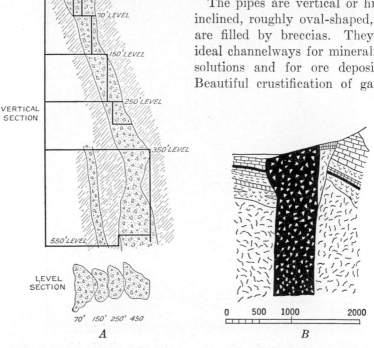

FIG. 5·4–25. *A*, Vertical section and plans of Bull Domingo gold-bearing volcanic breccia pipe in schist, near Lake City, Colo. (*Emmons, U. S. Geol. Survey.*) *B*, Vertical section of Ollie Reed pipe, Leadville, Colo. (*After Emmons-Irving-Loughlin, U. S. Geol. Survey.*)

and ore minerals is often seen around the rock fragments; and vugs partly filled by dainty crystals delight the mineralogist and indicate the sequence of mineral deposition.

The old Bull Domingo and Bassick breccia pipes (Fig. 5·4–25) of the San Juan area in Colorado are examples. The Bassick pipe is about 30 by 100 feet across and has been mined to a depth of 1,400 feet. It has yielded gold, silver, lead, and zinc. The great Braden copper deposit of Chile is considered by Lindgren and Bastin to be an old volcanic crater in part filled by breccia, but Brüggen holds that it is not a crater. The copper minerals in part fill cavities in the breccia.

Collapse Breccia Deposits. In the caving methods of mining, an excavation is started at the bottom of a block of ore; the roof rock then caves, and this caving extends itself upward until the block is a mass of jumbled, angular fragments of ore, with considerable open

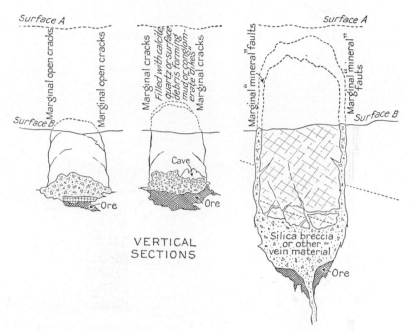

FIG. 5·4–26. Vertical sections of collapse breccia pipes produced by mine oxidation and " mineralization " collapse. (*Kingsbury-Locke, Econ. Geol.*)

space. It is an artificial breccia. Similarly, in nature, the roof of a solution opening may collapse and the collapse may extend upward to yield a mass of broken rock. Such subsidence may extend upward for hundreds of feet and may be bordered by crackled and fissured rock.

Breccias of this type (Fig. 5·4–26) have been described as cave-collapse breccias by Watson and Ulrich, as " crackle," and " founder " breccias, by Norton in the Appalachian zinc areas, and as mineralization-stoping breccias by Locke. The Cactus pipe in Utah (Fig. 5·4–27*A*), described by Butler, extends nearly 900 feet vertically. Locke believes that the subsidence was initiated by solution caused by the mineralizing solutions themselves, during the earliest stages of mineralization. He cites as other examples the Southwest pipe at

Bisbee, Ariz., the South Ibex and Cresson in Colorado (Fig. 5·4–27*B*),
and the Pilares, Catalina, and Duluth pipes in Sonora, Mexico.
Walker cites the Emma chimney, in the Little Cottonwood, Utah, as
an example of cave-collapse breccia; it is mineralized by silver-lead-

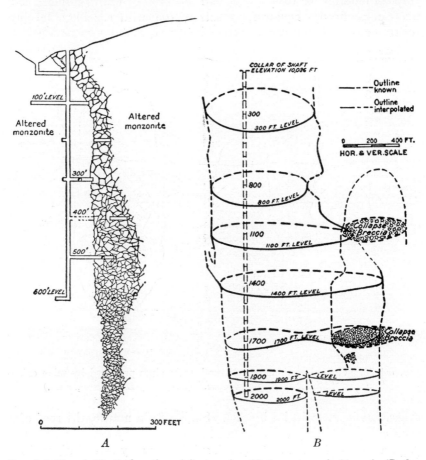

Fig. 5·4–27.   *A*, Vertical section of Cactus pipe, Utah, composed of breccia (*Butler-Locke, Econ. Geol.*).   *B*, Outline of Cresson pipe of basaltic breccia, Cripple Creek, Colo.   (*Loughlin, A. I. M. E.*)

zinc ore to a depth of over 1,000 feet.   The ore has been emplaced
chiefly by open space filling in the breccia cavities.

Wisser describes similar collapse breccias in Bisbee, Ariz., that were
initiated by shrinkage attendant upon sulphide oxidation, the effects
of which extend upward for over 1,000 feet.   Some reach the surface,
and such breccias and accompanying crackled areas are considered

to indicate the presence of post-breccia ore below them. The breccias themselves are not filled by ore, since they are post-ore features.

Whether the subsidence in all cases cited by Locke was initiated by hypogene solutions is open to question; it is not established for the Pilares mine in Mexico.

Tectonic Breccia Deposits. Breccias produced by folding, faulting, intrusion, or other tectonic forces have been referred to variously as

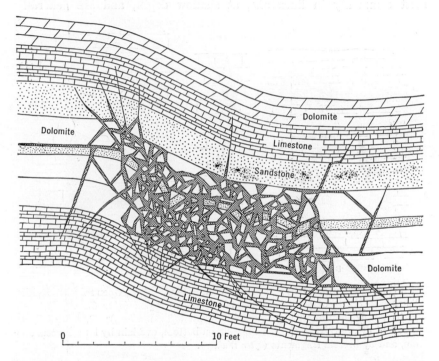

FIG. 5·4–28. Tectonic breccia formed by shattering of dolomite along axis of bend, forming an " ore run " in northern Arkansas lead and zinc district. (*After McKnight, U. S. Geol. Survey.*)

crush, rubble, crackle, and shatter breccias (Fig. 5·4–28). In the Appalachian zinc deposit region, Currier distinguishes rubble from shatter or crackle breccias. The rubble breccias are characterized by prominent relative displacement of fragments and by some rounding, whereas the shatter or crackle breccias are composed of angular fragments that show little rotation. The breccias localize the zinc ores, and in Currier's opinion, they were formed by warping or folding and by thrust faulting. Similar shatter breccias, according to Fowler, localize ore in the Tri-State zinc district. The zinc bodies at Mascot,

Tenn., which are examples of tectonic breccias, are limited to a stratigraphic range of about 200 feet, average 50 to 60 feet in thickness, are up to 150 feet wide, and are exposed for 1,000 feet in depth.

8. Solution Cavity Fillings

Various types of solution openings in soluble rocks have afforded receptacles for primary and secondary mineral deposits. They occur most commonly in limestone, at shallow depth, and are generally

Fig. 5·4–29. Solution caves and cavities in limestone. *A, B,* Open solution cavities lined with crusts of crystals, Wisconsin lead-zinc region (*after Chamberlin*); *C,* gash vein or solution enlargement along joint (*after Whitney*); *D,* solution cave occupied by ore (black) and cave breccia on bottom, overlain by later breccia and ore, and by breccia fragments (*after Walker*).

believed to have been dissolved above the water table by surface waters charged with carbon dioxide, although Davis has postulated their formation below the water table. Most solution cavities are modifications of pre-existing openings, such as joints, fissures, or bedding planes, resulting in caves, galleries, and gash veins.

Cave Deposits. Caves of various shapes and sizes (Fig. 5·4–29) are characteristic of limestone plateau areas that have undergone prolonged erosion. They are commonly accompanied by " sinkholes " that result from roof collapse or from surface solution of joint or fissure intersections. Joint solution gives rise to funnel-shaped sinks; collapse gives large irregular openings, and the floors of the underlying caves are strewn with broken pieces of rock and mud. Caves unaccompanied by sinkholes are generally floored with "cave earth," a residual accumulation of the insoluble materials of the limestone.

Small caves may be almost completely filled by ore minerals, but large caves generally contain only peripheral crusts of ore minerals, among which may be large and beautiful crystals. It was formerly thought that all large, massive, irregular ore bodies in limestone were cave fillings until replacement was realized. Caves are common containers of zinc and lead ores, such as those of Wisconsin and Illinois, and of oxidized ores of copper, lead, silver-lead, zinc, vanadium, and other metals. The caves of Bisbee, Ariz., have been particularly productive of oxidized copper ores in the past. Caves also yield nitrates, clays, fluorspar, barite, strontianite, guano, and other materials. In general, the individual deposits are relatively small and disconnected.

Galleries. Galleries are horizontal or gently inclined attenuated caves, which result from solution along fissures. They probably represent former underground streams. They yield ores similar to those found in caves.

Gash Veins. "Gash vein" is frequently erroneously employed to designate small wedge-shaped fissures in stressed, brittle rocks. The term as originally applied by Whitney, however, designates vertical solution joints in limestone (Fig. 5·4–29*C*). Solution is essential or they are not gash veins. They are confined to single formations, seldom reach 200 feet in depth, and widen and narrow conspicuously. In some respects they resemble fissure veins. They are common in limestone regions; those in the upper Mississippi Valley were the first described.

The fillings are characterized by crustification, large vugs, and beautiful crystals and consist of lead, zinc, silver-lead, copper, fluorspar, and barite. The deposits are mostly small, although they may be rather rich.

9. PORE-SPACE FILLINGS

Pore spaces in rocks may contain ores, in addition to oil, gas, and water. Copper ores occupying sandstone pores, known as the Red Bed ores, occur in Texas, New Mexico, Arizona, Colorado, and Utah. The best ones are in New Mexico, but none of them are of present economic importance. Similar sandstone copper ores occur in the Permian red beds of the Pre-Ural region, where large reserves average 2 to 2.5 percent copper; in the Don Basin and Djeskasjan, Russia; in Turkestan; in Nova Scotia; Germany; and the Midlands, England. Copper in pore spaces in Triassic arkose was formerly worked at Bristol, Conn. Similarly, lead ores have been mined in Rhenish

Prussia and silver ores at Silver Reef, Utah. Vanadium and uranium (with radium) ores occur as impregnations in sandstone in western Colorado and Utah, and quicksilver ore impregnates sandstone in California and Arkansas. For oil and gas occurrences see Chapter 16, and for subsurface water, Chapter 25.

10. VESICULAR FILLINGS

Permeable vesicular lava tops may serve as channelways for mineralizing solutions. In the Lake Superior region, vesicles in basalts are filled with native copper and have given rise to some of the greatest copper deposits in the world, which have been followed down-dip for nearly 9,000 feet. (See pp. 496–498.)

Somewhat similar, but at present, noncommercial deposits occur in the White River region of Alaska and the Yukon and around the Coppermine River in Arctic Canada. They also occur in Nova Scotia, New Jersey, Brazil, Faroe Islands, Norway, Germany, Siberia, and the Russian Arctic.

Selected References On Cavity Filling

Genesis of Ore Deposits. FRANZ POŠEPNÝ. A. I. M. E., New York, 1902. Pp. 54–188. *A classic account of various types of cavity-filling deposits.*

General Reference 8. Chaps. 12, 13, 15, 23. *Descriptions of various types of openings.*

Structural Relations of Ore Deposits. S. F. EMMONS. In Ore Deposits, A. I. M. E., 1913. *A general discussion of rock openings filled by ore.*

Principles of Economic Geology. 2nd Ed. W. H. EMMONS. McGraw-Hill, New York, 1940. Chaps. 12, 16. *Rock openings.*

Types of Ore Deposits. Edit. by H. F. BAIN. Min. and Sci. Press, San Francisco, 1911. Pp. 77–130 by H. F. BAIN and E. R. BUCKLEY. *Openings of lead-zinc ores.* Pp. 324–353, by R. A. F. PENROSE, JR. *Causes of ore shoots.*

Structural Control of Ore Deposition. C. D. HULIN. Econ. Geol. 24:15–49, 1929. *Kinds of openings that control ore deposition.*

Vein Systems of Anastre Basin, Colorado. W. S. BURBANK. Colo. Sci. Soc. Proc. 13, No. 5, 1933. *An interesting group of vein systems.*

General Reference 10. *Brief résumés of many hydrothermal deposits that exhibit localization of ore by structural features.*

General Reference 15. *This volume contains many descriptions of Canadian deposits that exhibit structural controls.*

Hydrothermal Differentiation. H. NEUMANN. Econ. Geol. 43:77–83, 1948. *Formation of hydrothermal solutions by differentiation.*

The Mechanism and Environment of Gold Deposition in Veins. W. H. WHITE. Econ. Geol. 38:512–532, 1943. *Control and environment of gold deposition.*

Solubility of Metal Sulphides in Dilute Vein-Forming Solutions. R. M. GARRELS. Econ. Geol. 39:472–483, 1944. *Dilute solutions cannot carry metal sulphides; colloids play important role.*

Factors in the Localization of Mineralized Districts. C. D. HULIN. A. I. M. E. Tech. Pub. 1762, 1945. *Igneous intrusion, structural controls, and mineralization.*

Factors in the Deposition of Hydrothermal Ores. C. D. HULIN. Econ. Geol. 45: 1950. *Stages of mineralization of hydrothermal ores.*

The literature on cavity fillings is very disperse. Descriptions occur in innumerable reports of individual mineral districts. Further references will be found in Part II, under the descriptions of the examples given in this chapter.

METASOMATIC REPLACEMENT

Metasomatic replacement, or simply replacement, as it is generally called, is the most important process in the emplacement of epigenetic mineral deposits. It is the dominating process of mineral deposition in the hypothermal and mesothermal deposits, and important in the epithermal group; the ore minerals of contact-metasomatic deposits have been formed almost entirely by this process. Likewise, it is the controlling process of deposition in supergene sulphide enrichment and dominates in the formation of most other supergene mineral deposits. In addition, it plays the major role in the extensive rock alteration that accompanies most epigenetic metallization.

Replacement may be defined as a process of essentially simultaneous capillary solution and deposition by which a new mineral is substituted for one or more earlier formed minerals. The process, however, is not so simple as the definition implies, as will be shown later. By means of replacement wood may be transformed to silica (petrification), a single mineral may take the place of another, retaining its form and size (pseudomorphs), or a large body of solid ore may take the place of an equal volume of rock. Thus, many mineral deposits originate. The replacing mineral (metasome) need not have a common ion with the replaced substance. The replacing minerals are carried in solution and the replaced substances are carried away in solution; it is an open circuit, not a closed one.

Replacement was early recognized in explanation of mineral pseudomorphs, particularly where the pseudomorph exhibited a diagnostic crystal form of an earlier unlike mineral. Naumann introduced the word metasomatism, meaning a change of body, to designate the process. It was vaguely realized as a mode of emplacement of mineral bodies by Zimmermann, who in 1749 ascribed the origin of lodes to the transformation of rocks into metallic minerals and veinstones by the action of solutions that entered through small rents and other openings in the rocks.

Hints concerning its application to ore formation were subsequently made from time to time, but only in the beginning of the present century was replacement realized to be a large-scale process of ore formation. Huge bodies of massive sulphide ore completely enclosed within limestone were early thought to be the filling of old solution caves in limestone. But, with the growing knowledge of the effectiveness and extent of replacement, and, particularly, of criteria for its recognition, it was realized that such bodies had been formed by replacement. As knowledge of this type of deposit has increased, especially in America, many deposits formerly thought to have originated by different processes are now ascribed to replacement. Even many formerly thought to be igneous injections are now conclusively recognized as replacement deposits. Likewise to this group now belong many deposits that formerly were thought to have been fissure fillings and impregnations of rock pores. In the case of supergene sulphide enrichment the supergene sulphides were formerly thought to have been merely deposited as coatings on earlier sulphides, but now it is generally known that they are deposited only by replacement of earlier sulphides.

THE PROCESS OF REPLACEMENT

If mineralizing solutions encounter minerals that are unstable in their presence, substitution may take place and replacement ensues. The exchange is practically simultaneous, and the resulting body may occupy the same volume and may retain the identical structure of the original body.

Mode of Interchange. If, in a brick wall, each brick were removed one by one, and a silver brick of similar size substituted as each brick in the wall was removed, the end result would be a wall of the same size and form, even to the minutiae of brick pattern, except that it would be composed of silver instead of clay. This is how replacement proceeds, except that the parts interchanged are infinitesimally small — of molecular or atomic size. Consequently, the shape, size, structure, and texture may be faithfully preserved even below the visible magnifications of the microscope.

If the replacement were molecule by molecule, a simple chemical equation would express the interchange, such as

$$ZnS + CuSO_4 = CuS + ZnSO_4$$

Here, dense covellite replaces less dense sphalerite, and a unit volume of sphalerite yields a smaller volume of covellite, with consequent shrinkage. Such interchange, however, takes place only with free-growing crystals or in incoherent materials where pressure is negligible.

Observations show that replacement in deep-seated, rigid rocks is not attended by change in volume. A cubic centimeter of sphalerite or calcite is replaced by a cubic centimeter of covellite or galena. This is the *law of equal volumes*. The calculable shrinkage, according to the chemical reactions, does not occur. It follows from the law of equal volumes that excess covellite or galena is deposited to make up for the shrinkage that would ensue if denser covellite or galena replaced lighter sphalerite or calcite according to the customary chemical reactions. Therefore, in volume-for-volume replacement, the interchange is not molecule for molecule. A single crystal of pyrite, for example, may cut across and replace some half-dozen different rock minerals, which proves further that volume-for-volume replacement cannot be expressed by any single chemical equation. It has been repeatedly demonstrated by large-scale field relations, by crystal pseudomorphs, and by microscopic observations that replacement is definitely a volume-for-volume interchange. Consequently, ordinary balanced chemical equations do not express what actually happens in volume-for-volume replacement, and the oft-quoted equations must be considered only as indicating the trend and end products of the exchange. The process is not as yet fully understood.

Procedure of Substitution. Although the procedure of substitution is imperfectly known, certain features are recognized. The simultaneous interchange must be by infinitesimal particles of molecular or atomic size. The growing mineral is in sharp contact with the vanishing substance; between them there must be a thin film of solution that supplies by diffusion the replacing materials and removes the replaced substances. In the case of liquids such a film will be supersaturated, facilitating reaction. The rate of reaction will depend upon the rate of supply of new material and the readiness of removal of the dissolved material. The instant that space is made available by solution, some of the replacing mineral will separate out from the supersaturated film. Thus, the metasome will continuously advance against the host and grow at its expense. In this, the growth pressure of the impinging crystals will also facilitate solution of the host. The replacing minerals thus present a constantly advancing front against the host as long as the supply of new material and disposal of dissolved material keeps up.

Where solution is supplied to a center, such as a pore space, growth may proceed outward in all directions from the center, giving rise either to discrete, shapeless grains, or to crystals with well-developed faces in sharp contact with the enclosing host. In this manner isolated, doubly terminated crystals may grow at the expense of limestone or other rock. It is obvious that such crystals could not have been formed by filling of pre-existing cavities with shapes coinciding exactly with the crystals. Therefore, it is concluded that they must have been formed by replacement, and such isolated, doubly terminated crystals are rightly considered diagnostic of replacement. Naturally, minerals with high power of crystallization, such as tourmaline, arsenopyrite, or pyrite, are the ones that most commonly assume euhedral forms. If the supply of material to feed a given crystal ceases for any reason, other minerals may continue to deposit at its margins, and eventually earlier formed euhedral minerals will be enclosed within later replacement minerals. Thus, pyrite crystals are common within copper, lead-zinc, and other ores. This explanation may also account for the interesting " polar " pyrite cubes of Ducktown, Tenn.; fringes of chalcopyrite at opposite pyrite faces probably are due to greater ease of replacement at these ends.

Entry and Exit of Solution. Replacement involves the necessity of continuing supplies of new material and removal of the dissolved material. How does the new material arrive at the point of deposition? This question becomes more pointed in the case of an unfractured pyrite cube that is undergoing supergene replacement by compact chalcocite from the outside toward the center, by means of a liquid solution at atmospheric temperature and pressure. First the faces of the cube are replaced, say to a depth of one-quarter of its diameter. Then, the replacing copper must penetrate a dense layer of chalcocite in order to arrive at the interior front of replacement, and the dissolved iron from the pyrite must escape through the same layer of compact chalcocite. The chalcocite may have no determinable pore space. Obviously, the necessary quantity of solution cannot flow bodily through the dense chalcocite layer.

Diffusion is probably the answer. This is the movement of molecules or ions in a solution from a point of supply to a point of deposition or from a place of higher to one of lower concentration. Particles of ionic or molecular size can move through a layer where a bodily flow of solution could not. But diffusion is known to be exceedingly slow and to act only over short distances. Therefore, it cannot be a means of transporting large quantities of replacing substances over long distances; it is incompetent by itself to build up large mineral deposits. However, it is an effective process for supplying and removing the products and by-products of replacement over the short distances at the actual front of replacement, whereas voluminous moving solutions could not gain entry.

Replacement will occur first along the major channelways referred to on page 100. The walls succumb, and further advances are fed through minor openings, down to capillary size, whose walls in turn succumb. If the wall rocks or early formed replacement minerals have been minutely fractured, this provides greater opportunity for replacement,

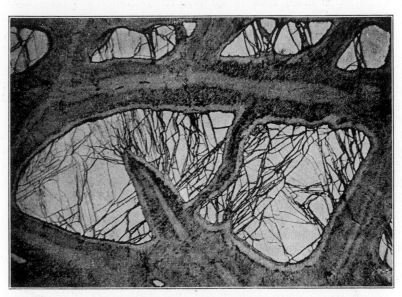

FIG. 5·4–30. Replacement extending outward from cracks. Residual grains of pyrite isolated by invading covellite and intersected by covellite veinlets. (*Photo by L. C. Graton.*)

and such openings may serve for the final stages of delivery of the replacing solutions without the necessity of having to call to any great extent upon the slow process of diffusion. However, the microscopic examination of polished ore surfaces discloses (Fig. 5·4–30) waves of replacement, whose growing fronts are nurtured by diffusion, extending outward from the smallest microscopic cracks.

The refuse of replacement is probably removed through conduits similar to those that permitted ingress to the new materials and is swept away to be dispersed within the mass of the ground water.

Stages of Replacement. Those who view polished ore surfaces through a microscope become aware that replacement ore deposits are commonly built up in stages, and that earlier replacement minerals are themselves replaced by later minerals. Definite successions are thus recognized. In the initial stages of the replacement of rocks, ferromagnesian silicates are attacked first, followed by the feldspars and quartz. Introduced gangue minerals replace all silicates. Likewise, all sulphides replace all rock minerals and also introduced gangue minerals; gangue minerals rarely replace sulphides.

The first formed metallic minerals, of which pyrite and arsenopyrite are common ones, may also be replaced by later sulphides. Pyrite, for example, is often seen to be veined by minute fractures along

Fig. 5·4–31. Camera lucida drawing of supergene chalcocite (white) replacing sphalerite (black), leaving unreplaced residuals; stippled, incompletely replaced sphalerite. Morenci, Ariz. ×40.

which the later sulphides have penetrated and replaced the walls. Little islands of unreplaced sulphides left between intersecting replacement veinlets are characteristic features (Fig. 5·4–31). Again, the margins may be attacked, and unreplaced residuals may remain enclosed within the later sulphides. Similarly, still later metallic minerals may, in turn, replace those of the second generation, and the process may be repeated until eight or ten successive stages have resulted. The sequence among some common hypogene metallic minerals generally is pyrite, enargite, tetrahedrite, sphalerite, chalcopyrite, bornite, galena, and ruby silver.

Growth of Replacement. In nonhomogeneous rocks the growth of replacement may be controlled by favorable beds, structural features, or chemical or physical properties of the host rock as already indicated.

In homogeneous rocks replacement may advance in one or more of three ways: (1) Starting from a fissure, the walls are first replaced and the replacement then advances outward with a bold face of massive ore against unreplaced country rock to the extreme limit of mineralization (Fig. 5·4–32A). It is a wavelike advance, and the end product is a massive ore body consisting mostly of introduced materials, such as the massive sulphide bodies of Kennecott, Alaska, or of Bisbee, Ariz. (2) The growth may take place with a bold front but preceding it, like skirmishers flung out in front of an advancing army, is a fringe of disseminated replacement where partial replacement is going on at many small centers (Fig. 5·4–32B). The latter may gradually coalesce to form a farther front of massive ore, beyond which still continues, however, the outward-flung fringe of disseminated

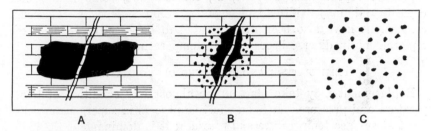

FIG. 5·4–32. Replacement advance. A, Bold face advance; B, outer fringe of disseminated replacement (black is ore); C, multiple center.

ore. The resulting deposit is a massive or high-grade ore body flanked by disseminated or lower-grade ore. The massive part of the ore body has resulted from the continued growth and coalescence of innumerable centers of replacement. (3) The third method may be termed *multiple-center* growth (Fig. 5·4–32C). If the country rock becomes permeated by mineralizing solutions, replacement may start simultaneously at innumerable closely spaced centers, such, for example, as at the loci of ferromagnesian minerals in a monzonite porphyry. With arrested growth, disseminated replacement deposits result. The individual metallic mineral grains are peppered throughout the rock and range in size from specks invisible to the naked eye to a half-centimeter in diameter; they commonly constitute only 5 to 10 percent of the rock mass. The porphyry coppers are examples of this mode of replacement, some of which contain several hundred million tons of copper ore.

AGENCIES OF REPLACEMENT

Replacement deposits are produced by both liquid and gaseous solutions, and in both of them water predominates. The liquid solutions occupy the major role. Most hypogene replacement deposits are considered to have been deposited mainly from hot alkaline solutions of igneous derivation. As shown in Chapter 4, such hydrothermal solutions may have left the magma chamber as alkaline liquids, or as acidic gaseous emanations that later condensed to liquids and generally became alkaline by reaction with the rocks through which they passed. The hot waters, at first entirely magmatic, may later have been diluted by intermingling with near-surface meteoric waters. The materials carried in solution largely came from the magma, but some were dissolved from the country rock.

Cold surface or artesian waters also produce both primary and supergene replacement deposits, as, for example, some manganese deposits and many oxidized and supergene sulphide deposits. Gaseous emanations have also been demonstrated to be effective agencies of replacement in diverse types of high-temperature deposits.

TEMPERATURE AND PRESSURE OF FORMATION

Replacement may take place under almost any condition of temperature and pressure. It is an effective process in supergene enrichment at surface temperature and pressure; it is dominant at the high temperatures and pressures that prevail during contact metasomatism. It is, of course, most effective at elevated temperatures, since heat tends to accelerate chemical activity.

The nature of replacement varies somewhat according to the conditions of temperature and pressure.

Atmospheric Temperatures. The formation of replacement deposits by cold meteoric waters is confined, mostly, to soluble rocks, such as limestones. These may be replaced by oxides of iron or manganese to form iron and manganese deposits, or by calcium phosphate to form phosphate deposits.

The most noteworthy replacement under atmospheric temperature and pressure is in the zone of oxidation and supergene sulphide enrichment. As shown in Chapter 5·8, near-surface limestone beds are extensively replaced by smithsonite and malachite-azurite, forming large and valuable zinc and copper carbonate deposits. Similarly, copper and zinc silicates form valuable surficial replacement deposits by means of cold solutions. In addition, vast tonnages of valuable metallic ores have resulted from the replacement of lean hypogene sulphides by supergene sulphides, such as has occurred in many of the great porphyry copper deposits. In general, the metallic minerals formed by replacement from cold solutions are of simple composition.

Some of the common ones are chalcocite, covellite, marcasite, wurtzite, native copper, native silver, and argentite.

Elevated Temperatures. With warmer solutions the ranges and intensity of replacement increase. More rocks are affected, and more extensive replacement occurs. The metallic minerals formed are mostly simple sulphides and sulpho-salts, and the gangue minerals are chiefly carbonates, quartz and simple silicates; many are hydrous.

With solutions at intermediate temperatures, extensive wholesale replacement of rocks may take place, and large replacement ore deposits may be formed. Many ore minerals replace carbonates, quartz, silicates, and metallic minerals.

At high temperatures hardly any rock may escape replacement. Even such relatively refractory rocks as granite may be almost completely altered to greisen. Silicates containing boron, chlorine, and fluorine are common. Sulphides and oxides (particularly magnetite) develop in coarse texture. Many tin, tungsten, and magnetite deposits have been formed by replacement at high temperatures.

Host Rocks

Every rock is susceptible to replacement, but, naturally, the readily soluble carbonate rocks are the most widespread hosts of replacement mineral deposits. Calcareous shales and sandstones are also congenial host rocks. Likewise, igneous rocks are readily susceptible to replacement and include some of the largest known ore deposits. Among the metamorphic rocks, field studies show that crystalline schists and marble yield most readily to replacement, and gneisses, phyllite, and quartzite the least. Even these, however, are known to succumb. Pure argillaceous rocks and quartzites are the least susceptible to replacement by ore.

In the granular igneous rocks, replacement is commonly selective and differential in that certain minerals, such as the dark minerals or feldspar, may be selectively replaced by ore minerals. This gives rise to one type of disseminated replacement deposit of which the porphyry coppers are outstanding examples.

Localization of Replacement

It is shown on pages 105–106 that various physical, chemical, and structural features serve to localize hydrothermal deposits. Most of these features operate in connection with replacement deposits. Also the remarks on pages 101 and 309 regarding the effect of the host rock are particularly pertinent to replacement deposits. Their chemical character alone may be the controlling factor in localizing ore, but generally structural features operate in conjunction. In addition, certain minerals of an igneous rock may localize replacement, as, for example, where ferromagnesian minerals are selectively replaced by sulphides. The chemical control of replacement is perhaps most clearly demonstrated in the selective replacement of disseminated hypogene sulphides by supergene sulphides, in which chalcopyrite is replaced in preference to pyrite, and adjacent rock silicates are untouched.

Structural Features. Of the various structural features mentioned previously, fissures are the most important ore localizers. Single fis-

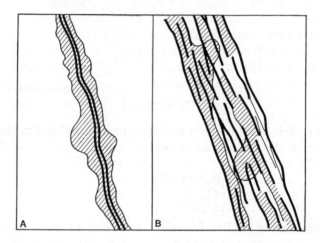

Fig. 5·4–33. Replacement lodes. *A*, Developed along single fissure; *B*, along shear zone.

sures may have their walls replaced along favorable beds to form tabular deposits known as *replacement veins* or *replacement lodes* (Fig. 5·4–33*A*). The fissure may intersect several favorable sedimentary beds, each of which may be replaced to form a succession of replacement bodies, of which the fissure is the major locus, and the beds the minor locus. Examples of such controls are numerous and form a working basis for the mining geologist in ore hunting.

Closely spaced *sheeted fissures* and *shear zones*, because of their large specific surface, give rise to still larger replacement lodes (Fig. 5·4–33*B*). Deposits along sheeted zones are likely to be more regular in width than those in shear zones, and replacement of the plates of rock may give rise to a banded structure that simulates crustification. Commonly, in shear zones, the replacement is more fitfully distributed within the zone, and an irregularly shaped deposit, although roughly tabular in form, results.

Intersections of fissures may yield a type of deposit shown in Fig. 5·4–34. These give rise to ore bodies of limited horizontal extent, but their vertical extent may be great. Intersections of two groups of closely spaced multiple fissures may divide the rock mass into many-surfaced polygonal blocks whose large specific surface offers opportunity for the development of large, irregularly shaped replacement deposits confined by the outlines of the intersection.

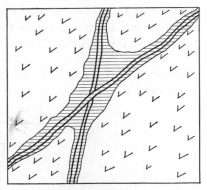

FIG. 5·4–34. Replacement ore localized by fissure intersections.

Pitching folds and drag folds have localized the great Homestake Mine (Fig. 5·4–35).

The tin-bearing pipes of eastern Australia and South Africa are another example of structural control.

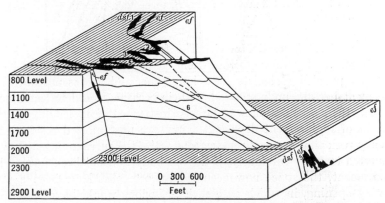

FIG. 5·4–35. Replacement ore deposit localized by folds: Homestake mine.
(*After McLaughlin.*)

Sedimentary Features. The features previously mentioned due to sedimentation are especially effective in localizing replacement bodies, as is shown in Fig. 5·4–36.

CRITERIA OF REPLACEMENT

Replacement deposits generally exhibit one or more diagnostic features characteristic of replacement. Some can be seen only in field relationships, some in hand specimens, and others only under the microscope. Some of the criteria afford positive identification, others negative. An extraneous mineral can gain entry into a rock only by filling a pre-existing cavity or by making way for itself by replacement. Cavity-filling deposition is distinguished by certain characteristic features discussed under Cavity Filling, which are lacking in replacement deposition. Therefore, their absence constitutes dependable negative criteria for replacement. The more important criteria, established by J. D. Irving and others, are enumerated briefly below.

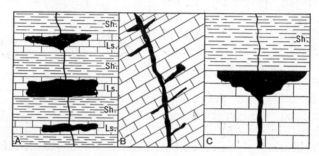

Fig. 5·4–36. Relation of replacement to sedimentary features. Black, ore. *A*, relation to intercalated limestone beds; *B*, to bedding planes; *C*, to overlying impervious bed.

Unsupported Residual Nuclei. The frequent observations of small islands of country rock entirely enclosed within mineral bodies (Fig. 5·4–37*A*) puzzled geologists for a long time, and it was not until replacement became understood that their presence could be explained. In fact, their elucidation was largely instrumental in leading to the concept of replacement deposits. They are unsupported residuals of country rock that escaped replacement while the surrounding rock was converted to ore. Obviously, if the ore body had not been formed by replacement, the space now occupied by ore must have been a pre-existing opening, in which case such inclusions would have rested on the floor of the opening. Where the residuals contain bedding or other structural features that are in alignment with similar structures of the wall rock (Fig. 5·4–37*A*) they constitute the strongest evidence of replacement.

Preservation of Rock Structure. When rock is replaced by ore, particle by particle, the structure of the replaced body is commonly faithfully preserved in the ore. Where such inherited structure is diagnostic of the pre-existing rock, it constitutes definite and usable criteria of replacement. Such features as stratification, cross-bedding, fossils, and dolomitization rhombs, if preserved in ore, indicate clearly that the ore has replaced the rock that formerly showed those features (Fig. 5·4–37B). Similarly, phenocrysts of igneous rocks and the schistosity of metamorphic rocks may be delicately preserved. In addition, larger structural features, such as folds (Fig. 5·4–37C), breccias, faults, and joints, are inherited. These various features are analogous to the preserved wood cells in petrified wood.

Intersection of Structural Features. Since many types of mineral deposits intersect rock structure, this criterion by itself is inconclusive.

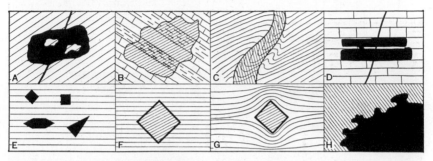

Fig. 5·4–37. Features illustrating criteria of replacement. A, Unsupported residual nuclei; B, preserved rock strata; C, preserved folded structure; D, ore abutting bedding; E, doubly terminated crystals; F, pyrite cube truncating bedding, in contrast to G, which has grown in yielding shale; H, irregular outlines of replacement contacts.

A massive sulphide body, for example, may abut the ends of thin limestone beds (Fig. 5·4–37A, B, D). This shows that the ore is later in age than the bedding of the rock, but the ore may have filled an open cavity that abutted the beds or it may have grown by replacement into an abutting position. If, however, the bedding extends uninterrupted from rock into ore, it indicates replacement.

Complete Crystals. Crystals alien to the original rock, that grow by replacement in homogeneous rocks, commonly display freely developed faces in contrast to the incompletely developed crystals that are attached to and grow outward from the walls of open spaces. Consequently, complete doubly terminated crystals tightly enclosed in rock (Fig. 5·4–37E) are diagnostic of replacement, particularly when they transect several individual rock grains. Delicate crystals of barite in a fine-grained limestone are an example. Such crystals may be of megascopic or microscopic size.

Care must be used in differentiating between these crystals and crystals which, through the force of crystallization, may have grown by pushing aside the adjacent rock. For example, cubes of pyrite in shale may have grown by pushing aside the shale laminae (Fig. 5·4–37G), which continue unbroken around the cube. However, if the cube face squarely intersects the shale laminae it is a sure indication of replacement (Fig. 5·4–37F).

Mineral Pseudomorphs. The presence of pseudomorphs of a mineral of one composition after another of quite different composition is evi-

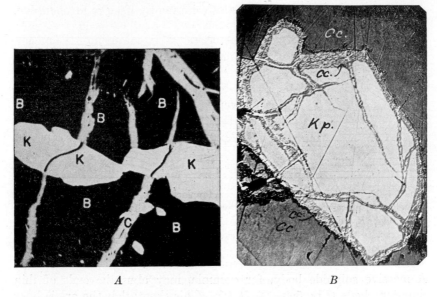

A   B

FIG. 5·4–38. Differential replacement. *A*, Veinlets of chalcocite *C* replace bornite *B* but do not replace mineral *K*, although veinlets cross it. ×260. (*Kreiger, Econ. Geol.*) *B*, Marginal replacement of klaprothite (*kp*) by chalcocite *cc*. (*Ray, Econ. Geol.*)

dence of replacement. A cube having the form and characteristic striations of pyrite may consist of chalcocite. The chalcocite must have replaced the pyrite crystal.

Outlines. The sharp outline of a compact mineral or body of ore against the host rock may indicate replacement. In Fig. 5·4–37H the wavy knobby outline of the ore, particularly the unusual-shaped protuberances and embayments into the rock, is not characteristic of solution cavities filled by ore but is typical of replacement. Even more characteristic are irregular outlines formed as a result of differential replacement whereby grains or bands of one mineral are more completely replaced than adjacent minerals or bands (Fig. 5·4–38).

Form. Extreme irregularity of form is characteristic of replacement deposits. Such shapes might be confused with those of solution cavities in limestone but not with cavities produced by fissuring or rupture. However, solution caves have characteristic walls marked by intersecting concavities and commonly have floor debris. Also, some deposits were formed at a geologic time when several thousand feet of uneroded cover overlay them — at a depth at which solution caves do not form.

Shrinkage Cavities. Irving noted irregular cavities or " volume vugs " at Leadville, Colo., and in the siliceous gold ores of the Black Hills, which he considered characteristic of replacement. They are, however, not diagnostic, since replacement proceeds without change of volume.

Transection of Different Crystals. Small wavy veins or veinlets of irregular width that transect several diversely oriented host crystals indicate replacement. Their irregularity of width and outline distinguishes them from somewhat similar veinlets formed by the filling of small fissures.

Absence of Crustification. Crustification is absent from replacement deposits, so its presence distinguishes cavity-filled deposits, but its absence does not prove replacement.

RESULTING MINERAL DEPOSITS

Mineral deposits formed by replacement may be divided into massive, replacement lode, and disseminated deposits. Except for deposits of iron ore, they are the largest and most valuable of all metallic mineral deposits.

Massive Deposits. Massive deposits are characterized by great variations in size and extremely irregular form. Bodies in limestone generally thicken and thin, display wavy outlines, and ramify irregularly in all directions. Others are great irregular massive bodies whose larger dimensions may be measured in thousands of feet. Generally, the deposit consists mostly or entirely of introduced ore and gangue minerals, and included rock matter constitutes only a small or negligible part. Some pyritic replacement deposits consist of stupendous masses of pure yellow sulphides, such as those at Noranda, Quebec (Fig. 5·4–39), Flin Flon, Manitoba, or Rio Tinto, Spain. More than 500 million tons of pyrite were originally introduced into the rocks at Rio Tinto.

These bodies result from the almost complete replacement of the host rock, and, commonly, massive ore terminates abruptly against the

country rock.   Although most common in limestone, they occur widely
in various kinds of rocks.

Replacement Lode Deposits.   Replacement lode deposits are localized along thin beds or fissures whose walls have been replaced.   Con-

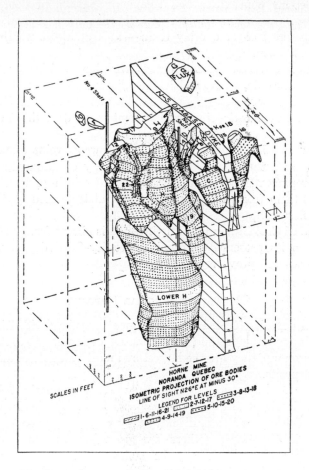

FIG. 5·4–39.   Massive replacement bodies of sulphide ore, Noranda mine, Quebec,
in isometric projection.   (*Peter Price, Bull. Can. Inst. Min. and Met.*)

sequently, they resemble fissure veins in form.   Many so-called fissure
veins are actually replacement lodes.   In general, they are wider than
fissure veins, and the width varies greatly along a single lode; it may
range from a few inches to several tens of feet.   The walls are commonly wavy, irregular, and gradational into the country rock.   The
ore may be massive or irregularly scattered in the rock, and commonly

it is flanked by a fringe of disseminated ore minerals, so that the ore boundary is a commercial one. The gold veins of Kirkland Lake, Ontario, the copper veins of Kennecott, Alaska, and the lead veins of the Coeur d'Alene, Idaho, are examples.

Disseminated Replacement Deposits. In disseminated replacement deposits, the introduced material constitutes only a small proportion of the ore. The ore minerals are peppered through the host rock in the form of specks, grains, or blebs, generally accompanied by small veinlets, and represent the multiple-center type of replacement. The amount of introduced gangue is small, and the ore consists of altered host rock and the disseminated ore grains. The total content of metallic minerals may be as low as 2 percent but commonly ranges from 5 to 10 percent of the mass. The ores are mostly low grade. The boundaries are vague; the metallized part fades gradually into waste rock, and the ore limits are determined by the workable grade of the ore.

Disseminated replacement deposits are generally huge, which permits large-scale mining operations and the utilization of low-grade ores. Since the workable parts of the deposits are fringed by zones of somewhat lower-grade material, a slight reduction in the grade of the rock that can be profitably treated permits the inclusion of the lower-grade marginal ore within the ore boundaries and thereby brings about a large increase in the size and tonnage of the deposits. A lowering of a workable-grade copper ore, for example by 0.25 percent, may increase the ore reserves by tens of millions of tons. The great "porphyry copper" deposits, many of which are mined from huge open pits, fall in this group of deposits. Some idea of the enormous size of disseminated deposits and the immensity of the operations may be realized from the following examples: The Chile Copper Co. mine at Chuquicamata, Chile, is reported to have reserves of 1035 million tons of copper ore, which to date has averaged 1.64 percent copper. The great Utah Copper mine at Bingham, Utah, has officially reported reserves of 640 million tons of ore containing 1.07 percent copper, though both reserves and grade are now lower, and has treated over 100,000 tons of ore per day. The Northern Rhodesia copper belt has estimated reserves of 550 million tons, averaging 4.11 percent copper. The Climax Molybdenum mine at Climax, Colo., has ore reserves of over 200 million tons averaging 0.7 percent molybdenum.

The Alaska Juneau gold mine at Juneau, Alaska, handled 12,000 tons of ore per day averaging about 0.035 ounce ($1.23) of gold per ton and has mined around 88 million tons yielding about 81 million dollars from ore averaging 0.043 oz gold per ton. Another example of ores and deposits that belongs to this group is the disseminated lead deposits of southeastern Missouri.

Form and Size. The form of replacement deposits is determined largely by the structural and sedimentary features that localize them.

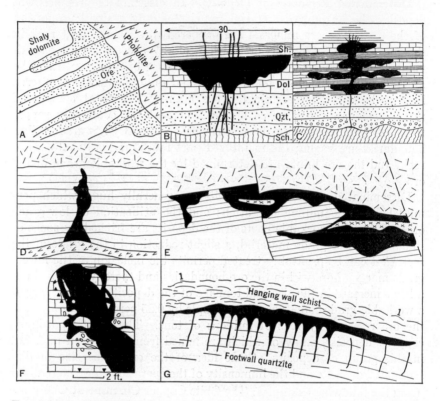

Fig. 5·4–40.   Forms of replacement deposits.   *A*, Along fissures and where phonolite dammed solutions, Mineral Farm, Black Hills, S. Dak. (*after Irving*); *B*, ore restricted below shale and abutting dolomite bedding, Union mine, S. Dak. (*after Irving*); *C*, Portland mine, S. Dak. (*after Irving*); *D* and *E*, cross and longitudinal sections of Iron Hill, Leadville, Colo. (*after Irving*); *F*, sketch of replacement vein, Jumbo mine, Kennecott, Alaska; *G*, relation of ore to fissures in quartzite and to overlying schist, Ferris Haggarty mine, Encampment, Wyo. (*After Spencer, U. S. Geol. Survey.*)

Accordingly, they are irregular, blanket-shaped, tabular, pipe-shaped, synclinal, or anticlinal, or they may be large irregular disseminated deposits.   Figure 5·4–40 shows representative forms.

The size varies greatly.   The deposits may be mere cracks containing high-grade ore, or they may have the dimensions of the massive

Henrietta-Wolftene ore body at Leadville, Colo., which was 3,500 feet long, 1,600 feet wide, and 200 feet thick. Lode replacements may reach several thousand feet in length and depth and as much as 200 feet in width. Disseminated replacement deposits range in size

Fig. 5·4–41. Sphalerite (dark) cutting across and replacing pyrite (light) along fractures in pyrite. ×90. (*Graton, Econ. Geol.*)

from small to the dimensions of the Chuquicamata deposit in Chile, whose maximum length is 10,500 feet and maximum width is 3,600 feet.

Texture of Replacement Ores. The texture of replacement ores varies considerably according to the conditions of temperature and pressure of formation and the degree of replacement. All replacement ores lack crustification, and drusy cavities are generally absent.

Dissemination ores are characterized by pepper-and-salt texture. The valuable mineral grains may be shapeless, as shown by copper ores, or crystals, as shown by disseminated lead deposits.

Massive ores may retain the texture and structure of rocks they replace, such as the texture of oölitic limestones, the rhombs of dolomite, or the phenocrysts of porphyry. Commonly, however, such original texture is wholly destroyed. With incomplete replacement residual rock particles may constitute a small or major part of the ore. With more or less complete replacement, the texture is generally holocrystalline. At high temperatures of formation, coarse texture is characteristic. Ores formed at intermediate temperatures are generally characterized by finer-grained texture, and those at low temperatures are mostly fine grained. Agatelike and colloform textures are not uncommon.

Under the microscope, crisscrossing and replacement of earlier minerals by later minerals are often observed (Fig. 5·4–41), or the replacing mineral may form rims at the expense of earlier minerals (Fig. 5·4–38B). Exsolution textures, such as dots, lenses, or plates, oriented along cleavage planes of the host mineral, may be present but are not diagnostic of replacement deposits.

Ores Formed, and Examples. Outside of some iron and some nonmetallic deposits, replacement processes have given rise to the world's largest and most important mineral deposits. Of first rank are the base metal and precious metal deposits, but the rarer metals and many nonmetals are well represented. The chief metals, the chief types of deposits, and some examples of important deposits are listed on page 157.

Selected References on Replacement

The Nature of Replacement. WALDEMAR LINDGREN. Econ. Geol. 7:521–535, 1912. *A fundamental discussion of the process of replacement.*

General Reference 8. Chaps. 9, 14, 26, 27. *Descriptions of some replacement deposits and processes operating toward their formation.*

Metasomatism. WALDEMAR LINDGREN. Geol. Soc. Amer. Bull. 36:247–261, 1925. *Development of ideas; processes involved and problems raised.*

Replacement Ore Deposits and Criteria for Their Recognition. J. D. IRVING. Econ. Geol. 6:527–561, 619–669, 1911. *Chiefly forms and criteria.*

Influence of Replaced Rock on Replacement Minerals of Ore Deposits. B. S. BUTLER. Econ. Geol. 27:1–24, 1932. *How the host rock affects the minerals deposited by replacement.*

The Mechanics of Metasomatism. G. W. BAIN. Econ. Geol. 31:505–526, 1936. *Relation of metasomatism to fine pores.*

Metasomatic Processes. W. T. HOLSER. Econ. Geol. 42:384–395, 1947. *Physicochemical processes of transport and reaction in fine-grained rocks.*

Specific references to districts and ore deposits are given under individual minerals in Parts II and III.

ORES	TYPE	EXAMPLES OF IMPORTANT DEPOSITS
Iron	Magnetite	Dover, N. J.; Lyon Mountain, N. Y.
	Hematite	Iron Mountain, Mo.
	Limonite	Oriskany ores, Va.
Copper	Disseminated	Utah Copper; Nevada Consolidated; Chino, N. Mex.; Ray, Ajo, Miami-Inspiration, and Morenci, Ariz.; Northern Rhodesia copper belt; Chuquicamata, Braden, and Potrerillos, Chile
	Lode	Kennecott, Alaska; Bingham, Utah; Magma, Morenci, and Bisbee, Ariz.; Britannia, British Columbia; Cerro de Pasco, Peru
	Massive	Bingham, Utah; Bisbee and United Verde, Ariz.; Rio Tinto, Spain; Noranda, Quebec; Flin Flon, Manitoba; Granby, British Columbia; Boliden, Sweden
Lead	Massive	Leadville, Colo.; Bingham, Utah; Sullivan, British Columbia; Santa Eulalia, Mexico; Broken Hill, Australia
	Lode	Coeur d'Alene, Idaho; Park City and Tintic, Utah
	Disseminated	Southeastern Missouri
Zinc	Massive	Leadville, Colo.; Bingham, Utah; Sullivan, British Columbia; Flin Flon, Manitoba; Silesia
	Lode	Broken Hill, Rhodesia; Park City, Utah; Franklin Furnace, N. J.; Trepca, Serbia
Gold	Massive	Noranda, Quebec
	Lode	Homestake, S. Dak.; Kirkland Lake, Ontario
	Disseminated	Juneau, Alaska; Witwatersrand, South Africa
Silver	Massive	Leadville, Colo.; Bingham, Utah
	Lode	Park City and Tintic, Utah; Coeur d'Alene, Id.; Cerro de Pasco, Peru; Santa Eulalia, Mexico
Tin		Transvaal; Australia
Mercury	Lode	Almaden, Spain
	Disseminated	New Almaden, Calif.
Molybdenum	Disseminated	Climax, Colo.; Utah Copper, Utah
Manganese	Massive	Leadville, Colo.; Potgietersrust, South Africa
Barite	Lode	Missouri
Fluorite	Lode	Illinois-Kentucky field
Magnesite	Massive	Manchuria; Washington; California

Ore Shoots

Ore deposits are rarely equally rich throughout. Generally, the valuable primary minerals tend to be concentrated in certain sections called *ore shoots*, which contrast with lean or barren portions of the deposits. Ore shoots may be present in most hydrothermal deposits, but they are most characteristic of fissure veins and replacement lodes.

An ore shoot may differ from the lean portions of a deposit by the presence of only as little as 0.0004 percent gold or 1 percent copper, or there may be a readily detectable mineralogical difference, such as the presence of galena in the ore shoot and its absence from the barren portion.

Terminology. The term *ore shoot* should apply only to concentrations of hypogene ore, and it is advisable to restrict the term to this usage in order to distinguish hypogene concentrations from supergene concentrations, because the latter are produced by quite different processes and may have been formed from either the rich or lean hypogene parts of veins. The terms *pockets, nests, bunches,* or *kidneys* are variously employed in different places to designate small, irregular concentrations of ore. They refer either to hypogene or supergene concentrations, and although they fall within the definition of ore shoots, the latter term is usually applied to larger bodies. *Bonanza* is commonly used to designate an exceptionally rich shoot or bunch of ore, particularly with reference to gold and silver. Generally, it refers to rich secondary masses, and Irving has suggested its restriction to that field. *Chimneys* or *pipes* are terms used to designate vertical or highly inclined elongated ore shoots that resemble huge smokestacks. These may occur within fissure veins, but the terms are more commonly applied to any body of pipelike shape.

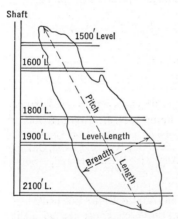

Fig. 5·4–42. S. Flurry ore shoot, Jumbo mine, Kennecott, Alaska, illustrating terminology of ore shoots.

Shape and Size. The outline of ore shoots may be irregular, but generally they tend to be elongated bodies that extend in a vertical or highly inclined position up and down a fissure vein. This inclination within a vein is called the *pitch* or *rake* (Fig. 5·4–42). Commonly, several shoots of more or less similar shape, size, and pitch occur in a

vein and tend to be spaced at approximately equal intervals; those in
nearby parallel veins tend to pitch in the same direction. The shape of
some typical ore shoots may be seen in Fig. 5·4–43.

In size, most ore shoots range in level-length from a few tens to a
few hundreds of feet, and in pitch length from a hundred or so to
several hundred feet; rarely they attain lengths of 2,000 or 3,000 feet.
Notable exceptions occur, however, as in the Mother Lode of California, where shoots attain pitch lengths of 5,000 feet or more, and at
Grass Valley, where the North Star shoot is 9,000 feet in pitch length.

Grouping. Ore shoots may be grouped as follows:

1. Open-space shoots — due to available open space.
2. Intersection shoots — due to vein intersections.
3. Impounded shoots — due to damming of mineralizing solutions.
4. Wall-controlled shoots — due to effect of wall rock upon precipitation.
5. Structure-controlled shoots — due to various structural controls.
6. Depth-controlled shoots — due to decrease of temperature and
   pressure.
7. Recurrent mineralization shoots — due to successive periods of
   mineralization.
8. Unsolved shoots — due to unknown factors.

The last group includes the greatest number of shoots.

Open-space shoots are localized by available open spaces in fissures,
such as are caused by the relative movement of opposite walls of a
curved fissure (Fig. 5·4–7). The ore shoots of the Mother Lode gold
veins of California are considered by Knopf to have been so controlled.
Similarly, Wandke ascribes many of the ore shoots of Guanajuato,
Mexico, to open-space control.

Intersection shoots are localized at vein intersections or cross fissures and are among the oldest known and the commonest types (Fig.
5·4–34). Intersections are particularly favorable because at such
places different solutions meet, also the walls are more shattered and
afford greater specific surface. Howe states that most of the ore shoots
of the Grass Valley gold veins are at vein intersections.

Impounded shoots result from the impounding of mineralizing solutions against impervious barriers, such as shales or fault gouge (Figs.
5·4–36, 5·4–40).

Wall-controlled shoots are those that occur adjacent to certain favorable wall rocks that presumably influence deposition from the mineralizing fluids. The well-known precipitating effect of carbonaceous
rocks upon gold is an example. A fissure may intersect several favorable beds and contain ore shoots adjacent to each, alternating with

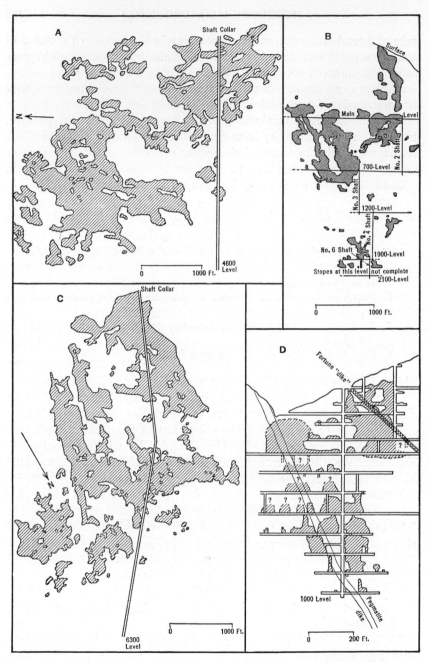

FIG. 5·4–43. Forms of ore shoots in longitudinal section. *A*, North Star vein, and
C, Empire vein, Grass Valley, Calif. (*after W. D. Johnston, Jr., U. S. Geol. Survey*);
B, Chichagoff mine, Alaska (*after Reed, A.I.M.E.*); *D*, Ingram vein, Gold Hill,
Colo. (*after Goddard, Colo. Sci. Soc. Proc.*).

barren stretches of vein opposite less favorable rocks. Wandke and Martinez point out that at Guanajuato, Mexico, bonanza silver ores occur in fissures opposite andesite or porphyry but not opposite schist. Similarly, at Porcupine, Ontario, good gold ore in greenstone gives way to barren gangue where the fissures pass into porphyry.

Structure-controlled shoots are localized by various structures. Places of change in strike and dip of fissures are favorable sites for ore shoots. Closely folded, competent beds alternating with incompetent beds form potential openings between the layers at the crests of anticlines and the troughs of synclines and localize ore deposition. Intermineral movement along veins may produce brecciation of one wall or a part of the vein, forming a locus for further mineral deposition, and thus give rise to ore shoots.

Depth-controlled shoots result from the control exerted by decreasing temperature and pressure upon deposition from solutions (Chap. 3). Rapid release of pressure by near-surface, shattered rocks may cause a sudden dumping of the minerals in solution, such as described by Turneaure for the rich tin shoots of Llallagua, Bolivia (Chap. 13). This type of shoot occurs particularly with " typomorphic " minerals (cinnabar, silver minerals), or those deposited within a narrow range of temperature and pressure. Similarly, near-surface chemical changes in the metallizing solutions, such as described by Ransome for Goldfield, Nev. (Chap. 12), and by Graton and Bowditch for Cerro de Pasco, Peru (Chap. 13), give rise to rich, near-surface ore shoots.

Recurrent mineralization shoots are due to reopenings by intermineralization movements, accompanied by recurrent mineralization during which certain parts of a vein, commonly along either wall, are enriched. Several stages of movement and recurrent mineralization may occur.

Unsolved shoots are numerous. Although many ore shoots may readily be placed in one of the above-described groups, a large number as yet defy interpretation. They do not appear to be localized by any of the conditions considered above. Probably, many are localized by chemical controls, such as the complexes described by R. M. Garrels.

Causes of Ore Shoots. Although separate causes of ore shoots are indicated above, a single shoot may be due to more than one cause. Its formation may involve the coincidence of two or more separate factors; and suitable conditions of temperature, pressure, and chemical character of the solutions must prevail. Once the cause is determined, however, a practical as well as a scientific achievement has been attained because a search for similar conditions may be rewarded by the discovery of other shoots.

Recognition and Search for Shoots. Shoots must be recognized before their cause can be determined or search directed for others. Often the two unfold themselves together. They may be recognized by visual observation of the mineralogy, by assays or analyses, and by

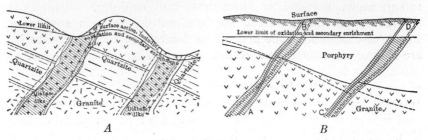

A   B

FIG. 5·4–44. Sections along veins showing geology plotted on both walls (solid lines, far wall; dotted lines, near wall), and ore shoots shaded. In *A*, the ore shoots coincide with the diabase dikes, indicating their control in localizing the ore and affording basis for search for other shoots. In *B*, intersecting fissures have localized the shoots.

plotting of data on maps, particularly on longitudinal sections. The plotting of visual observations is sufficient in the case of gross ores or minerals, but assays or analyses are necessary for invisible metals. The plotting will indicate if the valuable materials are uniformly or

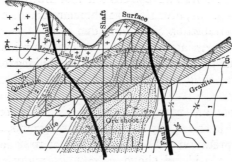

FIG. 5·4–45. Section along vein showing geology, and vein width indicated by contours, with superimposed ore shoot (stippled). Here the ore shoot occurs in disregard of geological features but is clearly related to the wider portions of vein.

haphazardly distributed, or if they are localized into definite shoots, whose outlines and size will also become evident.

If the geology is superimposed on a longitudinal section (Fig. 5·4–44), a relationship between ore shoot and kind of wall rock, or fissure intersection, or other controls, may become apparent. If the ore shoots occur in apparent disregard of such features, some other cause of ore localization must be sought. Any relationship between ore shoots and width of vein or former open space may be made apparent by plotting width contours, i.e., contour lines connecting parts of equal vein width, on the longitudinal assay plat (Fig. 5·4–45). If the vein is a fault fissure, the walls of which differ in their geology, longitudinal geological plats may be made for both walls, and the hanging wall tracing can be superimposed upon the footwall plat. An established ore-shoot control yields information for search for other shoots.

Selected References on Ore Shoots

Structural Control of Ore Deposition. C. D. HULIN. Econ. Geol. 24:15–49, 1929. *Excellent discussion.*

Outcrops of Ore Shoots. H. SCHMITT. Econ. Geol. 34:654–673, 1939. *Detection.*

Openings Due to Movement along a Curved or Irregular Fault Plane. W. H. NEWHOUSE. Econ. Geol. 35:445–464, 1940. *One type of ore-shoot formation.*

General Reference 10. *Innumerable examples and discussions of ore shoots.*

Certain Ore Shoots on Warped Plane Surfaces. W. H. EMMONS. A. I. M. E. Tech. Pub. 1545, 1943. *Comparison of ore shoots in veins; good bibliography.*

Formation of Primary Ore Shoots (Colo.). V. C. KELLEY. Colo. Sci. Proc. 14: 7:318–333, 1946. *Review of ore-shoot control in general.*

5·5 SEDIMENTATION (Exclusive of Evaporation)

The process of sedimentation as distinct from evaporation has resulted not only in the formation of common sedimentary rocks but also in valuable mineral deposits of iron, manganese, copper, phosphate, coal, oil shale, carbonates, cement rocks, clay, diatomaceous earth, bentonite, fuller's earth, magnesite, sulphur, and, less directly, uranium-vanadium deposits. These substances may be regarded as merely exceptional varieties of sedimentary rocks that happen to be valued because of their chemical or physical properties. Their mode of formation, therefore, is that of sedimentation with special variations to account for the special materials. They are composed of inorganic and organic materials; and their source, like that of any sedimentary rock, is from other rocks that have undergone disintegration, the ultimate source, of course, being the igneous rocks. Some of the materials, such as oxygen and carbon dioxide, have been obtained from the atmosphere, and a few have been derived from former deposits.

The formation of sedimentary deposits involves, first, an adequate source of materials; second, the gathering of the materials by solution or other processes; third, the transportation of the materials to the site of accumulation if that is necessary; and fourth, the deposition of the materials in the sedimentary basin. Subsequent compaction, chemical alteration, or other changes may take place.

The source materials, solution, mode of transport, and the nature of the deposition are generally similar for each product involved. The variations of some of the sedimentary cycles will be considered separately. The organic substances are considered in Chapter 16.

Source of Materials

The materials that enter into sedimentary mineral deposits have been derived chiefly from the weathering of rocks. Occasionally, materials have come from the weathering and oxidation of former mineral deposits such as iron, manganese, and copper. Others have passed through an intermediate organic stage. The rocks, however, constitute an adequate source for all of the sedimentary iron, manganese, and copper ore that is known.

Clarke and Washington have shown that in the earth's crust the average content of iron is 5.05 percent. Eckel has calculated that the portion of the earth's crust beneath the United States to a depth of 1,000 feet contains over 275,000 billion tons of iron, of which only about 0.01 percent has been concentrated into commercial deposits, in a ratio of about 80,000 to 1. Clearly, one need not look beyond the ordinary rocks for an adequate source of the iron in deposits. The iron comes from the weathering of iron-bearing minerals of igneous rocks such as hornblende, pyroxene, or mica, from the iron-bearing minerals of sedimentary and metamorphic rocks, and from the red coloring matter of sedimentary rocks.

Similarly, the manganese of sedimentary deposits has been derived chiefly from the weathering of manganese-bearing minerals in the rocks and, to a minor extent, from former sedimentary concentrations and epigenetic lode deposits. Hewett states that there are over 200 minerals that contain manganese as an essential element, and Clarke and Washington estimate that manganese makes up 0.09 percent of the earth's crust, there being about 56 times as much iron as manganese. Using this proportion, and applying it to Eckel's estimate for iron, there should be 5,000 billion tons of manganese beneath the United States to a depth of 1,000 feet — an adequate source of supply.

The source of sedimentary phosphate is phosphorus-bearing rock minerals, among which apatite is the most common. Some is also derived from the weathering of collophanite and dahllite in sedimentary rocks.

The constituents of sedimentary carbonate deposits such as the industrial limestones, dolomite, and magnesite are derived from the sea or saline waters to which they are largely supplied by rock weathering; likewise, the constituents of the numerous types of industrial clayey deposits such as clays, bentonite, and fuller's earth originate in rock weathering.

Solution and Transportation

Solution of the constituents of economic sedimentary deposits in large part goes on during weathering. This is true of iron, manganese, phosphates, carbonates, copper, and some other metals, but, of course, does not apply to clays. The chief solvents are carbonated water, humic and other organic acids, and sulphate solutions.

Carbonated waters are very effective solvents of iron, manganese, and phosphorus. Where iron is present in the ferrous state its solution offers no difficulty, since in that form it is unstable and soluble. But ferric iron is almost insoluble in most surface waters and to undergo solution must first be changed to the ferrous state. Organic matter aids this. The Precambrian iron ores that were formed before organic matter or vegetation came into existence were probably dissolved and transported as ferrous bicarbonate. Vast quantities of calcium carbonate, as well as other salts, are readily removed in solution and transported to bodies of standing water where precipitation may occur to form sedimentary deposits.

Humic and other organic acids derived from decomposing vegetation are considered effective solvents by Harrar. The hydroxy acids dissolve large quantities of iron, but the weak organic acids dissolve remarkable quantities and are the most effective of all natural solvents. Moore and Maynard's experiments on the solution and precipitation of iron indicate that carbonated water is the most effective solvent of iron and silica from norite and diabase and that peat solution is next. They concluded that iron is not carried as bicarbonate in surface waters high in inorganic matter but that the main part entering sedimentary iron formations was probably transported as ferric oxide hydrosol stabilized by organic colloids, although small amounts may have been carried as salts of organic acids or adsorbed by organic colloids. The idea of colloidal solution and transportation of iron is becoming more firmly established. The solution of manganese has not been investigated so thoroughly as that of iron, and knowledge of its solution by organic compounds and its transportation in colloidal form is inadequate. It will probably be learned, as in the case of iron, that organic compounds are important solvents and that manganese is removed largely in the colloidal state.

Sulphate solutions are effective solvents of iron and manganese, but they are rarely abundant enough to effect large-scale solution and transportation. The oxidation of pyrite yields sulphuric acid and ferric sulphate. At Rio Tinto, Spain, for example, the oxidation of the huge pyritic ore bodies yielded vast quantities of iron sulphate, some of which was precipitated nearby to form thick deposits of bog iron ore, but much of it has been carried down the Rio Tinto (hence the river's name) to the sea and deposited near the shore.

Most of the substances that make up sedimentary mineral deposits (coal excepted) are transported by means of rivers and subsurface waters. For the most part the substances reach the sea, but some are arrested en route or find a resting place in inland bodies of water or interior land basins.

The sedimentary substances will remain in solution as long as the solution does not undergo any appreciable physical or chemical change. Some or all of the iron or manganese, however, may be lost during transportation if the solutions traverse limestone or are subjected to other agencies of deposition. If the iron or manganese in solution escapes these hazards it may be transported to bogs, lakes, playas, or the sea, where quantity concentration can take place.

Deposition

The materials that form economic sedimentary beds are deposited mechanically, chemically, or biochemically. The manner of deposition depends upon the nature of the solvent and the place of deposition, as, for example, whether in the sea or in a swampy basin. The resulting products of deposition will also vary according to these conditions.

From Bicarbonate Solutions. Iron and manganese may be deposited from iron bicarbonate solutions by: (1) loss of carbon dioxide (Chap. 3); (2) oxidation and hydrolysis; (3) plants; (4) bacteria; (5) replacement of sea-bottom shells, in the case of iron, forming " fossil " hematite; or (6) reaction with sea-bottom colloidal silica and clay, yielding a gel that absorbs potassium from sea water to form glauconite, or similar reactions to form chamosite. Calcium phosphate is precipitated from solution in the presence of calcium carbonate.

From Sulphate Solutions. Iron and manganese ores may be deposited from sulphate solutions by: (1) reaction with calcium carbonate; (2) oxidation or hydrolysis; and (3) reaction of iron with silicates, giving greenalite.

From Organic Solutions. Precipitation of iron and manganese from organic solutions takes place by: (1) oxidation of ferrous and manganous carbonate to ferric and manganic oxides; (2) bacteria; (3) action by plants; (4) hydrolysis; (5) reaction with alkalies; (6) sea-water electrolytes acting on ferric oxide hydrosols stabilized by organic colloids, giving hydrous ferric oxide gel.

Bacteria and Catalytic Deposition. Harrar has long advocated a prominent role for bacteria in iron deposition. He divides the iron bacteria into three main groups: (1) those that precipitate ferric hydroxide from ferrous bicarbonate solutions, such as *Spirophyllum ferrugineum* and *Gallionella ferrugineum,* which thrive best in the absence of organic matter and the presence of carbon dioxide; (2) those that deposit ferric hydroxide from organic or inorganic iron solutions, such as *Leptothrix ochracea* and *Cladothrix dichotoma;* these need organic matter; (3) those that attack organic iron salts, using the organic radical for food, producing ferric hydroxide. Moore and Maynard also prove that ferric hydroxide is precipitated by soil bacteria of natural waters.

The chief manganese-precipitating bacteria are: *Crenothrix polyspora, Leptothrix ochracea, Cladothrix,* and *Clanothrix.* The first two forms were found in abundance by C. Zapffe in manganese precipitates at Brainerd, Minn., where the water mains, supplied by well water, were being rapidly choked by a precipitate of oxides of manganese and iron, the manganese predominating. In the well water iron predominated and had been largely removed by aeration, which, however, did not remove the manganese. After many experiments, Zapffe built a simple, effective, manganese-removal plant, in which the manganese was precipitated from the aerated water by using pyrolusite (MnO_2) as a catalyzer. Deposition of manganese dioxide in the water mains was started by bacterial action, and once started it was accelerated by the earlier deposition. By catalysis, the manganese bicarbonate in solution is broken up into carbon dioxide and manganous hydroxide and the latter is converted into manganic hydroxide and manganese dioxide.

Phosphorus is also precipitated by bacteria. Most of it, however, is probably removed from solution by vertebrates or shellfish, and deposition occurs best under reducing conditions and in shallow water. Most investigators agree that the concentration of the phosphate of sedimentary beds has taken place through organic agencies.

Bacteria are also effective in the deposition of sulphur from sulphates and sulphur dioxide.

Other Causes. Moore and Maynard showed that ordinary sea-water electrolytes cause almost instantaneous precipitation of hydrous ferric oxide from colloidal iron solutions. Manganese, undoubtedly, is similarly precipitated.

Products of Deposition. Iron is commonly precipitated as: (1) ferrous carbonate (siderite); (2) hydrous ferric oxide, goethite (limonite); (3) ferric oxide (hematite); and (4) minor basic ferric salts.

It seems probable that most of the marine hematites were deposited directly as ferric oxide. Glauconite, chamosite, and greenalite are less common forms. In the presence of air, ferric oxides form; and in the presence of organic matter, siderite, the ferrous carbonate, forms. Manganese is deposited largely as oxides. The carbonate may be deposited in the absence of air and in the presence of excess carbon dioxide.

Depositional Separation of Manganese and Iron. An interesting feature in connection with the formation of manganese ore is its separation from iron during deposition. In the case of chemical precipitation from carbonate solutions, the separation is due to the fact that manganese carbonate is more stable in solution than iron carbonate, hence, it is carried farther and is thus separated from the iron. Similarly, the ferric oxides are less soluble than manganese dioxide and so are deposited from solution first. The dioxide is the most stable manganese compound, is the easiest formed, and is the principal manganese ore mineral. Also, Vogt states that a point often overlooked by Americans is the difference between oxidic and neutral precipitation from carbonate solutions. Oxidic precipitation yields (1) iron hydroxide with little manganese, and (2) manganese hydroxide with a little iron, thus a separation; neutral precipitation gives simultaneous deposition of the carbonates.

Organisms also effect selective precipitation of manganese and iron, since some iron bacteria precipitate only iron, during an early stage, and some bacteria apparently exert a later selective precipitation on manganese.

Zapffe's investigations at Brainerd, Minn., also disclosed a selective precipitation of manganese by catalytic deposition.

Dunnington states that, from sulphate solutions, separation of manganese and iron is effected by their different reactions with calcium carbonate, which causes deposition of ferric hydroxide but has no effect upon the manganese sulphate until it is exposed to both air and calcium carbonate simultaneously.

CONDITIONS OF DEPOSITION

The conditions under which deposition occurs determine in large part the mineralogical composition of the resulting deposits, their size, purity, and distribution, both areal and stratigraphic. Sedimentary iron and manganese ores are deposited in both fresh and salt water, in bogs, swamps, marshes, lakes, lagoons, and in the ocean. Phosphates and sulphur form under marine conditions.

Bogs and Lakes. In restricted bogs and boggy lakes the resulting deposits are small and local. Iron is deposited as hydroxide or as ferrous carbonate, which in the absence of organic matter readily oxidize to ferric oxides. Impurities are generally present. In Quebec and Sweden iron ore is dredged from existing lake bottoms. The conditions for deposition of bog ores are widely present in glaciated regions.

Manganese may accompany bog iron or be deposited by itself. The resulting product consists of impure manganese wad. Hewett mentions that manganese oxide nodules have been found in lakes, such as those of Lake Tyne, Scotland, and Seller Sea, Austria. A bog manganese deposit 6 feet thick and with an area of 17 acres occurs at Hillsborough, New Brunswick.

Swampy Basins. In areas of brackish water and in marine swamps and marshes iron is deposited in the presence of plants. Precipitation takes place from ferrous bicarbonate or organic solutions chiefly through depletion of carbon dioxide. Decaying vegetation inhibits oxidation, and the iron is deposited as ferrous carbonate (siderite); if any were thrown down as hydroxide, reduction and carbonation to ferrous carbonate would take place.

The conditions of deposition are also those of coal accumulation; therefore sedimentary siderite occurs in coal measures. If the siderite is deposited along with accumulating vegetation, *black band iron ore* is formed. This occurs in coal-like beds, is associated with coal, looks like coal, and might even be mistaken for dull coal except for its weight. If the deposition takes place in coal measures along with clay but not with coal, *clay ironstone* results — an impure carbonate. Sedimentary carbonate ores generally contain impurities of organic matter, clay, sand, and other carbonates.

Manganese is similarly deposited, and Hewett states that carbonate deposition in sediments is more widespread than is generally realized. Residual enrichment (Chap. 5·7) of these carbonate beds has given rise to many economic oxide deposits.

Peneplain Depressions. Deposition might also take place in isolated basins on peneplains during eras of low seasonal rainfall, as suggested by Woolnough.

Marine Conditions. The majority of economic sedimentary deposits are formed in marine areas, mostly under shallow-water conditions but also as open-sea deposits. Manganese of uneconomic importance occurs in deep-sea sediments, and the iron silicates have been deposited abundantly in open-sea sediments. These only rarely constitute economic deposits, but some investigators hold that some iron oxide deposits have been derived from them.

Shallow-water areas such as marine lagoons or long, narrow, epeiric seas are the sites of the greatest deposition of sedimentary deposits of iron ores, manganese, phosphate rock, sulphur, commercial carbonate rocks, and clays.

The Cycle of Iron

Iron dissolved during rock weathering moves largely in streams to favorable sites of deposition.   The iron may be lost during transportation (1) if the solutions traverse limestone, where reactions cause deposition of ferrous carbonate or ferric oxides; (2) if the solutions come to rest in an enclosed basin undergoing evaporation; (3) by contact with organic matter; or (4) by decrease in carbon dioxide content of the solutions.   That which is precipitated in bogs gives rise only to small, impure, low-grade deposits rarely exceeding 45 percent iron. The ore consists chiefly of limonite with some ferrous carbonate and phosphate of iron mixed with sand and clay, as at Three Rivers, Quebec, and in Maine, Sweden, Russia, and England.   The iron that reaches coal basins is thrown down as impure and low-grade siderite deposits. For extensive oxide precipitation, however, the iron must reach the sea.

Open-Sea Deposition.   The iron that reaches the open seas becomes deposited in vast quantities as the hydrous iron silicates, glauconite, greenalite, chamosite, or thuringite.   " Greensands " are marine deposits of sand, glauconite granules, and some clay and shell matter. Eckel estimates that three Cretaceous beds in New Jersey contain some 250 thousand million tons of ferric oxide.   They are not iron ore, but their content of potassium and phosphorus make them useful for fertilizer and water purification.

Glauconite is deposited in deep-sea muds, and recent glauconite favors warm, slightly shallow mud bottoms, where some organic matter is present, and an environment that is neither strongly oxidizing nor reducing.   It is thought that glauconite is formed from pellets of colloidal silica and clay in which colloidal iron replaces the alumina and absorbs potash from the sea water.

Greenalite is thought by Van Hise and Leith to be the iron silicate from which the Lake Superior iron ores were derived by the action of alkaline silicates on ferrous salts.

Chamosite, a hydrous ferrous silicate, occurs mostly as oölites in the sedimentary iron ores of Newfoundland, England, and central Europe. With thuringite, it has been worked in Thuringia.

Marine Shallow-Water Deposition. The iron solutions that reach the shallow seas give rise to the largest iron-ore deposits of the world. Apparently, the optimum conditions are where sluggish streams enter from deeply eroded, low-lying, coastal areas, with gradients too low to permit abundant suspended matter to be transported. Consequently, little sediment accumulated with the iron ore. Shallow waters are indicated, where waves, alternating with periods of quiet water, gently churned the bottom, and macerated the fossils present. Ripple and current markings also show that the depth of water was not great, and mud cracks indicate occasional elevation and exposure to the sun. It is not necessary to assume an unusually high concentration of iron in the sea water. The marine life was not dwarfed, which gives evidence of no unusual conditions of environment. Limestone deposition occurred, alternating with the deposition of thin beds of shale and sandstone. The sluggish streams, relatively high in iron and soluble materials and low in suspended matter, introduced the iron, which was carried as bicarbonate or in colloidal form. Colloidal iron would be deposited almost instantaneously upon contact with oppositely charged electrolytes of the sea water; bicarbonate iron would be precipitated as shown above.

The iron was probably deposited mostly as ferric oxide rather than limonite because of the dehydrating effect of salt water. Some iron was deposited as oölites; some coated or replaced shell fragments on the sea bottom; some was precipitated as an iron mud, perhaps as a gel. A little calcium carbonate was precipitated at the same time, and variable amounts of clayey matter became admixed, relatively small amounts in the rich ores and relatively large amounts in the lean ores. Eckel points out that in the Clinton iron beds the associated clayey and shaly matter is higher in alumina and iron than normal shales, suggesting that the iron was derived from deeply weathered basic igneous rocks or limestones.

The marine character of these oxides is indicated by the contained fossils, the oölites, the nature of the sedimentation, and the size of the basins of deposition. The oölites are considered to be diagnostic of marine conditions and to have been formed by colloidal processes. The origin of these marine oxides, however, has not been unquestioned. Some geologists have contended that they have been formed by the replacement of limestone, and others that they are residual accumulations, or derived from glauconite beds. Such conclusions, however, are not substantiated by field evidence. The iron is considered to have been dissolved and transported as bicarbonate or as a colloid in organic solutions and to have been deposited directly as hematite or goethite. In some deposits, diagenetic replacement of fossil fragments is evident, and in others, such as those of Newfoundland, chamosite has been deposited in concentric shells with hematite to form the oölites; siderite is also relatively abundant here.

In places several iron-ore beds have been deposited of which from three to ten have proved commercial. A long period of continuous iron deposition over a wide area is indicated by the Big Seam of Clinton iron ore at Birmingham, Ala., which, according to Eckel, attains a thickness of 30 feet, averages over 10 feet, and has been tested over a width of 10 miles and a length of 50 miles. Outcrops of Clinton ore are almost continuous for a distance of over 700 miles, so the main basin of deposition must have been at least this long and 50 miles wide. Eckel states that in one part of it the enormous quantity of 5,000 million tons of iron oxide was laid down in continuous deposition and that the Wabana basin of Newfoundland probably had a continuous deposition of 7,000 million tons. Iron accumulation was on such a vast scale that purely local conditions cannot be invoked to explain it.

DEPOSITS FORMED, AND EXAMPLES

TYPE	CHARACTER	CHIEF LOCALITIES
Bog ores	Limonite	Quebec, Maine, Sweden, Russia
Carbonate	Black band ore, clay ironstone	Pennsylvania; Ohio; Middleboro, North Staffordshire, Lowmoor, Dowlais, Clyde Basin, Great Britain; Westphalia, Saarbrücken, Germany; Russia
Iron silicates	Thuringite	Thuringia, Switzerland, South Africa, Russia
	Chamosite	Cleveland Hills, England; Newfoundland
	Greenalite-glauconite	Widely distributed
Marine oxides	Hematite or limonite	Clinton oölitic hematites of United States; " Minette " oölitic limonites of Lorraine and Luxemburg; Newfoundland; Krivoi Rog, Krusch, Russia; Brazil

These various deposits are described in Chapter 14.

The Cycle of Manganese

The cycle of sedimentary manganese is very similar to that of iron, to which manganese has many chemical resemblances. The two metals have the same source, are dissolved by similar or the same solutions, may be transported together in similar chemical compounds, and are deposited as oxides and carbonates by the same agencies, although generally separately.

Manganese may be deposited as a minor constituent of iron ores, or it may be deposited separately as sedimentary manganese deposits relatively free from iron. Large deposits of sedimentary manganese, however, are relatively few but very important as compared with manganese deposits produced by other processes, and they are fewer and much smaller than sedimentary iron deposits.

Deposition of sedimentary manganese ore in the form of carbonate or oxides may occur both in fresh and salt water, in lakes or bogs, or in the sea. It parallels that of iron and so needs only brief mention. The two differ, however, in the degree of oxidation and in the oxides formed. Whereas ferrous oxide and ferrous hydroxide are unknown in nature, the equivalent manganese compounds are well-known minerals. Manganite ($Mn_2O_3H_2O$) and hausmannite (Mn_3O_4), the respective counterparts of goethite and magnetite, are common, but the dioxide (MnO_2), which is the chief ore mineral of manganese, has no iron counterpart.

Marine depositions, chiefly in the form of the dioxide, have been formed under shallow-water conditions and in deep-sea sediments, where it is widely distributed as nodules, as coloring matter, and as coatings on fossils; according to Hewett, it is greatest in amount in oceanic areas surrounded by basic rocks. Deep-sea deposits are of scientific, but not commercial, interest.

Under near-shore conditions, as in the case of iron, manganese oxide hydrosol, or bicarbonate solutions, precipitate oxides or carbonates or both. The oxides commonly form oölites, which are made up chiefly of psilomelane and pyrolusite. These, along with included marine fossils, indicate a marine origin of the manganese. The associated rocks are shales, limestones, and, less commonly, sandstones.

DEPOSITS FORMED, AND EXAMPLES

Carbonates. Hewett states that impure sedimentary manganese carbonate is widely distributed but is nowhere commercial. Beds of relatively pure sedimentary manganese carbonate are reported from Newfoundland, and others occur in Arkansas, Minnesota, South Dakota, California, the Appalachian states, Wales, Belgium, and Russia. The beds seldom exceed a few feet in thickness or contain more than 20 percent manganese. Their chief importance is in the light they throw on the origin of manganese and in the fact that they supply preliminary concentrations of manganese which, upon weathering, may yield marketable deposits of secondary oxides.

Oxide Deposits. The great manganese deposits of Russia at Chiaturi, Georgia, and at Nikopol, Ukraine, are not only the outstanding sedimentary oxide deposits but the greatest manganese deposits of the world. The ore occurs (see Chap. 14) in oölites and nodules in Tertiary beds of sand and clay from 3 to 12 feet thick. Somewhat similar deposits occur in the Urals, Siberia, and Milos, Greece.

The Cycle of Phosphorus

The sedimentary cycle of phosphorus is fascinating and puzzling. Dissolved from the rocks, some of it enters the soil, from which it is abstracted by plants, from them passes into the bodies of animals, and is returned via their excreta and bones to accumulate into deposits. These in turn may undergo re-solution, reach the sea, and there the phosphorus is deposited or accumulated by sea life, embodied in sediments, and returned to the land upon uplift, when a new cycle may start.

Phosphates are soluble in carbonated water and, in the absence of calcium carbonate, will stay in solution. The phosphate in limestones resists solution. Some phosphoric acid in solution reaches the sea, where it is extracted by organisms; some is redeposited as secondary phosphates, which may be redissolved; and some is retained in the soil. Swamp waters rich in organic matter also dissolve phosphates, and some phosphorus compounds are thought to enter solution as colloids. Phosphorus is probably transported by streams as phosphoric acid and as calcium phosphate. Some is transported by birds and animals.

Special Conditions of Deposition. Economic beds of phosphate are formed only under marine conditions in the form of phosphorite. The beds range in age from Cambrian to Pleistocene and extend with remarkable uniformity over thousands of square miles. They are interstratified with other sediments and grade laterally into them. The beds are sparingly fossiliferous and are interlaminated with marine fossiliferous beds. These features together with their own oölitic character prove a marine origin.

According to Mansfield, the accumulation of sufficient phosphates to produce sedimentary beds must have required unusual marine conditions. Other sedimentation must have ceased to permit sufficient accumulation of pure phosphatic materials to form the oölites that aggregated into phosphate beds. The presence of hydrocarbons and marcasite in these deposits indicates deposition under reducing conditions, and replacement of shells by phosphate proves some diagenesis. Mansfield thinks that the material represents a slow accumulation, shut off from the sea, of phosphatic debris under anaerobic conditions and that for a long time cool temperatures were frequent during climatic oscillations. Such conditions favored the growth of life in the shallow waters and reduced the activities of denitrifying bacteria, which curtailed the deposition of calcium carbonate and favored the concentration of phosphatic solutions from which the oölites were formed.

Mansfield considers that two features are essential to phosphate accumulation: (1) a combination of favorable paleogeographic circumstances, and (2) the presence of some agent to fix the phosphoric acid in relatively insoluble form. This agent, he suggests, may be fluorine, which enters (and is present in excess) in the fluorapatite of phosphate. He has observed a relation of phosphate deposition to unconformities, and particularly to volcanism, which would supply the needed quantities of fluorine. Volcanism combined with favorable paleogeographic conditions may, therefore, provide the unusual conditions necessary for marine phosphate deposition.

Sedimentary deposition has given rise to the great phosphate deposits of the world. Algerian, Tunisian, and Moroccan deposits together yield the largest production in the world, and the marine phosphate beds of Idaho, Montana, Wyoming, and Utah are the most widespread. Sedimentary phosphate beds also yield secondary " pebble " deposits. These and other deposits are described in Chapter 22.

The Cycle of Sulphur

Sulphur is abundantly distributed in the earth's crust in the form of sulphates and sulphides. It is a copious and important constituent of volcanic gases and magmatic emanations and is common in hot springs.

The sulphur of sedimentary deposits is derived from sulphates in rocks and from hydrogen sulphide obtained from volcanic emanations, anaerobic bacterial decay, and bacterial reduction of sulphates in solution. These substances are transported in solution to basins of sedimentation; sulphur may also be carried in solution as colloidal sulphur.

The sulphur is deposited from sulphates and hydrogen sulphide in bodies of water low in oxygen where reducing conditions and anaerobic bacteria prevail. Sulphates are reduced by bacteria to hydrogen sulphide, which, in turn, oxidizes to sulphur and water. Hydrogen sulphide becomes so highly concentrated in some waters lacking in oxygen that it inhibits marine organisms. Sulphur bacteria are also supposed to deposit sulphur from hydrogen sulphide; Trask has shown that native sulphur is not an uncommon constituent of marine muds.

Special conditions of deposition and accumulation are, however, necessary to give rise to commercial concentrations of sedimentary sulphur. The deposition of other sediments must cease, or the supply of sulphur must be unusually large, in order to permit the accumulation of layers of pure sulphur. Associated gypsum is thought to have been deposited during periods of high salinity induced by evaporation, which eliminates temporarily the sulphate-reducing bacteria and the deposition of sulphur.

Volcanic hydrogen sulphide, which oxidizes to sulphur, is considered by Kato to be the source of sulphur layers in a lake at Közuke, Japan, and by Sagui to have been fed by hot springs from underlying basalt into sulphur basins in Sicily. Murzaiev, likewise, attributes the sulphur of Kazbek, Gamur, and Kamchatka, in Russia, to a volcanic source.

Reduction of sulphates by sterile inorganic compounds met with no success by Bastin and Allen-Crenshaw-Merwin, but experiments by Thiel, Beyerinck, Van Delden, and others using microorganisms yielded hydrogen sulphide. Van Delden's results led to his estimate that 45 kg of sulphur per square meter would accumulate in 100 days from a zone of hydrogen sulphide water 10 m deep in a lake. Apparently, natural reduction of sulphates necessitates microorganisms.

Examples of Deposits. Important sedimentary sulphur deposits occur near Knibyshev, Sukeievo, and Chekur, in Russia. Those deposits at Knibyshev (Fig. 21–3) illustrate the sedimentary processes. They are thin gypsum beds with layers of pure sulphur, laminations of sulphur and calcite, or sulphur nodules in bituminous limestone. Celestite is an unusual associate. The sulphur is pure or bituminous; some is recrystallized, and some is oölitic. A lagoonal sedimentary origin is indicated. These deposits are of Permian age, and those of Chekur occur in upper Tertiary lagoonal clay beds. Hydrogen sulphide springs are numerous near the deposits.

The sulphur deposits of Sicily constitute another example of sedimentary sulphur. Sulphur-bearing formations lie in isolated basins up to 5 miles long and half a mile across. They consist of cellular limestone, interstratified with bituminous shale and gypsum, and overlain by beds of marl, clay, limestone, and sandstone, now folded and faulted. The sulphur is disseminated through the cellular limestone and occurs also in bands of pure sulphur up to 1 inch in thickness. The sulphur content ranges from 12 to 50 percent and averages 26 percent. It is considered to have been deposited largely by biochemical processes.

Other examples of sedimentary sulphur deposits occur in Persia, Rumania, Croatia, Galicia, and Upper Silesia. Other types of sulphur deposits are described in Chapter 21.

The Cycle of Copper

A syngenetic sedimentary origin is held for a few copper deposits, but not without dispute. It is, of course, known that copper also has a sedimentary cycle. Dissolving during oxidation, it moves to basins of fresh or salt water; 1.24 to 5.12 mg per oyster have been absorbed. It has been precipitated in sea muds as sulphides and native copper and has been thrown down by microorganisms.

A sedimentary origin has been maintained for the famed Kupferschiefer of Mansfeld, Germany, but an opposing school vigorously advocates a hydrothermal origin for it. These remarkable deposits (Chap. 13), mined since A.D. 1150, have been the chief copper reserves of Germany. The Permian Kupferschiefer basin has an area of 22,500 square miles. Above a basal conglomerate is the thin, black, cupriferous shale, 1 m thick, which Beyschlag describes as "one of the most remarkable products of the geologic ages." It is a shallow, marine, organic mud, full of land plants washed in from adjacent coasts. The syngenetic advocates believe that into this putrefying bottom there were swept cupriferous solutions, probably sulphate, derived from the oxidation of distant lodes, perhaps even in the Harz Mountains. The metals were precipitated as iron-copper sulphide gels by the organic matter or, as Schneiderhöhn believes, by sulphur bacteria. The chief minerals are copper sulphides, but there are also those of iron, lead, and zinc. Silver, nickel, cobalt, vanadium, molybdenum, and other metals are also present. Although the arguments for a syngenetic origin are strong, there are stronger arguments that support an epigenetic origin.

The "Red Bed" copper deposits of the southwestern United States have often been spoken of as sedimentary. However, the proof seems convincing that the copper minerals are epigenetic.

The Cycle of Uranium-Vanadium

The unique uranium-vanadium-radium deposits of the western Colorado Plateau are considered by F. L. Hess to represent the reworking of sedimentary deposits of these metals. The source and solution of the metals are unknown, but it is supposed that they were released from former veins by weathering. If so, such veins presumably contained pyrite, which by oxidation yielded sulphuric acid and ferric sulphate that would dissolve the metals as sulphates. Their association with gypsum lends color to this possibility.

The carnotite deposits occur in cross-bedded sandstone of Morrison and Entrada age in association with vegetable matter, logs, and saurian bones. The sandstones, according to Hess, indicate an origin in " shallow water with mobile islands, spits, and shores." He concludes that sulphate solutions of these metals entered the shallow waters where abundant vegetation was macerated by wind and waves. Logs were swept in by the rivers and stranded on the islands, spits, and shores, and around them were packed sands and macerated vegetation. The vegetation is supposed to have reduced the dissolved salts from the tenuous solutions and caused deposition of the uranium-vanadium minerals at these localized places. The logs, now petrified, are considered to have been replaced first by calcite and then by the uranium-vanadium minerals, or by both together. That the vegetable material and logs were important factors in the accumulation of the uranium-vanadium materials seems indicated by the amazing concentrations in some logs. Two petrified logs at the San Miguel River, one 100 feet by 4 feet, yielded 105 tons of ore containing $175,000 in radium, $27,300 in uranium, and $28,200 in vanadium. Two other logs, and the intervening sandstone, from the vicinity of Calamity Gulch yielded $350,000 — the most valuable logs ever known. (See also pp. 619–621.)

Later work casts doubt on these ores being of sedimentary origin. Fischer states that the vanadium deposits in the Morrison sandstone appear to occupy the thicker and coarser-grained parts of channel sandstone and were probably deposited by circulating ground-water solutions after the enclosing sands had accumulated but before regional deformation. (See p. 621.)

The Carbonate Cycle

The solution, transportation, and deposition of calcium and magnesium carbonate give rise to deposits of commercial limestones, dolomite, and magnesite.

Limestones are of marine or fresh-water origin, and magnesium may in part replace the calcium, giving dolomitic limestones. Impurities of silica, clay, or sand are commonly present, as well as minor amounts of phosphate, iron, manganese, and carbonaceous material. The calcium is derived from the weathering of rocks and is transported to the sedimentary basins chiefly as the bicarbonate, in part as carbonate, and abundantly as sulphate.

Calcium carbonate is deposited by inorganic, organic, and mechanical means. Carbon dioxide plays a dominant role in inorganic processes because the solution of the calcium carbonate in the sea is dependent upon it. If it escapes, calcium carbonate is precipitated, as in cave drip stones. The amount of carbon dioxide in the sea depends upon the water temperature and the amount in the air, which is in equilibrium with that in the water. More carbon dioxide is held in cold water than in warm water, a fact familiar to everyone who opens a warm bottle of soda water in contrast with an iced one. Similarly, warmed sea water loses carbon dioxide and, since it is practically saturated with calcium carbonate, precipitation ensues. Johnson and Williamson state that an increase of water temperature from 15° to 20° C causes deposition of 5.4 ppm calcium carbonate. Evaporation in lakes brings deposition of tufa. Other changes of equilibrium also bring about deposition.

Organic deposition is brought about by algae, bacteria, corals, Foraminifera, and larger shells. Calcium carbonate is also deposited by the photosynthesis of plants. Entire limestone beds may consist of Foraminifera, or nummulite shells, or of coral, or of larger shell forms (coquina).

Limestone may be formed mechanically through the deposition of comminuted shell matter and coral sand, which became cemented into compact limestone. Most limestones are deposited in shallow to moderately deep sea water, free from terrigenous sediments.

Marl, a friable, incoherent, pure limestone, is deposited in lakes from calcium carbonate supplied by streams or springs. It is common in glacial lakes because the glaciers which made the lakes supplied ground limestone and yielded cold water that held much carbon dioxide and therefore, calcium carbonate in solution. The cold melt waters lost their carbon dioxide content in the warmer lake waters and calcium carbonate was precipitated. Low aquatic plants, such as *Chara*, probably, however, deposit most of the marl.

Chalk, white earthy limestone, is deposited mainly in shallow waters and consists of a chemical precipitate of calcium carbonate and the minute shells of Foraminifera and other organisms.

Dolomite consists of the double carbonate of calcium and magnesium (54.35% $CaCO_3$ and 45.65% $MgCO_3$), but in dolomitic limestones the proportion of $MgCO_3$ is less than in dolomite. Most so-called dolomites are really dolomitic limestones; some of the magnesium may be replaced by iron or manganese. The three carbonates, with calcium, form isomorphous mixtures within certain limits. Many dolomites are not sedimentary but are epigenetic replacements of limestone. Dolomite is generally absent from sea oozes, and its origin is unsettled. Chemically, its direct precipitation has not been demonstrated. It is known, however, to be abstracted from sea water into the shells of organisms, and some coral reefs consist in part of dolomite. Sedimentary dolomites are generally considered to be sea-floor replacements of calcareous ooze. The staining of polished surfaces with Lemberg's solution shows that dolomite is intimately intercalated and contemporaneous with limestone.

Magnesite, the carbonate of magnesium, is an important industrial mineral. The sedimentary variety occurs in association with salt and gypsum, or shales and limestones, and is formed by deposition of magnesium carbonate, along with some calcium carbonate from concentrated waters of saline lakes. Apparently, the deposition has been brought about by chemical precipitation with subsequent dehydration. Presumably the magnesium was transported as magnesium sulphate by surface or underground waters and reacted with sodium carbonate to yield insoluble hydromagnesite, which accumulated as a relatively pure precipitate, and sodium sulphate, which with other soluble salts remained in solution. Examples of sedimentary deposits occur in Kern County, California, Nevada, Idaho, British Columbia, and Germany. (See Chap. 18.)

Varieties of sedimentary carbonate rocks of economic interest are listed herewith.

KIND	USE	REFERENCE CHAPTER
Building limestones	Building and structural	**18***
Cement limestones	Hydraulic cements	**18**
Siliceous limestones	Hydraulic limes	**18**
Silico-aluminous limestones	Natural cements	**18**
Limestone	Flux, fertilizer, chemicals	**19**, 21, 22
Lime rock	Quicklime	**18**, 19, 20, 21, 22
Chalk	Cements, powders, crayons, fertilizer	18, **19**, 22, 23
Marl	Cement, fertilizer	**18**, 22
Lithographic limestone	Fine engraving	—
Dolomite	Cement, refractory	18, **19**, 21
Magnesite	Cement, refractory, chemical	17, **18**, 19, 20, 21, 22, 23

* Bold-face indicates the chapter of chief description.

The Clay Cycle

The clay cycle differs from the preceding sedimentary cycles in that the constituents of clay are transported not in solution but in suspension, and their deposition is mechanical rather than by chemical or organic processes.

As shown in Chapter 5·7, the chemical decomposition of aluminous rocks yields clay minerals. These may become deposited *in situ* to form residual clay deposits (Chaps. 5·7, 17) or be transported and deposited as sediments. The sedimentary clays may be divided into marine, estuarine, lake, swamp, and stream clays.

Marine clays settle from mechanically transported suspensions in quiet water some distance offshore. The beds may be of great extent, of considerable thickness, and of fairly uniform composition, although lateral variation is to be expected because different streams supply the materials; vertical variations are similarly introduced. The beds are finely laminated. Marine clays are widely distributed in Paleozoic and Mesozoic formations.

Estuarine clays, since they are laid down in shallow ocean arms, are restricted in extent and commonly contain many sandy laminations that increase toward the source of supply. Marsh products are also interlaminated. Examples occur along the New Jersey side of the lower Hudson River and in the Chesapeake Bay region.

Lake clays occur in restricted basins and typically alternate with beds of sand. They are common in glacial lake beds. Consequently, many of them are varved, i.e., made up of thin bands of alternate coarse and fine material, each pair of bands representing seasonal accumulations of a year's growth. These clays are of Recent age and are abundant within the glaciated areas.

Swamp clays or *fire clays* may underlie coal seams, and upright stumps of ancient trees are found in them. The beds are small, lens-shaped, and exhibit little lamination. The clays are highly plastic, generally refractory, and relatively pure. Consequently they are eagerly sought for the host of uses to which they are put. They are thought to have originated from suspended matter carried by low-gradient streams into coal swamps whose early marginal fringe of vegetation filtered out the coarser sediment, allowing only the finest to reach and settle in the interior of the basin. The organic acids present have been held by some investigators to purify the clay sediment.

Stream clays are deposited in protected places on flood plains during periods of overflow. Consequently the deposits are pockety and grade laterally into sandy material. The pockets, however, yield a fine plastic clay, but different pockets vary greatly in composition. Streams may also deposit delta clays in isolated basins on deltas. Clays of stream origin are widely distributed.

Other Economic Products of Sedimentation

Substance	Composition	Reference Chapter
Coal (nonanthracite)	Carbonaceous materials	16*
Oil shale	Shale and bitumens	16
Diatomaceous earth	Chiefly diatoms	17, 18, **20**, 23
Fuller's earth	Clay minerals	17, **20**, 21, 23
Bentonite	Clay minerals	17, **20**
Tripoli	Silica	**23**
Sand and sandstone	Sand grains	**18**, 19, 20, 21, 23
Greensand	Sand, glauconite	21, **22**

* Bold-face indicates the chapter of chief description.

Selected References on Sedimentation

Principles of Sedimentation. W. T. TWENHOFEL. McGraw-Hill, New York, 1940. Chaps. 9, 10, 11, 13. *Comprehensive discussion of all phases of sedimentation.*

Iron Ores. E. C. ECKEL. McGraw-Hill, New York, 1914. *Treatment of important sedimentary iron-ore deposits of the world.*

Solution, Transportation, and Deposition of Iron and Silica. E. S. MOORE and J. E. MAYNARD. Econ. Geol. 24:272–303, 365–402, 1929. *Theoretical treatment of natural solution and deposition of iron that forms ore beds.*

Origin of Banded Iron Deposits — A Suggestion. W. G. WOOLNOUGH. Econ. Geol. 36:465–489, 1941. *Precambrian banded iron ores are epicontinental precipitates in closed basins on peneplains.*

Wabana Iron Ore Deposits of Newfoundland. A. O. HAYES. Econ. Geol. 26: 44–64, 1931. *Occurrence, mineralogy, and origin.*

Deposition of Manganese. C. ZAPFFE. Econ. Geol. 26:799–832, 1931. *Catalytic deposition of manganese from a water supply.*

Mineral Development in Soviet Russia. C. S. FOX. Min. Geol. and Met. Inst. India Trans. 34:98–210, 1938. *Manganese, gypsum, and iron ores.*

Cements, Limes, and Plasters. E. C. ECKEL. John Wiley & Sons, New York, 1928. Parts II–VII. *Good treatment of gypsum and limestones.*

General Reference 9. Chap. 10. *Deposits of uranium, vanadium, radium, copper, manganese, phosphate, and potash.*

Vanadium Deposits Near Placerville, Colo. R. P. FISCHER, J. C. HAFF, and J. F. ROMINGER. Colo. Sci. Soc. Proc. 15:117–135, 1947. *Ores cross stratification.*

The Geologic Role of Phosphorus. ELIOT BLACKWELDER. Am. Jour. Sci. 42:285–298, 1916. *Details of the phosphate cycle.*

Phosphate Deposits of the United States. G. R. MANSFIELD. Econ. Geol. 35:405–429, 1940. *Occurrence and origin of sedimentary phosphates.*

Potash Reserves of the United States. S. H. DOLBEAR. Amer. Potash Inst., Washington, D. C., 1946.

The Role of Fluorine in Phosphate Deposition. G. R. MANSFIELD. Am. Jour. Sci. 238:863–879, 1940. *Methods of transport and deposition.*

Genesis of the Sulphur Deposits of the U. S. S. R. P. M. MURZAIEV. Econ. Geol. 32:69–103, 1937. *Descriptions and origin of sedimentary sulphur.*

5·6 EVAPORATION

Evaporation has been important in producing many valuable types of nonmetallic mineral deposits. Underground waters have been drawn up to arid surfaces, there to be evaporated, leaving behind valuable minerals once in solution. Lakes have disappeared under relentless desert suns to form playas encrusted with various usable salts, or salt layers covered later by the shifting sands of arid regions. Or, the evaporation may not have gone to dryness but has yielded concentrated liquors from which are obtained much-used household salts. The searcher for useful substances eagerly avails himself of these rich liquors and extracts them for precipitation by more rapid natural or artificial evaporation. Natural brines are pumped from great depths and spread out to the sun or put in artificial evaporators, so that their salts may be won.

Great sections of the oceans have been cut off during slow oscillations of land or sea and gradually evaporated to the point that gypsum or common salt has been deposited in many parts of the world. Still greater concentration by evaporation has yielded rich and valuable potash deposits that have been a continuing source of wealth.

Thus, evaporation is a great mineral-forming process, simple and familiar in its operation, which supplies homely materials used by the householder, the farmer, the chemist, the builder, the engineer, the manufacturer, and even by the birds and beasts and plants.

Process of Mineral Formation by Evaporation

Evaporation proceeds most rapidly in warm, arid climates.

In evaporation of bodies of saline water, concentration of the soluble salts occurs, and when supersaturation of any salt is reached, that salt is precipitated. The least soluble salts are precipitated first and the most soluble last. The solubility of a given salt and, therefore, its deposition, is affected by temperature and the presence of other salts in solution. Time is also a factor in the formation of complex salts. Changes of temperature and salinity during evaporation, daily, seasonal, or cyclic, may cause reversals in the order of deposition. Also, overlapping of deposition occurs. Some salts undergo a change during and after precipitation; e.g., gypsum may be converted to anhydrite.

With sea water, which contains 3.5 percent by weight of salts, no deposition occurs until the volume has been concentrated by evaporation to nearly one-half of its original bulk; gypsum or anhydrite is deposited when the volume reaches about one-fifth, and common salt when it reaches about one-tenth. The composition of sea water, and the order of precipitation of sea salts and the reduction of volume at which precipitation starts, as determined by Usiglio, are given in the table. Below 0.0162, the bittern salts containing chlorides of K, Mg,

	Percent by Weight	Percent of Total Solids as Salts	Volume	Salt Precipitated
Water	96.2345	0	1.000	
NaCl	2.9424	77.758	0.533	Fe_2O_3
$MgCl_2$	0.3219	10.878		$CaCO_3$
$MgSO_4$	0.2477	4.737	0.190	$CaSO_4 \cdot 2H_2O$
$CaSO_4$	0.1357	3.600	0.095	NaCl
NaBr	0.0556			$MgSO_4$
KCl	0.0505			$MgCl_2$
K_2SO_4		2.465	0.039	NaBr
$CaCO_3$*	0.0114	0.345	0.0162	Bittern salts
Fe_2O_3	0.0003			
$MgBr_2$		0.217		

* Includes traces of all other salts.

and Na and some $MgSO_4$ yield different salts between night and day, and many double salts are thrown down.

The above sequence of salts is rarely obtained because evaporation seldom goes far enough to cause precipitation of the most soluble salts, although the Stassfurt deposits of Germany are fairly complete. The sequence given would not necessarily follow for other saline waters that contain different proportions of the same salts, because concentration and temperature determine the depositional sequence and the composition of the precipitates. Quite different products and sequences obtain in interior drainage lakes of arid regions where, instead of sodium chloride waters, sodium carbonate, sulphate, and borate predominate. Still different products result from evaporation of subsurface waters and hot springs. Consequently, each is considered separately.

Deposition from Oceanic Waters

MATERIALS INVOLVED

The salts of oceanic waters are mainly the contribution of land waters; small amounts are contributed by volcanism, and some are dissolved from the ocean basins. They are the soluble products of the weathering of rocks and of solution by subsurface waters. The ocean forms a great mixing pot for the diverse contributions of rivers. Those from northern North America contribute carbonates in excess of sulphates and chlorides; those from the western and southwestern United States contribute sulphates in excess of carbonates. The same difference exists between eastern and western South America. Carbonate waters, however, dominate over sulphate waters.

Ocean water, dropped over the land as rain, garners a new load of soluble materials and again reaches the ocean; the cycle repeats itself. Thus, salts have accumulated in the oceanic reservoir ever since streams first coursed over the earth. The total amount of salts in the ocean is calculated to be 21.8 million cu km, enough to form a layer 60 m thick over the ocean bottoms. Of this, common salt would constitute 47.5 m; $MgCl_2$, 5.8; $MgSO_4$, 3.9; $CaSO_4$, 2.3; and the remaining salts, 0.6. The rivers of the world are estimated to contribute annually about $2\frac{1}{2}$ billion tons of salts to the ocean.

It is noteworthy that the ocean is continually receiving from the present rivers water unlike itself. Clarke points out that the rivers contribute carbonates in excess of chlorides and calcium in excess of sodium. The ocean contains chlorides in excess of carbonates and sodium in excess of calcium. This is a complete reversal. Therefore, the composition of the ocean is slowly changing. Of course, the low content of calcium carbonate is due to its continual abstractions, but the great excess of chlorine is puzzling.

The composition of the ocean salts, according to Dittmars, is:

	PERCENT		PERCENT		PERCENT
Cl	55.292	Mg	3.725	CO_3	0.207
Na	30.593	Ca	1.197	Br	0.188
SO_4	7.692	K	1.106	Others	Trace

Ocean water also contains gold, silver, base metals, manganese, aluminum, nickel, cobalt, and radium, as well as iodine, fluorine, phosphorus, arsenic, lithium, rubidium, caesium, barium, and strontium. Of these, iodine concentrates in sea weeds and copper in shellfish. Except for iron, phosphorus, and iodine, none of these become concentrated into commercial deposits.

CONDITIONS OF DEPOSITION

The mean depth of the oceans is about 12,000 feet, and the surface would have to be lowered nearly 11,000 feet before salt could be deposited. Consequently, to attain the necessary concentration to induce precipitation, bodies of sea water must become isolated from the ocean in places where evaporation exceeds inflow. Such isolation may be effected by the formation of barriers, which, for a time at least, keep out ocean water and permit evaporation concentration. If the barriers are low, more sea water may be added from time to time. Natural cut-offs occur along coastal areas where bars isolate lagoons, or as a result of delta building or slow crustal movements.

Natural " salt pans " are numerous in coastal areas of arid regions. The Gulf of Karabugaz, on the eastern side of the Caspian Sea, is a well-known example of concentration of salt water behind a bar with a narrow, shallow inlet. Here, amid the surrounding deserts, evaporation is so rapid that a current flows continuously from the Caspian Sea into the Gulf, augmenting the Gulf waters with some 350,000 tons of salt daily. Gypsum is being deposited on the shallow shores of the Gulf and sodium sulphate toward its center. If its inlet should become closed, sodium chloride and other salts would be deposited eventually.

In a cut-off body of sea water under conditions of evaporation in excess of inflow, as the volume diminishes the original salt content becomes concentrated in the deeper portions of the basin. That which is deposited on the uncovered shores of the receding water body is mostly washed in again by infrequent rains. Consequently, the salt originally diffused throughout the entire body of water gradually becomes concentrated in a relatively small volume in a central depression. If the original cut-off body had a volume of 100 cu km of water (Karabugaz has 183 cu km) and this became concentrated to about 50 cu km, the iron oxide and calcium carbonate present would be precipitated. If evaporated to about 20 cu km the water would then contain about 3,500 million tons of salt, of which about 2,700 million tons would be common salt. At this point gypsum would be deposited. When the volume reaches about 10 cu km, common salt would be deposited. Subsequent evaporation would next bring about deposition of magnesium sulphate and chloride, followed by the bittern salts.

CALCIUM SULPHATE DEPOSITION

Calcium sulphate may be deposited either in the form of gypsum or anhydrite, depending upon the temperature and salinity of the solution. According to Posnjak, whose recent work supersedes that of Van't Hoff and Weigert, gypsum will be deposited from saturated calcium sulphate solutions below 42° C (not the 66° C of Van't Hoff), and anhydrite above that temperature. Posnjak shows also that, when sea water evaporates at 30° C, calcium sulphate will always be deposited as gypsum when the salinity reaches 3.35 times the normal concentration of sea water, and gypsum will continue to be deposited until the salinity reaches 4.8 times normal, beyond which anhydrite is deposited. Further, about one-half of all the calcium sulphate present will be deposited as gypsum before anhydrite is deposited. Somewhat lower temperatures would not greatly modify the above ratios. This means that, at the temperatures of evaporation of marine basins, much gypsum will always be deposited first and that marine beds of pure anhydrite imply either that the early deposited gypsum was converted to anhydrite or that deposition occurred above 42° C. If deposition was below 42° C the lower part of anhydrite beds cannot be primary anhydrite. After sodium chloride starts to precipitate, only anhydrite can be deposited. Changes in temperature could cause alternations of gypsum and anhydrite. Under hot, arid conditions anhydrite would be expected, and in temperate regions, gypsum. Either form may change into the other. The dehydrating effect of salt water may convert gypsum into anhydrite, and gypsum subject to the pressure and temperature accompanying deep burial is converted to anhydrite. Anhydrite is converted into gypsum under conditions of weathering, as shown by Newland for New York deposits.

Most salt deposits throughout the world consist only of calcium sulphate and common salt; the general lack of bittern salts means that the sea water has not been completely evaporated. Consequently, evaporation must have been interrupted, and the bittern liquors were drained off or diluted by influx of more sea water or fresh water. The common lack of bittern salts is puzzling. In fact, its commonness suggests that it may not be due to a coincidence of special conditions but rather to some general underlying principle as yet unknown.

Deposition of Thick Beds. The origin and conditions of deposition of thick beds of calcium sulphate, or calcium sulphate and salt, is a problem. Thicknesses of calcium sulphate of 250 feet are known in Nova Scotia, of 300 feet in Germany, and Darton reports a thickness of 1,325 feet in New Mexico. Since the evaporation of 1,000 feet of sea water yields only 0.7 foot of gypsum, the evaporation of 425,000 feet, or a depth of 80 miles, of sea water, would be required to yield 300 feet of anhydrite. As this is ridiculous, it follows that new supplies of sea water must have been added to the basin during evaporation, and the residual liquors became concentrated in subbasins. The depth of water in most closed basins should, upon evaporation, also yield a layer of common salt above the calcium sulphate. Therefore, in the case of gypsum beds with no overlying salt, unusual conditions must have prevailed toward the conclusion of deposition. Further complications are added by the occurrence of salt without gypsum. Any theory of origin of thick beds of gypsum or salt or both must therefore account for a method by which enormous volumes of additional sea water are added to the evaporating basins, for the cessation of deposition before the period of deposition of sodium chloride or bittern salts is reached, and must also account for the alternations of salt and gypsum.

Many theories have been proposed, few have survived, and none are entirely satisfactory. Most investigators adopt the " bar theory " of Ochsenius or some modification of it, which postulates partial evaporation of an arm of the sea isolated by a bar over which or through which new supplies of sea water are added. Under high evaporation, the surface layers supposedly become denser and sink, causing bottom concentration, and with reversal of flow the residual bitterns are removed. The Gulf of Karabugaz is used as an example. This theory does not account for the lack of sodium chloride or bittern salts in many occurrences. A modification by Branson postulates a series of semi-isolated subbasins separated by low barriers, so that new supplies of water first enter the outer basins and overflow into the inner basins, carrying there the salt brines of the outer basin; thus, gypsum would be deposited in the outer basins and salt in the inner basins. However, the influxes of new sea water should run over into the inner basins, and bittern salts should also be deposited in the inner basins. Other theories assume the influx of fresh supplies of salts from stream drainage.

Where salt is deposited without underlying gypsum the explanation probably is that the brines from which the gypsum has already been deposited have been decanted into another basin where later deposition of salt could occur. This could be brought about by coastal tilting.

Resulting Products. Calcium sulphate deposition gives rise to: (1) Beds of relatively pure gypsum or anhydrite from a few feet to a few hundred feet in thickness. The gypsum beds constitute one of the most important nonmetallic resources, but anhydrite finds little use because in plaster it absorbs water and swells. (2) Gypsum beds with impurities of anhydrite. (2) Alabaster, a softer and lighter variety of gypsum. (4) Gypsite, an admixture with dirt. The beds are generally interstratified with limestone or shale, and they are commonly associated with salt. For the distribution, occurrence, and uses of gypsum, see Chapter 18.

Common Salt (Halite) Deposition

The deposition of salt beds provides the source for about three-fourths of all salt used. Salt beds are generally lenticular and range in thickness from a few feet to a few tens of feet. However, beds 150 feet thick occur in Ohio, 325 feet thick in Michigan, and 630 feet thick in New Mexico. Thicknesses of thousands of feet have been inter-

sected in salt-dome regions of the Gulf Coast and Germany, where salt
beds have been squeezed and upthrust.   The salt is commonly dazzling
white and pure; gypsum, anhydrite, or shales may occur at the bottom.
The beds are of many geologic ages; in North America they occur
mostly in the Silurian, Permian, and Triassic.   Salt beds are extensive
in the eastern, central, and southern states.

Salt Domes.   Salt domes are unusual and puzzling.   They are known
in the Gulf Coast, Germany, Spain, Roumania, and Iran, but those of
the United States are of particular interest because of the prolific oil

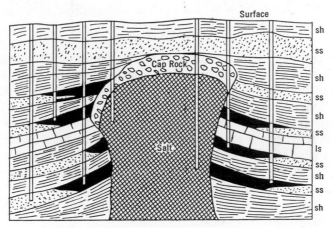

FIG. 5·6–1. Diagram of salt dome showing salt plug, anhydrite caprock, and
upturned strata containing oil and gas pools (black) tapped by wells.

pools associated with them.   Surface expression of them may be absent,
or they may be indicated by low mounds or depressions.   Most of
them have been located by geophysical methods and proved by drilling.
They are dome-shaped, pipe-shaped, or even mushroom-shaped (Fig.
5·6–1).   The plug consists of salt and anhydrite, and anhydrite or
gypsum generally constitutes a caprock 500 to 600 feet thick.   The
salt plug may be several thousand feet thick; one at Humble is at
least 5,400 feet.   The tops may be a few feet to a few thousand feet
beneath the surface.

No drill has yet penetrated to their bottoms.   The flanks are
bordered by upturned strata, which form the reservoir rocks of associated oil pools.   They give the impression of being gigantic plugs that
have been driven forcibly upward into the overlying strata, rending the
beds asunder and bending them upward to make oil traps.   That is
just what is considered to have happened.   The salt originally existed
in sedimentary beds (Mesozoic?) several thousand feet beneath the

surface. Under pressure, the yielding plastic salt was forced upward along some line of weakness, such as fault intersections. The great thickness of salt is thus not original but has been induced subsequent to deposition. Barton estimates that ten of the famous salt domes in the Gulf Coast region contain enough minable salt to supply the world for 4,000 years and that all known salt domes would supply the world demand for 300,000 years.

Salt domes are discussed in connection with petroleum in Chapter 16, and the production, distribution, and uses of salt are given in Chapter 21.

POTASH DEPOSITION

After the deposition of common salt, chlorides and sulphates of magnesium and potassium are the other chief salts deposited. The potassium minerals result from evaporation carried almost to completion, but only rarely are they deposited. Consequently they are not common, and important deposits are known only in Germany, Alsace, Russia, Spain, Poland, and United States. Most of the world's supply of potash is derived from marine evaporites, but a minor quantity is won from brines of salt lakes, vegetation, and other sources.

NAME	COMPOSITION	PERCENT K_2O
Anhydrite	$CaSO_4$	—
Bischofite	$MgCl_2 \cdot 6H_2O$	—
Carnallite	$KCl \cdot MgCl_2 \cdot 6H_2O$	16.9
Douglasite	$K_2FeCl_4 \cdot 2H_2O$	—
Glaserite	$3K_2SO_4 \cdot Na_2SO_4$	37.5
Halite	$NaCl$	—
Kainite	$MgSO_4 \cdot KCl \cdot 3H_2O$	18.9
Kieserite	$MgSO_4 \cdot H_2O$	—
Langbeinite	$K_2SO_4 \cdot 2MgSO_4$	22.7
Leonite	$MgSO_4 \cdot K_2SO_4 \cdot 4H_2O$	23.3
Polyhalite	$K_2SO_4 \cdot 2CaSO_4 \cdot MgSO_4 \cdot 2H_2O$	—
Sylvite	KCl	63.2

Potassium and Associated Salts. The famous Stassfurt deposits of Germany represent the only complete sequence of deposition of oceanic salts. There, some 30 saline minerals are known, but many of them have been formed by subsequent secondary reactions. Some of the more common salts of the deposits are shown in the table above. The important potash salts are italicized.

Sylvite is the preferred salt because it has the highest K_2O content, is a simple soluble chloride, and is the most usable mineral.

Special Conditions of Deposition. The deposition of potash salts from sea water requires special conditions of formation that have not occurred with the world-wide deposits of nonpotash gypsum and halite. Their abundance means not only almost complete evaporation but also a stupendous concentration of residual bitterns. This must mean a gradual drainage of bitterns, after removal of gypsum and much halite, into the lowest portions of large basins of evaporation. Thus a small area may come to contain the residual constituents formerly dispersed in a great volume of water spread over a large basin. In addition it seems probable that in most places there must have been some tilting of the land to drain off the bitterns into adjacent basins and aid in concentration.

Throughout the deposition of potash salts, halite is also deposited. Consequently beds of pure potash salts are rare; they also contain magnesium salts. Thus, certain beds are richer than others. Stratigraphic repetitions are not uncommon, indicating that the cycle of evaporation and deposition was interrupted and renewed many times.

The conditions of formation of potash minerals may be understood better by considering certain features of the Stassfurt deposits, of which the general description will be found in Chapter 22.

Features of the Stassfurt Deposits. The deposits of Stassfurt and the related areas are the outstanding illustration of potash deposition by evaporation of sea waters because they not only represent a complete sequence of oceanic salts but they have long been carefully studied. The stratigraphic sequence in the Stassfurt area is shown in the table on page 192.

Deposition of potassium salts closed with the older series (Zones 10–6), and a further influx of sea water started the deposition of the younger series (Zones 5–1) in which, however, no commercial potassium salts were deposited.

The potash zones also contain magnesium and calcium chlorides, and magnesium sulphate. Halite is present throughout all the zones from 10 to 2. Complex double salts also form as a result of resolution and reprecipitation due to slight changes in temperature and salinity during deposition. A yield of about 9 percent K_2O is obtained from the carnallite salts and about 13 percent K_2O from the sylvite-kainite salts.

Several of the Stassfurt salts have resulted from subsequent transformations due to increased temperatures attendant upon burial. Kainite and sylvite are alteration products of carnallite. Other salts represent retrograde metamorphism or reversals brought about by later rise of the beds after deep burials. Such potash minerals constitute delicate geologic thermometers. According to Van't Hoff and his associates, salts of the Stassfurt region had temperatures of formation above 10° C for loewite with glaserite and 60° with vanthoffite; and 72° for kieserite with sylvite. Recent synthesis by Ide suggests much higher temperatures for langbeinite, vanthoffite, and polyhalite. Investigations by Weber show that much replacement has gone on; polyhalite and kieserite replaced anhydrite; vanthoffite replaced kieserite and carnallite, and kieserite has been altered to sylvite. Some of

VERTICAL SECTION, IN METERS, OF SALTS AT STASSFURT, GERMANY

No.	Zones	Materials	Percent K_2O	Thickness	Depth from Surface
1		Surface drift		10	10
2	Red clay	Clays, sandstone, shales, salt, and anhydrite	0	100	110
3	Younger salt	Rock salt	0	100–150	210–260
4	Main anhydrite	Massive, widespread	0	30–90	240–350
5	Saline clay	Clay, salt	0	6–10	245–360
6	Carnallite	Carnallite, halite, kieserite, sand, clay, some sylvite, kainite	9.27	15–40	300±
7	Kieserite	Kieserite + carnallite and bischofite rock salt, anhydrite	2.17	20–40	350
8	Polyhalite	Rock salt, $MgCl_2$, gypsum, etc.	1.02	40–60	410
9	Older rock salt	With anhydrite in thin annular layers	Trace	300–500	800
10	Older anhydrite	Anhydrite and gypsum	Trace	60–100	1025
11	Underlying	Zechstein marine limestone		4–10	1030

these changes are considered to have occurred by downward percolation of potassium and magnesium chloride solutions through underlying anhydrite. Ahlborn has shown that where pressure was greatest "hartsalz" was converted to langbeinite, and where pressure was least, sylvite-rich or carnallite-rich beds were formed. It is evident that the formation of the complex salts has been a complicated geochemical process and not merely simple deposition through evaporation.

BORATE AND BROMINE DEPOSITION

Minor quantities of borates and bromine are obtained from marine salts. Although *borates* are mostly formed under other conditions, some are precipitated along with potassium minerals from marine residual liquors. Thus, boracite and other borates occur in the potash salt of Germany in association with carnallite and the overlying potash minerals. Magnesium borates are considered to be typical of marine conditions and calcium borates of lake-bed deposits. Most borates of commerce are obtained from lakes or lake-bed deposits. (See Chap. 21.)

Bromine is also deposited from residual liquors of sea water. The carnallite of Stassfurt contains 0.2 percent bromine, which is extracted in Germany during the refining of the potash salts. Bromine is remarkably concentrated in the Dead Sea to the extent of 0.4 percent compared with 0.0064 percent in ocean water. Most bromine, however, is a by-product of salt, from salt brines and sea water. (See Chap. 21.)

Deposition from Lakes

The salts of saline lakes are more diverse than those of the ocean since each lake, being an individual unit with no outlet, collects into itself the soluble salts of its own drainage basin. With the high rate of evaporation characteristic of arid interior drainage areas where saline lakes occur the salinity may be several times greater than that of the ocean. Great Salt Lake, Utah, in 1904 had a salinity of over 7 times that of the ocean, and its salts are of oceanic type. Others are characterized by high sodium carbonate, or sodium sulphate, or borates, depending upon their location with respect to areas of sediments, igneous rocks, or hot-spring discharge. Saline lake waters may be divided into salt or oceanic type, sodium carbonate and sodium sulphate, borate, nitrate, and potash. Some salt lakes also contain sulphates and the others contain more or less sodium chloride. Prolonged evaporation gives rise to waters of concentrated magnesium salts.

Deposition from Salt Lakes

Salt lakes contain the same salts as the ocean but generally in greater proportions. The salinity may range from less than that of the ocean to that of Great Salt Lake, which has varied, depending upon rainfall, from 13.79 percent in 1877 to 27.72 percent in 1904, and is now around 23 percent. Since Great Salt Lake is only a residuum of the greater Lake Bonneville, its salts have undergone considerable concentration by evaporation, and it has lost most of its calcium carbonate. Gilbert estimated it to contain 400 million tons of common salt and 30 million tons of sulphate. The Dead Sea is an example of a similar type that has been subject to extreme evaporation and is highly enriched in magnesium chloride.

Deposits. The deposits formed from the evaporation of salt lakes are similar to those obtained from ocean water. The relatively small size of lakes, however, makes them more responsive to climatic changes, with the result that they exhibit greater fluctuations of deposition. Evaporites formed during periods of desiccation may be redissolved during subsequent periods of expansion. Moreover, lakes constantly receive new supplies of fresh water, salts, and also sediments. The resulting saline deposits, therefore, are generally thin-bedded alternations of impure salts and clays. Also, on salt playas, desert winds distribute sands and silt, upon which later salts may be deposited during subsequent lake periods. This also gives alternations of salines with sand, clay, and minor calcium carbonate.

Examples of salt-lake deposits are common throughout the Basin and Range province of the United States. Miles of level glistening white salt lie west of Great Salt Lake and support a local salt industry. The Salton area of the Imperial Valley of California, before the last flooding of the Colorado, yielded considerable salt from a surface crust 10 to 20 inches thick. Examples of salt pans are numerous throughout central Asia and northern Africa. Huntington tells of the salt plain of Lop in Asia, where choppy waves of hard salt can be seen over a length of 200 miles and a width of 50 miles. Salt lakes of the past have been disclosed by drilling in Oklahoma and elsewhere.

Playa salt deposits throughout the world supply much of the local common salt used.

DEPOSITION FROM ALKALI AND BITTER LAKES

Lakes rich in sodium carbonate and sodium sulphate are common in the arid regions of North America where igneous rocks, or sedimentary rocks rich in sulphates, are found. Many igneous rocks upon weathering yield sodium and some potassium in greater abundance than do sedimentary rocks. Thus, the high content of sodium is explainable. The lakes are high in CO_3, Na, and SO_4, are relatively high in Cl, contain appreciable quantities of Mg and K and variable amounts of SiO_2, and are low in Ca and Fe. In the early stages of the formation of such lakes, calcium carbonate derived from weathering is introduced, but, with increasing concentration from evaporation, calcium is thrown down in the form of tufa, leaving sulphates and chlorides in solution.

Alkali Lakes. In alkali or soda lakes, sodium carbonate predominates, potassium carbonate may be abundant, and common salt is always present. Most of the sodium carbonate has been derived directly by decomposition of volcanic rocks, but some is also formed by slow and complex chemical reactions with other sodium and calcium salts; it may be formed also by the action of algae on sodium sulphate. The potassium carbonate is considered to be the indirect product of the work of organisms.

Examples of alkali lakes are Owens and Mono Lakes in California, the Soda Lakes of Nevada, Lake Goodenough in Saskatchewan, and the Natron Lakes of Egypt; they are numerous in the arid parts of the world. (See pp. 783–785.) Owens Lake has high salinity that varies with seasonal and climatic cycles. The Soda Lakes of Nevada contain about one-fifth each of sodium carbonate and sodium sulphate and three-fifths of sodium chloride. The deposits of hydrated salts, however, consist of nearly one-half sodium carbonate, one-third bicarbonate, and only a little sodium sulphate and halite. The Natron Lakes of Egypt are alternately wet and dry, and evaporation leaves a layer of natron and salt, bordered by sodium carbonate.

Bitter Lakes. In bitter lakes, sodium sulphate predominates, but carbonate and chloride are present. The sulphate may be derived from the decomposition of rocks that contain sulphates, or from the leaching of buried beds of sulphates. Such lakes are common in the arid regions of America and Asia. Examples are Verde Valley Lake in Arizona, Soda and Searles Lakes in California, and numerous lakes in New Mexico, Wyoming, and Saskatchewan. Lakes Altai and Domoshakovo in Russia are Asiatic representatives. The Soda Lake playa salts contain 88 percent anhydrous sodium sulphate, and the crude salts of the Downey Lakes of Wyoming contain over 95 percent sodium sulphate. The Gulf of Karabugaz deposits marginal sodium sulphate each winter, and this is harvested.

Deposits. The alkali and bitter lakes yield commercial sodium carbonate and sodium sulphate from brines, and from marginal and playa deposits.

Sodium carbonate is deposited around soda lakes as an efflorescence, known as " black alkali " because of its effect upon vegetation. The efflorescences also contain some chlorides, sulphates, or nitrate and borax. According to Chatard, fractional crystallization attendant upon evaporation yields (1) trona with some soda, (2) sodium sulphate, (3) common salt, (4) normal sodium carbonate. Gaylussite $(CaCO_3 \cdot Na_2CO_3 \cdot 5H_2O)$ is also deposited from soda lakes. The Natron Lakes of Egypt become desiccated during the dry season and leave a crust 10 to 27 feet thick of common salt and natron and an efflorescence of natural soda.

Sodium sulphate is obtained from concentrated brines of bitter lakes. Harvestable crops are also deposited as marginal efflorescences of "white alkali." It is deposited in the bottoms of dried-up lakes mainly as Glauber's salt or mirabilite ($Na_2SO_4 \cdot 10H_2O$), and as the anhydrous sulphate thenardite (Na_2SO_4). Minor amounts of halite, natron, and magnesium and calcium sulphates are generally present. Some of the associated sulphate salts are: glauberite, bloedite, and aphthitalite.

The deposition of sodium sulphate generally precedes the deposition of salt and may be controlled by the season of the year. Since it is much more soluble in warm than in cold water, whereas salt is not, it may be deposited as mirabilite in winter and washed up on the shores, as at Great Salt Lake and the Gulf of Karabugaz. From warm (above $32.4°$ C) concentrated waters, thenardite is deposited. The mirabilite in warm dry air loses its water and changes to thenardite. The bottom deposits may be interstratified with beds of epsom salts and sodium carbonate.

The crude salts as mined, mainly Glauber's salt or thenardite, are commonly relatively pure sodium sulphate (85 to 96 percent). Glauber's salt is harvested from the Saskatchewan lakes, Great Salt Lake, Downey Lakes, and Karabugaz; thenardite is characteristic of warm regions, and is the salt obtained from Searles Lake and Verde Valley.

For uses, production, occurrence, and extraction of sodium sulphate see Chapter 21.

DEPOSITION FROM POTASH LAKES

Some of the alkali lakes contain potash in amounts that permit commercial extraction. The potash lakes of Nebraska, which are just hollows in sand dunes, are of interest. The potash is believed to have come from the surrounding country that formerly was burned over by the Indians, releasing plant ashes. The evaporated salts are high in potassium sulphate and carbonate and contain soda, salt, and sodium sulphates; one crust contained 21% K_2O.

Searles Lake in California is the most important lake source of potash in the United States. The deposit is rich in saline minerals of which some 15 are unknown in the Stassfurt deposits. The list includes six sulphates, six carbonates, four triple salts, two chlorides, and one borosilicate. The potassium is confined to the residual brine. Potash is also present in Mono Lake, California, and Columbus Marsh, Nevada.

Deposition from Borate Lakes

Borate lakes are relatively uncommon, but several are known in California, Nevada, Oregon, Tibet, Argentina, Chile, and Bolivia. Formerly, most of the borax in the United States was obtained from lake waters in California and Nevada or from playas. Subsequently, borax was made less expensively from colemanite and ulexite, and later from kernite. At present, the only lakes yielding commercial borax are Searles and Owens, in California, where it is extracted in conjunction with other salts.

Boron is a constituent of some igneous minerals, and, in the form of boric acid and calcium and magnesium borates, it is well known in the volcanic exhalations of Tuscany, where steam jets have yielded much commercial boric acid. It also occurs in the exhalations of Vulcano. Borax is a constituent of some hot springs, and it occurs in some lakes in volcanic regions, as Borax Lake, California, where it constitutes 12 percent of the evaporated salts. The borax of the lakes is considered to have been leached from surrounding igneous rocks or to have been contributed by magmatic hot springs. Clarke thinks calcium borates are indicative of lake deposits.

Deposits. Borates are obtained from brines or playa deposits. The brines of Searles and Owens Lakes yield borax, as do also those of Borax Lake, where, in addition, borax crystals are embedded in the marginal mud. Incrustations on the Searles Marsh playa have also yielded much borax. The famed dazzling white surface crusts of borax in Death Valley, however, are no longer worked. Borax crusts also occur in Chile, Bolivia, and Argentina.

The chief boron minerals of playas and brines are borax $(Na_2B_4O_7 \cdot 10H_2O)$, colemanite $(Ca_2B_6O_{11} \cdot 5H_2O)$, and ulexite $(Na_2 \cdot 2CaO \cdot 5B_2O_3 \cdot 16H_2O)$; searlesite $(Na_2O \cdot B_2O_3 \cdot 4SiO_2 \cdot 2H_2O)$ is also found at Searles Marsh. Other boron minerals and their occurrence are discussed in Chapter 21.

Deposition from Nitrate Lakes

Nitrate lakes are common in arid regions but rarely contain commercial nitrate deposits. The nitrogen of nitrates originally came from the atmosphere, although some may have been magmatic. Sodium and potassium nitrates are characteristic of fertile soils, but, being readily soluble, they are removed in humid regions and become concentrated only in dry regions. They originate in the soil by oxidation of organic matter in contact with alkaline salts, aided by the action of " nitrifying " bacteria, which live in the roots of legumes.

Some may originate through electrical discharges in the atmosphere. The lakes of accumulation are mostly in volcanic regions, suggesting a possible connection between nitrate lakes and volcanoes. Soda niter is commonly associated with borates, which further suggests that nitrate, like borate, may have a volcanic origin.

Soda niter, along with ammonium compounds, is known in Searles Lake and in lake deposits in Death Valley, Utah, and Nevada. Saline lakes in Argentina contain liquors from which soda niter is obtained. It is also found in playa deposits in Colombia and Bolivia, again in association with boron compounds. It is reported also from Matsap, South Africa. Deposits occur near Winkelman, Ariz., and Glenrock, Wyo.

Deposition from Ground Water

Evaporation of ground water is universal, but in humid regions the evaporites are mostly redissolved and removed by rain water. In arid regions the evaporites may accumulate as long as the climate remains arid.

The ground water contains salts similar to those of lakes and oceans. However, the concentration is generally low, and the proportions of the individual salts vary with the character of the soil, bedrock, topography, and climate. Calcium carbonate is almost invariably present; magnesium, sodium, potassium, iron, and manganese compounds are common; silica is generally present, as is also phosphorus; and, locally, boron and iodine may be relatively abundant. The ground water in Chile has yielded vast quantities of commercial nitrates, and some deposits of copper are due to evaporation.

Deposition ensues when evaporation occurs at or near the surface, or in caves. If the site of evaporation is fed by fresh supplies of ground water, extensive deposits may eventually result. Evaporation will, of course, proceed most rapidly where ground water is supplied relatively close to the surface, such as valley bottoms, slopes where hills and valleys merge, and long hill slopes interrupted by gentler or reverse grades.

Deposits. The deposits of economic importance formed by evaporation of ground water are nitrate salts with iodine, some boron, calcium carbonate, sodium sulphate, and sodium carbonate.

The processes of formation of *nitrate deposits* are illustrated by the famed *Chilean nitrate deposits*, which are the only large deposits and, until the development of synthetic nitrates, formerly supplied the world with natural nitrate for fertilizer, explosives, and other chemicals. The deposits are found in the Atacama, Tarapaca, and Antofagasta deserts of northern Chile, where several discontinuous deposits occur at moderate elevations in a long narrow belt some 13 to 100 miles inland. The deposits, called *caliche,* lie on gentle slopes on the east side of the Coast Range. They consist of a thin bed of gravel cemented by sodium nitrate and associated salts at depths of a few inches to a few feet beneath the surface. About one-fourth is sodium nitrate, along with common salt, sulphates, borates, bromides, iodates, phosphates, and lithium and strontium compounds. They also yield most of the world's supply of iodine. The overburden contains a little nitrate, salt, gypsum, and sodium sulphate. (See also Chapter 22.)

The origin of these deposits is controversial. It is generally agreed that the nitrates were transported by ground water and were deposited by its evaporation, as advocated by Miller and Singewald. The source of the nitrate, however, has been attributed to (1) bird guano, subsequently leached (Penrose); (2) nitrogen fixation by electrical discharges from thunderstorms that sweep the Andes (Pissis, Rogers, and Van Wagenen); (3) the bacterial fixation of former nitrogenous vegetable matter; (4) a volcanic source via nearby Triassic and Cretaceous tuffs containing ammonium salts that were oxidized during weathering under unusual climatic conditions (Clark, Whitehead). The soluble nitrates were carried downslope to the site of evaporation. Whitehead estimates that the erosion of 1 m of the volcanic rock, carrying 1% NH_4Cl, over the nitrate areas would supply all the nitrate of the deposits. The correspondence between volcanic rocks and nitrates can hardly be casual. Moreover, such an origin is supported by the constant association of borates, a relationship noted previously in connection with the California borates.

Graham, however, states, without elaboration, that "it seems definite that nitrate was finally collected in an inland sea, which eventually deposited part of its load along its shore lines. Finally, the sea drained into the Pacific Ocean, but the deposits laid down on the eastern shore were either dissolved or washed away by the drainage from the Andes or covered with alluvial debris, leaving the present deposits on the western shore of the inland sea or the eastern slope of the Coast Range."

Somewhat similar although much less extensive nitrate deposits occur in northern Argentina, Cochabamba, Bolivia, and at Matsap, South Africa. A trivial amount occurs as cave depositions known as "cave niter" in the southern and western United States.

Natural brines result when sediments saturated with sea water are buried beneath the level of circulating ground water. These are called connate waters and become sources of commercial salts. Other natural brines result from underground solution of salt beds. The salt waters of oil pools are also natural brines. The brines yield salt, bromine, iodine, and calcium and magnesium chlorides.

Artificial brines are formed by pumping water down wells drilled into rock-salt beds, and the brines so formed are the main source of common salt production in the United States. For the occurrence and distribution of brines, see Chapter 21.

Calcium carbonate in solution in the ground water is also deposited by evaporation at or near the surface to form " caliche " deposits in the arid regions. These deposits commonly underlie arid valleys floored by detrital matter and serve to cement the detritus, but they are most abundant on piedmont slopes, where subsurface waters from higher regions approach the place where hills and valleys merge. Rarely such deposits are locally utilized for lime, fertilizer, or cement. Cave dripstones also result from evaporation.

Other depositions consist of crusts of common salt, Glauber's salt, soda, epsom salts, and borax, deposited in arid regions by evaporation of subsurface water. Silica is, likewise, deposited by evaporation, causing surface silicification or " case hardening " of altered rocks, gossans, or fault material, which have often been mistaken for outcrops of mineral bodies.

Deposition from Hot Springs

The substances contained in solution in hot-spring waters build up deposits about their orifices, partly as a result of evaporation. A few such deposits are of commercial importance; others are scenically beautiful. Their deposition, however, is not, in all cases, the result of evaporation alone; microorganisms help deposit some substances, and escape of carbon dioxide under the reduced surface pressure also causes deposition. The chief substances deposited in this manner are (1) calcium carbonate, in the form of tufa, travertine, or calcareous sinter, (2) silica in the form of siliceous sinter or geyserite, (3) iron oxide in the form of ocher, and (4) manganese dioxide in the form of wad. Many other nonmetallic and metallic substances are deposited from hot springs but in relatively small amounts.

The *calcium carbonate* deposited from hot carbonated waters forms deposits of tufa; the famed White Terrace of New Zealand and the Mammoth Hot Springs of the Yellowstone Park, composed of almost pure calcium carbonate, are noted for their beauty and scenic interest. Other deposits of tufa are locally utilized for cement, lime, and fertilizer. One form, known as *travertine,* is much prized as an interior decorative stone because of its pleasing texture and vuggy nature; the type locality is at Travertine, Italy, where it has been quarried since the time of the Romans.

Silica in hot-spring waters is deposited by evaporation and hot-water algae as a colloidal gel, which crystallizes to chalcedony, opaline quartz, agate, microcrystalline quartz, and quartz. The famous siliceous sinters or geyserite of the Yellowstone Park geysers consist of almost pure silica. Siliceous sinters of hot springs generally contain small quantities of metals, as at Steamboat Springs, Nev.; the Yellowstone geyserite contains arsenic and pyrite. A few spring deposits have been mined for metals, e.g., mercury in New Zealand. For other constituents of spring waters see Chapter 4.

Selected References on Evaporation

Data of Geochemistry. F. W. CLARKE. U. S. Geol. Surv. Bull. 770, 1924. *General chemical data.*

Principles of Sedimentation. W. H. TWENHOFEL. McGraw-Hill, New York, 1939. Chap. 12. *Good brief compilation of origin of evaporites.*

Geology of Nonmetallic Mineral Deposits. A. W. GRABAU. McGraw-Hill, New York, 1920. Vol. I, **Principles of Salt Production.** *Oceanic salts and their precipitation.*

Cements, Limes, and Plasters. 3rd Ed., E. C. ECKEL. John Wiley & Sons, New York, 1928. *Gypsum formation, occurrences, and uses.*

Geology and Origin of Silurian Salt in New York State. H. L. ALLING. N. Y. State Mus. Bull. 275, 1928. *Salt and gypsum deposits and review of theories of origin.*

Gypsum Deposits of the United States. R. W. STONE. U. S. Geol. Surv. Bull. 697, 1920. *Descriptions of gypsum occurrences.*

Origin of North American Salt Domes. E. L. DeGOLYER. A. A. P. G. Bull. Vol. 9:831–874, 1925.

Deposition of Calcium Sulphate from Sea Water. E. POSNJAK. Am. J. Sci. 238: 559–568, 1940. *Chemical controls of deposition of gypsum and anhydrite.*

Geology of Texas. G. R. MANSFIELD and W. B. LANG. Univ. of Texas Bull. 3401: 641–868, 1934. *Best description of U. S. Permian basin and potash deposits.*

Potash Deposits in Spain; in Alsace. H. S. GALE. U. S. Geol. Surv. Bull. 715, 1921. *Geologic occurrences.* **Salines in California.** U. S. Geol. Surv. Bull. 580, 1914. *Saline lake deposits.*

Potash Mining in Germany. J. H. EAST, JR. U. S. Bur. Mines Inform. Circ. 7405, 1947. *Description, production.*

Potash in North America. J. W. TURRENTINE. Am. Chem. Soc. Mon. Ser. 91, 1943. *General survey.*

Owens Lake: Source of Sodium Minerals. G. D. DUB. A. I. M. E. Tech. Pub. 2235, 1947. *Saline minerals and occurrence.*

5·7 RESIDUAL AND MECHANICAL CONCENTRATION

Under the slow, unrelenting attack of weathering, rocks and enclosed mineral deposits succumb to mechanical disintegration and chemical decomposition. As shown on page 34, weathering is a complex operation consisting of distinct processes that may operate singly or jointly. The minerals that are unstable under weathering conditions suffer chemical decay; the soluble parts may be removed and the insoluble residues may accumulate, and some of them may form residual mineral deposits. Stable minerals, such as quartz or gold, undergo little or no chemical change but may be freed from their enclosing matrix and undergo residual enrichment on the surface or be mechanically concentrated into placer deposits by moving water or air.

Weathering is generally considered to consist of mechanical and chemical action; usually both operate together. Mechanical disintegration, such as that by frost action, or expansion and contraction under temperature changes, does not create new minerals; it merely frees minerals already formed. However, it facilitates chemical decomposition by reducing the particle size and creating greater specific surface available for chemical attack. Chemical weathering, however, actually creates new minerals of which some remain stable under surface conditions.

WEATHERING PROCESSES

Mechanical disintegration by itself is confined largely to arid regions where pronounced annual and diurnal temperature changes occur, or to very cold regions where surface chemical changes proceed slowly and frost action is vigorous much of the time. Chemical weathering is most active in warm, humid regions where rainfall supplies moisture and nourishes plant life that yields humic and organic acids to assist chemical activity.

Deep and long-continued weathering is necessary to yield residual products. Hence, such products are generally absent from glaciated regions where residual products have been removed by ice and the time since glaciation has been too brief to permit much weathering under the unfavorable cold climates that prevail there.

The agents of decomposition that operate at the surface are water, oxygen, carbon dioxide, heat, acids, alkalies, vegetable and animal life, and some of the soluble products of the decomposition of the rocks themselves. Without water, little decomposition occurs; hydration is general and is an important factor in both decomposition and disintegration. Oxygen permits oxidation, a process almost universal in weathering. Carbon dioxide dissolved in water is a powerful solvent. Acids, such as sulphuric acid, and some sulphates, generated by oxidation of sulphides, are active agents of decomposition. Heat quickens many reactions. Vegetation supplies carbon dioxide and organic compounds that decompose rocks and dissolve materials. Bacteria are active in biochemical processes of decomposition and precipitation.

Under warm, humid climates most rock, ore, and gangue minerals succumb to weathering and many of their constituents are removed either in solution or mechanically with attendant decrease in volume. Obviously, the effects of weathering vary with the nature of the rock and with the climate. Some minerals, such as the carbonates, are readily dissolved in carbonated waters and are removed in solution. Vast quantities of limestone may thus vanish. Quartzites are little affected. Silicate rocks are readily decomposed, and some of the constituents go into solution and others form insoluble residues; commonly they yield soluble carbonates of alkalies, calcium, and magnesium, which, with considerable silica, are carried away. Aluminum silicates and the oxide compounds of iron and manganese tend to remain. Clay rocks are generally little changed. For the most part, quartz and zircon are little affected, and magnetite generally remains unchanged. Pyrite decomposes readily to form hydrous oxides of iron and sulphuric acid. During weathering, colloidal compounds are formed, some of which, e.g., colloidal silica, go into solution; others, such as aluminum compounds and the oxides of iron and manganese, may be precipitated as colloids.

Weathering commonly does not extend deeper than a few feet to a few scores of feet, but depths of 100 to 200 feet are known. Sulphides are known to oxidize to depths of more than 3,000 feet.

Temperate vs. Tropical Climate Weathering. There is a difference, not fully understood, in the nature of weathering in temperate climates as compared with that in tropical and subtropical climates. In temperate climates the silica of silicate rocks is not extensively removed but remains behind to form clay, along with hydrous oxides of iron and perhaps residual grains of quartz. Thus, clayey soils are the common products of weathering of such rocks in all temperate climates. Tropical and subtropical climates are characterized by alternate wet and dry seasons, by hot weather and warm surface waters throughout the year, and, generally, by luxuriant vegetation with an abundant supply of bacterial life and organic compounds. Under these conditions, rock decay is carried further, leaching is more complete, the silicates are more thoroughly decomposed, but, particularly, the surface waters readily and extensively remove silica in solution. This is a very important distinction. The result is a *laterite* soil, which is a mixture of hydrous oxides of aluminum and iron with some silica and other impurities. Instead of the hydrous aluminum silicate (clay) of temperate regions, it is hydrous aluminum oxide (bauxite). Laterites may be so high in iron that they constitute a laterite iron ore, or, conversely, they may be so high in alumina and low in iron and silica that they constitute an ore of aluminum; most laterites are neither.

Products of Weathering. Normally, weathering results in soils of either the clayey or lateritic variety. But special circumstances may exist in which the decomposing rock contains appreciable quantities of constituents that remain behind in the residue and may undergo concentration or purification to give rise to valuable mineral deposits. A silicate rock low in iron, for example, may, upon weathering, leave a residue of good-quality clay. One high in iron or manganese under temperate climates would leave a residue high in iron or manganese but also in silica, which would make it worthless as an ore of iron or manganese. However, under tropical weathering the removal of silica may leave a residue of workable iron or manganese ore. Consequently, the geographical location of weathering is important from the standpoint of the formation of economic mineral deposits.

The products of weathering may give rise to valuable residues, which, by concentration through residual accumulation, form one line of descent to *residual concentration deposits,* or, by concentration by means of water or air, form another line of descent to *mechanical concentrations* or placer deposits.

RESIDUAL CONCENTRATION

Residual concentration results in the accumulation of valuable minerals when undesired constituents of rocks or mineral deposits are removed during weathering. The concentration is due largely to a decrease in volume effected almost entirely by surficial chemical weathering. The residues may continue to accumulate until their purity and volume make them of commercial importance.

Process of Formation

The first requirement for residual concentration of economic mineral deposits is the presence of rocks or lodes containing valuable minerals, of which the undesired substances are soluble and the desired substances are generally insoluble under surface conditions. Second, the climatic conditions must favor chemical decay. Third, the relief must not be too great, or the valuable residue will be washed away as rapidly as formed. Fourth, long-continued crustal stability is essential in order that residues may accumulate in quantity and that the deposits may not be destroyed by erosion. Given these conditions, a limestone formation free from other impurities except minor iron oxide, for example, will slowly be dissolved, leaving the insoluble iron oxide as residue. As bed after bed of limestone disappears, the iron oxide of each bed persists and accumulates as an insoluble residue, until eventually there may be formed an overlying mantle of iron ore of sufficient thickness and grade to constitute a workable deposit.

The above example typifies one mode of formation wherein the residue is simply an accumulation of a pre-existing mineral that has not changed during the process. In another mode of formation, however, the valuable mineral first comes into existence as a *result* of weathering processes and then persists and accumulates. For example, the feldspar of a syenite decomposes upon weathering to form bauxite, which persists at the surface while the other constituents of the syenite are removed.

Both temperate and tropical climatic conditions yield residual concentrations, although tropical and subtropical conditions favor the formation of more kinds of deposits. Valuable deposits of iron ore, manganese, bauxite, clays, nickel, phosphate, kyanite, barite, ochers, tin, gold, and other substances accumulate as residual concentrations. The basic processes of weathering and concentration apply to each, but, inasmuch as the source materials, chemical changes, and other details of formation differ considerably for each substance, the cycles of the various products are considered separately below.

Residual Iron Concentration

Most rocks contain iron, and under favorable conditions a sufficient quantity may accumulate in the residue to form workable deposits of residual iron ore. This depends, in part, upon the nature of the rock that contains it, the form in which the iron occurs, and the amount of iron present. Also, the climatic and physiographic conditions must be favorable.

SOURCE MATERIALS

The iron of residual iron-ore deposits may come from:

1. Lode deposits of siderite or iron sulphide. The iron oxide residues of pyritic deposits, however, are only rarely used as iron ores.
2. Disseminated iron minerals contained in nonaluminous limestone.
3. Limestones that have been partly replaced by iron minerals, either before or during the period of weathering.
4. Basic igneous rocks.
5. Ferruginous siliceous sediments.

MODE OF FORMATION

Of the above source materials, epigenetic deposits of siderite, limestone, ferruginous chert, and basic igneous rocks are the chief sources of residual iron ores. In temperate regions, limestones and siderite lode deposits are the only source rocks of residual iron ores since all others leave a residue too high in silica or in alumina, or both, to constitute ore. The various source materials are considered below.

From Iron Carbonates. Bodies of siderite or ankerite readily weather to yield oxides of iron, which may accumulate as residual concentrations. The highly ferriferous carbonates yield good clean iron ore, such as the valuable deposits of Bilbao, Spain.

From Pyrite. Massive pyritic bodies yield extensive gossans of goethite and hematite (Chap. 5·8) that locally are used for iron ore, but their high sulphur content makes them generally undesirable. Limonite of this type has been mined at Rio Tinto, Spain, and in Shasta County, California. Tarr described sink holes 500 feet in diameter in Missouri, occupied by marcasite that weathered to hematite nearly 100 feet thick, which was mined as iron ore.

From Limestone. The weathering of relatively pure limestone containing a little iron carbonate will result in solution of the limestone and a residuum of iron oxide. If conditions favor accumulation and if a considerable thickness of limestone is weathered, as in weathering on plateaus, then workable iron deposits may result.

E. C. Eckel points out, however, that many of the limestones underlying residual iron ores contain only a slight amount of disseminated iron and much larger amounts of silica and alumina and that their weathering would yield a clay instead of an iron oxide residue. He concludes that the decay of such limestones cannot yield deposits of residual iron ores. If a limestone contains 4 percent of insoluble material, of which one-eighth is iron, it would be incorrect to conclude that the decay of 100 feet of limestone would yield a 6-inch bed of iron ore. On the contrary, it would yield, under ordinary weathering, a 4-foot bed of residue that would be a clay with only about 12 percent iron content. Consequently, residual iron-ore deposits resting on such a limestone must have been derived from overlying beds in which the iron was somewhat concentrated either before or perhaps during weathering. This is the conclusion of Eckel for most residual iron deposits resulting from limestone decay in temperate regions.

During weathering and accumulation of iron residue there was unquestionably some solution and redeposition of iron. This is indicated by the common radial and colloform structure of the limonite, characteristic of colloidal deposition.

Basic Igneous Rocks. Basic igneous rocks will not yield residual iron ores from normal weathering in temperate climates because clay accumulates in excess of iron oxide. Under tropical weathering, however, all the constituents except alumina and iron may be removed in solution, leaving a residue composed chiefly of either brown ore (limonite) or bauxite (aluminum hydroxide) or both. The weathering of serpentine has yielded extensive residual iron deposits in Cuba, and the formation of laterite in tropical and subtropical regions is well known. Since the weathering residuals in tropical regions are chiefly the hydrated oxides of iron and alumina, it follows that the iron ores of such regions are high in alumina and low in silica, phosphorus, and sulphur.

From Ferruginous Cherts. The rich hematite deposits of the Lake Superior district (see p. 565) are considered by Irving, Van Hise, Leith, Mead, and others, to have resulted from the weathering of ferruginous siliceous sediments. The unweathered source rock is the Biwabik formation, composed of ferruginous chert, greenalite (ferrous silicate), ferruginous slate, and other rocks. Weathering and oxidation are considered to have brought about the removal, by solution, of vast quantities of silica and some carbonates, and to have left a residue of high-grade iron ore. The greenalite alters to taconite or to ferruginous chert, which, in turn, becomes rich iron ore by the loss of silica. During the leaching process, magnesia, some lime, and alkalies are also dissolved to supply alkaline carbonates, which are supposed to aid in the solution of the silica. It should be mentioned that Gruner has forcefully advocated hydrothermal leaching of the silica for some of the Lake Superior deposits instead of leaching by surface waters.

Summary. There are many puzzling features about the formation of residual iron deposits that are not yet understood. In the Appalachian region, carbonates were removed and silica remained; in the Lake Superior region, silica and carbonates were removed and iron and a little clay remained; in Arkansas, silica, iron, and other substances were removed and aluminum hydroxide remained; in Cuba, silica was removed and iron remained. The large-scale removal of silica from the iron formation of the Lake Superior region has been questioned, but it is clearly recognized that silica has been removed in the formation of bauxite deposits.

CLASSES OF RESULTING DEPOSITS

Commercial iron-ore deposits of residual origin may be divided into:

1. Those derived from pre-existing masses of siderite (or associated ankerite). These are relatively few but locally important. The ores are of high grade, are valuable, and are relatively low in impurities.

2. Those derived from the oxidation of iron sulphides. They are small, impure, and rarely of even local importance.

3. Those derived from limestone that contain seams, bunches, or beds of iron carbonate. This group includes the greatest number of residual iron deposits and the most important of the limestone-derived ores. In general they contain medium-grade ores and impurities of clay and silica, but they are low in deleterious sulphur.

4. Those derived from disseminated iron minerals in beds of limestone. The ores commonly have a high clay content or consist of nodular and concretionary masses of hydrous ferric oxide embedded in clay. They are apt to be of low grade and difficult to treat.

5. Those derived from basic igneous rocks. They are formed only under tropical and subtropical climates. The deposits may be rather extensive but are generally high in alumina, although low in silica, phosphorus, and sulphur.

6. Those derived from ferruginous siliceous sediments. Extensive bodies of noncommercial banded ironstone may result, or, if the Lake Superior iron ores were derived from such materials, the resulting deposits are very high grade, large, and number among them the most important commercial iron-ore deposits of the world.

DISTRIBUTION AND EXAMPLES OF DEPOSITS

Residual iron ore deposits are widely distributed except in glaciated regions; they are particularly characteristic of warm, humid regions underlain by limestone or highly ferriferous silicate rocks. Thus, they are common in the southeastern United States, the West Indies, Brazil, Venezuela, southern Europe, and India. A few examples will illustrate residual concentration processes.

Valuable iron ores near Bilbao, Spain (see p. 573), have resulted from the weathering of large replacements of iron carbonate in limestone, yielding extensive surficial blankets of residual iron oxides of high purity that still contain some unweathered carbonate.

Southern States. The Southern States include numerous deposits of residual " brown ores," principally in three regions — the Appalachian, the Tennessee River, and northeastern Texas.

The best deposits occur in the Appalachian Valley, in lower Paleozoic strata near the Precambrian contact where there are several hundred million tons. The content ranges from 30 to 55% Fe, and the ores are high in Al_2O_3, SiO_2, P, and Mn, and low in S. They constitute an important future reserve. The ores are mostly irregular pocket deposits enclosed in clay and derived from the decomposition and solution of folded Cambrian-Silurian limestones and dolomites. The " valley ores " containing from 50 to 55% Fe, rest on limestone and are mostly shallow; the " mountain ores," containing 35 to 50% Fe, rest in residual material generally above the Lower Cambrian quartzite and may extend to depths of a few hundred feet. The ores consist of lumps and nodules in clay, from which they have to be " log-washed." The low-grade " Oriskany " ores of Virginia replace fracture zones in the folded calcareous Oriskany sandstone and reach depths of 600 feet.

The Tennessee River brown ores occur as nearly horizontal mantles in lower Carboniferous limestones. They also form pockets in clay and are of somewhat higher grade than most of the Appalachian ores.

FIG. 5·7–1.   Cross section of residual brown iron ores of East Texas, showing relation of limonite (black) in Weches greensand to perched water table, P (dashed line), which lies above permanent water table. (*After Eckel and Purcell, Tex. Bur. Econ. Geol.*)

The brown ores of northeast Texas (Fig. 5·7–1), according to Eckel and Purcell, are in Eocene beds. The Northern Basin ore occurs in nodules, concretionary masses, or as thin lenses in a 5- to 30-foot zone of weathered greensand. The washed ore contains from 48 to 57% Fe, 5 to 13% SiO_2, 2 to 7% Al_2O_3, and less than 0.12% P. The Southern Basin ore occurs as a solid, almost continuous horizontal bed 3 to 4 feet thick, resting on white clay that grades downward into weathered greensand. No washing is necessary. The ore contains from 42 to 48% Fe, 10 to 12% SiO_2, 8 to 12% Al_2O_3, and less than 0.24% P. Eckel and Purcell estimate that between 150 and 200 million tons of high-grade ore are available in this field. These ores illustrate the accumulation of alumina and silica in temperate-region weathering.

Cuba. Cuba contains extensive deposits of residual iron ores on the north coast, of which the chief ones are the Mayari, Moa, and San Felipe (see p. 576). The ores consist of goethite and hematite and occur as residual mantles overlying weathered serpentine, from which they have been derived. They rest on plateaus and are probably of Tertiary age.

The ore is a typical iron laterite from which all the magnesia and most of the silica content of the serpentine have been removed. Leith and Mead showed that 100 pounds of the original serpentine containing 1.5 pounds of alumina and 10 pounds of ferrous iron shrank during weathering to a residual of 17.5 pounds, consisting of 11.7 pounds of hydrous ferric oxide, 3.8 pounds of bauxite and kaolin, and perhaps 2 pounds of other constituents. The residue carries 44 percent iron and is an iron ore. Chromium and nickel, common constituents of serpentine, are also residually concentrated in the ore. This deposit is a clear example of the removal of silica under subtropical weathering.

Other Localities. Little-developed residual brown ores, generally similar to those of Cuba, occur in Colombia, Venezuela, and the Guianas. Extensive deposits of brown ores occur in the Kirtsch district of Russia. They are low-grade Pliocene hematites with 35 to 42% Fe, 1 to 8% Mn, and are high in phosphorus and silica. Several deposits of brown ores in Greece are similar to those of Cuba, with high chromium and some nickel content. In the provinces of Madras, Bombay, and Bengal, in India, are several deposits of low-grade brown ores, and smaller residual deposits occur in New Zealand and the East Indies.

Residual Manganese Concentration

Manganese accompanies iron and occurs under generally similar conditions but in smaller amounts. Its oxides likewise resist solution, and residues accumulate from the weathering of manganiferous rocks to form residual manganese deposits. The conditions of formation are similar to those of residual iron deposits, which commonly contain some manganese. Special conditions, however, are necessary to give rise to residual deposits in which the manganese dominates.

Source Materials. The sources of manganese that give rise to residual deposits are more restricted than those of iron. Manganese is always present in igneous rocks but in amounts too small to give rise to residual deposits; this is also true of many schists.

Most residual deposits of manganese result from the weathering of:

1. Limestones or dolomites low in alumina but containing disseminated, syngenetic manganese carbonates and oxides.

2. Limestones containing disseminated introduced manganese. Carbonate rocks precipitate manganese under conditions shown on page 173.

3. Manganiferous silicate rocks, such as crystalline schists or altered igneous rocks. Crystalline schists in Brazil containing rhodochrosite, spessartite, and tephroite, and in India and the Gold Coast containing rhodonite and spessartite, are sources of large residual deposits.

4. Lode deposits of manganese minerals, or ores high in manganese. Veins, replacement deposits, or contact-metasomatic deposits may

contain or consist of such minerals as rhodochrosite, rhodonite, manganiferous siderite and calcite, spessartite, tephroite, alleghenite, piedmontite, hausmannite, manganosite, and others. Hewett has shown that these minerals are much more widespread than previously realized.

Most residual manganese deposits, however, are derived from manganiferous rocks rather than from manganiferous lodes.

Mode of Formation.   Residual manganese deposits (Fig. 5·7–2), like iron, are formed by the accumulation of insoluble oxides released by

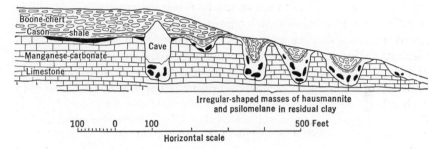

Fig. 5·7–2.   Sketch section at Clubhouse mine, Batesville, Ark., showing irregular bodies of residual manganese ore (black) in Fernvale limestone and masses in residual clay (white).   (*After Miser, U. S. Geol. Survey.*)

weathering of manganiferous rocks or lodes.   As shown on page 203, they result from silicate rocks only under tropical or subtropical weathering, during which process silica is removed.

Few large deposits have been formed from the weathering of limestone.   Since the amount of syngenetic manganese in limestone is, in general, much less than iron, a larger reduction in volume of limestone is necessary to yield a large deposit.   Consequently, more clay and silica accumulate.   Therefore, manganese deposits of limestone derivation are apt to be small, of low grade, and to have more clay per ton of ore.   The few large residual deposits of limestone derivation have come from limestone previously enriched in manganese minerals.

From silicate rocks, the silica, magnesia, lime, and other constituents are removed under tropical weathering, and a residue of manganese oxides and bauxite results.   Most of the important residual manganese deposits have been derived from crystalline schists.

Lode deposits containing manganese-bearing minerals, particularly the carbonates, yield upon weathering a residue of the oxides of manganese.   Carbonated solutions or humic acids dissolve the carbonate, and, in the presence of oxygen, the higher and nearly insoluble oxides of manganese are precipitated and may accumulate as a superposed residue. If sulphides are present in the lode, the manganese may be taken into solution as manganese sulphate which, upon neutralization, breaks down and in the presence of oxygen yields the higher oxides of manganese.

The formation of residual manganese ores is apparently accompanied by much solution and redeposition of the manganese — even more than with iron. The radiating and colloform structures and the complex admixtures so characteristic of manganese oxides suggest that most of the residual manganese oxide compounds have been deposited as colloids or as colloidal mixtures. The manganese hydroxide sol is negative and is precipitated by positive ions. Mixing of sols brings about precipitation of manganese compounds in bewildering complexity. Microscopic study of manganese ores shows that most so-called minerals are mixtures of two, three, or even four manganese minerals of different properties that occur in perplexing relationship to each other. Few manganese minerals in collections are correctly identified or labeled. Even excellent crystals of polianite, for example, are not internally what the external form indicates; they consist of intimate associations of several manganese minerals. One mineral changes readily to another, an action favored by the large number of oxide compounds in which manganese occurs. Microscopic examination of the ores suggests precipitation of colloidal mixtures and subsequent unmixing.

In general, the formation of residual manganese deposits closely parallels that of residual iron deposits just described. There is, however, need for much study before our understanding of the formation of these deposits approaches completeness.

Classes of Deposits and Distribution. Manganese deposits formed by residual concentration may be grouped as follows:

1. Those derived from crystalline schists in tropical regions.
2. Those derived from limestone previously enriched with manganese minerals.
3. Those derived from former manganiferous mineral deposits.

The first two groups are the most important and, together with the sedimentary deposits, yield most of the manganese of the world.

Residual deposits of manganese are found throughout the unglaciated parts of the world, but the best ones occur in tropical or subtropical regions. The large deposits of manganese in India, the Gold Coast, Brazil, Egypt, and Morocco are of residual origin. Other important residual deposits occur in Rumania, Japan, Malaya, and the Philippines; there are minor deposits in Hungary, Czechoslovakia,

Austria, Italy, Chile, Puerto Rico, East Indies, Belgian Congo, and United States.

Manganese deposits are described in Chapter 14.

Residual Bauxite Formation

Bauxite, the ore of aluminum, is formed in residual deposits at or near the surface under special and peculiar conditions of weathering. Although aluminum is the most abundant metal in the earth's crust and the third most abundant element, it occurs mainly in combinations that so far have defied commercial extraction. It is an important constituent of all clays and soil and of the silicates of common rocks, and yet until the beginning of this century it was necessary to travel to far-distant Greenland to obtain the mineral cryolite for a source of metallic aluminum, now displaced by extensive deposits of bauxite.

CONSTITUTION OF BAUXITE

Bauxite was formerly the mineral name of $Al_2O_3 \cdot 2H_2O$, which is no longer recognized as a mineral species. The present usage of the term, both mineralogically and in commerce, is to designate a commonly occurring substance that is a mixture of several hydrated aluminum oxides with considerable variation in alumina content. It is a hardened and partly crystallized hydrogel that consists of variable proportions of the minerals gibbsite or hydrargillite ($Al_2O_3 \cdot 3H_2O$), boehmite ($Al_2O_3 \cdot H_2O$), and its dimorphous form diaspore. Impurities are invariably present in the form of halloysite, kaolinite, nontronite, and iron oxides; rarely bauxite contains octahedrite. A typical bauxite contains 55 to 65% Al_2O_3, 2 to 10% SiO_2, 2 to 20% Fe_2O_3, 1 to 3% TiO_2, and 10 to 30 percent combined water. For aluminum ore, bauxite should contain preferably at least 50% Al_2O_3 and less than 6% SiO_2, 10% Fe_2O_3, and 4% TiO_2. For the chemical industry the percentage of silica is less important, but iron and titanium oxides should not exceed 3 percent each; and for abrasive use silica and ferric oxide should be less than 5 percent each. De Lapparent restricts the term bauxite to mixtures in which Al_2O_3 is greater than 40 percent, Fe_2O_3 is less than 30 percent; and the ratio of Al_2O_3 to SiO_2 is greater than 1.

Commercial bauxite occurs in three forms: (1) pisolitic or oölitic, in which the kernels are as much as a centimeter in diameter and consist principally of amorphous trihydrate; (2) sponge ore (Arkansas), which is porous, commonly retains the texture of the source rock, and is composed mainly of gibbsite; and (3) amorphous or clay ore. All three may be intermingled.

MODE OF FORMATION

General. Bauxite is not a product of normal weathering in temperate regions; it is almost entirely lacking from soils formed there. It is, however, a constituent of lateritic soils formed in tropical and subtropical regions.

Bauxite is an accumulated product of peculiar weathering of aluminum silicate rocks lacking in much free quartz. The silicates are

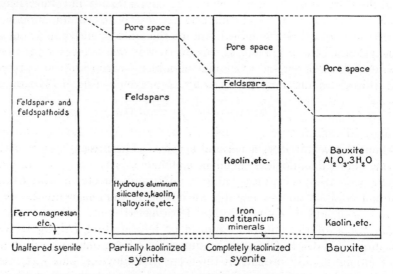

FIG. 5·7–3.   Diagram showing mineral and volume change in alteration of syenite to bauxite.   (*W. J. Mead, Econ. Geol.*)

broken down, silica is removed, iron is partly removed, water is added, and alumina, along with titanium and ferric oxide (and perhaps manganese oxide), becomes concentrated in the residuum (Fig. 5·7–3). This is the generally accepted view, but an origin by hydrothermal solutions has also been advanced.   Bauxite is not a normal constituent of the alteration halo that surrounds hydrothermal ore deposits, and an origin by hydrothermal solutions would require just as special conditions of formation as by surface waters.   Some investigators have thought that bacteria may have played a part in bauxite formation.

Since hydrous aluminum oxides, and not the stable hydrous aluminum silicates of clays, were formed, it means that special conditions peculiar to tropical weathering must have prevailed during bauxite formation.

The conditions necessary for bauxite deposits are: (1) humid tropical or subtropical climate; (2) rocks susceptible of yielding bauxite under suitable weathering conditions; (3) available reagents to bring about breakdown of the silicates and solution of silica; (4) surfaces that permit slow downward infiltration of meteoric water; (5) subsurface conditions that allow the removal of dissolved waste products; (6) time; and (7) preservation.

Climate. Dittler has shown that a temperature above 20° C favors chemical processes by which SiO_2 goes into solution and Fe_2O_3 and Al_2O_3 remain behind. Accumulation of free CO_2 on the surface is hindered during a wet season. The wet season of the tropics is one of formation of Al_2O_3 and Fe_2O_3, the dry season one of leaching of silica away from these oxides. This may supply the answer to why a tropical climate is necessary for bauxite formation. The large bauxite deposits in the temperate regions of Arkansas, Georgia, and France do not contradict this because they were all formed during warmer Tertiary climates.

Source Rocks. Bauxite deposits are formed from rocks relatively high in aluminum silicates and low in iron and free quartz. Thus, the Arkansas, Brazil, and French Guinea deposits have been formed from nepheline syenite, the Georgia-Alabama deposits from clay, the French deposits at Baux (the source of the name bauxite) from limestones or clays in limestone, the British Guiana deposits from residual clay derived from crystalline and metamorphosed rocks, the Gold Coast deposits from clay shales and other aluminous rocks, the Indian deposits from basalt, and Siam deposits from clay alluvium.

Chemical Reagents. The chemical processes involved in bauxite formation are imperfectly understood. Because sulphuric acid or sodium carbonate decomposes clays, these reagents have been mentioned as the active ones. Lindgren points out, however, that sulphuric acid leaching, so common in surficial alteration of sulphide deposits, produces kaolin from sericite, and not bauxite. It is more likely that the ordinary reagents available in tropical weathering have been the effective agents of alteration.

It seems probable, as concluded by Cooper for the Gold Coast deposits, that water, carbon dioxide, humic acids, and tepid rain water are the reagents involved. Carbonic and organic acids break up silicates, and the resulting alkali carbonates are competent solvents for silica. Cooper mentions that, in shafts sunk in bauxite in the Gold Coast, carbon dioxide and methane are so abundant that workers have to be safeguarded. Bacteria may also aid in bringing about solution and redeposition of alumina, as demonstrated by Thiel. He has shown that aluminum sulphate in solution is hydrolyzed to the hydrate and that natural reduction of sulphates may take place through biochemical processes.   It has also been suggested that peat or lignite in overlying beds may have contributed organic compounds that effected or aided the alteration.   However, not all deposits have overlying lignites, and further, the bauxite in most if not all cases was formed before the overlying lignites.   Probably the lignite resulted because of the conditions that gave rise to bauxite.   The strangeness of the alteration that yields

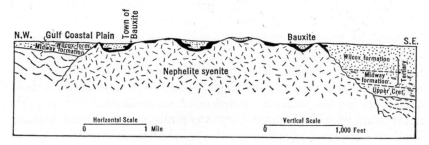

FIG. 5·7–4.   Cross section of bauxite beds in Arkansas.
(*After Branner, 16th Internat. Geol. Cong.*)

hydrous aluminum oxide instead of the usual aluminum silicate was early attributed to the action of hydrothermal solution — an origin with but few advocates at present.

Erosion Surfaces.   Bauxite deposits characteristically lie upon, or are part of, gently undulating surfaces of erosion.   Many are associated with old peneplains, as in the southern United States (Fig. 5·7–4) and the Gold Coast; most deposits mark unconformities.   Such gently sloping surfaces restrict waste removal and permit continual and slow downward seep of rain water and a seasonal high water table.   Apparently, for some unknown reason, this is one of the prerequisites for bauxite formation.   A flattish surface is also necessary to retain the accumulating residuum, which would otherwise be washed away as soon as formed.

Waste Disposal.   Subsurface drainage must permit underground withdrawal, commensurate with inflow, of the rain water charged with the silica and other substances removed during the formation of the bauxite.   If the ground water were stagnant, the waste products could not be removed, and alteration would be inhibited.

Time.   The formation of bauxite apparently requires a long period of time — perhaps equivalent to the amount involved in the formation of a peneplain.   Most deposits are post-Cretaceous in age; many are early or middle Tertiary.

Preserval. Probably countless deposits of bauxite have been formed throughout the broad tropical regions, but few have survived the ravages of subaerial erosion or destruction by encroaching marine invasions. Fortuitous circumstances were necessary. Burial during quiet sedimentation, followed by freedom from dissection, or preserval on undissected remnants of peneplains or old erosion surfaces, has been necessary. Many deposits occur on remnants of dissected peneplains.

RESULTING BAUXITE DEPOSITS

Classes of Deposits. Bauxite deposits formed by residual concentration occur as: (1) blankets at or near the surface and approximately horizontal; (2) interstratified bedded deposits lying on erosional unconformities and consisting of beds or lenses within sedimentary formations; (3) pocket deposits or irregular masses whose bottoms are sheathed in clay, occupying solution or erosional depressions in limestone or dolomite. Bauxite may also be moved from its site of formation and be redeposited in nearby sedimentary beds, or as rubble accumulations, giving rise to transported deposits.

Blanket and interstratified deposits are represented by the large deposits of Arkansas, British Guiana, Surinam, and many of the important southern European fields. Pocket deposits are common in parts of southern France, Dalmatia, Hungary, and Yugoslavia. More than one class may occur together, as, for example, in Arkansas and in the southern European fields.

The *blanket deposits* generally have some soil cover but are not interstratified. They are invariably associated with erosion surfaces or peneplains, as in the Gold Coast (Fig. 5·7–6) and Alabama-Georgia. Few are deeply buried, and most have a clay base.

The *interstratified deposits* lie on erosion surfaces and invariably occupy unconformities (Figs. 5·7–4, 13–33, 13–34). They rest on residual clay. Their bottoms are less regular than their tops, which are generally covered by sediments. The deposits of Guiana (Fig. 13–35) and Arkansas lie on erosion surfaces beneath partly eroded Tertiary sands, clays, and lignite. Those of France and Dalmatia lie on erosion surfaces beneath folded Upper Cretaceous or Eocene limestones.

Deposits derived from silicate rocks are, generally, fairly uniform beds that grade downward through clay into undecomposed rock. Those formed in limestone have a bottom sheath of residual clay and bottom protuberances that extend into the underlying limestone. Being residual deposits, they take the shape of an erosion surface on limestone, which, owing to solution, is generally irregular and cavernous.

The *pocket deposits* are restricted to limestone or dolomite, as in southern France, Hungary, Yugoslavia, Italy, and Georgia, U. S. A. They occupy solution depressions, or the cavernous openings of a karst topography (Fig. 5·7–5). They also rest on clay in association with

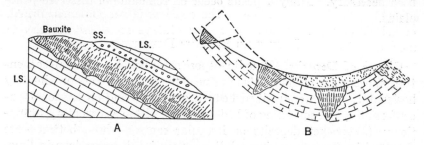

FIG. 5·7–5. Sections of French bauxite deposits. *A*, bedded deposits between compact limestones at Maigret; and *B*, pocket deposits in Jurassic limestone at Centre de Barjol. (*After de Lapparent.*)

erosion surfaces and unconformities. They are shaped like huge teeth with many projecting roots. They may be covered by thick consolidated sediments, as in Yugoslavia, or by a thin mantle of surface soil, as in Alabama.

The *transported deposits* may be bedded, as in Russia and Arkansas (Fig. 13–33), or they may consist of broken rubble, as in some of

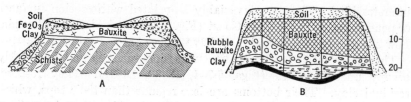

FIG. 5·7–6. Cross sections of Gold Coast bauxite deposits showing relation to erosion surfaces. *A*, Typical; *B*, Mount Ejuanema deposit. (*After Cooper, Gold Coast Geol. Survey.*)

the Gold Coast deposits. Probably most of the bauxite was washed away during formation, but some was collected and preserved.

Age. The optimum conditions for the formation of bauxite apparently existed between Middle Cretaceous and middle Eocene, when most of the commercial deposits of the world were formed. A few were formed earlier, and some in late Tertiary and Recent. The only Paleozoic bauxite known is that of Tikhvin, in Devonian rocks, and the only Triassic deposits are some in Croatia. The deposits of Arkansas are underlain by Cretaceous rocks and overlain by lower Eocene, and those of southern Europe mostly lie on Jurassic, Middle Cretaceous, and Upper Cretaceous formations and are overlain by Upper Cretaceous, lower Eocene, or middle Eocene. The Irish and Jamaica bauxite is Oligocene, and that of the Vogelsberg Mountains, Germany, is thought to be Miocene in age. The deposits of the Guianas, Brazil, French Guinea, Gold Coast, India, and Malaya are latest Tertiary to Recent.

It is thus evident that the conditions prevailing during the " most dramatic and puzzling event in the history of life on the earth " (Simpson) — that of the change from the age of reptiles to the age of mammals — was equally the optimum for the development of bauxite, the exact conditions of formation of which are no less puzzling than those of the paleontological change. Possibly the climatic conditions that affected the one likewise promoted the other.

Associations. The association of clay with bauxite is quite general. The limestone deposits rest on residual clay; the Arkansas deposits grade through clay into nepheline syenite; all the deposits of Alabama-Georgia, the Guianas, and the Gold Coast rest on clay. In most cases, the clay of the base is residual, derived from the underlying rocks.

The question arises whether the clay was formed first and the bauxite was later derived from it, or whether both clay and bauxite were formed at the same time, or whether clay was formed from bauxite (resilication). Bauxite certainly was not formed directly from limestone so must have been derived from its residual clay. If the clay was formed first, it would imply a change in conditions of weathering — from those that yielded clay to the special ones that yielded bauxite — and, moreover, the change was fortuitous enough to recur during different periods of bauxite formation, which seems improbable. The Arkansas occurrence, likewise, could have been formed from a previously formed clay. If, however, the clay was a transitory intermediate step in the continuous alteration of syenite to bauxite, would this reasoning apply to limestone deposits? It might, if surface conditions favored bauxite formation while contemporaneous underlying conditions favored clay formation.

The high silica content characteristic of the tops of many bauxite beds may be due to evaporation deposition of a capillary supply of the solutions removing the silica, and to some resilication.

Summary. The features of occurrence of bauxite deposits lead to the following conclusions:

1. They are all related to old peneplains or erosion surfaces. This implies the necessity of considerable time and of flattish drainage surfaces for their formation.
2. They have mostly been formed in the late Mesozoic or Tertiary time under climatic conditions different from those that prevail in the same places today. Possibly some of the conditions that affected the change from the age of reptiles to the age of mammals likewise promoted the formation of bauxite.
3. The association of bauxite with clay or claylike materials is too general to be a coincidence. Presumably the clay was formed as a transition stage in the alteration to bauxite. For the clay to have been formed as a distinct deposit first and then at a later period to have been altered to bauxite would necessitate an unlikely change-over of agencies of alteration.

Chapter 13 deals with aluminum production figures and uses and contains descriptions of the various bauxite deposits.

Residual Clay Formation

Clay results wherever aluminous rocks are chemically weathered, but most of it is carried off and deposited as ordinary argillaceous sediments. Under suitable conditions beds of pure sedimentary clay are laid down, as described in Chapter 5·5. With favorable climatic and topographic conditions and suitable source rocks valuable deposits of residual clays may result, and this is the most common mode of origin of all the high-grade clays. Clay can also be formed by hydrothermal action on feldspathic rocks.

Types of Residual Clays. Ries classified residual clays as follows:

1. Kaolins, white in color, and usually white burning.
    a. Veins, derived from weathering of dikes.
    b. Blanket deposits, derived from areas of igneous or metamorphic rocks.
    c. Replacement deposits, such as indianite.
    d. Bedded deposits derived from feldspathic sandstones.
2. Red-burning residuals, derived from different kinds of rocks.

For the various types of clays and their composition, uses, and occurrence the reader is referred to Chapter 17.

Source Materials. The chief source rocks of residual clays are crystalline rocks, more especially the silicic granular rocks that are rich in feldspars and low in iron minerals, such as granite and gneiss. These yield the purest residual clays, but they have to be washed to remove quartz and limonite. Some granites high in silica yield " lean " clays of low plasticity. Basic igneous rocks yield much ferric oxide, which stains the clay, often rendering it useless. Feldspar-rich pegmatites yield dikelike masses of high-grade white kaolin that is generally very low in iron and other impurities deleterious to chinaware manufacture. Syenites yield excellent clay. Limestones, after long-continued solution erosion, leave a mantle of insoluble clayey impurities that are used for brick clays. Shale, which is largely made up of clay minerals, is used as clay material as it is, but weathering often yields a purer product. Sericitized igneous rocks also yield clay.

Mode of Formation. Clay formation results from normal weathering processes. Vegetation supplies the necessary carbon dioxide, and it is noteworthy that good clays commonly underlie swamps or former swamps. Organic compounds serve to remove coloring materials and produce white clays; they change iron from the insoluble ferric to the soluble ferrous state, permitting its removal in solution and thereby bleaching the clay.

The formation of clay from silicate minerals is essentially a breaking down of the silicates to form hydrous aluminum silicates and the removal of the soluble silica and alkalies in solution. Free quartz will remain and must be extracted to obtain pure clay. The alteration of orthoclase, for example, yields kaolinite, potassium carbonate, and silica. The last two are removed in solution and the kaolinite persists. Colloidal substances are also formed. Noll has shown that a thorough removal of the alkalies gives rise to kaolinite, whereas less complete removal favors the formation of montmorillonite. The rate of removal of the alkalies is also important.

Kaolin is also produced by the action of sulphates and sulphuric acid, formed by the weathering of sulphides, upon sericitized rocks. Such surficial kaolinization is characteristic of all supergene-enriched copper deposits.

Many investigators have maintained that kaolin deposits result from hydrothermal action. Kaolinite has been produced experimentally by Schwartz, Noll, and others between $200°$ and $400°$ C. Kaolinite and montmorillonite occur in the halo of hydrothermal rock alteration that surrounds many hydrothermal ore deposits, particularly copper deposits. According to Noll, sericite forms in alkaline solutions at $300°$ C, but kaolin develops in acid solutions. J. W. Gruener, however, has shown that sericite also forms in acid solutions at temperatures of 350 to $525°$ C but not at $300°$. Sales and Meyer show that in the alteration halo at Butte, Mont., sericite with some dickite occurs next to the veins and kaolinite and montmorillonite farther out, the last representing the earlier formed, lower-temperature alteration products. They also found that sericite dominates in the intensely altered zone and kaolinite in the less intense zone.   Lovering has noted hydro-thermal kaolinite at East Tintic, Utah.   These occurrences prove con-clusively that kaolinite is a hydrothermal as well as a weathering alteration product.   Hypogene kaolinite is stated by Ransome to accompany alunite in the gold ores of Goldfield, Nev., and he postulates an origin by acid metallizing solutions.   The deep kaolin of the Corn-wall tin lodes has also been ascribed to hydrothermal action, as have the deep china clays of Cornwall.   The clays, however, can be traced downward into unweathered sericitized granite; and the kaolin may have resulted from the weathering of the sericite (see Chap. 17).

So many residual clays exhibit downward transitions through partly altered to fresh rocks, and so widespread are the mantles of clay over-lying feldspathic rocks, that there can be little doubt that most residual clays result from weathering.

A unique deposit described by Ries and others, known as indianite, from Indiana, is believed to have been formed by replacement of a sandstone bed by means of meteoric waters that were supplied with aluminum sulphate by the underlying Chester shale.

Occurrence and Distribution.   Residual clay deposits assume roughly the form of the source rock.   Dikelike deposits are derived from pegmatite dikes and are as wide as 300 feet and as deep as 120 feet.   Many of the kaolin deposits of the Southern States are derived from pegmatites.   Those derived from crystalline rocks occur as mantles that may be acres in extent and tens of feet in depth; the Cornish kaolins have been worked to a depth of 300 feet.   Some resid-ual clays are covered by later formations, such as the kaolins of Monroe County, Pennsylvania, which Mansfield states were residual on an old Pennsylvanian surface and are covered by Upper Cretaceous strata.

The color and purity vary greatly.   Those derived from granitic rocks range from 10 to 50 percent white clay minerals, the remainder being mineral fragments and decomposition products of the granite that must be removed by washing.   Residual clays from limestone are mostly highly colored by iron compounds.

Geologically, the residual clays are much more restricted in distribu-tion than the sedimentary clays.   Most of them are of Pleistocene age; they are nearly all post-Paleozoic, and few are of Mesozoic age.

Geographically, residual clays are widely distributed wherever suit-able source rocks crop out in favorable topography and humid climates. The best kaolins come from temperate humid regions and are lacking in arid regions.   The glaciated regions, although rich in sedimentary clays, contain no good residual deposits because of removal by ice and insufficient post-Glacial time.

In North America, residual clays are confined mostly to the Southern States and some of the Western States, although some kind of clay is produced in every state in the Union. England, China, France, Germany, and Czechoslovakia contain the most important deposits of high-grade clays.

The constitution, varieties, uses, mode of origin, and examples of deposits of all kinds of clays are given under ceramic materials in Chapter 17.

Nickel Deposits Formed by Residual Concentration

Many ultra-basic igneous rocks are known to contain very small quantities of nickel in some unknown form but presumably held in the silicate lattices. Under tropical and subtropical weathering such rocks become decomposed, lose silica, and yield hydrous silicates of magnesium and nickel. Formerly, several species were thought to exist, such as genthite, pimelite, nepouite, connarite, garnierite, and noumeite, but Caillère concluded (1937) that the nickel replaces magnesium in variable amounts, and he recognized only two names, garnierite or noumeite (formula approximately $H_4(Mg,Ni)_3Si_2O_9$). He considered them to be formed by laterization.

In several places, garnierite, derived from serpentinized peridotite, has undergone sufficient residual concentration on the surface to form workable deposits of nickel ore. Such deposits have been formed in New Caledonia, Cuba, Celebes, Brazil, and Venezuela.

The deposits of New Caledonia, formerly the world's chief source of nickel, illustrate the mode of formation. Part of the island is underlain by deeply weathered serpentinized peridotite of Tertiary age, and nickel is present throughout all the weathered mantle. Locally, the residual concentration has proceeded far enough to form hundreds of small deposits from 25 to 35 feet thick, mostly localized in six centers, three on each coast. The largest deposit contains 600,000 tons. The deposits lie on slopes rather than on plateau ridges, and from this fact it has been inferred that moving seepage waters have been important for their formation. In places the nickel silicates penetrate into cracks in the underlying serpentine, indicating some migration of the nickel during concentration. The composition of the ore treated gives an idea of the nature of the concentration; namely, 1.5 to 3.5% Ni, 6 to 10% H_2O, 40 to 45% SiO_2, 20 to 30% MgO, 13 to 20% Fe_2O_3, 0 to 1% Al_2O_3 + CaO, 0 to 2% Co, 0.1 to 0.8% Cr_2O_3. This indicates a high loss in Al_2O_3 and CaO, a reduction in SiO_2 and MgO, considerable concentration in Fe_2O_3, and a high concentration in Ni, Co, and Cr_2O_3. Some cobalt has been mined. In Brazil, a similar con-centration of cobalt with the nickel has taken place.

At Nicaro, Cuba, the minute nickel content of serpentine has been residually concentrated to 1.4 percent nickel in a flat ferruginous sur-face blanket 20 to 50 feet thick. Cobalt and chromium have also undergone concentration but have not been commercially extracted in the huge treatment plant.

The similar deposits of Celebes have been worked by the Japanese, and those of Venezuela are under exploration.

Other Products of Residual Concentration

Kyanite. High-quality kyanite is obtained from Singhbhum, Bihar, Orissa, and Ghagidih, India, in the form of massive residual boulders that have accumulated on the surface or are washed from the regolith. Their source is underlying Precambrian kyanite-quartz-granulite rocks.

In the United States at Baker Mountain, Virginia, Burnsville, N. C., and Habersham County, Georgia, kyanite occurs in kyanite schists formed by Precambrian metamorphism of aluminous rocks. Weath-ering of the schists has produced a deep clay regolith containing loosened kyanite crystals up to 3 inches long, and boulders of kyanite crystals, which are extracted from the clay and processed. Some residual concentration has taken place by the weathering processes, but the unweathered schist also contains abundant kyanite.

Barite. Barite ($BaSO_4$) is obtained mostly from residual lumps and masses embedded in clay, as in Missouri. (See Chap. 20.) The barite unquestionably came from the underlying rocks, and being resistant to weathering it persisted as accumulated fragments and masses in the overlying residual clay. It has been maintained that the barite nodules resulted from circulating ground water during weath-ering, but the evidence seems conclusive that the original barite was of hypogene origin and was deposited in the bedrock from igneous hydro-thermal solutions.

Phosphates. The "land pebble" phosphates of Bone Valley, Florida, which consist of pebbles and boulders in a matrix of sand and clay, are considered to have been weathered out of the underlying Alum Bluff phosphate formation. They are residual accumulations of weath-ering that were worked over by an advancing sea and included in marine sands and gravels. (Chap. 22.)

Tripoli. Tripoli (Chap. 23), a light, soft, porous, earthy substance that is nearly pure silica, is a residual product of weathering and is found in massive form, either blocky or in friable masses. It results from the weathering of chert, cherty limestones, or siliceous limestone, which leaves a residue of siliceous materials that change to tripoli by removal of included calcium carbonate. It is not known whether the chert accumulates first and is then converted to tripoli or whether tripoli is the direct decay of flint, or other forms of silica, during the initial weathering of the limestone. In North America the largest deposits are in Missouri, Oklahoma, and Illinois, but it is found also in Tennessee, Mississippi, Alabama, Georgia, and Arkansas. The Seneca, Missouri, deposits were derived from the Boone cherty limestone, and the western Tennessee Valley deposits from the weathering of the cherty Fort Payne and Warsaw limestones.

Mineral Paints or Ochers. Natural ochers or mineral pigments have been used since early man was first attracted by their colors and spreading ability. They consist of various colored earthy materials, generally iron oxide mixed with clay as a binder or filler. Minerals are chiefly goethite and hematite mixed in various proportions with clay or manganese oxides and are derived from the weathering of almost any iron-bearing rock. The brilliancy of color depends upon the proportion of clay and the proportions of hematite, goethite, and manganese oxide. Ochers occur mostly in regions of deep and thorough weathering. The well-known ones of Cartersville, Ga., overlie the Weisner quartzite and occur at the base of the residual zone. The source rock contains only 1.5% FeS_2, 0.5% Fe_2O_3, and 90% SiO_2, but the residual ocher contains 66 percent limonite and 25 percent clay. (See also Chap. 18.)

Zinc Ore. In Virginia and Tennessee, primary zinc sulphide deposits in the shaly dolomite have been oxidized to hemimorphite and smithsonite, which occur in nodules and minable masses in residual clay overlying the dolomite.

Tin Ore. Residual accumulations of tinstone occur in the Netherlands Indies in connection with the fluviatile and eluvial placers described under placers. Some remarkably rich pockets have been found.

Gold. Grains of gold released from their matrix during weathering have accumulated in places to form small residual deposits. Residual accumulations have been found in the Appalachian States, the Western States, the Guianas, Brazil, Madagascar, Tanganyika, and Australia. The gold occurs in angular particles, showing that it has not been transported, and is generally in loose ferruginous detritus. Derby described a residual gold deposit in Brazil, which is in the lower part of a bed of laterite iron ore that represents a residual concentration in place from the weathering of underlying schist.

Cobalt. In Katanga, there are residual accumulations of black oxide of cobalt associated with the oxidized copper ores (see also Chap. 13), presumably derived from underlying linnaeite.

Selected References on Residual Concentrations

General Reference 8. Chap. 21:344–378. *Residual deposition in general.*

Iron Ores, Their Occurrence, Valuation and Control. E. C. ECKEL. McGraw-Hill, New York, 1941. *Various residual deposits.*

Lake Superior Iron Region. W. O. HOTCHKISS and others. XVI Int. Geol. Cong. Guidebook 27, 1932. *Résumé of geology, origin, and references.*

Brown Iron Ores, Tennessee. E. F. BURCHARD. Tenn. Geol. Surv. Bull. 39, 1932.

Deposition of Manganese. C. ZAPFFE. Econ. Geol. 26:799–832, 1931. *Catalytic and bacterial deposition.*

Manganese. R. RIDGWAY. U. S. Bur. Mines Inf. Circ. 6729, 1934. *Statistics and résumé of occurrences.*

Manganese. 2nd Ed., A. W. GROVES. Imper. Inst., London, 1938. *General résumé of British and world occurrences; good bibliography.*

Aluminum Industry. EDWARDS, FRARY, and JEFFRIES. McGraw-Hill, New York, 1930. Vol. I, Chap. 4, by E. C. HARDER. *Résumé of bauxite deposits.*

Bauxite. C. F. Fox. London, 1929. *General.*

Bauxite Deposits of Arkansas. W. J. MEAD. Econ. Geol. 10:28–54, 1915. *Occurrence and origin by removal of silica.*

Relations of Bauxite and Kaolin in Arkansas Bauxite Deposits. M. GOLDMAN and J. I. TRACEY, JR. Econ. Geol. 41:567–575, 1946. *Claim that bauxite is derived from syenite and not from kaolin.*

Formation of Bauxite from Basaltic Rocks of Oregon. V. T. ALLEN. Econ. Geol. 43:619–626, 1948. *Alteration first to clay minerals, then to bauxite.*

Clays, Occurrence, Properties and Uses. H. RIES. John Wiley & Sons, New York, 1927. *General treatise; and* **General Reference 11.** Chap. 11. *Excellent summary and good bibliography.*

Clays and Other Ceramic Minerals. C. W. PARMELEE. Edwards Bros., Ann Arbor, 1937.

Geology of Some Kaolins of Western Europe. E. R. LILLEY. A. I. M. E. Tech. Paper 475, 1932. *Good description and discussion of origin of residual clays.*

Petrography and Petrology of South African Clays. V. L. BOSAZZA. Johannesburg, 1948. *Excellent treatise on origin and nature of clays.*

Residual Kaolin Deposits of Spruce Pine District, North Carolina. J. M. PARKER. N. Car. Dept. Cons. Dev. Bull. 48, 1946. *Occurrence and origin.*

Kyanite in Virginia. A. I. JONAS and J. H. WATKINS. Va. Geol. Surv. Bull. 39, 1932.

Kyanite Industry of Georgia. R. W. SMITH. A. I. M. E. Tech. Pub. 742, 1936.

Aluminous Refractory Minerals: Kyanite, Sillimanite, Corundum, in Northern India. J. A. DUNN. Geol. Surv. India 52: Part 2, 1929.

Zinc and Lead Region of Southwestern Virginia. L. W. CURRIER. Va. Geol. Surv. Bull. 43, 1935.

Introductory Economic Geology. W. A. TARR. McGraw-Hill, New York, 1930. Pp. 577–583.

Tripoli Deposits of the Western Tennessee Valley. E. L. SPAIN. A. I. M. E. Tech. Pub. 700, 1936.

Nickel-Silicate and Associated Nickel-Cobalt-Manganese Deposits, Goiáz, Brazil.
U. S. Geol. Surv. Bull. 935–E, 1944. *Residual nickel-cobalt deposits.*
See also references given under: " Iron," Chaps. 5·5, 14; " Manganese," Chap. 14;
" Aluminum," Chap. 13; " Clays," Chap. 17; " Mineral Pigments," Chap. 18;
" Barite," Chap. 20; " Tin," Chap. 13; " Nickel," Chap. 14; " Phosphate,"
Chap. 22.

MECHANICAL CONCENTRATION

Mechanical concentration is the natural gravity separation of heavy
from light minerals by means of moving water or air by which the
heavier minerals become concentrated into deposits called *placer
deposits.*   It involves two stages: (1) the freeing by weathering of the
stable minerals from their matrix, and (2) their concentration.   Concentration can occur only if the valuable minerals possess the three
properties: high specific gravity, chemical resistance to weathering,
and durability (malleability, toughness, or hardness).   Placer minerals that have these properties are gold, platinum, tinstone, magnetite,
chromite, ilmenite, rutile, native copper, gemstones, zircon, monazite,
phosphate, and, rarely, quicksilver.

Source Materials

The minerals that make up placer deposits may be derived from:

1. *Commercial lode deposits,* such as gold veins, e.g., Mother Lode gold
   veins of California.
2. *Noncommercial lodes,* such as small gold quartz stringers or veinlets
   of cassiterite, e.g., tin placers of Banka.
3. *Sparsely disseminated ore minerals,* i.e., minute grains of platinum
   sparsely disseminated in basic intrusives, e.g., Ural Mountains.
4. *Rock-forming minerals,* such as grains of magnetite, ilmenite, monazite,
   and zircon, e.g., ilmenite beach sands of India.
5. *Former placer deposits,* such as bench stream gravels or buried placers,
   e.g., California Recent gold placers.

Principles Involved

In the formation of placer deposits, nature has operated to produce
the results achieved by man when he mines, crushes, and concentrates
ores.   The placer minerals are released from their matrix by weathering.   The comminuted materials are washed slowly downslope to the
nearest stream or to the seashore.   Moving stream water sweeps away
the lighter matrix, and the heavier placer minerals sink to the bottom
or are moved downstream relatively shorter distances.   Waves and
shore currents also separate heavy minerals from light ones and coarse
grains from fine ones.   From thousands of tons of debris, the few
heavy minerals in each ton are gradually concentrated in the stream

or beach gravels until they accumulate in sufficient abundance to constitute placer deposits. Eventually, the little gold of countless thousands of tons of matrix is concentrated in relatively small volume. Rich placer gold deposits formed in this manner gave rise to the great California gold rush of 1849, to the Klondike stampede, and to the rich discoveries in Alaska, Australia, and other places and have yielded billions of dollars in gold.

The operation of mechanical concentration rests on a few basic principles involving chiefly the differences in specific gravity, size, and shape of particles, as affected by the velocity of a moving fluid. First, in a body of water, a heavier mineral sinks more rapidly than a lighter one of the same size. Moreover, the difference in specific gravity is accentuated in water as compared to air. For example, the ratio of gold (sp. gr. 19) to quartz (sp. gr. 2.6) is as follows:

$$\frac{\text{Gold in air, 19}}{\text{Quartz in air, 2.6}} = \frac{7.3}{1} \quad \text{whereas} \quad \frac{\text{Gold in water, 19-1}}{\text{Quartz in water, 2.6-1}} = \frac{11.2}{1}$$

Secondly, the rate of settling in water is also affected by the specific surface of particles. Of two spheres of the same weight but of different size, the smaller, with its lesser surface and, therefore, lesser friction in water, sinks more rapidly. Thirdly, the shape of a particle affects its rate of settling. A spherical pellet has less specific surface than a thin, platy disc of the same weight, and, therefore, will sink more rapidly. Thus, flaky specularite and molybdenite are difficult to concentrate by gravity despite their high specific gravity.

Now, add to these factors the effect of moving water. The ability of a body of flowing water (or air) to transport a solid depends upon the velocity and varies as the square of the velocity. A racing floodwater can carry substances that a quiet stream cannot. When the velocity doubles, the transporting power is increased about four times and stationary materials are moved. Conversely, if the velocity is halved, much of the transported load is dropped. Hence, placer minerals may be dropped where the current slackens. Faster-moving water accentuates the differences in settling rate based upon specific gravity, and if particles of gold and quartz were dropped into moving water the gold might drop directly to the bottom and the quartz be swept downstream. The specific surface again enters in; of two equal-weight particles, the one with the larger specific surface is increasingly rapidly swept away from the other with increased water velocity. Thus, flaky mica is readily separated from quartz, and fine materials are separated from coarse particles.

Again, add another factor, that a particle in suspension is more readily moved by a flowing fluid than one at rest. As one stirs up sugar from the bottom of a tea cup, so eddies in streams or shore currents raise light substances from the bottom and enable currents to swish them away. Separation of light minerals from heavy ones is thus aided in bringing about concentration. Also, the swirls and eddies of stream and wave action simulate the upward pulsations of jigs and tables in ore dressing by which lighter minerals are bounced higher than heavier ones so that they can be more easily moved away by flowing water. This jigging action enables gold particles scattered through bottom gravels to become concentrated on the bottom, even though the gravels are thick — the coarser gold on the bottom, the finer above.

The various factors enumerated above operate together to separate the light and fine minerals from coarse and heavy ones, and, with long-continued action, placer minerals may eventually become sufficiently concentrated to constitute workable deposits.

For the mechanical concentration of placer minerals it can thus be seen that the water velocity must be favorable. If too low, the lighter materials will not be removed from the heavier. If too great, the placer minerals will also be swept away and perhaps dissipated. A slackening of velocity, whether it be of a stream, a shore current, or undertow, causes deposition and accumulation. In a stream, a change of gradient, meandering, spreading, or obstructions all produce reduced velocity that permit heavier minerals to drop and accumulate.

The scraping and pounding of the placer minerals during jigging and transportation eventually crushes the nondurable ones to a powder; it rounds off sharp corners of the durable ones, and compacts and flattens the malleable ones. These features are, therefore, a criterion of the amount of concentration or the distance of travel they have undergone. Sharp, angular placer gold, for example, is not far from its source, and this principle is utilized by the wily prospector in his search for the " Mother lode."

An essential for any mechanical concentration is that a continuous supply of placer minerals be made available for concentration. This means that the most favorable regions are those of deep weathering and topographic relief — the weathering to free the placer minerals, and the relief to permit the debris of weathering to move streamward or beachward. A plateau or a peneplain cannot supply much debris. Still more favorable are areas of stream rejuvenation by recent uplift where new valleys are cut into older ones, causing rewashing and reconcentration of pre-uplift gravels. The more such reconcentration takes place, the greater is the degree of concentration.

When weathering yields debris on a hill slope, the heavier particles move downslope more slowly than the lighter ones, giving a rough concentration into *eluvial* placers. During water transportation, concentration may occur in streams, giving *stream* placer (or *alluvial*) deposits, or on beaches, giving *beach* placers. If concentration takes

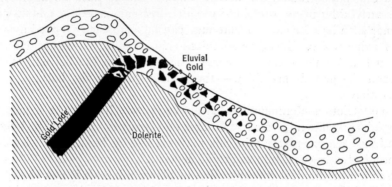

Fig. 5·7–7.   Eluvial gold ore at San Antonio vein, Chontales district, Nicaragua.   (*After Carter A.I.M.E.*)

place by wind, *eolian* placers result. The basic principles enumerated above apply to each of them, but special features contribute to the individual groups, and these are considered separately below.

Eluvial Placer Formation

Eluvial placers may be considered an intermediate or embryonic stage in the formation of stream or beach placers. They are formed, without stream action, upon hill slopes from materials released from weathered lodes that outcrop above them. The heavier, resistant, minerals collect below the outcrop (Fig. 5·7–7); the lighter nonresistant products of decay are dissolved or swept downhill by rain wash or are blown away by the wind. This brings about a partial concentration by reduction in volume, a process which continues with continued downslope creep. Fairly rich lodes are necessary to yield workable deposits by this incomplete concentration. The most important eluvial deposits are gold and tin; minor deposits include manganese, tungsten, kyanite, barite, and gemstones.

Eluvial gold deposits mined in Australia contributed much of the early placer gold production and several of the famous Australian nuggets. The first placer deposits worked in New Zealand were mainly eluvial deposits. Becker, describing early gold mining in the southern Appalachians, recorded the formation of eluvial gold deposits under secular decay 100 feet deep; at Dahlonega, the debris was hy-draulicked, and the enclosed quartz boulders were crushed and amal-gamated. Considerable eluvial gold was washed in California, Oregon, Nevada, and Montana, and Lindgren mentions that eluvial deposits in Eldorado County, California, gave rise to litigation to determine whether they were lode or placer deposits.   Eluvial gold deposits were formerly extensively worked in South America; some are now being mined in Sierra Leone, Tanganyika, Kenya, and British Guiana.

Tinstone is obtained in abundance from eluvial deposits in the Netherlands Indies and the Malay States; it is also being washed in Burma, Siam, French Indo-China, Belgian Congo, Nigeria, and Australia.

Wolframite is obtained as a by-product of the eluvial tin washing in the above-mentioned states, and minor amounts of eluvial manganese boulders are mined in the Gold Coast, also in India and Brazil. Eluvial kyanite boulders of high purity are gathered, or washed out, from the regolith in India, as are also barite masses in Missouri.

Stream or Alluvial Placer Formation

Stream placers are by far the most important type of placer deposits. They have yielded the greatest quantity of placer gold, tinstone, platinum, and precious stones.   Primitive mining, undoubtedly, started on such deposits. The ease of extraction and the richness of some deposits made them as eagerly sought in early times as in recent times. They have been the cause of some of the great gold and diamond " rushes " of the world.

Concentration. Flowing water is the most effective separator of light from heavy materials. Stream water, although ever flowing downstream, does so with irregularity. It rushes through canyons sweeping everything along with it; it slackens in wide places; it swirls around the outside of bends, creating back eddies on the inside; it laps up over bottom projections, forming quiet eddies on the lee side. In these slack waters the heavy substances drop to the bottom. In streams, jigging action is particularly effective in concentrating placer minerals in the bottom gravels. Further concentration is effected by the gradual abrasion of the bottom gravels and reduction in their thickness.

During dry seasons of low stream velocity the placer minerals re-main at rest, but in flood time they and the enclosing gravels may all be swept farther downstream and reconcentrated on bars, stream mar-gins of flood plains, or other favorable places. Minute particles of gold called " colors," however, may be carried far downstream.

Places of Accumulation. The lower sluggish reaches of streams are not favorable sites for placer accumulations, and neither are the upper headwaters because of the limited supply of source materials. The most favorable sites are the middle reaches. Where streams surge along polished canyon floors, placer minerals and gravels alike are rushed along with little opportunity for settling; but where they de-

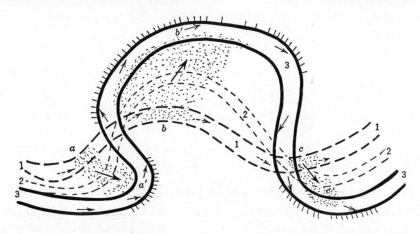

FIG. 5·7–8. Gravel deposition and formation of pay streaks in rapidly flowing meandering stream, in which meanders migrate laterally and downstream. Stream arrows represent point of cutting. *1*, Original position; *2*, intermediate position; and *3*, present position of stream. Deposits formed at *a, b, c,* or inside of meanders of stream *1*, become extended downstream and laterally in direction of heavy arrow growth to *a′, b′, c′* on present stream, and buried pay streaks result.

bouch into valley sections of gentler gradient the conditions are ideal for settling and concentration into deposits.

In a rapidly flowing meandering stream the fastest water is on the outside curve of meanders, and slack water is opposite (Fig. 5·7–8). The junction of the two, where gravel bars form, is a favorable site for deposition of placer minerals. With lateral migration of the meander, the "pay streak" becomes covered and eventually lies distant from the present stream channel. Obviously, placer deposits do not form in the downstream meanders of sluggish old-age streams, because the stream velocity is insufficient to transport heavy minerals.

Where streams cross highly inclined or vertically layered rocks, such as slates, schists, or alternating hard and soft beds, the harder layers tend to project upward and the softer ones to be cut away. This forms natural "riffles" (Fig. 5·7–9A), similar to the wooden riffles nailed in the bottom of a sluice box to arrest the gold in sluicing operations.

Such natural riffles are excellent traps for placer minerals and may give rise to *bonanzas* or exceptionally rich streaks.

The mode of entry of the placer minerals into a stream also determines the site of accumulation. Where materials are delivered by a swift tributary into a slower master stream they accumulate, under diminished velocity, as a pay streak down the near side (Fig. 5·7–9*B*).

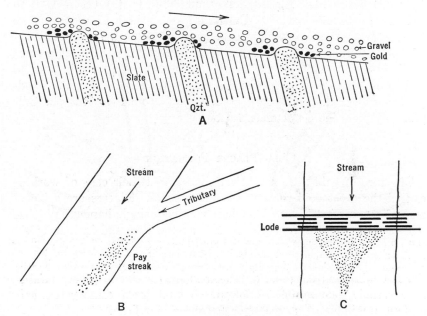

FIG. 5·7–9. *A*, Quartzite ribs, interbedded with slate, which serve as natural riffles for collection of placer gold (black); *B*, pay streak formed where fast tributary enters slow master stream; *C*, pay streak formed below gold lode crossed by stream.

If a stream crosses a mineralized lode, and through its own corrosion supplies the placer minerals, the pay streak will be spread across the stream channel on the downstream side of the lode (Fig. 5·7–9*C*). The early placer miners followed the trail of rich pay streaks or indications upstream to where they ceased, and they then looked for the "mother lode" nearby; they found many valuable bedrock deposits in this manner. Sometimes the search led to *bench* or *terrace gravels*.

The accumulation of placer minerals necessitated a nice, long-continued adjustment between stream velocity and gravel accumulation. The gravels could not be too thick; they had to be slowly moving downstream; and they had to be water-soaked so that jigging could occur. The presence or absence of these conditions meant rich pay streaks in some streams and sparsely disseminated gold in others.

Gradient of Streams. Placer accumulation requires well-graded streams, where a balance has been reached between erosion, transportation, and deposition. Examples of gradients in feet per mile of some placer streams are: (1) Klondike, "white channel," about 30 feet; (2) Fairbanks, Alaska, less than 30 feet; (3) other Alaskan streams, up to 150 feet; (4) California and British Columbia, 50 to 100 feet; (5) California high-level gravels, up to 150 feet; and (6) Netherlands Indies tin gravels, less than 30 feet. The high gradient of the high-level gravels of California has been shown by Lindgren to be in part due to a later westward tilting of the Sierra Nevada. Lindgren considers that moderate gradients of around 30 feet per mile probably yield the best concentration.

Gold Placer Formation

The lure of gold placers, their richness, and their ease of working caused the great gold rushes of modern times. Placers are a "poor man's" type of deposit. The hardy miner, single-handed, requires only a shovel and a gold pan to extract the gold and in a fortnight may accumulate fabulous riches. The great California gold rush of 1849, the rush to Australia in the fifties and the perilous stampede to the Klondike and Alaska in 1897 initiated ephemeral placer mining. More stable lode mining followed. Behind them came settlement, agriculture, industries; great new countries were born.

The California rush of 1849 amidst hardships of laborious overland travel and savage Indians, or around Cape Horn in inadequate ships, was, according to Rickard, but the modern counterpart of an ancient gold rush made classic by Greek legend. In the land of Colchis by the Euxine, ancient placer miners extracted alluvial gold by shoveling the gravels into sluice boxes made of hollowed trees. A lining of sheepskins entangled the gold particles; the coarse gold was shaken out, but the fine gold adhered to the wet wool, and the fleeces were hung in trees to dry so that the gold could be beaten out of them. The tale of a "gold strike" reached Greece, and Jason and his Argonauts set sail in the good ship *Argo* to seek the Golden Fleece! In Brazil today placer gold is extracted by using hollowed-out trees for sluice boxes and cowhides with the hair side up for riffles.

Today production of placer gold is small compared with its romantic past. The easily discoverable deposits have been found; fewer strikes or rushes can occur in the future. Only the slightly explored lands hold hope of yielding new placer discoveries of consequence. Russian rise in gold production is due largely to production of new placer deposits.

Source Materials. The gold in placers has come from primary workable lodes, lean lodes, or rocks sparsely traversed by gold-bearing veinlets. The source lodes need not necessarily be rich, as is popularly supposed; probably most of them are uneconomic. The rich placers of the Klondike, for example, never led to the discovery of workable lodes; some of those of Australia and California did. The richness of the placer deposit is more a result of abundant disintegration of auriferous rock and good concentration than of rich primary source material.

Pay Streaks. Gold tends to occur in concentrated pay streaks that are apt to be narrow and rich. The coarser gold is deposited in the upstream reaches of a run and the finer gold in the lower reaches. The pay streak may not lie in the present course of a stream, and, if it does, it is not necessarily in the central part. It is generally irregular in outline; it branches and splits and is absent in certain places. Although a definite pay streak may be present, some gold is also generally scattered throughout the main body of stream gravels. Thus, in many of the Alaskan gold placers the rich pay streaks were first mined by hand methods, and the low-grade gravels beyond the pay streaks are now being mined by means of large dredges.

Size, Shape, and Fineness of Gold. Most placer gold is in the form of fine specks called *dust*, but to the joy of the miner a few larger lumps called *nuggets* are found. They range from about the size of a pea or bean to that of a nut, but nuggets from 1 to 10 pounds are common. The largest nugget recorded is the Welcome Stranger from Ballarat, Australia, which weighed 2,280 ounces; the Blanche Barkley from Victoria weighed 1,752 ounces, and the Carson Hill, from California, weighed 1,296 ounces. A 1,050-ounce nugget was found in the Ural Mountains in 1936. The largest of the Australian nuggets were found immediately below the outcrop; therefore, they were not strictly stream placer nuggets. The largest nugget reported from the Yukon was 85 ounces. Rickard records nuggets of 47 ounces from Arizona and 40 ounces from Montana. All of these, of course, are exceptional. The general run of placer gold particles is roughly about the size of coarse sand.

An astonishing feature of placer gold is its divisibility into minute scaly particles called *flour* gold. Hite estimated that the flour gold of the Snake River averages about 5,000 *colors* to equal 1 cent in value and that discs 0.1 sq mm in area averaged about 17 million to 1 ounce Troy; the very smallest colors figured would require 7 to 8 millions to 1 cent. Such *flour* gold floats readily if exposed to air for an instant and may travel hundreds of miles. Its extreme fineness explains the wide distribution of " colors " throughout the entire thickness and width of stream gravels.

The shape of placer gold is generally disclike, owing to the continued pounding to which it is subjected. The nuggets have rounded but irregular outlines. The " dust " and " flour " gold consist of small flattened pellets and thin discs that may be only 2 microns in thickness. Crystallized gold is rarely present in placers, but the Latrobe nugget (23 oz) in the British Museum is a beautiful example of a crystal that escaped mutilation.

The quality or *fineness* of gold is usually expressed in parts per 1,000 (1,000 is pure gold). The fineness of placer gold varies from 500 to 999; that of veins, from 500 to about 850. Most placer gold is above 800 fine. Lindgren points out that the gold of the California veins averages 850 fine, whereas the Tertiary placer gold averages 930 to 950; also that the fineness of placer gold increases with the distance transported and with decreasing size of grains. McConnell showed that Klondike nuggets have greater fineness on the outside than on the inside. These observations suggest that some silver has been dissolved from the outside of the gold grains. The Snake River flour gold, according to Hite is 943 fine.

Associated Minerals. Gold-bearing gravels commonly consist largely of durable quartz pebbles; the other pebbles are comminuted or chemically destroyed. Alluvial gold deposits are, therefore, commonly marked by " white " runs of quartz gravel, as in the " white channels " of the Yukon, the " white bars " of California, or the " white leads " of Australia. " Black sand," which consists chiefly of magnetite and some ilmenite, is an intimate associate; and garnet, zircon, and monazite (yellow sand) are common. Any of the other placer minerals may also occur.

Relation to Bedrock. The character of the bedrock is important both in localizing gold placers and in effecting their recovery. Rocks that form smooth bedrock are less desirable than steeply dipping foliates or sediments that have irregular bottoms, or " riffles," to trap the gold. Clayey and decomposed rocks, and limestone with solution cavities make good traps. Crevices in the surface of the bedrock are generally intricately penetrated by the gold grains and a top layer has to be mined to obtain all the gold.

Coarse gold generally rests on or within a foot or so of bedrock; rarely it may lie within the lower 5 to 15 feet of gravels. Normally there is a decrease from the bottom up.   This is shown by McConnell's oft-quoted section of the " white channel " of the Klondike; the lower 6 feet contained $4.13 per cubic yard; the next 6 feet $0.18; and succeeding 6-foot sections contained 0.47, 0.04, 0.34, 0.32, 0.45, and 0.25.

In many places there is " false bedrock," which is generally a compact clay bed within the gravels.   It serves as a floor for placer gravels, and other gold-bearing gravels may or may not lie beneath it.

Growth of Gold in Gravels.   Do the large gold nuggets indicate some growth in the gravels?   The many proponents of accretion have claimed that large nuggets and coarse grains are larger than gold particles observed in veins and that crystals of gold have been seen on nuggets and pebbles.   However, the prevailing belief is that placer gold is entirely of mechanical derivation; nuggets polished and etched by Uglow showed crystalline structure typical of vein gold.

Bench Gravels.   Slight uplift or sudden decrease of stream load or increase of stream volume often causes stream incision and the stranding of *terrace* or *bench* gravels on one or both sides of the valley. Repetitions may form more than one set of terraces, such as " low " and " high " benches.   Their tops represent the former valley floors. They may consist entirely of alluvium or of alluvium-covered bedrock. Commonly the gravels from such terraces are side-washed again into the stream and undergo further reconcentration.   In the absence of side-wash the placer gold may persist in the benches, and much gold has been won from them.

High-Level Gravels.   Ordinarily, regional uplift causes rapid erosion of unconsolidated gravels.   However, in California, Victoria, and New South Wales, gold-bearing gravels have been preserved by later rock covers, uplifted, and re-exposed by cross-cutting valleys.   These are called variously "high-level gravels," " buried leads," and " high Tertiary gravels."

Those of California have been made classic by Lindgren.   He has shown that in the Sierra Nevada region during early Tertiary time gold-bearing gravels accumulated in stream channels that traversed a gently sloping country.   These were buried by thick deposits of rapidly accumulated lean gravels, and these in turn by tuffs and breccias that attained thicknesses of 1,500 feet.   The range was elevated; new consequent streams eroded canyons to depths of 2,000 to 3,000 feet, and these now form superimposed streams, many of which are out of adjustment with and transverse to the early Tertiary streams.   The new stream bottoms now lie on an average about 2,600 feet below the early Tertiary stream bottoms; the present valley walls disclose elevated cross sections of the former gold-bearing stream deposits, and these have been extensively mined.   Their eroded parts have supplied much of the Recent stream gold.

Submerged Placers.   In Alaska, California, Australia, New Zealand, and Siberia regional subsidence, stream overloading, or lava flows have buried placer gravels to depths of tens or hundreds of feet.   These are reached by dredging or through shafts, but they are difficult to mine because of excessive water.   In Alaska and Siberia, however, where the ground is frozen, these gravels are extracted by " drift-mining," or are excavated by dredges to depths of 20 to 30 feet; Lincoln records gold gravels at depths of 300 feet near Fairbanks, Alaska, and much drift-mining there has been carried on at depths of 50 to 150 feet. The thickness of the gold-bearing gravels ranges from 2 inches to 8 feet, and the gold content is from 1 to 8 dwt[1] per yard.   The most notable submerged placers are some of the " deep-leads " of Australia and New Zealand.   At Ballarat many rich " gutters " or channels have been found beneath 300 feet of cover.

Cement Gravels.   Consolidation of gold-bearing gravels into conglomerate gives " cement gravels," and these range in age from Cambrian to Recent.   A Cambrian conglomerate overlying schist in the Black Hills carries detrital gold, presumably derived from the Homestake Lode, and Cretaceous conglomerates in Oregon and California yield a little detrital gold.   Similar conglomerates of Permian age in Bohemia are described by Pošepńy; of Miocene age in New Zealand, by Park; and of Permo-Carboniferous age in Australia, by Wilkinson. The celebrated gold-bearing cements of Blue Spur, New Zealand, are reported by Park to have been profitably mined for over 30 years and were the source of the remarkably rich alluvial gold of Gabriel's Gully.

Distribution and Value.   Stream placer deposits have been mined in most parts of the world.   They were probably the chief source of the gold of the ancients and also account for a considerable part of the world's total gold supply.   The most notable placer gold deposits are those of California and other western states, Yukon-Alaska, Australia, New Zealand, Siberia, British Columbia, Peru, Bolivia, and Chile. Many others have been worked throughout Europe, Asia, and Africa. Gold placers are being mined at present on a large scale in Alaska, California, Colombia, Siberia, Central Africa, and New Guinea.

The tenor of gold placer deposits is referred to by the yield per cubic yard, by the running foot of channel, or by the square foot of surface. The first is generally used in connection with large-scale operations. Bonanza yields are in dollars per pan.   Individual small pockets have yielded several thousand dollars.   One area 40 by 40 feet at North

[1] One ounce troy equals 20 pennyweights (dwt).

Zachlan, New South Wales, yielded $38,000, or nearly $24 per square foot.

The yield of some of the better-known placer deposits is shown below.

Locality	Width of deposit (feet)	Yield	
		Per linear foot*	Per cubic yard*
Madame Berry, Victoria	450	$1,293	
Victoria, No. 2	1,000	443	
Victoria, general			$2.00–15.00
White Channel, Yukon		380	
Klondike			9.00–50.00
Nome Creeks, Alaska	50	100	
Seward, Alaska			2.00– 6.00
Red Point, Calif.	120	72	
Nevada County, Calif.	1,000	414	
New South Wales			0.24
Fairbanks, Alaska			0.17
California, general			0.25– 0.34
North Fork, Calif.			0.042
Columbia			0.15

* Gold at $20.67 per ounce.

PLATINUM PLACER FORMATION

Most of the world's production of platinum and platinum metals formerly came from stream and beach placers.

The source of placer platinum is ultrabasic rocks, such as dunite or peridotite, in which the metal, along with chromite, has undergone a preliminary concentration by magmatic differentiation. The enriched portions, although highly concentrated above the original magmatic content, are still too lean to constitute workable lodes (except in the Bushveld of South Africa), but through erosion the metal has been sufficiently concentrated to form commercial placer deposits.

Platinum is concentrated into placer deposits in the same manner as gold and has similar occurrence. In fact, it is commonly associated with gold placers. It occurs in the form of dust and little flattened pellets; nuggets are extremely rare. Associated minerals are chromite, magnetite, and gold. The platinum ranges in fineness from 700 to 850. Invariably one or more of the platinum metals — palladium, osmium, iridium, rhodium, ruthenium (and the minerals iridosmine and osmiridium) — are alloyed with the platinum, as well as small quantities of copper, iron, and other metals.

Most of the placer platinum of the world has come from the Ural Mountains, where streams transect Paleozoic intrusions of enriched peridotitic rocks. Although Russian production has diminished, it is still second in the world. Colombia has long yielded platinum metals from Recent gravels and Tertiary conglomerates and ranks fourth in world production. Peridotite in Tasmania and New South Wales yields platinum and osmiridium placers. Small amounts of placer platinum are also obtained in Ethiopia, Japan, Panama, Papua, Sierra Leone, and South Africa. About one-third of the world production of platinum metals comes from placers.

In North America a small but steady production of placer platinum is obtained from Goodnews Bay, in Alaska, and as a by-product of gold placers in Alaska, California, and Oregon, where streams cross peridotite or serpentine.

The production, uses, occurrence and deposits of platinum are given in Chapter 12.

TIN PLACER FORMATION

Cassiterite (SnO_2), because of its weight and resistance to chemical and mechanical disintegration, ranks second in value as a placer mineral; and most of the tin production now comes from placer deposits. The " stream tin," which is derived from the weathering of stockworks and lodes in granite and other rocks, has accumulated in alluvial and eluvial deposits in a manner similar to gold.

The stream tin is in the form of rounded pellets of shot, bean, or nut size. Associated minerals are similar pellets of limonite, hematite, magnetite, garnet, wolframite, tourmaline, and other placer minerals.

The greatest tin placer deposits are those of Malaya (p. 548), where extensive alluvial deposits containing only half a pound of tinstone to the cubic yard are worked by some of the largest dredges in the world.

The Netherlands Indies tin deposits (Chap. 13) resulted from deep tropical weathering of granite containing lean stockworks and sparse veinlets of quartz and cassiterite giving rise to eluvial concentrations of cassiterite (koelits) and of alluvial concentrations (kaksas) during Quaternary time. An eluvial origin is indicated by unworn cassiterite, in high concentration, along with angular fragments of other resistant minerals. These deposits contribute a considerable part of the Netherlands tin production. They are really a transient stage in the formation of the alluvial tin placers because eventually the eluvial material reaches the stream channels and there undergoes rigorous stream concentration to form the productive stream placers. During a slight subsidence of the region in Quaternary time, when the islands became separated from each other, the pay streaks of tinstone became covered by subsequent gravels and sands and the downstream parts became submerged beneath the sea, so that former stream gravels are now dredged from the sea bottom.

Small amounts of stream tin are also won in Nigeria, China, Belgian Congo, Bolivia, Siam, Burma, Victoria, New South Wales, and Tasmania. A trivial amount is obtained as a by-product from gold gravels in Alaska. About two-thirds of the world supply of tin is won from placer deposits.

See Chapter 13 for the history, production, uses, and deposits of tin.

Precious Stones Placer Formation

Diamond. The diamond is the chief precious stone won from placer deposits. Prior to 1871, when the first diamond in primary rock was found at Kimberley, South Africa, the entire world production came from placers in India and Brazil. About 95 percent of the world production has been won from placers in the Belgian Congo, Gold Coast, Angola, Sierra Leone, Tanganyika, Brazil, and South Africa. Small amounts come from French West Africa, Southwest Africa, British Guiana, Borneo, Australia, Nigeria, and Venezuela. The production of the different countries and description of the various deposits are given on pages 837–846.

The original home of the diamond, so far as known, is in pipelike intrusions of ultrabasic igneous rocks, such as kimberlite or peridotite. These occur in south and central Africa, and there is one occurrence in Arkansas; the original source of the South American diamonds is not definitely known. Since the diamonds are very sparsely scattered in the pipes an extremely high ratio of concentration must have taken place to yield the workable placer deposits.

The most remarkable placer deposits of South Africa are those of the Lichtenburg district, which witnessed the great " stampede " by 25,000 " runners " in 1926. The writer approached this district across a striking peneplain developed upon an ancient dolomite and was amazed to see that the stream gravels are sinuous, eskerlike embankments that project above the peneplain and stretch discontinuously across the country, marking the meandering course of a former stream. The stream was formerly entrenched in dolomite, and its valley contained diamantiferous gravels. The region developed a karst topography, the surface drainage disappeared, and the soluble dolomite plateau was slowly lowered by solution-weathering; but the protected dolomite beneath the gravels escaped solution and was left as a ridge capped by diamantiferous gravels. The diamonds now lie in the bottom gravels of the ridge. Even more unusual are sinkholes developed beneath the former valley bottom into which the diamond gravels have slumped; some of these pockets proved fabulously rich. The Lichtenburg gravels in 1927, the year of their greatest production, yielded over 2,100,000 carats valued at over $23,000,000 as against approximately 2,300,000 carats from the diamond pipes in the same year.

The diamantiferous gravels of the Belgian Congo, which yield about two-thirds of the world production, are former stream gravels, and the diamonds probably came originally from weathered kimberlite pipes, several of which are known within the Congo. The same is probably true of the adjacent Angola gravels. The diamonds of Minas Geraes, Brazil, are found in gravels derived from a conglomerate, which in turn represents a former lightly concentrated placer deposit whose stones came from an unknown source. The prized *carbonado* of Bahia, Brazil, desired for its use in diamond drilling, is obtained from present stream gravels.

Other Gemstones. Most rubies, sapphires, chrysoberyls, aquamarines, and spinels are washed from stream gravels. Their original site was in pegmatite dikes or in contact- or regional-metamorphosed rocks from which they have been released by weathering. The sapphires, aquamarines, and zircons of Ceylon and Kashmir also come from stream gravels. Formerly stream gravels from peridotite areas in North Carolina and Georgia supplied some fine gemstones of sapphire and ruby; sapphires also occur along the Missouri River near Helena, Mont.

Other Fluviatile Deposits. *Monazite* (see Beach Placers), an accessory mineral of igneous rocks, accumulates with other placer minerals. Stream deposits occur in the Carolinas and Idaho. Alluvial *quartz sands* are common. *Ilmenite*, which is won mainly from beach gravels, also occurs in stream gravels and is produced as a by-product of placer tin in Malaya. *"River pebble" phosphate* rock of alluvial origin occurs in streams in Florida (p. 814). Other minerals won in minor quantities from stream placers include zircon, garnet, native copper, chromite, quicksilver, wolframite, thorianite, thorite, and kyanite.

Beach Placer Formation

Beach placers are formed along seashores by the concentrating effects of wave and shore action. Shore currents shift materials alongshore, and the lighter materials are moved faster and farther than the heavy, thereby concentrating the heavy minerals. Wave action operates at the same time. Pounding waves throw up materials on the beaches; the backwash and the undertow carry out the lighter and finer material, which in turn is moved alongshore; and the larger and heavier materials are concentrated on the exposed beaches. A familiar manifestation of this is gravel or shingle beaches from which the sand has been removed. If placer minerals are available they likewise are similarly concentrated.

Placer minerals may be made available for beach concentration by (1) streams that debouch upon the coast; (2) wave erosion upon sea terraces or gravel plains; (3) wave encroachment upon former near-shore stream terraces; (4) wave erosion of rocky shores.

Placer minerals of beach deposits consist of gold, ilmenite, magnetite, rutile, diamonds, zircon, monazite, garnet, and quartz.

The beach *gold* is mostly very finely divided, ranging from 70 to 600 colors to the cent, giving evidence thereby of the travel and beating it has undergone. The deposits of Nome, Alaska (p. 435), illustrate beach placer gold. *Ilmenite* (p. 625), *rutile, and monazite* have undergone unusual beach concentration at Travancore and Quilon, India, that has resulted in commercial sands containing 50 to 70 percent ilmenite and 2 to 5 percent (in places up to 25 percent) monazite, along with zircon, rutile, garnet, and other minerals. Somewhat lower concentration has occurred in beach sands of Florida, Redondo Beach, Calif., North Carolina, Senegal, Ceylon, Argentina, Brazil, Evanshead and Ballina in Australia, and in New Zealand. *Zircon* has reached high beach concentration in Brazil where, according to H. C. Meyer, the sands will carry 51 percent zirconium oxide, along with titanium and cerium and yttrium earths. The beach sands in Australia contain up to 75 percent zircon. *Magnetite, or black sands,* containing also ilmenite, chromite, and other heavy minerals have been formed extensively on Oregon, California, Brazil, Japan, and New Zealand coasts.

One of the most remarkable beach concentrations is that of the diamantiferous marine gravels of Namaqualand, South Africa, which in 1928 yielded 50 million dollars. The diamonds occur in marine gravels that occupy wave-cut terraces 20 to 210 feet above sea level which extend intermittently for a distance of 200 miles or more along the desert coast south of the mouth of the Orange River at a maximum distance of 3 miles inland. The best deposit is near Alexander Bay in gravels 8 to 100 feet wide and 120 feet above sea level. After the finding of the first few stones, the mode of occurrence was not realized until Merensky and Reuning discovered that the diamonds were associated with certain diagnostic oyster shells that marked definite beach lines. These beach lines could then be traced along the coast. The diamonds in places occur in exposed coarse beach shingle from which the sand has been blown away. A cupful of excellent diamonds was said to have been picked up by hand from amongst the pebbles over a very small area. Other diamantiferous gravels are buried under surface deposits. The diamonds are thought to have been carried down by the Orange River and to have been distributed down the coast by the prevailing southerly shore currents, and by waves.

Eolian Placer Formation

Wind instead of water may act as the agent of concentration and give rise to placer deposits. Obviously this can occur only in arid regions. It is reported that some eolian gold placers have been formed in the Australian deserts from the disintegration of gold quartz lodes. The light decomposed materials have been blown away; the heavy gold particles, freed from their matrix, remained behind. The continuation of this process finally resulted in patches of surface accumulations of placer gold in a debris of sand and wind-worn pebbles. Some of the gold was recovered by " dry washing," utilizing wind instead of water to separate the debris from the gold after screening off the pebbles. Similar concentration has taken place at El Arco, in Lower California, Mexico.

Selected References on Placer Deposits

General Reference 8. Chap. XVII. *General discussion.*

Tertiary Gravels of the Sierra Nevada. W. LINDGREN. U. S. Geol. Surv. Prof. Paper 73, 1911. *A fascinating correlation of Tertiary gravel formation and physiographic development of the region resulting in remnants of elevated buried gravels.*

New Technique Applicable to the Study of Placers. O. P. JENKINS. Calif. Dept. Nat. Resources XXXI, 1935.

Klondike Gold Fields. R. G. McCONNELL. Can. Geol. Surv. Ann. Rept. 143, 1905; and **The Klondike High-Level Gravels.** Can. Geol. Surv., 1907. *Occurrence and origin of rich Klondike placers.*

Gold. J. M. MACLAREN. The Mining Journal, London, 1908. *Descriptions of important placer deposits of the world.*

Law of the Pay Streak in Placer Deposits. J. B. TYRRELL. Inst. Min. and Met. Trans., London, May 16, 1912. *Development of pay streaks in stream gravels.*

Textbook of Mining Geology. J. PARK. Griffin and Co., London, 1927. *New Zealand and Australian occurrences.*

Gold and Silver. W. R. CRANE. John Wiley & Sons, New York, 1908. *Descriptions of placer workings of the western United States.*

Diamond-Bearing Gravels of South Africa. A. F. WILLIAMS. 3rd Empire Min. and Met. Cong., S. Africa, 1930. *Descriptions of various types of placer gravels.*

The Fineness of Gold, Morobe Goldfield, New Guinea. N. H. FISHER. Econ. Geol. 40:449–495, 537–563, 1945. *Excellent treatment of placer gold pay streaks and source correlation.*

See also references under " Gold," Chap. 12, and " Gemstones," Chap. 24.

5·8 OXIDATION AND SUPERGENE ENRICHMENT

When ore deposits become exposed by erosion they are weathered along with the enclosing rocks. The surface waters oxidize many ore minerals and yield solvents that dissolve other minerals. An ore deposit thus becomes oxidized and generally leached of many of its valuable materials down to the ground water table, or to a depth where oxidation cannot take place. The oxidized part is called the *zone of oxidation*. The effects of oxidation, however, may extend far below the zone of oxidation. As the cold, dilute, leaching solutions trickle downward they may lose a part or all of their metallic content within the zone of oxidation and give rise to oxidized ore deposits—a familiar group of deposits readily accessible to mining and made colorful in the glamorous beginnings of many mining districts. If the down-trickling solutions penetrate the water table, their metallic content may be precipitated in the form of secondary sulphides to give rise to a zone of *secondary* or *supergene sulphide enrichment*. The lower, unaffected part of the deposit is called the *primary* or *hypogene zone*. This zonal arrangement (Fig. 5·8–1) is characteristic of many mineral deposits that have undergone long-continued weathering. In places the supergene sulphide zone may be absent, and in rare cases the oxidized zone is shallow or lacking, as in glaciated areas or regions undergoing rapid erosion. Special conditions of time, climate, physiographic development, and amenable ores are necessary to yield the results pictured above, but they are sufficiently common for oxidized and enriched supergene ores to occur in most of the nonglaciated land areas of the world.

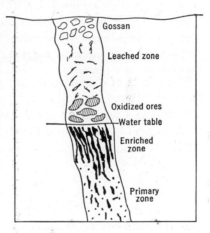

FIG. 5·8–1. Diagram of zones of a weathered vein, with oxidized, supergene enrichment, and primary zones.

The effect of oxidation has been either to render barren the upper parts of many deposits or to change ore minerals into more usable or less usable forms, or even to make rich bonanzas. The effect of super-gene enrichment has been more far reaching, for it has added much where there was previously little; leaner parts of veins have been made rich; unworkable protore (p. 274) has been enriched to ore grade. Indeed, many of the great copper districts would not have come into existence except for the process of supergene sulphide enrichment.

Supergene oxidation and enrichment go hand in hand. Without oxidation there can be no supply of solvents from which minerals may later be precipitated in the zones of oxidation or of supergene sulphides. The process resolves itself, therefore, into three stages: (1) oxidation and solution in the zone of oxidation, (2) deposition in the zone of oxidation, and (3) supergene sulphide deposition. Each is considered separately.

Oxidation and Solution in Zone of Oxidation

The effects of oxidation on mineral deposits is profound. The minerals are altered, and the structure is obliterated. The metallic substances are leached, or are altered to new compounds that require metallurgical treatment for their extraction quite unlike that employed for the unoxidized materials. The texture and the type of deposit are obscured. Compact ores are made cavernous. Ubiquitous limonite obscures everything and imparts to the gossan that familiar rusty color which from earliest times has attracted the curiosity of the miner. One can only infer what lies beneath, but in this inference one can be guided by the character of the oxidation products themselves.

Water with dissolved and entangled oxygen is the most powerful oxidizing reagent, but carbon dioxide also plays an important role. (Locally chlorides, iodides, and bromides play a part.) These substances react with certain minerals to yield strong solvents, such as ferric sulphate and sulphuric acid. Sulphuric acid, in turn, reacting with sodium chloride yields hydrochloric acid, which with iron yields the strongly oxidizing ferric chloride.

Chemical Changes Involved

There are two main chemical changes within the zone of oxidation: (1) the oxidation, solution, and removal of the valuable minerals, and (2) the transformation *in situ* of metallic minerals into oxidized compounds.

Most metallic mineral deposits contain pyrite. This mineral under attack readily yields sulphur to form iron sulphate and sulphuric acid;

pyrrhotite does the same. The following reactions indicate, without intermediate steps, their general trend:

[1] $$FeS_2 + 7O + H_2O = FeSO_4 + H_2SO_4$$

[2] $$2FeSO_4 + H_2SO_4 + O = Fe_2(SO_4)_3 + H_2O$$

Reaction 2 passes through intermediate stages during which S, SO_2, and $FeSO_4$ may form. The sulphur may oxidize to sulphuric acid. The ferrous sulphate readily oxidizes to ferric sulphate and ferric hydroxide:

[3] $$6FeSO_4 + 3O + 3H_2O = 2Fe_2(SO_4)_3 + 2Fe(OH)_3$$

The ferric sulphate hydrolizes to ferric hydroxide and sulphuric acid:

[4] $$Fe_2(SO_4)_3 + 6H_2O = 2Fe(OH)_3 + 3H_2SO_4$$

Ferric sulphate is also a strong oxidizing agent and attacks pyrite and other sulphides to yield more ferrous sulphate.

[5] $$Fe_2(SO_4)_3 + FeS_2 = 3FeSO_4 + 2S$$

Ferric sulphate, in addition, changes to various basic sulphates.

The above reactions indicate the importance of pyrite, which yields the chief solvents, ferric sulphate and sulphuric acid, and also ferric hydroxide and basic ferric sulphates. Moreover, ferric sulphate is continuously being regenerated not only from pyrite but also from chalcopyrite and other sulphides. The ferric hydroxide changes over to hematite and goethite and forms the ever-present limonite that characterizes all oxidized zones. The basic ferric sulphates, of which there are several, may be deposited as such, but generally limonite is the end product.

The part played by ferric sulphate as a solvent may be seen in the following equations. Although the end products are obtained, it is not established in all cases that the reactions given below are those that actually take place.

[6] *Pyrite* — $FeS_2 + Fe_2(SO_4)_3 = 3FeSO_4 + 2S$

[7] *Chalcopyrite* — $CuFeS_2 + 2Fe_2(SO_4)_3 = CuSO_4 + 5FeSO_4 + 2S$

[8] *Chalcocite* — $Cu_2S + Fe_2(SO_4)_3 = CuSO_4 + 2FeSO_4 + CuS$

[9] *Covellite* — $CuS + Fe_2(SO_4)_3 = CuSO_4 + 2FeSO_4 + S$

[10] *Sphalerite* — $ZnS + 4Fe_2(SO_4)_3 + 4H_2O = ZnSO_4 + 8FeSO_4 + 4H_2SO_4$

[11] *Galena* — $PbS + Fe_2(SO_4)_3 + H_2O + 3O = PbSO_4 + 2FeSO_4 + H_2SO_4$

[12] *Silver* — $2Ag + Fe_2(SO_4)_3 = Ag_2SO_4 + 2FeSO_4$

Similarly other minerals are dissolved yielding, except for lead, soluble sulphates of the metals. The sulphuric acid also attacks various sulphides, yielding sulphates of their metals. Chlorides, bromides, and iodides, chiefly of silver, are also formed.

Most of the sulphates formed are readily soluble, and these cold, dilute solutions slowly trickle downward through the deposit until the proper conditions are met to cause deposition of their metallic content.

If pyrite is absent from deposits undergoing oxidation, only minor amounts of the solvents are formed; little solution occurs, and the sulphides tend to be converted *in situ* into oxidized compounds, and the hypogene sulphides are not enriched. This is illustrated in the New Cornelia mine at Ajo, Ariz., where a deficiency of pyrite has resulted in chalcopyrite being converted to copper carbonate and supergene sulphides are negligible. This also happens where a supergene chalcocite zone lacking pyrite is oxidized; the chalcocite is not dissolved but is converted into copper carbonates, cuprite, or native copper.

A country rock of limestone tends to inhibit migration of some sulphate solutions; it immediately reacts with copper sulphate, for example, to form copper carbonates, thus precluding any supergene sulphide enrichment.

The general tendency of the chemical changes in the zone of oxidation is to break down complex minerals and form simple ones. In general, among the metallic minerals, those lacking in oxygen (sulphides, etc.) are most susceptible to oxidation, and most metallic oxides are little affected. Native metals may be attacked; quartz is resistant, but silica set free during oxidation is generally dissolved. Carbonates decompose readily, and most of the silicates are altered to a few stable minerals.

Solubility and Solution

Some constituents, like gold, generally resist solution and persist in the oxidized zone; others, like galena, are changed so slowly that they hang back in the oxidized zone while their more soluble companion minerals are removed; still others, like the minerals of copper, zinc, or silver, are readily soluble and so are generally removed. The oxides of iron are highly resistant to solution.

The simple sulphides are only slightly soluble in cold water and their order of solubility as determined by Weigel is shown on page 249. In dilute sulphuric acid, pyrrhotite, chalcopyrite, bornite, sphalerite, and galena are attacked; but arsenopyrite, pyrite, and silver compounds are attacked only slightly. Such minerals as covellite, chalcocite, and molybdenite are not attacked.   Ferric sulphate, however, readily dissolves most of them.

Gottschalk and Buehler found that the little electric currents generated in mixed sulphides accelerate and retard oxidation and solution of certain minerals.   Currents flowing from minerals of high potential to those of low potential accelerate solution of the latter and retard that of the former; e.g., sphalerite dissolves much more readily in the

COMMON SULPHATES	COMMON CARBONATES	COMMON SULPHIDES (*precipitated*)
1. $MnSO_4$	1. K_2CO_3	1. MnS
2. $ZnSO_4$	2. Na_2CO_3	2. ZnS
3. $MgSO_4$	3. $CuCO_3$	3. FeS
4. $Al_2(SO_4)_3$	4. $MnCO_3$	4. CoS
5. $FeSO_4$	5. $FeCO_3$	5. NiS
6. $CuSO_4$	6. $MgCO_3$	6. CdS
7. Na_2SO_4	7. $ZnCO_3$	7. Sb_2S_3
8. K_2SO_4	8. $BaCO_3$	8. PbS
9. Ag_2SO_4	9. $CaCO_3$	9. CuS
10. $CaSO_4$		10. As_2S_3
11. $PbSO_4$		11. Ag_2S
		12. Bi_2S_3
		13. HgS

presence of pyrite.   The sulphides can be arranged according to their differences in potential, which vary, however, according to the kind of solution.   The potentials of marcasite and pyrite are high, and of chalcocite and sphalerite low; therefore, the solution of the last two is accelerated in the presence of universal pyrite.

The carbonates are low in solubility and the sulphates are high, lead and barium sulphates being exceptions.   The order of solubility in cold water of the common sulphates and carbonates, as determined by Kohlrausch, and common sulphides as determined by Weigel (No. 1 being the most soluble), is shown in the lists above.

The oxidation behavior of the common metallic compounds of hypogene ore deposits is summarized in Table 8; their precipitation is considered later.

OXIDATION SEPARATION OF METALS

Oxidation of mixed ores commonly results in a separation of the contained metals, as in the lead-zinc-pyrite " manto " limestone deposits of Mexico (illustrated in Fig. 5·8–2).   The pyrite is largely removed; the galena becomes oxidized to anglesite and cerussite, which remain in place; the sphalerite is dissolved as zinc sulphate, which migrates, encounters limestone, and the zinc is deposited as ore bodies of zinc carbonate. Thus, complex metallurgy is simplified by nature, rendering the deposits of greater value than the original. At Sierra Mojada, Mexico (Fig. 5·8–2), for example, a body of rich lead cerussite

TABLE 8

OXIDATION BEHAVIOR OF COMMON METALLIC COMPOUNDS

Metal	Primary Compounds	Important Solutions Formed	Oxidized Compounds Formed in Oxidation Zone	Transported out of Oxidation Zone
Iron	Sulphides	Ferric sulphate, colloidal	Hematite, limonite, basic sulphates	Yes
	Carbonates	......	Limonite, ferric hydroxide	No
	Oxides	......	Hydrous ferric oxide	No
Copper	Sulphides, etc.	Copper sulphate	Carbonates, oxides, native, silicate	Yes
Lead	Sulphides, etc.	......	Sulphate, carbonate	No
Zinc	Sulphide	Zinc sulphate	Carbonate, silicate, oxide	Yes
Tin	Oxide, sulphide	?	Oxide	No
Aluminum	Silicates	Colloidal	Oxides, silicates	No
Silver	Sulphides, etc.	Silver sulphate, colloidal	Chloride, iodide, bromide, native	Yes
Gold	Au, tellurides	Acid ferric sulphate + Cl	Native	Yes
Nickel	Sulphide	Nickel sulphate	Nickel arsenate, Nickel silicate	Yes?
Cobalt	Sulphide, etc.	Cobalt sulphate	Cobalt oxides	Yes
Molybdenum	Sulphide, etc.	......	Oxides	?
Chromium	Oxide	?	Oxides	No
Tungsten	Oxides	......	Oxides	No
Manganese	Oxides, carbonates	Mn sulphate	Oxides	Yes; No
		Mn bicarbonate, colloidal	Sulphate	Yes
			Oxides	No; Yes
Vanadium	Sulphide, oxide	Sulphate	Oxides	?
Mercury	Sulphide	......	Oxides, chloride	No
Arsenic	As, arsenides	?	Arsenates, oxides, sulphides	Yes
Antimony	Sb, antimonides	?	Oxides	Yes
Bismuth	Sulphides	......	Oxides	No
Cadmium	Sulphide	Sulphate	Carbonate	No
Platinum	Pt, sulphide	?	Arsenide	Yes
Uranium	Oxides	?	Oxides, etc.	?

ore was underlain by large bodies of zinc carbonate and locally by areas rich in silver that had also undergone migration. Lindgren has described similar examples from Tintic, Utah.

In similar manner, gold is separated from sulphide matrix and rendered " free milling."

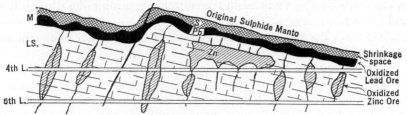

Fig. 5·8–2. Sketch of longitudinal section of part of ore manto of Encantada mine, Sierra Mojada, Mexico, illustrating separation of metals by oxidation. Original body (M) composed of pyrite, galena, and zincblende. Oxidation yielded body of oxidized lead-silver ore (Pb), overlain by open spaces (S) due to oxidation shrinkage, and underlain by large bodies of transported oxidized zinc ore (Zn) in iron-stained dolomite.

GOSSANS AND CAPPINGS

Gossans are signboards that point to what lies beneath the surface. They arrest attention and incite interest as to what they may mask. Most ore deposits, save in glaciated regions, are capped by gossans; hence the finding of one may herald the discovery of buried wealth. Noncommercial mineral bodies, however, also yield gossans. To distinguish between them is of vital importance; but it requires experience, knowledge, and careful observation. The distinctions involve delicate differences of color and form that are difficult to transmit by written word; they must be seen.

Gossan is a Cornish word used to designate the oxidized outcropping cellular mass of limonite and gangue overlying aggregated sulphide deposits. The present tendency is to use also the term *capping* or *cap rock* to designate the oxidized outcrops of disseminated deposits.

Materials of Gossans and Their Formation. The limonite universally formed during oxidation of iron-bearing sulphides persists in the oxidized zone and imparts to the gossan and capping its diagnostic color. Its many forms and colors are considered separately. Other metals whose oxidized compounds display characteristic colors are shown in Table 9.

Other persistent oxidized minerals in the gossan also indicate former sulphides, such as the sulphate and carbonate of lead, the carbonate and silicate of zinc, native silver, and horn silver. Gold generally persists in the croppings as native gold.

THE ROLE OF IRON IN GOSSANS

The importance of iron in croppings cannot be overemphasized. Not only is limonite universally present, but also it occurs in a variety of colors, structures, and positions, each of which is significant. Much information has been gained in recent years regarding the composition and interpretation of limonite.

Limonite. Posnjak and Merwin have shown that there is no mineral species " limonite " with a formula of $2Fe_2O_3 \cdot 3H_2O$. The only members of the hydrous iron oxide series are the ferric oxide hematite and the monohydrate $Fe_2O_3 \cdot H_2O$ which exists in the two forms *goethite* and *lepidocrocite,* both having the same composition but different crystalline forms. The so-called limonite is goethite or lepidocrocite

TABLE 9

SOME COMMON DIAGNOSTIC CROPPING COLORS OF COMMON METALS

MINERAL OR METAL	CROPPING COLORS	OXIDIZED COMPOUNDS
Iron sulphides	Yellows, browns, maroons, reds	Goethite, hematite, limonite, sulphates
Manganese	Black	Manganese oxides, wad
Copper	Green, blues	Carbonates, silicates, sulphates, oxides, native
Cobalt	Black, brilliant pink	Oxides, " bloom " (erythrite)
Nickel	Greens	Nickel " bloom " (annabergite), garnierite
Molybdenite	Bright yellows	Wulfenite, molybdite
Silver	Waxy greenish	Chlorides, etc., native
Arsenic	Orange, yellows	Oxides
Bismuth	Yellow	Bismite
Cadmium (in zinc)	Light yellow	Cadmium oxide

with or without hematite and with adsorbed water. " Limonite " is used in this book to designate a mixture of fine-grained, undetermined oxides of iron. Tunell has shown that most limonite of croppings consists of hematite, goethite, and jarosite (hydrous potassium iron sulphate); nontronite, the hydrous ferric silicate, may also be present.

The limonite of croppings is of two derivations — that from iron-bearing sulphides and that from iron-bearing gangue or rock silicates. The former is soluble in dilute hydrochloric acid, the latter is not. Also, the latter can usually be traced to its parent silicate and is generally negligible in amount.

Indigenous and Transported Limonite. In croppings, iron of sulphide derivation may either (1) become fixed as oxide at the site of the former sulphide, forming *indigenous limonite,* or (2) be dissolved, transported, and precipitated elsewhere, forming *transported limonite.* The indigenous is thrown down in the insoluble ferric state; the transported is in the soluble ferrous state. Posnjak and Merwin have shown that the oxidation of ferrous to ferric iron is retarded by free sulphuric acid and is accelerated by copper; also that the deposition of ferric oxides is likewise retarded by free acid and aided by copper. In plain words, this means that, when pyrite undergoes oxidation, the free acid generated tends to keep the iron in the soluble ferrous state and enables it to be removed in solution. The free acid, in addition, retards deposition as limonite. Also, it means that, if chalcopyrite undergoes oxidation which yields little or no acid, ferrous iron readily changes to the insoluble ferric and limonite is precipitated *in situ*. Thus, as pointed out by Locke and Morse, indigenous limonite indicates the former

FIG. 5·8–3. *A*, Sketch of indigenous limonite, (*c*), derived from oxidation of chalcopyrite and remaining within cavity; *a*, *b*, limonite fog surrounding cavity. Cactus mine, Utah, ×10 *B*, Transported limonite from oxidation of pyrite, with empty cavity *d*, surrounded by bleached zone *e* and by limonite halo *b*. Silverbell, Ariz., ×10. (*Blanchard-Boswell, Econ. Geol.*)

presence of copper, and transported limonite indicates a former high ratio of iron to copper or else lack of copper.

Indigenous limonite of sulphide derivation occupies the voids left by the former sulphides (Fig. 5·8–3*A*). It does not occur outside the voids. Its characteristic structure denotes the kind of predecessor sulphide. It is generally compact and fairly hard, and it has subdued colors.

Transported limonite may have been moved no farther than beyond the rim of the void, or it may have been transported tens or hundreds of feet from the site of its originating sulphides. The distance of transportation depends largely upon the precipitating power of the gangue or rock through which the solutions pass. Transported limonite thus may form halos around the empty voids (Fig. 5·8–3*B*), or it may thoroughly permeate the gangue, or enclosing rock. A little of it, like a drop of ink in a glass of water, goes a long way, and it makes, therefore, a conspicuous and exaggerated showing of iron oxide. It occurs in varied structural forms, colors, and positions. Paints, crusts, impregnations, and earthy mixtures are common. Other distinctions are discussed below.

Iron Migration.   The nature of the gangue affects the migration of iron.   Quartz, an inert gangue, exerts no precipitating effect upon iron in solution, but carbonate is a strong precipitant.   Locke divides capping gangues into four classes: (1) inert (quartz and barite); (2) slowly reacting (fine-grained mixtures of quartz, sericite, and adularia); (3) moderately reacting (coarse feldspar made porous by partial kaolinization); and (4) rapidly reacting (carbonates).   Some deposition of limonite is brought about by evaporation of solution, so that even inert quartz may become iron-stained.   In moderately to rapidly reacting gangues the iron is generally precipitated close to the sulphide voids, giving rise to the familiar limonite halos typical of transported limonite.   In slowly reacting gangues, the halos are fainter, farther out from the voids, and the rock tends to be more permeated throughout.

In massive gossans the iron generally migrates considerable distances, except in carbonate.   It may even migrate far beyond the ore body and give rise to a false gossan.   Large, coarse boxworks, such as those that characterize the outcrops at Ducktown, Tenn., are common. There is also much open-space deposition of limonite with botryoidal and reniform structure.

In general, indigenous limonite points to former presence of copper, or a few other sulphides, and transported limonite to a preponderance of pyrite or an absence of copper.   Field studies show that most sphalerite also gives rise to transported iron in inert gangue.

False Gossans.   Transported iron, precipitated by reacting rocks, may form an iron-stained area that resembles a true gossan, and the resemblance is closer if the limonitic area is reworked by weathering.   The false gossan is distinguished from the true by the lack of indigenous limonite, the lack of sulphide voids, and the nature of the transported limonite.   Such false gossans do not overlie ore deposits, but their presence may indicate former sulphides not far distant.

Copper, like iron, may yield false croppings beneath which there is no ore body.   Transported copper may move outward from the original deposit and be precipitated in the country rock as copper carbonates.   A disseminated copper ore body 140 feet wide observed by the writer in Kenya Colony proved to be a false cropping — a deposit without roots.   The copper had migrated laterally and presumably down structure from the eroded upward projections of narrow copper-bearing fissures.   The carbonate ore became exposed on a surface lowered by erosion.   It was recognized as transported copper because of (1) the lack of original sulphide voids; (2) no associated indigenous limonite; (3) no indigenous copper carbonate; (4) the occurrence of the copper carbonate in the form of paint, impregnations, carbonate soaked clay, veinlets, and as open-space fillings with banding and gel structure; and (5) the lack of hypogene rock alteration. An apparent large disseminated copper deposit actually turned out to be of little value. A similar occurrence of transported copper was observed in Lower California. Copper is rather inclined to form false croppings.

Some false croppings have no tie with any known ore bodies; they presumably have come from deposits above that have been completely eroded.

INFERENCES FROM GOSSANS AS TO HIDDEN DEPOSITS

Gossans and croppings are intriguing not only because of what may lie beneath them but also because they offer a fascinating study of rock and mineral interpretation, of which much field work yet remains to be done. They supply, however, many decipherable inferences as to size, character, and mineral content of the hidden ore deposits. To start with, a gossan indicates the site of former iron-bearing minerals, generally sulphides, which carries the possibility and hope that valuable minerals may also be present. Therefore, any clues that may be wrested from a gossan will aid in determining whether a commercial ore body lies hidden below and in evaluating and exploring it. Such clues may be won from the size of the gossan, from outcrop minerals, voids, rock alteration, and color, structure, and position of the limonite. Interpretation of cappings by aid of limonite has been advanced greatly by the painstaking work of Locke, Morse, Tunell, White, Boswell, Blanchard, and others. There is much need for further investigation, however, for many problems of croppings yet remain to be solved.

Form and Size. Gossans generally indicate the form and size of the underlying deposits. Tabular veins are faithfully reproduced. The width of the gossan, however, may be greatly exaggerated by limonite "mushrooming," as in Fig. 5·8–4. The writer recalls a large gossan in Oregon that under casual inspection appeared to cap a wide copper vein. Detailed examination, however, disclosed only a few narrow ribs of indigenous limonite enclosed in transported limonite, indicating a former deposit made up of a few sulphide veins of noncommercial width (5·8–4B).

In the case of cappings, the plan, outline, and size of the underlying disseminated sulphide body can be determined by careful mapping of the features diagnostic of copper.

Surface Subsidence and Crackling. Wisser, Locke, Kingsbury, and Walker have described collapse of rock overlying sulphide deposits occasioned by shrinkage of the ore attendant upon oxidation. In

Bisbee, Ariz., the collapsed area overlying shrunken oxidized ore is a breccia jumble cemented by supergene calcite or quartz. Above this, marginal cracks extend up to the surface, forming a diagnostic crackled area. (See Fig. 5·4–26.) Wisser states that such surface-crackled areas have indicated oxidized ore as much as 700 feet beneath the surface.

Relict Sulphides. Bits of sulphide that persist in the cropping give direct clues to underlying minerals. They survive because they are

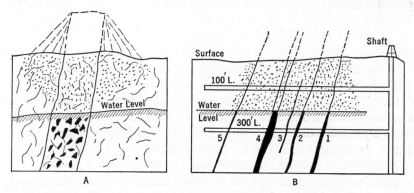

Fig. 5·8–4. Sketches illustrating exaggerated width or " mushrooming " of gossan over the original vein width by migration of limonite into the walls; *A*, broad gossan zone from a single vein; *B*, a broad gossan zone cut on the 100-level derived from several small noncommercial sulphide veins disclosed on the 300-level. Black is sulphide; dotted, limonite.

enclosed tightly in quartz, or have a protective oxidized coating, or are resistant to oxidation (e.g., galena).

Voids. Inferences of several kinds may be drawn from the voids of pre-existing minerals.

Lack of voids shows that no predecessor ore minerals existed.

Nonsulphide voids fall into three classes: (*a*) readily distinguishable vesicular openings in lavas; (*b*) rectangular voids of feldspar laths, which are most prominent in the surface skin but in Northern Rhodesia were observed in solid rock well below the surface; (*c*) those left by other minerals, such as carbonates, chlorite, amphibole, or pyroxene, which generally can be distinguished as such unless obscured by advanced tropical decay.

Abundance of voids of sulphide derivation indicates the abundance of the pre-existing sulphides. Their numbers permit rough estimates, based upon experience, of the sulphide content of the original ore.

Shape of sulphide voids may indicate characteristic predecessor minerals, such as molds of pyrite cubes or terraced cubic outlines of galena.

Other minerals that may leave diagnositc molds are marcasite, arsenopyrite (spear shapes), enargite (blades), molybdenite (plates), rhodochrosite (rhombs), and covellite (plates). Their identification may reward the painstaking observer.

Oxidized Cropping Minerals. In addition to limonite, the gossan minerals of Table 9 give clues as to former content of the deposit. Manganese stain, however, may indicate only the presence of manganese in the ore and not its mineral form. Many minerals, such as copper sulphides, may not be revealed in croppings because they have been completely removed. Oxidized copper compounds may also persist; if they are as abundant as the sulphide voids, no copper has been removed.

Colors of Limonite. The yellows, browns, maroons, and reds of cropping limonite are familiar colors not yet entirely understood. However, careful field correlation by many investigators between limonite and adjacent sulphide has provided satisfactory answers for some of them, and this has been furthered by the laboratory work of Morse, Tunell, Posnjak, and Merwin. Now certain colors are known to signify the presence of copper and certain sulphides.

Apparently the colors of limonite are contributed by the mineral composition, the influence of copper, impurities, porosity or compactness, and grain size. The pure iron oxides are stated by Posnjak and Merwin to have the following colors: hematite, deep red; goethite, orange-yellow; and jarosite, yellowish. In general, seal brown, maroon, and orange colors of cappings signify copper, and the yellows and brick reds indicate pyrite. Tunell states that, in cappings over disseminated ores, brick reds indicate much transported hematite; maroon, much indigenous hematite; deep brown, much indigenous goethite; yellowish-brown, much transported goethite; and yellow, much transported jarosite. Blanchard states that in his experience brick reds overlie pyrite areas low in copper and that dull reddish-brown to seal brown limonites overlie copper ore.

A striking example may be observed at Miami, Ariz., where capping of a brilliant reddish color, conspicuous to the eye, overlies what was then the noncommercial north ore body. A less conspicuous area to the south, characterized by sparse maroon and brownish limonite, overlies the high-grade disseminated ore body. At Ely, Nev., the good ore is overlain by capping spotted by maroon and brown hematite. In Northern Rhodesia the ore capping is likewise distinguished by specks of deep maroon and rich brown colors, whereas a few areas of pyritic material yield surface colors of bright reds and yellows.

Many individual minerals that contain iron, and some that contain no iron but have admixed pyrite, leave indigenous limonites of characteristic color. Brick reds (much hematite) indicate pyrite; deep browns and yellowish browns (much goethite) are characteristic of chalcopyrite; ocherous-orange to chocolate colors (much goethite and hematite) are characteristic of bornite; deep maroons (much hematite)

TABLE 10

TYPES OF LIMONITE BOXWORKS

Boxwork	Fig.	Character	Color	Derivation
Coarse cellular	5·8–5	Coarse, angular; thin, wide, rigid walls; blebs, masses	Ocherous	Chalcopyrite
Coarse cellular	5·8–6	Siliceous; thin, rigid, angular walls	Light brown	Sphalerite
Fine cellular	....	Thin, small, friable walls; specks, blebs	Yellow orange	Bornite-chalcopyrite
Fine cellular	....	Stronger cells; " shriveled " limonite jasper	Light brown	Sphalerite
Cellular sponge	5·8–12	Rounded, thick, rigid, empty cells; crisp; much silica	Brownish	Sphalerite
Triangular	5·8–13	Triangular cells, thick, fragile, crusted	Ocherous orange	Bornite
Triangular	5·8–7	Triangular, curved	Ocherous orange	Bornite
Contour	5·8–10	Long, narrow, angular rigid cells	Chocolate	Tetrahedrite
Relief	....	No boxwork; arrangement of sulphide grains; weak, porous, relief	Maroon	Chalcocite, covellite, bornite
Limonite pitch	....	Pitchlike; varnish; no cells	Dark brown	Chalcopyrite, bornite
Limonite crusts	....	Thin, fragile, flaky concentric foils	Dark brown to black	Chalcocite
Cleavage	5·8–9	Thin parallel cubic plates of limonite jasper	Ocherous orange	Galena
Diamond mesh	5·8–11	Diamond-shaped meshes	Ocherous orange	Galena
Pyramidal	....	Steplike arrangement	Ocherous orange	Galena
Foliated	5·8–8	Smooth, thin, rounded cells	Tan to maroon	Molybdenite

are characteristic of chalcocite. According to Blanchard and Boswell, ocherous orange is also indicative of galena; tan to brown, of sphalerite; and tan to maroon, of molybdenite.

Colors of limonite alone do not always signify specific minerals, but in conjunction with limonite structure they aid diagnosis.

Limonite Boxworks. The structure assumed by indigenous limonite is the most diagnostic feature of the predecessor minerals. Color and structure of the limonite combined are, in many cases, specifically diagnostic of certain minerals; some structural patterns that resemble each other closely may have distinguishing color differences.

When a grain of sulphide is oxidized and residual limonite remains in the cavity, the limonite assumes a honeycomb pattern, called *boxwork*, that persists in the cropping. Neither the goethite nor the hematite suffers alteration although the jarosite may do so. Thus, the original patterns are preserved. The boxwork patterns, like colors,

TABLE 11

OXIDATION RESIDUALS OF SOME COMMON SULPHIDES

Mineral	Voids	Limonite	Boxwork	Color	Composition
Pyrite	Empty	Transported; halos or flooding	None	Brick red	Hematite +, jarosite +
Pyrrhotite	Empty	Transported; halos or flooding	Coarse spongy masses	Brick red	Hematite +, jarosite +
Chalcopyrite	Occupied	Indigenous	Coarse cellular; fine cellular; limonite pitch	Ocherous	Goethite +, jarosite −
Bornite	Occupied	Indigenous	Triangular; sponge crusts; relief	Ocherous to orange	Goethite +, hematite
Chalcocite	Occupied	Indigenous	Relief; crusts	Maroon to seal brown	Goethite +
Tetrahedrite	Occupied	Indigenous	Contour; coagulated	Chocolate	Goethite
Sphalerite (inert gangue)	Empty	Transported	" Moss " limonite	Yellow to brown	Goethite +, silica +
Sphalerite (reacting gangue)	Occupied	Indigenous	Coarse cellular; fine cellular; cellular sponge	Yellow to brown	Goethite +, silica +
Galena	Occupied	Indigenous	Cleavage; crusts; diamond mesh; cellular sponge	Ocherous to orange, seal brown, dark chocolate	Goethite, hematite
Molybdenite	Occupied	Indigenous	Foliated; granular	Maroon	Goethite
Siderite	Occupied	Indigenous	Cleavage; mica plate	Yellow, deep brown	Goethite

+ means important constituent; − means minor constituent.

have been correlated in place with underlying sulphides, thus establishing their sulphide derivation. Several varieties have been described, chiefly by Locke, Morse, Blanchard, and Boswell, who recognize the 10 varieties of indigenous limonite shown in Table 10.

From Limonite Residues. Limonite colors, boxworks, and position, properly interpreted, can be resolved into many of the original sulphides. The diagnostic features of some common ones are listed in **Table 11.**

A   B   C

FIG. 5·8–5. Coarse cellular boxworks from chalcopyrite. *A*, Bagdad, Ariz., ×2; *a*, quartz veinlets; *b*, cell walls of limonitic jasper; *c*, *e*, interstitial limonite; *B*, quadrangular pattern, Duquesne, Ariz., ×5, with parallel cell walls; *b*, cross structure; *c*, *d*, thin angular web-work. *C*, average type, Creston Verde, Sinaloa, Mexico, ×5. (*Blanchard-Boswell, Econ. Geol.*)

A   B

FIG. 5·8–6. Coarse cellular boxwork from sphalerite, Spruce Mountain, Nev., ×2. *a*, Quartz veinlet; *b*, cell walls of limonitic jasper; *c*, interstitial limonite. (*Boswell-Blanchard, Econ. Geol.*)

FIG. 5·8–7. Triangular, curved boxwork from bornite with eye-shaped cells, ×5. *A*, Engel mine, Calif.; *B*, Black Mountains, N. Mex. (*Boswell-Blanchard, Econ. Geol.*)

A   B

FIG. 5·8–8. Foliated boxwork from molybdenite, containing limonite flakes and quartz veinlets. *A*, Nogales, Ariz.; *B*, Hodgkinson, Queensland, ×20. (*Blanchard-Boswell, Econ. Geol.*)

FIG. 5·8–9. Cleavage boxwork and sintered limonite crusts. "Cleavage boxes" follow galena cleavage. Lawn Hill, Queensland, ×1.5. (*Blanchard-Boswell, Econ. Geol.*)

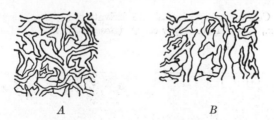

FIG. 5·8–10. Typical contour boxwork (like topographic contours) from tetrahedrite. *A*, Hachita, N. Mex., ×5; *B*, Patagonia, Ariz., ×5. (*Boswell-Blanchard, Econ. Geol.*)

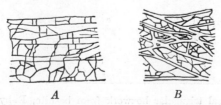

FIG. 5·8–11. Galena boxworks. *A*, Regular pattern. Lawn Hill, Queensland, ×3. *B*, Diamond mesh type from steely galena, Eureka, Nev., ×3. (*Boswell-Blanchard, Econ. Geol.*)

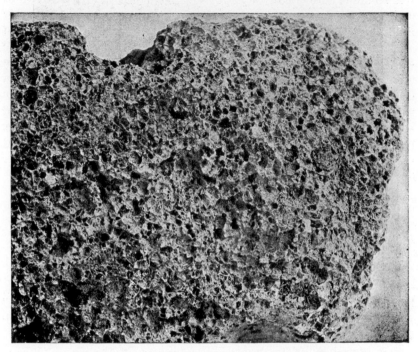

FIG. 5·8–12.   Cellular sponge of sphalerite derivation, Empire Zinc mine, Hanover, N. Mex., ×1½.   Note empty cells.   (*Boswell-Blanchard, Econ. Geol.*)

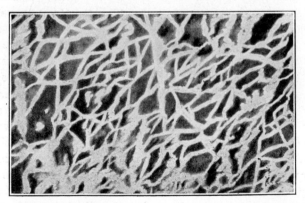

FIG. 5·8–13.   Typical triangular boxwork from bornite, Ruby mine, Calif., ×5. (*Blanchard-Boswell, Econ. Geol.*)

From Rock Alteration. Field observations disclose, particularly in disseminated deposits, that the solutions formed during oxidation of sulphides effect a pronounced kaolinization of hydrothermally altered country rock. The zone of supergene kaolinization coincides with the zones of oxidation and supergene sulphide enrichment. The degree of kaolinization indicates the intensity and extent of downward migration of sulphate solutions and of possible supergene sulphide enrichment. Thus, if, in copper cappings, intense kaolinization is associated with indigenous limonite, considerable supergene sulphide enrichment may be expected beneath such a kaolinized oxidized zone. It should be pointed out, however, that the oxidation of disseminated barren pyrite also yields solutions that produce kaolinization, but in such cases the limonite would be transported and not indigenous. Thus, kaolinization, plus indigenous limonite, superimposed upon intense sericitic alteration indicates hypogene sulphide mineralization that has undergone supergene sulphide copper enrichment.

Factors Controlling and Limiting Oxidation

It is obvious that oxidation cannot proceed forever nor to the lowest limits of metallization. Certain factors cause its cessation; others, such as the water table, rate of erosion, climate, time, rock structure, and burial, affect or regulate oxidation.

Part Played by Water Table. The position of the water table is a control of oxidation. Above it oxidation can take place freely; below it no free oxygen is ordinarily available for oxidation. Hence it normally constitutes the downward limit of the zone of oxidation. Since the water table ordinarily approximately parallels the surface of the ground, it follows, in such cases, that the bottom of the zone of oxidation also approximately parallels the ground surface. As the water table is generally a smooth, gently curved surface, the contact between the oxidized and sulphide zones is also generally smooth, and in many cases abrupt. It is common observation that at the water table oxidized minerals give way to sulphides. This is the normal condition. However, in many cases, there is a slight interpenetration of the oxidized and sulphide zones due to the oscillation of the water table, i.e., between the low stand of dry periods and the high stand of wet periods.

There are, however, deviations from the normal in which the contact between oxide and sulphide is not smooth, or where oxidation does not reach to the water table or extends far beneath it.

Oxidation Below Water Table. Since free oxygen is usually lacking in the ground water and the oxidized zone generally terminates at the water table, it is almost universally accepted that oxidation cannot take place to any appreciable depth below the permanent water table. It is generally assumed, where ores are completely oxidized beneath the water table, that the water table has risen.

Notwithstanding such evidence, field observations compel the conclusion that under special circumstances oxidation is possible and has taken place at scores of feet beneath the water table. Examples have been studied of oxide-sulphide contacts that are not flattish or gently curved surfaces but jagged like a vertical profile of a youthful mountain range (Fig. 5·8–14). No water table could have such sharp

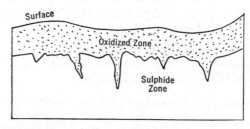

FIG. 5·8–14. Contact between oxidized and sulphide zones of an ore deposit. Prongs of oxidation extend down into the sulphides but inter-prong areas correspond to a water level, indicating oxidation beneath the water table. Sketch of conditions at Cananea, Mexico.

indentations, and that type of bottom of oxidation, therefore, could not be limited by a water table. Either the oxidation took place above the water table or below it. Since, in many cases, the deep pendants of oxidation are separated by flattish areas that conform to a water table, the pendants themselves must have been oxidized beneath the water table. Further, the pendants coincide with zones of high permeability, such as shear or fracture zones or faults.

One need not search far for an explanation. In flattish areas the water table is also horizontal, the water circulates slowly, and free oxygen is present only in the upper layer. In mountainous regions, however, where most ore deposits occur, relief imposes different conditions. Water that falls on interstream areas and joins the mass of the ground water may move rapidly down fractures to outlets at lower levels; a hydrostatic head exists. Oxygen thus carried deep beneath the water table would permit oxidation along the channelways. This may be an adequate explanation for such oxidation beneath the water table.

An occurrence at Morenci, Ariz., is illustrative (Fig. 5·8–15). A long pendant of oxidized material coincides with the hanging wall of a

large fault.   The oxidation away from the fault is clearly related to a
water table, *W. T.*   The water that travels down the fault presumably
has an outlet in the nearby Gila Valley that permits rapid downward
circulation.   This, in turn, has enabled available oxygen to be carried
to great depths, where it fostered deep sub-water-table oxidation.

There is no reason why some oxidation should not take place below
the water table.   It is a question of the relative rate at which oxygen
is supplied and is used up.

Changes of Water Table.   The position of the water table is not
permanent, and changes affect oxidation.   The normal change is
downward, owing chiefly to progressive valley deepening and erosion

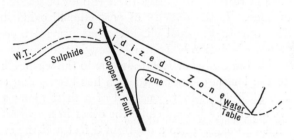

FIG. 5·8–15.   Sketch of oxidized sulphide zones at Copper Mountain, Morenci, Ariz.,
  showing deep penetration of oxidation along the hanging wall of an impervious
  fault, due to rapid downward circulation of ground water.

of the lands.   In its downward movement the oxide zone encroaches
upon the sulphide zone, which likewise progresses downward.   When
the water table comes to a standstill, so does the zone of oxidation; if it
remains stationary, oxidation eventually ceases.   A change of climate
from humid to arid also causes a lowering of the water table.   If the
water table is depressed rapidly, oxidation may not be able to keep
pace with it and the bottom of oxidation may become stranded above
the water table.

The water table may rise in response to opposite processes, i.e.,
deposition of sediments or a change of climate from arid to humid.
Valley filling, in places up to several thousand feet, is not uncommon.
A rising water level gradually submerges any oxidized zone that bottomed on it; further oxidation would be inhibited, and drowned oxidized zones would result.   Such submerged oxidized zones are common
to western United States.

Faulting may also bring about a relative rise of the water table; a
block from the oxidized zone may be dropped downward into the
ground water.

It is clear, therefore, that a study of oxidation must coincide with a study of the physiographic history of a region, which must be deciphered to unravel the conditions underlying oxidation and particularly to interpret oxidation during former geologic periods.

Rate of Erosion. The rate of erosion and the ensuing lowering of the surface and of the water table may be so rapid that oxidation cannot keep pace with it. Any former oxidized zone would then eventually be removed by erosion, and little or no oxidation would be evident on the surface. A very low rate of erosion would cause an almost stagnant position of the water level and little oxidation.

Climate. The important factors of climate that affect oxidation are temperatures and rainfall, but often these cannot be disentangled from the effect of time. Is the paucity of post-Glacial oxidation in glaciated countries due to the cool climates that prevail there or to the short time since glaciation? Is the thorough and deep oxidation of southwestern United States or Rhodesia due to the warm climate of the regions or to the length of time oxidation has been going on? In the Leonard mine at Butte, Mont., one end of a drift along a deep vein terminated in a very hot " dead end," the other in a cool airway; after 2 years the hot end oxidized to a depth of 3 feet and the cool end to a depth of only 2 inches. High temperature clearly accelerates oxidation; low temperature retards it. This agrees with laboratory results. Oxidation, therefore, is favored by warm much more than by cool climates.

The effect of seasonal vs. evenly distributed rainfall is little known, for this also is involved with temperature and time. Oxidation under the seasonal rainfall of warm central Africa is no greater or less than under the distributed rainfall of Montana. It is probable, however, that oxidation proceeds fastest in warm humid climates with evenly distributed rainfall. Obviously it is inhibited in regions of perennial ground frost.

Time. Studies indicate that thorough and deep oxidation is relatively slow and requires considerable time, geologically speaking. It is of the order of the length, say, of the Miocene or Pliocene periods rather than that of post-Glacial time. Many of the oxidized zones of the copper deposits of the southwestern United States have been formed during the Tertiary.

Rocks. Both physical and chemical properties of enclosing rocks affect oxidation. It proceeds fastest in porous rocks or brittle rocks that fracture or crackle easily. Consequently permeable sandstones, sericitized and silicified igneous rocks, and brittle schists favor oxidation. On the other hand, massive, dense, unfractured rocks, such as quartzite, impervious clay and shale, tough pliable greenstones, and kaolinized basic igneous rocks, do not favor oxidation.

The carbonate rocks develop porosity under oxidation and permit thorough penetration of surface waters with consequent deep and pervasive oxidation. Less conspicuously, sericitized porphyry lends itself to thorough oxidation.

Structure. Structural features exert control over the distribution and pervasiveness of oxidation. Faults influence oxidation in three ways: (1) they concentrate oxidizing waters within the fault zone and permit deep oxidation; (2) impervious faults impound and concentrate oxidizing waters on their hanging wall sides, causing concentrated oxidation (Fig. 5·8–15); and (3) impervious faults act as dams and protect underlying metallized rocks from down-seeping oxidizing waters.

Shear zones serve as permeable channelways and permit thorough and deep oxidation. Schistosity directs the flow of oxidizing waters, causing thorough oxidation along the schistosity. In places, underlying schistose formations may carry oxidation deep beneath overlying unoxidized materials. Similarly, inclined bedding planes, or interbedded permeable beds, may conduct oxidation down the dip beneath overlying unoxidized ores, as at the Mufulira and N'Changa mines, Northern Rhodesia, where vertical drill holes passed through sulphide ores into underlying oxidized ores at depths exceeding 1,000 feet.

Cessation of Oxidation

Since oxidation requires oxygen, it ceases when the supply of available oxygen ceases. This is brought about by the following:

Water Level. Since the ground water lacks free oxygen, oxidation ordinarily ceases at the water table.

Rise of Water Table. A raised water table submerges the oxidized zone, and oxidation ceases.

Refrigeration. Ground freezing, as in Arctic regions, stops all oxidation. At Kennecott, Alaska, pre-Glacial oxidation to depths of more than 2,000 feet ceased with the Glacial period; ground and ground water became frozen down to the 700-foot level, and further oxidation was prevented.

Oxygen Depletion by Abundant Sulphides. Abundant sulphides above a deep water table may exhaust the oxygen of down-seeping waters, and oxidation ceases above the water table. At the Utah Copper mine, Bingham, Utah, deep valley cutting rapidly lowered the water table and the sulphide zone became stranded high above it. The air in some abandoned workings within this sulphide zone contains no oxygen; it was used up in passing down through the sulphides.

Burial. Deep burial by sediments or volcanics stops oxidation, e.g., the deeply buried Precambrian oxidized and supergene sulphide zones of the United Verde Extension mine, Jerome, Ariz.

SUBMERGED AND STRANDED OXIDIZED ZONES

It was indicated under "Changes of Water Table" that oxidized zones may become out of adjustment with the water table or the topog-

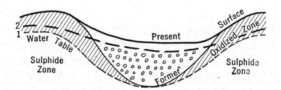

FIG. 5·8–16.   Sketch illustrating rising of water table, and submergence of a former oxidized zone, at Miami-Globe, Ariz.   1 represents former water table corresponding to former land surface (light line); with valley filling of Gila conglomerate the water table rose to 2 and part of the oxidized zone became submerged.

raphy.   Many are related to former water levels or topographic surfaces, and a physiographic study must be made to reveal the relationship.

Submerged Oxidized Zones.   Submerged oxidized ores may lie at depths of 2,000 feet or more beneath the water table, and one of four explanations may account for this phenomenon.

1. Oxidation below the water table; example, old Morenci, Ariz.

2. Water level raised by sedimentation (Fig. 5·8–16); example, Globe-Miami, Ariz., where desert valley filling by 2,000 feet of Gila conglomerate (Recent) deeply submerged the oxidized ores of the adjacent Miami Copper Co. and Old Dominion mines.   Volcanic extrusions will, of course, produce the same result.

3. Faulting has dropped the oxidized zone; example, East Butte mine, Butte, Mont., where an oxidized zone was down-faulted by the Continental Fault and the Continental Divide was relatively upraised. Impounded drainage and lake-bed deposition deeply buried the oxidized ores, and the water table was raised.

4. Change of climate; example, Northern Rhodesia copper belt, where a change from arid to humid climate raised the water level and submerged oxidized ores to depths of 2,000 feet.   Several lines of evidence converge to indicate that this region was formerly arid and that the Kalahari Desert once extended over the area.

Stranded Oxidized Zones. Stranded oxidized zones are temporary features because most of them have subsequently been lowered to the water table and eventually all will be. One example — Bingham, Utah — has already been given, and similar conditions exist at Bisbee, Ariz., at Rio Tinto, Spain, and in other localities. At Morenci, Ariz., rapid down-cutting by the Gila River has left sulphide ores stranded above the water table. In Kenya Colony, East Africa, the sinking of the surface of the Lake Victoria Nyanza has lowered the surrounding water level and left sulphides stranded above the water table.

DEPTH OF OXIDATION

Several instances have been observed of oxidation to depths greater than 2,000 feet. At Kennecott, Alaska, the depth of oxidation is over 2,000 feet; at Tintic, Utah, according to Lindgren, it is 2,200 feet; at Bisbee, Ariz., it is 1,400 feet; in the Northern Rhodesia copper mines it is known to a depth of more than 2,000 feet; and at the Lonely mine, Southern Rhodesia, it reaches a depth of 3,000 feet. In general it ranges in depth from a few tens to a few hundreds of feet.

Where there is no evidence of a post-oxidation rise of water level, broad generalizations can be made regarding the depth of oxidation: (1) In humid regions of low relief it is generally shallow, but with high relief it will probably be shallow to medium. (2) In glaciated regions, post-glacial oxidation is negligible or lacking; in areas sheltered from deep glacial erosion, pre-glacial oxidation may be present and is likely to be shallow in flattish areas and deep in areas of high relief. (3) In arid regions it is apt to be very deep and is generally rather deep under old or mature topography and shallow under youthful topography.

Ore Deposition in the Zone of Oxidation

Oxidized ore deposits may result within the zone of oxidation if the down-trickling oxidizing solutions encounter precipitants above the water table.

There is abundant evidence that such solutions are present in the zone of oxidation. In copper mines, trickles and pools of copper sulphate eat into the steel rails and the nails of miners' boots. A story is recorded of the early days of Butte, Mont., that a " tin " can fell into discharged mine water draining through a miner's yard, and became converted to copper, whereupon the miner unobtrusively obtained a lease upon the water and became wealthy producing " cement "[1] copper. Since then, copper has been commercially precipitated by iron from the copper sulphate mine water. At Rio Tinto,

---

[1] Copper precipitated from copper sulphate solution on scrap iron.

Spain, Ruth, Nev., Chino, N. Mex., Inspiration, Ariz., and Bingham, Utah, commercial heap leaching yields copper sulphate solutions from which the copper is precipitated by iron. In addition, efflorescences of sulphates of copper, iron, zinc, magnesium, manganese, cobalt, and other metals are commonly found on the walls of underground workings, indicating seeping sulphate solutions.

The economic minerals redeposited within the zone of oxidation are chiefly native metals, carbonates, silicates, and oxides. The metals of greatest commercial importance are copper, zinc, lead, silver, vanadium, and uranium, of which large and valuable deposits are known. Less valuable deposits of cobalt, manganese, iron, and other substances are similarly formed. For the most part, the ores are deposited in the lower part of the zone of oxidation.

METHODS OF PRECIPITATION

Various processes, acting singly or together, operate to bring about precipitation of the metals in the zone of oxidation.

Evaporation and Saturation. By simple saturation attendant upon evaporation, particularly in arid climates, metallic compounds are deposited as veinlets, blebs, and " paint." Such compounds are chiefly sulphates. This process has operated at Chuquicamata, Chile, the largest copper deposit in the world, where all the production to date has come from the oxide zone. The chief ore minerals are various copper sulphates. Antlerite is the dominant ore mineral, but brochantite, chalcanthite, and krohnkite are common. Other sulphates of iron, copper, magnesium, sodium, and potassium also occur.

Oxidation and Hydration. Oxidation and hydration yield oxides and hydrous oxides of iron, manganese, copper, and other materials. The formation of ferric hydroxide and finally goethite is an example already mentioned.

Reactions between Solutions. The sulphate solutions formed in the zone of oxidation may encounter various other solutions that bring about precipitation within the oxidized zone.

Chloride Solutions. Sodium chloride, a widespread constituent of the soil and ground water of arid regions, reacts with sulphates and other compounds to form chlorides of silver, copper, lead, and other metals. Silver chloride, or cerargyrite, the ore of most commercial importance, is supposedly formed as follows:

$$Ag_2SO_4 + 2NaCl = 2AgCl + Na_2SO_4$$

Similarly embolite, bromyrite, and iodyrite, the bromides and iodide of silver, are formed in less abundance. Most of the silver early mined in the western United States was won from silver chloride, and numerous other examples of such ores occur in Mexico and South America.

Carbonated Solutions. Bicarbonate solutions react with sulphates and sulphate solutions to form carbonates of the metals. Thus, reaction with copper sulphate yields the common basic carbonates of copper, azurite, and malachite; with zinc sulphate, the zinc carbonate, smithsonite; with lead sulphate, the lead carbonate, cerussite, is formed, an ore mineral of great commercial importance. Also native copper and cuprite are attacked to yield copper carbonate.

Other Solutions. Cuprite (Cu_2O) and tenorite (CuO), common ore minerals of the oxidized zone, are possibly formed by reactions between copper sulphate and ferrous and ferric sulphate. Native copper is also formed, among other ways, by reaction between cupric and ferric sulphate solutions. Likewise, native silver is formed by reaction between silver sulphate and ferrous sulphate.

Reaction with Gangue or Wall Rocks. Solutions of copper or zinc sulphate react with carbonate minerals to form malachite and azurite, the carbonates of copper, or smithsonite, the carbonate of zinc. These solutions may also react with available silica to form chrysocolla, the copper silicate, or hemimorphite, the common zinc silicate. Hydrozincite and monheimite may also be formed. Iron carbonate is commonly precipitated along with zinc carbonate and generally becomes oxidized, adding iron impurity and limonite color to the zinc ore.

The formation of lead sulphate, which is practically insoluble, results, not in a solution, but in the deposition of indigenous anglesite, and the carbonate, cerussite, is formed from the sulphate. No silicate of lead is known, but oxides of lead form in the zone of oxidation.

Silver sulphate solution does not give rise to natural carbonates or silicates of silver by reactions with wall rocks. Either the silver solutions formed in the zone of oxidation form silver salts as shown on page 270 or pass through the oxidized zone into the supergene sulphide zone.

The availability of carbonate minerals, such as common calcite or limestone, to form the copper or zinc carbonates offers no problem; but the availability of silica to form copper or zinc silicates is less simple. Quartz is inert to solutions of the oxidized zone, and the answer is probably colloidal silica set free during weathering. For example, at Globe, Ariz., the weathered dacite was found to have lost about one-fifth of its silica content, and beneath it in the Old Dominion mine was one of the largest known bodies of chrysocolla ore. Similarly, at the Miami and Live Oak mines at Miami, Ariz., weathered porphyry yielded silica to form depositions of chrysocolla, some of it in alternating layers with opaline silica. Many valuable deposits of zinc carbonate, some remote from the parent sulphide, have been so formed at Tintic, Utah, Leadville, Colo., and in many Mexican deposits.

Copper carbonates formed as described above have given rise to large and rich ore bodies at Bisbee, Morenci, and Globe in Arizona and in Northern Rhodesia and the Belgian Congo.

DISTINCTIVE OXIDIZED ORE-FORMING MINERALS

Many of the ore minerals formed in the zone of oxidation are also formed by hypogene processes. It is necessary to distinguish between them, since broad, practical deductions may be drawn if it is determined that the valuable minerals are oxidation products.

Ore-Forming Minerals Diagnostic of Oxidation. (See also Table 12.) The carbonates, silicates, and sulphates of copper and zinc originate only in the zone of oxidation. This is also true of the oxides of copper, cobalt, antimony, molybdenum, and bismuth, of the chlorides, iodides, and bromides of silver, and of anglesite, cerussite, psilomelane, garnierite, and many nonindustrial minerals.

Ore-Forming Minerals of Dual Formation. Other ore-forming minerals that are formed both by oxidation and by primary processes include native gold, silver, and copper. Consequently, ore deposits composed of these minerals may have resulted either through oxidation or hypogene processes and carry the generalizations that pertain to either. Likewise, hematite, limonite, pyrolusite, braunite, arsenic and antimony sulphides, and other less common compounds may have resulted from either oxidation or primary processes. Consequently, they are not necessarily diagnostic of oxidized ores.

Criteria of Distinction. Oxidized deposits whose value lies in native silver, gold, or copper, for example, can generally be distinguished from hypogene or secondary enrichment deposits of the same metals by the mineral associations. Should the accompanying minerals consist of sulphides or other minerals that are definitely of hypogene occurrence, the presumption is justified that the deposit is not oxidized ore. On the other hand, if the associated minerals are diagnostic oxidation minerals, and native silver occurs only with cerargyrite, or native copper occurs only with cuprite or malachite or indigenous boxwork limonite, it is probable that the native metals have also originated by oxidation. Manganese minerals offer greater difficulty.

GENERALIZATIONS REGARDING OXIDIZED ORES

When it is established that given ores are oxidized, it follows that: (1) such ores will change in nature in depth; (2) there is likely to be a

pronounced change in tenor in depth; (3) in most cases, only shallow
depth may be expected; (4) different metallurgical treatment will be
required for underlying ores; (5) extraction plants should not be erected
until the volume of the oxidized ore is delimited; and (6) more adequate

EXAMPLES OF IMPORTANT DEPOSITS OF OXIDIZED ORES

Metal	Deposit or District	Location	Type	Chief Minerals
Gold	Mount Morgan	Australia	Stockwork	Au
Silver	Potosi	Bolivia	Veins	Ag, cerargyrite
	Mapimi	Mexico	Veins	Ag, cerussite
	Horn Silver mine	San Francisco, Utah	Vein	Cerargyrite
Copper	Copper Queen	Bisbee, Ariz.	Replacement	Carbonates
	Old Dominion	Globe, Ariz.	Vein	Carb. chrysoc.
	New Cornelia	Ajo, Ariz.	Porphyry	Carbonates
	Tintic	Tintic, Utah	Veins	Carbonates
	Chile Copper Co.	Chuquicamata, Chile	Porphyry	Antlerite
				Brochantite
				Chalcanthite
	Union Minière	Katanga, Congo	Surficial	Carb. chrysoc.
Lead	Santa Eulalia	Mexico	Replacement	Cerussite
	Sierra, Mojada	Mexico	Replacement	Cerussite
				Anglesite
	Leadville	Colorado	Replacement	Cerussite
	Silver King	Park City, Utah	Veins	Cerussite
Zinc	Mammoth	Tintic, Utah	Replacement	Smithsonite
	Leadville	Colorado	Replacement	Smithsonite
	Goodsprings	Nevada	Replacement	Hydrozincite
	Austinville	Virginia	Replacement	Smithsonite
Manganese	Philipsburg	Montana	Veins	Oxides
	Batesville	Arkansas	Veins	Oxides
Iron	Bilbao	Spain	Replacement	Limonite
Nickel	New Caledonia		Residual	Garnierite
Cobalt	Union Minière	Katanga	Replacement	Oxides
Molybdenum	Mammoth	Arizona	Vein	Wulfenite
Vanadium	Broken Hill	N. Rhodesia	Replacement	Descloizite
	Minasragra	Peru	Residual	Complex
Uranium-	Union Minière	Katanga	Veins	Pitchblende
radium	Colorado Plateau	Colorado	Beds	Carnotite, etc.
	Great Bear Lake	Canada	Veins	Pitchblende

transportation is generally necessary for oxidized than for unoxidized
ores, because most are shipped directly to smelters.

It should be determined if the oxidized ores are indigenous or transported. If indigenous, sulphide ores may be expected to underlie
them; if transported, they may be remote from their source and so
may have no sulphide roots — their source is more likely to be upward
and lateral, as has been found in many zinc and copper deposits. If
transported, the size of the ore body may have no relation to the size
of the source body.

Supergene Sulphide Enrichment

The changes in ore deposits produced by oxidation are so clear and prominent that the major features have long been recognized, although cropping interpretation is relatively recent. Secondary sulphide enrichment, however, although mentioned by DeLaunay, was not clearly understood until the appearance of the classical papers by S. F. Emmons, Van Hise, and Weed in 1901; it was later clarified by Ransome in 1910. The next 15 years saw the greatest advance, led by L. C. Graton, W. H. Emmons, and others, in the knowledge of supergene sulphide deposition — a study greatly aided by the development of the ore microscope and the use of reflected polarized light.

The metals in solution that escape capture in the oxidized zone trickle down to where there is no available oxygen, generally the water table, and there undergo deposition as secondary sulphides. The metals removed from above are thus added to those existing below, thereby enriching the upper part of the sulphide zone. This forms the *zone of secondary enrichment* or, as it is now more generally referred to, the *supergene sulphide zone*. It in turn is underlain by the *primary* or *hypogene zone* (Fig. 5·8–1). Progressive erosion permits deeper oxidation, and after a time the supergene sulphides themselves become oxidized, and their metal content is then transferred to the downward-progressing enrichment zone. The primary ore may thus be enriched to as much as ten times its original metal content. Rich ores are made richer, lean ores are made valuable, and noncommercial primary material or *protore* is built up to commercial grade. Many great copper camps, such as Utah Copper, Ray, Miami, Inspiration, and Santa Rita (Chino), owe their start to supergene enrichment of protore. The process is therefore not only of great scientific interest but also of far-reaching economic importance to the mineral industry.

Favorable conditions must exist for supergene sulphide enrichment to take place, but they are sufficiently common that sulphide-enriched deposits are widely scattered over the nonglaciated areas of the earth. The process is of greatest importance with copper and silver deposits, and the following pages deal mainly with them.

Requirements for Supergene Enrichment

Oxidation. Oxidation of ore deposits may occur without attendant sulphide enrichment, but enrichment cannot take place without accompanying oxidation. Therefore, oxidation is the primary requisite, and the factors that have been shown to favor oxidation must be present.

Suitable Hypogene Minerals. The ore deposit must contain primary minerals which upon oxidation yield the necessary solvents. Iron sulphides are essential, and deposits lacking them rarely contain supergene sulphide zones. Chalcopyrite alone does not yield sufficient solvents, and deposits like Ajo, Ariz., where this mineral predominates in the primary ore, have negligible supergene enrichment. The primary ore also must contain metals that can undergo supergene enrichment. Lead or zinc, for example, do not ordinarily yield supergene sulphides; copper and silver do, tin apparently does not, and little is known of the supergene enrichment of nickel and the less common metals; much research still remains to be done in this field.

Permeability. Permeability of the ore deposit is essential to enable trickling solutions to penetrate beneath the zone of oxidation. If the walls are quite pervious and the deposit only slightly so, the solutions may enter the walls and in the absence of suitable precipitants become dissipated. Impervious inclined faults may bar the entry of the downward-moving solutions into the underlying deposit.

Absence of Precipitants in Oxidized Zone. The oxidized zone must be free from precipitants, such as carbonate rocks, that fix the metal content of the enriching solutions in the form of oxidized compounds.

Zone of No Available Oxygen. Supergene sulphides can be deposited only where oxygen is excluded. Generally this is below the water level. Under certain circumstances, as previously pointed out, oxygen may also be lacking within the oxidized zone, and sulphide deposition may then take place there.

Precipitants. It is widely and mistakenly thought that supergene copper sulphides are deposited beneath the water table upon quartz, gangue, or any other minerals. This is erroneous. They are deposited from sulphate solutions only at the expense of other sulphides, arsenides, or similar minerals, and by replacing them. Hydrogen sulphide also effects precipitation. Thus, underlying ore minerals, generally hypogene ores, are essential for deposition of supergene sulphides. If the down-trickling sulphate solutions do not encounter sulphides in zones lacking available oxygen, the sulphates become dissipated, and no supergene sulphide enrichment occurs.

CAUSE AND METHOD OF DEPOSITION

Schürmann determined that certain metals in solution are deposited as sulphides in the presence of other sulphides and in order as follows: *mercury, silver, copper, bismuth, lead, zinc, nickel, cobalt, iron,* and *manganese.* He showed that any one of these metals in solution is precipitated as a sulphide by any one lower in the series. He attributed it to their relative order of affinity for sulphur. Later, Wells showed that Schürmann's series corresponded essentially with Weigel's order of solubilities of the sulphides, and that deposition is determined by their relative solubilities.

The list above is arranged with the least soluble first and the most soluble last. This simple law clarifies the cause and order of the deposition of supergene sulphides and why one supergene sulphide replaces another. For example, if silver sulphate solution encounters in the supergene sulphide zone any sulphide lower in the series than itself, silver sulphide (argentite) will be deposited at its expense. Similarly, copper sulphide from copper sulphate solution will replace galena, zincblende, and pyrite. The farther apart in the series they are, the more complete will be the replacement. Manganese sulphide, being the most soluble, could not be precipitated by any sulphides of Schürmann's series; this may be the explanation of its rarity in nature. Lead sulphate should yield lead sulphide, but it is insoluble. The soluble zinc sulphate, however, should yield zinc sulphide in contact with nickel, cobalt, or iron sulphides. Its general absence as a supergene sulphide is, therefore, striking. It is probable that the relative concentrations of ferrous and ferric ions in the solutions also play a part in the sequence of deposition.

In copper deposits, both chalcocite and covellite are found to replace bornite, chalcopyrite, enargite, galena, sphalerite, pyrrhotite, pyrite, tetrahedrite, and tennantite, and other more complex minerals. Chalcocite is the most common supergene sulphide, but covellite is locally dominant, as, for example, at Bingham, Utah. The cupric sulphide (covellite) tends to be precipitated where the ratio of ferric to ferrous iron is relatively high and the cuprous sulphide (chalcocite) where the reverse ratio exists. Chalcocite is also replaced by supergene covellite, and hypogene covellite is replaced by supergene chalcocite.

Examination, by the writer, of hundreds of mines and thousands of polished surfaces under the microscope, has never disclosed deposition of any supergene sulphides except by replacement of other metallic minerals, mostly sulphides. One exception was observed of a deposit of supergene chalcocite upon quartz in Morenci, Ariz., apparently brought about by escaping hydrogen sulphide. Instances have been recorded of the replacement of wood cells by supergene chalcocite in the Red Bed type of copper deposits. The chalcocite has clearly inherited the wood cell structure, but some specimens examined by the writer under the microscope show clear evidence that the woody material was first replaced by hypogene bornite and later by chalcocite, which inherited the cell structure from the bornite.

Supergene sulphide replacement is invariably volume for volume and is not, therefore, a molecular interchange between replacing and replaced substances. If the replacement of pyrite by chalcocite, for example, were a molecular interchange, a grain of denser chalcocite should occupy only a portion of the volume of a grain of pyrite. This is not the case in nature. Hence, customary balanced chemical equations depicting such changes do not correctly represent what has taken place. As used here, they are given to indicate the trend of the change rather than the exact chemical interchange in all instances.

Chemistry of Supergene Sulphide Deposition. Since the chemistry of copper enrichment has been more thoroughly investigated than that of other metals, it will be used mainly to exemplify the changes that take place. Under proper conditions, copper sulphate will react with other ore minerals to form copper sulphides, as is indicated in the following equations, some of which have been established in the laboratory.[1]

[1] $PbS + CuSO_4 = CuS + PbSO_4$ (*Covellite*)

[2] $ZnS + CuSO_4 = CuS + ZnSO_4$ (*Covellite*)

[3] $5FeS_2 + 14CuSO_4 + 12H_2O = 7Cu_2S + 5FeSO_4 + 12H_2SO_4$[2]   (*Chalcocite*)

[4] $4FeS_2 + 7CuSO_4 + 4H_2O = 7CuS + 4FeSO_4 + 4H_2SO_4$ (*Covellite*)

[5] $CuFeS_2 + CuSO_4 = 2CuS + FeSO_4$ (*Covellite*)

[6] $5CuFeS_2 + 11CuSO_4 + 8H_2O = 8Cu_2S + 5FeSO_4 + 8H_2SO_4$ (*Chalcocite*)

[7] $Cu_5FeS_4 + CuSO_4 = 2Cu_2S + 2CuS + FeSO_4$ (*Chalcocite, covellite*)

[8] $5Cu_5FeS_4 + 11CuSO_4 + 8H_2O = 18Cu_2S + 5FeSO_4 + 8H_2SO_4$ (*Chalcocite*)

[9] $5CuS + 3CuSO_4 + 4H_2O = 4Cu_2S + 4H_2SO_4$ (*Chalcocite*)

Copper sulphate is also known to react with marcasite, pyrrhotite, enargite, tennantite, tetrahedrite, and other minerals. Supergene chalcopyrite and bornite are also formed as transition products in the alteration of pyrite or chalcopyrite to chalcocite or covellite. As far as the author's experience goes, they are formed only in microscopic amounts, and, except for one instance, he has never observed either bornite or chalcopyrite as supergene minerals in amounts large enough to be seen by the naked eye. Certain reactions also yield native copper, and this mineral is common in the deep chalcocite ores of the Chino mine in New Mexico. In general, among the copper minerals, those

---

[1] See preceding paragraph regarding chemical equations.
[2] Stokes' reaction.

highest in copper content are more readily replaced by supergene sulphides than those of low copper content.

Silver sulphate reacts with some sulphides to form silver sulphide as, for example, $ZnS + Ag_2SO_4 = Ag_2S + ZnSO_4$. Silver sulphide is also precipitated directly, presumably by H_2S released by attack of H_2SO_4 on pyrrhotite, sphalerite, or galena. Native silver is also commonly deposited in association with the sulphide, as in $Cu_2S + 2Ag_2SO_4 = Ag_2S + Ag + 2CuSO_4$. Rich silver sulphantimonides and sulpharsenides also are deposited by supergene replacement of many of the common hypogene ore minerals. It was formerly thought that these compound silver minerals were all supergene, but Bastin showed that they occur as both hypogene and supergene minerals.

Although sphalerite is not a normal supergene sulphide, Brown has advocated a supergene origin for sphalerite, ilvaite, and other minerals at Balmat, N. Y.

Gold may be dissolved by ferric sulphate in the presence of H_2SO_4, NaCl, and MnO_2. If the ferric sulphate is reduced to ferrous sulphate, the gold can no longer be maintained in solution and is deposited. Consequently, gold could be deposited in the supergene sulphide zone, thus bringing about supergene gold enrichment; none is definitely known to be so redeposited, however.

Little is known of the chemistry of supergene sulphides, antimonides, or arsenides of the other metals. Both nickel and cobalt form soluble sulphates from which sulphides should be precipitated, and the sulphates of both are not uncommon in mine waters. Unfortunately, most of the nickel sulphide deposits of the world lie in glaciated countries that lack supergene sulphide zones. Lindgren and Davy suggest that millerite and violarite have been deposited as supergene sulphides at Key West, Nev., and Michener and Yates suggest that violarite has replaced pentlandite during oxidation at Sudbury, Ontario. Manganese sulphide is so low in Schürmann's scale that alabandite would hardly be expected as a supergene sulphide — even the hypogene sulphide is rare. Mercury minerals resist solution by sulphuric acid or ferric sulphate, although one occurrence of supergene cinnabar is reported in Germany. Platinum minerals are generally resistant to oxidation, but Wagner and Schneiderhöhn report supergene platinum in the footwall part of the platinum reef at Pilansberg, Transvaal. The enrichment of tin ores has been the subject of controversy, particularly in connection with the Bolivian tin deposits. Koeberlin has vigorously supported a supergene origin, and Singewald, after careful review of the evidence, concluded that it was improbable — a finding with which most geologists are in agreement. The uncommon cadmium sulphide, greenockite, is a supergene mineral. No established supergene sulphides of lead, bismuth, molybdenum, tungsten, uranium, or vanadium have been recorded.

DEGREE OF ENRICHMENT

The degree of enrichment is the extent to which the hypogene minerals have been replaced by the supergene sulphides. This can some-

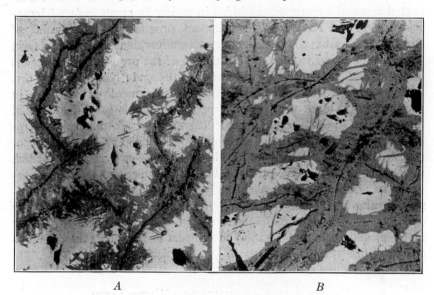

A   B

FIG. 5·8–17. Photomicrographs showing two stages of sulphide enrichment. *A*, Incipient enrichment of chalcopyrite (white) by feathery crystals of covellite that are growing outward from minute cracks (black lines). ×180. *B*, Partial stage in which the chalcocite-covellite veinlets have widened and coalesced, leaving only scattered residuals of unreplaced chalcopyrite; original cracks still visible. ×180.

times be determined by the eye, but generally a reflecting microscope is necessary. Microscopic examination reveals, from specimen to specimen, the various stages of enrichment from an initial stage to almost complete replacement. For convenience, three stages may be designated — *incipient, partial,* and *complete.*

Incipient enrichment (Fig. 5·8–17*A*) indicates a beginning stage and represents either weak enrichment or the bottom of the zone of enrichment, depending upon where it occurs in the zone. It is characterized by a tarnish or thin coating of supergene sulphides on the outside of hypogene grains and microscopic veinlets within them. Partial enrichment (Fig. 5·8–17B) represents replacement of about one-quarter to two-thirds of the hypogene ore minerals. The coatings are thicker, and the intersecting veinlets are wider and in part coalesced, leaving islands of unreplaced hypogene residuals. Complete enrichment is reached when most of the hypogene minerals are replaced by supergene sulphides. The intersecting veinlets have widened or coalesced and have merged into the enlarged concentric coatings to form a connected mass of supergene sulphides in which lie small isolated hypogene residuals. Rarely has enrichment progressed to the point of elimination of all vestiges of the hypogene minerals; some residuals can generally be found to indicate the nature of the pre-existing hypogene minerals and, therefore, the character of the underlying primary ore.

Selective vs. Pervasive Enrichment. Enrichment is selective in two ways: (1) only veinlets of hypogene minerals are affected, and disseminated grains are unaffected; (2) only certain minerals are replaced (Fig. 5·4–38); supergene chalcocite may replace grains of bornite, leaving contiguous grains of chalcopyrite or pyrite unattacked. In pervasive enrichment, on the other hand, supergene sulphides pervade the entire ore body, relentlessly attacking veinlets and disseminated grains and chalcopyrite and pyrite alike. It is a vigorous, strong enrichment. Selective enrichment, on the contrary, is generally only incipient and is characteristic of the bottom of a strong zone of supergene enrichment or of weak, partial, or incipient attack.

RELATION OF ENRICHMENT TO WATER LEVEL

Sulphide enrichment starts at the water table and extends far below it. The position of the top of the zone is thus controlled by the water table and is similarly related to the topography. The upper surface may be sharply separated from the oxidized zone, but generally there is an interpenetration of the two; the surface may be gently curved or highly irregular and deeply penetrated by long roots of oxidized material. The bottom of the enrichment zone is highly irregular and is a gradual transition to the primary ore. In places it terminates abruptly downward against impervious faults, or it may send roots downward along faults or other structural features.

Sulphide Enrichment above Water Table. In masses of sulphides stranded above the water table the supply of oxygen may be consumed by the exteriors, and supergene sulphides may be deposited in the oxygen-free interiors. This has occurred at Bingham, Utah, and at Bisbee, Ariz.

Subsiding Water Table. Under normal, slow-sinking water level, coincident with erosion, the upper part of the sulphide enrichment zone is progressively enriched. Rapid uplift with rejuvenated erosion and canyon cutting may depress the water level more rapidly than oxidation can keep pace, resulting in lessened enrichment or a stranded sulphide zone. A rising water table causes cessation of sulphide enrichment. Many zones of sulphide enrichment are out of adjustment with the present water table; they are either elevated or depressed, and their tops are related to former water tables, which generally can be correlated with older erosion surfaces.

Factors Influencing Sulphide Enrichment

Water Level. A high, stationary water table, such as occurs in flattish, humid regions, means a thin zone of oxidation, little metal dissolved, and a thin but well-enriched supergene sulphide zone, as at Ducktown, Tenn. A deep water table, especially one being slowly depressed, favors a thick and well-enriched supergene sulphide zone. The ideal condition for sulphide enrichment is active erosion with progressive depression of the water table at a rate such that oxidation and enrichment keep pace with it.

Primary Ores. Supergene enrichment is dependent upon the enriching solutions making contact with primary ore precipitants. In disseminated copper deposits, which are permeable, there is little chance of the enriching solutions escaping contact with hypogene sulphides. In inclined vein deposits the enriching solutions may escape vertically downward and not encounter any hypogene sulphides beyond the footwall; the solutions are wasted, and little sulphide enrichment results. In steep veins with pyrite-impregnated wall rocks, the wall-rock pyrite may become replaced by chalcocite and give rise to an enrichment zone much wider than that of the hypogene vein. In general, the higher the grade of the primary ore the greater the degree of enrichment.

Wall Rocks. In disseminated deposits, the character of the enclosing rock controls enrichment. Carbonate rocks generally inhibit enrichment by precipitating the metals in the oxidized zone. Crackled brittle rocks favor complete enrichment, and pliable rocks restrict it. At Morenci, Ariz., porphyry and quartzite had thorough enrichment, adjacent shales had little, and adjacent limestones no enrichment.

In veins, nonmineralized, shattered wall rocks permit escape and dissipation of the enriching solutions, and impervious walls confine them to the metallized vein.

Structure. Faults have the same effect on enrichment as on oxidation. Large fault zones generate deep roots of enrichment, and in-

clined gouge faults prevent underlying blanket enrichment or form
lateral ore boundaries, as at Ray, Ariz. In veins, gougy strike faults
that dip crosswise may deflect enriching solutions from the vein out
into barren walls and thus prevent enrichment. Strike faults may
also shatter veins, rendering them more pervious.

Topography. Although the top of the enrichment zone conforms in
general with topography, its conformation is with the topography that
existed at the time of enrichment. In many cases, the correlation is
with an old erosion surface whose recognition helps predict the location of probable enriched areas. For example, at Morenci, Ariz.,
the enrichment is related to a mature topographic surface into which
youthful canyons were cut with such rapidity that oxidation could not
keep pace, and they cut across the zones of oxidation and enrichment
and penetrate the primary zone. The sweeping generalization was
drawn that no enriched sulphide ore could be expected beneath the
youthful topography but that good enrichment may be expected beneath the undissected areas of mature topography. This broad conclusion was of economic importance because it delimited at once the
areas where exploration might succeed in disclosing ore.

An enrichment zone related to a buried erosion surface would also be
unrelated to the present surface. If a physiographic study indicates a
long erosion interval preceding the covering, then hidden supergene
sulphide ore might underlie the cover, as was disclosed at Ely, Nev.

Climate. The climatic factors that favor or retard oxidation also
apply to sulphide enrichment.

Erosion. Rapid erosion may outstrip oxidation and little enrichment
result; instead, the enrichment zone may be destroyed. Very slow
erosion results in an almost stationary water level and a thin but rich
supergene sulphide zone. The optimum condition of erosion for rich
and thick supergene sulphide zones is a rate just about equal to the
rate of oxidation and solution. This results in a continuous supply of
enriching solutions.

Time. Considerable geologic time is necessary for extensive sulphide
enrichment. It is of the order of a good part of a geologic epoch or the
equivalent of the time necessary to produce a mature erosion surface.
Post-Glacial time is probably too short to yield considerable enrichment. The important feature in regard to time is not when a given
deposit was first exposed to oxidation but the length of time it continued to be exposed. A porphyry copper deposit may first have
become exposed to oxidation in early Miocene, which means an
intervening time more than sufficient for abundant enrichment, but,
if the erosion surface was buried by middle Miocene lava flows and

exhumed only in late Pleistocene, there has been insufficient exposure to yield much enrichment. Many enriched deposits are related to late mature and old-age erosion surfaces; some are related to peneplains.

CESSATION OF ENRICHMENT

The bottom of the zone of sulphide enrichment is ever moving downward, coincident with erosion, and it will continue to move downward as long as fresh supplies from above are received and hypogene ores lie beneath. Obviously, however, it must cease somewhere. Cessation is brought about by more than one means, some of which are the same as for oxidation, e.g., climatic changes.

Burial. Burial beneath a thick cover of sediments or volcanics prevents further oxidation and enrichment. The United Verde Extension mine, Arizona, is an example of a buried Precambrian enriched zone. A buried enriched zone may later become exhumed and again be subject to enrichment.

Submergence. Submergence of the oxidized and enriched zones beneath the water level causes enrichment to cease. Later valley scouring may again lower the water level and re-expose a deposit to further enrichment. Some deposits have been subjected to at least three such cessations and renewals of enrichment.

Base Leveling. If a region becomes base leveled and is undisturbed by uplift, the water table becomes essentially stationary. Under this condition no more sulphides become oxidized and enrichment ceases.

Bottoming of Ore. When sulphide enrichment in its downward progression reaches the bottom of the hypogene metallization, no more precipitants are available and enrichment ceases. This occurs rarely with veins but is common with horizontal bedded deposits and contact-metasomatic deposits.

Complete Enrichment. When enrichment becomes complete, the solvent-yielding sulphides have practically been replaced by low-sulphur supergene sulphides. Further erosion and depression of the water table then raises into the zone of oxidation enriched ores lacking in minerals whose oxidation yields solvents, and further enrichment ceases.

If a copper deposit reaches this stage the chalcocite zone will then become oxidized to indigenous copper carbonate, cuprite, or native copper, or all three, and the result will be either a deposit of mixed oxidized and sulphide ore or of oxidized ore. This appears to be the explanation of many deposits of carbonate ores, such as the " mixed " and oxidized part of the Miami-Inspiration porphyry copper deposit in Arizona, the oxide zone at Bagdad, Ariz., and parts of the Bisbee, Ariz., carbonate ore bodies. It also explains the oxidation to indigenous carbonate without sulphide enrichment in the pure, hypogene, copper sulphide ore of Kennecott, Alaska, where there are no hypogene minerals whose oxidation products could yield ferric sulphate solvent.

THICKNESS OF SULPHIDE ENRICHMENT ZONES

The following examples illustrate the thickness in feet of some supergene sulphide enrichment zones.

LOCALITY	THICKNESS OF CAPPING	MAXIMUM THICKNESS OF ENRICHED ZONE	AVERAGE THICKNESS OF ENRICHED ZONE
Ducktown, Tenn.	100	8	3–8
Bingham, Utah	—	1,400+	
Miami-Inspiration, Ariz.	250	300	200–300
Ray, Ariz.	250	400	150
United Verde Extension, Ariz.	850	450	
Magma, Ariz. (minor)	450		350
Morenci, Ariz.	—	1,000	450
Butte, Mont.		1,000	

RECOGNITION OF SULPHIDE ENRICHMENT

The recognition of sulphide enrichment is not always simple because many single features characteristic of it are also common to hypogene metallization. Therefore, combined rather than single criteria should be used. Some of these have already been discussed in detail.

Zoning. The three superposed zones — oxide, supergene sulphide, and primary — characterize sulphide enrichment. One or the other may be minor, but they are rarely lacking.

It should be realized that hypogene processes also give rise to ore zones, such as the rich near-surface zones at Cerro de Pasco, Peru, and in the tin veins of Bolivia. Therefore, one must be able to distinguish between supergene and hypogene zoning. The following features aid in distinguishing the two. A supergene sulphide zone is generally related to a water table and, therefore, conforms to a present or past erosion surface, perhaps one of considerable relief. A rich near-surface hypogene zone probably resulted from a rapid temperature drop near the surface that existed when the deposits were formed, but this surface has probably been destroyed long ago. Consequently, there would be no conformation to an existing surface. Also supergene-enriched zones are marked by a pronounced difference in mineralogy from the underlying zones, whereas rich hypogene zones differ from the underlying leaner zones by a difference in amount rather than in kinds of minerals present.

Gossans and Cappings. Gossans and cappings are indicative of pre-existing ore that may or may not have been enriched. Enriched ores, however, generally leave some indication in the gossan such as (1) gossan evidence of metals that undergo sulphide enrichment, e.g., copper; (2) limonite boxwork evidence of supergene sulphides, e.g., chalcocite; (3) removal of metals from gossan; (4) kaolinization, and (5) association of 1 and 4. Evidence from gossans opposed to sulphide enrichment would be (1) rich oxidized ores in oxidized zone; (2) transported rather than indigenous limonite; (3) lack of voids to account for pre-existing sulphides; (4) lack of kaolinization of minerals susceptible of being kaolinized; and (5) limestone country rock.

Erosion. The amount of erosion must have been sufficient to release enough metal to produce enrichment. No appreciable sulphide enrichment could take place from the erosion of a few tens of feet of low-grade primary ore. In certain cases, as at Ray, Ariz., the upper limit of the primary metallization can be determined; the present lowered surface indicates removal of enough protore to produce appreciable enrichment.

District Habit. Supergene enrichment is indicated if a group of similar sulphide deposits display a district habit of a rich zone beneath the oxide zone that gradually becomes impoverished with depth. This is especially true if such rich zones are related to an erosion surface of some relief, and particularly if the rich zones are of shallow depth.

Mineralogy. The mineralogy alone is not conclusive evidence of supergene enrichment. Sooty chalcocite is the only mineral absolutely diagnostic of supergene sulphide origin. Some of the minerals that are commonly of supergene origin are given in Table 12. The association of minerals is more diagnostic than single minerals. For example, chalcocite associated with native copper, limonite overlain by some malachite or native silver, or argentite and chalcocite overlain by malachite and cerargyrite indicates supergene origin.

Chalcocite, the most common supergene sulphide, is often considered to be diagnostic of enrichment. This is not the case. High-temperature chalcocite (see Chap. 3) excludes a supergene origin. The low-temperature orthorhombic chalcocite, if it had a high-temperature ancestry (and this can generally be determined), could not be supergene; if the ancestry was not high-temperature, the chalcocite may have had either a low-temperature hypogene origin or a supergene origin (e.g., Bristol, Conn., chalcocite is late hypogene). In general, the occurrence of chalcocite and/or covellite in abundance in the upper

TABLE 12

COMMON HYPOGENE, SUPERGENE, AND OXIDIZED ORE MINERALS

Metal	Minerals Generally Hypogene	Minerals Generally Supergene	Minerals Generally Originating in Oxidized Zone
Copper	Chalcopyrite Bornite Enargite* Tetrahedrite* Tennantite*	Chalcocite Sooty chalcocite* Covellite	Native copper Malachite* Brochantite* Antlerite* Atacamite* Azurite* Chrysocolla* Cuprite* Tenorite*
Silver	Tetrahedrite* Tennantite*	Native silver Argentite Pyrargyrite Proustite Stephanite Polybasite Pearcite	Cerargyrite* Embolite* Bromyrite*
Gold	Native gold Gold tellurides*	Native gold	Native gold?
Zinc	Sphalerite Willemite*	Wurtzite	Smithsonite* Hemimorphite* Hydrozincite*
Lead	Galena*		Cerussite* Anglesite* Pyromorphite* Leadhillite
Iron	Pyrite* Marcasite Pyrrhotite* Arsenopyrite* Magnetite* Hematite Specularite* Siderite	Marcasite	Goethite* Iron sulphates* Hematite
Manganese	Rhodochrosite* Rhodonite* Manganite(?) Alabandite*		Psilomelane Pyrolusite Braunite
Nickel	Millerite Pentlandite* Niccolite*	Bravoite (?)	Garnierite*

* Always hypogene or supergene or oxidized according to the column in which they are placed.

part of the sulphide zone and diminishing with depth is a safe indication of supergene enrichment.   The same is true of argentite and some sulphantimonides.

Texture.   The texture, where decipherable, is more diagnostic than the minerals, but supergene texture is often difficult to distinguish from late hypogene; experience and a knowledge of the microscope are essen-

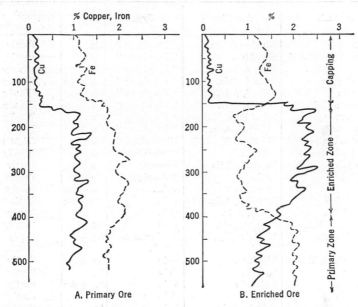

Fig. 5·8–18.   Drill-hole assays of copper and iron plotted in percentage against depth.   In *A*, unenriched ore, the curves of copper and iron are approximately parallel; in *B*, enriched ore, the copper content jumps up at the water table and gradually decreases toward the bottom of the enriched zone; the iron content, being replaced by copper in the enriched zone, drops down but its curve crosses the copper curve again in the primary zone.

tial.   Coatings of sooty chalcocite are absolutely diagnostic.   Under the microscope, supergene sulphides are seen to be the latest minerals, to replace sulphides and not gangue, and the replacement is along late cracks or around grain boundaries (Fig. 5·8–17).   Selective enrichment is characteristic of supergene sulphides; veinlets of chalcocite will expand in chalcopyrite and terminate abruptly at pyrite boundaries.

Metal Ratios.   In copper deposits, supergene enrichment can be determined from drill-hole records by plotting curves of the copper content against depth.   In Fig. 5·8–18 are shown typical curves for hypogene and well-enriched ore.   If iron and sulphur curves are also plotted, the zone of supergene enrichment is even more strikingly shown. In Fig. 5·8–18*A* is shown by curves the relation between copper and iron in hypogene ore, and in Fig. 5·8–18*B*, enriched ore. Note that the curve for iron crosses that of copper at the top and the bottom of the enriched zone; in the middle zone, copper has largely

EXAMPLES OF IMPORTANT SUPERGENE ENRICHMENT DEPOSITS

Deposit	Location	Type	Metal	Dependence upon Enrichment
Utah Copper	Bingham, Utah	" Porphyry "	Cu	Partly
Nevada Consolidated	Ely, Nev.	" Porphyry "	Cu	Partly
Ray Consolidated	Ray, Ariz.	" Porphyry "	Cu	Wholly
Miami Copper	Miami, Ariz.	" Porphyry "	Cu	Wholly
Inspiration	Miami, Ariz.	" Porphyry "	Cu	Wholly
Sacramento Hill	Bisbee, Ariz.	" Porphyry "	Cu	Wholly
Chino Copper	Santa Rita, N. Mex.	" Porphyry "	Cu	Wholly
Clay Ore Body	Morenci, Ariz.	" Porphyry "	Cu	Partly
Rhodesian Copper Belt	Northern Rhodesia	Disseminated	Cu	Slightly
Morenci (old camp)	Arizona	Replacement	Cu	Wholly
Braden Mine	Braden, Chile	Disseminated	Cu	Partly
Potrerillos	Chile	" Porphyry "	Cu	Partly
Butte	Montana	Veins	Cu + Ag	Partly
Magma Mine	Superior, Ariz.	Vein	Cu + Ag	Little
Old Dominion	Globe, Ariz.	Vein	Cu	Partly
United Verde Ext.	Jerome, Ariz.	Vein	Cu	Mostly
Comstock Lode	Virginia City, Nev.	Replacement lode	Ag + Au	Partly
Granite-Bimetallic	Montana	Vein	Ag + Au	Partly
Rochester	Nevada	Vein	Ag	Partly
Georgetown	Colorado	Vein	Ag + Au	Partly
Chanarcillo	Chile	Vein	Ag	Partly
Potosi	Bolivia	Vein	Ag	Partly
Broken Hill	Australia	Replacement lode	Ag + Au	Partly

replaced the sulphide iron. Sulphur is negligible in the oxidized zone, less than copper in the enriched zone, and its curve crosses the copper curve at the bottom of the enriched zone.

Selected References on Oxidation and Supergene Enrichment

The Enrichment of Ore Deposits. W. H. EMMONS. U. S. Geol. Surv. Bull. 625, 1917. *Older comprehensive treatment of supergene sulphide enrichment.*

Copper Resources of the World. XVI Int. Geol. Cong., Washington, 1935. Vols. 1 and 2. *Brief details of individual deposits.*

General Reference 8. Chap. 32. *Résumé.*

General Reference 9. Chap. 9, Supergene Enrichment by W. H. EMMONS. *Later developments.*

Leached Outcrops as Guides to Copper Ore. A. LOCKE. Williams and Wilkins, Baltimore, 1926. *Outcrop characteristics of oxidized and enriched deposits.*

The Secondary Enrichment of Ore Deposits.   S. F. Emmons.   A. I. M. E. Trans., Vol. 30, 1900.   *The original concept.*

Some Reactions Involved in Secondary Copper Sulphide Enrichment.   Zies, Allen, and Merwin.   Econ. Geol. 11:407–503, 1916.   *Fundamental chemical data.*

Fractional Precipitation of Some Ore-Forming Compounds.   R. C. Wells.   U. S. Geol. Surv. Bull. 609, 1915.   *Some causes of precipitation.*

The Hydrated Ferric Oxides.   E. Posnjak and H. E. Merwin.   Am. Jour. Sci., Vol. 47, 1919, and Jour. Am. Chem. Soc., Vol. 44, 1922.   *Iron oxides, their composition and formation.*

Oxidation Products from Sulphides — Group of papers by R. Blanchard and P. F. Boswell.   Econ. Geol., Vols. 22, 25, 29, 30, 1927–1935.   *Various types of limonite boxworks.*

Oxidation at Chuquicamata, Chile.   O. W. Jarrell.   Econ. Geol. 39:251–286, 1944.   *Excellent discussion of processes of oxidation of large copper deposits.*

Oxidation and Enrichment in the San Manuel Copper Deposit.   G. M. Schwartz.   Econ. Geol. 44:253–277, 1949. *Good example.*

5·9   METAMORPHISM

Metamorphic processes profoundly alter pre-existing mineral deposits and form new ones. The chief agencies involved are heat, pressure, and water. The substances operated upon are either earlier formed mineral deposits or rocks. Valuable nonmetallic mineral deposits are formed from rocks chiefly by recrystallization and recombination of the rock-making minerals.

Metamorphism of Earlier Deposits

When rocks are metamorphosed, enclosed mineral deposits may also be metamorphosed. However, unlike rocks, which undergo both textural and mineralogical changes, ores rarely suffer mineral recombinations. Textural changes, however, are pronounced. Schistose or gneissic textures are induced, particularly with sectile minerals, and flow structure is not uncommon. Galena, for example, becomes gneissic, as in the ores of Coeur d'Alene, Idaho. It may also be rendered so fine grained that individual cleavage surfaces cannot be discerned with a hand lens. It " flows " around hard minerals, such as pyrite. Other minerals, such as chalcopyrite, bornite, covellite, or stibnite, behave similarly. The result is that ores may exhibit streaked, banded, smeared appearances with indistinct boundaries between minerals of different color. The original texture and structure may be so obscured that it is impossible to determine to what class the original deposits belonged. Such deposits are then classified as " metamorphosed."

Formation of Mineral Deposits by Metamorphism

Several kinds of nonmetallic mineral deposits are formed as a result of regional metamorphism. The source materials are rock constituents that have undergone recrystallization or recombination, or both. Rarely water or carbon dioxide has been added, but other new constituents are not introduced as they are in contact-metasomatic deposits. The enclosing rocks are wholly or in part metamorphosed; it is the rock metamorphism that has given rise to the deposits. The chief deposits thus formed are asbestos, graphite, talc, soapstone, andalusite-sillimanite-kyanite, dumortierite, garnet, and possibly some emery.

Asbestos Formation

There are two main groups of asbestos minerals — serpentine and amphibole. The serpentine ones are hydrous magnesium silicates, chrysotile and picrolite, and are of the same composition as serpentine; the fine, silky chrysotile is the most valuable. The amphiboles are silicates of calcium, magnesium, iron, sodium, and aluminum. They comprise the minerals amosite, crocidolite, tremolite, actinolite, and anthophyllite. For descriptions of these minerals see Chapter 20.

Serpentine Asbestos. Chrysotile asbestos occurs in serpentine that has been altered from (a) ultrabasic igneous rocks, such as peridotite or dunite, or (b) magnesian limestones or dolomite; the first yields about 90 percent of the world's asbestos supply.

In the ultrabasic occurrences, the fiber is in lenslike veinlets enclosed in serpentine and has three modes of occurrence: (1) *cross-fiber*, with fibers normal to the walls, their length being the width of the veinlet, or less if they contain " partings "; (2) *slip-fiber*, parallel or oblique to the walls, and long but of poor quality; (3) *mass-fiber*, composed of a mass aggregate of interlaced, unoriented, or radiating fibers. The three modes of occurrence are found in a single deposit. Chrysotile fibers range up to 4 or 5 inches in length, rarely 8 inches; most of them are less than 1 inch. Chrysotile may make up from 2 to 20 percent of the rock.

The fiber veinlets are commonly short and discontinuous and crisscross in all directions, forming a network (Fig. 5·9–1). Where numerous and closely spaced, they constitute a workable deposit. Less commonly they occur in parallel veinlets, as in the " ribbon rock " in the Transvaal. The deposits generally constitute only a part of the ultrabasic igneous masses; the deposits in Quebec, the most valuable in the world, are as much as 800 by 200 feet in dimension.

In the deposits in magnesian limestone, cross-fibers in discontinuous bands of serpentine develop within the limestone beds, parallel to the bedding. Several parallel bands may alternate with unserpentinized limestone. The chrysotile veinlets are discontinuous and lie en echelon within a serpentine band. This type of asbestos is very pure,

FIG. 5·9–1. Veins of asbestos at depth of 500 feet. Thetford Mines, Quebec. (*Bowles, U. S. Bur. Mines.*)

and its freedom from included magnetite makes it desirable for electrical insulation.

Amphibole Varieties. The amphibole varieties, of which crocidolite and amosite are the most important, are inferior in quality to chrysotile. These two minerals are found in slates, schists, and banded ironstones over an extensive belt in the Transvaal and Cape Province of South Africa. They occur as cross-fiber in greater lengths than chrysotile; some of the amosite attains lengths up to 12 inches and averages around 6 inches. The crocidolite deposits are said to be the most extensive asbestos deposits in the world. They are in part associated with dolerite sills.

The other asbestiform minerals occur largely as mass fiber with some slip-fiber. The most important is anthophyllite, which is mined in the United States. It occurs as lenses and pockets in peridotite and pyroxenite, and the best quality is from weathered portions. The fibers are harsh and break to lengths of less than 1/4 inch. Fibrous material may make up 90 percent of the rocks. Tremolite and actinolite varieties, except for the Italian tremolite, are commercially unimportant.

ORIGIN

Chrysotile. Chrysotile asbestos is confined entirely to serpentine and, strictly speaking, is a fibrous variety of serpentine. Serpentinization is an autometamorphic process, and in the ultrabasic rocks, such as dunite, serpentinization has proceeded along fractures; in Quebec, Dresser noted a rather constant ratio of width of serpentine bands to chrysotile of 6–1–6. Chrysotile is not formed except where there is serpentinization; but serpentine may occur without chrysotile. In some occurrences the entire rock mass may be serpentinized. Cooke recognizes two stages of serpentinization, a first general stage whereby 40 to 60 percent of the rock mass is converted to serpentine, and a second stage during which portions of the partly altered rock along fractures are completely altered to serpentine. This alteration was probably accomplished by hot residual solutions that emanated from within the intrusives. Since granitic intrusives may have come from the same reservoir, the solutions might possibly have emanated from them. In the alteration, the magnesium silicate olivine is converted to the hydrous magnesium silicate serpentine, only water being added.

The puzzling problem of origin, however, is how chrysotile, having the same composition as serpentine, was formed and how it became emplaced. Various theories have been advanced: (1) The veinlets are fissure fillings (a) in openings of hydration expansion from serpentine solutions of short distance transportation (Cirkel); (b) in fractures produced by dynamic stresses, by means of hydrothermal solutions of remote source (Keith and Bain). (2) Replacement and recrystallization of serpentine walls outward from tight cracks (Dresser, Graham). (3) Serpentine extracted from rock and deposited as asbestos in tight fractures, the walls of which are pushed apart by the force of growing crystals (Taber, Cooke). Cooke, the most recent investigator of the subject, concludes that for the Quebec deposits replacement is untenable and ordinary fissure filling is impossible and he, therefore, favors the theory that the fiber commenced to crystallize in tight fractures, the walls of which were pushed apart by the fiber growth, aided by tension set up by deformation movements.

It would seem that any adequate explanation advanced to account for chrysotile in serpentinized peridotite must also account for chrysotile in serpentinized limestone. The occurrence of chrysotile in serpentine in both rocks can hardly be accidental. Therefore, a necessary and preliminary step must be the formation of serpentine, and the formation of the chrysotile must be closely related to that of serpentine. In the limestone the magnesia was largely introduced into nonmagmatic rocks from diabase intrusions; layers were thus serpentinized, and in these serpentine layers horizontal lenslike veins of chrysotile were formed, which are clearly recrystallization of parts of the serpentine bands. As the horizontal limestones are overlain by other sediments, tension could hardly have been a factor in such cases. Likewise, in the ultrabasic rocks of the New Amianthus mine, Transvaal, dozens of nearly horizontal bands of chrysotile lie one above another with the fiber vertical. A heavy column of sediments and lava overlies it. Tension could not have been a factor here either. Also, if due to the force of growing crystals, the growing fibers would have had to lift up the enormous weight of the overlying rocks, which does not seem likely. The bedding of the limestone above and below chrysotile lenses is undisturbed, horizontal, and continuous with bedding in places where chrysotile is absent. This shows that the horizontal walls were not pushed apart. Consequently, tension and crystal growth do not explain the limestone occurrences. If tension and crystal growth are inadequate to explain the limestone occurrences it seems probable they are likewise inadequate to explain the ultrabasic occurrences as well. The Quebec occurrences with many crisscross veinlets suggest that open fissures did not form. Fissure filling also inadequately explains the limestone occurrences, since flat fissures could hardly stay open under superimposed weight, and, moreover, the undisturbed bedding planes above the chrysotile lenses show that no fissures opened. The most tenable explanation is that certain bands of limestone were converted to serpentine by circulatory solutions and that some slight change in the character of the solutions caused the serpentine to undergo molecular rearrangement into fibrous form. May not the same hypothesis apply to the peridotite occurrences, such as those of Quebec or Africa?

It is commonly stated that serpentinization involves considerable volume expansion because of the addition of water. Serpentine replaces olivine, and it is the experience of economic geologists that replacement is a volume-for-volume interchange. Therefore, the alteration to serpentine may not be attended by volume increase. Actually, there may possibly be a decrease in volume, thus accounting for the numerous fractures.

Amphibole. The crocidolite is thought by Peacock to have originated by molecular reorganization, without essential transfer of materials or constituents of the enclosing banded ironstones. Deep burial is thought to have supplied heat and pressure that resulted in the metamorphism of the rock constituents into the blue asbestos. This is strongly suggested by its wide distribution in similar rocks generally unassociated with igneous intrusions. The amosite is chemically dissimilar to the enclosing rocks, and its occurrence around the contact aureole of the Bushveld Complex suggests contributions from solutions of Bushveld origin in addition to static metamorphism.

For production, uses, occurrences, and references on asbestos, see Chapter 20.

Graphite Formation

Graphite or " black lead " is a form of carbon that occurs in two varieties, *crystalline*, consisting of thin, nearly pure black flakes, and *amorphous*, a noncrystalline, impure variety. It is soft, black, has a greasy feel, and marks paper; hence the term " graphite " (to write). It is debatable that the material of graphitic slate, which yields " amorphous graphite," is really graphite or amorphous carbon. True graphite yields graphitic acid when treated with nitric acid; amorphous carbon does not.

Occurrence. Graphite occurs chiefly in metamorphic rocks produced by regional or contact metamorphism. It is found in marble, gneiss, schist, quartzite, and altered coal beds; it also occurs in igneous rocks, veins, and pegmatite dikes. Most of the crystalline variety occurs in minute flakes disseminated through metamorphic rocks. The amorphous variety is in dustlike form. The deposits may be of large size, and the graphite content may be as much as 7 percent. Associated minerals are quartz, chlorite, rutile, titanite, and sillimanite. Disseminations and fissure veins are the most important types of deposits.

Origin. Graphite originates by (1) regional metamorphism; (2) original crystallization from igneous rocks as shown by its occurrence in granite, syenite, and basalt; (3) contact metamorphism, as at Calabogie, Ontario, where it occurs with contact metamorphic silicates in limestone adjacent to an igneous intrusion; and (4) introduction by hydrothermal solutions, which accounts for vein deposits and, as Beverly considers, for deposits in pegmatites and shear zones in schist in the San Gabriel Mountains, California. The graphite of 2, 3, and 4 is considered of magmatic derivation; that of 3 and 4 may have resulted from gaseous carbon compounds given off by the magma, as believed by Weinschenk; or the carbon may have been derived from intruded sediments, and later deposited. The coal beds that have been altered to graphite in Sonora, Mexico, and Raton, N. Mex., are clearly the result of igneous metamorphism. The volatile materials of the coal have been driven off and the residual carbon converted into the crystalline condition carrying 80 to 85 percent graphite.

Two views exist for the deposits resulting from regional metamorphism: one, the graphite is altered organic matter formerly present in the sediments, and the other that it results from the breakdown of calcium carbonate. Black, carbonaceous limestones, when metamorphosed, yield white marbles with disseminated graphite. Either the original hydrocarbons have been broken up, causing direct precipitation of the carbon, or they have been converted into carbon monoxide and carbon dioxide, which in turn were deoxidized and the carbon precipitated. In either case the distributed carbon has been moved into concentration centers.

The other idea, advanced by Winchell and others, is that carbonates are broken down, yielding their Ca, Mg, or Fe to form silicates and releasing CO and CO_2, which in turn become deoxidized to form graphite. A. N. Winchell suggests two possible reversible reactions:

$$C + 2H_2O \rightleftarrows CO_2 + 2H_2$$

$$C + CO_2 \rightleftarrows 2CO$$

If they are reversible, either could account for free carbon.

The occurrence of graphite in Precambrian rocks suggests an inorganic rather than an organic origin for the carbon. Under either hypothesis, the carbon has come from the sediments. It is also possible that the carbon of the graphite found in igneous rocks, dikes, and veins was picked up from underlying carbonate rocks.

The uses, distribution, occurrence, related references, and examples of graphite deposits are given in Chapter 19.

Talc, Soapstone

Talc, a product of metamorphism, is a hydrous magnesium silicate $[H_2Mg_3(SiO_3)_4]$, which, when finely ground, forms the familiar talcum powder. The pure, soft mineral is known in trade as *talc; steatite* describes a massive, compact variety; *agalite* is a special name applied to fibrous talc from New York State. Soapstone is a soft rock composed essentially of talc but also containing chlorite, serpentine, magnesite, antigorite, and enstatite, and perhaps some quartz, magnetite, or pyrite. It is a massive impure talcky rock that can be quarried and sawed into large blocks. *Pyrophyllite,* sometimes included among soapstones, is a hydrous aluminum silicate that serves some of the same uses as soapstone.

Occurrence. Commercial talc and soapstone deposits occur in metamorphosed ultrabasic intrusives or dolomite limestones. They are thus restricted to metamorphic areas and are largely confined to the Precambrian. The best quality talc comes from metamorphosed dolomitic limestones and is generally associated with tremolite, actinolite, and related minerals. These deposits are generally lens-shaped, in beds, and reach widths up to 125 feet. The important deposits of Ontario, New York, North Carolina, Georgia, California, Bavaria, and Austria are of this type. Talc also occurs in Europe intercalated in schists and gneisses, of which they are supposed to be replacements. However, Gillson suggests they may be replacements of included magnesian limestone beds.

The deposits in, and associated with, ultrabasic masses are more numerous but smaller than those in altered limestones. They occur with serpentine, whose formation preceded that of the talc, as in the case of the soapstone deposits of Virginia. The pyrophyllite deposits of North Carolina occur in slates and tuffs with interbedded volcanic breccias and flows, all of which have been metamorphosed. The pyrophyllite is chiefly in acid tuff.

Origin. Talc is an alteration product of original or secondary magnesian minerals of rocks. It results from mild hydrothermal metamorphism, perhaps aided by simple dynamic metamorphism, but never from weathering. It is rare in ore deposits. It is pseudomorphic after minerals such as tremolite, actinolite, enstatite, diopside, olivine, serpentine, chlorite, amphibole, epidote, and mica. Lindgren states that it may be formed from any magnesian amphibole or pyroxene acted on by CO_2 and H_2O according to the reaction:

$$4MgSiO_3 + CO_2 + H_2O = H_2Mg_3Si_4O_{12} + MgCO_3$$

It thus originates in (a) regionally metamorphosed limestones, (b) altered ultrabasic igneous rocks, and (c) contact-metamorphic zones adjacent to basic igneous rocks. Talc is always late in the mineral sequence. It is formed largely from other minerals that in turn represent alteration products of original minerals. Where present in serpentine, it was not formed as a result of the serpentinization but, according to Hess, by subsequent, unrelated processes by means of which the serpentine was replaced by talc. Gillson states that talc in serpentine rocks is pseudomorphic after actinolite, or after chlorite that replaced biotite. Its random orientation suggests that it cannot have formed from dynamic metamorphism alone as is believed by Harper. The magnesia is largely, if not entirely, derived from the rocks in which the talc occurs. According to Stuckey, the pyrophyllite deposits are hydrothermal replacements of silicic volcanics.

The uses, production, distribution, examples of deposits, and selected references are discussed under Talc in Chapter 20.

Sillimanite Group — Andalusite, Kyanite, Sillimanite

The four interesting minerals andalusite, kyanite, sillimanite, and dumortierite withstand high temperature, change over to mullite, are sought for high-grade refractories, and are used for similar ceramic purposes. The first three have identical composition ($Al_2O_3 \cdot SiO_2$) but differ in crystallization, andalusite and sillimanite being orthorhombic and kyanite triclinic. Dumortierite is a basic aluminum borosilicate (orthorhombic). At high temperatures ($1,100°–1,650°$ C) these minerals change over to mullite ($3Al_2O_3 \cdot 2SiO_2$) and vitreous silica, which is considered to be cristobalite. This material, rare in nature, remains stable up to $1,810°$ C, therefore is heat resistant, is a good high-temperature insulator, and is particularly resistant to shock.

Occurrence and Origin

Kyanite is a common mineral of metamorphic rocks, but commercial deposits are few. The commercial deposits consist of disseminated crystals or small masses in gneiss or schist. Kyanite also occurs as lenses in pegmatite dikes and as bunches in quartz veins. It is considered to have been formed from mica schists or other aluminous silicate rocks by dynamothermal metamorphism, perhaps accompanied by magmatic emanations.

Andalusite occurs in argillaceous crystalline rocks and also in pegmatites. Its common associates are tourmaline, garnet, corundum, topaz, quartz, and mica. The largest deposit is at White Mountain, Calif., where it occurs in irregular segregations in a quartz mass enclosed by sericite schist. The minable portions, averaging 70 to 80 percent andalusite, occur in a zone 10 to 50 feet thick. According to Kerr, the deposit was formed by a sequence of metamorphic processes during which aluminous rock (volcanic or sediment) was converted to andalusite segregations as a result of pneumatolytic action by a nearby intrusive. Another occurrence near Hawthorne, Nev., is a veinlike deposit 2 to 4 feet thick, about 3,000 feet long, and explored to a depth of 100 feet.

Dumortierite occurs at Oreana, Nev., in pegmatites or quartz veins that cut aluminous rocks, where a quartz mass in a sericite schist contains irregular lenses of andalusite altered, or partly altered, to dumortierite. Kerr recognizes three generations of dumortierite. He thinks that igneous emanations caused " crystallization of the andalusite and later dumortierite by metamorphism along the boundaries of the quartz mass adjoining the schist. The earlier phase was probably pneumatolytic when andalusite and earlier dumortierite were formed.

Fig. 5·9–2. Andalusite zone in quartz mass and mining openings. White Mountain, Calif. (*Kerr, Econ. Geol.*)

Later, presumably hydrothermal metamorphism resulted in the formation of late dumortierite."

Sillimanite occurs as slender prisms in aluminous crystalline rocks and results from high temperature metamorphism.

Distribution and Examples of Deposits

The United States, India, and Kenya are the only countries that produce these refractory minerals. In the United States a desire for better spark plugs and refractories led to a search for workable deposits of these minerals, and, since 1920, production has been steadily increasing. During a 10-year period the Champion Spark Plug Co. produced 350 million spark plug cores from andalusite and dumortierite.

Kyanite occurs in commercial deposits in North Carolina, Virginia, Georgia, California, India, and Kenya Colony. The Indian occurrences, mostly in the Singhbhum district, and those of Kenya are the largest. In India kyanite occurs in kyanite schist or quartz-kyanite schist, and is in disseminated crystals or in bunches. Most of the material shipped consists of residual boulders of kyanite collected from the regolith which have been weathered out of underlying kyanite rock. The deposits are extensive, and the unweathered material has hardly been touched. In the Bhandara district both kyanite and sillimanite are found exclusive of each other. It is thought that the two minerals have been formed from chlorite-muscovite schists by pneumatolytic and hydrothermal metamorphism resulting from granitic intrusions. The Appalachian occurrences (Chap. 5·7), and those near Ogilby, Calif., and Kenya have been formed from aluminous schists.

Andalusite occurs at White Mountain, Calif. (Fig. 5·9–2), Oreana and Hawthorne, Nev., and in small amounts in the Black Hills of South Dakota, the Transvaal, and Switzerland.

Dumortierite is known in a number of localities, but the only commercial deposit of dumortierite is at Oreana, Nev.

Sillimanite comes from India, where it is recovered in boulders from the weathered parts of sillimanite schist. Large inaccessible deposits are reported by Dunn in India at Khasi Hills, Assam, and Pipra, Rewa. The mineral occurs in sillimanite schist enveloped in granite and associated with corundum.

Possible sources of commercial kyanite are reported from Russia, western Transvaal, Nyasaland, Western Australia, and Black Mountain, N. C.; possible sources of sillimanite are reported from South Africa.

Selected References on Sillimanite Minerals

Formation of Mullite from Kyanite, Andalusite, and Sillimanite. J. W. GREIG. Jour. Am. Ceramic Soc. 8:465, 1925. *Chemical and physical changes.*

Andalusite and Related Minerals, White Mountain, Calif. P. F. KERR. Econ. Geol. 27:614–643, 1932. *Occurrence and origin.*

Dumortierite-Andalusite, Oreana, Nev. P. F. KERR and C. P. TENNEY. Econ. Geol. 30:287–300, 1935. *Occurrence, origin, production.*

Kyanite Deposits of North Carolina. J. L. STUCKEY. Econ. Geol. 27:661–674, 1932. *Geological features.*

Occurrences of Sillimanite in North Carolina. C. E. HUNTER and W. E. WHITE. N. Car. Dept. Cons. Devel. Inf. Circ. 13, 1946. *General occurrence.*

General Reference 11. Chap. 42. Sillimanite Group, by F. H. RIDDLE and W. R. FOSTER. *Brief summary of mineralogy and geology, distribution, and uses. Complete bibliography.*

Mining and Treatment of the Sillimanite Group of Minerals and Their Use in Ceramic Products. F. H. RIDDLE. A. I. M. E. Tech. Pub. 460, 1932. *Technology and uses.*

General Reference 5.

Other Metamorphic Products

Garnet. There are seven varieties of garnet (see p. 826), of which two are of commercial importance — almandite and rhodolite. Garnets occur as accessory minerals in many rocks, but their common home is in gneisses and schists. They are formed by regional metamorphism, contact metamorphism, and as original constituents of igneous rocks; but metamorphic processes are responsible for all the commercial deposits.

Miscellaneous Materials. *Emery,* which is a mixture of corundum and magnetite with hematite or spinel, is formed by metamorphic processes, mostly contact metamorphism. The Peekskill deposits, New York, occur in the Courtland igneous series where it contains inclusions of mica schist. The Virginia deposits occur in schist bands within quartzite and granite. At Chester, Mass., pod-shaped pockets occur in amphibolite. The Grecian deposits are pockets in crystalline limestone cut by granite as are the Turkish deposits of Asia Minor. (See Chap. 23.)

Metamorphism also gives rise to many varieties of abrasive stones (pp. 827–829).

5·10 SUMMARY OF ORIGIN OF MINERAL DEPOSITS

In this lengthy chapter an attempt has been made to follow the processes of origin of mineral deposits from the source of the materials to their final resting place as mineral deposits. Magmas are the source of essentially all the ingredients of mineral deposits. A few constituents, such as oxygen, carbon dioxide, or water, are derived from the atmosphere or the oceans, but even these are in part of magmatic derivation.

The initial stages of magma crystallization are attended by separation of certain metallic oxides, sulphides, and native metals. Some of these crystallize and become segregated by crystal separation into mineral deposits early in the magmatic stage; others solidify later than the rock crystals and either become segregated at the original site of accumulation or are injected into the cooled intrusive or the surrounding rocks, forming late magmatic mineral deposits.

During the progressive crystallization of the magma, the abstraction of the early crystallizing rock minerals leaves a residuum liquid, generally silicic, which gradually becomes enriched in volatiles and gases. These contain and in part consist of compounds of the metals and other valuable substances formerly sparsely distributed throughout the magma; they tend to collect in the upper part of the magmatic chamber. If the pressure becomes relieved they escape into the enclosing wall rock, and, under favorable circumstances, contact-metasomatic mineral deposits result.

Toward the close of the solidification of the magma, some of the accumulated, highly silicic, mother liquors may be withdrawn to form pegmatites, accompanied or followed by aqueous solutions containing valuable mineral compounds from which economic minerals are deposited by replacement of the pegmatitic minerals. Upon final consolidation of the magma the residual aqueous solutions, in the form of gases, liquids, or both, are ejected toward places of less pressure. These constitute the mineralizing solutions from which nearly all epigenetic mineral deposits are formed. The gases upon cooling condense to form magmatic liquids, which as they near the surface mingle with meteoric waters, and many hydrothermal solutions consist of both. The hydrothermal solutions in their journey outward from the magma chamber undergo chemical change by reaction with the wall rocks and thus become the alkaline solutions from which most economic minerals are considered to be deposited. The solutions seeking lines of easiest flow follow cracks, joints, bedding planes, rock pores, and other openings. The mineral substances in them may replace the rock substances, giving rise to replacement deposits, or they may be precipitated from solution and fill up the rock openings, forming cavity-filling deposits. The deposition may occur at high, medium, and low temperatures and pressures, forming respectively the hypothermal, mesothermal, and epithermal groups of Lingdren (p. 94), each characterized by distinguishing minerals and textures. The magma is thus the parent and mineral deposits the offspring. The offspring may be deposited in sufficient concentrations to constitute economic mineral deposits or they may be sparsely deposited, requiring other means of concentration to render them valuable.

Secondary processes may then operate upon the previously formed mineral deposits or upon rocks to form yet other types of economic mineral deposits. Weathering releases many valuable mineral substances that are transported in solution or mechanically to sedimentary basins and there are deposited as sediments giving rise not only to the common sedimentary rocks but also to economic deposits of metals and many industrial nonmetallic minerals and products. Organic processes also take part in the growth of plants and animals from which coal and oil are formed. The inorganic substances were derived originally from igneous rocks and magma, although they may have passed through previous sedimentary cycles. Other soluble substances released during erosion are concentrated in bodies of water of the oceans, of enclosed basins, or of the ground water, from which they are deposited by evaporation, giving rise to numerous valuable saline deposits. The circulating ground water is considered by some to be effective in dissolving, transporting, and redepositing mineral substances in more concentrated form.

Weathering, combined with the sorting action of water and air, effectively garners heavy, insoluble, and durable minerals from their enclosing rock matrix and concentrates them into valuable placer deposits of both metallic and nonmetallic minerals. Weathering alone, in its relentless attack upon the rocks, deliberately sorts out valuable and nonvaluable materials. Soluble waste products are removed from desired insoluble substances, which persist and accumulate *in situ* as important residual mineral deposits. Other products, such as clays or bauxite, are created during weathering and persist, while associated undesirable substances are removed in solution, leaving accumulated masses as residual deposits of economic importance. Surficial oxidation profoundly modifies most ore deposits, rendering barren the upper parts of many deposits or changing the ore minerals into more usable or less usable forms. Metals are dissolved within the zone of oxidation, and are then carried down below the water table, where they are reprecipitated. The metals removed from above are thus added to those existing below, thereby bringing about a supergene enrichment of the upper part of the sulphide zone. Leaner parts of veins have been made richer, and worthless protore has been made workable. Thus have many large and rich ore deposits been created.

Metamorphism not only drastically changes the form and texture of pre-existing mineral deposits but it also creates new ones. Under high pressure and temperature, aided in some cases by hot waters, metamorphic minerals that are stable under the new environment are produced. The change may consist only of recrystallization, or of recombinations of materials to form the new minerals. Generally nothing, except perhaps water, is added during the metamorphism.

It is evident that the types of economic mineral deposits are many and varied and that numerous and unrelated geologic processes must be invoked to explain their origin. Four major processes operate to produce mineral deposits, namely, igneous activity, sedimentation, weathering, and metamorphism. In certain cases more than one process is involved, and these may overlap or operate at different periods of time. In the deciphering of the genesis of mineral deposits, and in the application of the deductions therefrom, it is imperative that multiple working hypotheses be utilized. The student of economic geology must not only have at his command a knowledge of the other geologic sciences, but he must apply them to the problem of ore genesis.

CHAPTER 6

CONTROLS OF MINERAL LOCALIZATION

The question of *why ore is where it is* has stimulated thought to the cause of localization of economic mineral deposits, and this in turn has brought forth much information regarding structural control of ore deposition. In fact, so much information has appeared that there is a growing tendency to regard structural control as synonymous with ore localization — a common swing of the pendulum that in time will right itself. However, it must not be overlooked that stratigraphic, physical, chemical, and other factors also control ore localization, either alone or in cooperation with structural factors. A structural feature, such as a fissure, for example, may guide the flow of mineralizing solutions, but ore deposition may not occur within the fissure unless a favorable wall rock is encountered. Structural control of ore localization is now being used extensively to guide ore-finding.

Controls of ore deposition, particularly structural controls, obviously are of most importance in the localization of epigenetic ore deposits, which comprise the bulk of the ores of the American cordillera. To a limited extent they determine the location of certain types of syngenetic and residual concentration deposits and may determine if and where oxidized ores and supergene sulphides are deposited. A number of ore controls have already been referred to or discussed in Chapter 5.

Structural Control of Ore Localization

Structural features are by far the most important loci of ore and may be divided broadly into regional and detailed. The regional features determine the broader localization of ore belts or mineral districts within wide areas barren of economic minerals. The relationship is clear in a broad way but vague in detail. The detailed features determine the immediate localization of ore; they are the ones that come under the daily observation of the mining geologist; they are well exposed in mining operations, and naturally they are the ones that have received most attention.

Orogenic Movements. In general, most epigenetic mineral deposits occur in regions that have undergone orogenic or mountain-building movements, although the mountains themselves may have been re-

moved by erosion and only their roots remain. They are sites of
thick sedimentation, crustal movements, dislocations, and igneous
intrusions that yield ore-forming fluids; abyssal rocks have been elevated thousands of feet, and erosion has revealed deep-lying mineral
deposits. Mountain building also produces many kinds of openings
that serve as channelways for mineralizing solutions. Thus, mountains, existing or eroded, are the home of epigenetic ore deposits.

Igneous Intrusions. Broadly speaking, belts of igneous intrusions
are belts of ore deposits. Since igneous intrusives are a source of ore-
forming fluids, they constitute ore loci on a regional scale. Commonly,

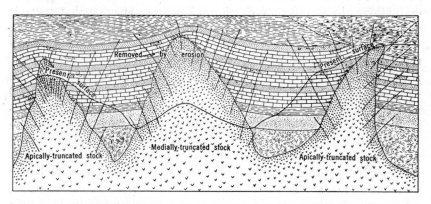

Fig. 6–1. Ideal section through intrusive showing loci of differentiation, places
where veins are formed, and relation of vein outcrop to depth of erosion of intrusive.
(*After B. S. Butler, A.I.M.E.*)

mineral deposits occur within, or are clustered about, stocks and
batholiths, as the Bingham, Utah, monzonite stock or the Boulder,
Idaho, and Sierra Nevada granodiorite batholiths (Fig. 6–1). In
numerous places the ores are zonally arranged with respect to the
intrusive. Few epigenetic deposits of the American cordillera are far
removed from intrusives; the rich deposits of the Canadian shield are
intimately related to intrusives; those without genetically associated
intrusives are rare.

Major Faults. Large regional faults serve as master channelways
along which mineralizing solutions move up from the depths, thence
into subordinate connecting channelways. Thus, the great Keewee-
naw fault in Michigan is believed to have served as a major channelway for the 300-mile Lake Superior copper belt. Similarly, Quentin
Singewald and R. Butler consider the London fault in Colorado to have
been a major locus for the ore deposits of that district.

Spurr, B. S. Butler, and others have shown that, in a broad way, folding, igneous intrusion, faulting, and metallization are related and occur in definite sequence. First, folding, accompanied by low-angle thrust faulting, has been followed by large-scale igneous activity and then by faulting, which regionally and locally exerted control over the metallization that immediately followed.

Detailed Structural Controls. Of the smaller detailed structural controls, the various types of rock openings are of prime importance. They are for the most part the direct and immediate cause of ore being in one place rather than in another. They are features readily observable and worthy of the careful study of the geologist. Many of them owe their origin to structural processes and constitute, therefore, structural controls of ore localization. The part played by various types of openings in rocks in localizing ore is discussed in detail under Hydrothermal Processes in Chapter 5·4.

Stratigraphic Controls of Mineral Localization

Stratigraphic controls are of greatest importance in the localization of gas, oil, and water, as shown in Chapter 16. Obviously, they are the chief factors in the localization of sedimentary mineral deposits such as coal, iron, manganese, phosphate, salines, and other deposits described in Chapter 5. They also play a part in the localization of other types of deposits.

REGIONAL STRATIGRAPHIC CONTROLS

Geosynclines. In geosynclines, thick accumulations of sediments are followed by uplift, folding, faulting, and batholithic invasion, which provide channelways for mineralizing solutions. Thus, the association of ores with mountains is also an association with former geosynclines. In addition, geosynclines are also sites for deposition of sedimentary ores, rocks, and fuels of economic value, and the geosynclines delimit the occurrence of such products.

Basins of Deposition. Many kinds of sedimentary deposits are laid down in basins of deposition of lesser scale than geosynclines. They include beds of coal, iron ore, manganese, phosphate, sulphur, salines, various earths, clays, and industrial rocks. Such basins delimit the area of possible occurrence of these substances. They range in size from small swamps to areas as large as a state or country. The potash basin of New Mexico and Texas has an area of 60,000 square miles.

Plateau Margins. As previously mentioned, igneous activity and mineralization are associated with belts of strong folding and faulting, and these are greatest in basins of extensive sedimentation. Further, belts of sedimentation surround plateau areas. B. S. Butler has shown that the Colorado Plateau has existed as a positive area for much of pre-Tertiary time, around which are marginal areas of extensive sedimentation that later were strongly folded, faulted, intruded, and mineralized. The barrenness of the plateau itself and the concentration of mineral districts surrounding it are strikingly shown in Fig. 6–2. The influence of the plateau and the localization of the metallization within the plateau margin is undeniable.

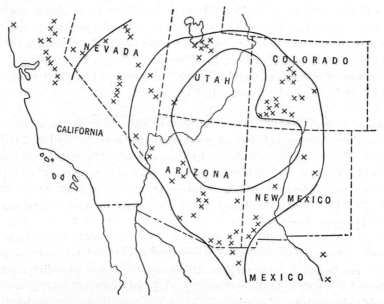

Fig. 6–2. Barren Colorado Plateau and concentration of mineral deposits and districts (marked by x) surrounding it. Heavy lines denote boundaries of positive and negative elements. (*After B. S. Butler, A.I.M.E.*)

Unconformities. Surfaces of erosion indicated by unconformities are favorable sites for accumulation of residual concentration deposits (Fig. 5·7–4). The unconformities control the location of the deposits, which lie neither above nor below them, and may be coextensive with them. Bauxite deposits, for example, lie on unconformities, notably those of southern Europe. Residual iron and manganese ores likewise are generally associated with unconformities. Some deposits of oxidized ores lie just beneath unconformities; buried placer deposits also lie on them.

Because the weathered zones beneath unconformities are generally more porous than the contiguous rocks, they serve as channelways for mineralizing solutions, permitting deposition and replacement. The important function of unconformities in localizing oil, gas, and water is shown in Chapter 16.

DETAILED STRATIGRAPHIC CONTROLS

Bedding. Bedding facilitates the movement of underground fluids through sedimentary rocks; and oil, gas, water, ores, and nonmetallic minerals are concentrated along certain beds. Replacement deposits become extended along bedding, ores are limited at bedding planes, and contact-metasomatic deposits may be localized by them. Cross-bedded strata are favorable sites for metallization, and basal beds in places contain buried placer deposits. Intercalations in bedding affect the flow and accumulation of petroleum.

Lenses. Included permeable sandstone lenses or strata localize oil, water, and ores. Carbonate lenses, but not the surrounding rocks, may be replaced by ores, forming bedded ore lenses.

Impervious Covers. Impervious covers, such as shale, constitute effective barriers to ascending mineralizing solutions (Figs. 5·4–36, 5·4–40). The result is that in many districts, such as Ouray, Colo., mineral deposits are localized just beneath the shales. Irving has shown that deposits of Tertiary gold ores in the northern Black Hills spread out underneath shales. They also form caprocks for oil pools.

Impervious Bases. Impervious base rocks similarly impound descending fluids. Subsurface water may thus be localized to emerge as springs elsewhere; oil may be trapped in synclines. Descending supergene mineralizing solutions are dammed, and their metallic load is deposited in the form of oxidized ores or supergene sulphides on such impervious strata. Pitching synclines and anticlines with impervious troughs and arches (Fig. 5·4–35) control the flow of descending and ascending mineralizing solutions and thus localize ore deposition.

Physical and Chemical Controls

Physical and chemical properties of host rocks influence ore deposition and thereby help to localize mineral bodies. However, the exact properties that aid or cause mineral deposition are seldom evident; they are mostly surmised. In general, they operate along with structural features. A fissure that traverses sedimentary beds, for example, acts as a channelway for solutions, but a congenial host rock is also necessary to induce deposition (Fig. 5·4–36); the formation of the ore only where limestone is the wall rock is an observable fact, but the cause of the deposition is uncertain.

Permeability. The mineralizing solutions of epigenetic deposits require openings to reach the site of deposition. In general, induced openings are utilized as main channelways, but connecting rock pores are necessary to permit passage of the solutions or the solutes through the wall rock to the sites of deposition. In some deposits main channelways are absent and permeability alone seems to have been the main or only factor of localization as, for example, the Roan Antelope and N'Kana deposits of the Northern Rhodesia copper belt, where a bed 25 feet thick has been almost uniformly metallized by copper for the remarkable length of several miles.

Permeability is also a factor in the localization of the great disseminated copper deposits. The crackled enclosing rock and porosity together permitted wholesale permeation by the ore fluids. In the localization of many other types of epigenetic deposits permeability has been the controlling or assisting factor. In the formation of oil pools and ground-water supplies it is essential.

Permeability also controls the localization of many oxidized and sulphide enrichment bodies. Sulphide enrichment of disseminated deposits cannot occur without it; impervious sections are unenriched.

Brittleness or Toughness. Brittle rocks tend to crackle readily under slight stress such as has generally been present in regions of metallization. Their resulting high permeability is favorable for ore deposition (Figs. 5·4–14, 5·4–15A). Rhyolite, quartzite, limestone, and silicified and sericitized rocks come under this class. " Tough " or pliable rocks do not crackle and are consequently less permeable. Shale, schists, chloritic rocks, greenstones, and some basic igneous rocks come under this class.

The relative effects of the permeability resulting from these physical properties can often be observed clearly in secondary enrichment deposits.

Chemical Properties. The chemical character of wall rock has long been ascribed a dominant role in the localization of epigenetic deposits and ore shoots. Deserving as is this role, it has also been overrated, occult attributes having been assigned to it. Much unnecessary exploration has been undertaken on the *assumption* of the part played by the wall rock. Some rocks unquestionably are more congenial hosts to ore than others, and often for no observable reasons.

Carbonate rocks are especially favorable localizers of ore. In district after district in North America, fissures carry ore in limestone or dolomite and not in other rocks (Fig. 5·4–40), a selective replacement due to chemical control, of which some examples are discussed on p. 101. At Kennecott, Alaska, the ore is confined to dolomitic limestone and terminates against limestone; there are no observable differences in the two rocks; the porosity is about the same in both, but one contains magnesium and the other does not — a chemical difference. This preference of ore for dolomite or limestone is as yet not understood.

Many other kinds of rocks display relative preferences for ore. Ore may occur in a vein opposite dolerite and not opposite granite, or in rhyolite and not in andesite. In the Porcupine gold district, Ontario, ore " makes " in greenstones but not generally in the Pearl Lake quartz porphyry; in the Kirkland Lake district, the gold ore favors syenite over the large porphyry mass; at the Noranda mine, Quebec, the ore is confined almost wholly to rhyolite and avoids andesites; at the Homestake mine in South Dakota the gold ore is localized in cummingtonite schist. Innumerable other examples of similar nature exist. In general, highly silicic rocks are not as favorable for ore formation as those less silicic, and ultrabasic igneous rocks and highly aluminous rocks, such as shale, are not generally favorable host rocks. In rocks like monzonite, the dark minerals, or feldspar, are commonly preferentially replaced by ore minerals.

The relatively high content of ferric iron in the upper parts of the Michigan lava flows is considered by B. S. Butler and others to have localized the deposition of the native copper of the Lake Superior deposits, and a similar localization has occurred at Corocoro, Bolivia.

In supergene alteration the chemical character of the wall rock may control directly the loci of oxide and sulphide deposition. Reactive limestone causes immediate precipitation of carbonates of copper or zinc. The large copper carbonate deposits of Bisbee, Ariz., are confined to limestone, and the great oxidized copper deposits of Katanga, Belgian Congo, are restricted to dolomitic beds. Rocks that yield colloidal silica upon weathering cause precipitation within themselves of bodies of copper and zinc silicate. Limestone inhibits supergene sulphide deposition, and nonreactive rocks favor it.

Igneous Rocks and Associated Ores

Even before the genetic connection between most hypogene mineral deposits and magmas became widely realized, certain associations between specific igneous rocks and specific kinds of mineralization had become generally known. The fuller realization of the genetic associations stimulated field and laboratory investigations, with the result that the group of recognized associations has been broadened. These associations have been summarized by Buddington, whose work supplies some of the following remarks.

A genetic connection between ores and volcanic rocks is often impossible to establish because both may have emanated from undisclosed reservoirs at depths. Further, ores in extrusive rocks may have come from hidden intrusives. Consequently only intrusives will be considered.

Even with intrusives, a genetic association is often difficult to establish, and commonly mere conjecture is considered evidence. In the absence of direct connection, the tendency is often to assume that a given intrusive in a mineral district must be responsible for the ores present. In many cases a genetic connection is impossible to establish. In others, it is often difficult to determine to which of several separate intrusives the ores may be related, assuming that they are related. Many intrusives are composite, and ores may be related to one facies and not to another. Or, they may be connected with no specific single facies but with an underlying magmatic reservoir that yielded both facies and ores. In the case of small intrusions, the question arises as to whether the ores are related to that mass or to an underlying body of which it is an offshoot. Mining men are commonly inclined to assume a relation between ores and dikes, and much inadvisable exploration has been undertaken upon such an assumption. Generally there is no direct genetic connection, but rather both dikes and ores have sprung from the same reservoir. If assumptions are to be made, and direct evidence is lacking, it is generally safer to infer that both hypogene ores and minor intrusives have emanated from an unknown underlying magmatic reservoir since each is a phase of igneous activity.

Ores may be genetically related to specific intrusives as, (a) magmatic concentrations, and (b) magmatic emanations given off from the magma during its consolidation. The genetic relationship is generally obvious in the case of magmatic concentrations (Chap. 5·1). Magmatic injections, however, are less obvious. The products of magmatic fluids offer the greatest difficulty in establishing genetic relationship to specific intrusives. The following criteria aid in establishing genetic associations:

1. Deposits of magmatic concentrations enclosed within a specific intrusive; e.g., chrome deposits in peridotite.

2. Magmatic injections within or nearby specific intrusives; e.g., magnetite deposits in and about anorthosite.

3. Deposits of contact-metasomatic origin confined to the intruded rocks adjacent to a specific intrusive; e.g., copper-magnetite-garnet deposits adjacent to monzonite.

4. Deposits confined to the periphery but absent from the interior of a specific intrusive; e.g., porphyry coppers.

5. Deposits in or near roof pendants of older rocks surrounded by an intrusive; e.g., Mackay, Idaho.

TABLE 13

Igneous Rocks and Associated Ores

Rock Type	Associated Ores and Minerals	Important Examples
Kimberlite-eclogite	Diamond	African occurrences
	Garnet (pyrope)	Diamond pipes
Peridotite-pyroxenite	Chromite	World-wide occurrences
	Platinum metals	Bushveld, South Africa
	Chrysotile asbestos	World-wide occurrences
Norite	Nickel-copper sulphides	Sudbury; South Africa; Norway
Gabbro-anorthosite	Titaniferous magnetite	Bushveld, South Africa; New York; Iron Mountain, Wyo.; Sweden
	Ilmenite	Norway; Ivry, Allard L., Quebec
	Native copper	Lake Superior; Norway; Japan
Dolerite (diabase)	Silver-cobalt-nickel	Cobalt, Ontario; Germany
Diorite-monzonite	Magnetite	Banat, Hungary
	Copper	World-wide
	Gold	World-wide
Granodiorite-quartz monzonite-quartz diorite	Magnetite-hematite	Fierro, N. Mex.; Ely, Nev.
	" Porphyry " coppers	United States; Chile
	Base metals	World-wide
	Gold-silver	World-wide
	Molybdenum	Climax, Colo.; Bingham, Utah
	Tin-tungsten	Nevada; California; Bolivia
Syenites	Magnetite	Kiruna, Sweden
	Gold	Canadian Shield
Nepheline syenites	Corundum	Ontario; Russia
Granite and granite pegmatites	Tin	World-wide
	Tungsten	Burma; Nevada; Bolivia
	Uranium and radium	Katanga; Great Bear Lake; Joachimsthal

6. Spatial distribution of deposits with respect to a specific intrusive or to several similar intrusives; e.g., tin deposits of Cornwall, England.

7. Coincidence of time of formation of deposits and of an intrusion of an igneous mass or its differentiates; e.g., Bisbee, Ariz.

8. Zonal distribution of mineralization with respect to a specific intrusive; e.g., copper-lead-zinc ores about porphyry intrusive, Bingham, Utah.

9. World-wide association of specific ores and specific kinds of intrusives; e.g., nickel-copper ores with norite.

In general, the basic rocks have associated with them more limited but more definite groups of ores than the silicic rocks, and the latter have a greater variety but less specific association of ores. Certain ores occur with more than one rock type as, for example, copper, which occurs abundantly with rocks of intermediate composition and also with basic rocks.

Some consistent associations between specific rocks and specific ores are shown in Table 13.

Localization of Ore Deposits with Batholiths

W. H. Emmons has published at length on the relationship between ore deposits and batholiths. Such relationships, if established, would indicate important ore controls. He points out a feature long recog-

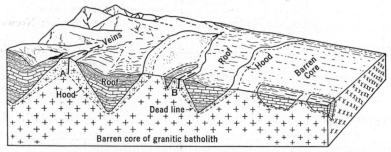

FIG. 6–3. Idealized batholith showing barren core, dead line, hood, roof, roof pendants, and location of veins. (*Emmons, Gold Deposits of the World.*)

nized, that many ore deposits tend to be concentrated in roof pendants, in the upper parts of stocks and batholiths, and in the surrounding invaded rocks but that the central parts of batholiths are generally barren of deposits. He concludes that the magmatic fluids tend to accumulate in cupolas in the upper parts of batholiths, from which they stream upward into the outer frozen shell and into the invaded rocks. The cupolas would thus be the centers of distribution and, in that way, the localizers of deposits.

Emmons recognizes (1) the roof, composed of the invaded rocks; (2) the hood, or the upper frozen part of a batholith; (3) the core, generally barren of minerals; and (4) the dead line, below which economic deposits rarely form. (Fig. 6–3.) The depth of erosion determines which of these features is exposed on the surface. The economic deposits form in the roof and above the dead line in the hood, within 1 mile of the contact in deeply eroded areas. He also recognizes three types of cupolas — summit cupolas, intermediate cupolas, and trough cupolas. Few deposits occur around the trough cupolas, and most are associated with the summit cupolas, from which they commonly extend outward in zonal arrangement. Contact-metasomatic

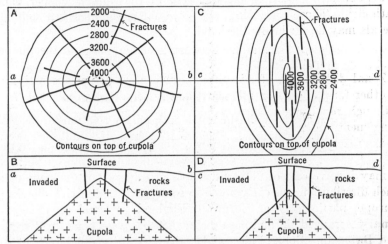

FIG. 6-4. Conical cupola with radial fractures (A, B), and more common elongated cupola with elongated fractures (C, D). (Emmons, Gold Deposits of the World.)

deposits may lie close to the summit cupolas. The batholith itself may not be exposed, but the cupola may project upward as a stock. Two types of stocks are recognized, circular and elliptical (Fig. 6-4).

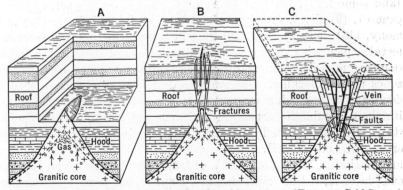

FIG. 6-5. Cupolas and associated hood and roof fractures. (Emmons, Gold Deposits of the World.)

The rare circular ones are likely to be accompanied by radial fissures and the common elliptical ones by parallel fissures. Fingers of the batholith also give rise to parallel fissures (Fig. 6-5). Thus Emmons considers that the form of the batholith has controlled the orientation of the fracture systems, which in turn serve as loci of ore deposition. It should not be overlooked that cupolas would probably solidify early and thus may not be ore loci.

Pegmatites containing metals form in the hood and roof of the batholith. Those containing feldspar, quartz, and mica but lacking in metals may lie far below the hood.

Zonal Distribution of Mineral Deposits

Zonal distribution of minerals about centers of igneous activity is another form of ore control, sometimes referred to as the *zonal theory* although *zonal arrangement* would be a better term. In a zonal arrangement, the ores or minerals formed at high temperatures lie closest to the magmatic source and the lower-temperature minerals lie in more distant zones. The zoning may be vertical or horizontal or both, and it may involve ores of different metals or minerals of the same metal. Such distribution was early noted in some of the old mining districts of Europe, particularly at Cornwall, England. Several occurrences of zonally arranged tin and base metal deposits were noted by Ferguson and Bateman. Spurr, Sales, and many others have pointed out such distributions in specific districts in America, and Emmons has assembled much information indicating widespread occurrence.

An ideal sequence of metals presented by W. H. Emmons is, from the parent intrusive outward, (1) barren zone with quartz, (2) tin, (3) tungsten, (4) bismuth and molybdenum, (5) gold, (6) copper, (7) zinc (and some lead), (8) lead (with some zinc, silver, copper, and manganese), (9) silver, (10) barren zone, (11) gold and silver, (12) antimony, (13) mercury, and (14) upper barren zone. Of course no such metal sequence is to be found in any one mineral district. Parts of the sequence may occur in one district and other parts elsewhere, or some zones may be lacking. For example, in Cornwall, England, in some deposits tin lies below copper, and copper below zinc and lead; at Bingham, Utah, the arrangement is concentric about the Bingham stock, with copper ores in and about the intrusive, copper and lead-zinc ores farther out, and silver-lead ores still farther out. At Zeehan, Tasmania, tin lies in granite; a zone of magnetite, tungsten, and copper is outside of it; a lead, zinc, stannite, copper zone is next outside; and farthest is a siderite zone with lead, zinc, silver, nickel, antimony, and manganese. At Butte, Mont., Sales has clearly demonstrated a pronounced zoning of copper minerals and metals, even within individual veins. In the central and hottest zone are copper ores consisting chiefly of quartz, pyrite, enargite, bornite, and chalcocite; in the next or intermediate zone enargite and chalcocite are less, chalcopyrite and tennantite are important, and toward the margin sphalerite is common — and galena and rhodochrosite are present; the peripheral zone is characterized by zinc, silver, and manganese, and the dominant minerals are sphalerite, galena, rhodochrosite, and calcite, with subordinate rhodonite, ankerite, and barite. The zonal arrangement is both horizontal and vertical.

A zonal arrangement of the minerals is presumed to be a function of their temperature of deposition. As distance from the igneous source increases, the temperature drops and successively lower-temperature minerals are deposited, the least soluble ones being deposited near the source. However, it should be pointed out that factors other than temperature are involved in the deposition of minerals. Pressure, concentration, relative concentrations, reactions with wall rock, reactions within the solution causing progressive precipitation, chemical complexes, and other factors enter in. The relative concentration of certain substances, which may cause precipitation of one or other minerals, is given too little consideration. Garrels has shown that chemical complexes produce zoning in reverse order. Nevertheless, there are many examples of zonal arrangement according to temperature gradient. In many districts or deposits an orderly sequence is lacking, and gold and silver have been deposited near the surface. Spurr called these " telescoped " deposits.

It should be noted that there are several contradictory facts in connection with zonal distribution. In Bolivia some of the inner high-temperature minerals, such as tin, also occur high up among lower-temperature minerals. Tungsten behaves similarly. On the other hand, some of the large lead deposits (Sullivan, British Columbia, Broken Hill, Australia) occur in the deep zone along with high-temperature minerals. Emmons, however, states that zoning is to be expected only with summit cupolas in the roofs of batholiths and not in hood deposits deep in roof pendants. Lindgren points out that the Mississippi Valley lead-zinc deposits occur high up, where only barren gangue would be expected, and that " in the epithermal veins there is no zoning except a common impoverishment in depth." Also zinc may occur below copper and lead below zinc. These are called reversals, which Emmons thinks can generally be explained. He considers some hypogene reversals are due to exceptionally high concentrations of certain salts of metals, causing a precipitation of the more soluble salts ahead of the less soluble. He explains other reversals by the action of supergene enrichment. There are, however, many that are not adequately accounted for, and so many discrepancies occur as to cast doubt on the zonal arrangement being more than ideal.

Metallogenetic Epochs and Provinces

In a broad way metallogenetic epochs have acted as large-scale localizers of ore. Most mineral deposits of igneous affiliations represent but one event within a period of igneous activity. They are formed in regions and periods of igneous intrusion, and these in turn are generally associated with periods of crustal disturbance and orogenic revolutions. Consequently, such mineral deposits are related in time and place to periods of crustal and igneous activity that have ocurred at definite periods in the earth's history. These constitute metallogenetic or minerogenetic epochs. The areas within which specific types of mineral deposits have been relatively concentrated constitute metallogenetic provinces. Other types of mineral-forming processes not associated with igneous activity, such as sedimentation, occur in periods and places of quiet sedimentation, and these likewise are metallogenetic epochs and provinces. Weathering that gives rise to surficial concentrations of economic minerals by mechanical, residual, and chemical processes has operated at all times and places, and the resulting deposits, except bauxite, do not fall within definitely recognized metallogenetic epochs or provinces.

The concept of metallogenetic epochs was first developed by De Launay and furthered by the studies of Lindgren, Spurr, and many others. Considerable confusion exists in the literature between metallogenetic provinces and epochs.

Metallogenetic Epochs

Not all mineral deposits of the world fall into definite metallogenetic epochs. Nevertheless on each continent the major number of deposits do fall within clearly defined, if rather broad, periods.

Precambrian. The Precambrian the world over was a period of great and varied mineral formation, perhaps because of the great length of time involved. It embraces many epochs of ore formation. Periods of igneous activity and accompanying ore formation alternated with periods of deep weathering, sedimentation, and accompanying mineral deposition. Many large deposits or iron, chromium, gold, silver, copper, nickel, tungsten, and some lead and zinc were formed in this period. Three-quarters of the world production of gold comes from Precambrian deposits. All the great Precambrian shields are areas of intense mining activity.

In *North America* the Precambrian was characterized by the formation of iron, titanium, nickel, copper, gold, silver, and zinc. Examples in the West are the gold deposits of the Black Hills, Wyoming, New

Mexico, and Arizona; the gold-lead-zinc deposits of the Pecos mine, N. Mex.; the copper deposits of Jerome, Ariz., and southern California, and the iron ores of Wyoming. In eastern North America there are the extensive iron and copper formations of the Lake Superior region, the magnetite deposits of New York, New Jersey, Quebec, and the ilmenite deposits of New York and Quebec. In the Canadian Shield are the famous gold deposits of Porcupine, Kirkland Lake, Rouyn, Little Long Lac, and others; the great nickel-copper deposits of Sudbury, Ontario, and Lynn Lake, Manitoba, the silver-cobalt deposits of Cobalt, Ontario, the pyritic copper-gold deposits of the Noranda district, Quebec, the heavy sulphide deposits of zinc, copper, silver, and gold of the Flin-Flon district of Manitoba, and the greatest ilmenite deposits of the world in Quebec. These are but large examples of many similar deposits.

In *Europe*, the Precambrian epoch was also marked by mineralization of iron, copper, gold, and some lead and zinc, the most notable being the great magnetite deposits of Kiruna. Most of the deposits occur in the Fenno-Scandia Shield. In *Asia* again, Precambrian deposits of iron, gold, copper, and manganese are found in Siberia and India, and tungsten in China.

The great Precambrian platform of southern, and central *Africa* includes vast deposits of gold, chromium, iron, manganese, asbestos, and also copper, tin, tungsten, nickel, and platinum. *Australia* was marked by wide gold metallization during the Precambrian. In *South America* gold, iron, and manganese are typical of the Precambrian of Brazil and Colombia.

Middle Paleozoic Era. Paleozoic metallogenetic epochs are less well defined than the Precambrian and later ones. There is a distinct middle Paleozoic epoch marked by important sedimentary deposits. Sedimentary iron deposits of this period are widespread in north and central Europe, in the Clinton of the United States, and in Newfoundland. Salt and gypsum deposits of this epoch occur in the eastern United States, Russia, and southern Europe. Middle Paleozoic metallization is less well marked. There are the Caledonian deposits of copper, nickel, and titanium, and chromium in Norway; gold and copper-gold in Australia and New Zealand, and weak gold metallization in Nova Scotia.

Late Paleozoic-Hercynian. The Hercynian folding between the Carboniferous and Triassic, and its attendant granitic intrusions, resulted in metallization of tin, copper, silver, gold, zinc, lead, and platinum in Europe. To this period belong the tin lodes of Cornwall and Saxony, the silver-gold lodes of Spain, France, and Germany, and the great pyritic deposits of Rio Tinto, Spain. Deposits of silver, gold, copper, zinc, and lead of this epoch are found in Russia, China, Japan, Burma, and Malaya, and platinum in the Urals. To the same epoch belong the tin, tungsten, gold, and bismuth deposits of Australia and New Zealand. Most of the deposits are of the deep seated type. There are hardly any representatives in North America.

Permo-Triassic Epoch. This epoch embraces the widespread salt, gypsum, and potash deposits of Germany, Russia, Persia, Texas-New Mexico, and other places. Also to it belong the disseminated copper deposits, possibly of sedimentary origin, in Germany and Turkestan. In general it was a period of aridity with widespread deserts and evaporation pans. In western North America extensive marine phosphate beds were laid down.

Jurassic Period. In Europe the Jurassic was marked by broad incursions of the sea and the deposition of the oölitic iron ores of central Europe and of England. In America there was extensive igneous activity in the Jurassic, and copper was associated with these intrusions in California, British Columbia, and Alaska.

Late Mesozoic Epoch. This was one of the great mineral epochs of western North America. Batholithic and stocklike intrusions of Jurassic and Cretaceous age occurred along the Pacific Coast from Mexico to Alaska, including the Sierra Nevada, Idaho, and Coast Range batholiths. Other intrusions took place in Idaho, Washington, Montana, Nevada, and Arizona, accompanied by widespread metallization. Deposits of gold, silver, copper, zinc, and lead lie in the marginal parts of the intrusives and in the nearby invaded rocks. The gold belts of California and Alaska fall in this epoch, and the many deposits that flank the Coast Range batholith belong here also. To this epoch also belong the gold, silver, copper, lead, and zinc ores of the outer zones of the Japanese islands and of parts of Russia.

Early Tertiary Epoch. This was a world-wide epoch of extensive and rich mineralization. Intrusions of intermediate character, such as granodiorite and monzonite porphyry, broke through the Cretaceous and older formations in stocklike bodies and brought with them a train of precious and base metals.

In *America* these intrusions occurred in the Cordillera, mainly in the Eocene as a part of the Laramide revolution, and gave rise to contact-metasomatic, replacement, and cavity-filling deposits of copper, silver and gold, lead and zinc, and molybdenum. The large porphyry copper deposits of southwestern United States as well as the large molybdenum-bearing deposits of Colorado, Utah, and New Mexico belong to this epoch. Other examples are Butte, Coeur d'Alene, Rossland,

Tintic, Leadville and adjacent regions of Colorado, and Chihuahua, Mexico. There are also many rich gold-silver veins of this epoch in the Central American countries.

In southern *Europe* extensive metallization accompanied the Alpine intrusions of granite rocks and gave rise to deposits of many metals. Examples are the lead-zinc deposits of the Alpine Trias, southern Spain, Sardinia, Serbia, and perhaps of the Harz; the iron deposits of Banat and Elba; the gold-silver deposits of central France, Spain, Hungary, and Transylvania; the silver-cobalt-nickel ores of Germany; the chrome deposits of Greece and Turkey; the emery deposits of Turkey; the uranium-radium of Joachimsthal; and the mercury deposits of Almaden, Austria, and Italy.

In *Asia* representatives are gold, silver, copper, lead, zinc, and antimony deposits in the East Indies, Philippines, and the inner Japanese zones.

In *South America* are disseminated and lode replacements, and fissure veins, mostly of mesothermal or rarely hypothermal types related to Tertiary intrusives of intermediate composition. The ores yield mainly copper and silver but also contain gold, tungsten, tin, lead, zinc, antimony, and mercury. Examples include the three great copper deposits of Chile, each characterized by enargite, and the silver treasures of Potosi and Cerro de Pasco.

Africa has a few early Tertiary representatives of lead, zinc, antimony, and quicksilver deposits in the Mediterranean section. In northern *New Zealand* are many rich gold-silver deposits of this epoch.

Late Tertiary Epoch. After the early Tertiary intrusions, orogenic disturbances accompanied by vast extravasations of lava occurred around the Pacific belt. They were particularly evident in the Americas, notably in the western United States and Mexico, where they gave rise to rich epithermal gold-silver veins, accompanied by tellurium and antimony but rarely by lead, zinc, or copper. Representative examples are Comstock, Tonopah, Goldfield in Nev.; San Juan and Cripple Creek in Colorado; De Lamar in Idaho; and Pachuca, Guanajuato, San Luis Potosi, and El Oro in Mexico. Probably some of the deposits assigned to the early Tertiary epoch belong here also.

The occurrence of mineralization in distinct epochs, many of which were of short duration and separated from each other by long intervals of time, is further proof of the association with igneous activity.

METALLOGENETIC PROVINCES

Certain regions characterized by relatively abundant mineralization dominantly of one type are referred to as metallogenetic provinces.

Such provinces may contain mineralization of more than one epoch, each superimposed upon the other, but essentially of the same type. In Arizona, for example, copper metallization took place in the Precambrian and recurred in the Permian, the early Tertiary and the late Tertiary epochs. It has been dominantly a copper province throughout geologic times.

North America. This continent affords many fine examples of metallogenetic provinces. An outstanding one is the Canadian Shield gold belt (Precambrian) that extends for 2,000 miles from Great Slave Lake to eastern Quebec. The gold deposits, consisting of replacement lodes and veins are generally similar and lie in or near belts of altered sediments and lavas close to or in granitic and syenitic intrusives. Gold quartz is the chief ore; sulphides or arsenides are scarce, and some tellurides are present. This province, which includes all the well-known gold camps of Canada, makes Canada the third or fourth ranking gold-producing country of the world.

The Mississippi Valley region constitutes a large zinc-lead metallogenetic province where zinc and lead ores occur in the 500 miles from Wisconsin to Oklahoma. The deposits are cavity fillings and replacements and are not in direct association with known igneous rocks, although present knowledge indicates magmatic affiliations. Minerals other than sphalerite and galena, with lesser pyrite or marcasite, are rare. This province is the largest lead-zinc producing area in the world. Most of the production comes from the Tri-State, Joplin, and southeastern Missouri areas. The Tri-State area has for the last 50 years yielded 50 to 80 percent of the United States production and 20 to 30 percent of the world production.

The southwestern copper province includes the southern parts of Arizona, New Mexico, adjacent Mexico, and extends into parts of Utah and Nevada. This province for many years provided about one-third of the world's copper production. The deposits are mainly " porphyry coppers," replacements in limestone, and contact-metasomatic, mainly in association with monzonitic porphyry intrusives of the early Tertiary epoch. Important Precambrian deposits occur also. The ores are dominantly copper, but they contain some gold and silver, molybdenum, and a little lead and zinc. The important mining districts within the province are Bisbee, Globe, Ray, Miami, Inspiration, Chino, Ajo, Jerome, Morenci, Burro Mountains, Cananea, Nacozari, and to the north, Ely and Rio Tinto in Nevada, and Bingham, Park City, and Tintic in Utah.

A northern copper province centers in the rich district of Butte, Mont., in association with the Boulder batholith, surrounded by a bevy of lesser mining camps. The ores are dominantly copper, marked by the presence of abundant enargite, and contain considerable gold and silver. The amount of silver, zinc, lead, and manganese increases outward.

The Pacific Coast gold belt constitutes a metallogenetic gold province, embracing the Mother Lode belt of California, the Grass Valley region, and extends northward into Oregon and British Columbia. The deposits are gold quartz veins accompanied by few sulphides and rarely by tellurides.

A Basin and Range gold-silver metallogenetic province embraces parts of Nevada, southern California, Utah, northern Arizona, and New Mexico. This province includes the late Tertiary gold-silver deposits that yielded the precious metal treasures of the middle and late nineteenth century.

A silver-lead metallogentic province lies in Idaho and adjacent parts of Montana, Washington, and British Columbia. The center is the rich Coeur d'Alene district. The ores are galena with included silver minerals, pyrite, and sphalerite, with quartz and carbonate gangue. The deposits are fissure veins and replacement lodes.

Another important base metal-precious metal province lies in part of Colorado and extends into Utah. This includes deposits of both the early and late Tertiary epochs, which consist chiefly of mesothermal and epithermal fissure veins and replacement deposits in limestone. The ores yield silver and gold in association with lead, zinc, and copper. Leadville, the San Juan, and many of the other well-known Colorado mining camps lie within this province.

A native copper metallogenetic province centers in the Lake Superior part of Michigan and embraces a part of Ontario and Wisconsin. The deposits are associated with basic lavas and intrusives and consist chiefly of native copper, with subordinate chalcocite and silver and zeolites.

North of this region lies the nickel-copper metallogenetic province of the Sudbury basin, Ontario. It includes some 50 deposits about the margin of the basin and a few outside, all in close relationship to norite. Their mineralogy is monotonously similar, the assemblage consisting chiefly of pyrrhotite, pentlandite, chalcopyrite and containing nickel, copper, platinum metals, and gold. In the center of the basin are some gold and lead-zinc deposits.

A silver-cobalt-nickel province centers around Cobalt, Ontario, including also South Lorrain, Gowganda, Elk Lake, and other minor localities, all in association with dolerite sills. The deposits are mesothermal fissure veins consisting of generally similar ores yielding silver and cobalt and containing native silver, cobalt and nickel arsenides, and bismuth, in carbonate gangue.

A mercury province embraces the Coast Range of California and Mexico, with representatives also in Nevada, Oregon, Washington, and British Columbia. These belong to the late Tertiary epoch and consist of veins and impregnations. The ores contain cinnabar and metacinnabarite. Examples are the New Almaden, New Idria, and Great Eastern mines. A generally similar belt heads at the Terlingua mine in Texas and extends southward into Mexico as far as San Luis Potosi.

A large and rich silver-lead province lies in north-central Mexico where there are widespread, silver-bearing, lead-zinc ores of similar character and great richness. They occur as replacements in limestone and in fissure veins. This province includes also the well-known manto-deposits. Representative districts are Santa Eulalia, Sierra Mojada, Mapimi, Parral, and San Francisco del Oro.

A rich silver-gold province lies farther south, embracing parts of Guanajuato, San Luis Potosi, Queretaro, and Hidalgo. The deposits belong to the late Tertiary epoch and contain quartz and minor sulphides. Representative examples are Pachuca, Guanajuato, and El Oro. Similar provinces are found in other parts of Mexico and Central America.

Other smaller metallogenetic provinces occur in other parts of North America, such as the copper-pyrrhotite ores of the southern Appalachian region, the gold-quartz deposits of the Appalachians, the Clinton sedimentary iron ore province of the East, the saddle-reef gold deposits of Nova Scotia, the asbestos province of Quebec and Vermont, the molybdenum centers of Colorado and Arizona-Sonora, the tungsten area of Nevada-Idaho-California, the copper area of the Kennecott region, Alaska, the Juneau, Alaska gold belt, the north Pacific Coast copper belt, and the Texas-New Mexico-Kansas potash basin.

South America. The great Chilean copper belt, extending northward into Peru, includes three great mines and supplies a considerable part of the world's copper output. The copper ores are sulphide replacements (in part oxidized) containing chalcopyrite, bornite, and chalcocite and marked by enargite. They are associated with Tertiary monzonitic intrusives.

A rich silver province that has yielded vast treasure includes parts of Bolivia, Peru, and Argentina. Silver dominates, but gold is present and sulphides are sparse. The deposits are Tertiary fissure veins, and Potosi, Cerro de Pasco, Morococha, and Oruro are examples.

A smaller metallogenetic province of tin-silver veins, marked by the presence of the uncommon tin sulphide stannite, and tungsten and bismuth, includes part of Bolivia. Examples are Potosi, Oruro, Chocaya, Llallagua, and Huanuni.

Europe. Metallogenetic provinces are less well defined in Europe than in America and are of small size. *Scandinavia* contains a nickel province that is a small counterpart of Sudbury, Ontario. There is also a magnetite-hematite province in the North, centering at Kiruna, where the largest magnetite ore body of the world occurs. A tin province with zonally arranged copper and lead-zinc deposits centers in Cornwall, *England*. *Germany* has a cobalt-silver province in the Erzgebirge embracing Annaberg and Schneeberg and extending into Joachimsthal in Bohemia; the classical silver-lead district around Freiberg; and the lead-zinc lodes of the Harz. In *Spain* there is a silver-lead-zinc province of similar mineralogy, centering around Linares, the pyritic copper area of massive sulphide bodies in the Huelva district, and the restricted quicksilver area around Almaden. The Alpine intrusions were accompanied by scattered lead-zinc deposition in southern Europe, constituting a distinct metallogenetic province.

Other provinces are the gold-telluride deposits of the Transylvanian Carpathians and the sedimentary iron ore areas of Luxemburg and France.

Africa. Africa boasts of several distinct metallogenetic provinces. The most outstanding is the Witwatersrand and Orange Free State gold belts, provinces of identical gold occurrence and mineralogy that extend for distances of many scores of miles.

The copper belts of Northern Rhodesia and Katanga constitute one of the great copper provinces of the world. It stretches 400 km in a northwesterly direction. The Rhodesian copper sulphide ores, with the distinctive cobalt sulphide linnaeite, are similar throughout and apparently are the same materials from which the Katanga oxidized copper ores with cobalt were derived.

The diamond province is unique. Kimberlite pipes occur in numbers in the Cape Province, Orange Free State, and Transvaal. Many more are known than are worked, and the widespread distribution of alluvial and beach diamonds indicate a former more widespread distribution of pipes than at present. Similar pipes are known in Tanganyika, Angola, and central Africa. The diamonds are of different ages; the pipe deposits are post-Karoo, and those in the Rand banket are Precambrian. The diamond province is unusual in that the containing rocks represent the deepest known materials of the earth's crust.

Another larger metallogenetic province of Africa includes the Bushveld and Great Dyke areas of ultrabasic rocks carrying chromite, magnetite, platinum, and asbestos. Occurrences extend for a length of 280 miles in the Transvaal but reach the greatest economic importance in Southern Rhodesia.

Other smaller but distinct metallogenetic provinces of Africa are the asbestos province of the Transvaal and Rhodesia, the manganese provinces of Postmastburg and the Gold Coast, and the gold and tin provinces of the Belgian Congo, Nigeria, and Uganda. The various African provinces are noteworthy for the removal of great thickness of cover by erosion.

Asia. Asia contains several metallogenetic provinces. The intrusions accompanying the Ural disturbances made that region richly metalliferous. There is a platinum province, which for many years yielded most of the world's supply of that metal. There is also a copper province made up of pyritic copper replacement deposits and a lead-zinc province made up of massive replacement and vein deposits composed chiefly of galena and sphalerite. Another lesser copper province is a small counterpart of the German Kupferschiefer. In addition there are smaller provinces of chromite and of asbestos.

The Siberian Shield area has many resemblances to the Canadian Shield and includes a large gold province of widely scattered quartz lodes, many of which have been deeply eroded to form extensive gold placers. This province is largely responsible for the gold that makes Russia the second largest gold producer of the world.

Other Asiatic metallogenetic provinces are the manganese province of Georgia; the extended antimony province of Hunan and the tungsten province in China; the inner gold-silver zone and the outer copper zone of the Japanese Islands; the great tin province of Malaya; and the gold province of the Island arc, including the Philippines.

Australia. Despite its large mineral production, Australia has fewer well-defined metallogenetic provinces than the other continents. Of outstanding interest is the great gold province of Victoria, New South Wales, and Tasmania that has yielded over one and one-half billion dollars in gold, from lodes, saddle reefs, and placers. There is also a distinct tin-tungsten-molybdenum province and a copper-gold province in Tasmania. The silver-lead-zinc area of Broken Hill, one of the greatest lead-producing areas of the world, is a lead province of small area. Many of Australia's numerous mineral deposits do not appear to fit into distinctive metallogenetic provinces.

General. The fact of metallogenetic provinces is undisputable but their cause is a problem. Why should the Arizona province be so rich in copper or the Canadian Shield province in gold? It cannot be advanced that the Canadian gold province is such because prolonged erosion has cut down to the horizon of a gold zone of mineral deposition. This presupposes a former zonal arrangement, which may or may not have been there, but also presupposes the impossibility that throughout this great area all the deposits, if zoned, had their gold zones at the same horizon. Individual deposits exhibit no evidence of zoning. The exposures of the Arizona copper province also belie such a possibility, since their altitudes vary with respect to the intrusions. It can only mean that the magmas of this region were rich in copper, those of northern Canada and the Sierra Nevada in gold, and those of Idaho in silver and lead. The vagaries of erosion and depth of exposure can be eliminated. But why certain intermediate magmas are rich in copper and others in gold is unknown. One can understand that a constant association of tin with silicic granites and of chromium with ultrabasic rocks is a matter of differentiation whereby the chromium collects with the basic differentiate and the tin with the silicic residual magma. But the predominance of such diverse metals as gold, copper, lead, silver, zinc, molybdenum, and others in genetic association with identical host rocks cannot be a matter of simple differentiation associates. The underlying magmas must have been richer in one or other of the metals. It is possible that metals, under the conditions prevailing in the depths where magmas originate, may undergo some unknown atomic change by means of which electrons are removed or added, thereby creating one or other of the metallic elements.

Selected References on Mineral Controls

General Reference 9. Chap. 6, Structural and Stratigraphic Controls, by B. S. BUTLER et al.; *broad features*. Chap. 7, Ores and Batholiths, by W. H. EMMONS. Chap. 8, Rocks and Associated Ores, by A. F. BUDDINGTON; *relationships*. Chap. 12, Utilization of Geology by Mining Companies; *mine methods, structural controls*.

Structural Geology. B. STOČES and C. H. WHITE. Van Nostrand, New York, 1935. *An elementary textbook with many mining examples of ores and structures.*

Structural Control of Ore Deposition. C. D. HULIN. Econ. Geol. 24:15–49, 1929. *Relation of mineralization in time and place to structural deformation.*

Structural Control of Ore Deposition in the Uncompahgre District, Colo., with Suggestions for Prospecting. W. S. BURBANK. U. S. Geol. Surv. Bull. 906–E, 1940. *An excellent discussion.*

Physical Chemistry and Stratigraphic Problems. G. R. MANSFIELD. Econ. Geol. 32:533–549, 1937. *Stratigraphic controls of nonmetallic mineral deposits.*

Structural Environment of Bendigo Goldfield. J. B. STONE. Econ. Geol. 32: 867–895, 1937. *Saddle-reef type of structural control of gold ores.*

Gold Deposits of the World. W. H. EMMONS. McGraw-Hill, New York, 1938. Chap. 1. *Relation of ores to intrusives.*

Gites Minéraux, I. L. DE LAUNAY. Paris, 1913. Chap. 13. *Examples of structural and wall rock controls.*

The Ore Magmas, I. J. E. SPURR. McGraw-Hill, New York, 1923. Chap. 9. *Several structural controls.*

General Reference 8. Chap. 33. *Metallogenetic epochs.*

Copper Deposits of the World. XVI Int. Geol. Cong., Washington, 1936. *Many fine examples of structural control.*

Structural Control of Ore Deposition in the Red Mountain, Sneffels, and Telluride Districts of the San Juan Mountains, Colo. W. S. BURBANK. Colo. Sci. Soc. Proc. 14, No. 5:141–260, 1941.

Factors in the Localization of Mineralized Districts. C. D. HULIN. A. I. M. E. Tech. Pub. 1762, 1945. *Larger structural controls for ore.*

General Reference 10. *Excellent examples of structural controls.*

Some Physical Characteristics of Certain Favorable and Unfavorable Ore Horizons. OLAF N. ROVE. Econ. Geol. 42:57–77; 161–193, 1947. *Testing of carbonate host rocks for physical features.*

General Reference 15. *Comprehensive treatment of all Canadian mining districts or mines from control standpoint.*

Mining Geology. H. E. McKINSTRY. Prentice-Hall, New York, 1948. Chaps. 11, 12, 13. *Stratigraphic and structural controls as ore guides.*

CHAPTER 7

FOLDING AND FAULTING OF MINERAL DEPOSITS

A rock mass that is folded or faulted is of scientific interest, but when a rich and valuable vein is abruptly cut off by a fault, the economic interest transcends the scientific. The life of a mine may be jeopardized, large expenditures may be necessary to find the faulted portion, and important legal aspects may also be involved. Geologic principles must be utilized in unraveling the complexities of faulting and folding and in the search for faulted portions of ore bodies. The broader features of folding and faulting as affecting large rock masses are not considered here; this chapter deals only with their relation to mineral deposits.

Folding

The folding of rocks affects mineral bodies that may be contained in them. It is particularly noticeable in sedimentary beds, and interbedded mineral deposits, such as iron ore or coal, may become intricately folded. Many deposits are rendered uneconomic. Many deposits in metamorphic rocks are also closely contorted by folding. Close folding commonly passes into faulting.

KINDS AND TERMINOLOGY

A simple bend is a *monocline;* an up-arched fold is an *anticline* (Fig. 7–1); a down-arched one is a *syncline;* and a hat-shaped one is a *quaquaversal* fold. The *limbs, axis,* and *axial plane* are shown by Fig. 7–1. A fold may be *open* (Fig. 7–1), *closed, upright, inclined,* or *recumbent* (Fig. 7–2), or *symmetrical* (Fig. 7–1) or *unsymmetrical* (Fig. 7–2). The axis or crest of an anticline may be horizontal or *pitching;* the *pitch* is shown in Fig. 7–3. A fold with horizontal axis yields parallel outcrops; a pitching fold yields converging outcrops (Fig. 7–3).

Folds are *similar* or *parallel, concentric, fan, box,* or *isoclinal* (Fig. 7–4). *Competent* folds are composed of strong beds capable of supporting the overlying beds; *incompetent* folds are composed of weak beds that collapse under pressure. Two *competent beds* may include an intervening *incompetent bed,* with resulting contortion, as some

327

coal seams. The minor folds of an incompetent bed resulting from slipping between folded competent beds are called *drag* folds.

DECIPHERING OF FOLDS

Folded mineral beds must be deciphered in order to (1) correlate deposits, (2) determine whether separate deposits arc parts of one or

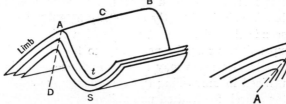

FIG. 7–1.   Anticline *A* (*symmetrical*), syncline *S*, with crest *C*, trough *t*, axis *A–B*, axial plane *A–D*, and limb.

FIG. 7–2.   *A*, Closed, asymmetrical anticline; *B*, overturned anticline.

more single beds, (3) determine the probable position of undisclosed portions of beds, and (4) indicate areas that may or may not be underlain by a given mineral bed. In addition, the crests of anticlines may indicate mineral loci, such as accumulations of petroleum or gold-bearing saddle-reefs.

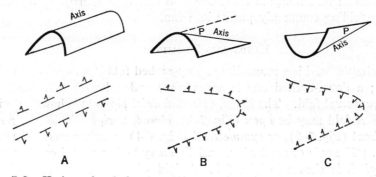

FIG. 7–3.   Horizontal and plunging anticline and syncline; dips and strikes plotted in plan.

Folding can be deciphered best in sedimentary rocks, particularly where a " key " bed is available. The simplest method is by observing and plotting dips and strikes, as in Fig. 7–3.

Reconstruction may be made by means of a cross section. Whether the point *x* of Fig. 7–5*A* is an anticline, or is a syncline, as in Fig. 7–5*B*, rests on whether the bed x is older or younger than the bed P. In gentle folding, the assumption is usually sound that an overlying bed is younger than an underlying bed, but in complex folding this may not be the case.

Care must be exercised to decipher correctly a sequence of dipping mineral beds. For example, in Fig. 7–6, A represents three outcrops of

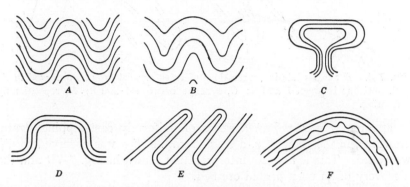

FIG. 7–4. Folds. *A*, Parallel; *B*, concentric; *C*, fan; *D*, box; *E*, isoclinal; *F*, incompetent and drag.

sedimentary iron-ore beds that may have resulted from: (*a*) three separate ore beds; (*b*) a single bed folded and reconstructed as in B; (*c*) a single bed folded as in C; or (4) a single bed repeated by faulting as in D. If the bed y contains basal pebbles of iron ore, the interpretation must be as in B because y would then be the younger bed; but if

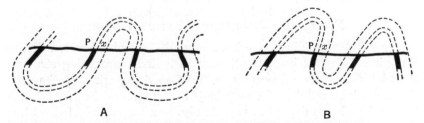

FIG. 7–5. Reconstruction of folded strata. Observed dips in solid black. *A*, reconstruction as an anticline; *B*, as a syncline.

bed z contains iron-ore pebbles then the interpretation must be as in C. If folding has caused the repetition, then bed y will appear on the hanging wall side of ore bed 1 and ore bed 3, but on the foot wall side of ore bed 2. However, if the repetition is due to faulting, bed y will appear on the hanging wall side of all three ore beds. If the relative age of beds x, y, and z cannot be ascertained, a drill hole d will determine whether the folding is as shown in B or C.

The deciphering of folding often rests on slight clues which may be overlooked by the unobservant but which if found bring reward. For example, in Fig. 7–7A, a sedimentary bed of iron ore c was being mined.

FIG. 7–6. Three possible interpretations of three parallel beds as observed in A; B, anticline between 1 and 2; C, syncline between 1 and 2; D, repetition by faulting.

It apparently occurred in a series of uniformly east-dipping strata. However, in the river at r, a single stratum of a was observed to dip westward. This led to the interpretation shown in Fig. 7–7B and the discovery of an unsuspected ore bed.

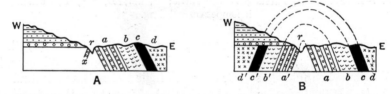

FIG. 7–7. A, Cross section showing iron-ore bed; B, interpretation based upon discovery of stratum x of formation a, in bed of creek, with resulting discovery of hidden ore bed at c'.

EFFECTS OF FOLDING ON MINERAL DEPOSITS

The effect of folding on mineral deposits may be direct, affecting the mineral deposit itself, or it may be indirect, producing structure that later localizes mineral deposits. The deciphering of the folds that have localized deposits may be just as important in mineral finding as though the deposits had actually been folded. An oil pool, or the continuation of the Homestake ore body, for example, could not be determined if the folding were not first known.

Thickening and Thinning. Folds commonly thicken or thin and include beds of soft, plastic, or brittle materials, such as coal seams, beds of salt, gypsum, potash or graphite, bedded bauxite, clay, marble, and beds of iron and manganese ore. A thinned portion may appear to be of uneconomic thickness, or a thickened portion may give an exaggerated impression of the average thickness. This change in

ancient rocks. Anticlinal folding causes loss of economic beds by exposing them to erosion (Fig. 7–14).

Physical Effects on Beds. Folding subjects beds to limb stresses, giving rise to cleavage or crushing; a building stone may be rendered unfit for quarrying, or commercial slate is formed. This effect tends to pulverize coal, commonly rendering it unfit for use or making it difficult to mine. The same effect, on the other hand, causes bedded iron ore to be more cheaply broken and mined.

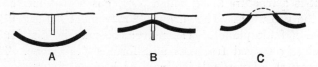

A   B   C

FIG. 7–13. Folded potash beds: *A*, Deeply buried; *B*, made accessible to mining by anticlinal fold; *C*, made to outcrop.

Creation of Favorable Structure. Folding creates many structural features that localize minerals, such as petroleum traps, saddle reefs, or drag folds. (See Chap. 6.)

Effect on Mining Operations. Folding affects mining and quarrying operations, rendering them easier or more difficult. A bed inclined by folding is generally more readily and cheaply mined than a horizontal one; in an inclined bed the ore can be drawn and loaded by gravity; ventilation is simpler, and timbering is less. On the other

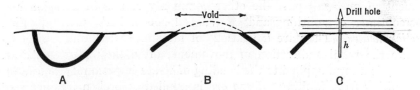

A   B   C

FIG. 7–14. Preserval of Roan Antelope copper bed in a syncline (*A*); loss of ore bed by folding (*B*); *h*, barren drill hole sunk to find ore bed (*C*).

hand, close or complex folding makes exploration and development more costly, renders haulage more difficult, necessitates more expensive mining methods and timbering, and often renders the deposit unworkable.

Faulting

Faulting plays a more important role than folding in connection with mineral deposits. Few mineral districts are free of faults, which are commonly a scourge to the mine. Many ore deposits would not exist except for them, and many do not exist because of them. Faults serve as channelways for mineralizing solutions; they localize ore; they repeat themselves along former fractures now mineralized by ore; they cut and displace mineral bodies; they repeat and cut out beds; and they form and drain petroleum reservoirs. They play a manifold part.

TERMINOLOGY AND COMPONENTS

A fault is a fracture along which there has been relative displacement of the two walls. The surface of movement is the *fault surface* (miscalled a fault plane). The movement may be distributed over several closely spaced fractures and form a *fault zone*. The fault surface may stand vertical, inclined, or horizontal; most faults are inclined. The wall above the surface of an inclined fault is the *hanging wall*, that below is the *foot wall*. The inclination of the fault surface with respect to a horizontal plane is the *dip* (*hade*, rarely used, is the complement of the dip), and the *strike* is the direction of its intersection with a horizontal plane. The *apex* of a fault is its top part, and if this reaches the surface it forms an *outcrop* or *fault line*. A fault may be *open* if its two walls are separated or *closed* if they are not. If the two walls have rubbed against each other, a clayey attrition product called *gouge* is commonly formed. The space between the walls may be filled by broken rock fragments or *fault breccia*. A large slab of rock caught between the two walls is a *horse*. The movement of one wall past the other commonly results in abrasion and smoothing, or polishing and grooving of one or both surfaces, called *slickensides*. They are indications of fault movement and may indicate the relative direction of movement; often they are misleading. *Drag* is a term applied to the bending of strata in proximity to a fault, resulting from faulting. *Drag ore* is applied to broken fragments of ore disrupted from the faulted ends of an ore body and contained within the fault between the faulted portions. Drag ore often leads the miner along a fault toward the lost portion of the ore body. The dislocation of a geologic body on either side of the fault, measured along the shortest horizontal distance, is called the *offset*. The wall that moves upward relative to the other is called the *upthrown* side, and the opposite one, the *downthrown* side; these terms are strictly relative since it can rarely be told which side moves up or down.

Components. A point on the hanging wall of a fault relative to one on the foot wall may move up or down, or right or left, or any combination of those movements. It is rare to find either vertical or horizontal movement alone; both occur together. Consequently, fault movement is tridimensional and resolves itself into three components. Four terms are employed to designate this movement. In Fig. 7–15, the movement along the fault surface from the point o to o' is the *displacement* (or net *slip*), d, measured along the fault surface. But the point o' has also moved to the right, and this is the *strike-slip, ss*, measured along the fault surface. The amount of vertical drop of o' is the *throw, t*, and the horizontal increase of distance between o and o', measured at right angles to the fault, is the *heave h*. The throw, heave, and strike-slip are the three components of the displacement. Displacement is also commonly used as a general term referring to disloca-

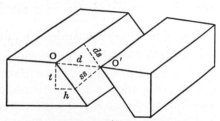

FIG. 7–15. Components of faulting. d, Displacement; t, throw; ss, strike slip; h, heave; ds, dip slip.

tion or movement. Slip is always measured on the fault surface. *Dip-slip* (*ds*, Fig. 7–15) is another term sometimes used, referring to the slip, up or down, measured on the fault surface normal to the strike of the fault.

On the surface it is generally the outcrop or fault line that is observed, but in mines it is the fault surface itself that is seen and commonly the apex is not visible. Thus, it is the slip that is generally determined. In mines also, the terms *vertical* and *horizontal displacement* are often used. The former refers to the vertical dislocation of a geologic body as disclosed in the workings and is generally equivalent to the throw. The horizontal displacement, a common term in mines, refers to the dislocation of a geologic body (i.e., vein) as actually measured on a mine level.

KINDS OF FAULTS

Faults are as long as 100 miles or more, and displacement ranges from a fraction of an inch to several thousand feet. Faults commonly terminate in flexures, or they may abut other faults.

The kind of fault is determined by the direction and character of movement. There are three types of movement on faults: (1) *parallel* movement or block movement, in which straight lines on opposite sides of a fault remain parallel after movement, as in Fig. 7–15; (2) *rotary* or *pivotal* movement, like that of scissors; (3) movement which is a combination of (1) and (2), i.e., there is a rotary movement, but the hinge line is also translated along the surface of the fault (sometimes called *trochoidal*).

It is often assumed that fault motion is generally parallel, but mining experience indicates that commonly it is rotary. Consequently, estimates of throw based on one location will not apply to the same fault in another location. Strictly speaking, all those faults that terminate by a gradual dying out of movement are rotary. The two sides of any short section of a rotary fault, however, are essentially parallel.

Normal and Reverse Faults. Faults are generally classified as normal or reverse according to the character of the movement. A *normal*

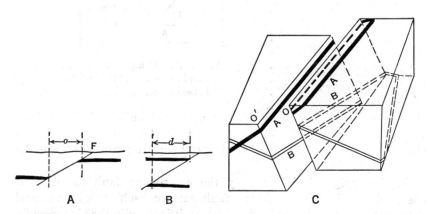

FIG. 7–16.   *A,* Cross section of a normal fault, eliminating ore bed at *o; B,* reverse fault duplicating ore bed at *d; C,* apparent normal fault with reference to stratum *A,* and same movement gives an apparent reverse fault with respect to *B,* but a point *o* indicates that it is a normal fault.

fault is one in which a point on the hanging wall has moved downward relative to the same point on the foot wall (Fig. 7–16A); the fault surface dips toward the downthrown side. Most high angle faults encountered in mines are of this movement, hence the name. A *reverse* fault has the opposite movement (Fig. 7–16B). The terms are commonly applied with respect to strata, and apparent reverse faulting can result from faults of horizontal movement, or normal faults cutting obliquely dipping strata, as in Fig. 7–16C. However, if the terms are applied with respect to a *point,* as *o* in Fig. 7–15, and not to *strata,* the definitions hold and no confusion results.

A normal fault increases the width of a given block of ground, (*o,* Fig. 7–16A) and a reverse fault decreases it, as *d* in Fig. 7–16B. Thus, in an area underlain by a horizontal coal seam, normal faulting will produce a blank area and reverse faulting will produce a surface area underlain by a double thickness of coal.

Slip Faults. Faults are often named according to the direction of slip on the fault surface, as *dip-slip, strike-slip,* and *oblique-slip* faults. These are useful terms because oftentimes one can determine that there has been movement up or down the dip of the fault surface, or

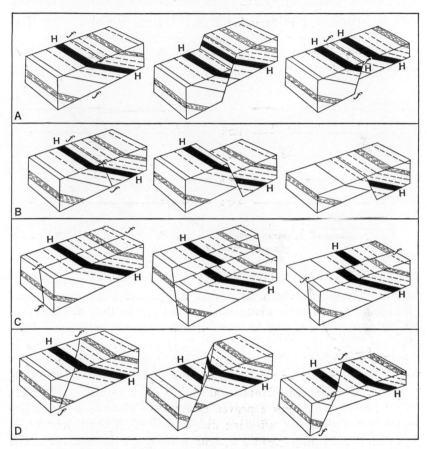

Fig. 7–17.   Faulting in relation to sedimentary strata. *A,* Strike fault with repetition of beds after erosion and, *B,* with elimination of beds after erosion; *C,* dip fault, with parallel offset beds; and *D,* oblique fault with offset beds.

obliquely along it, without being able to tell whether it was normal or reverse movement.   Also, a strike-slip fault, with horizontal movement parallel to the strike, is a definite dislocation that is neither normal nor reverse.

Thrust Faults. These are low-angle reverse faults, not often found in direct connection with mineral deposits but commonly encountered on a large scale in areal geology.   The dip may be only a few degrees;

the length is as much as a few hundred miles, and the horizontal displacement as much as a few tens of miles. Complete mountain ranges have ridden over adjacent territory along low-angle thrusts. The movement is spoken of as *overthrusting* and *underthrusting,* a distinction generally difficult to determine.

Faults in Relation to Stratified Rocks. Faults are also grouped as subclasses according to their relation to stratified rocks. A *strike fault* is one whose strike is parallel to the strike of the strata and whose dip crosses the dip of the strata (Fig. 7–17*A*). A *dip fault* is one

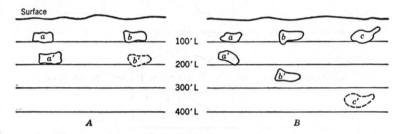

FIG. 7–18.   Effects of *A,* parallel movement, and *B,* rotational movement on displaced portion of an ore body.

whose strike is approximately normal to the strike of the strata (Fig. 7–17*B*). Extended over a large area it is called a *transverse fault.* An *oblique fault* is one whose strike is oblique to that of the strata (Fig. 7–17*C*). A *bedding fault* is one that coincides with the bedding of strata. This type is common but rarely evident.

Rotary Faults. Detailed underground mapping on a large scale has indicated that rotary or rotational faults are much more common than is generally supposed. Rotational rather than parallel movement should be suspected unless proven otherwise. The important difference between them, as affecting dislocations of mineral deposits, is that with parallel movement a measurement of the throw made at one point will indicate the expected throw at any other point on the same fault; but with rotary movement, the throw would be different at each point along a fault.   For example, on Fig. 7–18*A,* a longitudinal section parallel to the surface of a fault, a replacement ore body in a favorable limestone bed is shown on the foot wall side at *a* and its faulted position on the hanging wall side at *a′,* a throw of 100 feet. With parallel movement, the faulted end of another known ore body, *b,* in the same horizon, may be expected at *b′.* In Fig. 7–18*B,* a similar ore body *a* is found at *a′,* a throw of 100 feet, but a companion body, *b,* in the same horizon was encountered at *b′* with a throw of 200 feet.   Therefore, the faulted end of a parallel body, *c*, would be expected at *c'* on the 400-foot level.

There are three varieties of rotary faults: (1) *Hinge faults*, which are hinged at one end, and the movement, increasing outward, is con-

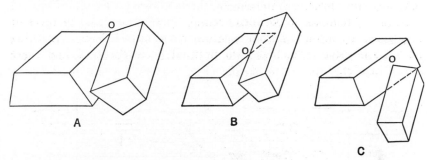

FIG. 7–19. Varieties of rotary faults: *A*, Hinge fault; *B*, pivotal fault; *C*, trochoidal fault.

fined to *one* side of the hinge line (Fig. 7–19*A*). (2) *Pivotal faults*, which also rotate about an axis perpendicular to the fault surface, but with movement increasing outward from *both* sides of the pivot (Fig. 7–19*B*). (3) *Trochoidal faults*, which resemble hinge faults except

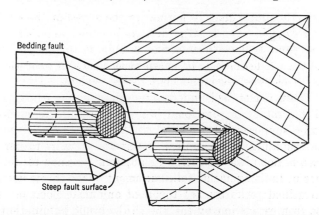

FIG. 7–20. Interbed fault, Kennecott Mines, Alaska. Ore body (shaded) displaced by a fault which is parallel to bedding and which does not show on level above.

that the hinge point also slips along the fault surface, as in parallel faulting (Fig. 7–19*C*). Their decipherment is generally baffling.

Other Types of Faults. *Step faults* are closely spaced faults of parallel strike and dip forming blocks tilted in one direction and resembling a giant flight of steps. Generally they are normal faults. *Block* *faults* bound blocks of ground. *En echelon faults* are fault zones in which individual short faults are staggered, and strike around 45° to the trend of the major zone. *Branch faults* are minor branches of a major fault. *Graben faults* are paired faults of parallel strike that generally dip toward each other and have allowed a trough of ground to subside (relatively) between them. *Horst faults* are the reverse; they generally dip away from each other and have left an intervening elevated block of ground. Two unusual occurrences of faults are shown in Figs. 7–20 and 7–21.

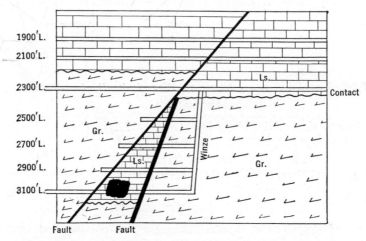

Fig. 7–21. Complicated faulting in Jumbo mine, Kennecott, producing a block of deeply buried, unsuspected (from the 2,300-foot level) favorable limestone (Ls) and included ore body. Gr, greenstone.

EVIDENCE OF FAULTING

Evidence of faulting, as applied to mineral deposits rather than to large areas, is not always present or apparent or easy to distinguish. Indications of faulting are as follows:

1. In stratified rocks by (*a*) repeated or omitted beds (to be distinguished from repetition by folding), (*b*) abrupt termination of beds along their strike, (*c*) angular contacts of strata, and (*d*) strata out of place stratigraphically.

2. Slickensides, grooves, and polish on the terminating body, which indicates that it has moved but does not show if movement is pre- or post-mineral. Such evidence, therefore, must be used in conjunction with other lines of evidence.

3. Fault gouge along the structural surface. The remarks under (2) apply here also.

4. Slickensided and polished ore, when present, is positive evidence of post-mineral faulting.

5. Drag ore within the fault is generally positive evidence.

6. Drag or bending of structures within the ore body, or parallel to it, adjacent to the structural surface; this is not always positive evidence.

7. Brecciation or fracturing of the ore adjacent, and particularly parallel, to the structural surface.

8. A post-mineral fault may displace a vein and later be mineralized. The later ore generally contains inclusions of the earlier ore.

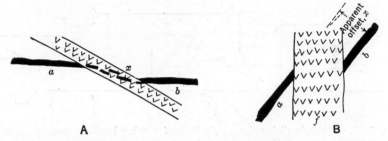

Fig. 7–22. *A*, Refraction of fissure vein across dike producing apparent displacement; *B*, apparent displacement of fissure vein produced by dike widening; no actual faulting.

9. A line of springs or underground water courses indicates a permeable fissure that may be a fault.

It is the unhappy experience of most mining operators to find that a paying ore body abruptly terminates underground at some structural surface. The customary conclusion is that it has been faulted. Because of the expense of trying to find the displaced portion of the ore body it is essential to determine if it actually has been faulted. Positive evidence of faulting should, therefore, be sought.

Apparent Displacement. The features enumerated above often give the appearance of faulting where no faulting has occurred.

1. The termination of ore against a pre-mineral fault is very deceptive, and post-mineral faulting can be assumed only after positive evidence is gained. Still more deceptive and more common is ore terminating against a pre-mineral fault that has had just enough post-mineral renewal of movement along it to shatter or shear the abutting ore without displacing it.

2. Deflection of a fissure vein that crosses a dike gives an apparent displacement of the vein. In Fig. 7–22*A* the parts *a* and *b* appear to be offset by an amount *x*, which is due only to the refraction of the

vein by the dike.   Deflection is more likely to occur if the angle of incidence (Fig. 5·4–15) is not large; reflection, may occur if the angle of incidence is small.

3. A mineralized fissure of an intersecting cognate system (whose members do not offset each other) may abut another member of that system and appear to be faulted by it.   Positive evidence of such faulting is lacking.

4. A fissure vein, No. 2, may abut a vein of earlier mineralization, No. 1, and appear to be faulted by it.   Positive evidence of such conditions is also lacking.   Apparent dislocation is even more pro-

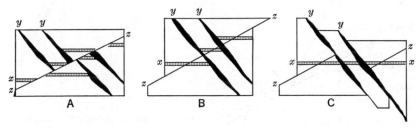

FIG. 7–23.   Multiple dislocations.   A, Part of a mine level as mapped showing fault systems x, y, z;   B, after restoration of movement on z;   C, after restoration of movement on two faults of system y.

nounced if vein No. 1 has post-No. 2 vein renewed movement along it.   In such cases vein No. 1, even though it has shattered vein No. 2, will contain no inclusions of vein No. 2; it is only apparent faulting by vein No. 2.

5. Apparent displacement is also produced by the formation of a wide fissure or a dike that intersects a vein at a low angle.   In Fig. 7–22B the dike or fissure, f, has without faulting separated the two ends of vein a–b, producing an apparent offset x.

Multiple Dislocations.   In many mineral districts, such as Butte, Mont., multiple dislocations form complicated patterns.   Faults of several ages displace earlier faults that in turn displace earlier veins. Fig. 7–23 represents a relatively simple example showing mapped effects and the resolution of double faulting.   Fig. 7–24A shows a vertical section of two veins and two faults.   In A there would appear to be two veins displaced by fault z; B shows the restoration before such faulting, and C before faulting on y.

Recurrent Displacement.   It is common experience to find that movement recurs along old faults.   A fault represents an adjustment to accumulated strain, and subsequent stress in the same direction naturally finds relief along the earlier formed lines of weakness.   Some faults show several such recurrent movements. Recent gravels and even mine timbers have been found displaced along older faults. In an Alaskan mine an ice-filled fault fissure was observed to contain slickensided ice resulting from post-glacial movement. Such recurrent

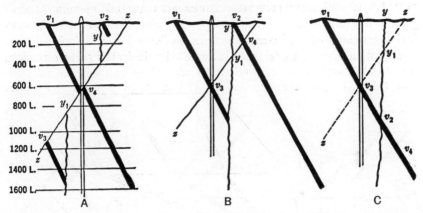

FIG. 7–24. *A,* Apparently two veins disclosed in mine workings as mapped in section; *B,* restoration before faulting on *z; C,* restoration before faulting on *y.*

movements are indicated by progressively greater displacement of older formations (Fig. 7–25).

Puzzling patterns of intersecting fault veins are sometimes encountered, which lead to paradoxical deductions. These, however,

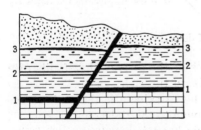

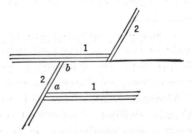

FIG. 7–25. Recurrent movement resulting in progressively greater displacement of older beds.

FIG. 7–26. Vein 1 displaced by vein 2 at *a,* and vein 2 displaced by vein 1 at *b.* Which is the earlier? Produced by recurrent movement along vein 1.

clarify when recurrent movement is taken into consideration. For example, Fig. 7–26 is a plan of two veins, the two segments of each of which can be positively correlated. Vein 1 is older than vein 2. At *a,* vein 1 is definitely cut off and displaced by vein 2; at *b,* vein 2 is definitely cut and displaced by vein 1. Each vein, therefore is apparently older and younger than the other — a paradox. However, consideration of recurrent movement shows that vein 1 was displaced by vein 2, and vein 2 was later displaced by recurrent movement on vein 1. If still later recurrent movement occurred on vein 2, complexity would result.

Displacement in Different Directions on Same Fault. Each member of a system of faults offsets an intersected vein in the same direction.

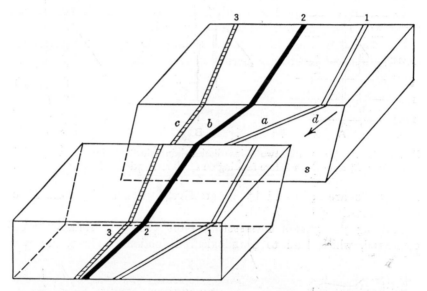

Fig. 7–27.   Opposite displacements of different veins by a single fault

In many mining camps there are rules of the district that faults of a given direction always throw the vein to the right, or to the left, as the case may be. This has come to be so axiomatic that, where one vein is thrown to the left and another vein to the right on disconnected parts of apparently the same fault, doubt has arisen as to the correlation of the two disconnected parts. This feature is entirely a function of the relation of the dip of the veins to the direction of displacement. A normal fault with no strike-slip will cause opposite displacements of different veins as shown by Fig. 7–27.

Two or three such veins with throws in different directions are helpful in determining the amount and direction of the throw and strike-slip of a fault.

Offset and Direction of Displacement. The direction of offset of a vein bears a direct relation to the angles between the trace of a vein on a fault surface and the direction of displacement. For example, in Fig. 7–27, the trace *b* of vein 2 on fault surface *s* is parallel to the direction of displacement *d;* therefore there will be no offset; if the trace *a* of vein 1 and the direction of displacement *d converge* upward, then the offset is *opposite* to the direction of the strike-shift; if the trace *c* of vein 3 and the direction of displacement *d diverge* upward, then the offset is in the *same direction* as the strike-shift.

EFFECTS ON FAULTING OF MINERAL DEPOSITS

Faults exert both pre- and post-mineral effects on ore bodies; the former are mostly helpful, the latter mostly harmful.

Pre-Mineral Effects. Since faults localize ore they exert an important pre-mineral effect on mineral deposits. Most faults that become

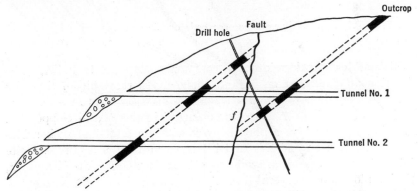

Fig. 7–28. Repetition by reverse faults. Apparently two veins, as disclosed in drill hole. Tunnel No. 2, expected to cut the two veins disclosed in Tunnel No. 1, where fault *f* was overlooked.

mineralized are of small displacement. Faults of large displacement may serve as feeding channelways. Faulting also has given rise to broad fault zones that have become large mineral bodies. Pre-mineral faults may displace beds that subsequently become mineralized, and in order to follow the ore from one fault segment to another it is necessary to decipher the faulting just as though the faulting had been post-mineral.

Faulting, however, is most evident in its post-mineral effects upon mineral deposits.

Repetition and Omission. The commonest effect of faulting is the repetition and omission of beds and veinlike bodies.

Horizontal or inclined coal seams and mineral beds are duplicated by reverse faults (Fig. 7–28) and may be dropped below mining depths by normal faults (Fig. 7–16A). Inclined beds are repeated, or are eliminated in outcrop by strike faults (Fig. 7–17B). Dip faults produce offset outcrops of the same bed with the same strike (Fig. 7–17C), and oblique faults produce similar results (Fig. 7–17D). Also reverse faults may conceal a deposit that might easily be overlooked unless the stratigraphy is known (Fig. 7–29). Step faulting by faults of the same system produces a series of repetitions of a dipping deposit.

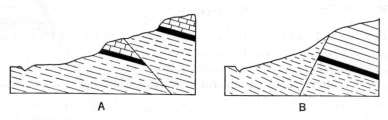

FIG. 7–29. *A*, Apparently two iron-ore beds on hillside; repetition by faulting. *B*, Iron-ore bed concealed and overlooked if stratigraphic column is overlooked.

Slickensides. Slickensides are commonly supposed to indicate the direction of movement on a fault surface. In some places they do; generally they do not. Small slickensides may represent only one part of the fault motion. A case was observed in the Miami Copper Co. mine in Arizona where a plate of rock within a fault had slickensides in one direction on one side and in another direction on the opposite side. Slickensides on a well-exposed fault surface at Kennecott, Alaska, plotted over several levels, yielded an S-shaped track; any one of the readings would be misleading. Large grooves or "mullions" several inches or feet from trough to trough are fairly reliable indications of direction of movement.

It should be realized that, in faulting, the path of motion from starting point to finishing point need not have been a straight line connecting the two but may have been a rather circuitous path, leaving variable directions of slickensides. Ordinarily the slickensides, or even grooves, represent only the direction of the last part of the movement on a fault surface.

Brecciation. Faulting is commonly accompanied by brecciation of the walls, giving rise to fault breccia. This is rarely present with small single faults but does occur with step faults and large faults. The breccia may fill the fault from wall to wall and its width may be many feet. The material is generally angular and has come mostly from the adjacent walls. It is not uncommon, however, to find fragments that can be correlated with distant formations. Where ore bodies are intersected, fragments of ore occur in or make up the breccia.

The cut-off portion of the ore body is also commonly brecciated or shattered.

Drag and Drag Ore. Faulting commonly is accompanied by a *drag* or bending of strata, dikes, or veins against the fault surface (Fig. 7–30). The drag generally bends in the direction of the slip on the opposite wall and is thus used as an indicator of the direction of move-

ment. This is not infallible, however, since " reverse drag " occurs; i.e., the drag instead of bending toward the direction of slip, bends in the opposite direction. This may be caused by a later fault movement in the opposite direction to that which just caused the drag or by a normal fault that cuts through an inclined anticline or syncline.

FIG. 7–30. Drag or bending of strata by faulting.

A fault that cuts off an ore body generally contains *drag ore*, or brecciated fragments of ore, within the fault between the two displaced ends of ore (Fig. 7–31). Drag ore indicates the direction in which to turn along a fault to find the displaced part of the ore body. In Butte, Mont., a post-mineral fault was mined for the large amount of drag ore that was incorporated in it from the intersection of several veins. Drag ore is generally discernible as ore

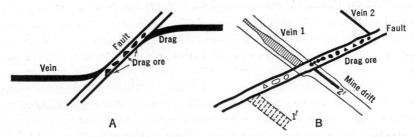

FIG. 7–31. *A*, Drag ore, from faulted portion of vein; *B*, mine drift following vein 1 encountered fault and drag ore, and mistakenly followed vein supposed to be vein 1, but actually was vein 2, and vein 1 lay undisclosed to right. (*Adapted from description by Knopf.*)

fragments, but tests have shown that it may be finely comminuted and incorporated in the gouge, from which it may be released by washing and panning.

Drag ore may also be deceptive as to direction of movement. For example, vein 1 at Rochester, Nev., was followed to fault (Fig. 7–31*B*). Drag ore was encountered on the left side and followed into vein 2′, which was thought to be a continuation of vein 1; actually 2′ proved to be the faulted end of vein 2 indicating that the faulted end of 1 should lie to the right.

Effects of Oxidation and Enrichment. As shown in Chapter 5·7, faults play an important role in oxidation and supergene enrichment in serving as channelways and barriers. The crushed zones accompanying faults facilitate complete and extensive enrichment.

Legal Effects. Legal considerations also result from faulting. Under extra-lateral mining laws the owner of a vein apex is entitled to follow the vein downward on its dip, within his end lines, beneath the land of another; if the vein is faulted, the owner of the apex is entitled to that portion of the displaced part within his end-line extensions, even though it lies in another's ground and provided further that the displaced parts can be proved to have been originally one and the same vein. Faulted veins do not carry the legal rights of junctions. In extra-lateral mining law, if a vein branches upward and each branch has a different surface owner, the prior locator of a branch is entitled to the rest of the vein below the junction, within his end-line extensions. However, if the junction is caused by a fault and is not a natural split the prior locator has no right to the lower part of the vein below the fault junction and this part belongs to the one who holds the apex of the continuous vein. Other, more complicated legal considerations are involved by faulting.

Effect on Mining Operations. Faulting produces both harmful and favorable effects on mining operations. Horizontal seams and beds may be broken into upthrown and downthrown blocks, making continuous level workings impossible, with consequent greater mining and haulage expense. Coal seams become crushed and explosive gases are released, creating hazards.

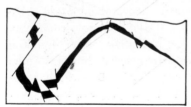

Fig. 7–32. Iron deposit of Nucice, Czechoslovakia, bent and broken by folding and faulting. (*After Stočes and Šusta.*)

Inclined deposits are also broken into discontinuous blocks necessitating crooked workings, pipe lines, and tracks, and requiring more raises and winzes, thereby increasing mining costs (Fig. 7–32). Veins may contain long blank spots as a result of faulting, thus decreasing the ore reserves in a given tract of ground and requiring a much greater development footage of workings per ton of ore developed. Such faulting introduces uncertainty into the estimation of possible ore reserves.

Heavy water flows may be encountered in faults in underground workings. In many places water under considerable pressure may be impounded behind faults, which, when punctured by mine workings, may permit the water to burst forth with violence, hazarding life and mine, and adding greatly to pumping expense. Such conditions necessitate drilling in advance of mine headings to test for hidden water pressures.

Faulting along mineral bodies may crush the ore and mix in waste rock, causing ore dilution. More harmful, however, are faults that crush the walls of mineral bodies, thereby allowing caving of waste rock and causing great dilution of ore during ore extraction, causing hazards from rock falls in stopes, and necessitating expensive timbering. This feature is worst with (a) faults that closely parallel mineral deposits, (b) those that are parallel in strike but of opposite dip, and (c) those that intersect mineral bodies at acute angles in strike and dip. Caving of crushed walls may be so extensive as to destroy entire stopes. Faults often prove bothersome in quarrying or in opencut mining by providing planes of weakness along which large masses of rock tend to slip into the excavations, and by causing minor rock falls.

Faults also cause disturbances of ground that affect mine workings away from ore bodies. They cause ground caving and swelling and necessitate expensive timbering or other support. Cases are known where workings could not be kept open against such swelling and caving, the largest timbers being crushed like matchwood. If shafts penetrate large fault zones the shaft alignment is often disturbed, necessitating costly maintenance and sometimes abandonment. It is generally difficult and expensive to keep open and maintain drifts and crosscuts that follow large faults. Rarely, narrow noncaving faults cheapen underground drifting by providing easier breaking ground and a " free face " to break to. In general, however, mine workings, particularly shafts, should be laid out to avoid fault zones.

Miscellaneous Effects. Faulting also produces many minor effects on mineral bodies. Downfaulting of parts of mineral bodies may preserve them from erosion, or upfaulting may permit their erosion. Deposits inaccessible because of depth may be faulted up within reach of exploitation, and, conversely, accessible deposits may be faulted down beneath the depth of economic exploitation. Normal faulting may cause the omission of bedded or inclined deposits beneath a given tract of land, and reverse faulting may double the mineral reserves. Faulting also conceals deposits by preventing their outcropping.

Faults make petroleum traps, and if they cut oil pools they allow the oil to become dissipated. They give rise to lines of surface springs, and impound and make water resources. If they intersect aquifers they may cause artesian springs or may allow stored water to become dissipated.

Search for Faulted Ends

The first step in the search for the faulted ends of mineral bodies is the positive establishment that 1, faulting has taken place; 2, it is of post-mineral age; and 3, a part of the mineral body has been displaced. Working out displacement of faults in three dimensions has been most carefully considered in mining camps. The following methods have been utilized: geologic, graphic, geometric, geophysical, and empirical rules.

Geologic Methods. The best way to find a faulted mineral body, if the necessary evidence is available, is by straight geologic methods. This is simplest in stratified rocks with recognizable horizons. An examination of the geologic column on the far side (i.e., displaced side of mineral body being followed) of a fault may disclose whether that side has moved relatively up or down. Fissures, dikes, shears, joints, or contacts may give a clue to the amount of strike-slip. This information may be obtained by either underground or surface mapping. If underground mapping does not disclose the needed clues, the surface mapping may do so. If the fault is not visible at the surface, it may be projected to the surface, and by means of structure contours the curving trace of the projected fault (and projected vein also) can be drawn upon a topographic map. Mapping of the strata on either side of the projected fault apex may then indicate the relative motion on the two sides. In unstratified rocks other faults, fissures, mineral bodies, or other structural surfaces must be used as key markers. Lacking identifiable markers, other veins of different dip cut off by the same fault may afford the necessary clues.

The direction of the movement may perhaps be inferred from drag ore, drag, and large slickensides, but, as previously pointed out, the last two factors are not always dependable; they must be used in conjunction with other data.

The district habit of other faults may indicate the direction, if not the amount, of displacement. For example, in Butte, Mont., the east-west veins are always offset to the left by the northwest faults. An undetermined northwest fault in such an area therefore, would be expected to behave similarly.

Graphic Methods. When definite geologic markers are not available, as often happens, various graphic methods are helpful. A simple and

expedient one is to make a sheet of paper represent the foot wall
surface of a fault upon which is plotted all known foot wall geologic
features, and then have a superimposed sheet of tracing linen representing the hanging wall fault surface upon which is also plotted all
known hanging wall geologic features.   The linen may be moved over
the paper until geologic features coincide, thus giving the amount and
direction of movement of the hanging wall with respect to the foot wall.

A somewhat similar graphic method used by Stočes and others is
shown in Fig. 7–33, where X–Y represents a fault surface in vertical
projection; the heavy lines represent veins on the hanging wall side, and
the light lines those on the foot wall side; solid lines represent known,

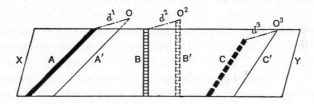

FIG. 7–33.   Graphic method of solving for faulted portion of vein; X–Y, fault
surface; solid lines, known veins; dotted lines, unknown parts; heavy lines, hanging wall veins; and light lines, foot wall veins.

and dashed lines unknown, veins.   The direction of displacement d
has been determined by reliable mullions or slickensides.   Veins A
and B are known on the hanging wall side, and A' and C' on the foot
wall side.   It is desired to find B', the faulted end of B, on the foot
wall side, and C on the hanging wall side.   Draw d^1 parallel to d until
it meets the projection of A' at O.   Then draw d^2 similar to d^1; from
O^2 draw B' parallel to B, and this indicates where B', the faulted end
of B, should lie on the foot wall.   Similarly draw d^3 from C, and from
O^3 make C' parallel to C, which gives the hanging wall position of
vein C'.

Another simple and effective method is to plot on a cardboard a plan
of all geologic features and workings, and, after cutting the card along
the line of the fault, to move one side back and forth until like features
coincide; a vertical section may be similarly used (Figs. 7–23, 7–24).
A similar method also effectively used by the author for more complex
cases is to construct light three-dimensional models of thin celluloid
with a double fault surface so that one part of the model can be shifted
over the other along the fault surface until like features coincide.
Stereograms or isometric projections, representing in three dimensions
the geologic features on each side of a fault, are also helpful at times.

The construction of detailed large-scale mine models has often proved useful, but expensive.

Geometric Analyses. The utilization of solid geometry and trigonometry in mathematical analyses of faults sometimes helps to locate the lost portion. Commonly, the number of variables involved does not permit such analyses or else the information available is insufficient. In simple cases, where the necessary information is available, these methods generally yield results, but then a mathematical solution is generally unnecessary. In the author's experience, so many mathematical solutions have been proved incorrect by subsequent actual finding of the lost portion that he has shaken confidence in such results and utilizes them only as a last resort.

Geophysical Methods. Geophysical methods, which have been utilized by the author, often can and often have found lost portions of mineral bodies. The electric and magnetometer methods are the most satisfactory (Chap. 10), but they all have limitations and their results require careful geological interpretation. Geophysical methods are expensive, relatively slow, and often yield geophysical " indicators " but not the desired lost ore body, and have to be tested by subsequent expensive underground exploration.

Empirical Rules. In the mining industry, where fortunes are lost by ore bodies being cut off, it is natural that the finding of the lost portions has received careful attention since the inception of mining. In consequence, before much of the present knowledge regarding faults was attained, empirical rules were formulated to guide the search for lost portions. Most of these were formulated for simple normal faults, for which they generally hold. Others apply to both normal and reverse faults, and do hold within certain limitations. There are, however, many classes of faults for which they do not hold. Consequently, the results cannot always be depended upon. They are to be considered only as a last resort for faults that have baffled other methods of solution, or by those not conversant with other methods. Since some of these rules are widely accepted they are mentioned briefly below. Some of them duplicate each other.

Schmidt's Rule. To find the faulted end, turn toward the obtuse angle. This does not apply to reverse or vertical faults or to normal faults when the vein dips at a greater angle than the component of the dip of the fault, and in the opposite direction.

Parks' Rules. (1) If the vein dips left and fault is toward one, then the lost vein is to left, or if away from one, lost vein is to the right; (2) if the vein dips right and fault toward one, then lost vein is right, or left if fault dips away; (3) if vein is horizontal, and fault dips toward one, then lost vein is higher, or lower if fault dips away.   These rules of course apply only to normal faults.

Carnall-Schmidt Rule.   After cutting through a fault, prospect on that side of the line of intersection of vein and fault indicated by the relative motion of the opposite block of ground.   The rule was formulated for normal faults of small strike-slip.

Zimmerman Rule for Flat Beds.   When a fault is struck on its foot wall side, the lost portion lies down the dip of the fault; if struck on its hanging wall side the lost portion lies above.   This was formulated for flat beds and normal faults.   The opposite procedure applies to reverse faults.   This rule works well for flat beds if it is known whether the fault is normal or reverse.

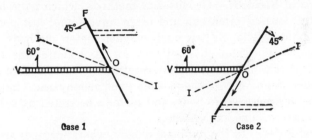

Case 1   Case 2

Fig. 7–34.   Two graphic solutions (in plan) of finding faulted portions of veins, according to Zimmerman's rule.   Heavy line is vein; light line is fault; dotted line is projection of line of intersection of the two.

Zimmerman Rule for Inclined Bodies.   (Fig. 7–34.)   If in driving along a vein V a fault F is met, make a plan map; plot the strike of the fault and the line of intersection of the planes of the vein and fault I; then the displaced vein lies on that side on which the plan of the intersection I makes the largest angle (angle FOI) with the plan of the fault.   This is a much-used popular rule and holds for a great many normal faults and some reverse faults.   It also holds for many hinge faults, both normal and reverse.   It has limitations, however. It does not hold with normal faults (1) when the trace of the movement is at a less angle with the horizontal than the trace of the vein; or (2) when fault and vein are parallel.   It holds with reverse and strike-slip faults only if the direction of displacement is the same as that of the trace of the vein on the fault, and at a less angle to the horizontal.   Consequently it does not hold for any reverse fault without strike-slip, or for any kind of reverse fault whose far wall has shifted to the right.   This means that the rule is practically useless as applied to most reverse faults.

Selected References on Faulting and Folding

Principles of Structural Geology. 4th Ed. C. M. NEVIN. John Wiley & Sons, New York, 1949. Chaps. 3 and 4. *General features of folding and faulting.*

Structural Geology. B. STOČES and C. H. WHITE. Van Nostrand, New York, 1935. Pp. 109–260. *Folding and faulting of ore deposits and mineral beds; excellent illustrations.*

Structural Geology. MARLAND BILLINGS. Prentice-Hall, New York, 1942. Chaps. 3, 8–11. *Excellent discussions and illustrations of folding and faulting.*

Dislocated Ore Bodies. Chap. 14 of **Mining Geology.** H. E. McKINSTRY. Prentice-Hall, New York, 1948. Pp. 343–363. *Good discussion and illustrations.*

Classification of Faults. M. L. HILL. Bull. Am. Assn. Petrol. Geol. 31:1669–1673, 1947.

General References 10, 15.

CHAPTER 8

CLASSIFICATIONS OF MINERAL DEPOSITS

Classifications attempt to arrange related subjects in logical order or sequence and thus help clarify a diverse assemblage. Classifications of natural phenomena, however, are seldom so comprehensive that all phenomena fit exactly into their own pigeonholes. This is the case with classifications of mineral deposits, which consist of substances that vary greatly in metal and mineral content, and in form, size, mode of origin, and value.

A classification should be logical, orderly, and permit clean-cut separation, as far as possible. It should not allow one group to fit equally well into two or more pigeonholes. A classification of people, for example, set up according to color, size, and head shape would be illogical, since tall men might fit under either color or head shape. So, in the early classifications of mineral deposits, material, form, texture, position, and origin were hopelessly given equal rank. Ever since Agricola first classified ore deposits, successive writers have attempted classifications of mineral deposits, none of which have attained unanimous endorsement. In any classification, uncertainty always arises as to which division certain deposits belong. Some investigators tend to consider classification as an ultimate objective; others refuse to worry much over it. In this book no formal classification of mineral deposits has been used. However, a review of classifications that have been used is desirable.

Early Classifications. Early schemes of classification, reviewed by Kemp, appeared in the middle of the nineteenth century and involved veins alone. Among them are those of von Wissenbach, von Cotta, and LeConte; but each illogically used for equal major divisions, form, origin, and position or materials; obviously veins could not be so classified. The latter half of the century saw a more logical development of classifications. First a group appeared based partly on form and partly on origin without subdivisions. Next came a more logical group based upon form alone, by von Cotta, Prime, Koehler, Callon, and Lottner-Serlo. Von Cotta divided deposits into I, Regular, with *A*, Beds, *B*, Veins; and II, Irregular, with *C*, Segregations and *D*, Impregnations. This group although logical was too simple and not sufficiently inclusive. It was followed by a more logical group based chiefly upon origin, by Grim, Von Groddeck, Pumpelly, and Phillips. Some of these classifications interspersed form and origin but none of them were sufficiently inclusive for the then-existing knowledge of ore deposits.

The end of the century saw the first group of classifications based entirely upon origin, such as those by Monroe, Wadsworth, Pošepńy, and Kemp. They were the forerunners of later genetic schemes, but they lacked sufficient subdivision to permit desired separation of deposits.

Later Classifications. Around the beginning of the twentieth century much attention was directed to schemes of classification, several of which were based upon origin. Most of them utilized terms that are still widely used today. Examples of them are:

Beck, 1904	Bergeat — Stelzner, 1904	Irving, 1908
I. Primary.	I. Protogene.	I. Bedrock deposits.
A. Syngenetic.	A. Syngenetic.	A. Syngenetic.
1. Magmatic segregations.	1. With eruptive rocks.	1. Igneous.
2. Sedimentary ores.	2. With sedimentary rocks.	2. Sedimentary.
B. Epigenetic.	B. Epigenetic.	B. Epigenetic.
1. Veins.	1. Cavity fillings.	1. Cavity fillings.
2. Epigenetic deposits not veins.	2. Replacements.	2. Replacements.
		3. Contact-metamorphic deposits.
II. Secondary.	II. Secondary.	II. Disintegration deposits.
A. Residual.	A. Residual.	A. Mechanical.
B. Placers.	B. Placers.	B. Chemical.

These schemes, like the accompanying ones, separated primary from secondary deposits and divided the primary into syngenetic, or those formed at the same time as the enclosing rock, and epigenetic, or those formed later than the enclosing rock. This is an advantageous distinction that permits broad scientific and practical conclusions to be applied to each group. The epigenetic groups are those formed by gases or liquids dominantly of igneous derivation, and are divided into subgroups based upon processes of origin. The cavity fillings in turn are subdivided according to form into fissure veins, saddle-reefs, amygdaloidal fillings, etc. The above schemes are genetic and simple;

the terms are everyday words and are readily usable by the geologist in the field and by the mining profession.

Beck in 1909 changed his classification as follows:

1, Magmatic segregations; 2, contact-metamorphic deposits; 3, fissure veins; 4, bedded deposits; 5, stocks; 6, secondary alterations; 7, sedimentary ore deposits; 8, detrital deposits. This is inconsistent in that process and form are ranked equally; genetically different types must go in the same group and the same deposit could be placed in more than one group.

A genetic scheme formerly used by the author was a modification of the 1908 classification by J. D. Irving, as follows:

I. Bedrock deposits.
    A. Syngenetic deposits: (1) igneous; (2) sedimentary.
    B. Epigenetic deposits.
        1. Cavity fillings: (*a*) fissure veins, (*b*) shear zones, (*c*) ladder veins, (*d*) stockworks, (*e*) saddle-reefs, (*f*) tension-crack fillings, (*g*) solution cavity fillings (caves, channels, gash veins), (*h*) breccia fillings, (*i*) pore-space fillings, (*j*) vesicular fillings.
        2. Replacement deposits: (*a*) massive, (*b*) lode, (*c*) disseminated.
        3. Contact-metamorphic deposits.

II. Disintegration deposits.
    A. Mechanical. *B*. Residual. *C*. Chemical.

Although the above classification has many advantages, it has some disadvantages, among which is the difficulty common to all classifications, that of determining into which group certain deposits should be placed. The distinction between replacement and cavity filling is at times difficult, although generally it can be determined if the deposit has been formed dominantly by one or the other process. The advantages of such a classification are that it is genetic, it is readily usable in the field by the geologist or mining man, and it employs familiar terms that are themselves descriptive of a deposit. The assignment of a deposit to a division is not always dependent upon subsequent laboratory work. Further, in order to classify a deposit, it directs attention in the field to the cause of ore localization, such as structural control, and the subdivisions take cognizance of the form of the deposit, which is one of the most striking observable features of an individual deposit. In this classification, once a deposit is fitted into its proper niche, broad geological conclusions affecting origin, occurrence, and search for other deposits immediately apply. It is, therefore, a valuable classification for the working geologist. For example, if an epigenetic deposit is determined to be a cavity filling of tectonic breccia, it indicates that mineralization by circulating solutions was later than the rock shattering and that other similar deposits might be found in the region where similar brecciation is known to occur or might be expected. Also, the distribution, shape, and continuity of the deposit may be inferred.

Lindgren's Classification. The desirable emphasis on genetic schemes of classification culminated in Lindgren's, which first appeared in 1911. He considered that most deposits have been formed by physicochemical reactions in solutions, whether liquid, igneous, or gaseous, which constitute one large class as distinct from those formed by mechanical concentration. These two subdivisions, therefore, constitute the main divisions of his classification, and practically all deposits fall under those produced by chemical concentrations. The main outlines of his classification are condensed below.

I. Deposits by Mechanical Processes.

II. Deposits by Chemical Processes.

    A. In surface waters.

	TEMPERATURE °C	PRESSURE
1. By reactions.	0–70	Medium to high
2. Evaporation.		

    B. In bodies of rocks.

1. Concentrations of substances contained within rocks:		
a. By weathering.	0–100	Medium
b. By ground water.	0–100	Medium
c. By metamorphism.	0–400	High
2. By introduced substances.		
a. Without igneous activity.	0–100	Medium
b. Related to igneous activity.		
(*a*) By ascending waters.		
(1) Epithermal deposits.	50–200	Medium
(2) Mesothermal deposits.	200–300	High
(3) Hypothermal deposits.	300–500	High+
(*b*) By direct igneous emanations.		
(1) Pyrometasomatic deposits.	500–800	High+
(2) Sublimates.	100–600	Low to medium

    C. In magmas by differentiation.

1. Magmatic deposits.	700–1500	High+
2. Pegmatites.	575 ±	High+

This is the most advanced genetic scheme that has yet been produced, and it has now become the most widely adopted classification. Its terminology is extensively used and parts of it are incorporated in many books. It is genetic, logical, and sufficiently complete that most deposits can be assigned to their proper pigeonholes. However, as in all other classifications, it is not always possible to determine into which group certain deposits should be placed.

Some deposits find no place in any of the classification groups but are treated by Lindgren under separate headings, such as "deposits of native copper," "regionally metamorphosed sulphide deposits," and "deposits resulting from oxidation and supergene sulphide enrichment." The last certainly deserve a place in any classification since many of the great copper deposits, for example, owed their initial operation to the process of supergene enrichment.

The chief basis of distinction between groups in the classification is the temperature and pressure of formation of deposits. Unfortunately, this is often determinable only after laboratory work or not at all, and many anomalous features appear.

Separation of deposits on the basis of temperature and pressure of formation, if determinable, should offer fairly precise lines of demarkation, but it is also obvious that it may be quite impossible to draw a distinction between deposits at the upper part of the hypothermal group, limited by an upper temperature of 300° C, from those of the lower part of the mesothermal group limited below by a temperature of 300° C.

All good classifications have some defects, and this also applies to Lindgren's. Some objections that may be pointed out are: (1) The classifications, divisions, and subdivisions do not themselves constitute appropriate names by which the deposits within them may be designated; for example, deposits of class IIA1, or "Deposits produced by chemical processes of concentration in bodies of surface waters by interaction of solutions," could be more satisfactorily and briefly called sedimentary deposits, which, except for bog iron ores, are all that are described under this grouping. The term sedimentary deposits at once connotes process of formation, geologic occurrence, shape, distribution, and other features. (2) The terminology of class IIB2b (a), or "deposits produced by hot ascending waters of uncertain origin, but charged with igneous emanations," which are more generally known as hydrothermal deposits, and which constitute most of the deposits of the American Cordillera, is insufficiently descriptive of the deposits themselves. For example, to state that a given deposit is mesothermal does imply the approximate temperature and pressure of formation and indicates the mineral assemblage, but it gives no hint as to the exact mode of formation or as to the localization, form, size, or continuity of the deposit, which are desirable features to convey. In general, most geologists unconsciously use the terminology of other classifications along with Lindgren's terminology to avoid this difficulty. Thus, instead of stating a deposit is mesothermal, they tend to say it is a mesothermal disseminated replacement deposit, using the term mesothermal as a qualifying adjective and the term disseminated replacement deposit to give a word picture of the deposit. (3) The Lindgren classification does not apply to many zoned deposits. The Butte, Mont., deposits, for example, are classed as mesothermal, yet the Gagnon vein extends from the central or higher-temperature zone, through the intermediate and into the outer or lower-temperature zone. The central, with its characteristic enargite, might be classed as mesothermal, but the outer part characterized by the absence of enargite and copper minerals and by the presence of sphalerite, galena, manganese, and rhodochrosite gangue would be epithermal. Thus the same vein would be both mesothermal and epithermal. Further, the mesothermal section contains abundant chalcocite and bornite and considerable covellite, minerals which are late in the paragenetic sequence and which are more typical of the epithermal than the mesothermal zone. Thus, even the central part of the deposit is a composite of epithermal superimposed upon mesothermal. Consequently, classifying these deposits as either mesothermal or epithermal carries little significance and gives an incomplete picture of them. Similarly, Loughlin and Behre have pointed out that the Leadville, Colo., deposits indicate transitions from hypothermal to mesothermal to epithermal, and even to the " telethermal " class of Graton. Some of the deposits have been built up by additions of minerals representing three of Lindgren's stages. One would, therefore, have to state that in a single deposit the magnetite is hypothermal, the quartz-pyrite, sphalerite, and chalcopyrite are mesothermal, and dolomite, silver, lead, and gold are epithermal. To which class would it belong? (4) Certain deposits, such as Cerro de Pasco, Peru, from their mineralogy would be classed as mesothermal but according to Graton and Bowditch they have been formed, not at great depth and high pressure, but at shallow depth, equivalent to that of epithermal deposits. High-temperature minerals actually are found at shallow depths and low-temperature minerals at great depths. (5) In general, most hydrothermal deposits are formed in successive stages with later minerals replacing earlier ones; and earlier minerals of epithermal deposits commonly appear to have formed at higher temperatures than later minerals of mesothermal deposits, so temperature alone is not the controlling factor. The mineral sequence of earlier to later minerals is generally also that of higher- to lower-temperature forms, and commonly minerals characteristic of the epithermal zone replace or are later than those characteristic of the mesothermal zone.

It seems clear, therefore, that depositional factors other than temperature and pressure play a part in the formation of mineral deposits. Thus, structural control, the physical and chemical effects of wall rocks, the relative ratios of concentration of different ions in solution, and chemical complexes all play a part in determining the position and mineralogic content of mineral deposits.

Graton proposes two additions to Lindgren's classification of deposits formed by hot ascending waters, namely, the " leptothermal " group, lying above the mesothermal, and the " telethermal " group, above the epithermal. The separation of " leptothermal " from upper mesothermal or lower epithermal involves difficulties, and the " telethermal " group is receiving slow acceptance.

Lindgren in 1922, recognizing the disadvantage of concise terminology in his classification, proposed additional terminology as follows:

DEPOSITS OF ORIGIN DEPENDENT UPON THE ERUPTION OF IGNEOUS ROCKS

    A. Hydrothermal deposits.
        a. Epithermal.
        b. Mesothermal.
        c. Hypothermal.
    B. Emanation deposits.
        a. Sublimates.
        b. Exudation veins, surface type.
        c. Pyrometasomatic deposits.
        d. Exudation veins, deep-seated type.
    C. Magmatic deposits.
        a. Orthotectic.
            1. Differentiation *in situ.*
            2. Injected.
        b. Pneumotectic.
            1. Differentiation *in situ.*
            2. Injected.

This arrangement met with objection because it distinguished emanation deposits formed from magmatic vapors from deposits formed from liquid solutions. Unfortunately, conclusive criteria for recognition of deposits formed by vapors are not available. This revision was not adopted by Lindgren in the later editions of his *Mineral Deposits.*

Recent Classifications. Beck and Berg (1922) presented a classification somewhat similar to that of Beck (1909). It is: (1) magmatic ore segregations; (2) contact deposits; (3) ore veins; (4) epigenetic ore stockworks; (5) epigenetic sulphide deposits; (6) gossan formations. This classification groups incongruous deposits together, contains insufficient groupings (as for massive replacement deposits in limestones), and permits the same deposit to be classed in more than one group.

Niggli in 1925 introduced a new major separation on the basis of "plutonic" and "volcanic," similar to that made for igneous rocks. His classification is: I. Plutonic: (A) hydrothermal, (B) pegmatitic-pneumatolytic, (C) ortho-magmatic. II. Volcanic: (A) exhalative to hydrothermal, (B) pneumatolytic, (C) ortho-magmatic. It appears logical to state that if rocks from a magma can be separated into plutonic and volcanic, ore deposits could similarly be separated; unfortunately mineralizing solutions do not congeal *in situ;* they are mostly mobile, and those in and associated with volcanic rocks may have come from a deep underlying magma reservoir that at depth produced hidden plutonic rocks. The separation into plutonic and volcanic deposits can hardly stand.

A more extended genetic classification was introduced by Schneiderhöhn in 1932, as follows:

A. Magmatic rocks and ore deposits.
    (*a*) Intrusive magmatic.
        I. Intrusive rocks and liquid magmatic deposits.
        I-II. Liquid magmatic-pneumatolytic.
        II. Pneumatolytic.
            1. Pegmatite veins.
            2. Pneumatolytic veins and impregnations.
            3. Contact pneumatolytic.
        II-III. Pneumatolytic-hydrothermal.
        III. Hydrothermal.
    (*b*) Extrusive magmatic.
        I. Extrusive-hydrothermal.
        II. Exhalation.

B. Sedimentary deposits.
    1, Weathered zone (oxidation and enrichment); 2, placers; 3, residual; **4,** biochemical-inorganic; 5, salts; 6, fuels; **7,** descending ground water deposits.

C. Metamorphic deposits.
    1, Thermal contact metamorphism; 2, metamorphic rocks; **3, metamor**phosed ore deposits; 4, rarely formed metamorphic deposits.

This classification has many good points but is too subjective to find much favor. Also, most geologists will balk at the usage of intrusive and extrusive magmatic — an almost impossible distinction. Many deposits would be difficult to classify under the intrusive magmatic group, and the sedimentary grouping is not entirely logical.

Proposed Classification. The following classification is proposed as a simple genetic classification which is usable for the field or laboratory and which carries familiar, well-established terminology. It is proposed primarily as a working classification for the use of the beginning student in economic geology and for the mining geologist and mine operator.

PROPOSED CLASSIFICATION

PROCESS	DEPOSITS	EXAMPLES
1. Magmatic concentration.	I. Early magmatic:	
	A. Disseminated crystallization.	Diamond pipes.
	B. Segregation.	Chromite deposits.
	C. Injection.	Kiruna magnetite?
	II. Late magmatic:	
	A. Residual liquid segregation.	Taberg magnetite.
	B. Residual liquid injection.	Adirondack magnetite, pegmatites.
	C. Immiscible liquid segregation.	Insizwa sulphides.
	D. Immiscible liquid injection.	Vlackfontein, S. Africa.
2. Sublimation.	Sublimates.	Sulphur.
3. Contact metasomatism.	Contact metasomatic: Iron, copper, gold, etc.	Cornwall magnetite, Morenci (old), etc.
4. Hydrothermal processes.		
A. Cavity filling.	Cavity filling (open space deposits):	
	A. Fissure veins.	Pachuca, Mexico.
	B. Shear-zone deposits.	Otago, New Zealand.
	C. Stockworks.	Quartz Hill, Colo.
	D. Ladder veins.	Morning Star, Australia.
	E. Saddle-reefs.	Bendigo, Australia.
	F. Tension-crack fillings (pitches and flats).	Wisconsin Pb and Zn.
	G. Breccia fillings:	
	a. Volcanic.	Bassick pipe, Colo.
	b. Tectonic.	Mascot, Tenn., Zn.
	c. Collapse.	Bisbee, Ariz.

PROPOSED CLASSIFICATION — *Continued*

PROCESS	DEPOSITS	EXAMPLES
4. Hydrothermal processes (*Continued*)	*H.* Solution-cavity fillings.	
	a. Caves and channels.	Wisconsin-Illinois Pb and Zn.
	b. Gash veins.	Upper Mississippi Valley Pb and Zn.
	I. Pore-space fillings.	" Red bed " copper.
	J. Vesicular fillings.	Lake Superior copper.
B. Replacement.	Replacement:	
	A. Massive.	Bisbee copper.
	B. Lode fissure.	Kirkland Lake gold.
	C. Disseminated.	" Porphyry " coppers.
5. Sedimentation (exclusive of evaporation)	Sedimentary: Iron, manganese, phosphate, etc.	Clinton iron ores.
6. Evaporation.	Evaporites:	
	A. Marine.	Gypsum, salt, potash.
	B. Lake.	Sodium carbonate, borates.
	C. Ground water.	Chile nitrates.
7. Residual and mechanical concentration		
A. Residual concentration.	Residual deposits: Iron, manganese, bauxite, etc.	Lake Superior iron ores, Gold Coast manganese, Arkansas bauxite.
B. Mechanical concentration.	Placers:	
	A. Stream.	California placers.
	B. Beach.	Nome, Alaska, gold.
	C. Eluvial.	Dutch East Indies tin.
	D. Eolian.	Australian gold.
8. Surficial oxidation and supergene enrichment.	Oxidized, supergene sulphide.	Chuquicamata, Chile. Ray, Ariz., copper.
9. Metamorphism.	*A.* Metamorphosed deposits.	Rammelsberg, Germany.
	B. Metamorphic deposits.	Graphite, asbestos, talc, soapstone, sillimanite group, garnet.

Selected References on Classification of Mineral Deposits

Ore deposits of the United States and Canada. J. F. KEMP. Eng. and Min. Jour., London, 1906. Appendix I. *Review of earlier methods of classification.*

General Reference 8. Chap. 16. *The Lindgren classification.*

General Reference 9. Chap. 2, by G. F. LOUGHLIN and C. H. BEHRE, JR. *A review of the Lindgren classification.*

Classification of Magmatic Mineral Deposits. ALAN M. BATEMAN. Econ. Geol. 37:1–15, 1942. *A new classification of magmatic deposits and processes.*

CHAPTER 9

RESOURCES, INTERNATIONAL RELATIONS, AND CONSERVATION IN MINERALS

The rise of the industrial age has so accelerated the demand for minerals that the world exploited more of its resources between World War I and World War II than in all preceding history, and mineral resources have become industrial power. Industrial power in turn is dependent upon a liberal endowment of mineral resources, and this is purely fortuitous. The insatiable demand for minerals has made sources of supply, formerly thought adequate, now look relatively trivial, and sources capable of meeting large supply are becoming fewer. It is now realized, as pointed out by Leith, that adequate supplies are concentrated in relatively few places on the globe and in relatively fewer hands, and this feature became a major factor in world unrest and aggression.

Leith states that 85 percent of the world's minerals originate in the countries bordering the North Atlantic Ocean and that their preponderant position in world affairs is due more to these natural resources than to the initiative and energy of their peoples — a factor that assures leadership for centuries to come.

The concentration of large supplies in few countries and under fewer owners gives rise to large international movements of minerals and to disregard of boundaries in seeking new supplies. It has also given rise to nationalization of mineral resources, to prevent inroads upon supplies, and to political control of weak nations that own important reserves. Thus, mineral resources raise international problems of trade, industry, national policy, peace, conquest, and war.

The Mineral Background

In this modern industrial age few people stop to realize the extent to which we have become dependent upon materials of the mineral kingdom. In the home are the conveniences of spring beds, gas, electricity, running water, plumbing, heating, cooking, and labor-saving devices. Outside are paved streets, facilities of transport in auto, train, ship, or plane, and the vast array of machines. The modern fireproof building, from cellar to shingles, save for a little wood, is a structure of minerals. All these are dependent upon mineral supplies.

Without them, there would be primitive agriculture and handicraft. These are the contrasts between nations that have and utilize minerals and those that do not.

A glance over the history of large industrial nations reveals that their rise coincided with the utilization of their mineral resources. Where coal and iron occurred together industry arose. Coal was the motive power, and iron the substance. It is no accident that great manufacturing cities sprang up in the United States around the Great Lakes, in Pennsylvania and in Alabama, and in England, the Ruhr, northern France, and Belgium. There coal and iron met, and the products of their union reached the far places of the globe. Those countries attained predominance above other nations. They became the great manufacturing and trading nations, and economically and industrially aggressive. Political prominence followed, and armaments arose for the protection of country and trade, and also, unfortunately, for the acquisition of extra-national mineral resources.

Leith has emphasized that self-sufficiency of mineral supplies is now regarded as one of the chief goals to economic nationalism and sufficiency. Those nations lacking an adequate supply of minerals face either a sharp curtailment of industrialism or must find means of gaining access to them. This is normally accomplished by peaceful trade, or investment in foreign resources. Some strong nations, however, have exercised forcible control over the mineral resources of weaker nations, and some minor nations have found abundant mineral resources as much a liability as an asset.

MINERAL RESOURCES AND THEIR DISTRIBUTION

A host of minerals enter into modern industry, and a single minor one may constitute an essential cog in a wheel. No single nation or empire is adequately supplied with all the minerals necessary for its industrial life, or for armament, or war. Many countries are devoid of most of them. Even the United States, despite the fact that it is richly endowed with the most bountiful mineral supply of all nations, could not continue as an industrial nation if its essential mineral imports were cut off. The predicament of other nations is worse. The distribution of mineral resources becomes, therefore, of great importance.

Chief Industrial Minerals. At the conclusion of the Dark Ages the chief mineral substances in use were iron, copper, lead, tin, gold, silver, mercury, precious stones, clays, and building stones. Today, more than 75 minerals enter international trade. Certain minerals are necessary for essential industries, some contribute only to unessential industries, or to luxuries.   Others are absolutely vital for peacetime development or armament or war.   Many assume an importance far in excess of the volume used.

The chief industrial minerals may be grouped as follows: mineral fuels — coal, oil, gas; iron ores; iron-alloy metals — chromium, manganese, molybdenum, nickel, tungsten, vanadium; nonferrous metals — copper, lead, zinc, tin, aluminum; minor metals; metallurgical minerals; chemical minerals; fertilizers; ceramic minerals; abrasives.   The more important industrial minerals and their distribution are given in Table 14.   Gold because of its dominant monetary use is not considered as an industrial mineral, but it plays an important part in affording means of purchasing needed mineral supplies.

Resources and Distribution.   Minerals, unlike products of the soil, are but " one crop " materials; they are vanishing assets, most of which once removed are largely gone forever.   Therefore their reserves and distribution command interest.   Table 14 indicates, horizontally, the major sources among important countries, and vertically, the resources of these same countries in these vital minerals.

In the table, A designates an adequate supply, or a supply which could be made adequate if required, or an actual or potential surplus available for export.   This designation in the case of nonindustrial or low-consumption countries implies availability for export.   D designates an important deficiency, which has to be made up by imports; for the countries of low consumption it implies minor quantities available for export; i.e., minor molybdenum in North Africa could be obtained by those desiring it.   O designates absence or unimportant supplies; few countries are entirely lacking in these minerals, however, and Brazil, for example, although given O for chromite, does export small quantities.

Table 14 is based upon the numerous tables of individual mineral production figures given in Parts II and III of this book, where the percentage of world production of each country for each mineral may be found, and also approximate unit production figures in many cases. The uses of each of the minerals of Table 14 are given there also. What Table 14 does not disclose is the very important factor of the concentration distribution of each mineral. It shows that copper occurs in surplus in 14 countries or areas, but it is much more revealing to learn that 86 percent of the world's copper comes from only five countries and that 55 percent is concentrated in two countries which are the chief owners of 82 percent of the production and 92 percent of the reserves.   Further, three important nations consume 22 percent of

TABLE 14

DISTRIBUTION OF IMPORTANT MINERALS AND STATUS OF IMPORTANT COUNTRIES

[A, adequate or surplus; D, important deficiency; O, no appreciable supply.]

Important Minerals	United States	Canada	Great Britain	British Empire	France	French Empire	Belgian Colonies	Russia	China	Germany	Italy	Japan	Scandinavia	Poland	Spain	Netherlands Empire	Balkans	North Africa	South Africa	Mexico	Brazil	Venezuela	Chile	Peru	Argentina	Bolivia
Iron	A	A	D	A	A	A	D	A	A	D	D	D	A	D	A	O	D	A	A	D	A	A	A	O	D	O
Copper	A	A	O	A	O	D	A	D	O	D	O	D	A	O	A	O	D	O	A	A	O	O	A	A	O	O
Lead	D	A	D	A	O	A	D	A	D	D	D	O	D	A	A	D	D	A	D	A	O	O	O	A	A	A
Zinc	O	O	D	A	O	D	D	A	D	D	A	D	O	A	A	D	D	A	D	D	O	O	O	A	A	A
Tin	O	O	D	A	O	D	A	O	A	O	O	A	A	O	O	A	O	O	O	D	O	O	O	A	O	A
Aluminum ore	D	O	D	A	A	A	O	D	O	O	A	O	O	O	O	A	A	A	A	O	A	O	O	O	O	O
Manganese	A	O	O	A	O	A	O	A	A	O	O	O	O	O	O	A	O	A	A	O	A	O	A	O	O	O
Chromium	O	O	O	D	O	A	O	A	O	O	O	O	O	O	O	O	D	O	D	O	O	O	A	O	O	O
Nickel	A	A	O	A	O	A	O	D	O	O	O	O	A	O	O	O	D	O	D	A	O	O	O	O	O	O
Molybdenum	D	D	O	A	O	D	D	D	O	O	O	O	O	O	O	O	D	O	A	A	O	O	A	D	D	A
Tungsten	D	D	O	D	O	O	O	D	A	O	O	O	O	O	A	O	D	O	O	A	O	O	O	A	O	O
Vanadium	A	O	O	A	O	D	O	D	O	O	A	O	O	O	O	O	A	A	A	A	O	O	A	A	O	O
Cobalt	D	A	O	A	O	O	A	D	O	O	O	O	O	O	A	O	A	O	A	O	O	O	O	D	O	D
Mercury	A	A	O	A	O	O	O	D	A	O	O	O	O	O	D	O	O	O	O	A	O	A	O	A	A	A
Antimony	D	D	O	A	O	O	A	D	O	O	D	D	A	O	O	O	O	D	O	D	A	O	O	D	A	O
Gold	A	A	O	A	O	A	A	A	O	O	O	O	O	O	O	O	A	O	A	O	O	O	O	A	O	O
Platinum	O	D	D	A	O	O	O	A	O	O	O	O	D	O	O	O	O	A	A	O	O	A	O	O	O	D
Tantalum	D	O	O	A	O	O	A	O	O	O	O	O	D	O	O	O	O	O	A	O	A	O	O	O	O	O
Coal	A	A	A	A	A	A	D	A	A	A	O	D	O	A	O	A	A	O	A	O	D	O	A	A	A	D
Oil	A	D	O	A	O	D	O	A	O	D	O	D	O	D	O	A	D	O	O	D	O	A	D	A	D	O
Sulphur	A	O	O	A	O	O	D	A	O	D	A	A	O	O	A	O	D	O	D	A	O	O	A	A	O	O
Potash	A	D	O	O	A	A	D	A	O	A	O	O	A	D	A	O	O	O	O	A	O	A	A	A	O	O
Phosphates	A	O	A	A	O	D	O	A	O	D	D	O	D	O	O	A	O	A	D	O	O	O	A	O	O	O
Fluorspar	D	D	O	A	A	O	A	A	O	A	O	O	D	A	O	O	O	O	D	A	O	O	O	A	O	D
Asbestos	D	A	O	A	O	A	O	A	A	O	D	A	O	O	A	O	O	O	A	O	O	O	D	O	O	A
Graphite	A	O	O	A	O	D	O	D	O	O	O	A	O	O	O	O	A	O	O	O	O	D	O	O	O	O
Magnesite	D	D	O	O	O	A	O	A	O	A	A	A	A	O	O	O	A	O	D	A	A	O	O	O	O	O
Mica	O	O	O	A	O	D	O	D	A	O	O	O	D	O	O	A	O	O	A	O	A	O	O	O	O	O
Diamonds	O	O	O	A	O	A	A	O	O	O	O	O	O	O	O	O	O	O	A	O	A	D	O	O	O	O
Quartz Crystals	D	O	O	D	O	O	O	O	O	O	O	O	O	O	O	O	O	O	D	O	A	O	O	O	O	O
Important Minerals: Surplus	12	14	2	25	4	10	5	16	6	4	4	3	5	4	8	5	4	7	13	12	8	2	5	7	4	5
Important Minerals: Deficient	14	6	6	4	1	13	8	10	3	10	6	12	8	3	2	5	13	1	9	5	2	3	4	5	4	3
Important Minerals: Absent	4	10	22	1	25	7	17	3	21	16	20	15	17	23	20	20	13	22	8	13	20	25	21	18	22	22

the world's copper but produce only 6 percent. Some figures of concentrations of normal production are given in Table 15.

TABLE 15

PRODUCTION DISTRIBUTION OF SOME IMPORTANT MINERALS

Mineral	Percent of Total	In Number of Countries	Chief Producing Countries
Oil	82	4	U. S., Venezuela, Russia, Middle East
Coal	80	4	U. S., Britain, Germany, Russia, S. Africa
Iron	64	4	U. S., France, Russia
Iron	78	5	U. S., France, Russia, Sweden, Britain
Copper	85	5	U. S., Chile, Rhodesia, Canada, Congo
Copper	82	2	Ownership: U. S., Britain
Copper	99	3	Reserves owned by U. S., Britain, Belgium
Lead	73	5	U. S., Australia, Mexico, Canada, Russia
Lead	70	2	Ownership: U. S., Britain (38)
Zinc	78	5	U. S., Belgium, Canada, Mexico, Russia, Poland
Zinc	62	2	Ownership: U. S. (44), Britain (18)
Tin	79	3	Malaya, D. E. Indies, Bolivia
Tin	64	2	Ownership: Britain (47), Holland (17)
Bauxite	64	5	France, Surinam, Hungary, U. S., Yugoslavia, British Guiana, Brazil
Manganese	82	5	Russia, India, S. Africa, Gold Coast, Brazil
Manganese	90	2	Russia (45), Britain (45)
Chromium	66	5	Rhodesia, Russia, Turkey, S. Africa, Cuba
Chromium	67	2	Ownership: Britain (50), Russia (17)
Nickel	97	3	Canada, New Caledonia, Russia (Finland)
Molybdenum	95	2	U. S., Chile, Mexico
Tungsten	62	3	China, Burma, U. S., Bolivia, Brazil
Vanadium	83	3	Peru, U. S., S. W. Africa
Mercury	73	4	Italy, Spain, U. S., Mexico, Canada
Antimony	81	3	China, Mexico, Bolivia
Gold	74	4	S. Africa, Russia, U. S., Canada
Platinum	85	3	Canada, Russia, S. Africa
Sulphur	76	5	U. S., Spain, Italy, Norway, Russia
Asbestos	78	3	Canada (53), Russia (18), Rhodesia (7), S. Africa
Potash	95	4	Germany, France, Russia, U. S.
Fluorspar	90	5	U. S., Germany, Russia, Britain, France
Graphite	80	5	Russia, Japan, Germany, Austria, Ceylon

Sufficiency and Deficiency of Leading Nations. Table 14 discloses, in the bottom summary, enlightening information of excesses and deficiencies in mineral resources.

The United States lacks only 4 of the minerals listed, is deficient in

14, has an actual or potential surplus in 12, and is the world leader in 8 essential minerals and many minor ones.

The contrast between the figures for Great Britain and the British Empire is illuminating from a political and naval standpoint, by forcibly bringing attention to her own paucity and to her dependence upon the Empire for vital minerals. It indicates that her industrial existence depends upon a protected and uninterrupted flow of minerals over long ocean voyages. The British Empire leads all nations in mineral sufficiency, since of these minerals it lacks only 1, is deficient in 4, and has a surplus of 25. Moreover, it is the world leader in such internationally critical minerals as gold, tin, lead, nickel, chrome, asbestos, mica, and diamonds and is a close second in manganese, copper, zinc, and some minor minerals.

France is likewise notably deficient, but with her colonies is well provided. However, she also is dependent either upon a strong navy or alliance with a strong naval power to insure receipt of her overseas supplies.

Russia now is extremely well provided and has few important deficiencies. She leads the world in manganese and is second in iron ore, gold, chrome, platinum, and asbestos, has abundant oil reserves, and is probably self-contained in most of the nonferrous metals. Her mineral resources, combined with a large population, make her potentially a colossal industrial giant of the future.

Germany, Italy, and Japan are notably lacking in most critically important minerals. They are the least provided of all large nations and are embarrassingly short of oil, iron ore, copper, tin, bauxite, iron-alloy metals, gold, and asbestos; and Italy also lacks coal. It is no coincidence that they became aggressor nations and challenged the unequal distribution of minerals. It is also significant that Great Britain and the United States together own three-fourths of the entire mineral wealth of the world, and large sources of supplies can come only from them.

Continental Europe, outside of Germany, Italy, and Russia, also has a surplus of few minerals and is strikingly deficient in most minerals. Several of the smaller countries, however, each possess surplus supplies of one or two important minerals that are absent from the strong nations. The most striking ones are iron ore, copper, and sulphur (pyrite) in Sweden; zinc-lead and potash in Poland; copper, iron, mercury, and sulphur in Spain; lead-zinc, copper, and bauxite in Yugoslavia; oil in Roumania; bauxite in Hungary; and magnesite and chromium in Greece. Holland and Belgium in their colonies have surpluses of oil, copper, tin, diamonds, and bauxite, of world importance.

Another striking feature revealed by Table 14 is that many of the nonindustrial countries have large resources of critical minerals of international and perhaps political importance. China has the largest supply of tungsten and antimony and large amounts of iron and tin. French colonies in north Africa have important deposits of iron, phosphates, lead-zinc, and antimony. South and central Africa have vast deposits of gold, copper, chrome, manganese, vanadium, platinum, asbestos, and diamonds. Mexico has seven important minerals; Venezuela and the Middle East countries have large reserves of oil; Chile has great copper and nitrate deposits; Brazil has iron, manganese, beryl, quartz, tungsten, and diamonds; Peru has nonferrous metals and vanadium; and Bolivia has tin, tungsten, and antimony. It is rather important that each of these nations has essential minerals desired by others.

MINERALS IN INTERNATIONAL RELATIONS AND WAR

Since no nation possesses all the minerals necessary to modern industry, each must obtain certain essential ones from others. Deficient nations become absolutely dependent upon others for needed supplies. Large international movements of minerals take place, and expanding mining companies become international in scope, with large investments in foreign lands. This brings on political entanglements, alliances, protectorates, and even forceful domination. During periods of peace the international flow of minerals supplies exchange for the purchase of other goods; the gold and nickel of Canada permit her purchase of needed manufactured goods from the United States.

Essential minerals of which the domestic supply is inadequate, or potentially inadequate, are known as *strategic minerals*, but what is a strategic mineral to one nation is not to another. Every nation, however, has some strategic minerals. Nationalistic independence in minerals is impossible, but there are enough of the essential and strategic minerals to supply the needs of all nations, and no one need go without. This means, however, that, if industrial nations are to continue to be industrial nations, they must have access to mineral supplies of other lands in normal trade. If access is denied them by nationalistic tendencies, or export or import restrictions, conditions are bred that become threats to world peace.

For armament and war, minerals play an even more important part. They become absolute necessities. The World Wars have demonstrated the extent to which modern armies and navies are dependent upon minerals. The modern mechanized and motorized units are entirely made of and propelled by minerals, and there must be abundant supplies to maintain them. Mineral supplies, therefore, may determine the course or cessation of war; they may determine peace no less than they sustain industry in times of peace. Any warring nation cut off from supplies of strategic minerals during a long struggle must inevitably face defeat if its stocks become exhausted, because minerals cannot be grown like wood or cotton. Minerals can serve to control war, and to enforce peace.

The most essential minerals for modern mechanized war are oil, iron, nonferrous metals, iron-alloy metals, mercury, antimony, platinum, sulphur, asbestos, graphite, magnesite, mica, quartz, fluorspar, beryl, tantalite, uranium, and industrial diamonds. These must move without interruption in large volume over long sea routes, which involve open sea lanes.

As minerals are the basis of modern industry, their deficiency in certain countries has been an important factor in world unrest. Permanent world peace and uninterrupted flow of international trade depends in considerable part upon all nations having access to mineral supplies and being allowed to pay for them.

Conservation of Mineral Resources

Conservation of mineral resources arises particularly because they are irreplaceable, wasting assets which have often been wantonly exploited. They do not re-grow like forests or wheat. There is no mineral " crop." The older concept of conservation against possible future shortage of minerals by withdrawal of mineral resources from exploitation has given way to conservation by prevention of waste during production; by wise use; by utilizing lower-grade materials; by the development of substitutes and of preservatives; and by conserving strategic minerals in favor of importation during times of peace. These questions concern both public interest and private ownership. It is commonly assumed that measures of conservation must be imposed upon owners of mineral lands, since they are motivated only by selfish interests, and that public action alone can bring the desired results. This is unquestionably true with regard to such minerals as oil and gas. It is also largely true of the small producer whose sole interest is concentrated in one property, without expectation of continuance elsewhere. Most large organizations, however, propagate their resources by developing new properties to replace exhausted ones, utilize research in mining, ore dressing, and chemistry, and practice conservation for long-time economy. In general, conservation acts that prohibit development of minerals that they may be conserved for posterity do more harm than good because they inhibit the development of industries that give growth to other more permanent industries, particularly in regions devoid of other opportunities for development. Enforced conservation of petroleum and natural gas, however, has proved beneficial to the public and the petroleum industry.

Conservation as applied to the prevention or elimination of loss or waste of mineral resources during mining is highly desirable. Too often a thin seam of coal is lost during the mining of a thick one; or a layer of marginal ore is left behind in the mining of adjacent profitable ore; or pillars of coal or ore may be left behind to support overlying ground because they are cheaper than artificial supports. These are unrecoverable losses of minerals that are made to permit a higher profit per ton of the materials extracted, and such practices should not be condoned. Again, too often rock below economic grade is sent to waste dumps and intermingled with barren material, thus eliminating its possible future use, instead of segregating it against the time when higher prices or lower costs might permit its utilization.

Improvements in mining methods, and mineral beneficiation, such as oil flotation, have greatly aided mineral conservation, by permitting utilization of lower-grade materials and thereby increasing reserves. Much still remains to be done in increasing extraction of tin ores, in utilizing lower-grade chrome and manganese ores, in decreasing slag losses in zinc and other metals, and in better processing recovery of nonmetallic minerals.

Other needed conservation measures are the better utilization of scrap metals — little zinc is recovered from scrap brass, for example — and the wiser use of scarce minerals. The use of scarce lead, chrysotile asbestos, metallurgical chrome, and others should not be permitted for secondary purposes for which other minerals or synthetic substitutes could be utilized. These materials could then be conserved for uses for which there are no known substitutes.

These and other conservation practices will help make unreplaceable metals and minerals go further and conserve rapidly wasting mineral resources. Conservation is generally practiced by the large mining and oil companies in their goal toward economy and efficiency. Conservation aimed toward withholding limited resources of strategic minerals in favor of importation has justification in national security preparedness.

Selected References on Mineral Resources

Strategic Mineral Supplies. G. A. ROUSH. McGraw-Hill, New York, 1939. *Discussion of United States strategic minerals.*

World Minerals and World Politics. C. K. Leith. McGraw-Hill, New York, 1931. *Broad survey of international mineral distribution.*

Mineral Economics. Ed. by F. G. Tryon and E. C. Eckel. McGraw-Hill, New York, 1932.

Political Control of Mineral Resources. C. K. Leith. Foreign Affairs, July, 1925. *Essay on mineral geography.*

Mineral Resources and Peace. C. K. Leith. Foreign Affairs, April, 1938. *Use of mineral sanctions to prevent war.*

World Minerals and World Peace. C. K. Leith, J. W. Furness, and C. Lewis. Brookings Institute, Washington, D. C., 1943. *Survey of production, resources, geographic distribution, cartels, and policy.*

General Reference 13. *Economics, distribution, occurrence, technology, and production of 16 critical minerals.*

Wartime Dependence on Foreign Minerals. Alan M. Bateman. Econ. Geol. 41:308–327, 1946. *Sources, amounts, and problems of obtaining some 60 strategic minerals for wartime industry.*

General Reference 7. *Comprehensive survey of domestic mineral resources, production, consumption, and imports.*

Our Natural Resources and Their Conservation. 2nd Ed. A. E. Parkins and J. R. Whitaker. John Wiley & Sons, New York, 1939. *Good chapters on conservation of minerals, soils, and water.*

America's Stake in World Mineral Resources. Alan M. Bateman. Min. Engr. 1:23–27, July, 1949. *Survey of domestic and foreign mineral resources and policy problems.*

CHAPTER 10

GEOPHYSICAL PROSPECTING, EXPLORATION, DEVELOP-
MENT, AND VALUATION OF MINERAL PROPERTIES

It becomes more and more difficult to discover new mineral deposits because most of the easily discoverable ones have already been found, and unprospected areas are gradually eliminated. Scientific prospecting has now become the vogue, and geology is called upon to play an increasing part in it. Geophysical prospecting has advanced by leaps and bounds, but geology must enter into the interpretation of results. In the exploration and development of known mineral bodies, geology is playing a more and more important role, and in the valuation of mineral deposits, knowledge of the occurrence, forms, and behavior of mineral bodies is essential. In these functions the field of the geologist crosses or coordinates with that of the mining engineer and geophysicist. Since these fields are extremely broad and the subject matter voluminous, only brief reference is made to them here.

Areal Prospecting

Homage is due the old-time prospector who braved the hardship of wilderness and desert in the search for minerals. His discoveries in the past have constituted the lion's share, and many famous mines perpetuate his memory. His industry has eliminated areas for obvious discovery, and his opportunities have diminished thereby. It is now common practice to supplement or supplant him with trained geologists or mining engineers for areal prospecting. Formerly one or two prospectors traveled afoot, with pack animal, or canoe, but modern prospecting is more organized. A practical prospector and a geologist are often paired together; or groups of paired, practical prospectors operate in different localities, under the direct supervision of one or more geologists who check over their results at intervals or move them to more promising sites. In places, all prospecting is done by groups of trained geologists. Such prospecting groups often are serviced by trucks, as in open African regions, or by airplane, as in Canada and Siberia.

In field work on ore deposits, examination of outcrops is facilitated by study of geologic maps and reports, if such exist. Work is directed to regions where the geologic conditions are favorable for mineralization. The mineral habit of the region can be utilized. Features that direct structural control of mineralization are checked up and followed. Intrusive bodies attract attention, particularly if they intrude limestones. Careful attention is given to introduced minerals and rock alteration, and oxidized areas are studied to see if croppings give a clue to the character and abundance of pre-existing sulphides.

Much information can be gained from careful examination of float and deducing its source, or direction of source. The tracing of glacial erratics containing copper-gold ore followed in the direction of the glacial striae led to the discovery of important mineral deposits in Sweden. From earliest time panning of alluvial gravels has been utilized to trace minerals upstream to their source. It has led to the discovery of rich diamond deposits in central and south Africa, and even helped in the finding of manganese deposits on the African west coast.

If no geologic maps exist, the formations may be broadly outlined by reconnaissance trips. If the bedrock geology is unknown, reconnaissance geologic maps are made, as was done in large prospecting concessions in Rhodesia with beneficial results.

For petroleum areal prospecting, different information and methods are utilized. Geological knowledge is essential. Surface expressions of petroleum, as oil and gas seeps, asphalt, and other evidences, are seldom present, although soil analyses may reveal hydrocarbons. The geologic column must be learned, the stratigraphy deciphered, possible source beds located, and the rock structure unraveled. Areal prospecting eliminates areas that cannot be petroliferous and directs attention to areas that may be petroliferous, although by itself rarely results in the direct finding of oil. Such areas are then checked or supplemented by geophysical work and by test drilling; the latter is necessary in either case.

Geophysical Prospecting

As alchemy was the forerunner of chemistry, so was the witch stick or divining rod the forerunner of modern, geophysical instruments. From the time of the ancients the quest for buried treasures has stimulated man's inventiveness to find them, and this has culminated today in the development of scientific instruments and methods whose foundation, unlike that of the divining rod, is physics and mathematics. They are used, and with great success, to locate mineral substances that lie within the earth. Rapid advances are being made in geophysical prospecting, particularly for oil, but also for metals, and it has grown to be a science in itself with a voluminous literature.

These methods are also used for determining the depth to bedrock for bridges, dams, and buildings, and for determining the nature of excavation and available sands and gravels for highway construction. The principal methods and adaptations used for oil and metals are briefly outlined in principle only; details of instruments, procedure and interpretation will be found in the extensive literature on this subject.

The various geophysical methods in use may be grouped as follows:

1. Magnetic   3. Electromagnetic   5. Seismic
2. Electrical   4. Gravitative   6. Radiometric

Magnetic Methods

The principle of the well-known horizontal and vertical control of the earth's magnetic field upon the compass needle is utilized in certain types of geophysical prospecting. It was early learned that magnetic bodies within the earth caused local deviation of the compass needle. Conversely, a local deviation of the needle indicates a disturbing magnetic body.

Magnetic Needle. Magnetic surveys are made with horizontal and vertical dip needles, along parallel lines, with readings at specific intervals. The position of attracting bodies is thus indicated and can be plotted.

Magnetic needle methods are usable only for substances that themselves exert magnetic attraction, or that are enclosed in rocks which, owing to their mineral composition, exert a greater magnetic influence than surrounding rocks lacking such minerals. Thus, bodies of magnetic iron ore, pyrrhotite, nickel, cobalt, and other substances may be detected; also rocks high in iron, such as basalt flows, that may contain copper.

A refinement of the common magnetic dip needle is the Hotchkiss Superdip, which is a more delicate detector of magnetic substances.

Magnetometer, Variometer. Magnetic work is now more commonly utilized for structural studies because different rocks show different magnetic susceptibilities that can be measured by delicate instruments. Magnetic permeable rocks such as basic igneous ones (paramagnetic substances) apparently increase the earth's magnetic field; others such as many sediments (diamagnetic substances) decrease the strength of the earth's magnetic field from place to place. Thus, the variations in the strength of the horizontal and vertical components of the earth's magnetic field are measured instead of the absolute values, as with the dip needle. Two types of balance are used: the one a vertical variometer, which measures the vertical variations; and the horizontal variometer, which measures the variations in the horizontal component of the earth's magnetic field. The Askania Magnetometer is one popular instrument. Magnetometer surveys are now carried on successfully by " flying magnetometers " in airplanes.

The magnetic susceptibility of basic igneous rocks is 10 to 100 times stronger than that of granite or most sediments, and that of some

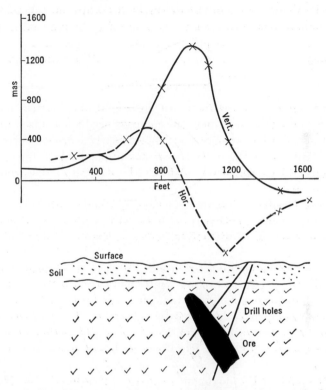

FIG. 10–1. Profile of vertical and horizontal components of magnetic field across a pyrrhotite sulphide body in northern Quebec. (*Private survey, 1941.*)

dolomites is some 20 to 50 times less than that of some shales. These relative differences can be detected deep beneath the surface.

Measurements are made along profiles or in a squared network. The field work is generally simple but requires great care and considerable calculations. The determined local vector is resolved into horizontal and vertical local vectors and a vector map prepared. The features that cause the determined local anomalies lie mostly at depths of 2,000 to 15,000 feet, and the horizontal effect is felt within a distance of about twice the depth. A magnetic profile across a pyrrhotite body in Quebec is shown in Fig. 10–1.

ELECTRICAL METHODS

Electrical methods are chiefly valuable for locating metallic mineral deposits, but they are also used in finding oil, deciphering structure, and in electrical coring or logging wells. Several methods are based upon the electrical conductivity and low specific electrical resistance of metallic minerals. In some methods, currents are introduced into the

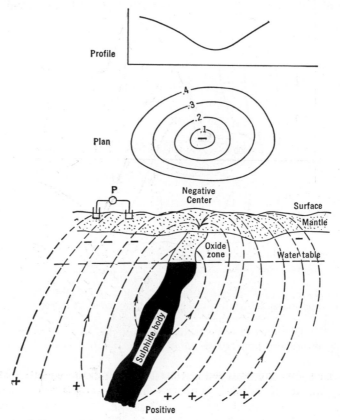

FIG. 10–2. Self-potential currents produced by oxidizing sulphide body, with lines connecting equal current readings (plan), indicating center of body.

ground and differences in potential measured; in others, natural currents generated by metallic minerals in the ground are measured, or differences in resistance between points are recorded.

Natural Current Methods. This method, also called self-potential and spontaneous polarization, requires no outside energizing force. The ore itself produces currents of electrical energy that can be measured and the presence of hidden ore bodies thereby detected. In an ore body beneath a soil mantle (Fig. 10–2), the upper part is chemically more active than the lower part. In consequence, differences in potential exist, and electrical currents flow naturally downward through the ore body and return upward through the enclosing rock (Fig. 10–2), spreading out to great distances because of the high resistivity of the country rocks. These currents may be traced out on the surface by placing two porous pots in watered holes and connecting them

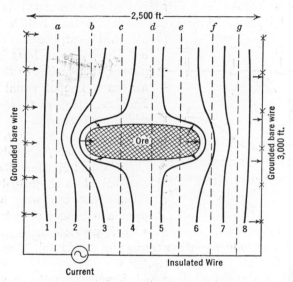

FIG. 10–3. Equipotential method using parallel grounded wires connected to current supply. *a, b, c* are normal equipotential lines in absence of a conductor; lines 1, 2, 3 are measured equipotential lines distorted by underground conductor. (*From survey of Beatson mine, Alaska.*)

with a sensitive galvanometer, thus indicating the negative center toward which currents flow. Curves connecting points of equal potential give an equipotential diagram indicating the position of the ore body. This method has been particularly developed by Schlumberger and his followers. It is not so well suited to arid regions or to regions where the subsoil is very deep. Also there must be present in the ore body 5 percent or more of conductive metallic sulphides.

Equipotential Method. This is one of the simpler electrical methods, of which there are several variations. A procedure perfected by Hans Lundberg is as follows: Two parallel bare copper wires about 3,000 feet long and 2,000 to 3,000 feet apart are pegged into the ground at intervals (Fig. 10–3). The ends of the bare wires are connected by insulated cable to a generator or batteries with a buzzer attached.

Current passes between the two grounded electrodes.. If the intervening ground is of uniform conductivity, then lines of equal electrical pressure or potential will be parallel to the grounded wires, as shown by dotted lines a, b, c, etc., of Fig. 10–3.   However, if there is a hidden ore body that is a better conductor than the surrounding rock, the lines of equal potential will be distorted by the conductor, as shown by the solid lines, 1, 2, 3.   Therefore, the lines of equal potential are determined.   An assistant sticks a steel rod into the ground; it is connected by a 30-foot insulated wire to an operator's steel rod, to which are

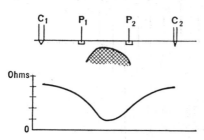

attached earphones. When the operator hears no buzz in the earphones, then he knows the two rods are on a line of equal potential. Numbered pegs mark these points, which are surveyed.

This method has been successful in many places, notably Sweden, Canada, and Newfoundland. It is best suited to shallow deposits, in regions not too wet, and where prominent aquifers interbedded with impervious beds are absent.

Fig. 10–4.   Resistivity arrangement with four electrodes and curve showing drop in resistance through a good conductor.

The point electrode variation consists of single points or groups of points instead of the parallel wires.

The leapfrog method of Eve and Keyes is another variation by which the drop in potential between parallel wires is measured by means of two porous pots inserted in the ground and connected by a 100-foot line.   First one pot is placed on the bare wire and the other in the ground 100 feet away and the drop in potential noted.   Then the pot from the wire is planted beyond the first pot and the fall in potential noted.   Repetition of this gives lines of equal potential.

Resistivity.   Different rocks and ores have characteristic electrical resistivities whose differences can be determined to yield subsurface information.

The earlier and simpler methods were by the use of a " Megger." The procedure is rapid, simple, and requires no special training.   Four (or three) ground electrodes are equally spaced in a straight line (Fig. 10–4).   The depth of subsurface penetration is about equal to the distance between electrodes.   Current is introduced at one end electrode and is taken out at the other end and measured.   The potential difference between the other electrodes is also measured, and the resistance of the earth followed by the current is calculated.   Shallow readings are obtained by close spacing of the electrodes and deep readings by wide spacing. The survey is generally made along a grid, the spacing of the lines being determined by the purpose of the survey.

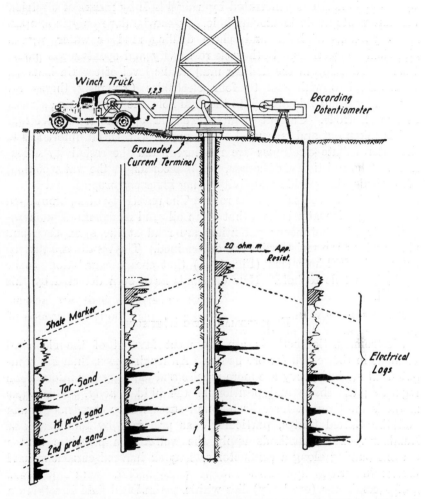

Fig. 10–5. Electrical coring or logging arrangement (schematic). (*Heiland, Geophysical Exploration.*)

This method is suitable for shallow determinations, such as the depth of bedrock, depth to water table, construction foundations, gravel fills, etc., and is also satisfactory to depths of several hundred feet for indicating geologic structure, strata, and mineral bodies.

For deep work more complicated methods and instruments are required. Subsurface structure as much as several thousand feet in depth has been successfully worked out in California.

Electrical Coring. This procedure relates to the determination of the geologic horizons penetrated by a drill hole by means of electrical measurements made inside the hole. It can be done only in the un-cased portions of holes and where drilling mud or water permits electrical connections. Both rock resistivity and spontaneous potentials that develop in the drilling mud at the level of certain beds are used. Devices with one to four electrodes (generally three) are lowered by insulated cables into the drill hole, the upper casing serving as one electrode. Current is introduced through the lower electrode and is taken out and measured at the upper. Potential differences at the intermediate electrodes are measured, and the resistivity calculated. The resistivity, of course, varies according to the water content of the beds, the specific resistivity being inversely proportional to the amount of dissolved salts in the rock. The resistivity of a porous sand containing salt water is low; that of an oil sand is higher. The measurements of spontaneous potentials are read at the same time and places where the resistivities are determined. The results are plotted as graphic drill-hole logs (Fig. 10–5) that enable correlations to be made from hole to hole. New oil sands have been detected by this method.

ELECTROMAGNETIC METHODS

Electromagnetic methods are the most favored of the electrical methods in the search for ore bodies. Although slower than the equi-potential methods, they are more precise and yield greater information regarding the shape and position of the hidden body. It is these methods mostly that have been found successful in Sweden, Canada and the United States, particularly as practiced by Sundberg and Lundberg. These methods apply the well-known principle that a current passing along a conductor sets up an induced current around the conductor.

If a conductor (ore body) lies within the induced field it sets up a secondary induced current around it, which can be measured. Two variations are common.

Loop Method. In this procedure there is no contact with the ground. For surface work, a rectangular loop of insulated cable is laid out, as in Fig. 10–6, to which an alternating current of 500 to 1,000 cycles is supplied by a gasoline-driven generator. Lines are surveyed normal to the longer axis of the loop and normal to the expected structure. The observers, by means of portable horizontal and vertical coils and suitable instruments, take readings along these lines at intervals of about 50 feet.

The charged loop sets up an induced current within the surrounding ground, which diminishes with distance from the loop. If a conductor, such as an ore body, is present, a secondary induced field is set up about the conductor, and the detection of such a secondary field indicates a hidden conductor. However, both the primary and secondary fields are present at the same place. Consequently the total strength is measured, from which is subtracted the calculated amount due to the

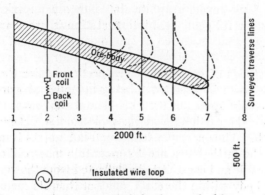

FIG. 10–6. Scheme for electromagnetic induction method by use of insulated wire loop on ground. Lines 1 to 8 are surveyed lines along which coil readings are taken and vertical (dotted curves) and horizontal components observed.

primary field. A residue indicates the presence of a secondary induced field, which, if large, shows that the disturbing body is a large one or else close to the surface. Thus, the observers with their two connected coils walking outward from the loop normally meet with decreasing measurements, but if a hidden conductor is approached the measurements increase and fall off again as it is passed. Similar measurements or parallel lines will indicate the extension of the conductor, and thus its size and boundaries may be approximately outlined; the approximate depth may also be determined.

This inductive method is the most precise of the electrical methods. It is especially valuable as a check on other, faster electrical methods since it gives with greater precision the shape, size, and position of conductors. It is adaptable to rocky ground, barren mountainous regions, dry sand, ice-covered lakes, or deep snow. For underground work, the loop may be laid on the surface and readings taken underground along crosscuts and drifts.

Single-Cable Method. An alternative method is to use a long straight single cable, well grounded at both ends, and measurements are made along straight lines normal to this. This method is faster for surface work since the cable may be 1 or 2 miles long. The single cable can also be used underground. It may be strung along an inclined shaft, and measurements taken along crosscuts from the shaft, or it may be strung along a crosscut and measurements made along drifts. Rails, pipes, and wires must be disconnected. This method gives less penetration but is more sensitive than the loop method and indicates more weak conductors that may not be ore bodies.

Indicated conductors may be ore bodies, but also water-bearing strata, damp fault gouge, graphite slates, or other conducting bodies. Consequently careful geological interpretation is necessary.

GRAVITY METHODS

Gravity methods of geophysical prospecting utilize the well-known principle of gravitational attraction. The force of gravitation that holds the heavenly bodies in their orbits and that causes a free object to drop toward the earth also exerts an attraction between masses on or in the earth. Thus a suspended plumb bob normally points toward the center of the earth, but a nearby mountain mass will exert a gravitative pull upon it and cause it to be deflected from the vertical. Similarly, a fixed body within the earth, heavier than its surroundings, will attract and deflect a free swinging mass of metal by an amount that depends upon the size of the heavy body and its distance from the metal. Lighter bodies will appear to repel the plumb bob because their more dense surroundings provide greater attraction. Measurements determine the direction and size or distance of the attracting body.

There are three general methods for measuring gravitative attraction. The oldest is by means of the pendulum. but the torsion balance and gravimeter are more widely used commercially. Each of these gravitative methods is employed in petroleum prospecting.

Pendulum. A pendulum oscillates faster where gravity is greater, or vice versa. Therefore, if a heavy body of rock or ore lies beneath the surface, the pendulum will oscillate faster above it than away from it. The exceedingly small differences are detected by special apparatus. Thus an anticline or a salt dome may be indicated. This is the method employed by the Coast and Geodetic Survey for its regional gravity determinations over the United States. In commercial oil work generally two or more pendulums are employed simultaneously. Stations are occupied along traverses and the gravity determinations are plotted to give a gravity contour map, of which the high and low closed contours indicate heavier or lighter subsurface bodies.

Torsion Balance Method. This successful method utilizes measurements of differences of gravity, or the horizontal rate of change of gravity, rather than the direct measurements of gravity itself. When weights are placed north of a heavy mass, the deflection will be to the south, when to the south of the heavy mass the deflection will be north, and when directly above the mass there will be no horizontal deflection.

The instruments used are modifications of the Eötvos torsion balance. The principle is shown in Fig. 10–7. An aluminum bar suspended on

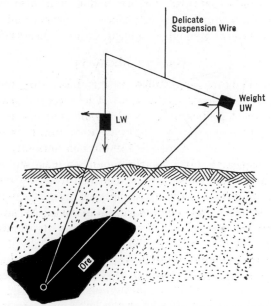

Fɪɢ. 10–7. Schematic representation of torsion balance whose suspended weights are attracted toward a heavy body, thus twisting the suspension wire.

a delicate platinum-iridium or tungsten torsion wire carries a weight, UW (gold or platinum) on one end and a similar suspended weight, LW, from the other end. The resistance of the wire to torsion is carefully calibrated. The gravitational pull of a buried mass upon the weights creates a deflection, which by means of a mirror on the arm can be read visually, or photographed. The fragile apparatus is housed in a triple aluminum casing. It is so sensitive that if a man stands near it it is deflected, and by the amount of its deflection the man's weight can be calculated. Moreover, it twists horizontally toward the man's mass and so indicates his position. It can detect a weight of 15 pounds at a distance of 6 feet from the instrument. It will detect changes of the order of one part in a million million. Its sensitivity requires that it be shielded from air currents, from temperature changes, and from the attraction of nearby ground protuberances. The mass of nearby hills has to be calculated and their attractive pull deducted from its readings. Consequently, it is best adapted to flattish regions.

In practice the instrument is set up in a temporary hut on flat terrain and the ground leveled off around it. When unclamped, the delicate

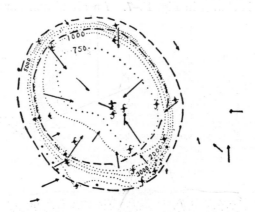

FIG. 10–8. Torsion balance gradient maps of Nash salt dome. Length of arrows indicates horizontal gradient of gravity and direction points to maximum gradient. (*Barton, A.A.P.G.*)

instrument requires about 40 minutes to come to rest, and about 4 hours is necessary for the three readings of a set-up. A series of stations are occupied, and the results are plotted as shown in Fig. 10–8.

The instruments detect the difference between limestone and sandstone, salt and rock, ore and gravel, and bedrock. Since they are best adapted to flat terrain they are most useful for petroleum work. They are unsuited to underground work because close wall and floor protuberances affect them too greatly. They have been extensively used for petroleum work in North America and have revealed many hidden structures and salt domes associated with oil. They are also used to check seismic findings. The drawbacks are that the method is slow, costly, and is not suited to all terrains.

Gravimeter. The gravimeter method has almost displaced the torsion balance method in gravitational surveying (Fig. 10–9). This is a static method in which gravity is compared with an elastic spring force. An object is weighed with great precision in several localities. Since a greater gravitational force is exerted by a heavy than a light mass, an object will weigh more over dense rocks than over lighter rocks. The observed changes of weight indicate the density of sub-surface materials (Fig. 10–10). These changes are of the order of one part in ten thousand or one part in ten million of the total value of gravity. Many types of gravimeters are available, and they yield results faster and with less expense than the torsion balance.

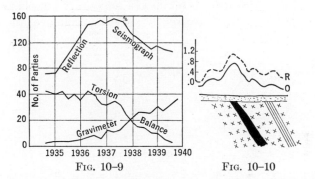

FIG. 10–9.   FIG. 10–10.

FIG. 10–9. Plat of results of geophysical prospecting for oil around Gulf of Mexico. (*After Eckhardt, Explor. Eng.*)

FIG. 10–10. Gravimeter survey over zinc-pyrite ore body, Skellefte district, Sweden. (*After Hedstrom, A.I.M.E.*)

SEISMIC METHODS

Seismic methods of prospecting utilize earthquake knowledge. A quake in Mexico sets up waves that travel through the earth with high velocity and are registered on seismographs in Washington, Ottawa, and London. The exact time of arrival is automatically recorded. Some waves travel faster than others and therefore arrive at the reading instrument earlier. Dense rocks transmit the waves faster than less dense ones, and deeper rocks are denser than shallow ones, hence the relative time of arrival of the waves indicates the character of the material through which they travel. The depth of penetrations of recorded waves depends upon the spacing between the point of shock and the recording instrument.

In seismic prospecting the same principles are employed. An artificial earthquake is made by setting off a charge of buried explosive; the waves set up are recorded on portable seismographs. The time of explosion is transmitted by radio, and the arrival time is recorded by an oscillograph on moving photographic film upon which time marks are projected at intervals of 1/10 or 1/100 of a second.

There are two seismic methods called reflection and refraction. In the first, the reflected wave or echo is used; in the second, the refracted

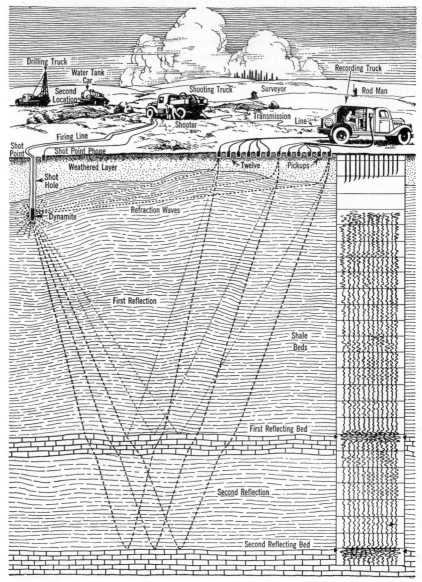

FIG. 10–11. Wave paths, records, and arrangements for seismic reflection method.
(*Heiland, Geophysical Prospecting.*)

or deeper penetrating waves are used. Of course both waves are produced at the same time, but either one may be screened out.

Reflection Method. In the reflection method, the reflection wave or the echo from a reflecting surface is used. By timing the interval between explosion and reflection record, the depth to the reflecting bed can be calculated when the average velocity of wave travel is known. Limestones overlain by shales form excellent reflecting beds. With a number of receivers in use, the reflected wave can be distinguished from the refracted wave because the former arrives simultaneously at almost all points and the latter arrives at different times (Fig. 10–11), depending upon the distance of the recorder from the explosion. The principle is the same as in echo sounding from ships, where a compression wave is transmitted downward from a ship's hull to the ocean bottom and back to a receiver. The speed of travel of the wave through water being known, the ½ time interval indicates the depth. In geophysical prospecting, however, it is more complicated because rocks vary in their transmitting speed and both compression and transverse waves are generated and received. Further, more than one reflecting surface is generally present, giving interfering waves, particularly within the weathered zone. Consequently, complicated equipment and procedure are necessary to screen out or minimize the undesirable reflections. Shot distances are mostly less than 3,000 feet from the recorder.

This method is the most widely used in oil exploration in the United States. It has been highly successful in accurately revealing subsurface structure and salt domes.

Refraction Method. This method simulates earthquake recording more closely than reflection. The shot generates waves that travel along the surface and through the deeper layers. The surface waves are slow waves, the deeper ones travel faster, their speed depending upon the medium through which they pass. Consequently, they are the first to arrive, if the recorder is not too close to the shot. By spacing recorders at equal linear intervals a travel-time curve can be established. The velocity in overburden, for example, may be 1,600 feet per second and in a lower layer 9,000 feet per second. With a travel-time curve set up, the nature of an underlying stratum can be inferred from the speed with which it transmits the waves (Fig. 10–12).

A common field practice is to set up recorders in a fan arrangement at different distances from the shot point. A network of fans covers the area to be explored. The time of the first arrival is recorded at each recorder and plotted on a time-distance chart. Thus, abnormally short time paths do not fall on the mean time curve and are indications of the presence of salt domes or other fast-transmitting mediums between the shot and the observation point. This area is then tested in detail, generally by a linear traverse method to determine the exact position and outlines of the salt dome. The observation points may also be distributed in an arc of a circle the center of which is the shot point. With normal strata the shot will be recorded approximately simultaneously upon each seismograph, but an abnormal speed will be at once evident and will indicate an abnormal stratum worthy of further investigation.

The refraction method has been found successful not only for petro-

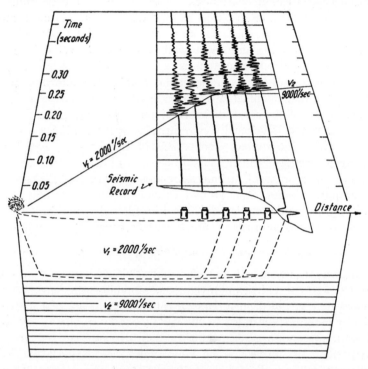

Fig. 10–12.   Wave path, record, travel-time curve in single-layer refraction method.
(*Heiland, Geophysical Prospecting.*)

leum work but also for bedrock determinations, gold and diamond gravels, ground water, and other purposes.

Seismic prospecting has reached an expenditure up to ten million dollars a year in the United States, and vast quantities of oil have been discovered by means of it.

RADIOACTIVE AND OTHER METHODS

In addition to radioactive ores and minerals such as uranium and radium, most rocks are measurably radioactive, the acid rocks more so than basic ones. Soils overlying areas of high radioactivity, and surrounding ground water, also may contain radon, the gaseous radioactive emanation. The radioactive emanations may be detected by various devices utilizing ionization chambers. The Geiger-Müller counter detects surface or underground emanations, which, however, are blanketed by a relatively few feet of soil. A more sensitive supercounter utilizes gamma rays and detects rock contacts readily. An airborne type yields a continuous record and detects rock differences and various common ores. Modern petroleum practice records gamma ray and neutron differences through well casings and indicates rock changes and fluid contents. Such well logs parallel electric coring and serve for rock correlation and porosity and fluid indications.

Geothermal gradients, established by careful temperature readings in drill holes, in some places give an indication of subsurface structure. Temperatures in excess of the normal regional gradient may be due to frictional heat developed by folding, faulting, fluid flow, chemical reactions, or radioactivity. Reflections of radio waves from subsurfaces have also been tried.

Geochemical determinations may indicate traces of elements in soils, foliage, streams, and ground waters, thereby affording clues to underlying hydrocarbons or metallic mineral deposits. Zinc deposits yield measurable quantities of zinc in certain types of overlying shrubs.

Exploration, Development, and Mining of Mineral Properties

After mineral deposits have been discovered, geology again is utilized in connection with their exploration and development. Mere discovery does not indicate whether a deposit is of value. That must be determined by careful exploration to outline the character, shape, size, grade, and prospective tonnage of a deposit. Few deposits survive this stage; mostly they are proved to be noncommercial. Exploration is also undertaken to find other deposits in the same area. Upon the completion of an exploratory program, the property is developed for commercial production. Again, geology plays a part. Development is followed by mining operations, during which geology also enters in.

Ore Deposits. In the exploration and development of ore deposits, in distinction to oil deposits, the geologist is commonly called upon to outline, and follow closely, the exploratory work. His detailed knowledge of the rock column and structure, the origin of the deposits, the causes of their localizations, their shape, behavior, and probable extent permits him to choose the location of surface workings, shafts, adits, drill holes, and underground workings so that they will yield the maximum results with the minimum expenditure. The mapping and study of folds, faults, intrusives, congenial host rocks, ore outlines, and numerous other features supply data that indicate the type of deposit, expected continuity, features controlling the distribution of values, and the most likely places to find ore continuations, dislocated ore bodies, and new deposits.

In mine development the geologist's knowledge of ore bodies, the factors controlling their location, their behavior and their expectable continuation, enable him to recommend permanent workings that will reach and develop the deposits from level to level with the minimum of footage and with the least hazard to workings under mining operations. For example, an inclined working shaft may be so inclined and pointed as to be kept in ground which will not be affected by stoping and which will necessitate the minimum cross-cut distances to the ore. Many an inclined shaft has been so badly pointed and inclined that, at lower levels, long haulageways with expensive upkeep are necessary to reach the ore body. Others, located without due regard to the ore loci, have penetrated through the ore or into the hanging wall, where " heavy ground " occasioned by subsequent mining operations necessitated their abandonment. Workings can be directed to avoid heavy ground of weak formations or faulted areas. Assistance can be given in the planning of mining methods by which the vagaries of rock formation and ore behavior can be largely anticipated. Thus, caving stopes, or stoping methods that tend toward loss of ore in walls or inclusion of waste rock in the ore, can be avoided, and " ore dilution " may be cut down, as has been successfully done at the Homestake mine in South Dakota.

During mining operations the geologist is continually in demand for routine and special problems. It is his task to map and interpret the geology, to find dislocated ore bodies, to see that ore is not overlooked, to direct the sampling, to plan the location of new workings, to avoid heavy ground, to make estimates of ore tonnage and grade, to direct exploration for new ore bodies, and to carry on investigations and research leading to successful and continued mining operations.

Oil. The aid of geology in the development and production of oil is quite different from that in metal mining and quarrying, and is largely subsurface geological work.

Once an oil discovery is made the pool must be developed, and for this it is necessary to learn in greater detail the kind of oil structure, the nature of the reservoir rock, and the shape and extent of the pool. This information must be gained largely from well cuttings or cores. Fossils and microfossils have to be determined for the purpose of locating the oil horizon and correlating it from hole to hole. Analyses of the heavy minerals in sandstones and of the silica residues from acid-treated limestones are made for the same purpose. Structure contour maps of the oil sand are drawn and locations for new holes are spotted with respect to the subsurface structure. Studies are made of porosity, permeability, resistivity, drill-hole temperatures, pressure, and water salinity.

During the production stage, the problems of well spacing, bottom hole pressures, and salt-water encroachment, and their bearing on the rate of well production, have to be continually checked up.

More and more the petroleum geologist is concentrating attention on various kinds of stratigraphic traps that bring about the accumulation of oil and on deep subsurface structures underlying unconformities and not reflected in the overlying rocks. In this " layer-cake " geology, two or three distinct subsurface structures may be worked out underlying successive unconformities, providing data for deep drilling.

Valuation of Mineral Properties

Since the valuation of most mineral properties depends not only upon the amount of mineral actually disclosed but also upon its probable extent beyond the disclosures, the economic geologist is called upon for a valuation because of his special knowledge of such features. Valuations may be desired to determine the size of plant to be installed, rate of production, justifiable initial expenditures, or for purposes of sale, purchase, merger, taxation, or investment.

Mining Properties. In a partly developed mining property there is, in addition to the ready calculable " blocked out " ore, expectable ore extensions. These may be indicated by sparse workings, scattered drill holes, or merely by observation that ore lies under one's feet or beyond the limit of the proven ore. Commonly, the value of the property rests more upon the probable ore extensions than upon the blocked-out ore. The probable extensions in turn depend upon the behavior of the ore body, its known persistence, the ore controls, delimiting formations along its projections, and other geological features, which must be evaluated. Most mines keep but a few years' ore supply blocked out at any one time and yet may have many years of life ahead of them. Judgment and experience are, therefore, important factors in valuation.

In its simplest terms the gross value of a mining property is the amount of metal or mineral it is estimated to contain multiplied by the market price of the product. Such a figure, however, is meaningless; a property might be stated to have a gross value of $5,000,000, but this might be arrived at by estimating 5,000,000 tons of ore with 10 pounds of copper per ton, worth $1.00 per ton, which would be valueless. What is desired is the net worth of a property, which depends upon the assay value, the cost per unit for extraction, the price of the product, the net profit per unit, the rate of production, and the life of the property. A simple example is as follows:

> Equipped gold mine: Estimated tonnage, 1,000,000 tons.
> Annual production, 100,000 tons; life, 10 years.
> Grade, $10.76; extraction, 93%; recovery, $10 per ton.
> Costs per ton, $7.00; profit per ton, $3.00.
> Annual profit, $300,000; gross profit, $3,000,000.

What is desired, however, is the present value. Since a profit of $3,000,000 accumulated 10 years hence is not worth that today, it must be discounted to the present. Thus, on the basis of a 6 percent return and amortization at 4 percent, the annual income is multiplied by a factor of 6.98 (from Inwood's interest tables), which equals $2,094,000, which represents the present value of the gross profit at the end of 10 years. The mine would return this purchase price and interest at 6 percent at the end of 10 years. If $500,000 is necessary to equip the mine for production, the present value is $1,594,000. If 100,000 shares are issued, the present value per share is $15.94. Other refinements of interest on equipment cost could be included. The fundamental basis of such a valuation is the estimated tonnage and grade of ore, which comes within the realm of the geologist.

Oil Properties. In the valuation of oil properties, similar principles are applied. Estimates are made of the yield of oil in barrels per acre, which in turn are based on the thickness of the oil sand, its porosity, and recovery yield. Estimates are made of the cost of development and extraction, which give an estimated profit per barrel of oil. The determined rate of extraction will give the life of the pool, and the annual profits can be reduced to present worth. In such valuations geology plays a greater role than in evaluating mining properties.

Selected References

General Reference 9. Chap. 12. *Geological methods employed by mining companies.*

Handbook for Prospectors. M. W. Von Bernewitz. McGraw-Hill, New York, 1931. *Prospecting and mining methods.*

Mineral Valuations of the Future. C. K. Leith. A. I. M. E. Series, New York, 1938.

Mine Examination and Valuation. C. H. Baxter and R. D. Parks. Mich. Coll. Mines and Tech., Houghton, Mich., 1939. *Methods of sampling and valuing mineral properties.*

Oil Property Valuation. P. Paine. John Wiley & Sons, New York, 1942. *Theory and methods.*

The Science of Petroleum. Oxford Univ. Press, Oxford, 1937. Vol. I. *Comprehensive treatise of petroleum geology, occurrence, and technique.*

Geophysical Prospecting. Ed. by BROUGHTON EDGE and T. H. LASKY. Cambridge Univ. Press, Cambridge, 1931. *An abridged summary of some methods.*

Applied Geophysics. 3d Ed. A. S. EVE and D. A. KEYES. Cambridge Univ. Press, Cambridge, 1938. *A brief, readable book.*

Geophysical Exploration. C. A. HEILAND. Prentice-Hall, New York, 1940. *A comprehensive technical textbook, with Part I, 64 pp., a summary of the subject in nontechnical language.*

Exploration Geophysics. J. J. JAKOSKY. Times Mirror Press, Los Angeles, 1940. *A comprehensive technical treatise.*

Choice of Geophysical Methods in Prospecting for Ore. H. T. F. LUNDBERG et al. Min. & Met. 26: No. 464, 1945. *How to choose the most suitable method.*

Evaluation of New Geophysical Methods. W. M. RUST, JR. Bull. A. A. P. G. 29:865, 1945.

Sampling Ore and Calculating Tonnage; Chap. 2 of **Mining Geology.** H. E. McKINSTRY. Prentice-Hall, New York, 1948. *Methods of sampling and ore calculation.*

Some Practical Aspects of Radioactive Well Logging. W. J. JACKSON and J. L. P. CAMPBELL. A. I. M. E. Trans. 165: 241–267, 1946. *Gamma ray and neutron well logging.*

Geochemical Prospecting for Ores: a Progress Report. H. E. HAWKES. Econ. Geol. 44:706–712, 1949. *Summary of geochemical methods.*

Aeromagnetic Survey of Allard Lake District, Que. (Ilmenite.) W. BOURRET. Econ. Geol. 44:732–740, 1949. *Airborne magnetometer survey of ilmenite deposit.*

CHAPTER 11

EXTRACTION OF METALS AND MINERALS

The winning of metals and industrial minerals involves one or more steps within the sphere of mining and metallurgical operations, such as mining, milling or ore dressing, smelting, and refining. The number of such steps necessary is determined by the nature of the ore and the form in which the mineral substance is desired. Thus, a copper ore must be mined, milled, smelted, and refined to obtain the desired pure metallic copper, but mica need only be mined and processed, and petroleum extracted (mined) and refined. These operations mainly fall under mining and metallurgical engineering, but geology may enter into them in a minor role and sometimes in an important one. However, the economic geologist should have some knowledge of them since geology is so closely allied with mining and to a lesser extent with metallurgy. Often the mining methods imposed because of the shape or character of an ore body may determine whether a given deposit may have economic value. Similarly, the treatment made necessary because of the mineralogical character of the ore may be so costly that the deposit cannot be profitably worked. For example, a low-grade gold ore consisting of free gold and quartz can be treated at nominal cost by amalgamation or cyanidation. However, if the gold is finely dispersed in arsenopyrite and minor chalcopyrite, the ore is not amenable to such simple treatment but must be concentrated and the concentrates roasted or shipped to one of the few arsenical smelters, and all at high cost. A brief outline of mining and metallurgical operations, omitting details, follows.

Mining Methods

Mining methods are broadly divided into two large groups — surface and underground. Surface methods are preferred where practicable because they are generally much lower in cost.

SURFACE METHODS

Surface methods are determined by the volume of ore, ore tenor, ore width, and depth of overburden. Many low-grade bodies cannot be profitably mined except by large-scale, low-cost surface methods.

Pitting and Trenching. Small shallow excavations are sometimes used for removal of scattered surface products, such as residual deposits, pegmatite minerals, and placer gravels.

Quarrying. Quarrying is generally employed for removal of stone, pegmatite minerals, and residual deposits. Quarries may be retreating benches open on two sides, or they may be nearly enclosed openings that extend horizontally into a hill. Commonly the ground is blasted down in benches. Operations may be on a small hand scale or they may be highly mechanized, using power shovels and railroad trains for rock removal.

Stripping. Many near-surface, flat sedimentary beds, such as coal, clay, pebble phosphate, or bauxite, are excavated by stripping opera-

FIG. 11–1. Stripping Pittsburgh coal in Jefferson, Ohio. Shovel rests on coal. (*Campbell, U. S. Geol. Survey.*)

tions. The overburden must not be too deep, otherwise the value of the mineral cannot sustain the cost of stripping. In coal stripping, which is a typical example, a large, long-boom, power shovel excavates a trench through the overburden to the coal, upon which the shovel rests. The exposed coal is then power-shoveled out into railroad cars. A parallel swath is then cut out, the overburden being dropped directly into the coal excavation (Fig. 11–1). Thus each new swath parallels and adjoins the last one mined, the coal being removed in progress and the spoil filling in the excavation behind. Some of the largest power shovels made are employed for stripping.

Open Cutting. Open cutting is employed for deposits of large volume and low grade or for low-value materials, most of which are surficial. It is commonly used for deposits of iron ore, " porphyry " copper, and other large deposits. The open cuts may attain lengths or diameters of several thousand feet, and the scale of operations may be many thousand tons per day. At Bingham, Utah, 125,000 tons of ore

and 200,000 tons of waste rock can be removed in a day.   Open cuts
are generally depressed beneath the surface level (Fig. 11–2) ; a few,
such as at Bingham, Utah, rise above it.   Examples of some large
open-cut mining operations are the great pits of the Lake Superior
iron region; the porphyry copper deposits of Bingham, Utah, Ely, Nev.,
Chino, N. Mex., Ajo, and Morenci, Ariz., and Chuquicamata, Chile;

FIG. 11–2.   Airplane view of Ruth pit of Nevada Consolidated Copper Co., Ely,
Nev., looking down 800 feet into bottom of pit.   Black lines are ore trains.

the bauxite pits of Arkansas; the asbestos pits of eastern Quebec; the
pyrite pits of Rio Tinto, Spain; the clay pits of Cornwall, England, and
the huge diamond pits of South Africa.

In open-cut mining the overburden is stripped off and carried to
waste dumps; it may be soft enough to be shoveled directly or it may
be hard rock that has to be blasted.   The ore is generally broken in
benches that spiral downward.   In large operations, power shovels
on the benches load ore and waste into railroad cars that are hauled
upward around the spiraling benches to the surface level.   Less commonly the ore is hoisted out on sloping cableways or by " horizontal
and drop " cableways.   As mining proceeds and an open cut becomes
deeper, the sides must be extended outward to give a safe slope and
prevent rock infalls.   Thus much waste rock in proportion to ore

must be removed. Open-cut mining is a low-cost operation; mining may be done for as little as 40 cents per ton of ore delivered at the mill.

A variation of open cutting, combined with underground work, is known as "glory-holing." In this, the rock is broken in open cuts and is dropped into ore and waste passes where it is drawn off underground, trammed to a shaft and hoisted to the surface. This is more costly than open cutting but is necessary when pits become very deep or the slopes cannot be cut back. There are many deposits which yield a profit by open cutting but which by underground methods would be unprofitable.

Hydraulicking. This is used for placer deposits. Overlying, unconsolidated gravels are excavated and swept by powerful jets of water into sluiceways; the gold gravels are sluiced through troughs, the particles of gold being trapped by bottom riffles and the barren gravel being stacked by other water jets at the exits.

Dredging. This is a more elaborate and larger-scale method of placer mining. Huge dredges are built in the gravel area and dig themselves into artificial ponds. They contain mechanical excavators for mining subaqueous gravels, and washing, screening, pumping, and propulsion machinery. The dredge digs on one side of the pond and rejects the washed tailing on the other side, thus carrying the pond with it. Dredges excavate to 100 feet beneath the water line. The excavator is a continuous bucket elevator passing over rollers on a long boom that is let down to the excavating face. A dredge handles from 300 to 500 cubic yards of gravel per hour at costs as low as 3 cents per cubic yard.

Underground Mining

Underground mining is as variable as the types of ore deposits (Figs. 11–3, 11–4). Individual methods are devised for each deposit, so details here will be omitted. In essence, entry is made into the ground by vertical or inclined shafts or, in mountainous regions, by adits (tunnels). Levels are driven off from the shaft at intervals of 100 to 200 feet vertically. Horizontal workings consist of drifts that parallel the ore, and cross-cuts that cross the ore trend. Vertical workings consist of raises (upward), winzes (downward), and stopes where the ore is excavated. Ore above a level is drilled, blasted downward, and drawn off through chutes into cars, trammed to the hoisting shaft, dumped into ore pockets, loaded into skips, hoisted to the surface, and automatically dumped into ore bins. The skips hold from 1 to 5 tons of ore and are hoisted at the rate of 1,000 to 2,000 feet a minute.

Underground mines produce up to 20,000 tons of ore per day. The deepest mines have reached depths of more than 9,200 feet.

Veins. In vein mining, chute raises are driven up on the vein from the level drifts, and at 10 to 20 feet above the level are joined to form the bottom of a stope. This is then drilled overhead and the ore is

FIG. 11-3.   An ancient method of underground mining with two levels.   (*Agricola*.)

blasted down onto the ore pillar and dumped into the chutes, where it is drawn off into cars. If the walls are strong, broken ore is allowed to fill the stope, only the excess being drawn off, and the miners thus stand on the broken ore to drill into the vein above them (Fig. 11-6). When the stope is completed, the broken ore is then all drawn off. This is called shrinkage stoping. The empty stope may be left standing open, or it may be filled (Fig. 11-7) with waste rock, or sand, or in narrow veins the walls may be supported by timber to prevent ground movement. In many narrow veins, all the ore is drawn off as broken, and cross timbers (stulls) support the walls and provide working platforms.

Where the ground is heavy (i.e., the walls will not stand up readily), shrinkage stoping is impossible because the walls cave into the ore and dilute it or block off the ore when it is being drawn. Then *square*

FIG. 11–4. The medieval method of hoisting ore. (*Agricola.*)

setting (Fig. 11–8) may be necessary. This consists of erecting square or rectangular sets of timbers from wall to wall and reaching upward as ore is excavated; these support the walls and back (roof). The hollow squares are filled with waste rock as soon as erected. Square setting is expensive. Other methods called *top slicing* may be used for similar progress to avoid the high cost of square-set timber. In this, successive slices are cut off the top of the ore, which is dumped down shoots, and a mat of timber provides support above. *Rill stoping,* or steplike stoping, using fill, is another variation.

Large Irregular Deposits. Large irregular-shaped ore bodies, particularly if low-grade, are generally *caved*. There are many variations of caving methods, the details of which are beyond the scope of this summary. The general principle is that a caving block or stope, oriented with respect to rock structure, is undercut and is penetrated and flanked by branching raises, which are drilled up and blasted (Fig. 11–9). The whole mass of the delimited block then caves, and the

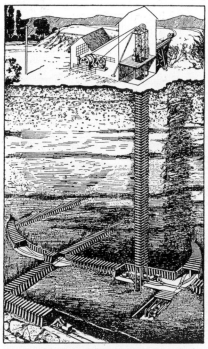

Fig. 11–5. A later method of underground mining. (*Fehlmann, Eizenerzengung.*)

caving automatically crushes the ore, which descends into underlying chutes in solid rock where it is drawn off. This allows the mass to cave further and the surface is allowed to subside as the ore is withdrawn. Most of the rock breaking is thus accomplished without much blasting. Large pieces of ore that may escape crushing are blasted (bulldozed) in entries above the ore chutes. Once started, the ore caves by itself, and the ore is drawn off at the bottom until the overlying barren caprock makes its appearance. The caving block is then finished and another one is similarly started. This method permits extremely cheap mining on a large scale with a minimum of drilling, blasting, and ore handling. It is the method used in all of the large

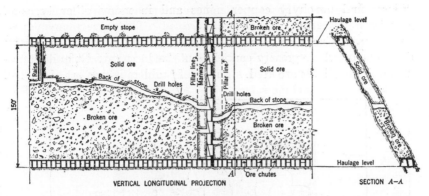

FIG. 11–6. Example of shrinkage stope, longitudinal section. (*Jackson-Hedges, U. S. Bur. Mines.*)

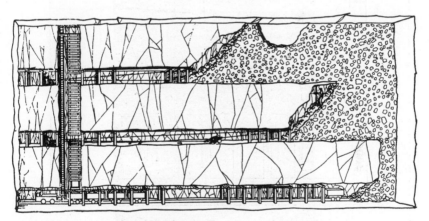

FIG. 11–7. Example of horizontal cut and fill stoping. (*Works Prog. Admin.*)

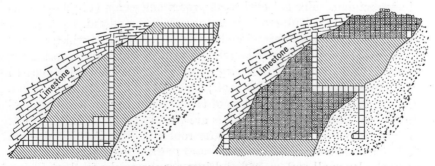

FIG. 11–8. Example of square-set stoping. (*U. S. Bur. Mines.*)

underground porphyry copper mines and in many other types of deposits.

Flat Ore Beds or Coal Seams.   For flat-lying ore beds and coal seams (Fig. 11–10) where gravity cannot be utilized extensively, quite different mining methods are used.   The chief problem is support of the roof

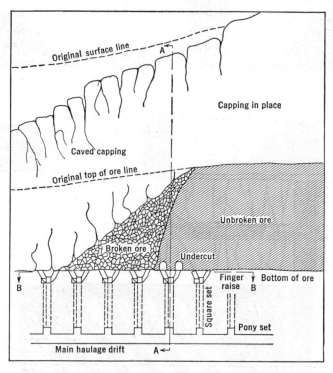

Fig. 11–9.   Example of caving used at Ruth mine, Nevada.   (*Wright, U. S. Bur. Mines.*)

during mining.   For flat coal seams, *room and pillar* (Fig. 11–11) and *longwall* (Fig. 11–12) methods are generally employed.   The former consists of dividing an area into a series of rooms, from which the coal is excavated, separated by a series of pillars, which serve to hold up the roof while the rooms are being mined.   The face of coal is undercut and the roof pressure cracks it off; thus the coal face retreats toward the end of the room.   Cribbings of timber, stone, or masonry are built to support the roof, and the pillars are then in part excavated, smaller pillars of coal generally being left for roof support.   Considerable coal is lost in this method.

In the longwall system, radial drives are run to the farthest limit of mining, and from a peripheral drive a coal face is retreated toward the shaft. Timbers, small coal pillars, or stone cribs support the roof immediately in front of the working coal face, and as the coal face

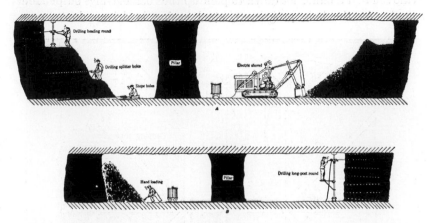

FIG. 11–10. Mining a horizontal coal seam. (*Anderson, U. S. Bur. Mines.*)

retreats, the roof slowly settles down over the excavated area. A coal pillar is left around the shaft to support it. This method results in a small loss of coal.

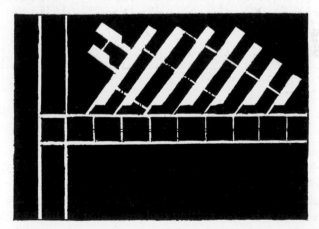

FIG. 11–11. Room and pillar method of mining coal. The rooms are excavated; intervening pillars are left for support and are later withdrawn. (*Moore, Coal.*)

Oil. Oil is removed by drilling oil wells, properly spaced, over the property or oil pool to the oil producing zones. One or more oil sands may be present and the depths may be greater than 15,000 feet. Formerly drilling was accomplished by a heavy wedge-shaped cutting bit suspended on a steel cable and jerked up and down by a reciprocating pivoted boom (*cable-tool method*).   After a round of drilling the tool is removed and a bailer let down to pick up the rock cuttings suspended in mud.   As the hole progresses, steel casing is driven down, new

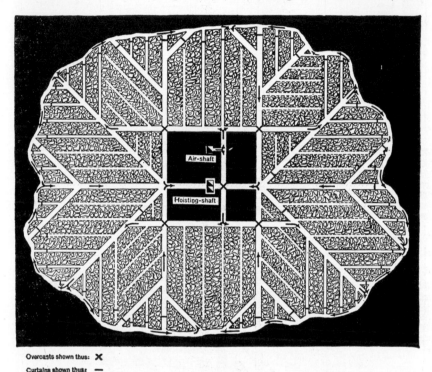

Overcasts shown thus: ✕
Curtains shown thus: —

FIG. 11–12.   Longwall method of flat-seam coal mining, where faces retreat from periphery toward central shaft pillar. Arrows indicate ventilation control. (*Swift, U. S. Bur. Mines.*)

lengths being added at the top.   Water flows are sealed off by casing and concrete, and perforated casing permits inflow of oil.

The cable-tool method is now largely supplanted by the *rotary* method, in which a bottom cutting tool suspended by tubing is rotated from the surface and cuts in a medium of " drilling-mud."   Casing follows the deepening of the hole and prevents caving.   The top of the hole is capped and equipped with high-pressure valves to regulate oil and gas flow.

The oil generally flows out naturally because of the gas pressure, which is transmitted by hydrostatic or rock pressure.   The bottom-hole

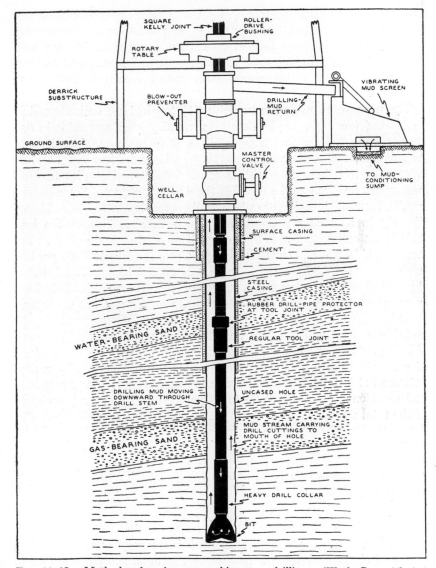

FIG. 11–13.   Method and equipment used in rotary drilling.   (*Works Prog. Admin.*)

pressure may exceed 5,000 pounds to the square inch.   Where gas pressure is lost, or is absent, or in declining or old wells, the oil may have to be pumped out.   These are called "pumpers."   In certain localities, gas or water may be forced under pressure into some holes, made to flow through the oil sands and be withdrawn from other holes, thus flushing the beds of residual oil.

Milling

Once ore is mined it generally must be milled, treated, or processed to separate the desired minerals from the undesired ones. The varieties of ore and desired end products necessitate several types of milling processes.

The purpose of milling is twofold: (1) to yield the desired product in finished form; (2) to yield a concentrated product amenable to further treatment. The first applies to many nonmetallic minerals, such as asbestos, fluorspar, and diamonds. Gold and silver are also obtained directly from free-milling ores. The second applies to most base-metal ores, where the metal is the desired product. Since milling results in some loss, high-grade metallic ores commonly are shipped directly to a smelter without intervening milling, to save the loss. This is not economical with low- or medium-grade ores, however. For example, if freight and smelting charges on copper ore are $5.00 per ton, ore containing $4.00 per ton in copper could not be so treated, but if it is milled so that the copper in 15 tons of ore is concentrated into 1 ton of concentrates, then the freight and smelting charges are equivalent to only about 33 cents per ton of original ore. Consequently most metallic ores are milled. Multimetal ores may have to be milled, even if high grade, in order to separate the metallic minerals. The amount of concentration, termed the *ratio of concentration,* generally ranges between 5:1 to 30:1. The economy of milling is expressed as *recovery,* and commonly ranges from 80 to 95 percent, meaning the proportion of the total metal recovered during milling.

Practically all milling processes have one step in common, namely, preliminary crushing, which is divided into two stages, yielding coarse and fine lumps. Further reduction is called grinding, the degree of fineness depending on the nature of the ore and the process used, the resulting product being a powder.

AMALGAMATION

This is a simple, low-cost process usable only for free-milling native gold and silver ores and for placer gold. The crushed ore is ground wet in ball mills, rod mills, or stamp mills until the particles of gold are released. The ground pulp is then passed over amalgamating plates (metal plates coated by quicksilver), which absorb the gold, forming amalgam. This is scraped off, and the gold is removed from the amalgam by retorting. In places quicksilver is also introduced into the grinding mills, or the pulp may be introduced into rotating amalgam barrels instead of passing over plates. Where coarse gold is also present, it is commonly removed before amalgamation by passing the pulp over concentrating tables or blankets. Amalgamation is not effective where the gold is not " free," i.e., where it is not released by grinding. Placer gold is collected mainly by amalgamation.

CYANIDATION

This process is the one generally used for the reduction of simple gold and silver ores but cannot be used for the extraction of other metals. Neither is it adapted to gold and silver ores that contain such metals as copper, which cause high cyanide consumption.

In simple cyanidation the ore is generally finely ground and the pulp is dewatered and placed in large tanks through which is passed a solution of potassium cyanide to dissolve the gold and silver. The gold solution is passed through a precipitating filter containing zinc that precipitates the bullion, which is then purified and cast into bricks. The silver content is later extracted, leaving the gold. Where the gold is contained in sulphides, these may be concentrated and treated separately. If some of the gold is associated with sulphides, or is present as tellurides, in a form not amenable to cyanidation, these may be concentrated from the ore, roasted to drive off the volatile elements, and then cyanided or amalgamated.

CONCENTRATION

Concentration processes are used for obtaining dispersed minerals and for separating desired from undesired minerals. They are not applicable for rich or solid sulphide ores or for those ores that consist largely of the desired mineral, such as hematite iron ore or bauxite.

There are two main stages in concentration. One is the freeing of the desired minerals from the enclosing gangue or undesired minerals, which is accomplished by crushing and grinding, with or without accompanying roasting. The other is the collection or concentration of the desired minerals (called concentrates), with the minimum of loss, and the rejection of the undesired minerals (called tails). Several methods are employed.

Hand Sorting. This is a simple, crude method used mostly in small operations for coarse ores, or necessitated for coarse nonmetallic minerals that should not be ground. In many metallic ores the ore minerals may occur in coarse chunks easily separable from the enclosing gangue by coarse crushing. The crushed rock travels on a moving belt or revolving circular table from which the ore pieces are removed by hand. In other ores the gangue may be sorted out, leaving only the ore minerals, which are thus enriched for further concentration or are shipped directly.

With many nonmetallic products, hand sorting is the only concentration step necessary. Some are too fragile or too easily harmed to be treated by any other method as, for example, long asbestos, mica, feldspar, and gemstones.

Gravity Concentration. This older method is particularly adapted for simple ores containing heavy and light minerals. Formerly, it was the only milling method used for base-metal minerals but is now largely supplanted by flotation. It is still used, however, for simple ores of lead, zinc, copper, tin, tungsten, fluorspar, and others. It is carried on in stages, some ores needing only one stage, others more.

In a simple ore, after crushing, the classified " heads " flow into jigs and rest on a screen through which a column of water is pulsated vertically. The heavy minerals settle to the bottom and are drawn off; the lighter ones flow into a lower screen where the process is repeated, the final tails being ejected. Different-sized fragments are placed on different-sized screens. This stage may complete the concentration cycle for some coarse ores.

Ores with fine mineral particles may be jigged first to remove the coarse materials, and the tailing discharge reground, or they may all be finely ground, classified as to grain size, and passed with water over concentrating tables, which consist of tilted tables, containing wooden riffles, subjected to quick pulsations. The heavier particles gravitate toward the higher end and side, and lighter ones to the lower. The discharge is reground and passed over other tables; the concentrates are dewatered and shipped.

Sink and Float. In this method, crushed ores are dropped into a heavy solution; the heavy particles sink, the lighter ones float, and separation is thus effected.

Gravity concentration is the only method for obtaining concentrates of many of the heavy nonmetallic minerals.

Spiral Concentration. Heavy particles are separated from light particles by downward passage through a vertical " Humphrey " spiral.

Flotation. This process is now used for most metallic minerals. The equipment occupies much less space than that for gravity methods; the method is faster, more economical, and gives a higher extraction. The ore is ground to a fine powder, is classified as to grain size, receives a little oil and chemicals, and flows to different flotation cells. Here a froth of air bubbles, coated by oil, is created. Sulphides and some other metallic minerals adhere to such bubbles, and are carried upward and swept into a collecting trough. Passage through several cells removes most of the metallic minerals. Some nonmetallic minerals can be pre-treated so that they can be made to float. The concentrates are dewatered and sent to a smelter.

Dry Concentration. Strongly and weakly magnetic minerals are concentrated from nonmagnetic gangue minerals by magnetic and electrostatic processes. Ground ore travels on a conveyor belt past magnets that remove the magnetic minerals. Asbestos fibers, after being crushed and torn are concentrated by being blown by air currents from the rock matrix.

LEACHING

Some minerals are not amenable to other treatment processes but can be recovered by leaching. This applies to certain minerals soluble in common solvents, such as some of the copper minerals, nitrates, or halite.

Acid Leaching. Sulphuric acid leaching has formerly been employed extensively on a large scale for the treatment of carbonate copper ores at Ajo, Ariz. It is also used for leaching oxidized copper ores at Inspiration, Ariz., and for sulphates and oxychloride of copper at Chuquicamata, Chile, where the crushed ore is subjected to an acid bath in asphalt-lined tanks; the liquor is drawn off and the copper is electrolytically precipitated.

Natural leaching is utilized at Rio Tinto in Spain, Ohio Copper in Utah, Ray, Ariz., Chino, New Mex., and in other places. In this process the solvents are self-generated by the action of water on oxidizing sulphide ores. At Rio Tinto, pyritic ore with copper is placed in surface heaps and water is added, which generates the sulphuric acid and ferric sulphate that dissolve the copper as sulphate. The liquor then passes over scrap iron, which precipitates copper. At Ray and Ohio Copper water is made to seep through low-grade sulphide ore in place; the solutions are collected in drainage tunnels and precipitate their copper on scrap iron.

Acid leaching is not possible where limestone is abundant because of the high acid consumption.

Ammonia Leaching. Leaching by ammonia solution was employed for copper carbonates enclosed in limestone gangue at Kennecott, Alaska. The jig discharge was subjected to an ammonia bath in covered tanks. The liquor was pumped into stills and heated to drive off the ammonia, which caused precipitation of the copper in the form of the black oxide.

Water Leaching. Water leaching is commonly employed for the extraction of underground salt, being admitted and withdrawn through drill holes.   It is also used to leach Chilean nitrate from the enclosing impurities, for potash extraction and for other substances.

Smelting

Smelting is the process of melting ores in blast or reverberatory furnaces to obtain metals.   Concentrates, hand-sorted ore, or run-of-mine ore constitute the *feed*.   Coke and natural gas are the normal fuels. *Flux* must be added, according to the nature of the ore, to permit melting, to make a fluid *slag*, and to allow the molten metal to settle to the bottom of the furnace, where it is tapped off.   Iron ores require much limestone flux; siliceous ores require iron flux; and pyritic concentrates require siliceous flux.   Gold-silver ores or concentrates must have a *collector*, such as copper or lead, to carry down the molten droplets of gold.   Zinc becomes volatilized, so it must be a minimum in other ores. Sulphur and arsenic are driven off as gas, but the gas must be collected because of its poisonous nature.   The sulphur may be converted into sulphuric acid.   Smelting dust is trapped and recovered.   Practices vary with different metals.   In copper smelting, for example, the feed, perhaps pre-roasted, is introduced with flux and fuel into the furnace, slag is drawn off at the top, and the metals with impurities are tapped off at the bottom as *matte*.   This is transferred to converters, where air is blown through, forming more slag and purifying the copper, which is drawn off as blister copper.   Lead concentrates are roasted and smelted to lead bullion, which is drawn off and cast into pigs. Zinc is distilled in retort furnaces or else electrolytically obtained. Copper ores with high pyrite may be treated by pyritic smelting, where the sulphur of the ore supplies all or most of the melting heat.   Quicksilver is volatilized from its ores in furnaces and condensed in retorts.

Refining

The purpose of refining is (1) to recover the included precious and other metals, (2) to obtain pure metal for commercial use, and (3) to get rid of deleterious ingredients.   It consists of fire and electrolytic refining.   Most metallic ores, such as copper and lead, yield smelter products that contain additional metals and so need refining.

Electrolytic Refining.   With copper, for example, the blister from the smelter is cast into anodes, which are placed in tanks containing acid electrolyte through which an electric current is passed.   Copper is dissolved from the anode and precipitated as pure copper on the cathode, the impurities dropping out as sludge, which is recovered for its gold and other metal content.   The cost of refining is generally less than the value of the other metals obtained.   At the same time deleterious elements are removed. Most copper is sold as electrolytic.

Fire Refining. Copper matte lacking gold and other metals is fire refined for use as high-grade blister copper by means of converters, as in Braden, Chile, Rhodesia, and elsewhere. Lead bullion is also fire refined and desilverized.

Selected References

Mining Engineers Handbook. Ed. by ROBERT PEELE. John Wiley & Sons, New York, 1941. *General reference to all phases of prospecting, mining, milling, and metallurgy.*

Principles of Mining. HERBERT C. HOOVER. McGraw-Hill, New York, 1912. *Mining practice and valuation.*

Principles of Mineral Dressing. A. M. GAUDIN. McGraw-Hill, New York, 1939. *A standard textbook on concentration methods.*

Metal Mining Practice. C. F. JACKSON and J. H. HEDGES. U. S. Bur. Mines Bull. 419, 1939. *Best general treatment of exploration, sampling, ore reserve estimates, mining methods, concentration, and ore dressing.*

PART II
Metallic Mineral Deposits

In Part II the discussions of the various metals are necessarily brief. However, much of the related theoretical matter has been covered in Part I, and for details of form, occurrence, and origin reference should be made to the corresponding sections in Part I. A knowledge of mineralogy is presupposed.

CHAPTER 12

THE PRECIOUS METALS

Gold

Since antiquity gold has been prized as an ornament, as a concentrated form of wealth, and for monetary use. Man's urgent desire for it in ancient days led to barter, invasion, conquest, colonization, and exploration in India, Asia, Africa, and Iberia. The first gold rush started under Jason in the *Argo*. Gold was a strong incentive in the discovery of America, and its greedy acquisition by the Conquistadores was accompanied by treachery, robbery, and murder. Man has roamed the world in search of it, exploring far-off parts of the earth and undergoing almost unbelievable hardships and privations. It has little use other than decorative or monetary, but no other substance has been the cause of so much horror and misery, or has done so much good. For over 2,000 years its value has generally increased, rarely decreased. It is still eagerly sought and its discovery is attended by the springing up of new communities, the pushing back of frontiers and the rising of accompanying agriculture and industry; it has been the forerunner of civilization in many distant lands.

HISTORY OF PRODUCTION AND DISTRIBUTION

History. The first noteworthy rise in gold production followed the discovery of America, when Mexico, Peru, Bolivia, and Chile poured forth streams of yellow and white metal to enrich the capitals of Europe. The United States, however, suffered a lag of 300 years, until payable gold was discovered in North Carolina in 1801 and in Georgia in 1829. Then came the sensational discovery of placer gold in California in 1848 that culminated in the great gold rush of '49. Production increased tenfold, and by 1853 the United States had become the leading gold producer of the world, a position maintained for 50 years. A similar discovery in Australia in 1851 boosted the world output to over 6 million ounces in 1850–1860. This flash production quickly declined, but it was offset by other discoveries in the western United States, culminating in Cripple Creek, Colo., in 1891, which started another sensational gain in United States output, which was surpassed

419

by the even more sensational rise in South Africa, following the discovery of the great "Rand" in 1886.

With the romantic and tragic Klondike rush of 1896 the world production in 1890–1900 exceeded 15 million ounces annually and by 1915 had reached a peak production of nearly 23 million ounces — a peak that stood until the change in the price of gold.

Since 1905 South Africa has assumed dominant world leadership, seconded by the United States until 1931, when Russia and Canada pushed ahead.

Prior to 1930 it was generally considered that the world would soon be facing a serious shortage of gold. Peak production had apparently been reached, and it was estimated that the great Witwatersrand would begin to decline about 1933 and have a production of only $2\frac{1}{3}$ million ounces by 1947. All this was suddenly changed, however, with the increase in the price of gold. Former marginal ores became included in reserves, low-grade deposits became commercially important, and production was greatly stimulated. World production has again decreased, owing to stable price and increased cost of production.

Current Production. The highest world production of 41 million ounces was reached in 1940. The average since then has been about 30 million ounces. The total world production since gold was first known is estimated to be 50,000 tons. The chief producing countries and their average percent of world production over a number of years are as follows:

	PERCENT OF WORLD		PERCENT OF WORLD
South Africa	35	Japan and Chosen	4.0
United States	15	West Africa	2.5
Canada	13	Mexico	2.4
Russia	12	Congo	2.2
Australia, New Zealand	5	Rhodesia	2.0
South America	4	British India	0.7
		Others	2.0

The United States average production of 100 to 200 million dollars is ranked as follows:

1. California	5. Colorado	9. Washington
2. South Dakota	6. Nevada	10. Idaho
3. Alaska	7. Arizona	11. Oregon
4. Utah	8. Montana	

For yearly production of gold by countries consult the annual volumes of *Minerals Yearbook*.

Distribution. Gold deposits range in age from Precambrian to late Tertiary and are found throughout the world, mostly where igneous

activity has occurred.   They show a preference for intrusives of intermediate to silicic composition.

In *North America* the greatest gold belt lies within the Precambrian
Canadian Shield.   Another is associated with the batholiths of the
Pacific Coast region extending, with interruptions, from California to
Alaska.   A central Rocky Mountain belt contains numerous gold-
bearing lodes in association with smaller intrusives and is flanked by
two eastern outliers — an important one in the Black Hills and an
insignificant one in the Appalachians.   An extensive Tertiary volcanic
belt containing shallow lodes projects into Arizona, Nevada, and southern California and extends through central Mexico toward Nicaragua.

In *South America,* small, scattered gold veins and replacements
occur throughout the Andean chain in association with Tertiary and
older intrusives.   Outliers occur in Brazil and the Guiana Highlands.

In *Africa* is the great Witwatersrand of the Transvaal which dominates the world, and its new counterpart in the Orange Free State.
Also numerous smaller deposits occur in northern South Africa, Rhodesia, and central and east Africa.

In *Australasia* many important lodes and placer deposits occur in
Australia, New Guinea, the Netherlands Indies, and New Zealand.

In *Europe* meager occurrences lie in Scandinavia and the mountainous parts of central and southern Europe, but numerous and important deposits lie in Russian Europe, chiefly in the Urals.

In *Asia*, Siberia boasts many important placer and lode deposits, and
a prominent belt of small deposits extends from the mainland through
Japan, the Philippines, Indo-China, and Burma.   The Kolar district
of India is one of the world's great goldfields.

MINERALOGY, TENOR, TREATMENT, AND USES

Ore Minerals.   The economic gold minerals consist only of native
gold and minor amounts of gold tellurides, electrum, and amalgam.
The tellurides include calaverite, sylvanite, krennerite, and petzite.
Nearly all gold contains some silver but if silver is present in considerable amount the result is pale yellow to white electrum.   Natural
amalgam occurs in a few deposits.

Gangue Minerals.   The common gangue mineral of gold is quartz,
but carbonates, tourmaline, fluorspar, and a few other nonmetallics
may also be present.   Gold is commonly intimately contained in base-
metal sulphides and related minerals, or in their oxidation products.

Association and Treatment.   Gold is so commonly associated with silver that gold and silver deposits are generally considered together.

Where the gold is in the native form and easily amalgamated or cyanided it is said to be *free milling;* where it occurs in minerals difficult to treat it is called *refractory.* Its common associate, pyrite, generally offers no trouble in treatment except for the necessity of fine grinding or roasting. Associated copper minerals make cyaniding difficult, often necessitating roasting or concentration and smelting. Associated arsenical minerals make treatment difficult and expensive.

Most gold ores are treated by cyaniding, or amalgamation, or both together, with or without flotation and roasting (see Chap. 11). Refractory ores are mostly concentrated and smelted. Massive sulphide ores are generally smelted directly; if copper is absent they may be ground and cyanided.

The tenor of gold ores varies widely. Formerly the lowest-grade ores treated were those of Alaska Juneau, which averaged about 0.04 ounce gold (= 0.8 dwt or $1.40) per ton. The South African mines average about 4 pennyweight per ton or $7.00. The Canadian ores range from about $6.00 to $22 per ton. United States gold ores average 0.239 oz per ton. The tenor of gold ore changes with the cost of production; rising post-war costs have necessitated utilizing higher-grade ores.

Uses of Gold.   The foremost use of gold is for monetary purposes, most of it being kept as bullion in reserve for notes issued. United States gold coins (0.9 fine) contain 13.714 grains of gold per dollar.

The next most important use for gold is for ornamentation, chiefly jewelry, using " white," " green," and " yellow " gold. Because of its softness, jewelry gold is generally alloyed with copper, silver, nickel, or palladium. Its purity or fineness is designated in carats; one carat means 1 part gold in 24. It is also used for gold plating, glass and china inlays, gold lace, gilding, book binding, lettering, and interior decoration. For these purposes gold leaf is generally employed, which is made by pounding out gold to 1/200,000 of an inch in thickness; 1 ounce of gold will cover about 100 square feet of surface and 1 grain of gold ($0.73) will cover 52 square inches. Gold is also used for dentistry, glass making, and in the chemical industry.

KINDS OF DEPOSITS AND ORIGIN

The principal classes of gold deposits, and some examples of each, are:

1. Magmatic deposits (?): Waarkraal, South Africa (quartz gabbro); Golden Curry, Mont.; Gold Hill, Utah.
2. Contact metasomatic: *Nickel Plate* mine, British Columbia; Cable mine, Mont.; Ouray, Colo.

3. Replacement deposits:
   a. Massive — *Noranda*,[1] Quebec; Rossland, British Columbia; *Morro Vehlo*, Brazil; northern Black Hills, South Dakota.
   b. Lode — *Kirkland Lake, Porcupine*, and Little Long Lac, Ontario; *Homestake*, S. Dak.; *Kolar*, India.
   c. Disseminated — Alaska Juneau; *Witwatersrand;* Beattie mine, Quebec.
4. Cavity filling:
   a. Fissure veins — *Mother Lode*, Grass Valley, Calif.; *Cripple Creek* and Camp Bird, Colo.; *Porcupine*, Ontario; *El Oro*, Mexico; Kalgoorlie, Australia; Benguet, Philippines.
   b. Stockworks — Quartz Hill, Colo.; Victoria, Australia; Treadwell, Alaska; Lake Athabaska, Saskatchewan.
   c. Saddle reefs — *Bendigo*, Australia; Nova Scotia, Canada.
   d. Breccia deposits — Bull Domingo and Bassick, and Cresson, Colo.; Cactus, Utah.
5. Mechanical concentrations: Placer deposits of *California*, Yukon, *Alaska*, *Russia, Australia*, and other places.
6. Residual concentrations: Brazil, Madagascar, Australia, western United States, Mexico.

[1] Italics in this and subsequent similar tables indicate important deposits of each type.

Origin. Most gold deposits originate through igneous emanations or surficial concentrations. A few gold deposits have been formed by contact metasomatism, but most lodes have been formed by hydrothermal solutions. Emmons has shown a world-wide association between gold lodes and intrusive rocks that clearly indicates kinship between the two.

Mechanical concentration is responsible for the vast wealth of placer gold throughout the nonglaciated regions of the world. Supergene enrichment of gold is negligible.

Examples of Gold Deposits in the United States

The United States has produced about 20 percent of the world's gold, mainly from the following provinces: (1) Sierra Nevada, (2) Great Basin, (3) Rocky Mountain, (4) Black Hills, (5) the Southwest, (6) the Appalachians, (7) southeastern Alaska, (8) central Alaska, and (9) the Philippines. The bedrock deposits are dominantly fissure veins and replacements. Placer deposits account for about one-quarter of the total production. About 12 percent of the gold produced in the United States is a by-product from base-metal ores. Most base-metal ores carry some gold.

Homestake Mine, Black Hills, S. Dak.

The Homestake mine at Lead, S. Dak., is the ranking gold producer of the Western Hemisphere. Since 1879 it has produced about 470 million dollars in gold, from ores that ranged in tenor from $4.05 per ton in 1905 to $9.00 in 1933 and since then up to $16.25 per ton. It can produce about 5,000 tons of ore daily and carries greater ore reserves than any other American gold mine. It illustrates structurally controlled hydrothermal replacement of folded beds.

The ore deposits occur in altered Precambrian sediments, with included basic sills. The beds have been folded into a major anticline and syncline about 6,000 feet from crest to trough, and each contains prominent minor folds and crenulations; the folds plunge 35° to 40° to the southeast. Large dikes and sills of Tertiary rhyolite intersect the beds. The ore bodies follow closely the crests of the major anticline and minor folds and plunge with the structure.

The ores are localized in Homestake formation. Originally this was a bed about 60 feet thick consisting of iron-magnesium carbonate with included thin bands of shale and quartz sand, but the folding metamorphosed it to cummingtonite and chlorite, and squeezed the 60-foot stratum into thicknesses ranging from a knife-blade to 300 feet. The Homestake formation is overlain by arches of impervious Ellison slate, which may have channelized the rising mineralizing solutions.

McLaughlin states that the major loci of ore are: (1) the Homestake formation; (2) the plunging folds into which the Homestake formation is crumpled, thickened, and sheared beneath the impervious Ellison slate; and (3) zones of fracturing and shearing within the Homestake formation.

The ore bodies are inclined pods, veins, saddles, and lenses that lie within the Homestake formation along the crests of the plunging folds, each following its own minor fold (Fig. 5·4–35). In plan the main deposit is 300 to 800 feet long and from 50 to 150 feet wide, but it attains widths up to 300 feet in areas of compressed minor folds. Some bodies extend down the plunge over 4,500 feet. Another lower-grade ore body lies about 1,000 feet away from the main deposit. The ore bodies are cut and displaced by the Tertiary intrusives and faults.

The ore consists of coarse white quartz, cummingtonite, chlorite, ankerite, and minor amounts of garnet and mica, with pyrrhotite and arsenopyrite. The ratio of silver to gold is exceptionally low (0.25). The ore is generally uniform in character from top to bottom of the mine. Sericitization and silicification are lacking and the ore abuts sharply against barren walls.

The deposit has been formed by hypothermal replacement of the cummingtonite schist and is generally considered of Precambrian age, although a Tertiary age related to other Black Hills Tertiary gold mineralization has also been held.

PACIFIC COAST GOLD DEPOSITS

Mother Lode, California. The Mother Lode of California is not a mine but two-score mines within a belt 120 miles long and a mile wide, on the western foothills of the Sierra Nevada. The total production is about 300 million dollars, one-half of which has come from a 10-mile segment in Amador County. Mining has progressed over 90 years and has reached a vertical depth of 5,912 feet in gold quartz ores that have averaged around $7.00 per ton.

The gold belt traverses northwesterly trending, steeply dipping slates, schists, and greenstones of Jurassic and Carboniferous age. Some included green schists are favorable host rocks for many veins; they contain chlorite and an amphibole and are altered tuffs or lavas. The stratified rocks were invaded, according to Knopf, by a succession of plutonic rocks beginning with peridotite — now serpentine — and ending with granodiorite. This epoch of igneous activity ended with the formation of the gold deposits.

The deposits consist of (*a*) quartz veins up to few thousand feet in length, generally parallel, branching, or arranged en echelon along the strike, and (*b*) bodies of mineralized country rock. The veins are fillings of reverse faults with displacements up to 375 feet. Knopf points out that the veins cut the slate cleavage at an acute angle and pass from slate to greenstone or vice versa with marked deflection or refraction which he computed at 1:4. The veins pinch and swell; the swells consist of quartz and the pinches of quartz and slate, or of fault gouge. The ore is localized in short shoots that pitch steeply and persist in depth. The shoots not only contain more quartz but also more gold per ton than the thin frayed parts. Most of the shoots crop out boldly; some apex 3,300 feet beneath the surface. The bodies of mineralized rock consist of " gray ore " or mineralized schist and occur adjacent to quartz veins, between intersecting veins, or in broad zones of fissuring.

The wall rocks have been intensely altered. The slates, schists, and serpentine have been altered to ankerite, with sericite, albite, pyrite, and arsenopyrite, and some gold. Much carbon dioxide has been added and silica removed, which Knopf thinks was more than enough to supply the quartz of the veins.

The vein ore, formerly worked as low as $2 to $3 a ton, averaged above $7 per ton prior to 1942, but since World War II most of the mines have been forced to close. The ores consist of quartz and 1 to 2 percent of metallic minerals made up of base-metal sulphides, minor arsenopyrite, and, in places, petzite. The "gray ore" consists of ankerite, sericite, albite, quartz, pyrite, and some arsenopyrite, with

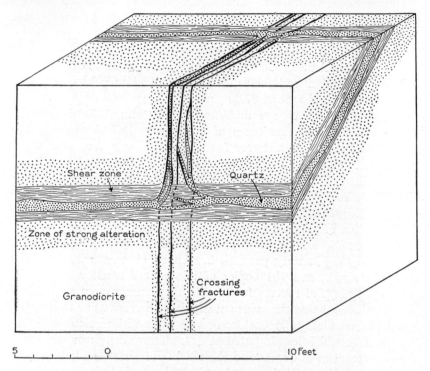

Fig. 12–1. Intersection of vein and crossing without displacement. Grass Valley, Calif., North Star mine, No. 3 vein, 8,600-foot level. (*Johnston, U. S. Geol. Survey.*)

gold up to $13 per ton. The mineralized schist ore contains ramifying ankerite-quartz veinlets, up to 6 percent sulphides, and $2 to $5 in gold per ton.

The veins are hydrothermal fissure fillings, with successive enlargement due to reopening by renewed movement along the faults. The solutions are thought to have been derived in part from a deep-seated consolidating granitic magma of late Jurassic or early Cretaceous age, and in part by mingling meteoric waters. Knopf thinks that the magma supplied the gold, sulphur, arsenic, carbon dioxide, and some other constituents but that the quartz was supplied from the wall rock — a view in which not all geologists concur.

Grass Valley-Nevada City, Calif. This district has produced about 145 million dollars from ores that have averaged about $11 per ton. The veins are associated with granodiorite plutons that intrude altered Paleozoic to Jurassic rocks.

The lodes at Grass Valley are conjugated fissure veins that parallel the long axis of the pluton; " crossings " (Fig. 12–1), or steeply dipping

Fig. 12–2.   Sheeted zone at intersection of two veins.   Grass Valley, Calif., North Star mine, 8,200-foot level.   (*Johnston, U. S. Geol. Survey.*)

joints, traverse the long axis.   The veins are only 3 to 10 feet wide, but they have been traced for 8,000 feet along the strike and over 9,000 feet down their flat dip (35°).   The ore is localized in shoots from a few hundred to 5,000 feet long.

The Nevada City veins occur in the granodiorite and the invaded rocks.   Some, according to Johnston, occupy thrust faults with displacements up to 1,000 feet.   The veins occupy fracture zones up to 40 feet wide, in which the quartz follows one and then another of the fractures.

The ores consist of quartz (Fig. 12–2) with as much as 20 percent pyrite, gold, and minor base-metal sulphides. Molybdenite and pyrargyrite have been noted by Johnston at Nevada City; tellurides and scheelite are rare. Ankerite and calcite are later than the quartz. The gold-silver ratio is about 3½ to 1. There are three stages of vein filling, according to Johnston, an earlier one of quartz, pyrite, and arsenopyrite followed by chalcopyrite and sphalerite, and then one of later quartz, tetrahedrite, galena, gold, and carbonates. Vugs,

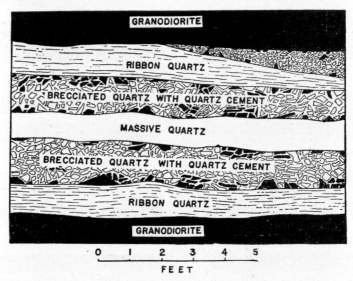

FIG. 12–3. Three episodes of quartz deposition. Murchie mine, Nevada City, Calif. (*Johnston, Geol. Soc. Amer.*)

comb structure, and other features indicate cavity filling by hydrothermal solutions. The quartz is ribboned by repeated opening of the fractures (Fig. 12–3).

Johnston accounts for the wide fillings by growth by accretion accompanying recurrent deposition and movements, of which he recognizes six at Grass Valley and four at Nevada City. These are ingeniously illustrated in Fig. 12–4. Few agree with Farmin's conclusion of vein-dike formation.

The nearby *Allegheny district* is noted for its spectacular free gold and for its massive quartz, which Ferguson thinks represents extended deposition, probably accompanied by recrystallization.

Sierra Nevada Placer Deposits. About two-thirds of the Pacific Coast gold production has come from "high-level" Tertiary and Quaternary gravels. The Sierra Nevada belt has yielded, since 1848, around 1.3 billion dollars in placer gold from an area about 150 miles long and 50 miles wide in the middle and lower reaches of the western slope of the Sierras. This is a region of gentle undulation passing upward into rugged slopes with canyonlike valleys, 3,000 to 4,000 feet deep.

The *Tertiary placers* represent the accumulations from erosion of hundreds or thousands of feet of quartz veins. The geologic history of

MOVEMENT   DEPOSITION

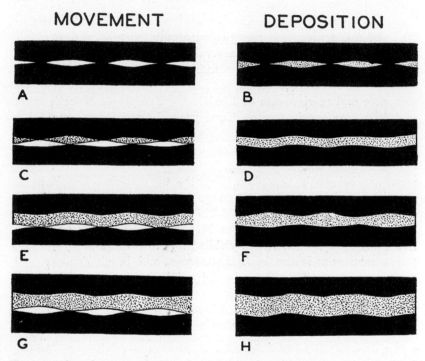

FIG 12–4. Development of thick quartz veins by alternate movement of vein walls (with continuous support), and deposition of quartz. Nevada City, Calif. (*Johnston, Geol. Soc. Amer.*)

these interesting deposits is given in Chapter 5·7. The buried Tertiary gravels that escaped erosion now lie in old valley remnants high up on the Quaternary valley slopes. Their gold is won by drift mining and hydraulicking from a lower layer of "deep gravel" and an upper layer of "bench gravel" within 2 to 3 feet of the bottom. Drift mines, taking only the richer parts, yielded up to $30 per cubic yard; hydraulic mines averaged about 10 cents. Platinum metals and rarely diamonds have been found with the gold. Remnants of gravel deposits in Tertiary channels are estimated to aggregate 500 miles in length and to contain 30 million ounces of gold.

The *Quaternary gravels* contain much gold reconcentrated from the Tertiary gravels from the Feather, Yuba, and American Rivers. The richest gold, which was in the channels, has now been removed, but smaller channel deposits and larger dredging areas, aggregating 19,000 acres, remain. These gravels are 18 to 65 feet in depth and yield 9 to 15 cents per yard. Some gold is also won from the residual mantle.

Klamath Mountain Placers. In the Klamath Mountains from Redding, Calif., to Coos Bay, Ore., is a placer belt 200 miles long and 50 to 80 miles wide that has yielded around 80 million dollars since 1852. This region underwent several oscillating movements recorded in erosion surfaces and elevated marine terraces. A Pliocene uplift, which caused valley incision, was followed by a Pleistocene submergence and re-elevation of 1,500 feet. The placer deposits consist of Cretaceous conglomerates, Tertiary and Quaternary gravels, and residual mantle deposits.

The *Cretaceous gold-bearing conglomerates* are considered by Pardee to be marine shore deposits, 170 to 800 feet thick, that lie on or near the Klamath peneplain. Weathered material has been hydraulicked, yielding $2 to $3 in gold per ton. *Uncapped Miocene gravels,* cemented, occur up to 1,000 feet thick in the Trinity River basin. Weathered portions have been mined, and unweathered material has been hydraulicked, yielding 2½ to 3 cents per yard. The *Quaternary gravels* are in flats, terraces, and beach deposits. The flats yield 12 to 14 cents per yard by dredging. The beach deposits are ancient and modern. Ancient ones are worked at elevations of 150 to 800 feet above sea level, where " black sands " with fine gold, buried beneath 20 to 100 feet of barren sand, are exposed by streams. The present beach sands contain fine gold and platinum in " black sand."

The Nevada Gold Province

This province embraces Nevada and parts of adjacent Arizona, California, Utah, and Idaho. It contains a great many gold and silver deposits that belong to the metallogenetic province that extends into Central America. In general the deposits are relatively shallow veins, of epithermal to mesothermal type, mostly of Tertiary age, and some of them are very rich in gold and silver.

Goldfield, Nev. Since 1902 the Goldfield district has produced 87 million dollars from ores with a gold-silver ratio of 3:1, but the production is now small. Here, a succession of Eocene to Pliocene volcanic flows resting on granite and shale are overlain by 1,000 feet of lacustrine beds containing flows of rhyolite and basalt. Deformation and low domal uplift followed. A dacite sill 700 feet thick intrudes andesite and contains most of the ore bodies, which are of late Miocene or Pliocene age.

The ore bodies are small rich shoots with indefinite boundaries, in silicified and fractured country rock called "ledges." The "ledge" matter consists of quartz, pyrite, and unique alunite and kaolinite. The bonanza ores contain gold, tellurides, pyrite, marcasite, wurtzite, sphalerite, bismuthinite, famatinite, and tennantite. The tellurides include petzite, hessite, sylvanite, and the uncommon goldfieldite. Tolman and Ambrose conclude that the first stage was the development of alunite, kaolin, silica, and pyrite along channels that became the "ledges." Next, ore solutions deposited tennantite, famatinite, sphalerite, and bismuthinite, followed by tellurium, copper, and antimony minerals, with gold and silver later, and quartz last. Deposition was by successive precipitations in open spaces, with some late mineral replacement.

The hypogene alunite indicates acid solutions. Ransome considers that magmatic alkaline waters containing hydrogen sulphide were oxidized near the surface, supplying sulphuric acid which caused the alunitic alteration; the mingling of surficial acid waters and ascending waters caused near-surface precipitation of the gold. The ore shoots, although very rich, did not persist downward. Much ore averaged $400 per ton; $30 ore was common; but more recently $8 to $10 ore has been treated.

Other Deposits. The area includes many other similar but smaller gold and gold-silver deposits. *Aurora* and *Tuscarora* in Nevada and *Oatman,* Ariz., have yielded in excess of 30 million dollars each from fissure veins in Tertiary volcanics carrying gold, argentite, sulphides, and silver sulpho-salts, with quartz and adularia. Oatman had five stages of vein filling with progressive increases in gold-silver ratios. *Jarbidge* and *Round Mountain,* Nev., contain generally similar deposits, except that the ores of Round Mountain carry realgar and alunite.

Rocky Mountain Region

Cripple Creek, Colo. Cripple Creek, with a production of over 375 million dollars is a famed mining camp. Its maximum production was reached in 1900 and a secondary peak occurred in 1915, when the rich Cresson pipe was discovered. Individual mines have produced from 10 to 30 million dollars. There are or were 64 mines in the district, among them the Portland, Cresson, Golden Cycle, Stratton-Independence, and Vindicator.

The deposits lie at an elevation of 10,000 feet mostly within a complex breccia plug of Tertiary volcanics some 2 to 3 miles across, generally referred to as the Cripple Creek volcano. It breaks through a basement of Precambrian granite, gneiss, and schist, in which some of the deposits occur. According to Loughlin the plug tapers downward into nine roots or subcraters (Fig. 12–5) through which latite-phonolite, syenite, phonolite, and alkaline basaltic rocks were successively intruded. Water-laid sediments containing fossil leaves of probable Miocene age occur within the plug. A late explosive eruption in the central part of the area formed the Cresson breccia pipe.

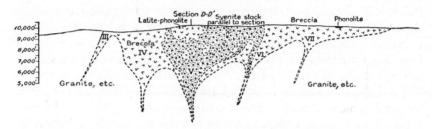

Fig. 12–5.   Cross section through the Cripple Creek " crater " showing five of the subcraters.   (*Loughlin-Koschman, Colo. Sci. Soc. Proc.*)

Deposits. The ore bodies are veins, sheeted zones, irregular bodies, and breccia fillings. Some veins occupy fissures that coincide with older fissures that margin the volcano. They resulted mainly from regional compression and shearing, aided by local settling of the breccia within the plug, and the upward thrust of intruding lava. The most persistent vein zones are along master zones in granite, and upward extensions of pre-volcanic fissures. The lodes are chiefly cavity fillings but there is local replacement of the wall rocks.

The veins range from mere cracks filled with rich ore accompanied by narrow bands of mineralized wall rocks to sheeted zones up to 40 feet in width. Most of the veins contain ore shoots formed at intersections of cross fissures or " flats," some of which are persistent horizontally and vertically, but few exceed 500 feet. The veins converge downward toward the roots of the subcraters; few were profitable below 1,500 feet; the maximum depth reached is 3,100 feet. The upper levels contained many short but extremely profitable veins; in the lower levels they are fewer and of lower grade.

The Cresson " blowout," 700 feet in diameter and 2,000 feet deep, is fractured basaltic rock within the crater and has a pipelike mass of collapse breccia on its eastern side (Fig. 5·4–27B).

The ores are noted for their paucity of native gold and abundance
of tellurides and fluorite. The principal mineral is calaverite
($AuAgTe_2$). Other tellurides include sylvanite, krennerite, petzite,
hessite, and a silver-copper telluride. Druses lined with calaverite are
common. Phenomenally rich veins and cavities were found. One
" vug " in the Cresson mine, lined with calaverite, yielded $1,200,000;
the Cresson pipe yielded about 35 million dollars.

Origin. Loughlin and Koschmann recognize three stages of ore
formation: 1, a first stage of intense corrosion of the wall rocks and
deposition of jaspery and porous quartz, adularia, fluorite, and pyrite;
2, a repetition of the first stage followed by dolomite, celestite, roscoelite, sphalerite, galena, tetrahedrite, and lastly tellurides; and 3, cavity
deposition of smoky quartz, chalcedony, pyrite, calcite, and locally
cinnabar. The rich telluride ores are considered to have been formed
from hot alkaline solutions that rose from the subcrater vents, particularly those of the Portland, Golden Cycle, and Vindicator mines.
The solutions rose along trunk channels and nearer the surface spread
out into numerous branch fissures.

San Juan Region, Colorado. The interesting San Juan region includes many important mineral districts, most of which, however, are
base-metal deposits with silver and minor gold. Large gold districts
are few, but there are many small ones. Within the region are such
well-known districts as Telluride, Ouray, Silverton, Creede, Bonanza,
Lake City, and Needle Mountains.

The region has a complex geologic history of sedimentation, folding, erosion, stupendous volcanic eruptions, intrusions, and faulting.
Paleozoic and Triassic sediments were folded and eroded, and then
received Mesozoic sediments, culminating in crustal disturbances and
volcanic activity. Eocene monzonitic intrusions, accompanied by
early Tertiary mineralization, were revealed by deep erosion, and the
late Eocene Telluride conglomerate was deposited on the peneplaned
surface. After further erosion the great volcanic eruptions began,
reaching maximum intensity in the Miocene and continuing into the
Pliocene, resulting in many thousands of feet of lavas and tuffs. The
San Juan tuff is 3,000 feet thick; the overlying Silverton volcanics,
4,000 feet, above which are the Potosi and Hinsdale volcanics. Another period of intrusion was also accompanied by faulting and late
Tertiary mineralization.

The *Ouray district* embraces older deposits in sedimentary rocks and
more important younger late Tertiary ones in the volcanics, such as
the Camp Bird ($30,000,000) and Virginius. The older ones are
replacement blanket deposits and fissure fillings, with some spread-

ing outward by replacement. These deposits carry base-metal sulphides, native gold and tellurides. The American Nettie mine is a
replacement in quartzite. The younger deposits are remarkably long,
continuous fissure veins carrying gold and silver. They are open-space
fillings exhibiting banding, crustification, and colloform structures.
The famous Camp Bird vein in tuff and andesite is a continuous
sheeted fissure zone about 5 feet wide and followed 3,000 feet below its
outcrop. The ore consists of native gold, minor sulphides, and sparse
tellurides in a gangue of crustified quartz, carbonates, and fluorite; the
gold is late. The ores undergo a striking change with depth. Toward
the lower levels specularite appeared as an early stage vein mineral
in association with quartz and gold, indicating " telescoping." It
appears to indicate a gradual impoverishment with depth, which has
necessitated closing the mine.

In the *Telluride district,* across the divide from Ouray, are other
similar late Tertiary fissure veins with beautiful crustification. They
are long and narrow; the Smuggler Union vein was 8,000 feet long and
2,300 feet deep. Other gold-copper ore veins occur in the *Silverton-
Anastra* districts in the *North Star, Shenandoah-Dives,* and *Sunnyside*
mines.

APPALACHIAN REGION

Although the Appalachian belt of the eastern United States has
yielded some 30 million dollars since 1800, most of this has come from
placers. Lenticular gold quartz veins occur in the Carolinas, Georgia,
and Alabama in ancient crystalline rocks. They consist of quartz,
accompanied by carbonates, apatite, chlorite, albite, tourmaline, and
garnet. The metallic minerals are gold, common sulphides, and molybdenite; some tellurides, enargite, and tetradymite have been reported.
The values are low — from $2 to $8 per ton, rarely $20 — and occur in
shoots. The deposits are now chiefly of historical interest. Only a
few deposits have been worked, as, for example, the Haile, Dahlonega,
and Hog Mountain.

ALASKA

Since the beginning of this century, Alaska has produced nearly 600
million dollars in gold, of which two-thirds was from placers. The
annual production of around 26 million dollars comes from a few large
deposits and many small ones.

Placer Deposits. Alaska's placers include stream, bench, buried
and beach gravels. The most productive areas are the Fairbanks
region, the Upper Yukon, and the Seward Peninsula. Other areas are

the Kuskokwim, Kenai, Endicott Range, and Susitna. About 80 percent of all placer gold is produced by dredges.

The *Upper Yukon River*, below Dawson, had its heyday during the great Klondike rush of 1898. The gold came mostly from phenomenally rich bottom and bench gravels that carried much coarse gold and nuggets.

The *Fairbanks district* is the main producing region of Alaska. The richer gravels have long since been extracted, and the present yield is largely from low-grade stream and bench gravels by dredging. The average dredging season is about 260 days, and it takes 3 years to prepare a tract for dredging and to thaw the gravels. Although the region lies well north of the glaciated area, all the ground is frozen to considerable depth. The gold gravels occur at various depths; many are 100 to 200 feet beneath the surface; they are up to 80 feet thick and yield about 60 cents per yard.

In the *Seward Peninsula* some 23 dredges are operated for about 120 days a year with an annual yield of about 4 million dollars from stream and bench gravels of a dozen districts. The *Nome* district has yielded over 70 million dollars from present and elevated beach gravels. On the present beach, gold gravels about 8 feet thick and as much as 300 feet wide in places lie on a clay layer 45 feet above bedrock. There are six beaches, two submarine and four elevated ones. The Second beach, at an altitude of 47 feet above sea level, lies on a false bottom and carries a 6-inch layer of gravel yielding 50 cents to $1.00 per yard. The submarine beach lies 56 feet below the Second with a pay streak from 1 to 3 feet thick carrying $4 to $10 per yard. The Third and richest beach is 70 feet above sea level with a pay streak 5 to 10 feet thick and 100 to 600 feet wide, lying under 20 to 120 feet of largely frozen overburden.

Lode Deposits. Alaskan lode deposits prior to World War II yielded about 8 million dollars annually, chiefly from the Juneau district, which includes the noted Alaska Juneau mine and two mines on Chichagoff Island. The Willow Creek district, north from Seward, is the next most important district, having contributed about 8 million dollars since 1909 from a group of small lodes. Minor lodes occur in the Fairbanks and Cook Inlet regions.

The *Alaska Juneau mine,* closed at present, is noted for being the largest tonnage gold mine in the world, and having the lowest-grade ore and lowest costs. Some 13,000 tons of ore per day were mined, of which 46 percent was discarded by hand sorting. The annual pre-war production was 5 to 6 million dollars, and the total production exceeds 70 million dollars from a little over 70 million tons of ore. The average gold content of the ore since 1893 is 0.0449 ounce per ton, and the pre-war content was 0.04 ounce, or a recovery of $1.15 per ton, mined at a total cost of 72 cents per ton.

The deposit lies at tidewater in a belt of schist and altered Triassic volcanic rocks intruded by sills and masses of gabbro. Wernecke states that these in turn were intruded by Coast Range granodiorite accompanied by dikes, and again by albite-diorite toward the Middle Cretaceous. The invaded rocks were profoundly altered, and mineralizing solutions then penetrated a zone of weakness in greenstone

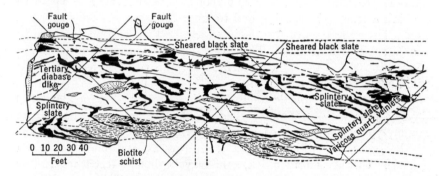

FIG. 12–6. Quartz stringers (black) shown in geologic sketch of part of No. 200 stope, south ore body, Alaska Juneau mine. (*After Wernecke, Eng. Min. Jour.*)

and slate 1,000 to 2,000 feet wide and 3 miles long, depositing stringers of auriferous quartz and producing a halo of intense wall rock alteration.

The deposits contain innumerable bunches and stringers of quartz from an inch or so to a few feet across (Fig. 12–6). It is a gigantic stockworks. The vertical range of the known part of the zone is 2,200 feet. The mineralized zone as a whole is mined by large-scale caving methods.

The gold is in the quartz along with minor silver, base-metal sulphides, and ankerite.

Wernecke thinks the ore fluids were derived from the depths of the Coast Range batholith and rose as hot liquids into a zone of structural weakness where minor irregular fractures had been formed as a result of physical variations in the rocks. The quartz was deposited in part in open spaces and in part by replacement of slate. The main part of the gold was late introduction into fractures in the quartz.

Formerly similar gold deposits were worked across the channel at the Treadwell mine, on Douglas Island, until a submarine cave-in occurred.

Chichagoff Island near Juneau contains two mines — the *Chichagoff* and the *Hearst Chichagoff* — which have produced about 16 million dollars in gold. According to Reed a satellitic dioritic mass has intruded deformed graywackes that are also cut by dikes and sills of aplite. Two long, prominent fault zones that intersect the bedding at

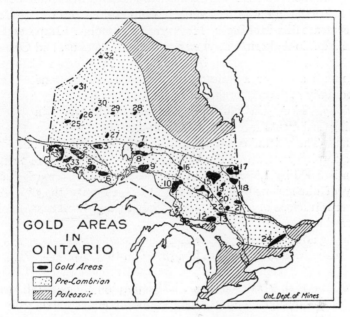

FIG. 12–7. Geologic sketch map of Ontario, showing distribution of 33 known gold areas. (*Ont. Dept. Mines.*)

about 45° localized the ore to form tabular bodies of rich gold quartz ores in distinct ore shoots that have greater depth than horizontal distance.

CANADIAN GOLD DEPOSITS

Canada contributes about 13 percent of the world's gold supply, or 3 to 5 million ounces annually. Of the total production, Ontario has produced approximately 57 percent, British Columbia 20 percent, Northwest Territories 14 percent, and Quebec 19 percent. Ontario now produces about 60 percent and Quebec 20 percent. There are some 140 gold mines in Canada.

The chief area is the Canadian Shield (Fig. 12–7), where gold quartz lodes are found from western Quebec to the Northwest Territories. Most of the lodes yield only gold; by-product silver is minor, and placer gold is negligible. The lodes are mostly hypothermal veins and replacement deposits.

Porcupine District, Ontario. Porcupine is the premier gold district of Canada, having produced about 30 million ounces of gold and paid nearly 200 million dollars in dividends; it has an annual production of about 1 million ounces. It includes 36 mines, of which the Hollinger for a period was the world's leading gold mine. The major producers are the Hollinger, McIntyre, and Dome. Important ones are the Buffalo-Ankerite, Paymaster, Delnite, Pamour, and Coniarum. Other producers are the Hallnor, Aunor, and Preston East Dome. Some mines are over a mile deep, and about 12,000 tons of ore are crushed daily with an average tenor of $8.25 per ton.

Geology. Keewatin volcanics unconformably overlain by Temiskaming sediments are intricately folded to form the Porcupine syncline. The volcanics were intruded by pipelike bodies of Algoman quartz porphyry, of which the principal one, the Pearl Lake porphyry, in plan is 5,000 by 1,500 feet and plunges 45° E with remarkable continuity of dimensions and shape for 1½ miles down the dip. There are later intrusions of albitite, quartz diabase, and Keweenawan olivine diabase dikes.

Shearing and fracturing formed loci for the veins, which occur mainly in the volcanics and sediments and rarely in the porphyry. The main Hollinger and McIntyre veins are grouped around the west end and flank of the highly altered Pearl Lake porphyry. Hurst thinks the fractures are tension openings produced by successive movements along lines of structural weakness.

Deposits. The ore zones lie on both sides of the Porcupine syncline, the North Break includes the Hollinger and McIntyre mines, and the South Break includes the Dome and other mines. The total production from the Hollinger-McIntyre zone is impressive, about 600 million dollars.

The deposits consist of (*a*) quartz veins, (*b*) irregular replacement lodes, (*c*) zones of roughly parallel quartz lenses, and (*d*) pipelike deposits. The veins are long and deep, as in the McIntyre and Hollinger, or form a linked system of short, closely spaced veins, so remarkably shown in the upper levels of the Hollinger (Fig. 12–8). Irregular quartz lodes, with the values mainly in the wall rock, characterize one Hollinger type, and wide replacement zones of quartz stringers and mineralized country rock typify the Pamour and Hallnor mines. In the Dome mine are zones of en echelon lenses of quartz and intervening mineralized country rock; the new East-end deposit is a pipelike mass of ore.

The average width of the veins is around 10 feet; the width of the irregular lodes is about 100 feet. Langford states that the McIntyre " long " veins range from 500 to 2,000 feet in length and the " short " veins are less than 500 feet long; the long veins occur in dacite contacts, the short ones in basalt; both avoid the porphyry.

Ores. The ore consists of quartz and mineralized wall rocks (Fig. 12–9); but in the McIntyre mine there are also sulphide ore bodies in

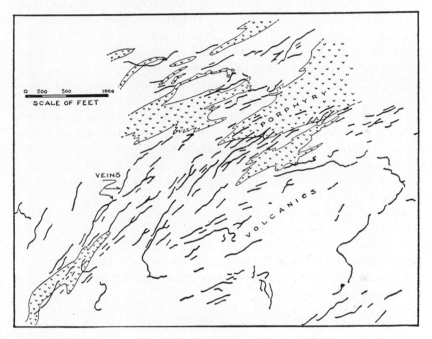

Fig. 12–8. Vein systems (linked) of upper levels of Hollinger mine, Porcupine district, Ontario. From mine plans of Hollinger mine. (*Bruce, A.I.M.E.*)

altered dacite, of which pyrite forms 5 to 15 percent, averaging 0.5 ounce gold.

The quartz ores carry gold, pyrite, sericite, tourmaline, and carbonates. Sporadic minerals are base-metal sulphides, tellurides, scheelite, chlorite, albite, siderite, dolomite, selenite, and anhydrite. The gold is with the pyrite and is largely free, in places forming spectacular gold, much subject to " high-grading." Considerable carbon is present in the McIntyre ore.

In the Hollinger mine Graton and McKinstry recognize a central zone of quartz-ankerite and quartz-calcite, with albite and an outer zone of quartz-calcite and minor clinozoisite. Langford notes similar zones concentric to the Pearl Lake porphyry in the McIntyre mine with the gold values concentrated in a central ankerite zone and no gold in the outer calcite zone. Hurst and Langford recognize three periods of mineralization: (1) barren quartz tourmaline veins; (2) main quartz deposition with the mineral assemblage noted above; and

FIG. 12–9.   Banded quartz lode of No. 7 vein, McIntyre mine, Porcupine district, Ontario.   (*Dougherty, Econ. Geol.*)

(3) barren quartz-calcite, probably overlapping No. 2. Graton and McKinstry think that the quartz of the Hollinger ore replaced ankerite and that ore deposition was largely replacement, while Hurst considers that open space filling dominated.

The approximate average gold recovery per ton is as follows: Hollinger, $8.25; McIntyre, $10.50; Dome, $9.50.

The ores are hypothermal fissure fillings and replacements of the walls. The solutions were probably derived from an underlying magmatic source, of which the porphyry bodies are one offshoot.

Kirkland Lake District, Ontario. This productive district ranks closely with Porcupine, yielding annually about a million ounces of

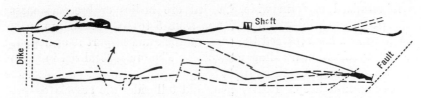

FIG. 12–10. 2,200-level, Lake Shore mine, Kirkland Lake, Ontario, showing vein fractures. Workings omitted. Black, ore; dash lines, vein fractures; dotted lines, faults. (*After Robson, Bull. Can. Min. Met. Inst.*)

gold. It includes 7 contiguous producers in a length of 3½ miles consisting of Lake Shore, Wright-Hargreaves, Teck-Hughes, Kirkland Lake Gold, Macassa, Sylvanite, and Toburn. A depth of 7,300 feet has been reached. The average tenor of the ores is $10.80 per ton. The Lake Shore has been the second largest gold producer in North America.

The deposits lie within a synclinal belt of metamorphosed Temiskaming sediments, underlain by Keewatin volcanics and intruded by syenites and basic dikes; together they make up most of the ore zone. Parallel reverse faults localized the ore, the main fault having a throw of 2,000 feet (Fig. 12–10). Post-ore movement produced diagonal cross faults and strike faults that displace the veins. There are two chief ore breaks, the North and South (Fig. 12–11), each being made up of several parallel fractures joined by diagonal fractures, resulting in a diamond-shaped mesh. In addition, larger diagonal veins connect the north and south zones.

The great producer is the North vein, 2¼ miles long, which in the Lake Shore property is continuously mineralized. In other mines, faults break the ore body into segments. The

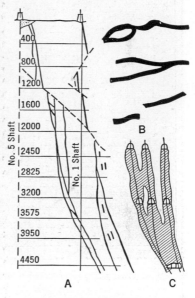

FIG. 12–11. *A*, vertical cross section of Lake Shore mine from surface to 4,450-level, showing vein fractures. Dashed lines are faults; workings omitted. *B*, plan of part of 800-level, showing parallel ore shoots; workings omitted. *C*, cross section of North vein, 2,200- to 2,450-level. (*After Robson, Bull. Can. Min. Met. Inst.*)

fracturing is narrow in the syenite porphyry but widens to 100 feet west of the porphyry. In places the ground between parallel fractures is mineralized to form wide, banded ore bodies or breccia masses (Fig. 12–12). Mining widths range from 5 to 100 feet or more.

The quartz has replaced the country rock and in part is open space filling, especially in the diagonal veins. The ore is localized in shoots that range from a few tens of feet to 2,000 feet in length. Some shoots are over a mile deep. Native gold and tellurides are associated with

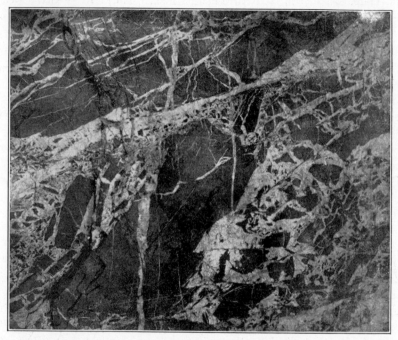

Fig. 12–12. Stope face No. 2 vein, Lake Shore mine, Kirkland Lake, Ontario, showing breccia type of quartz ore. (*Courtesy of W. T. Robson.*)

pyrite in quartz, calcite, and altered rock; sparse base-metal sulphides occur. The tellurides are altaite, calaverite, petzite, and coloradite. A little tourmaline is present.

The ore fractures occur in the brittle and readily fractured syenites. The intense silicification of the ore zone and carbonatization of the wall rocks, and the ore minerals, suggest a genetic relationship to a deep-seated granitic magma whose differentiation supplied the intrusives and the ore solutions. The Kirkland Lake ores are classed by Lindgren as mesothermal, but the differences between them and those of Porcupine are surprisingly small. Both probably belong to the same metallogenetic epoch and province, and similarities are more striking than differences.

Little Long Lac District. This district, which lies about 50 miles north of Lake Superior, has five producing mines. The rocks consist of Keewatin volcanics overlain unconformably by a sedimentary series and cut by diorite, feldspar porphyry, and diabase dikes. The gold veins, which occupy shear zones, consist of lodes made up of parallel quartz stringers and intervening altered and mineralized country rock. The veins range from 3 to 10 feet in width. They are localized near drag folds, and at Little Long Lac the ore is confined to arkose, ending at overlying and underlying graywacke. The introduced minerals are quartz, gold, minor amounts of base-metal sulphides, stibnite, tetrahedrite, bournonite, and berthierite. The ore ranges from $12 to $25 per ton.

Related Deposits. Numerous other gold districts occur within the Precambrian Shield, chiefly in areas of Keewatin volcanics cut by silicic intrusives. They contain quartz veins and replacement deposits generally similar to those described above. The more important districts, each including a number of mines, are: Yellowknife, Southeast Manitoba, Red Lake-Patricia, Matachewan-Larder Lake, and the large Rouyn-Chibougamou district.

WITWATERSRAND, SOUTH AFRICA

The Rand since 1886 has yielded over 16,000 tons of gold worth over 12 billion dollars; for years it produced over 50 percent of the world's gold. Its scores of mines, centering around Johannesburg, hoist annually around 53 million tons of ore, and to date they have hoisted nearly 1.7 billion tons. The annual production is about 12 million ounces, worth about 420 million dollars. The average recovery of gold is about 0.21 ounce per ton, and the average working cost about $5.25 per ton. The greatest depth penetrated is nearly 9,000 feet, vertically, and there are about 6,600 miles of underground workings. The main district is about 90 miles long and in places 25 miles wide. A new Rand-type district is being developed in the Orange Free State.

Geologic Setting. The gold of the Rand occurs in thin conglomerate beds, called reefs, that lie within the Precambrian Witwatersrand system, a 5-mile thickness of conformable sediments that rest unconformably upon ancient granite and schist. The sediments have been folded into a broad syncline and have been traced 180 miles east and west and are 90 miles across. Most of the mines are concentrated within a 50-mile stretch on the northern limb, but later explorations have extended the field eastward and westward where payable ore has also been discovered at depth beneath younger formations. In the central part of the field the beds dip steeply; toward the margins the dip is gentle. Considerable faulting has displaced the beds.

The Lower Witwatersrand (3 miles thick) consists of argillaceous zones alternating with quartzites and including some iron formations and volcanics; the Upper is dominantly quartzites with conglomerate bands. The system has been invaded by dikes and sills of basic to intermediate igneous rocks.

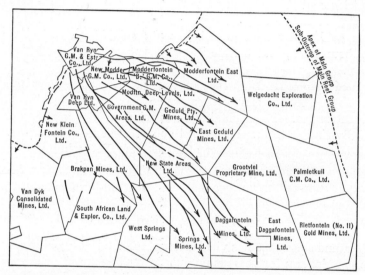

Fig. 12–13. Plan showing positions and directions of " pay streaks " in the Main Reef Leader of the East Rand, South Africa. (*After Reinecke, Trans. Geol. Soc. S. Africa.*)

The Reefs. There are eight principal conglomerate zones within the Witwatersrand system. All of them contain some gold, but the payable deposits are confined mainly to the Main Reef group; the Kimberley and Bird Reefs yield ore in the West and Far East Rand, and the Bird and Livingstone Reefs in the Central Rand.

The Main Reef group, lowermost member of the Upper Witwatersrand, is a zone from 20 to 150 feet thick including several conglomerate bands within quartzites and grits. There are three payable members — the Main Reef, Main Reef Leader, and South Reef — each from 1 to 10 feet thick. Each in places is split by quartzite intercalations. Of the three, the Leader is the most persistent and is the chief gold producer.

In the Main Reef the pebbles are small and uniform, and the distribution of the gold is uniform. Eastward it merges into the Leader and the South Reef to form the "Composite Reef." The Main Reef Leader, which contains larger pebbles, has been mined continuously for 23 miles, but it thins eastward. Its base is well marked and cuts across underlying filled channels; in places it lies almost on top of the Main Reef. The South Reef is separated from the Leader by 70 to 110 feet of quartzites in the Central Rand; eastward the two are in contact. The Bird Reef lies about 1,600 feet above the Leader. The Kimberley Reefs are composed of a dozen or so conglomerate beds characterized by high pyrite content lying about 5,000 feet above the Main Reef. The individual reefs are not continuous throughout the Rand, and the spacing between them varies from place to place.

Ore Shoots. The gold occurs in elongated ore shoots or pay streaks (Fig. 12–13) as much as 5,000 feet long and 1,000 feet wide; in the Central Rand they range from 50 to 450 feet in width. The longer axes, parallel to the longer axes of the pebbles, trend roughly in the same direction, or they are braided. Reinecke found that the gold distribution coincided with well-graded pebbles, large pebbles, aligned pebbles, and wide sections of the reef. These features he interpreted as originating in stream channels and concluded therefrom that the gold is of placer origin.

Nature of Ore. The ore is mineralized conglomerate, and the matrix carries recrystallized quartz grains, sericite, and minor chlorite, chloritoid, rutile, tourmaline, carbon, zircon, and calcite. The gold is mostly in minute grains (0.07 to 0.1 mm) that are ragged in outline. Veinlets of gold pass from the matrix into and across quartz pebbles, which they replace. The gold contains 8 to 9 percent silver and 3 percent base metals, mostly copper; the fineness varies from mine to mine. The gold is in two or more generations.

Pyrite is rather abundant, in places making up 10 to 20 percent of the ore; it averages 3 percent of the Leader. It replaces matrix and pebbles; gold occurs intergrown in veinlets crossing pyrite. Other minor metallic minerals are pyrrhotite, galena, sphalerite, chalcopyrite, and chromite, and minute quantities of cobalt-arsenide and uraninite. Some minerals, such as chromite, corundum, diamonds, and some of the tourmaline, are clearly detrital; others were introduced after the consolidation of the sediments. Numerous secondary minerals occur. Hydrothermal minerals are present throughout the reefs.

Origin. The origin of the gold of the Banket has been controversial between the placerists and the infiltrationists. The early thought was that the gold was of direct placer origin, but later microscopic examinations revealed features inconsistent with a simple placer origin *in situ* and caused a swing toward introduction of gold by hydrothermal solutions.   Reinecke's finding that the ore shoots coincide with thick lenses of large, aligned, and sized pebbles and his conclusion that this indicated a placer origin impelled another swing back toward a placer origin.   This was strengthened locally when Reinecke's predictions of ore-shoot prolongations, based upon his mapping, were proved by subsequent exploration to be correct; but Reinecke's predictions would have been equally correct based upon a hydrothermal origin.   A swing back again to a hydrothermal origin was later given impetus by Graton. It is striking that many of the arguments advanced for one side apply with equal force to the other side, and the criteria are not so convincing as to compel universal acceptance of either view.

A placer origin for the gold involves also the origin of the host conglomerate, and this is far from being settled.   The prevalent hypotheses are deposition from stream currents over flood plains, delta, and marine deposits.   Johnson's suggestion of a piedmont alluvial fan deposit has much to commend it.   This problem is not so important in the case of the hydrothermal hypothesis.

The modified placer theory, favored by most Rand geologists, supposes the gold to have been deposited with the gravels, along with iron ore, other detritals, hydrocarbons, and organic matter.   Later solutions introduced sulphur (to make sulphides) and dissolved and slightly redistributed the gold.   DuToit's statement is: (*a*) solution and redeposition of gold almost *in situ;* (*b*) conversion of iron oxides into ferric sulphate, which dissolved the alluvial gold; (*c*) conversion of ferric to ferrous sulphate and sulphide; (*d*) precipitation of gold by the latter compounds and hydrocarbons; (*e*) conversion of iron ore into pyrite and hydrocarbons into graphite; (*f*) recrystallization of silica, silicates, and gold.   Few geologists could accept this sequence of improbable geological and chemical processes.

The chief positive arguments advanced for the placer origin include the wide occurrence of gold in gravels that contain some undoubted detritals, combined with its occurrence in pay streaks corresponding to stream-gravel distribution and a tendency to be concentrated in the lower parts of reefs, as in placer deposits.   Many negative arguments, such as the lack of a known source for hydrothermal solutions and the difficulty of accounting for the wide distribution of gold by infiltration, are given more credence than negative arguments deserve.   Volumes have been written on this subject.

The chief arguments for a hydrothermal origin are that the gold is ragged, replaces pebbles, is contained within and replaces pyrite that has replaced pebbles, and is accompanied by pyrite associated with a suite of hydrothermal ore minerals and rock alteration products (sericite and chlorite) typical of hydrothermal deposits the world over. Gold is also contained in cross-cutting quartz veins. Fisher shows that the carbon, formerly considered to be a precipitant of the gold, was introduced by replacement along with, and later than, other metallic minerals. He gives the order of deposition as pyrite, carbon, pyrite, gold. Volumes have also been written about these arguments.

To a detached observer the weight of evidence favors a hydrothermal origin. The pebble arrangements and trends shown by Reinecke, would also afford ideal channelways for hydrothermal solutions and would constitute excellent stratigraphic ore loci. The wide distribution of metal is likewise a problem in many deposits of undisputed hydrothermal origin, such as the long mineralized copper beds of Northern Rhodesia.

RUSSIAN GOLD DEPOSITS

Although Russia ranks second or third in world gold production, detailed information regarding the deposits worked at present is extremely meager. Most of the gold apparently is won from placer deposits.

Lode Deposits. The most important lode deposits are in the Trans-Baikal region, including such districts as Beleï, Darasun Titagara, and Minusinsk. Other important lodes occur in the Altai and Ural mountains. These are mostly narrow quartz veins containing base-metal sulphides, arsenopyrite, some tourmaline, and 0.2 to 0.5 ounce of gold per ton. They are mostly associated with granitic rocks intruded at the end of the Paleozoic. The Altai lodes are pyritic quartz veins and polymetal veins associated with Paleozoic quartz keratophyres. They contain base-metal sulphides, many rare metals, and rarely tellurides, arsenopyrite, and pyrrhotite. The gangue consists of quartz, barite, and carbonates.

The Trans-Baikal lodes consist of (1) pyritic quartz veins, (2) quartz veins with base-metal sulphides and sulphantimonides, (3) quartz-tourmaline veins, and (4) quartz veins with tellurides and bismuth. These lodes occur in metamorphic rocks in association either with Paleozoic granodiorites or post-Jurassic granitic rocks.

The Beleï mine is the most productive. Other smaller lodes are widely scattered throughout Siberia in the drainage basins of the Amur, Yenisei, and Lena Rivers, and in the Okhotsh regions.

Placer Deposits. Some 85 or more gold dredges have been reported to be in operation in sixteen districts; about 27 are employed in the Urals, 22 in the Yenisei River region, and 8 in the Yakut region.

The *Lena* district, of first rank, has produced about 500 million dollars in gold from placers derived mainly from small gold quartz stringers in folded, peneplaned Precambrian rocks. The pay gravels are perpetually frozen and lie at depths of 20 to 300 feet under glacial drift, recent gravels, and muck. The *Ural* gravels are mostly of low grade, are small, and lie at depths of 30 to 120 feet. They occur all along the eastern slope of the range. The *Yenisei* district has long yielded the richest gravels of Siberia, derived from small Precambrian gold quartz veins. The *Amur* and *Yakut* districts have yielded much placer gold.

AUSTRALASIA GOLD DEPOSITS

Australasia occupies fifth place in world gold production. West Australia is the main source, with considerably less production from Queensland, Victoria, and New South Wales.

West Australia. West Australia gold is won almost entirely from lodes that average 13 dwt per ton. The region is underlain by Precambrian metamorphic and sedimentary rocks intruded by gneissic granitic rocks. Most of the deposits lie in minor islands of invaded rocks. The lode deposits consist of (a) fissure lodes, (b) auriferous sediments, (c) quartz reefs, and (d) auriferous dike rocks.

The lodes are mostly replacements along shear zones in basic metamorphic rocks; they attain several thousand feet in length and 3,800 feet in depth. The associated minerals are pyrite, base-metal sulphides, and minor tellurides, scheelite, vanadinite, and bismuthite. The auriferous sediments are replacements by silica with considerable pyrite, carrying low gold values. The quartz reefs are fissure fillings of quartz, base-metal sulphides, and minor rare minerals, with the gold localized in short, shallow, but rich ore sheets. The auriferous dike rocks are alaskites and aplites that consist dominantly of quartz with low gold content, and minor iron sulphides, molybdenite, wulfenite, and crocoite in the alaskites, and of silicates, pyrrhotite, and gold in the aplites.

Kalgoorlie became famous by the "Golden Mile," which for a time made this district one of the richest gold camps of the world, with a yield of over 22 million ounces. The deposits, according to Gustafson and Miller, consist of gold-pyrite-telluride replacement lodes, mainly in a quartz dolerite sill of the regionally folded "younger greenstone." A cross fold helped localize the controlling fractures. Some of the rich bodies are pipes at fracture intersections. There are two lode systems. The Western, confined to the Golden Mile proper, consists of few but large, persistent lodes, with regular ore bodies and even distribution of gold. The Eastern System consists dominantly of swarms of smaller veins and pipes with irregular distribution of gold and locally rich pockets. The famous Oroya shoot, which yielded 2 million ounces, is a freak flat pipe of ore, 4,600 feet long, localized by two nearly flat shears connected by a thin, nearly flat shear in a minor fold on the flank of the main Kalgoorlie syncline.

The gold occurs in structurally controlled ore shoots. The tenor has declined from $40 per ton near the surface to $7 to $10 per ton at depth. The shoots bottom in the underlying, unfavorable "calc schists," and are lenticular in shape, range from 2 to 82 feet in width, and still persist at a depth of 3,800 feet. The ore minerals include yellow sulphides, native gold, and tellurides (calaverite, kremerite, petzite, hessite, and coloradoite). Accessory minerals include galena, sphalerite, enargite, pyrargyrite, löllingite, specularite, magnetite, fluorspar, tourmaline, carbonates, and roscoelite. Spectacular seams a few centimeters across are filled by gold and tellurides.

Other Lodes. The Leonora Center district contains the Sons of Gwalia mine, which is a replacement lode in schistose dolerite, containing quartz, gold, and base-metal sulphides. The Murchison field contains the Wiluna and Great Fingall quartz lodes, and these as well as the lodes of the Meekatharra gold field are of generally similar type. Quartz hematite schists occur in the Yilgam district and auriferous sediments in the Yilgam and Peak Hill gold fields. Similar lodes, but containing arsenopyrite, occur at Coolgardie. During 1938 the Big Bell mine at Cue became an important gold producer.

Victoria. Victoria leads in gold production, being credited with some 72 million ounces. It contains the celebrated districts of Bendigo and Ballarat. Bendigo is noted for its unique saddle-reef deposits.

At *Bendigo,* Ordovician slates and sandstones have been acutely compressed into sharp anticlines and synclines, intruded by granitic batholiths, and deeply eroded. The axes of the folds are gently undulating, and the axial planes, or "center country," are steeply inclined or vertical. Fifteen parallel anticlinal lines, or "lines of reef," spaced 700 to 1,200 feet apart have been mapped.

The ore deposits are saddle reefs localized mainly at the crests of the anticlines between slate and sandstone beds (Fig. 12–14). The individual saddles lie mostly within 100 feet of "center country" and seldom exceed 20 feet across. The leg depth may reach 300 feet; most of them are less than 100 feet. Most legs are of unequal length or one may be lacking. The crests have many "spurs" and protuberances and are connected by steeply inclined angling faults (Fig. 5·4–19) which Stone thinks have been the main solution channelways. One saddle has been followed for 9,000 feet along the strike. Saddles underlie each other, and mining has reached a depth of 4,600 feet.

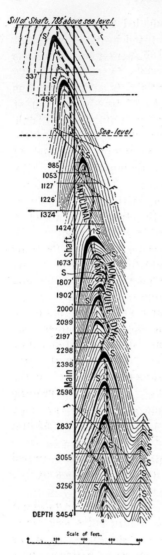

Sill of Shaft, 788 above sea level.

337'
498'
Sea-level
985'
1053'
1127'
1226'
1324'
1424'
1673'
1807'
1902'
2000'
2099'
2197'
2298'
2398'
2598'
2837'
3055'
3256'
DEPTH 3454'

ANTICLINAL

Shaft.

GIANT MONCHIQUITE DYKE

Main

Scale of Feet.

FIG. 12–14. Bendigo, Australia. Saddle reefs shown in Great Extended Hustlers shaft. (*Baragwanath, Gold Res. World.*) For sections see Fig. 5·4–19.

Five anticlinal lines have yielded most of the 300 million dollar production, the principal ones being New Chum, Garden Gully, and Hustlers. " Inverted saddles " and " leg reefs " also occur.

The ore consists chiefly of gold-bearing quartz with small amounts of pyrite, arsenopyrite, and pyrrhotite. Minor galena, sphalerite, stibnite, molybdenite, and bournonite are present, but tellurides are absent. Dolomite and ankerite are generally present.

Evidently the small saddlelike openings created by the folding have been the important loci of deposition, the metallizing solutions being fed through the connecting faults. Some open-space deposition is clearly indicated by crusts of quartz and ankerite; but the chief process of deposition, according to Stilwell, has been by replacement from hydrothermal solutions, which were probably yielded by the magmatic reservoirs that supplied the intrusive granitic rocks.

Similar saddle reefs are found at Castlemaine, also at Ballarat, Clunes, Beringa, Scarsdale, Blackwood, and other Victoria localities.

Ballarat ranks about equally with Bendigo in gold production, more than half of which, however, has come from rich placer deposits. The lodes consist of saddle reefs, and rich quartz bodies or " leather jackets " associated with the Indicator, a thin band of black slate replaced by quartz and pyrite, lying on the east limb of an anticline. Where intersected by west-dipping strike faults, rich concentrations of gold occur; masses of gold weighing 444 ounces have been extracted from the Indicator.

The Placer deposits of Victoria rank with those of California and have yielded over 700 million dollars. The earlier worked placers were surficial gravels in streams, benches, and hill cappings. Later, the rich "deep lead' gravels became very productive. These consist of: (1) *Pre-early basalt gravels* found at elevations over 2,000 feet above present stream bottoms and overlain by sediments, and early basalt. (2) *High-level gravels,* also overlain by early basalt and capping or flanking hills above present stream bottoms. They have been regarded as marine but are probably lacustrine or flood-plain deposits. (3) *Pre-later basalt gravels.* These consist of three groups: (*a*) stream gravels underlying the younger basalt and including the highly productive deep leads of Ballarat and Loddon Valley. These lie mostly below the present stream levels and contain coarse gold and many nuggets; (*b*) stream gravels underlying sand and clay beds to depths of 350 feet below the present surface. Such deep leads of Ballarat, Tarnagalla, and Dunolly were famous for the number and size of nuggets found in them; (*c*) buried coastal gravels underlying basalt or sediments. (4) *Post-later basalt gravels,* formed by a late Tertiary uplift that resulted in the rapid erosion of rich reefs and former placers and their concentration in present-day streams.

Queensland. This state has produced around 500 million dollars in gold, mainly from lodes of gold quartz or auriferous sulphides, and includes such well-known camps as Mount Morgan, Mount Isa, Charters Towers, Tympie, and Croydon. At *Charters Towers* productive fissure veins in Permian granodiorite have been mined to depths of 3,000 feet and yielded 112 million dollars. The gold, averaging about $15 per ton, is associated with quartz and sulphides in crustified fillings. Rich ore shoots lie at intersections with basic dikes. *Mount Morgan,* a great copper mine, yielded 110 million dollars in gold from ore averaging 4.7 dwt per ton. The ore body is a huge beet-shaped mass, 1,150 feet long, 750 feet wide, and 950 feet deep, and is a pyritic stockwork in tuffs carrying copper and gold.

New South Wales has produced over 300 million dollars from many productive gold districts. It has gold quartz fissure veins associated with granitic intrusives and carrying base-metal sulphides, as in the *Wyalong* and *Adelong fields;* rich lenticular lenses, as at *Hillend,* where $3,300,000 was recovered from 10 tons of quartz; fissure veins in sediments carrying stibnite, scheelite, and arsenopyrite; saddle reefs at *Mount Boppy;* large replacement lodes along shear zones carrying gold in massive sulphides of iron, copper, lead, and zinc at *Cobar;* and impregnations of auriferous arsenopyrite, pyrite, and pyrite in sediments.

New Zealand has produced about 470 million dollars in gold, but its productivity is largely past. The *Hauraki* district contains gold quartz veins in andesite, carrying arsenic, antimony, mercury, copper, lead, zinc, and iron. Rich bonanzas at *Thames* contained 1 to 6 ounces of electrum per pound of ore. This remarkable concentration has been attributed to supergene enrichment, but the evidence does not appear convincing. The *Otago* field has yielded about one-third of the production of New Zealand, most of it from Tertiary placers. The lodes are stockworks, rich at the surface, and carrying gold, pyrite, arsenopyrite, and scheelite.

New Guinea has become an important gold producer from placers of the *Bulolo* River. Seven dredges have been operated, and the recovery has been about 140,000 ounces annually.

Japan and Chosen

Japan and Korea together, prior to 1942, produced some 55 million dollars a year from 13 mines in Chosen and 18 in Japan. The gold deposits of Japan are dominantly fissure fillings carrying free gold, or gold with argentite associated with intrusions of rhyolitic and andesitic stocks and early Tertiary lavas. The ores are mostly crustified and contain base-metal sulphides, argentite, and some pyrargyrite and stibnite.

In Korea auriferous quartz fissure veins occur in sediments or igneous rocks ranging in age from Precambrian to Mesozoic. The gold is included in sulphides, with cinnabar and rarely molybdenite and wolframite. There are also several typical contact-metasomatic gold deposits.

The Kolar Gold Field, Mysore, India

The production of this great field is derived from a single reef — the Champion Lode — whose thickness averages between 3 and 4 feet. This deposit has been mined in 5 mines for 4½ miles in length and to a depth of over 8,500 feet. The ore has averaged 13 dwt per ton and it has yielded over 450 million dollars in gold. The lode lies in a schist belt 50 miles long composed of Precambrian Dharwar conglomerates, hornblende schists, ferruginous quartzites, and older porphyritic granite, intruded by younger granite. The walls are well defined, and the lode dips about 55° conformable with the enclosing schists but is vertical at 8,000 feet in depth. It lies on the limb of a major syncline. The ore is concentrated in shoots that pitch about 45° within the vein. The rich shoot of the Mysore mine had a stoping length of 800 feet, a maximum width of 35 feet, an average width of 4 feet, and a pitch length of 4,000 feet.

The ore consists of quartz with minor base-metal sulphides; some tourmaline, pyroxene, albite, actinolite, and biotite also occur in the altered walls. The fissuring is presumed to be due to faulting, and the localization of the ore shoots to be controlled by drag folds. The magmatic source that gave rise to the later granite is also considered to have supplied the mineralizing solutions.

Selected References on Gold

Gold Deposits of the World. W. H. EMMONS. McGraw-Hill, New York, 1937. *Gives location of all gold deposits.*

General Reference 8. *Descriptions of many gold deposits.*

General Reference 9. *Thumbnail sketches of many western gold districts.*

Economic Geology. H. RIES. John Wiley & Sons, New York, 1937. *Good bibliography.*

Gold Resources of the World. XV Int. Geol. Cong., South Africa, 1929. *A summary of gold resources and deposits.*

Summarized Data of Gold Production. R. H. RIDGWAY. U. S. Bur. Mines Econ. Paper 6, 1929. *Statistics of past production.*

Location of Payable Ore Shoots on the Witwatersrand. L. REINECKE. Geol. Soc. South Africa Trans. 30:89–119, 1927. *Coincidence of gold with sedimentary features of the conglomerates, and arguments for a placer origin.*

Hydrothermal Origin of the Rand Gold Deposits. L. C. GRATON. Econ. Geol. 25: Supp. to No. 3, 1930. *A lengthy discourse upholding a hydrothermal origin.*

Symposium on Gold Deposits of the Rand. Geol. Soc. South Africa 34:1–93, 1931. *Group of papers on occurrence and origin; chiefly arguments against Graton's advocacy of a hydrothermal origin.*

The Gold Mines of Southern Africa. D. LETCHER. Johannesburg, South Africa, 1936. *Tabular data.*

Gold Occurrences of Canada. H. C. COOKE and W. A. JOHNSTON. Can. Geol. Surv. Econ. Geol. Ser. No. 10, p. 61, 1932. *Brief summary of occurrences.*

Mineral Deposits of the Canadian Shield. E. L. BRUCE. Macmillan Co., Toronto, 1933. *Contains descriptions of important gold deposits.*

Genetic Relations of Gold Deposits and Igneous Rocks in the Canadian Shield. E. S. MOORE. Econ. Geol. 35:127–139, 1940.

General Reference 15. *132 papers; geologic details of all Canadian gold mines.*

Geology of the Homestake Mine, Black Hills, South Dakota. D. H. McLAUGHLIN. Eng. and Min. Jour. 132:324–329, 1931. *Occurrence, structure, and Precambrian origin.*

The Mother Lode System of California. A. KNOPF. U. S. Geol. Surv. Prof. Paper 157, 1929.

Mines of the Southern Mother Lode Region. C. E. JULIHN and F. W. HORTON. Mineral Industries Survey of the United States. U. S. Bur. Mines Bull. 424, 1940.

Geology of Gold. E. J. DUNN. London, 1929.

U. S. Geol. Surv. Prof. Papers and Bulletins on various mining districts of the Western United States.

Mineral Resources of the Philippines, 1934–1938. W. F. BOERICKE and N. N. LIM. Manila, 1939. Part I, Gold. *A summary.*

A Contribution to the Published Information on the Geology and Ore Deposits of Goldfield, Nev. FRED SEARLES, JR. Univ. of Nev. Bull. 42, No. 5, Oct. 1948.

Mechanism and Environment of Gold Deposition in Veins. W. H. WHITE. Econ. Geol. 38:512–532, 1943.

Host Minerals of Native Gold. G. M. SCHWARTZ. Econ. Geol. 39:371–411, 1944. *Ore and gangue mineral hosts of gold.*

Structural Principles Controlling Occurrence of Ore in Kolar Gold Field, India. J. W. BICHAN. Econ. Geol. 42:93–136, 1947. *Localization of gold ore by drag folds.*

Structure of a Part of the Northern Black Hills and the Homestake Mine, Lead, South Dakota. J. A. NOBLE, J. A. HARDER, and A. L. SLAUGHTER. Bull. Geol. Soc. Amer. 60:321–352, 1949.

Origin of the Bendigo Saddle Reefs and Formation of Ribbon Quartz. F. M. CHACE. Econ. Geol. 44:561–597, 1949. *Excellent discussion of origin; fine illustrations.*

General References 3, 5.

Silver

Silver, along with gold, has been prized and sought since the time of the ancients. It has found use continuously as a monetary metal, although in diminishing quantities. Its call for adornment, once widespread, is now more restricted, but it does find increasing use in the arts and industries. It has fluctuated in price from a high of $1.29 to a low of 24 cents in 1932.

Production. The production of silver has paralleled gold. It was mined early in Mediterranean Europe, but great production followed the discovery of America when streams of silver metal, robbed from the Incas and Aztecs, flowed to Europe from Peru, Bolivia, and Mexico. This increase from 1520 to 1620 reached a peak in 1800 but fell off again until the rich discoveries of the western United States in the 1860's. Since then there was a steady increase to 1912, and a subsequent high of 275 million ounces. The chief producing countries and their normal ratio to world production are:

	PERCENT OF WORLD		PERCENT OF WORLD
Mexico	33	Australia	7
United States	16	Bolivia	5
Canada	10	Belgian Congo	5
Peru	9	Honduras	2
		Others	13

The western part of North America yields over half of the world's silver.

The chief producers of the United States (rank of *total* silver production in parentheses) are:

1. Idaho	(5)	4. Arizona	(6)	7. Nevada	(4)
2. Utah	(3)	5. Colorado	(2)	8. New Mexico	(8)
3. Montana	(1)	6. California	(7)		

Of the total silver produced in the United States, the relative percentage proportions supplied by the different classes of ore are:

Dry and siliceous ores	24	Lead ore	7
Copper ore	24	Zinc ore	2
Zinc-lead-copper ore	41	Placers	0.3

MINERALOGY, TENOR, TREATMENT, USES

Much of the silver won is a by-product of gold, lead, copper, and zinc. It is the silver that makes many such deposits profitable.

Ore Minerals. The chief ore minerals of silver are:

MINERAL	COMPOSITION	PERCENT SILVER
Native silver	Ag	100
Argentite	Ag_2S	87.1
Cerargyrite	AgCl	75.3
Polybasite	$Ag_{16}Sb_2S_{11}$	75.6
Proustite	Ag_3AsS_3	65.4
Pyrargyrite	Ag_3SbS_3	59.9

Other economically important ore minerals of silver are the tellurides, argentiferous tetrahedrite, stromeyerite, stephanite, and pearcite. There are 55 well-known silver minerals. Silver also occurs in solid solution with gold and base-metal sulphides. The common gangue minerals of silver deposits are quartz, calcite, dolomite, and rhodochrosite, and oxidation products.

Silver is most commonly associated with lead, gold is rarely free from it, and most copper and zinc ores carry some silver; cobalt is also an associate.

Tenor and Treatment. The tenor of silver ores varies with the price of silver and upon other economic considerations. In general, straight silver ores need to contain about 15 ounces to the ton. Mexican ores are of lower tenor. In the western United States the general silver content of the different classes of ores is as follows:

ORE	SILVER (ounces)	ORE	SILVER (ounces)
Gold	0.3	Lead-copper	37
Gold-silver	3.3	Zinc	0.2
Silver	17.5	Zinc-lead	1.5
Copper	0.1	All ores	0.3
Lead	6.6		

The ores of the large lead-silver deposits of the world average 1 ounce silver to the ton and 1.7 percent lead.

The recovery of silver from its ores depends entirely upon its association. Straight silver ores, if free-milling, may be cyanided, or concentrated and cyanided and the silver precipitated from the cyanide solution by zinc or aluminum. If not free-milling, the ores are concentrated and smelted to bullion. Silver-lead ores are concentrated and smelted, and silver is carried down with the lead, and the alloy is desilverized. Where silver is present in copper ores, the concentrates are smelted, the silver (and gold) is carried down with the molten copper, and is then separated out by electrolytic refining. About 98 percent of all American silver is recovered by smelting and 1 percent by cyanidation.

Uses of Silver. Of the total world supplies of silver, about 70 percent is used for monetary purposes. This includes new production and demonetized silver. Much silver is hoarded; the remainder is used in the arts and industries, of which the largest amount is consumed for sterling and plated silverware. The photographic industry with its annual manufacture of millions of feet of film is using increasing amounts. Smaller amounts are used in the chemical, soldering, bearing, and electrical industry.

KINDS OF DEPOSITS, ORIGIN AND DISTRIBUTION

The silver that is a by-product of gold, copper, or other deposits is not considered separately. The principal classes of straight silver deposits and silver-lead deposits and some examples of them are:

Contact-metasomatic deposits: Zimapan, and *Velardeña*, Mexico; Magdalena, N. Mex.

Cavity fillings:
  a. Fissure veins — Cobalt, Ontario; Mayo, Yukon; *Pachuca*, Guanajuato, *San Francisco*, and *Fresnillo*, Mexico; *Rosario*, Honduras; San Juan, Colo.; *Sunshine*, Idaho.
  b. Stockworks — Quartz Hill, Colo.; Schneeberg, Germany; *Fresnillo* and Guanajuato, Mexico.
  c. Breccia fillings — Emma, Utah.
  d. Pore-space fillings — Silver Reef, Utah.

Replacement deposits:
  a. Massive:
    1. Silver deposits — Shafter, Tex.
    2. Silver-lead deposits — *Bingham, Tintic, Park City*, Utah; *Sullivan* mine, British Columbia; *Leadville*, Colo.; *Santa Eulalia*, Mexico.
  b. Lode — *Comstock Lode*, Tonopah, Nev.; *Potosi*, Bolivia.
  c. Disseminated — Coeur d'Alene, Idaho.

Supergene Sulphide Enrichments: *Parral*, Mexico; Chañarcillo, Chile.

Origin. Most silver deposits have been formed as replacements or cavity fillings by hydrothermal solutions. Massive and lode replacements of silver-lead ores are numerous, but most of the world's silver is won from fissure veins of the mesothermal and epithermal type.

World Distribution. Most of the silver of the world comes from the *North American* Cordillera, where it is associated with Tertiary intrusive volcanic rocks. A belt prolific in silver extends from Utah-Nevada, through Mexico, and down to the Rosario mine in Honduras and includes such famous districts as the Comstock, Tonopah, Tintic, Pachuca, Guanajuato, Real del Monte, San Luis Potosi, Zacatecas, Fresnillo, Mapimi, Parral, Sierra Mojada, and Santa Eulalia. A minor silver belt embraces the Coeur d'Alene and projects into southern British Columbia. Another silver-lead belt lies in Colorado. In eastern North America the unique Cobalt district was formerly a large producer of native silver.

In *South America* the Andean region of Peru, Bolivia, Chile, and Argentina is a silver belt which, together with Mexico, long supplied most of the world's silver. These epithermal deposits are associated mainly with Tertiary andesites. The famed mountain of Potosi contained the greatest concentration of silver in the world.

EXAMPLES OF UNITED STATES SILVER DEPOSITS

The United States is the second largest producer of silver, but it contains few large silver mines and few straight silver deposits. Most of its silver is won from silver-lead deposits and from gold and base-metal ores. The eight ranking silver producers of the United States, according to the Minerals Yearbook, are:

MINE OR COMPANY	LOCATION	NATURE OF ORE
1. Anaconda Copper Co.	Butte, Mont.	Copper ore
2. Utah Copper Co.	Bingham, Utah	Copper ore
3. Sunshine	Coeur d'Alene, Idaho	Silver ore (siliceous, dry)
4. Darwin	Darwin, Calif.	Lead ore
5. Polaris mine	Idaho	Silver ore
6. U.S. Smelting, Refining and Mining Co.	Utah	Lead, zinc-lead ore
7. Bunker Hill and Sullivan Co.	Idaho	Lead, zinc-lead ore
8. Phelps Dodge Corp.	Arizona	Copper ore

Sunshine Mine, Coeur d'Alene District, Idaho. This, the premier silver mine of the United States and the second largest silver producer in the world, is the only large, straight silver deposit north of Mexico.

The deposit consists of quartz veins in a broad zone in quartzite which has been mined for a length of about 1,500 feet and a depth exceeding 2,700 feet. In the upper part of the mine the ore occurs

only in bunches and splits that approach each other on the 500 level
and coalesce on the 1,500 level, below which the vein becomes wider
and richer and has yielded most of the silver production.

The ore consists of massive siderite and quartz with abundant
argentiferous tetrahedrite and minor galena. The annual production
has ranged between 3 and 12 million ounces silver, from ore that has
averaged 28 to 48 ounces silver to the ton. The adjacent Polaris mine
is similar to the Sunshine.

Silver is also won from lead and zinc ores of the adjacent disseminated replacements in the Bunker Hill and Sullivan, Federal, and
Hecla mines.

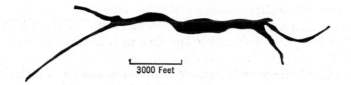

3000 Feet

FIG. 12–15. Plan of the Comstock Lode, Nevada. (*After Becker,*
U. S. Geol. Survey.)

Comstock Lode, Nevada. The famous Comstock Lode was the scene
of early western mining and since 1859 has produced over 400 million
dollars in silver and gold (1:40 ratio), from a group of mines scattered
along its 3 miles of length. The bonanza period was in the 1870's,
since which time production has declined steadily, although the change
in the price of gold has brought some rejuvenation.

The Comstock Lode (Fig. 12–15) is a fault of 3,000 feet throw,
separating hanging wall Tertiary and older volcanic rocks from foot
wall Mesozoic rocks. It branches upward and toward the bottom.
The dip is about 40°; the width attains several hundred feet, and it
has been followed 3,000 feet vertically, although the bonanza ore
ceased at 2,000 feet. The wall rocks have been intensely propylitized
and sericitized.

Most of the ore occurred in small bonanza ore shoots and in hanging
wall branches or chambers and consisted of crushed quartz, some
calcite and minor sulphides, with gold, electrum, argentite, polybasite,
and other silver sulpho-salts, which have replaced early quartz, pyrite,
sphalerite, and galena. A fine-grained later quartz is present. Bastin
recognized supergene silver, argentite, and polybasite above the 500
level.

At 3,000 feet the lode was flooded with hot calcium sulphate waters of 170° F temperature. The lower-level rock temperatures reached 114° F, suggesting an underlying mass of uncooled igneous rock. The early output of this mine made Nevada a great silver state.

Tonopah, Nev. The Tonopah district since 1900 has produced over 150 million dollars from ores with a gold-silver ratio of 1:100. Nolan, in his noteworthy survey, recognizes a Tertiary sequence as follows: Older rhyolitic flows, breccias, tuffs, and water-laid sediments of volcanic materials, called the Tonopah formation, followed by the " Mizpah trachyte " andesitic flows, both of which interfinger. Next came an intrusion of the Extension breccia, in the western part of the district, between the Tonopah and Mizpah formations. The same contact was followed by a sill of West End rhyolite. These formations were broken first along the Halifax fault zone and then by the remarkable mineralized compound Tonopah fault, convex eastward in plan and convex upward in section, and with a total throw of 500 to 1,500 feet.

The mineralization was probably contemporaneous with the faulting. The first stage was wholesale albitization of the andesite, followed by quartz-adularia-sericite. The metallic minerals are electrum, pyrite, argentite, pyrargyrite, polybasite, and minor chalcopyrite, galena, blende, selenium minerals, scheelite and huebnerite with " chalcedonic " quartz, barite, and rhodochrosite. Supergene cerargyrite, iodyrite, and embolite occurred in the upper levels. The ore bodies are replacement veins along faults.

The chief control of mineralization was the Tonopah fault and its branches; locally other faults served as ore channels. The ore deposition was limited to a relatively thin shell that domes upward in the central part of the district and the upper ore boundaries transgress both faults and formation boundaries. Nolan believes that the coincidence of the high point of the shell with the zone of intense rock alteration may indicate isogeotherms and that the shell represents the temperature range within which deposition could take place; this is further suggested by the gold-silver ratio.

The principal mines are the Tonopah, Tonopah Belmont, Tonopah Extension, and West End, whose ores averaged more than $100 per ton prior to 1904; around $15 between 1911 and 1930; and around $10 per ton or less thereafter.

Tintic Standard Mine, Utah. This is the second largest straight silver mine in the United States. Workings below the 1,000 level broke into the great " Tintic Standard Pothole," which is one of the greatest ore centers in the western United States, having already yielded ores worth over 80 million dollars. The total value of the entire Tintic district output is over 350 million dollars, of which one-half has been silver and the remainder copper and gold.

The rocks of the district consist of about 13,000 feet of Paleozoic sediments, unconformably overlain by volcanics and intruded by mon-

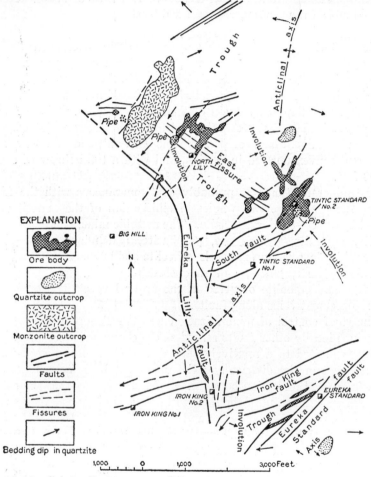

Fig. 12–16. Subrhyolite structure and fissure systems. East Tintic, Utah. (*Billingsley, 16th Int. Geol. Cong.*)

zonites and quartz porphyry of Tertiary age. The sediments are folded into a broad, asymmetric, pitching syncline. The folding was accompanied by numerous transverse faults, some of which have horizontal displacements of 200 feet, and by many bedding faults. The faults have been important localizers of ore (Fig. 12–16).

The ore bodies are massive replacements of the crumpled and faulted Ophir limestone (Fig. 12–17). Post-crumpling, northeast fissures have yielded successively monzonite, porphyry, pebble dikes, and ore solutions. The ore solutions ascended through trough conduits at critical intersections and formed massive replacement of the limestone (Fig. 12–18) that Billingsley states extends up through "vein roots, pipe, expanding funnel, and finger-tip extensions." Ore has been mined down to the conduit on the 1,400 level.

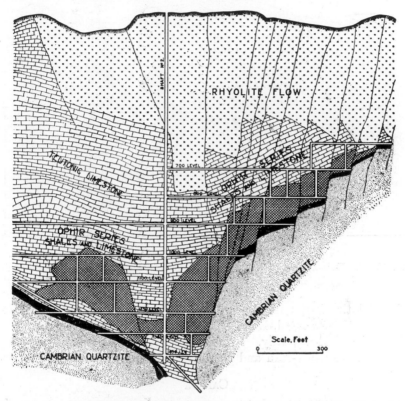

Fig. 12–17.   Northeast section through Tintic Standard mine, Tintic, Utah, show-ing relation of ore (cross hatched) to limestone and to faults.   (*After Wade, U. S. Bur. Mines.*)

The mineralization occurred in stages the first of which is repre-sented by quartz and barite with minor pyrite, enargite, and tetra-hedrite, and in places, marcasite, occurring chiefly in the fissure roots. A later stage deposited quartz, galena, and pyrite.   The ores range from siliceous silver ores to lead-silver ores, to pyritic silver ore.   The metallization does not extend upward into the overlying rhyolite, but a surrounding halo of rock alteration does penetrate to the surface. Above the ore bodies are surface depressions caused by underlying shrinkage attendant upon metallization and oxidation.

Oxidation is complete to the 900 level and partial to the 1,250 level. The upper oxidized ores are locally excessively enriched in silver, but normally they are about twice as rich in silver as the underlying sulphide ore.

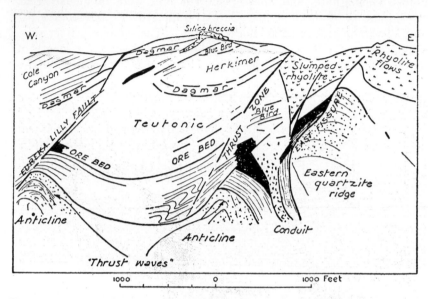

Fig. 12–18. East-west section through Tintic Standard trough showing ore conduit and relation to structure. (*Billingsley, 16th Internat. Geol. Cong.*)

The Tintic district includes several other important mines, such as the Chief Consolidated, Eureka, Centennial, Mammoth, Gemini, Iron Blossom, and North Lily. Most of these are replacements in limestone and are silver-lead and argentiferous copper, zinc, or gold ores.

Canada

Canada, the third largest silver producing country, obtains its metal almost entirely as by-product silver. The most important producers of silver in order of rank are:

Mine or Company	Location	Nature of Ore
1. Consolidated Mining and Smelting Co.	British Columbia	Lead-zinc
2. Treadwell-Yukon	Yukon Territory	Silver-lead
3. International Nickel Co.	Sudbury, Ontario	Nickel-copper
4. Hudson Bay Mining and Smelting Co.	Flin Flon, Manitoba	Zinc-copper
5. Noranda mine	Noranda, Quebec	Copper-gold
6. Premier mine	Premier, Brit. Columbia	Gold-silver

Cobalt, Ontario. This unique silver camp of northern Ontario, now exhausted, has produced around 370 million ounces of silver since 1904, a quantity exceeded by only three other silver camps, namely, Potosi, Guanajuato, and Zacatecas. The ores were fabulously rich, the shipments of the first few years averaging over 1,000 ounces per ton; low-grade ores ran 200 ounces per ton. The famous " silver sidewalk " of the La Rose was almost solid native silver for a length of 100 feet and to a depth of 60 feet yielded 658,000 ounces of silver.

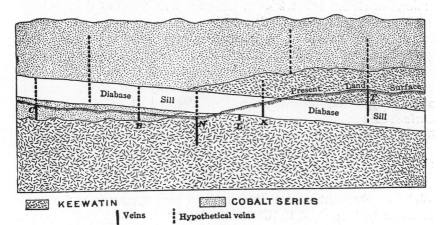

FIG. 12–19. Generalized section through productive part of Cobalt area, Ontario, showing relation of veins to diabase sill and overlying Cobalt series and underlying Keewatin greenstones. (*Knight and Miller, Ont. Dept. Mines.*)

Great slabs of almost pure silver were extracted; one famous specimen 5 feet long weighs 1,640 pounds and contains 9,715 ounces silver. The deposits illustrate low-temperature hydrothermal fissure filling.

The rock formations consist of Keewatin greenstones overlain by Cobalt conglomerate (Huronian) and intruded by a tilted Keeweenawan diabase sill, 1,000 feet thick (Fig. 12–19). The productive veins occur in the conglomerate beneath the sill.

The deposits were steep joint or fault fissure veins 4 to 5 inches wide, a few hundred feet long, and mostly less than 300 feet deep. About 100 veins are known.

The ore consists of an unusual assemblage of minerals. The gangue is calcite or dolomite and may be absent. The chief ore mineral is native silver. Dyscrasite, argentite, tetrahedrite, stromeyerite, polybasite, stephanite, and ruby silver also occur. Bismuth, galena, pyrite, chalcopyrite, pyrrhotite, and sphalerite are known, and arsenopyrite is abundant. The striking feature of the mineralogy is the unique group of nickel and cobalt arsenides of which the following have been checked by Thomson:

1. Smaltite	CoAs$_2$	5. Rammelsbergite	NiAs$_2$	9. Cobaltite	CoAsS
2. Chloanthite	NiAs$_2$	6. Skutterudite	CoAs$_3$	10. Glaucodot(CoFe)AsS	
3. Löllingite	FeAs$_2$	7. Arsenopyrite	FeAsS	11. Niccolite	NiAs
4. Safflorite	CoAs$_2$	8. Gersdorffite	NiAsS	12. Breithauptite	NiSb

Of these, numbers 9, 4, 7, and 11 are the most abundant. Later R. J. Holmes discarded 1 and 2 as mineral species. Some of the minerals are isomorphous mixtures. Dendritic forms are common, the centers of which are commonly occupied by native silver or calcite. The general mineral succession, according to Thomson, was first the arsenides and sulpharsenides of nickel and cobalt, followed by the iron compounds, and then by the monoarsenides; later some silver, argentite, and bismuth; and lastly the rare sulphides and silver sulpharsenides. Surface oxidation products of " cobalt bloom " (erythrite) and "nickel bloom " (annabergite) lent pink and green colors to the outcrops.

The ores are cavity fillings with minor replacement. They have been formed by hydrothermal solutions, and the presence of bismuth shows that the last minerals were deposited below a temperature of 271° C. The solutions are considered to have sprung from the same reservoir that gave rise to the diabase intrusion. The Cobalt deposits have counterparts at Annaberg and Schneeberg in Saxony.

Great Bear Lake. Rich silver ores associated with uranium-radium ores occur in veins in Precambrian sediments in the Great Bear Lake district. The ores contain native silver, pitchblende, cobalt and nickel arsenides, copper sulphides, and bismuth, along with quartz and rhodochrosite.

MEXICO

Mexico has long been the leading silver-producing country of the world, and camps may still be seen where silver mining has been almost continuous since the early 1500's. Its annual production ranges from 40 to 80 million ounces. About 10 companies exceed 1 million ounces annually. The most important producers are:

MINE OR DISTRICT	LOCATION	NATURE OF ORE
1. Fresnillo Co.	Fresnillo, Zacatecas	Oxidized and sulphide silver ore
2. Howe Sound Co.	Santa Eulalia, Chihuahua	Silver-lead-zinc
3. San Francisco	San Francisco del Oro, Chihuahua	Silver-lead-zinc
4. Inversiones Co.	El Oro, Zacatecas	Silver ore
5. Pachuca	Pachuca, Hidalgo	Silver ore

Fresnillo, Zacatecas. This famous mine, discovered in 1570 and worked almost continuously since then, is stated to have produced around 175 million ounces silver, chiefly from oxidized ores. Some lead, zinc, and copper are also won from deeper sulphide ores. Gold is present in both. The oxide ores average about 6 oz silver and 0.3 dwt gold. The oxide ore is cyanided; the sulphide ore is concentrated.

The deposits occur in slate, graywacke, and altered volcanics in two forms — fis-
sure veins and stockworks. Main veins from below spray out into minor veins and veinlets above, forming a stockwork several hundred feet across (Fig. 12–20) that has yielded some 15 million tons of ore from an open cut. Seven main veins oc-
cur below and numerous small ones extend outward from the stockwork. The ores within a shallow depth display a vertical zoning with 2 to 6 percent copper ores

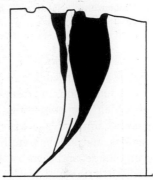

FIG. 12–20. Cross section of South Branch and Catillas stockwork ore bodies, Fresnillo, Zacatecas, Mexico. (*Livingston, mine records.*)

beneath, grading upward into lead-zinc ores poor in copper, and into silver ores, free from base metals, above. The primary ore minerals are base-metal sulphides, quartz, and calcite.

El Potosi Mine, Santa Eulalia, Chihuahua. The El Potosi mine of the Howe Sound Co., famed since 1703 as a rich silver-lead producer, ranks second in present Mexican silver production. Its 250 miles of workings disclose mantos, chimneys, irregular replacements, and fissure veins, all in gently dipping Cretaceous limestone. The Santa Eulalia district is credited with a metal production of over 300 million dollars from 500 miles of underground workings.

The limestones have been folded into a gentle anticline along which the ore bodies are localized. Rhyolite flows and tuffs of pre-mineral age cap the limestones, which are cut by numerous dikes and sills and by fractures that have helped localize the ores.

The deposits are massive replacements of limestone along fracture zones in several favorable beds (Fig. 12–21). The ore bodies may be mantos following along single beds, or chimneys that cut across sev-
eral beds and connect with mantos. The almost horizontal mantos attain lengths of 1,000 or more meters and cross sections from 100 to 5,000 square meters. The great "P" chimney is 310 meters high and 1,100 square meters in cross section. It throws off mantos into upper and lower beds. Other bodies are fissure replacements that extend as

flanges into favorable beds. Limestone is metamorphosed by selective replacement to dolomite.

The ore is of three classes: (1) high-grade sulphide mill ore; (2) low-grade sulphide silicate ore; and (3) oxidized ores, now largely exhausted. The massive sulphide ore consists of carbonate and base-metal sulphide with minor silver minerals, carrying 9 to 10 ounces

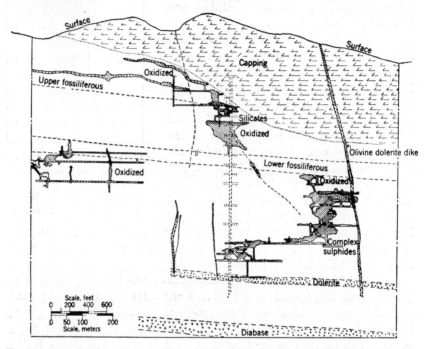

Fig. 12–21. Cross section of part of ore bodies of Potosi mine, Santa Eulalia, Mexico, showing mantos and pipes in limestone (white). Ore is shaded. (*Walker, U. S. Bur. Mines.*)

silver and 9 to 10 percent each of lead and zinc. The oxidized counterpart of this carries 15 ounces silver, 13 percent lead, and 2.5 percent zinc. The unusual silicate ore consists of quartz, silicified limestone, ilvaite, actinolite, hedenbergite, fayalite, magnetite, and hematite, with minor base-metal sulphides and a silver content of 20 ounces in the unoxidized and 33 ounces in the oxidized portions. Zinc and lead are low. Oxidation is complete to the 10th level, partial to the 13th, and absent below that. The reserves are running low.

These interesting ores indicate both intermediate and fairly high-temperature conditions of formation combined in one mine. The silicate ores overlap, grade into, and lie above the sulphide ores, and there are no indications of separate periods of formation. They need further investigation.

San Francisco del Oro, Chihuahua. These deposits lie northwest of the Parral district and have been highly productive in precious and base metals from strong fissure veins. There are nine veins, of which the San Francisco is the most important. The ore averages about 6 to 7 ounces silver, 7 percent lead, 9 percent zinc, and a little gold.

Pachuca, Hidalgo. This famous district, including the Real del Monte, since its discovery in 1534 has been an almost continuous producer of silver, being credited with around 400 million dollars. Filled fissure veins, crustified and brecciated, occupy faults in Lower Cretaceous sediments and Tertiary extrusives and intrusives. Wisser states that after intrusion, warping, gentle folding, and intrusion, a great series of northwest fault-fracturing followed. These northwest fault-fractures were cut by a late north-south system closely associated with mineralization. Hulin states the first deposition was barren quartz, then later, rhodonite, bustamite, base-metal sulphides, and lastly silver minerals, chiefly argentite and minor polybasite and stephanite; gold occurs in traces. The andesite country rock is widely propylitized and the vein walls are chloritized and silicified. The veins are 2 to 5 meters wide and mostly less than 600 meters in depth. The ore is in structurally controlled ore shoots affected by fault movements contemporaneous with metallization. Silver deposition took place during reopening. Zoning is indicated; the silver zone is overlain by a barren zone up to 400 meters thick, which in places is entirely removed by erosion. Oxidation extends to 700 meters, and supergene sulphide enrichment is absent. The veins are hydrothermal fissure fillings of epithermal type.

Guanajuato. This great district since 1548 has yielded a half billion dollars in silver. According to Wandke and Martinez, pre-Cretaceous shales, sandstones, and conglomerate are cut by rhyolite, andesite, granite porphyry, and monzonite and are overlain by later volcanics. Faulting then occurred on a large scale. One fault, the Veta Madre, extends 25 kilometers. The faults became metallized and the resulting veins have been dislocated by post-ore faulting.

The ore bodies are of three types, (1) the great Veta Madre vein, (2) smaller fault-fissure veins, (3) stockworks.

The famed Veta Madre (mother lode) vein is a rotational fault with an average displacement of 800 meters, a maximum width of 20 meters, and an average width of 6 to 8 meters. Some 5 kilometers of its 25-kilometer length contains most of the ore bodies. The ore is concentrated in shoots 200 to 400 meters long, where enlargements have re-

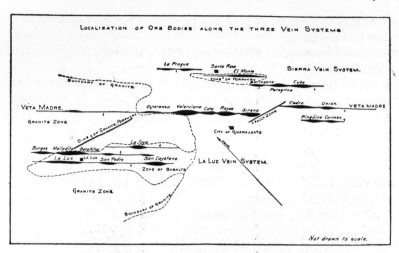

FIG. 12–22.   Guanajuato, Mexico, showing location of ore bodies along three vein systems.   Note pinches and swells.   (*Wandke, Econ. Geol.*)

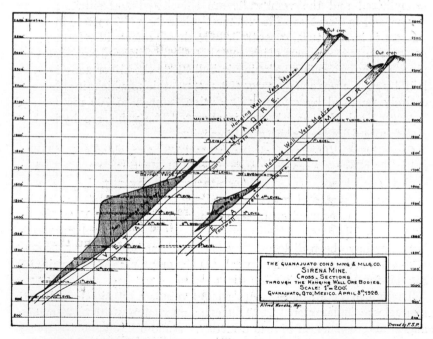

FIG. 12–23.   Hanging wall ore bodies (stockworks) of the Veta Madre, Guanajuato, Mexico.   (*Wandke, Econ. Geol.*)

sulted from fault motion (Fig. 12–22). Other large fault fissures have
been traced for 1½ to 7 kilometers. The vein matter is in part open
space deposition and in part replacement of brecciated rock that
filled the fissure.

The stockworks lie in the hanging walls of fault fissures where their
downward movement over a hump in the dip induced shattering.

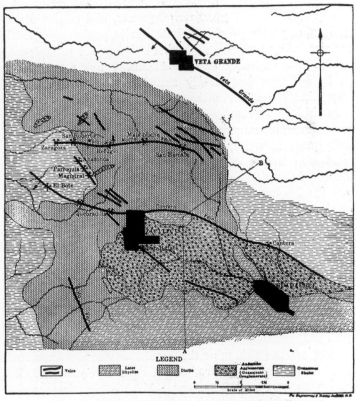

FIG. 12–24.   Veins of the Zacatecas silver district, Mexico.
(Bastin-Botsford, Econ. Geol.)

They are most common along the Veta Madre, where they form large
ore bodies 100 meters wide, 200 meters long, and 200 meters deep
(Fig. 12–23).

The ores carry silver and gold in the ratio of 100 to 1. The ore
minerals are disseminated or in bands in crustified quartz-carbonate
gangue and consist of numerous silver sulphides and antimonides,
base-metal sulphides, along with quartz, carbonates, and adularia, with
some rare fluorite, barite, and zeolites. Oxidation has extended only

20 to 30 meters and the ore minerals are characteristically hypogene and typical of epithermal fissure veins.

Parral-Santa Barbara, Chihuahua. These two camps, 20 kilometers apart, are of interest because of silver ores in similar geological environment but in contrasting mineralogy. The deposits are filled fissures, and the ores are dominantly silver with minor gold and low copper; lead and zinc are present in both areas. At Parral, Schmitt recognizes quartz, chalcedony, fluorite, garnet, and calcite containing blende, galena, argentite, proustite, chalcopyrite, pyrite, and specularite. These are classed by Lindgren as epithermal deposits (note specularite and garnet). Considerable silver enrichment has occurred. Santa Barbara, in addition, has closely associated high-temperature ores classed by Lindgren as hypothermal. They are both siliceous and massive sulphides. Accompanying the silver, lead, zinc, and copper minerals, however, are surprising gangue minerals, including, according to Schmitt, quartz, fluorite, garnet, pyroxene, orthoclase, epidote, and zoisite and also ilvaite and fayalite. These minerals denote high temperature. According to Barry and Schmitt, the high-temperature phase cuts the low-temperature phase, and vice versa; they are both phases of the same mineralization in the same mineral district.

Other Deposits. Rich silver ores have been mined from the many veins (Fig. 12–24) of Zacatecas since 1548, and numerous other camps have produced tens to hundreds of millions of dollars in silver.

SOUTH AMERICA

The former glory of South America as a silver region has diminished until now South America produces only 15 percent of the world's silver. Nearly half of this amount is derived from the copper ores of Cerro de Pasco in Peru, making this mine the world's largest producer of silver. Of the total South American production, Peru now yields 57, Bolivia 30, Argentina 10, and Chile 3 percent. Many of the famous districts of the past are now largely exhausted.

The present ranking silver producers are:

MINE OR COMPANY	LOCATION	NATURE OF ORE
1. Cerro de Pasco Co.	Cerro de Pasco, Peru	Copper-base metal
2. Huanchaca	Huanchaca, Bolivia	Silver
3. Aramayo	Chocaya, Bolivia	Silver-tin
4. Colquijirca	Colquijirca, Peru	Silver-lead
5. Potosi	Potosi, Bolivia	Silver-tin

Potosi, Bolivia. This district, famed as the richest silver hill on earth, is said to have produced well over 2 billion ounces of silver since

its discovery in 1544. Its conical volcanic peak (Fig. 12–25) rising
2,500 feet above its surroundings to an elevation of 16,000 feet consists

FIG. 12–25.   Cerro Potosi, seamed by mining since 1544 and containing the greatest
concentration of silver in the world. (*Evans, Econ. Geol.*)

of a core of rhyolite surrounded by Pliocene sediments and tuffs that
dip outward (Fig. 12–26). Numerous steep veins cut the mountain
both in rhyolite and sediments. The veins are sharp, well defined,

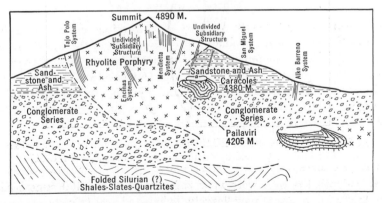

FIG. 12–26.   Section of Cerro Rico de Potosi showing rock types and main vein
systems. (*Evans, Econ. Geol.*)

and frozen, and the vein filling is banded and drusy. The walls are
bleached and altered and in places shattered and impregnated by ore.
The veins (Fig. 12–27) branch upward, are closely spaced, and inter-

secting, so that most of the mountain is mineralized. They range in
thickness up to 4 meters. The veins are open-space fillings and replacements of sheeted zones. Oxidation extended to 300 meters, and

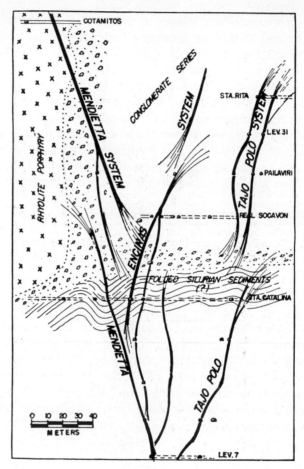

FIG. 12–27. Typical section of Potosi facing north, showing echelon character,
vein relationships, and upward branching. (*Evans, Econ. Geol.*)

to this depth the ores were very rich, many above 100 ounces silver
per ton. These ores, now largely exhausted, contained cerargyrite,
native silver, and some argentite and ruby silver. No enrichment in
tin has been noted.

The primary ores average 10 ounces silver and 1–4 percent tin. They
are fine-grained aggregates consisting of quartz, alunite, cassiterite,
stannite, base-metal sulphides, and silver minerals. The silver min-

erals consist of argentiferous tetrahedrite, andorite, ruby silver, matildite, and jamesonite. Chalcocite and covellite are supergene; and
oxidation products described by Lindgren and Creveling are jarosite,
halloysite, hematite, goethite, sideronatronite, and voltaite.

The ores apparently are related to the rhyolite porphyry intrusion
and were formed after it by ascending hydrothermal solutions, which
Lindgren thinks may have been colloidal and from which deposition
took place between 300° and 100° C.

Huanchaca, Bolivia. This district produces 4 to 5 million ounces of
silver annually. The Pulacayo for some time has been the largest
silver-producing company of Bolivia. The deposits are veins occurring
in a mountain composed of red shale and conglomerate cut by a
central mass of trachytic porphyry surrounded by granite porphyry.
The veins are known for a length of 2,350 meters and a depth of 750
meters. They range in width from 1 meter to 4 meters and branch
upward.

The veins are crustified fissures with marginal crusts of quartz and
pyrite, followed by bands of pure pyrite, sphalerite and tetrahedrite,
galena-chalcopyrite-quartz, and in the middle by blende, tetrahedrite,
bournonite, and boulangerite. Rare ruby silver, stibnite, wurtzite,
jamesonite, and bismuth occur. The galena and tetrahedrite are high
in silver. Galena decreases, and blende and pyrite increase, with
depth. The rich tetrahedrite shoots carry 180 to 600 ounces silver
per ton. The average ore runs 50 ounces silver, 5 percent copper, and
5 percent zinc.

Chocaya, Bolivia. This district worked by the Aramayo Company
contains complex silver-tin ores and is now the second largest silver
producer of Bolivia. The deposits are filled fissures up to 1 meter
wide carrying base-metal sulphides, tetrahedrite, ruby silver, and
cassiterite in quartz, barite, and calcite gangue. They are shallow-
seated cavity fillings. Similar deposits occur in the old Oruro and
Colquiri districts in Bolivia.

Colquijirca, Peru. This mine yields ores averaging 60, 40, and 20
ounces silver, respectively, from three mantos in folded limestone beds.
Lindgren states that three beds have been replaced to form three
mantos, separated by a few meters of shale. The upper manto (5
meters) contains rich silver-copper ores. Underneath this is a manto
higher in lead, and a lower pyritic copper manto with enargite lies
nearer a monzonite intrusion. In places the three merge. The different minerals in them are noteworthy.

The silver manto consists of chert, barite, dolomite, tennantite,
stromeyerite, pyrite and some wittichenite, galena, pearcite, stern-

bergite, galena, and sphalerite; enargite is rare. Magnificent specimens of supergene native silver occurred in the upper zone. The mineralization closed with a late specularite-marcasite phase. The richness of the ores has made this single mine one of the most productive of the world.

McKinstry considers that cross fracturing helped localize the ores and that the mineralogical difference of the mantos is small. The general mineral sequence was (1) rock alteration; (2) early sulphides: pyrite, marcasite, blende, tennantite and enargite, galena; (3) later botryoidal sulphides: blende, galena, pyrite; (4) stromeyerite; (5) specularite-marcasite; (6) supergene native silver. Here is a peculiar intermingling of mesothermal and epithermal mineralization. Most of the deposition is distinctly low-temperature, and telescoping is absent. McKinstry thinks there is a zonal " gradation northward from the intrusive from copper to silver-lead mineralization."

Other Districts. Other smaller silver mines are scattered throughout Peru and Bolivia and also in northern Chile, such as the camps of Oruro, Colquiri, Colquechaca, and Negrillos in Bolivia; those of Morococha in Peru; and Chañarcillo, Caracoles, Huantajaya and Tres Puntos in Chile. In the Argentine, which ranks third in South American production, most of the silver is a by-product of lead-zinc and copper deposits, such as those of Famatina, Iglesia, Hoyada, Los Cobres, and Aguilar.

AUSTRALIA

Australia is the fifth ranking country of the world in silver production, New South Wales accounting for 65 percent of it, Queensland 25 percent, and Tasmania 8 percent. Most of the silver is by-product metal from lead, zinc, and copper ores. In the following ranking properties the ore is the lead-zinc-silver type; all the properties are located near Broken Hill, New South Wales, except Mount Isa, Queensland: (1) Mount Isa, (2) North Broken Hill, (3) Broken Hill South, (4) Sulphide Corporation, (5) Zinc Corporation. Since these deposits are dominantly lead-zinc, and silver is secondary, they will be described under " lead " and " zinc."

Selected References on Silver

General Reference 9. *Brief references to silver deposits.*
Tintic District. PAUL BILLINGSLEY. XVI Int. Geol. Cong. Guidebook 17:101–124. *An outstanding example of geological field work and structural control of ores.*

Gold Deposits of the World. W. H. EMMONS. McGraw-Hill, New York, 1938. *Annotations only.*

Tonopah District, Nevada. T. B. NOLAN. Nevada Univ. Bull. 24, Pt. 4:35, 1930. *A fine illustration of careful field work, and structural and geothermal ore localization.*

Mineral Deposits of the Canadian Shield. E. L. BRUCE. Macmillan Co., Toronto, 1933. *Occurrence of Cobalt deposits.*

Pachuca Silver District, Mexico. E. WISSER. A. I. M. E. Tech. Pub. 753, 1937. *A good study of detailed structure.*

Santa Eulalia District, Mexico. BASIL PRESCOTT. A. I. M. E. Trans. 51:57–99, 1916. *An unusual district of manto deposits.*

Guanajuato Mining District, Mexico. A. WANDKE and J. MARTINEZ. Econ. Geol. 23:1–44, 1928. *Details about the great Veta Madre vein.*

Mineral Deposits of South America. B. L. MILLER and J. T. SINGEWALD, JR. McGraw-Hill, New York, 1919. *Various brief descriptions of silver districts.*

Silver Deposit at Colquijirca, Peru. H. E. McKINSTRY. Econ. Geol. 31:618–635, 1936. *A fine mineralogical and structural study.*

The Ores of Potosi, Bolivia. W. LINDGREN and J. G. CREVELING. Econ. Geol. 23:233–262, 1928. *Remarkable sequences of mineralization.*

Tin-Silver Veins of Oruro, Bolivia. F. M. CHACE. Econ. Geol. 43:333–383; 435–470, 1948. *Excellent discussion of occurrence and origin.*

Mineral Relationships in the Ores of Pachuca and Real del Monte. E. WISSER. Econ. Geol. 43:280–292, 1948. *Chiefly structural control.*

Rock Alteration as a Guide to Ore — East Tintic District, Utah. T. S. LOVERING. Mon. 1, Econ. Geol., 1949. *Details of a silver district.*

Platinum

Platinum may be considered one of the precious metals since it is more costly than gold. About 60 percent of that consumed in the United States is for jewelry purposes. It was once used for coinage in Russia until its value exceeded that of the coins. Its industrial use is increasing.

Platinum is only one of a group of related metals consisting of osmium, iridium, palladium, rhodium, and ruthenium. They are not only associated together but also are generally alloyed, and are called, therefore, the " platinum metals." They are very heavy, insoluble in most acids, melt at temperatures of 1,549° to 2,700° C, and range in hardness from 4.8 to over 7. Iridium is the heaviest metal and osmium the hardest.

Platinum is a modern metal and for long its production was small. Formerly 92 percent of it came from Russia and 7 percent from Colombia. World production, ranging between 500,000 and 900,000 ounces, comes mainly from Canada, Russia, South Africa, and Colombia. The United States production is minor and comes largely from Alaska and as a by-product in metal refining.

Mineralogy, Treatment, and Uses

Most platinum used today is won as a by-product in the refining of other metals, chiefly nickel.

Ore Minerals. The chief mineral is the native metal, but platinum also occurs as *sperrylite* ($PtAs_2$), *cooperite* ($PtAsS$), *stibio-palladinite* (Pd_3Sb), and *braggite* ($PtPdNiS$). There are probably also arsenides of the other platinum metals. The platinum metals form natural alloys with one other, such as *osmiridium* or *platiniridium*, also with iron to form *ferroplatinum* (16 to 21% Fe) and *polyxene* (6 to 11% Fe), and with copper to form *cuproplatinum* (8 to 13% Cu). Native platinum is never pure platinum.

Treatment. By-product platinum is separated from the containing metal during electrolytic refining. Placer platinum is concentrated by delicate gravity concentration. Lode platinum is recovered by combined gravity and flotation concentration. The individual metals of the platinum group are separated by complex refining methods.

Associations. Platinum is invariably associated with basic igneous rocks and with the ore minerals characteristic of those rocks. Most of the platinum of the world is intimately associated either with chromite or nickel. Even platinum placers are derived from basic rocks rich in chromite. The platiniferous nickel ores also contain copper and appreciable quantities of gold and silver.

Uses. The chief uses for platinum are in the jewelry, electrical, and dental industries. Its white color and hardness make it a desirable setting for diamonds. In the electrical industry it is used for resistance and contacts in the more delicate instruments, such as telephones and radios. Platinum finds wide uses in the chemical industry, such as for containers, wire, electrodes, coils, acid making, X-ray equipment, and as a catalyst. Large acid stills utilize considerable quantities.

Kinds of Deposits, Origin and Distribution

The classes of deposits that yield platinum metals, and some examples, are:

Magmatic concentrations:
  1. Early magmatic:
      (a) Disseminations — sparse disseminations with chromite in dunites, the erosion of which yields placers — *Urals*, Alaska, Colombia.
      (b) Segregations, by fractional crystallization — *Merensky Reef, Rustenburg, South Africa.*

2. Late magmatic:
    (a) Immiscible liquid segregations — Vlackfontein, South Africa.
    (b) Immiscible liquid injections — possibly Frood mine, Sudbury, Ontario.
Contact-metasomatic deposits — Tweetfontein, Potgietersrust, South Africa.
Hydrothermal — *Sudbury*, Ontario; Waterburg, South Africa.
Placer deposits — *Urals, Colombia*, Alaska.

Origin. The home of platinum is in ultrabasic igneous rocks, where it has been concentrated by magmatic processes. The erosion of disseminated magmatic concentrations has yielded the placers of the Urals, Colombia, and Alaska. Richer magmatic concentrations have formed lode deposits in South Africa. The by-product platinum from the nickel ores of Sudbury, Ontario, may have been formed either by magmatic or by hydrothermal processes. Only one deposit of contact-metasomatic origin is known (South Africa), and only one freak deposit has been formed as a fissure vein deposit by hydrothermal solutions.

Distribution. There are three main centers of platinum in the world, namely, the Ural Mountains, Sudbury, Ontario, and the Bushveld of South Africa. Less important localities are the Choco district of Colombia and Goodnews Bay, Alaska. Platinum is also won as a by-product from Alaska, Oregon, and California gold placers. Minor production comes from Ethiopia, Sierra Leone, Katanga, New South Wales, Victoria, and Tasmania. Small amounts have also been found in Papua, India, Sumatra, New Zealand, and the Philippines.

EXAMPLES OF DEPOSITS

Sudbury, Ontario. The main source of platinum in Canada is the Frood mine at Sudbury, where it is concentrated to an unusual degree in the massive copper-nickel sulphide body of the lower levels. Although small amounts of by-product platinum had formerly been obtained from the other nickel-copper mines, it was not until the rich Frood ores were discovered that Canada became an important producer of platinum. The platinum content is stated to average 0.05 ounce per ton and about three-quarters of that amount of the other platinum metals, chiefly palladium. A high gold and silver content is also associated with the high platinum ores.

Russia. For 100 years after the discovery of platinum in the Ural Mountains in 1819, Russia was the leading and almost the only source of platinum. The metal is won entirely from placers, but primary deposits in dunites and pyroxenites have been worked on an experimental scale. The placers occur on both sides of the northerly 350 miles of the Ural Mountains, where 10 small oval-shaped masses of dunite contain magmatic concentrations of platinum in irregular pockets, veins, and bands of chromite. The dunite is richer than the pyroxenite. These may eventually become workable lode deposits.

The erosion of these masses and of additional pyroxenite and serpentine bodies has yielded the placer deposits, which are of two kinds: (1) old river beds, and (2) later river beds that cut the older. The productive sands are up to 5 feet thick, under 35 feet of overburden. The chief centers are on the eastern slopes in the basins of the Iss and Tura Rivers and in the Nizhni Tagil district on both slopes. About 80 percent of the production is recovered by large electric dredges.

South Africa. South Africa is the only place where straight platinum ores are mined. There are two main groups: magmatic concentrations and contact-metasomatic deposits. The former consist of magmatic concentrations with (*a*) hortonolite dunite, in pipes, (*b*) chromitite, and (*c*) nickel-copper sulphides.

The *hortonolite-dunite pipe deposits* are unique. The rock itself is the heaviest silicate rock known and the richest in ferrous iron. It consists of hortonolite (olivine) with subordinate phlogopite, hornblende, diallage, ilmenite, chromite, and magnetite, and is a differentiate of the Bushveld Igneous Complex. Some 60 occurrences are known in the lower part of the norite zone; only 3 are of commercial importance, and these are all pipes. They are carrot-shaped bodies, up to 60 feet in diameter and as much as 1,000 feet deep. The bodies, which transgress the pseudostratification of the norite, consist of a central zone of platinum-bearing hortonolite dunite with local segregations of chromite, surrounded by envelopes of olivine dunite, and pyroxenite, grading outward into norite (Fig. 12–28). The central zone only is mined. The platinum content varies considerably, assays up to 1,200 dwt being obtained, the mining averages ranging between 4 and 20. At the Onverwacht pipe, Wagner states that the upper levels average 20 dwt, the 250-foot level 18.4 dwt, and the 750 level 9.5 dwt; the Mooihoek pipe averaged between 6 and 7 dwt. These peculiar ultrabasic pipes, crossing the stratification, with the most basic part in the center, and intrusive into the norite, suggest that the most basic portions remained fluid longest and are an enigma; possibly they may be pegmatitic.

The *chromitite segregations* are in the lower part of the differentiated norite zone and occur within the *Merensky Horizon* or " Reef " — a zone that has been traced scores of miles. In places the platinum is sufficiently concentrated to be workable, as at Rustenburg, Potgietersrust, and Lydenburg. At Rustenburg the zone dips gently, conforming with the pseudostratification of the differentiation zones in the Bushveld Complex. The platinum band is about 12 inches thick and averages 10 to 12 dwt platinum. This lies directly and sharply on anorthositic norite and is overlain by bands of pyroxenite, spotted norite, mottled anorthosite and again by anorthositic norite, which in

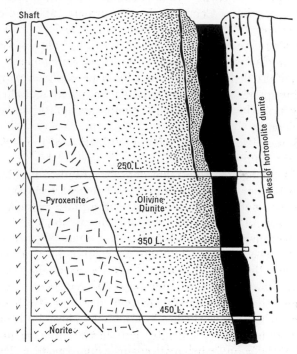

Fig. 12–28. Mooihoek platinum pipe, South Africa. Center, hortonolite dunite with chrome, carrying platinum, surrounded by olivine dunite, then by pyroxenite, all enclosed in country rock of norite, repeated on other side. (*Based upon Wagner, modified by sketches made by author at mine.*)

turn underlies the Bastard Reef 40 feet above the main reef. The ore consists of unusual pyroxenite or harzburgite with chromite, native platinum, magnetite, and specks of pyrrhotite, pentlandite, chalcopyrite, cubanite, millerite, and nickeliferous pyrite. Considerable palladium accompanies the platinum, and minor quantities of rhodium, osmiridium, gold, and silver. Some supergene platinum enrichment is claimed by Schneiderhöhn.

The *magmatic nickel-copper sulphide deposits* of platinum ore occur at Vlackfontein, Rustenburg. They are isolated masses of irregular shape composed of disseminated or massive sulphides in bronzitite.

The ore minerals are pyrrhotite, pentlandite, and chalcopyrite. Only the oxidized portions are worked for platinum.

The *contact-metasomatic* deposits occur northwest of Potgieters-rust, where Bushveld rocks rest on dolomite and ironstones. The dolomite has been silicated to a diopside-grossularite rock impregnated by platinum-bearing yellow sulphides, with unusually large crystals of sperrylite. Cooperite is also present, and the ores carry nickel and copper. The platinum tenor is 7 dwt.

Other platinum occurrences in South Africa include some pegmatitic bodies, minor placers, osmiridium in the Rand conglomerates, and the freak Waterburg deposits, where platinum occurs in a vuggy quartz vein carrying quartz, chalcedony, specularite, pyrite and chrome mica, presumably a shallow-seated, high-temperature hydrothermal or pneumatolytic deposit.

United States. The small platinum production of the United States comes mainly from Goodnews Bay, Alaska, where high-grade placers are worked in nonglaciated stream deposits near two peridotite intrusions. The best gravels lie within 2 or 3 feet of bedrock, and chromitic magnetite and ilmenite occur with the platinum. All six platinum metals are present, platinum constituting 68 to 75 percent, iridium 6 to 13 percent, and palladium is low.

Colombia. The waning platinum production of Colombia comes from the Choco district, where platinum sands are dredged from streams that cross ultrabasic intrusions; the sands contain up to 0.5 ounce platinum and the same quantity of gold per ton. Both stream-bed and terrace gravels are worked. As elsewhere, chromite is an intimate associate, and magnetite, ilmenite, and heavy silicates are present. Most of the platinum is in small grains, but nuggets up to 1 pound have been found. All six metals are present, platinum constituting about 85 percent. The dredging gravels yield 60 to 90 cents per cubic yard. Native lead has been reported in the platinum sands.

Selected References on Platinum

Platinum Deposits and Mines of South Africa. P. A. WAGNER. Oliver and Boyd, Edinburgh, 1929. *The pipes and " reefs " in the Bushveld Igneous Complex.*

Strategic Mineral Supplies. G. A. ROUSH. McGraw-Hill, New York, 1939. *Occurrence, distribution, and technology of platinum.*

Platinum and Allied Metals. 2nd Ed. Imperial Inst. Mon., London, 1936. *British and other deposits; uses and technology.*

Platinum at Work in 1942. E. M. WISE. Min. and Met. 23:421–425, 1942. *Chiefly properties and uses.*

CHAPTER 13

THE NONFERROUS METALS

The nonferrous metals include copper, lead, zinc, tin, and aluminum. Next to iron they are the chief metals of industry and war, and are indispensable in modern civilization. They are used in large volume and enter extensively into international trade. Their use is increasing faster than new supplies are being discovered, and their development, like that of gold, is pushing back frontiers and establishing large industrial centers in distant uncivilized regions. Unlike gold mining, whose output can be transported in an airplane package, the mining of these metals entails adequate transportation, movements of bulk freight, large treatment and power plants, and vast expenditures for their development.

Copper

From early times until 1800 copper was widely produced in small quantities. From 1801 to 1810 the annual world production was only 18,200 tons, equivalent to less than one month's production of some present-day mines. England was the world's leading producer until 1850, when Chile assumed first place, which she held until 1883, since when the United States has been the world leader.

In the Western Hemisphere copper was mined in Chile before the Conquistadores, but the first Spanish mining was in 1601 and the first modern smelting in 1842. Although copper was discovered in the United States in 1632, none was worked until 1705, when the Simsbury mine in Connecticut was opened. Similar ores were worked in the Schuyler mine, New Jersey, in 1719, and at Gap, Pa., in 1732. Considerable copper was mined in 1800 by the Spaniards from what is now the Chino mine, at Santa Rita, N. Mex. Records show that the Lake Superior copper was known in 1771, but the first real mining did not take place until 1830. The chalcocite ores of Bristol, Conn., were discovered in 1836, and workings reached a depth of 240 feet by 1853. Mining started at Ducktown, Tenn., in 1843. By 1874 the mines of Butte and Bisbee became prominent and by 1905 the " porphyry copper " era began in the United States and Chile. Subsequently large production came from Canada and central Africa.

PRODUCTION AND DISTRIBUTION

The tremendous growth in the use of copper is indicated by the fact that of the total world copper produced in the last 100 years about 80 percent was mined in the last 25 years and about one-third of it in the last dozen years. Annual production ranges from about 2 to 2.8 million tons of metallic copper. The chief producing countries, and their approximate percentage of current world production, with approximate percentage of world reserves in parentheses, follows: United States (20) 36, Chile (40) 20, Rhodesia (30) 10, Canada (5) 9, Belgian Congo (5) 7, Russia 7, Mexico 3. Lesser production comes from Japan, Germany, Yugoslavia, Peru, Finland, Sweden, Turkey, South Africa, and Australia.

The order of production in the United States is: Arizona, Utah, Montana, New Mexico, Nevada, Michigan, Tennessee, California, and Washington.

Distribution. Although copper is widely distributed, 80 percent of the world supply emanates from four regions, the southwestern United States, the Andean belt, the Canadian Shield, and the central African belt.

In *North America* the greatest concentration of copper in the world centers in Arizona and the adjacent parts of the United States and Mexico, and includes all the well-known North American " porphyry coppers " and a host of other famous districts. All the ores are associated with monzonitic types of intrusions. A compact, productive Montana province centers around a granodiorite massif at Butte. Other smaller provinces include the Appalachian, the fruitful Lake Superior district, and the sprawling Cascadian-Coast Range belt extending from northern British Columbia to Washington. A lone but rich center is at Kennecott, Alaska.

The *Canadian Shield,* from Manitoba to Quebec, includes the Hudson Bay, Sudbury, Noranda, and other deposits, with minor representatives in Newfoundland.

The *Andean* belt, from Chile to Peru, includes the large Chuquicamata, Braden, Potrerillos, and Cerro de Pasco deposits, also associated with monzonitic intrusives.

The *central African* province constitutes the most concentrated copper belt in the world and includes the prolific mines of Northern Rhodesia and adjacent Katanga.

Other minor belts are the Uralian province of Russia, the outer Japanese Island arc, and isolated European centers in Spain-Portugal (Rio Tinto), Bor in Yugoslavia, Mansfeld in Germany, Outokumpo in Finland, and Boliden in Sweden. Australia includes several minor copper centers such as Mount Lyell, Mount Morgan, and Mount Isa.

Mineralogy, Tenor, Treatment, and Uses

The copper of commerce occurs in several forms and in a variety of minerals. Mineralogically, copper ores are divided into four large groups, namely, native, sulphide, oxidized, and complex. Each requires a different metallurgical treatment to obtain the copper, and different ore tenor to form economic deposits.

The sulphide ores are the most valuable; the complex ores contain copper admixed with lead, zinc, gold, and silver minerals.

Mineralogy. Some 165 copper minerals are known, but the chief economic ones are those listed below.

MINERAL	COMPOSITION	PERCENT OF COPPER
Native:		
Native copper	Cu	100.0
Sulphides:		
Chalcopyrite	$CuFeS_2$	34.5
Bornite	Cu_5FeS_4	63.3
Chalcocite	Cu_2S	79.8
Covellite	CuS	66.4
Enargite	Cu_3AsS_4	48.3
Tetrahedrite	$Cu_8Sb_2S_7$	52.1
Tennantite	$Cu_8As_2S_7$	57.0
Oxidized:		
Cuprite	Cu_2O	88.8
Tenorite	CuO	79.8
Malachite	$CuCO_3Cu(OH)_2$	57.3
Azurite	$2CuCO_3Cu(OH)_2$	55.1
Chrysocolla	$CuSiO_3 \cdot 2H_2O \pm$	$36.0\pm$
Antlerite	$Cu_3SO_4(OH)_4$	54.0
Brochantite	$Cu_4SO_4(OH)_6$	56.2
Atacamite	$CuCl_2 \cdot 3Cu(OH)_2$	59.4

The chief gangue minerals of copper ores are rock matrix, quartz, calcite, dolomite, siderite, rhodochrosite, barite, and zeolite. Contact-metasomatic silicates may also be present in ores of that type. Most copper ores are free from other associations, except for gold, silver, and molybdenum. A few zinc, lead, and silver ores may carry copper as a by-product, and in complex ores copper may also be associated with zinc and lead. In general, copper sulphide ores are associated with intrusions of quartz monzonite and related rocks; less commonly with basic intrusives.

Tenor and Treatment. The type of copper ore and its tenor generally determine the method of treatment. The lowest-grade ores are the simple and easily treated native copper deposits, which may run as low as 0.6 percent. Sulphide ores run as low as 1 percent or less for the lowest-grade ores; high-grade ores may range from 5 to 30 percent. Most oxidized ores range from 2.5 to 10 percent, since they are more difficult to treat. Ores carrying 4 percent or more copper are generally smelted directly to avoid concentration losses; ores with 1 to 4 percent are generally concentrated and the concentrates are smelted. In the United States the average copper content of all smelting ores has been 3.52 percent and of all concentrating ores 0.90 percent; the average of all ores has been 0.91 percent.

Smelting ores are smelted with siliceous or base ores, flux, and coke, and the matte is refined in a converter to blister copper. This may be sold as " blister," or fire-refined, but generally it is electrolytically refined to remove harmful substances and to recover the precious metals.

Low-grade sulphide ores are floated, and the concentrates are smelted. Selective flotation permits the separation of undesired pyrite or pyrrhotite and of zinc and lead minerals. Concentrates range from 10 to 40 percent copper. About 95 percent of all copper ores mined in the United States are concentrated.

Oxidized ores are smelted directly or are leached. Dilute sulphuric acid leaching is used for carbonate ores in nonreacting gangue. Water is used for leaching sulphate and chloride ores and for " heap leaching " sulphide ore dumps and ore in place. The copper is precipitated from such liquors by electrolysis or scrap iron.

Uses of Copper. Copper is one of the very essential minerals in modern industry. It is normally a prosperity metal, i.e., used when electrical expansion takes place, but it is also an essential metal of war. The United States, the world's largest consumer, uses between 1 and 1.5 million tons annually, as follows:

Use	Percent of Total	Use	Percent of Total
Electrical manufacture	24.6	Radios	2.7
Automobiles	13.2	Refrigerators and air	
Miscellaneous wires	11.8	conditioning	2.3
Light and power lines	9.6	Ammunition	1.7
Buildings	8.2	Manufacture for export	5.2
Telephone and telegraph	4.6	Other uses	16.1

Most wires and electrical equipment are made of pure copper, but considerable alloy copper is used, chiefly as brass and bronze. The brasses are copper-zinc alloys (55 to 99% Cu), and the bronzes are copper-tin-zinc (88% Cu, 10% Sn, 2% Zn). There are also nickel, aluminum, and steel alloys of copper; and minor special alloys utilize arsenic, beryllium, cadmium, chromium, cobalt, iron, lead, magnesium, manganese, and silicon.

Electrolytic copper is the form normally distributed, but the use of fire-refined copper is growing. Low-grade blister is commonly utilized for making copper sulphate. Considerable scrap copper, or copper and brass cuttings and reclaimed metal, is used along with primary copper; the proportion may reach 0.8 pound scrap to 1 pound primary copper.

The world reserves of copper are large, and as economies in production are achieved, the reserves will increase.

KINDS OF DEPOSITS AND ORIGIN

The principal classes of copper deposits and some examples of each are:

1. Magmatic: Copper-nickel deposits — Insizwa, South Africa; Merensky Reef, Transvaal; possibly Frood mine, Sudbury, Ontario.
2. Contact-metasomatic: Older deposits of *Morenci*, Ariz., Bingham, Utah, and Cananea, Mexico.
3. Hydrothermal:
    A. Cavity Filling:
        (1) Fissure veins — *Butte*, Mont.; Walker, Calif.
        (2) Breccia filling — *Nacozari*, Mexico; Bisbee, Ariz.; *Braden*, Chile.
        (3) Cave fillings — Bisbee, Ariz.
        (4) Pore-space fillings — " Red bed " deposits of New Mexico and Arizona; Urals, Russia.
        (5) Vesicular fillings — *Lake Superior* district.
    B. Replacement:
        (1) Massive — *Bisbee*, United Verde, Ariz.; *Bingham, Tintic*, Utah; Ducktown, Tenn.; *Noranda*, Quebec; *Flin Flon*, Manitoba; Granby, British Columbia; *Rio Tinto*, Spain; Boliden, Sweden; *Cerro de Pasco*, Peru.
        (2) Lode — Kennecott, Alaska; *Magma*, Globe, Bisbee, Ariz.; Butte, Mont.; Britannia, British Columbia.
        (3) Disseminated — The " porphyry coppers ": *Bingham*, Utah; *Ely*, Nev.; *Ray, Miami*, Inspiration, *Ajo*, and *Clay* Orebody, Morenci, Ariz.; *Santa Rita*, N. Mex.; the *Rhodesia copper belt; Braden, Chuquicamata* and Potrerillos, Chile.
4. Sedimentary: Kupferschiefer, Mansfeld, Germany?
5. Surficial Oxidation: Bisbee and Globe, Ariz.; *Chuquicamata*, Chile; *Katanga*, Belgian Congo.
6. Supergene Enrichment: Some " porphyry coppers "; Cananea " porphyry, " Mexico; United Verde Extension, Ariz.

Origin. Most copper deposits, and all large ones, have been formed by hydrothermal solutions, with replacement dominant over cavity filling. Contact metasomatism accounts for a few. There are few, if any, large representatives of magmatic origin. There are no sedimentary copper deposits free from controversy. Most copper deposits in unglaciated regions have undergone oxidation and some supergene enrichment. Many deposits have been almost completely converted to oxidized compounds, and there are numerous deposits whose economic importance is due solely to supergene enrichment. It is thus evident that copper deposits have originated by diverse processes, but practically all of them are either the direct result of igneous activity or of weathering processes.

Examples of Deposits in the United States

The " Porphyry Coppers "

Mining of " porphyry copper " deposits of the western-southwestern province sprang into prominence between 1905 and 1910 and now constitutes the backbone of American copper production. One mine, the Utah Copper at Bingham, Utah, is the greatest copper-producing mine of the world. The others of the group are shown in Table 16.

All these deposits have similar characteristics: they are of low grade and are operated on a large scale by low-cost methods; they are associated with stocklike intrusions of monzonitic porphyries; they are disseminated replacements in porphyry or intruded schists; they are of blanket shape, with greater horizontal than vertical dimensions. Their primary mineralogy is generally similar and is accompanied by intense sericitization and, in places, silicification of the host rocks; all are overlain by leached cappings and have been subjected to more or less supergene sulphide enrichment; and all have similar modes of origin. The differences between them are in details of host rock, shape, size, tenor, oxidation, and degree of supergene enrichment. They are mined either by large-scale open-cut methods or underground caving, and the ores are readily amenable to flotation, with high ratios of concentration. Some of their common features are summarized in Table 16.

Intrusives. The porphyry coppers are all closely associated with intrusions of monzonite, quartz monzonite, or diorite porphyry of Mesozoic or early Tertiary age. These are stocks or chonoliths a few thousand feet across and of irregular shape, which have entered along major tectonic breaks. All are bared by erosion. The ore occurs in their upper or outer crackled parts or in the intruded rocks. The host rocks have been intensely altered during the early phase of mineralization and are largely converted to sericite, clay minerals, and chlorite with minor silicification; the dark minerals have disappeared,

TABLE 16

DATA OF UNITED STATES PORPHYRY COPPERS

Mine	Ore Tenor (percent ±)	Daily Production (tons ore)	Annual Production Capacity (tons Cu ±)	Average Depth of Capping (feet)	Average Thickness of Ore (feet)	Degree of Enrichment	Host Rock
Utah Copper	0.9	50–100,000 +	320,000 +	115	2,500	Medium	Monzonite
Nevada Copper	1.4	10– 15,000	50,000	150	600	Low	Quartz monzonite
Consolidated Copper mines	1.4	4– 6,000	20,000	100	200	Low	Quartz monzonite
Ray	1.6	15,000	16,000	250	150	High	Schist
Chino	1.22	15– 20,000	75,000	150	700	Medium	Granodiorite
New Cornelia	0.7	28,000	150,000	55	425	None	Quartz monzonite
Miami	0.9	8– 10,000	26,000	250	300	High	Monzonite schist
Inspiration	1.3	5– 10,000	30,000	250	300	High	Monzonite schist
Clay	0.99	50,000	150,000	220	850	Medium	Diorite porphyry
Bagdad	1.4	. . . .	. . . .	150	100	High	Monzonite porphyry
San Manuel	0.78	25,000	. . . .	600	900	Low	Quartz monzonite
Castle Dome	0.7	12,000	25,000	80	135	Medium	Quartz monzonite

the outlines of the feldspars are commonly indistinguishable, and the rock is bleached to a creamy color. Superimposed upon this hypogene alteration is a supergene kaolinization, which extends as deep as the supergene sulphides.

Character of the Deposits. The deposits are huge blankets roughly parallel to the topography and extending downward in irregular pro-

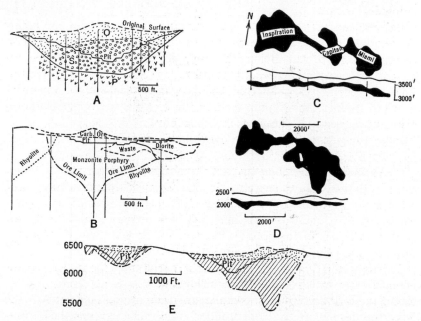

FIG. 13–1. Cross section of typical porphyry coppers. A, north-south section across Pit ore body, Ely, Nev.; B, cross section of New Cornelia deposit, Ajo, Ariz. (after Ingham-Barr, U. S. Bur. Mines); C, plan and section of eastern ore bodies of Inspiration-Miami district (after Ransome, U. S. Geol. Survey); D, generalized plan and section of ore bodies of Ray mine, Arizona (after Ransome, U. S. Geol. Survey); E, generalized north-south section across Chino ore body, Santa Rita, N. Mex. Stippled area is leached zone.

tuberances (Fig. 13–1). Their tops are generally smoothly undulating and are sharply demarked from the overlying leached capping, marking present or former positions of the water level. The capping ranges in depth from 50 to over 400 feet, and the thickness of the deposits is measured in hundreds of feet, rarely thousands. Areally, they range from 1,000 to 10,000 feet across.

The bodies are a combination of disseminations and stockworks in crackled porphyry and invaded rocks. Closely spaced veinlets of quartz and sulphides ramify in all directions; and discrete grains of sulphides, in places hardly visible to the eye, give a pepper-and-salt effect to the altered host rock. The lower boundary of the ore is gradational, merging into primary yellow sulphides. The total sulphide content amounts to 5 to 18 percent of the rock, and the copper minerals around 3 or 4 percent.

Mineralogy. The mineralogy is extremely simple. The primary sulphides consist of pyrite, chalcopyrite, and bornite with minor sphalerite and molybdenite. In the capping these sulphides are largely or wholly removed, leaving voids occupied by limonite of diagnostic colors and patterns. Below the capping the yellow sulphides are coated, or partially or wholly replaced by chalcocite and covellite. The molybdenite is unaffected and is won in considerable quantity at Utah Copper, and Chino, making the Kennecott Copper Corporation the largest producer of molybdenum in the world. At Utah Copper the present percentage distribution of ore minerals is: chalcopyrite 80, chalcocite 9, covellite 7, bornite 4. Pyrite is in excess.

Oxidation and Enrichment. All the porphyry coppers have undergone oxidation, and all except Ajo and Copper Mountain have been secondarily enriched. The capping is leached of most of its copper, except at Ajo, where the sulphide has been converted *in situ* to carbonate. At Miami-Inspiration, Bagdad, and San Manuel, the capping also carries some carbonate copper.

It was formerly thought that all the porphyry coppers were made commercially by supergene enrichment, but this is not true of all of them; the primary ore is of commercial grade in places, as at Ajo and Ely; the deeper ore at Utah Copper is only slightly enriched. Secondary enrichment is rather complete at Miami-Inspiration, Ray, and Chino but is lacking at Ajo. The chief supergene sulphide is chalcocite, but covellite is important at Utah Copper.

Mode of Formation. These huge deposits result from the intrusion of plugs of monzonite porphyry along old faults or other lines of weakness. The upper and outer margins became crackled, probably due to adjustments or to shrinkage either from cooling or from mineralization. This crackled area and enclosing schist permitted penetration by pulsations of uprising hydrothermal solutions given off from the magma chamber that supplied the porphyry. First, widespread rock alteration occurred, which further increased the rock permeability. During the alteration some iron, magnesia, and soda were extracted, and potash and silica were added. The cracks became filled by quartz and sulphides, and little grains of sulphides replaced silicates in intercrack areas and penetrated in places into the adjacent invaded rocks, mineralizing them also. Thus a great volume of rock became metallized with sufficient copper to constitute commercial ore. Subsequent deep erosion and weathering of the metallized portions released copper, giving rise to a zone of supergene sulphide enrichment.

Individual Properties. The porphyry coppers together produce about two-thirds of the total United States copper. Their similarities are greater than their differences.

Utah Copper is a steep mountain about 1,600 feet high, composed of a porphyry mass that intrudes Pennsylvanian sediments. Replacement deposits of copper, zinc, and silver-lead are zonally arranged in limestones outward from the intrusive. The disseminated ore is confined to the altered porphyry and is an oval body 6,000 feet long, 4,000 feet wide, and 2,500 or so feet deep (Fig. 13–1). The shape of the intrusive is unknown in depth. The last reported ore reserves are 640,000,000 tons averaging 1.07 percent copper. Present mining tenor is 0.9 percent. Formerly 40 percent of the copper was in the form of supergene sulphides but this has decreased to 16 percent. Molybdenite is separated from the concentrates. The ore is shot down in benches, loaded by huge electric shovels into trains, and is transported to concentrating mills of 100,000 tons daily capacity. Some 80,000 to 150,000 tons of waste is removed daily. Over 5 million tons of metallic copper have been mined.

At *Ely, Nev.,* are funnel-shaped monzonite porphyry intrusions, aligned along a major fault system. Several of these are metallized and four are being mined, namely in the Ruth underground mine, the Pit mine, and the Consolidated mines. The Ruth body is 1,600 by 1,200 feet across and 200 feet thick, and the Pit body is 4,000 by 2,500 feet across and 600 feet deep. These two mines have produced about 80,000,000 tons of ore averaging 1,4 percent copper. The upper parts were well enriched; the lower parts are primary chalcopyrite ore. The mill handles 18,000 to 20,000 tons per day (Fig. 11–2).

The *Chino mine* at Santa Rita, N. Mex., is a circular ore body 4,000 feet across divided by a barren island in the northern part. It is worked as an open pit. The ore is fairly well chalcocitized, but parts of the body consist of oxidized ores containing native copper, cuprite, and malachite.

At *Ray, Ariz.,* Precambrian Pinal schist has been metallized over an area of 3,000 by 2,000 feet to form a flat blanket. The lean protore has been built up to commercial grade by secondary enrichment, which is fairly complete.

At *Ajo, Ariz.,* the ore, a body 3,600 by 2,600 feet in area, is in monzonite porphyry and some intruded andesite-rhyolite flows. Deficiency in pyrite has prevented sulphide enrichment. The upper 55 feet is oxidized to carbonate without loss of copper. The ore from this part of the deposit was treated by sulphuric acid leaching. The underlying primary ore, consisting of chalcopyrite, minor bornite, and trivial pyrite, extends to a maximum depth of 1,000 feet and an average depth of 425 feet. It is worked by shovel in an open pit.

The *Miami-Inspiration* deposits occur both in porphyry and in Pinal schist. The highly enriched ores occur as an inclined blanket whose upper end has been converted to carbonate at the surface; this blanket dips beneath a later cover of Gila conglomerate. The ore bodies are about 2 miles long and 1,500 feet across. They are mined by underground caving methods.

The *Clay mine* at Morenci, Ariz., one of the newest porphyry copper mines, is an open-cut mine that contains about 400,000,000 tons of 1 percent chalcocite ore lying beneath a capping of variable thickness. It yields 50,000 tons of ore per day.

The *San Manuel mine* at Tiger, Ariz., is the newest American porphyry copper deposit. It contains some 460,000,000 tons of 0.78 percent copper ore with some molybdenum and minor gold and silver in quartz monzonite. The primary ore contains pyrite, chalcopyrite and bornite with a chalcocite layer and occurs as a blanket that is mostly buried by Gila conglomerate up to depths of 1,700 feet. The water table is 600–700 feet deep. About one-third of the tonnage is oxidized ore consisting of chrysocolla.

The *Castle Dome mine,* near Miami, Ariz., is generally similar to the San Manuel mine except that it is smaller and of lower grade.

Butte, Mont.

The famed Butte district, hardly more than 2 by 4 miles in area and three-quarters of a mile deep, and with over 1,000 miles of underground workings, has produced since 1879 over 2½ billion dollars in copper with considerable silver, gold, zinc, lead, and some manganese. More than a score of mines are operated. Butte is called the richest hill on earth and is exceeded only by the Rand gold deposits in metallic wealth extracted.

The district lies near the exposed margin of the granodiorite Boulder batholith, which intrudes Cretaceous andesites. Barren aplite and quartz porphyry dikes cut the granodiorite, and are in turn cut by seven systems of fissures, and by later rhyolite dikes.

The Ore Deposits. The deposits are steeply dipping fissure veins of the first three systems, the oldest being largely lode replacements, and the later ones filled faults, all with generally similar fillings. The deposits range in widths from a few feet to a few tens of feet and are as

much as 7,000 feet long. Since each later system faults the earlier,
the result is a complex parquetry of vein or fault segments, cleverly
resolved by Reno Sales and his staff.

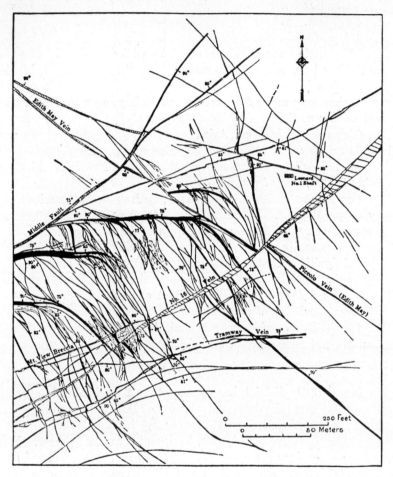

FIG. 13–2. Part of 1,200-level, Leonard mine, Butte, Mont., showing "horsetail"
   structure in the Colusa-Leonard vein. Shows also, E–W, NW, and NE veins and
   fault relations. (*Sales, A.I.M.E.*)

   The older, or Anaconda, are east-west, nonfault tension fissures of
southerly dip. They are long, wide (up to 100 feet), deep (over 4,000
feet), rich, and fairly continuously mineralized and have been the
great producers of the district. Eastward, where faults are numerous,
they exhibit unique "horsetail" structure (Fig. 13–2). The individual "hairs" of the tails are also accompanied by disconnected en

echelon segments. The result is an intensely fractured and metallized area about 2,000 feet long and 300 to 500 feet wide. Well-known veins of the Anaconda system are the Anaconda-St. Lawrence-Original-Steward, Rainbow-Black Rock, State, Badger, and Mountain Con.

The Blue or Northwest veins are strike-slip faults of steep southwest dip that fault to the left (Fig. 13–3). The movement is mainly horizontal and reaches 300 feet. The veins contain thick fault gouge, their walls are crushed, and they include drag ore of the Anaconda

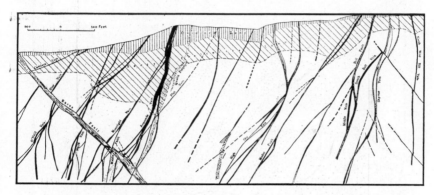

FIG. 13–3. N–S section through Butte, Mont., copper district near Anaconda shaft. Note fault relations of Anaconda and Moonlight veins and Rarus fault. (*After Sales, A.I.M.E.*)

veins. Their ore is identical with that of the Anaconda system, except that it occurs in shoots. They are long and productive.

The Steward or Northeast fault veins dip steeply south and displace veins of the two earlier systems. They carry similar ores in minor quantity. They commonly strike-fault the East-West veins and are themselves faulted by other northeast faults.

These three systems contain identical ores and apparently all were formed during the period of metallization, which was dying out while the Steward system was being formed. Although each system displaces its predecessor veins, all may have been formed essentially contemporaneously as a result of rotational stresses.

The subsequent systems are post-mineral and contain only drag ore, but they are troublemakers because they cut and displace members of all preceding systems, forming disconnected vein segments.

Blind veins are common; some of them do not extend above a depth of 2,000 feet.

Mineralization. The chief copper minerals are chalcocite, enargite, bornite, chalcoyprite, tennantite, tetrahedrite, and covellite. Pyrite

and quartz dominate. Sphalerite is locally dominant; galena is rare;
huebnerite occurs as specimens; and " colusite " (tin-bearing tetrahedrite?) is known. Rhodochrosite, calcite, rhodonite, barite, and
fluorite also occur. Both steely and " sooty " chalcocite are present,
and digenite is locally abundant in the lower levels.

The metals and minerals are zonally arranged outward from the
central zone (Fig. 13–4) and are identical in the different vein systems
within the same zone. A single long vein at Butte may contain min-

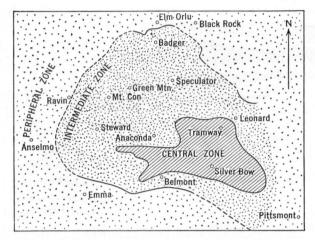

FIG. 13–4. Zonal arrangement of ore minerals at Butte, Mont., at
4,600-foot elevation. (*After Sales, A.I.M.E.*)

erals characteristic of each zone (see Zonal Distribution, Chap. 6).
The ores average 1 to 4 percent copper and 1 to 3 ounces silver; the
central zone is the richest in copper. The mineralization of each vein
system began with quartz-pyrite and ended with chalcocite-covellite
(Fig. 13–5).

The accompanying rock alteration is intense in the central zone,
where little fresh granodiorite can be seen, and diminishes in the intermediate zone to bands paralleling the walls, and in the peripheral zone
is much less. It is also less in the veins of the later systems than in
the earlier. Sales and Meyer have shown that the rock alteration
sequence is, from the veins outward, sericitic, argillic (kaolinite next
to sericite and montmorillonite farthest out) and fresh granodiorite.
This sequence is significantly the same for all veins, regardless of size
or age, and indicates a continuous mineralization by the same solutions.

Oxidation and Enrichment. The depth of the oxidized zone averages
about 300 feet. The copper and most of the silver have been com-

pletely removed from it, the galena is oxidized *in situ,* and the manganese carbonate has been altered to oxides, forming some residual deposits. A shallow zone of supergene enrichment, characterized by "sooty" chalcocite, in places reaches a depth of 1,000 feet. Within this zone, the pyrite of the wall rocks has been enriched to form wide stopes. The deep-level steely chalcocite, formerly considered to be supergene because it was chalcocite and the latest mineral formed, has been demonstrated by Sales to be hypogene, a conclusion amply sub-

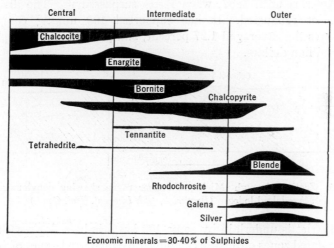

Fig. 13–5. Zonal range of economic minerals at Butte, Mont.
(*After Hart, Copper Res. of World.*)

stantiated by subsequent microscopic studies of polished ore surfaces. Some supergene silver enrichment has occurred in the peripheral zone.

Origin. The Butte deposits are replacements and fissure fillings formed by hydrothermal solutions emitted from the magmatic reservoir that supplied the intrusives. These solutions must have continued through the first three periods of faulting. It is striking that the first minerals in veins of *each system* are quartz and pyrite, slightly preceded by wall-rock alteration, followed by copper minerals ending with chalcocite, and not the earlier minerals in the first fissures and later minerals in the later fissures, as one might expect. This repetition of sequence in each of the later fissures suggests that the solutions underwent changes in composition while traversing the rocks. This is also shown by the sequence of rock alteration outward from each vein.

The central and intermediate zone mineralization would be classed as mesothermal; the peripheral as epithermal; thus one vein may contain both mesothermal and epithermal ores.

Lake Superior Copper District, Michigan

The Lake Superior district is the second greatest copper district of the world and the outstanding deposit of native copper. It forms a belt in Keweenaw Peninsula 2 to 4 miles wide and 100 miles long, of which 26 miles has been highly productive. First mined by a prehistoric race, it was discovered by the Jesuits in the seventeenth century, and modern mining started in 1845. It was the premier district of North America until 1887, when Butte surpassed it. The district is now almost exhausted. It has supported 100 mining companies, yielded ores that averaged 1.27 percent, and paid dividends approaching 350 million dollars.

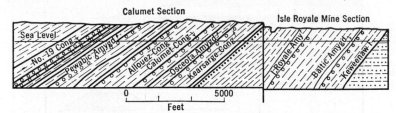

FIG. 13–6. Section through Michigan copper range showing conglomerate and amygdaloid lodes. (*Broderick, 16th Internat. Geol. Cong.*)

The district is underlain by some 400 basaltic lava flows with 20 to 30 intercalated zones of felsitic conglomerates and sandstones of Keweenawan age, totaling 25,000 feet. The copper belt beds lie in the south limb of a huge syncline that dips 40° northwest beneath Lake Superior and reappears in Ontario. To the south, the beds are faulted against Cambrian sediments. Many of the flows are amygdaloidal. The copper lodes occur near the middle of this Keweenawan series.

The Lodes. The copper lodes (Fig. 13–6) are of three types: 1, conglomerate lodes; 2, amygdaloidal lodes; and 3, fissure veins. The chief mineral is native copper. More than 90 percent of the production has come from six lodes, namely, the Calumet conglomerate, and the Baltic, Isle Royale, Kearsarge, Osceola, and Pewabic amygdaloids.

The *conglomerate lode* at Calumet is a reddish conglomerate lens from 5 to 20 feet thick composed of pebbles in a sandy matrix. Native copper fills the interstices and replaces matrix and pebbles; the ore portions are bleached. There is a little adularia, epidote, calcite, and quartz. The ore is localized in an ore shoot shaped like an upright funnel. It has been mined for 18,000 feet in length and for 9,000 feet down the 38° dip. It is opened by 10 shafts and 200 miles of workings. The tenor has gradually decreased with depth.

The *amygdaloidal lodes* occur in the upper, permeable parts of flows classed as coalescing, or fragmental. These tops are reddish owing to included hematite formed by oxidation during solidification; where mineralized, they are bleached. Native copper occupies the vesicles along with quartz, calcite, epidote, chlorite, adularia, sericite, pumpellyite, ankerite, and zeolites. The lodes average 13 feet in thickness and contain from 0.6 to 1.5 percent copper but average about

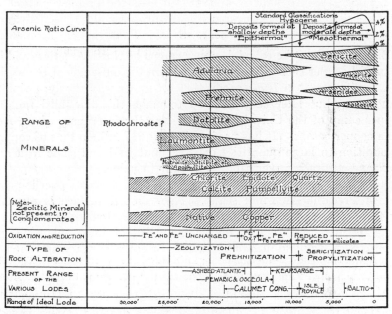

FIG. 13–7.   Zonal ranges of lode deposits and their mineralogical features.
(*Broderick, Econ. Geol.*)

0.8 percent. The Quincy lode has been followed 10,000 feet down the dip. The amygdaloids have furnished about half the total copper.

The few veins were noted for their "mass" copper — one mass weighed 500 tons; they also yielded beautiful specimens of native copper and native silver within clear calcite, interesting sulphides, and arsenides.

Origin. Butler, Broderick, and other Calumet and Hecla geologists consider that these unique deposits were formed by uprising hydro-thermal solutions given off from underlying basic intrusives. They believe that the structure is pre-ore and that the relations of copper beneath impervious cappings indicate rising rather than descending solutions (Fig. 13–7). The solutions, rich in copper, sulphur, and arsenic, were guided by the permeability of the conglomerate and amygdaloidal tops. The hematite of the lava tops yielded oxygen to combine with the sulphur and arsenic, causing reduction of copper to the native metal, and bleaching of the reddish flow tops. This concept is in direct contrast to an origin by descending, oxidizing, chloride waters, formerly held. The origin is likened to that of Corocoro, Bolivia, where native copper was deposited in red sandstones, and to Bristol, Conn., where chalcocite instead of chalcopyrite was deposited in red arkose.

Bisbee, Ariz.

Bisbee is one of the colorful copper camps of North America, and its beautiful copper carbonate specimens adorn the museums of the world. It was discovered in 1877; production got under way slowly, owing to remoteness, but later surged ahead to make this the third ranking copper camp of the United States. It was the forerunner of the Arizona copper camps. It has produced about $2\frac{1}{4}$ million tons of copper, in addition to large quantities of silver and gold and some lead and manganese, totaling over a billion dollars in value.

Some 5,000 feet of Paleozoic limestones and quartzites overlie peneplaned Precambrian Pinal schist. Their eroded surface is overlain by 4,000 to 5,000 feet of Cretaceous sediments bottomed by the Glance conglomerate. A laccolith, dikes, and sills of quartz monzonite to granite porphyry and of late Cretaceous or early Tertiary age, intrude the sediments. The mineralization is closely associated with this igneous activity. The beds have been lightly folded, tilted gently eastward, and broken into blocks by numerous normal faults. The prominent pre-mineral Dividend fault, which dropped Paleozoic sediments down against Precambrian schist exposed upon the northeast, was an important locus of metallization.

The Ore Deposits. The primary deposits are hydrothermal replacements of limestone and porphyry, along with minor contact-metasomatic ore bodies. The former are of four classes: 1, massive tabular sulphide replacements of limestone beds; 2, pipes of massive sulphide replacement in limestone; 3, silica breccia pipes; and 4, disseminated replacements in porphyry (porphyry copper) (Fig. 13–8).

The ore deposits are clustered around independent centers of metallization that are separate porphyry intrusives, of which the Sacramento Hill porphyry, itself a small porphyry copper deposit, is the most important. The deposits radiate outward from this center but are particularly localized along the Dividend fault. Certain favorable limestone beds within a 700-foot thickness have been selectively replaced by ore. The Campbell ore body, a pipe of massive sulphide, lies a mile east of the Sacramento Hill porphyry and is separated from its nearest satellitic ore body by 2,000 feet of unaltered limestones.

Mineralization. Near the porphyry centers the pure limestone beds are marmorized or altered to lime-silicate minerals. The impure beds are altered to calcium-aluminum-iron silicates with some magnetite,

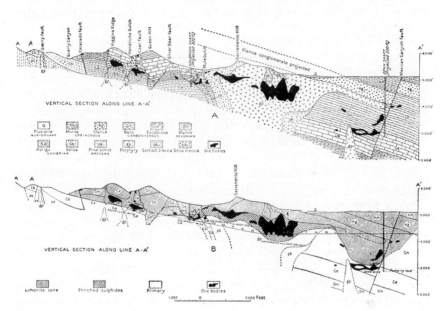

FIG. 13–8. E–W vertical section through Bisbee district, Arizona. *A*, geologic section; *B*, oxidized, enriched, and primary ores. (*Tenney, 16th Internat. Geol. Cong.*)

typical of contact metasomatism, and quartz, or chalcedony. No alteration accompanies porphyry that is not itself metallized. Replacement bodies occur in both altered and unaltered limestone.

The pipes are mainly massive pyrite masses with included chalcopyrite and bornite, and rare tennantite, zincblende, and galena, with an average grade of 4 percent copper. Farther out from the centers more blende and galena appear, and still farther out are silver-lead deposits with quartz and calcite. The disseminated ores consist of pyrite, chalcopyrite, bornite, and chalcocite.

Oxidation and Enrichment. The Bisbee ores have been deeply oxidized. For 25 years most of the copper came from oxidized ores to a depth of 2,200 feet. Rich supergene sulphide ores occur, and primary ores lie at greater depths eastward (Fig. 13–8*B*). The water table slopes gently eastward, and above it all the ores are oxidized or enriched, but oxidized ores also occur hundreds of feet beneath the water table. This is due to valley filling in the adjacent Sulphur Springs Valley, which caused a rising of the water level and drowning of oxidized ores. Also, supergene sulphide bodies persist above the water table, because of a protecting envelope of oxidation and an arid climate. Owing to the tilt of the host beds, the western ore bodies are all oxidized ores. In the unaltered limestones, sulphide enrichment was inhibited by the calcium carbonate; in altered limestones, where little carbonate remains, supergene sulphide enrichment took place.

Gossans envelop bodies of oxidized ore and cap the ore bodies. Oxidation collapse by removal of underlying sulphides has given surface depression markers.

Origin. The ores originated by replacement of altered limestone, fresh limestone, or porphyry by ore fluids that emanated from several porphyry apices. The resulting ores vary somewhat according to the center and to the beds replaced. The first and hottest mineralizing wave, nearest the Sacramento center, produced some contact-metasomatic minerals. The succeeding wave carried much iron, sulphur, and copper and little silica and alumina, replacing the earlier formed minerals and the limestone and porphyry. The last solutions also carried zinc, lead, and silver, which were deposited around the copper bodies or farther out. The various stages of mineralization grade imperceptibly into each other and represent contact-metasomatic to hypothermal, mesothermal, and even epithermal types of metallization.

Other Important United States Districts

Jerome, Ariz. The Jerome district includes the United Verde and United Verde Extension mines and is the seventh ranking copper district in the United States, with a total production of copper and precious metals of about half a billion dollars.

Paleozoic sediments have been stripped from an area exposing Precambrian greenstone schists cut by a diorite stock that forms the hanging wall of the United Verde ore body. The Verde fault, separating the two mines, has about 1,700 feet throw and drops the post-Paleozoic rocks against the schists.

The *United Verde deposit* is a cylindrical pipe about 800 by 700 feet across, and over 3,000 feet deep, enclosed in schist on the foot wall side of the Verde fault. It consists of massive iron and copper sulphides with quartz, dolomite, and chlorite that have replaced the schist. The average tenor is 5 to 6 percent copper with considerable silver. The upper part was oxidized to rich copper-silver ores to a depth of 160 feet, and supergene chalcocite extended to 600 feet.

The *United Verde Extension mine*, now closed, was discovered 41 years after Verde by following a geologic " hunch " and searching at depth for the down-faulted top of the United Verde pipe. The geologic history is revealed in Fig. 13–9. After Precambrian faulting and erosion, both segments were oxidized and enriched before being covered by Devonian sediments. Renewed post-basalt faulting buried the Extension deposit still deeper and permitted erosion of the up-thrown Verde portion to below the former Precambrian peneplain, giving rise to a new oxidized and enriched zone. The top of the Verde

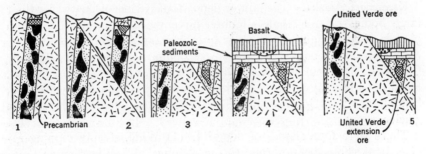

FIG. 13–9.   Stages in the faulting, erosion, and enrichment of the United Verde and United Verde Extension ore bodies, Jerome district, Arizona.   (*After Ransome, 16th Internat. Geol. Cong.*)

Extension deposit thus represents a " fossilized " pre-Devonian (Pre-cambrian) gossan and supergene enrichment zone. The top of the gossan was found at a depth of 700 feet and its bottom at about 1,150 feet. The gossan was abruptly underlain by bonanza supergene chalcocite ore that contained 25 percent copper to a depth of 150 feet, below which was partly enriched sulphide ore for another 300 feet before primary sulphides were reached. The vertical dimension of the unfaulted pipe must have been around 5,000 feet.

The deposits are hydrothermal replacements.

Magma Mine, Superior, Ariz.   The Magma mine, halfway between Ray and Miami, since 1914 has produced about 700 million pounds copper and 13 million ounces silver from ore that averages about 5 percent copper and 4 ounces silver.

The Magma " vein " occupies a fault zone 5 to 40 feet wide that cuts through inclined Paleozoic sediments and into two underlying diabase sills at least 2,500 feet thick. Quartzite and limestone constitute the walls in the upper levels and diabase in the lower levels.

The "vein" had an insignificant leached outcrop. The main ore body apexed at a depth of 450 feet, and to 800 feet it consisted of rich supergene chalcocite ore, below which is rich primary copper sulphide. In the upper levels the "vein" was about 500 feet long and 5 feet wide, but in the underlying diabase it is longer, wider, and richer. An eastern ore shoot extends from the 450 to the 1,900 level; a western one from the 1,200 to the lowest level. Its apex is sphalerite-galena ore which gives way shortly to bornite ore. Both shoots are sheathed by zinc-lead ore and sericitized wall rocks. The horizontal workings are 8,000 feet long, and the vein has been explored below 4,800 feet in depth. The deposit is a replacement lode with no open-space filling.

Kennecott, Alaska. The unique deposits of Kennecott, noted for their large bodies of rich chalcocite ore, have yielded around 1.2 billion pounds of copper and much silver from four mines, now closed.

The ores are confined to the lower dolomite belts of the Triassic Chitistone limestone, which overlies 5,000 feet of altered basaltic flows, and is overlain by thick shales. All beds dip gently northeastward, being part of a major anticline whose center is excavated by Kennecott glacier.

There were four classes of deposits: (1) Wide, steeply dipping replacement veins strike normal to the bedding. They bottom on limestone bedding planes, are widest at the base, and pinch stratigraphically upward from 200 to 500 feet above their base. Their vertical and horizontal dimensions are thus small, but they extend down dip for 4,000 feet. Few veins outcrop, and some do not reach within 1,000 feet of the surface. (2) "Flat" ore or tabular bedding replacements are localized by fissures restricted within certain beds. (3) Glacier ore, consisting of pieces of chalcocite eroded from the Bonanza vein outcrop, is incorporated as lateral moraine within a small glacier during its growth. The country rock is ice. (4) Slide ore, consisting of fragments of outcrop chalcocite, is incorporated within large talus slopes.

The ores consist of hypogene digenite and orthorhombic chalcocite, covellite, very minor bornite, rare traces of enargite, luzonite, chalcopyrite, and tennantite, and an occasional speck of sphalerite and galena. Pyrite, quartz, and other introduced gangue minerals are absent. The mineral assemblage is thus unusual. The sulphides constitute from 10 to 100 percent of the ore; pure chalcocite in masses of tens of thousands of tons are common. The Jumbo Main vein contained one chalcocite mass 80 feet wide, 150 feet long, and 400 feet down dip that averaged over 60 percent copper. The average tenor of all ores mined was 12.4 percent copper. Supergene sulphide enrichment is absent, but pre-Glacial partial oxidation to copper carbonate extends to a depth of 2,500 feet. The ores represent low-temperature, hydrothermal replacements.

Ducktown, Tenn. The Ducktown district yields copper and sulphuric acid made from the smelter fumes.

The deposits are massive sulphide replacement lenses in folded and metamorphosed Cambrian schists and graywackes. W. H. Emmons considers that the deposits were localized by former limestone lenses now replaced by ore, but the evidence for this appears inconclusive; fault zones appear to have been important localizers of ore. The deposits generally conform to the bedding of the schist but cross it in places. The ore itself has been slightly metamorphosed. The ore lenses are up to 300 feet wide and 3,000 feet long. The Burra Burra mine discloses its lens to a depth of 1,800 feet.

The primary ore consists of pyrrhotite, pyrite, and chalcopyrite, with minor zincblende, bornite, specularite, magnetite, quartz, chlorite, and iron-magnesium-lime silicates; the copper content averages about 1.5 percent. The deposits outcrop as strong gossans which were shipped for iron ore. A thin, rich, supergene sulphide zone from 3 to 8 feet thick and carrying 5 to 25 percent copper underlay the gossan. The deposits are high-temperature hydrothermal replacements.

Other Deposits. The *Holden* mine at Lake Chelan, Wash., is a large sulphide lens in Precambrian sediments associated with silicic Mesozoic intrusives. It consists of quartz, iron sulphides, chalcopyrite, and minor zincblende, galena, and gold. The *San Juan district*, Colorado, includes several epithermal quartz veins and limestone replacements that yield considerable copper from mixed base-metal ores. The *Rio Tinto* mine, Mountain City, Nev., contains rich supergene sulphides developed in a replacement lode in shale. The *Walker* and *Engels mines*, Plumas district, California, are high-temperature replacement lodes. The Walker carries iron and copper sulphides, with tourmaline, actinolite, and barite in schist. The Engels is at the contact of gabbro and quartz diorite and in igneous roof pendants and carries iron and copper sulphides, with exsolved ilmenite and magnetite.

CANADIAN COPPER DEPOSITS

Although Canada ranks third in copper production, most of it comes from ores in which other metals dominate. The Canadian Shield is the chief source. There are no supergene enrichment ores. The important districts or mines in order of rank are:

DISTRICT OR MINE	LOCATION	CHIEF ASSOCIATED METALS
1. International Nickel Co.	Sudbury, Ontario	Nickel-copper-platinum-gold
2. Noranda	Rouyn, Quebec	Gold-copper
3. Hudson Bay	Flin Flon, Manitoba	Copper-zinc-silver-gold
4. Howe Sound	Britannia, British Columbia	Copper
5. Sherritt-Gordon	Flin Flon, Manitoba	Copper-zinc-silver
6. Granby Consolidated	Copper Mountain, British Columbia	Copper
7. Waite-Amulet	Rouyn District, Quebec	Copper-gold
8 Aldermac	Rouyn District, Quebec	Copper-sulphur

Sudbury, Ontario. The famed Sudbury deposits, including those of International Nickel and Falconbridge mines, are primarily deposits of nickel ore but do produce about 1½ pounds of copper per pound of nickel, having a value of about one-third of that of the nickel. Over 200 million pounds of copper is produced annually. The deposits are described under Nickel.

Noranda, Rouyn District, Quebec. The Rouyn district of northern Quebec is dominated by the large Noranda mine, which illustrates massive replacement deposits. The enclosing rocks consist of altered Keewatin rhyolite and dacite extrusives cut by dikes of quartz diorite, syenite porphyry, and later gabbro. These rocks have been drag-folded over a width of 1,900 feet. The deposits are massive sulphide bodies in dacite breccias or tuffs, formed by hydrothermal replacement. There are also rich siliceous gold-quartz replacement veins. Some 24 sulphide lenses are known, dominated by the large H and lower H bodies (Fig. 5·4–39). The large H is 150 feet wide and extends to 1,300 feet; the lower H extends from the 1,600 level to the 2,600 level and is 600 feet across (Fig. 13–10). The ores are solid pyrrhotite-pyrite bodies with cores and fringes (Fig. 13–11) of chalcopyrite and minor zincblende and magnetite. The massive pyrrhotite of the upper levels has given way to pyrite in the lower levels. The ores average around 2 percent copper and $5.00 gold per ton, and the reserves amount to about 30 million tons. Profound alteration, consisting of silicification and chloritization, accompanied ore deposition.

Generally similar but smaller deposits occur in the nearby *Waite-Amulet, Normetal,* and *Aldermac* mines. At the Aldermac mine, mas-

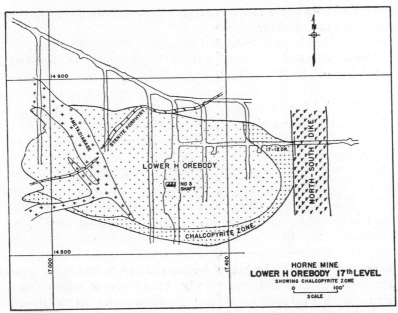

FIG. 13–10.   Plan of lower H ore body, 17th level, Horne mine, Noranda, Quebec, showing relation of chalcopyrite zone to sulphide mass and to diabase dike.   (*Suffel, Econ. Geol.*)

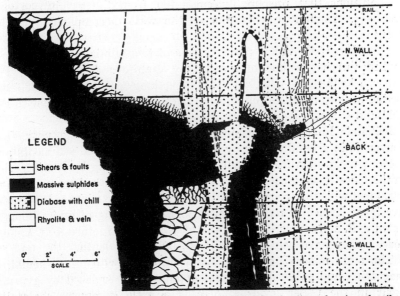

FIG. 13–11.   Part of 17th level, Horne mine, Noranda, Quebec, showing detail of sulphide contacts and relation to diabase dike.   (*Suffel, Econ. Geol.*)

sive pyrite lenses carry low copper and gold, and the pyrite was formerly used as sulphur ore.

Flin Flon District, Manitoba-Saskatchewan. This copper district in the Canadian hinterland includes the *Flin Flon* mine of the Hudson Bay Mining and Smelting Co., the *Sherritt-Gordon*, and the exhausted *Mandy* mine. The Flin Flon deposit reached production in 1930 and is now the third largest copper mine and the second largest zinc mine in Canada. It lies in altered Keewatin basic lava flows, tuffs, and coarser fragmental beds intruded by hornblende, lamprophyre, and younger quartz porphyry.

The ore body is a large, solid, sulphide lens, bifurcated in places, 2,600 feet long on the surface and up to 400 feet wide. At a depth of 900 feet it is 1,000 feet long and 35 feet wide, but at greater depths (3,500) it is still longer and wider. The core of the body is massive sulphide, and this is flanked by disseminated sulphides. The ore consists of an intimate mixture of iron, zinc, and copper sulphides, with some quartz, calcite, and schist residuals and rare arsenopyrite, galena, and magnetite. It averages 2.4% Cu, 5.2% Zn, 0.09 ounce Au, and 1.3 ounces Ag. In addition, cadmium, selenium, and tellurium are recovered. The reserves are listed at 22 million tons of 2.9 percent copper, and the daily production exceeds 6,000 tons. The deposit has been formed by hydrothermal replacement of a large shear zone by solutions that presumably emanated from the granitic magma reservoir. Intense sericitization of the walls accompanied ore deposition.

The *Sherritt-Gordon* deposit consists of two tabular sulphide bodies in Precambrian sediments invaded by granite and pegmatite. The two ore bodies are 4,200 and 5,200 feet long and 15 feet wide, consisting of base-metal sulphides carrying 2.6% Cu and 2% Zn, with low gold and silver.

Britannia Mine, British Columbia. The Britannia mine, near Vancouver, is in a band of steeply dipping metamorphosed sediments and igneous rocks that form a roof pendant 7 miles long by 2 miles wide in the Coast Range batholith. The deposits are in a shear zone up to 2,000 feet wide that occurs mainly in quartz-sericite schist. It includes five large lenticular replacement deposits consisting of schist impregnated with and replaced by ore and localized by structures of the shear zone. They have been followed to 2,500 feet in depth.

The main minerals are base-metal sulphides with minor hematite, magnetite, and gold. Wide lenses of anhydrite, replaced by gypsum, occur in the upper levels. There is a little supergene chalcocite, covellite, and marcasite. The copper content averages about 1 percent, but gold and silver add to the value. The reserves are large and the mill can treat 6,000 tons per day.

Copper Mountain, British Columbia. This deposit is a variety of small hypogene "porphyry copper." It consists of disseminated bornite and chalcopyrite with a little pyrite and magnesite in hydrothermally altered Copper Mountain porphyry and fragmental volcanic rocks. Most of the ore minerals are in little veinlets along with pegmatite minerals. Other minerals are hematite, galena, and sphalerite, and introduced gangue minerals include feldspars, pyroxene, apatite, epidote, zoisite, scapolite, garnet, and chlorite. The ore averages 2 percent copper, with some gold and silver.

MEXICAN COPPER DEPOSITS

Mexican copper production comes mainly from the extension of the Arizona belt into Sonora. Lower California accounts for about one-fifth, and the remaining production is by-product copper obtained from base-metal ores carrying silver and gold. Cananea and Nacozari in Sonora are the outstanding copper deposits.

Cananea. This notable district has yielded over 1.5 billion pounds copper, 25 million ounces silver, and total minerals exceeding 230 million dollars from porphyry copper deposits, brecca pipes, and minor limestone replacements and contact-metasomatic deposits. The region has a complex geologic history involving Paleozoic sediments, later extrusives, profound deformation, and porphyry intrusions, accompanied by widespread mineralization.

The *pipelike* bodies, of which the rich *Colorada* (Fig. 13–12) is outstanding, are the most important. They may have been the feeder funnels to overlying deposits. The unique Colorada ore pipe, according to Sales and Perry, is goblet shaped, apexing at the 500 level as a quartz-sulphide ring 600 feet by 500 feet across. The ore ring converges into the goblet stem of solid ore pipe at the 10th level, and the base of the goblet is pegmatitic quartz and phlogopite with sulphides. Also, stringers of quartz and sulphides penetrate the walls. The ore minerals are bornite, chalcopyrite, and molybdenite with minor pyrite, tennantite, covellite, and chalcocite. The sequence of events was a pre-porphyry breccia pipe intruded by quartz porphyry, which divided into upper and lower elements with the ore fluids trapped in the lower one and later injected along the sides and into the fractured walls of the upper element; subsequently the sulphides were fractured by later slumping. The unusual features suggest to Sales that the deposit may exemplify an ore magma injection.

The possibility of a replacement origin, characteristic of the overlying deposits, is not eliminated, however.

Other ore pipes are the *Capote,* extending from the 400 to the 1,600 level and underlying the Capote " porphyry copper " deposit, and the *Duluth* breccia pipe, an oval-shaped pipe 1,200 by 200 feet extending from the surface to 1,400 feet in depth. A minor pipe at the *Veta* mine extends upward into enriched " porphyry copper " ore. A low-grade porphyry copper deposit was opened up in 1944.

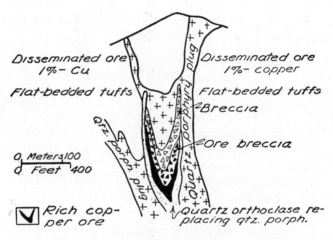

FIG. 13–12.   Section of Colorada pipe, Cananea, Mexico.
(*Emmons-Billingsley, A.I.M.E.*)

Nacozari. The Pilares mine at Nacozari, now largely exhausted, has yielded over 600 million pounds of copper and 4 million ounces of silver. Some 3,000 feet of volcanic rocks, intruded by monzonitic porphyry, have been dropped into a synclinal trough 2,000 feet by 1,000 feet across, bounded by circumferential faults (Fig. 13–13). Some call it a pipe. Apparently, the pipelike body has slumped, and the rock mass became intensely brecciated, particularly around the periphery and along cross faults. Hydrothermal solutions produced intense rock alteration, filled the breccia spaces, and replaced breccia fragments by ore minerals to form a circumferential deposit of iron and copper sulphides with specularite and scheelite. Supergene chalcocite enrichment occurred to a depth of 200 feet. The ore averages 2.5 to 3 percent copper.

El Boleo. The peculiar Boleo deposits in Lower California are mined like coal seams. Three clay ore beds, 1 to 4 meters thick, occur in gently sloping Pliocene marine tuffaceous sediments. According to

Touwaide they are the alteration products of tuff and are composed of 30 percent water, clay minerals, and calcite impregnated with pin points of chalcocite and minor bornite and chalcopyrite in a capillary matrix. Chalcocite veinlets cross clay layers and occur as free crystals. The ore averages 2½ percent copper and carries a little cobalt, nickel, and molybdenum. Some native copper, covellite, pyrite, zeolites, barite, and carbon (wood) are present, and the ore

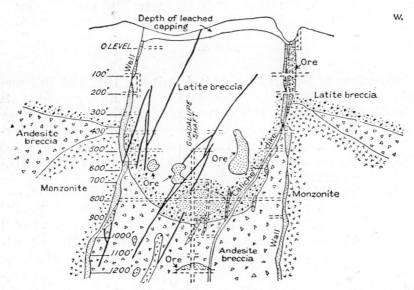

FIG. 13–13. Section across Pilares mine, Nacozari, Mexico, where marginal ore bodies are narrow. A section at right angles would show ore bodies many times wider. (*Wade-Wandke, A.I.M.E.*)

looks like a seam of dark clay. Its origin is a bit mystifying. A sedimentary origin is untenable, and a hydrothermal origin lacks support. Touwaide thinks the copper was extracted from the tuffs by subsurface waters and penetrated the clay beds where it was precipitated by hydrogen sulphide or organic matter in the clays.

SOUTH AMERICAN COPPER DEPOSITS

Chile and Peru are the two chief copper countries of South America, with minor production from Bolivia and Argentina; Chile is the second largest copper-producing country of the world. Its copper belt includes three of the large deposits of the world and a group of smaller ones, all lying on the western flank of the Andes. The Peruvian deposits are also on the western flank; those of Bolivia are in the heart of the Andes;

and the Argentina occurrences are on the eastern slopes. The important deposits are shown below.

MINE OR DISTRICT	LOCATION	PERCENT Cu	ANNUAL CAPACITY (1,000 tons Cu)	ORE RESERVES (million tons)
1. Chuquicamata	Chuquicamata, Chile	2.15	260	1,035
2. Braden Copper Co.	Rancagua, Chile	2.11	160	200
3. Andes Copper Co.	Potrerillos, Chile	1.28	70	65
4. Cerro de Pasco	Cerro de Pasco, Peru	—	25	
5. Corocoro	Corocoro, Bolivia	3.0		

Chuquicamata Mine, Chile. This deposit, the largest of all " porphyry coppers," lies in Antofagasta, 90 miles from the Pacific, at an elevation of 9,300 feet. It is a large, open-pit, shovel deposit of low-grade oxide ores which is noted for its huge ore reserves and the diversity of its ore minerals. Steam shovel operations started in 1915, but large production did not begin until 1923, when it became an Anaconda subsidiary.

According to A. V. Taylor diorite and granodiorite porphyry intrusions accompanied the intense folding and faulting of Mesozoic sediments and volcanics at the end of the Cretaceous. This was followed by andesitic eruptions, prolonged erosion, Pliocene uplift and volcanism, erosion, gravel formation, and recent uplift and aridity. Mineralization, localized by strong fissuring, accompanied the porphyry intrusions, and the Pliocene erosion exposed the deposits and oxidized them to soluble minerals that persisted in the extreme arid climate.

The deposit is a pear-shaped porphyry mass 2 miles long by 3,600 feet wide and over 1,500 feet deep (Fig. 13–14). Drilling has penetrated sulphides to a depth of 1,920 feet. The host rock is intensely sericitized and silicified, five types of alteration being recognized. The ore minerals occur in stringers and specks throughout the altered rock. Operations have been confined to the oxide ores, but the underlying sulphide ores are now being mined.

The primary mineralization apparently was by hydrothermal replacement of the porphyry, resulting in quartz, sericite, hematite, pyrite, and enargite, with rare tetrahedrite, chalcopyrite, zincblende, bornite, and molybdenite. Considerable covellite is present, but it and late alunite are probably supergene.

Weathering yielded an upper leached zone, not everywhere present, an underlying oxide zone that now constitutes the ore body, and an

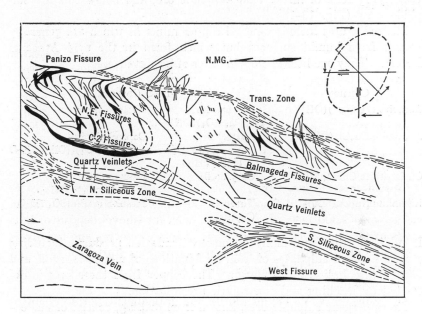

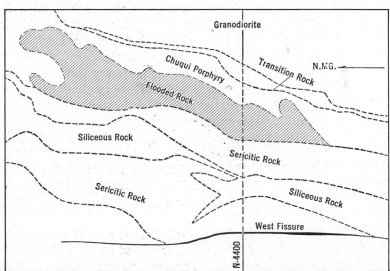

FIG. 13–14. Sketch map of structure pattern at Chuquicamata, Chile. (*Lopez, Econ. Geol.*)

underlying zone of mixed oxide-sulphide ore, and below this are sulphide ores (Fig. 13–15).

The oxide ores include unusual copper minerals which are generally absent from humid surfaces but which form in the arid Atacama Desert. They are listed below in order of importance.

COMMON		LESS COMMON
Antlerite $Cu_3(SO_4)(OH)_4$	Brochantite $Cu_4(SO_4)(OH)_6$	Native copper
Atacamite $Cu_4Cl_2(OH)_6$	Natrochalcite $Na_2Cu_4(SO_4)_4$- $(OH)_2 \cdot 2H_2O$	Cuprocopiapite $CuFe_4(SO_4)_5(OH)_4 \cdot 7H_2O$
Chalcanthite $CuSO_4 \cdot 5H_2O$	Chrysocolla	Lindgrenite $Cu_3(MoO_4)_2(OH)_2$
Krohnkite $Na_2Cu(SO_4)_2 \cdot 2H_2O$	Turquois Cuprite	Pisanite $(CuFe)SO_4 \cdot 7H_2O$

In addition, there are many uncommon oxidation products, including several hydrous sulphates of iron, combinations of magnesia, soda, and potash, also halite, and gypsum. The copper from these ores is extracted by leaching and electrolytic precipitation.

The supergene sulphide ores, known from drilling, consist of chalcocite-covellite ore with unreplaced remnants of the original sulphides. Between the oxide zone and the mixed zone there may be an intervening leached zone caused by a rising of the water level, which resulted in the leaching of the soluble oxide minerals.

The primary mineralization attended the porphyry intrusion. Oxidation and leaching during the damp middle Tertiary climate produced a leached capping and a deep supergene sulphide zone. The Pliocene uplift initiated erosion, removed most of the capping, and during the arid climate converted the upper part of the sulphide zone to the oxide ores.

Potrerillos Mine, Chile. This subsidiary of the Anaconda Copper Co. lies in the Atacama Desert at an elevation of 10,600 feet, 94 miles from the Pacific port of Barquito, and is a smaller replica, of the Chuquicamata mine.

According to W. S. March, Jr., an early Tertiary uplift accompanied by folding and faulting of Mesozoic marine sediments and volcanics was followed by erosion and early Tertiary andesite volcanics. Next came an intrusion of the Cobre quartz-diorite porphyry, accompanied by metallization of a fractured zone.

The ore deposit consists of stockworks and disseminations in shattered porphyry. The veinlets of the stockworks have been filled, the intervening rock has been partly replaced, and the host porphyry has been intensely sericitized and chloritized. There are three separate bodies: the central body (Fig. 13–16) is 2,000 by 1,115 feet in area and tapers downward to a depth of 1,510 feet; the south body is 1,650 by 295 feet in area and also tapers down to a depth of 1,950 feet. The north body is known only by drilling.

Weathering has yielded leached capping, oxide ore, and supergene sulphide ore. The primary mineralization in order of sequence was

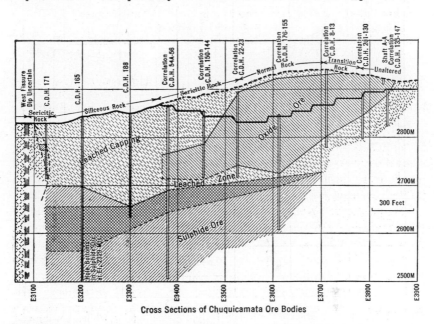

Cross Sections of Chuquicamata Ore Bodies

FIG. 13–15. Cross section of ore body at Chuquicamata, Chile, showing ore zones, open-cut levels, and drill holes. (*After Taylor, Copper Res. of World.*)

as follows: (1) rock alteration, (2) quartz, chalcopyrite, and pyrite, (3) quartz, pyrite, enargite, with minor tetrahedrite, zincblende, and galena, followed by calcite, dolomite, and enargite, and (4) quartz and molybdenite. The shallow leached zone contains diagnostic limonite, some oxidized copper minerals, and traces of residual arsenic and silver.

The oxide ore zone is 260 and 360 feet deep and averages 1.28 percent copper. The chief ore minerals are brochantite and antlerite with some copper oxides, carbonates, and silicates. The tenor is higher than that of the sulphide ore because it represents the oxidation of the richer upper part of the supergene sulphide zone. Below the oxide ore is another leached zone, up to 165 feet thick, formed by a fluctuating water table during its depression in response to oncoming aridity. The supergene sulphide zone averages 1.43 percent copper and is from 300 to 325 feet thick. The upper part is highly enriched to chalcocite ore with a little covellite and residuals of chalcopyrite and pyrite. The origin is similar to that of Chuquicamata.

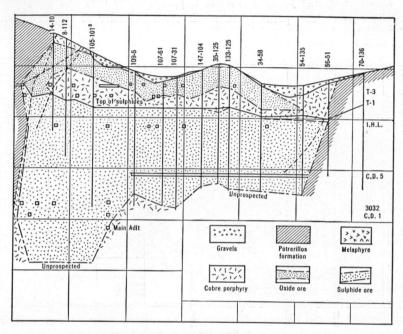

FIG. 13–16.   Cross section of Central ore body, Potrerillos mine, Chile, showing oxide and sulphide ore bodies.   (*After March, Copper Res. of World.*)

Braden Copper Deposit, Rancagua, Chile. This large Kennecott copper subsidiary lies east of Santiago at an elevation of about 9,000 feet.

Lindgren and Bastin consider that the Braden deposit (Fig. 13–17) lies in and around an explosive volcanic vent through igneous masses and that ore fluids followed the same paths. Their interpretation involves three periods of intrusion, three periods of hypogene metallization, uplift, erosion, and supergene enrichment. Tertiary volcanics intruded by andesite, followed by high-temperature mineralization, was interrupted by an explosion that produced a vent, later filled by fragmental material. Solutions rose around the periphery, producing copper protore with 0.5 to 1.5 percent copper. Then dacite and latite porphyry and breccia (Teniente breccia) penetrated the plug and walls, accompanied by tourmalinization of the breccia. A third upwelling of ore fluids produced the " bornite body," augmented the pyrite and chalcopyrite, and added tennantite, molybdenite, rare galena, and zincblende, huebnerite, and enargite, by replacement. Carbonates, barite, and anhydrite also came in, and the final deposition was minor sulphides, crystals of barite and quartz, and mammoth crystals of gypsum.

This complicated sequence of events does not carry the plausibility of Brüggen's simpler explanation of no explosion crater, downfaulting of regional breccias, and one long, single period of mineralization.

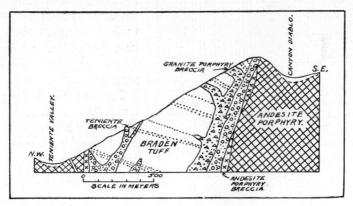

Fig. 13–17. Section across so-called Braden crater, Chile. (*Lingren-Bastin, Econ. Geol.*)

Uplift, tilting, and erosion initiated supergene enrichment that raised the ore grade of the upper levels by the addition of chalcocite and covellite, but in the deeper ores only a trace of enrichment is visible under the microscope.

The official ore reserves are given as 226 million tons of 2.18 percent copper. Molybdenite is also commercially extracted.

Cerro de Pasco, Peru. The Cerro de Pasco mines are at Cerro de Pasco, Morococha, Yauricocha, and Casapalca, the first being the main center of copper production. All are reached from Lima, Peru, Morococha being 75 miles distant and at an elevation of 14,800 feet and Cerro being 180 miles and at an elevation of 14,200 feet. The deposits are heavy sulphides that also yield lead, zinc, silver, gold, bismuth, cadmium, and arsenic.

At *Cerro de Pasco,* for 200 years after 1630 there was active production of silver. Copper mining was desultory from 1890 to 1915, when large-scale operations commenced.

McLaughlin and others describe the geologic setting as that of metamorphosed sediments (Silurian?) overlain by Carboniferous and Mesozoic sediments, dominantly limestones, and buried by thick early Tertiary extrusives. Intense folding and faulting was followed by silicic intrusions accompanied by mineralization. Faulting and erosion reveal Triassic limestone lying against early schists and later volcanics, cut by the Cerro quartz monzonite porphyry.

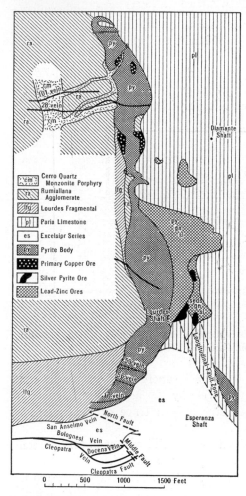

The copper deposits are of three classes (Fig. 13–18): (1) copper bodies in a massive pyrite deposit, (2) transverse copper-silver veins, and (3) copper mantos in limestone. The first is a massive pyritic body 5,000 by 400 feet, and 1,400 feet deep, that contains irregular bodies and transverse veins of copper ores carrying abundant enargite and some bismuth, silver, and gold. Other portions of the pyrite mass carry silver-lead-zinc ores and pyritic-silver ores. The transverse veins are curving fault-fissure, copper-silver veins near the pyrite body that are up to 3,000 feet long and 10 to 15 feet wide.

FIG. 13–18. Generalized geologic plan of ore bodies at 600-level (elevation 13,620 feet) of Cerro de Pasco mines, Peru. (*McLaughlin et al., Copper Res. of World.*)

The replacement mantos in limestone are all oxidized and carry mainly lead-zinc-silver ores with subordinate copper.

All the ore bodies formerly yielded rich supergene chalcocite ores that extended to a maximum depth of 500 feet beneath the oxidized zone. Oxidation also extends several hundred feet below the water table. Only trivial supergene silver enrichment is discernible. The silver remained in the oxidized zone, forming a thick mantle of siliceous silver-bearing ground called "pacos," of which parts are mined as low-grade silver ore.

The mineralization solutions were part of the early Tertiary igneous activity that gave rise to the volcanics and to the Cerro porphyry. They were localized along major fractures from which replacement

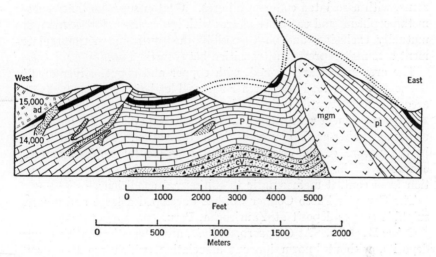

FIG. 13–19. Mantos on east-west projection, Morococha Mine, Cerro de Pasco, Peru. (*D. H. McLaughlin, Min. and Met., A.I.M.E.*)

proceeded. A rough zoning is evident, with the pyrite body on the vent margin, copper-gold ores being close, lead-zinc and pyritic-silver ores lying outside, and copper-free, lead-zinc-silver ores lying most remote from the igneous center.

For each 10 pounds copper, the ores yield 4.4 pounds lead, 1.14 ounces silver, and 2 cents in gold.

The *Morococha* deposits were mined for silver in pre-Spanish and colonial times but large-scale copper production started only in 1905.

The ores occur in altered Triassic limestone cut by Tertiary quartz monzonite intrusives and consist of sulphides of iron, copper, zinc, enargite, magnetite, and lead, sulpho-salts, tennantite-tetrahedrite, iron oxides, and rare aikenite, bournonite, cubanite, jamesonite, luzonite, and matildite. Quartz, carbonates, and contact-metasomatic silicates are associated.

The deposits, according to Graton and others, consist of veins, mantos, and "ore clusters." The veins are mineralized faults in altered limestone, volcanics, and porphyry, and occur in intersecting systems, each with distinct ore shoots. The mantos (Fig. 13–19) are replacements of limestone beds in the form of mantos, pipes, sill bodies, and contact bodies, all connected with feeding fissures. The ore clusters are localized groupings of veins and mantos.

The limestones near the monzonite display contact metamorphism, which grades into hydrothermal alteration of the monzonite and limestone, with associated commercial ores. Graton suggests that contact metamorphism, and commercial ores with hydrothermal alteration, are mutually exclusive, and that possibly the pyrite-magnetite replacement ores may represent a transition between the two.

The ores are notably zoned, with copper and copper-silver ores in the center, silver-lead ores farther out, and lead-zinc ores more remote, accompanied outward by decreasing intensity of rock alteration. The zonal arrangement and other features clearly point to the monzonite magma chamber as the source of the metallization. The rapid change in the nature of the ores vertically suggests telescoping, and the character of the mineralization indicates a complete range of formation from contact-metasomatic to epithermal conditions.

The Cerro de Pasco Copper Co. has developed a large and massive sulphide copper deposit at Yauricocha, Peru.

Other Deposits. At *Corocoro, Bolivia,* are interesting native copper deposits, without known igneous association, which are thought to have been deposited by the reducing action of ferric oxide in red Pliocene sediments. These grade upward into chalcocite-covellite-domeykite ores. At *Famatina, Argentina,* and near by, shallow fissure veins carry rich copper-silver ores of enargite, famatinite, and sulphides of iron, copper, zinc, and lead. Small deposits in Chile contribute about 5 percent of Chile's copper production.

African Copper Deposits

The outstanding development of copper in Northern Rhodesia has made this country the third ranking producer and the Rhodesian copper belt the most important single copper district in the world. This belt and the adjacent Katanga belt constitute a central African copper province. The deposits of both areas occur in parts of the same formation, but oxidized ores dominate in Katanga and sulphide ores in Rhodesia. Small copper deposits are also found elsewhere in North-

ern Rhodesia, in the Transvaal (Messina), in Namaqualand (O'okiep), in Southwest Africa (Tsumeb), Angola, and Kenya (Macalder).
The important deposits are shown below.

MINE OR DISTRICT	LOCATION	PERCENT Cu	ANNUAL CAPACITY (1,000 tons Cu)	ORE RESERVES (million tons)
Rokana	Northern Rhodesia	3.44	90	114
Roan Antelope	Northern Rhodesia	3.26	60	95
Mufulira	Northern Rhodesia	4.05	90	86
N'Changa	Northern Rhodesia	4.6	50	143
Chambishi, Baluba	Northern Rhodesia	3.4	—	46
Union Minière	Katanga	6.4	180	85
O'okiep Copper Co.	O'okiep, Namaqualand	2.0	21	—
Messina Copper	Messina, Transvaal	2.1	12	—
Tsumeb	Tsumeb, Southwest Africa	4.0	10	—

The Rhodesian Belt. This district has emerged, since 1927, from a forest wilderness to a center of industrial activity. Its unique disseminated copper deposits are more dissimilar than similar to the "porphyry coppers." The area is 140 miles long by 40 miles wide adjacent to the Belgian Congo, and includes four large producing mines, one developed mine, and several smaller deposits, all generally similar. They represent large-scale disseminated replacement deposits.

The formations include some 5,000 feet of continental sandstones, arkoses, and shales of the Mine series, which overlies a basement complex of ancient Precambrian schists intruded by granite. Above them is 13,000 feet of Kundelungu sediments. All are supposedly cut by granites, probably of two ages. The Mine series has been folded into pitching anticlines and synclines, of which erosion has left only synclinal remnants on a peneplaned surface. Post-folding basic sills occur in the Mine series.

Deposits. The chief deposits all occur in the Lower Roan group of the Mine series in a narrow belt 70 miles long. There are eight proved deposits, consolidated into four operating units — the Roan Antelope, Rokana, Mufulira, and N'Changa.

The deposits consist of one or more beds impregnated by specks of copper sulphides. Their outstanding features are the persistency and uniformity of the metallization, over a uniform width for great horizontal distances. Being bedded deposits, their shapes are those of the parts of the synclines that are metallized. At Roan Antelope (Fig.

13–20) the metallized bed laps around the nose of a plunging syncline and extends along both limbs for a total proven length of 5 miles and an indicated length of another 4 miles. The ore bed is 25 to 30 feet thick and has been tested to a depth of 2,000 feet. The N'Kana ore bed, in the north limb of another syncline, has been developed for a continuous length of 8,000 feet and, after a break, for another several thousand feet. Its width is 31 feet and tested depth is 2,700 feet. The Mufulira contains three dipping ore beds (Fig. 13–21) that coalesce to form a deposit 100 feet thick. It is 7,000 feet long and is known

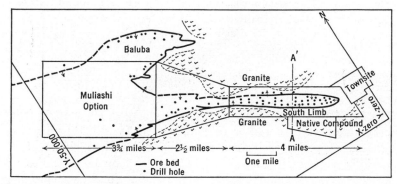

FIG. 13–20. Sketch of outcrop of Roan Antelope ore bed, northern Rhodesia (*Bateman-Sharpstone, Econ. Geol.*)

to extend 3,000 feet deep. The other limb of this syncline crops out in Katanga, 15 miles distant, and is ore-bearing there also. The other deposits are similarly located in synclines.

Ores. The ore consists of minute disseminated specks and rare veinlets of sulphides in feldspathic sandstones and "shales." The only gangue is the host rock. The minerals are chalcocite, bornite, and chalcopyrite; there are traces of pyrite, covellite, linnaeite, zincblende, carrollite (?), and two unidentified minerals. Both hypogene and supergene chalcocite are present. The iron content is extremely low. The order of hypogene deposition was of diminishing iron and sulphur, with chalcocite the latest mineral. At the Roan Antelope, chalcopyrite dominates in the deeper part of the syncline and chalcocite dominates nearer the nose. In general, chalcopyrite and bornite are more abundant at depth and chalcocite less so.

Oxidation and leaching is universal above the water table, which attains a depth of 200 feet. The oxidized zone generally contains some oxidized copper minerals and limonite "boxworks," diagnostic of former sulphides. There is also erratic deep copper oxidation to depths of 2,000 feet beneath the water table, formed during an earlier

period of aridity, when the water table stood lower than today.   Two
deposits consist of mixed oxides and sulphides.   There is some supergene sulphide enrichment, but much hypogene chalcocite is present.
The paucity of pyrite, which upon oxidation yields solvents for copper, was a deterrent to widespread, pervasive sulphide enrichment.

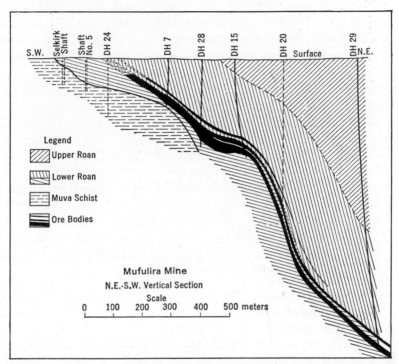

Fig. 13–21.   Vertical section across Mufulira ore body, northern Rhodesia, showing
three ore beds.   (*Gray, Econ. Geol.*)

The remarkable uniformity and wide distribution of the metallization, sharply confined to specific beds, early led to a suggestion of a
syngenetic origin.   Field and microscopic study, however, shows that
the metallization was later than the folding and was a hydrothermal
replacement of the rock silicates.   The sulphides also show a definite
paragenetic sequence.   The ore presumably was localized by certain
permeable beds and emanated from the magma chambers that supplied the later intrusives.

Katanga. The rich Katanga belt, just across from Rhodesia, follows
the regional structure for 300 kilometers.   The geology is similar to
that of Rhodesia, except that the folding has been closer, faulting more
abundant, and nappe thrusting important.   In addition, erosion has

penetrated less deeply, and anticlines and more of the higher formations are exposed.

Deposits. There are some two score commercial copper deposits, of which all except two are oxidized, and a radium-uranium deposit at Shinkolobwe.

The oxidized deposits lie at or near the surface and range in size from small bodies to large power-shovel pits. Most of them are confined to dolomitic beds of the Upper Roan group of the Mine series, about 200 meters thick. The oxidized minerals are malachite and chrysocolla, with subsidiary cuprite and native copper, and these form films, veinlets, and replacement masses. The oxidized minerals are transported; i.e., few have been formed *in situ* from pre-existing sulphides. Presumably the copper has come from higher levels now largely eroded.

Most of the deposits have no sulphide roots, and in most of those that do have them, the oxide ores pass directly into primary sulphides without an intervening supergene sulphide zone. The dolomitic host rock tends to precipitate the migrating copper as carbonate and prevent supergene enrichment. However, a little supergene chalcocite and covellite do occur in siliceous rocks. The few primary sulphides visible are disseminated chalcopyrite, bornite, pyrite, and linnaeite, similar to the Rhodesian sulphides.

The Prince Leopold mine, Kipushi, is a sulphide pipelike replacement deposit in limestone, extending up into Kundelungu beds. Rich ores consist of bornite, chalcopyrite, galena, considerable supergene chalcocite and silver, but no cobalt or gold. Some disseminated sulphide ores, similar to the Rhodesian ores, have been developed near Tshinsenda. The reserves are given as 5 million tons of copper.

The Shinkolobwe uranium deposit, which carries copper and cobalt sulphides and also nickel, is considered to be a higher-temperature phase of the copper mineralization.

Other African Deposits. In *Namaqualand,* the O'okiep and East O'okiep mines contain low-grade copper sulphide ores disseminated in norite, for which both a magmatic and replacement origin have been claimed. At *Messina, Transvaal,* hypogene chalcocite, bornite, and chalcopyrite, with minor pyrite and specularite, and quartz, feldspar, prehnite, delessite, and zeolites, occur in granite gneiss altered to zoisite quartz. The ores extend to 2,300 feet in depth. At *Tsumeb, Southwest Africa,* are two hydrothermal replacement lenses in dolomite adjacent to aplite, which consist of massive aggregates of chalcocite, galena, and sphalerite with some enargite, germaninite (?), stibioluzonite, and vanadium compounds. Limited oxidation has yielded an astonishing variety of oxidized minerals. At *Insizwa, South Africa,* are small magmatic copper sulphide deposits carrying cubanite; and similar low-grade nickel-copper deposits occur in the Bushveld rocks at *Pilansberg.* Massive gold-copper lenses carrying pyrrhotite, chalcopyrite, and pyrite have recently been disclosed at *Macalder, Kenya.*

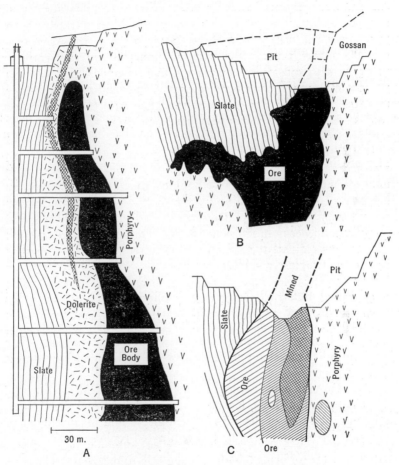

FIG. 13–22. Cross sections of Rio Tinto ore bodies, Huelva, Spain. *A,* Perrunal deposit; *B,* San Dionisio Lode, central section; *C,* South Lode. (*After D. Williams, Inst. Min. and Met.*)

OTHER IMPORTANT DEPOSITS

Rio Tinto (Huelva) Spain. The pyrite deposits of the famed Rio Tinto or Huelva district (Fig. 13–22) of Spain-Portugal have been mined for 3,000 years, first for gold and then for copper and sulphur.

They are the largest pyritic copper deposits of the world, having produced over 200 million tons of pyrite ore and 5 million tons of copper, with as much still in the ground.

The ores are associated with sheared quartz porphyry intrusives in Paleozoic slates, and both are cut by diabase dikes. The deposits lie in porphyry, in slate, or at contacts. Some 50 ore bodies have been worked, mostly in huge open cuts. Eight large ones are concentrated near Rio Tinto, and a similar group at Tharsus.

The deposits are collossal lenses of solid sulphide flanked by disseminated sulphides and sericitized wall rock, merging into fresher rock with diminution of sulphides. The San Dionisio (Fig. 13–22B), Eduardo, and South Lode (Fig. 13–22C) are connected bodies 3,000 meters long, up to 2,500 meters wide, and 500 meters deep. As the deposits were formed at different levels, erosion has removed some, disclosed others, and not reached others.

The ore is massive pyrite with 1 to 5 percent silicates, carrying chalcopyrite, minor sphalerite and galena, traces of tetrahedrite, enargite, and pyrite, and uncommon arsenides, selenides, and antimonides. A little chalcocite and covellite occur in a shallow supergene zone. The ores yield sulphur, copper, a little lead, zinc, gold, and silver, and small amounts of nickel, cobalt, and other metals. They consist of smelting ores, leaching ore, copper-sulphur ore, and sulphur ore.

Their origin is controversial. A magmatic origin has been maintained with vigor, and a hydrothermal origin has been strongly advocated. The associated hydrothermal rock alteration and the nature of the ore, however, indicate that these interesting deposits do not differ from other massive pyritic deposits of undisputed hydrothermal origin.

Russia. Russian copper comes from many widely scattered small deposits, chiefly in the Urals, Kazakstan, central Asia, and the Caucasus. The important types of deposits are sediments and magmatic copper-nickel, contact-metasomatic, and disseminated replacements.

Magmatic copper-nickel deposits in norite occur in Norilsky and Kola Peninsula. Pyrrhotite, chalcopyrite, and pentlandite form erratic low-grade disseminations and veinlets in peridotite, or form small aggregated bodies of 1 to 3 percent copper ore carrying 0.5 percent nickel and some platinoids.

Contact-metasomatic deposits associated with granodiorite intrusions in the Urals, central Asia, and Kazakstan yield ores carrying 1 to 4 percent copper in chalcopyrite-magnetite associations. This type constitutes 3 to 4 percent of the known Russian reserves.

Massive pyritic replacement bodies of the Rio Tinto type yield most of the Russian copper. These bodies range from 50 to 600 meters in length, are 50 meters wide, and carry 1.4 to 2.7 percent copper. Some deposits are secondarily enriched to depths of 100 to 250 meters. The reserves are large.

The *Djeskasgan deposits* in Kazakstan, one of the two largest copper reserves of Russia, are impregnations in Carboniferous sandstone, of the Rhodesian type, forming gently dipping lodes extending outward from feeder faults. Forty square miles of these beds have been surveyed and 0.9 mile has been explored by drilling, disclosing 37 ore bodies. The tenor of the workable ore is stated to be 2.17 percent copper, and 3.25 million tons of copper reserves are claimed.

Porphyry-copper type deposits have been developed in Asia, particularly around Kounrad. They consist of disseminated yellow sulphides and supergene chalcocite in small porphyry masses and invaded rocks. Only shallow, enriched zones, averaging around 1.1 percent copper, are workable. The protore carries 0.5 to 0.7 percent copper. Several zones have been explored and are claimed to aggregate 7.9 million tons of metallic copper, but noncommercial protore is included in these estimates. The Kounrad deposits are stated to contain 2.6 million tons of copper in an area of one-quarter square mile, which must mean that a vast amount of unprofitable protore is included. Similarly 3.5 million tons of copper is claimed to be in the Tashkent deposits. The Transcaucasian porphyry deposits of Agarak contain copper-molybdenum ore.

The *Permian red beds* of the Urals and Don Basin contain small sandstone layers 0.1 to 0.4 meter thick, carrying disseminated copper minerals with an average tenor of 1.5 to 1.9 percent copper. Similar deposits occur in the Silurian red beds of the Lena region.

Yugoslavia. The important deposits of Bor consist of three massive bedlike bodies in hydrothermally altered andesite porphyry. The largest consists of pyrite and enargite, with minor luzonite and famatinite, and supergene chalcocite and covellite, carrying 5 to 6 percent copper and 1 to 4 grams gold. The deposit has yielded over 4 million tons of ore from an open cut 350 feet deep. The second ore bed yields 2,000 tons of 5 to 6 percent ore daily, and a third body is pyrite-chalcopyrite ore, carrying 1 to 2 percent copper. The pre-war annual production was about 45,000 tons of copper metal.

Germany. Since A.D. 1150 most of Germany's copper has come from the Kupferschiefer of Mansfeld, where a thin but widespread cupriferous bed of black shale 40 centimeters thick has been mined for its 2 to 3 percent copper content. It occurs in folded Permian beds between underlying " Rothliegendes " and overlying limestone, gypsum, and salt. The ore minerals constitute about 12 percent of the ore and organic substances 15 percent. The chief ore minerals are bornite and chalcocite with some chalcopyrite, galena, sphalerite, and tetrahedrite. The ores also contain small quantities of silver, nickel, cobalt, selenium, vanadium, and molybdenum. The bed is cut by faults along which occur pyrite, chalcopyrite, niccolite, bornite, gypsum, and calcite.

The origin of the ore bed is controversial. Advocates of a syngenetic sedimentary origin hold that the ore was deposited in a shallow sea with decaying vegetation into which cupriferous surface waters were discharged and deposited and that the other metals present were supplied from the desert shores. However, the mineralogy and mode of occurrence of the minerals give most weight to a hydrothermal origin.

Finland. Glacial erratics containing copper were traced 50 kilometers and led to the discovery of the Outokumpo deposit beneath a swamp. The ore body is in brecciated quartzite and conforms to its structure. It is the shape of a long plate 6 to 14 meters thick, 300 to 400 meters wide and at least 3,700 meters long. The breccia fragments are cemented by chalcopyrite and pyrrhotite and impregnated by pyrite and chalcopyrite. The ore consists of 30 percent pyrite, 15 to 16 percent pyrrhotite, 11 to 12 percent chalcopyrite, 1 percent zincblende, and 42 percent quartz or quartzite. The tenor is 4% Cu, 0.8% Zn, 25% S, 0.8 gram Au, 12 grams Ag, and some Ni and Co. The reserves are 20 million tons of ore with 800,000 tons of copper, which makes it one of the largest individual copper deposits of Europe outside of Russia. Yearly production is 400,000 tons of ore. The pyrite concentrates are electrolytically smelted in the largest electric furnaces in the world.

Japan. Japan receives its annual production of about 55,000 tons of copper from 110 small mines. The ores are mainly chalcopyrite, rarely tetrahedrite or enargite. There are three main classes of deposits: (1) replacement (Besshi and Hitachi mines; " black ores "), (2) fissure veins, and (3) contact-metasomatic.

Other Deposits. At *Boliden, Sweden,* is a huge lens of gold-silver-copper-arsenic-sulphur ore, and at *Falun* are other pyritic masses carrying some copper. At *Skouriotissa, Cyprus,* a large pyrite lens 1,850 by 800 by 140 feet, containing 2.1% Cu, is being mined. *Australia* produces a little copper from gold-copper deposits, such as *Mount Morgan, Queensland,* and from huge pyritic lenses at *Mount Lyell, Tasmania.* The *Matahambre mine, Cuba,* contained a large replacement lode with fat bulges, carrying chalcopyrite and pyrite with a little galena and blende. Copper deposits opened up at *Ergani,* *Turkey,* now yield about 8,500 tons of metal per year.

Selected References on Copper

Copper Resources of the World. 2 Vols. XVI Int. Geol. Cong., Washington, 1935. *Vol. 1 covers the geology of all important United States occurrences, and Vol. 2 covers the rest of the world. Bibliography.*

Preliminary Study of Vein Formation, Butte, Mont. R. H. SALES and C. MEYER. Econ. Geol. 44:465–484, 1949. *Theories of formation.*

The Story of Copper. W. DAVIS. Appleton-Century, New York, 1924. *Popular descriptions of occurrence, mining, smelting, refining.*

Sulphide Enrichment. W. H. EMMONS. U. S. Geol. Surv. Bull. 625, 1917. *Supergene enrichment in various copper deposits.*

The Porphyry Coppers. A. B. PARSONS. A. I. M. E., New York, 1933. *A story.*

XVI International Geological Congress, Washington, 1933. Guidebooks: 2, **Ducktown;** 14, **Chino;** 23, **Butte;** 17, **Bingham;** 14, **Bisbee;** 27, **Lake Superior;** 19, **Colorado.** *Good summaries.*

Mineral Deposits of the Canadian Shield. E. L. BRUCE. Macmillan, Toronto, 1933. *Brief descriptions of Canadian deposits.*

Recent Contributions to the Geology of the Michigan Copper District. T. M. BRODERICK, C. D. HOHL, and H. N. EIDEMILLER. Econ. Geol. 41:675–725, 1946. *Data later than in the Copper Resources of the World.*

The Ajo Mining District, Arizona. J. GILLULY. U. S. Geol. Surv. Prof. Paper 209, 1946. *Detailed geology of a "porphyry copper."*

Hydrothermal Alteration in the "Porphyry Copper" Deposits. G. M. SCHWARTZ. Econ. Geol. 42:319–352, 1947. *Characteristic rock alteration accompanying disseminated copper ores.*

Foreign Ore Reserves of Copper, Lead, and Zinc. W. P. SHEA. Eng. and Min. Jour. 148:53–58, 1947. *A statistical review.*

General References 1–6, 8–10, 16; 7, pp. 93–99; **12,** pp. 61–71; **13,** pp. 263–278.

Lead and Zinc

Despite chemical dissimilarity, geological conditions favor the formation of lead and zinc together. The world over, the companion sulphides galena and zincblende, or their oxidation products, yield most of these important metals of modern industry. They are also associated with copper and other base sulphides; less commonly, lead and zinc occur separately. The association is so general, however, that the two are considered together.

History. Lead has been known since the beginning of history. Lead water pipes found in Pompeii show that its present-day use for plumbing was known to the Romans. The Chinese used it for money and debasing coins before 2000 B.C. despite Pliny's statement that it was discovered by Midas, King of Phrygia, about 1000 B.C. Ancient silver-lead deposits were worked in the Mediterranean countries, India, China, Persia, and Arabia. The famed Laurium deposits of Greece were worked in 1200 B.C. Lead was used by the ancients for ornaments, coins, solder, bronzes, vases, and pipes. Later it was extensively mined in Spain, Greece, the Pyrenees, the Harz, and Silesia.

Zinc was discovered as a metal in 1520, but bracelets filled with zinc were found in the ruins of Cameros, burnt in 500 B.C. The Greeks and Romans unknowingly used it to make brass, as they found that copper melted with smithsonite resulted in a metal more yellow than bronze. In the sixteenth century zinc was imported into Europe from India and China, and mining of zinc in Europe began in 1740. In the Americas, lead was mined by the pre-Spaniards, and then by the Spaniards in Bolivia and Peru. The first real mining in the United States was in Missouri in 1720. The first production of zinc was at the Washington arsenal in 1838 from New Jersey ores.

Production. The large industrial demand for lead and zinc has caused exhaustion of the earlier mined small deposits throughout the world, and large steady production now comes from relatively few regions. The following production estimates are of interest.

	LEAD	ZINC
Production since 1800 (million tons)	80	57
Percent production, 1800–1900	31	22
Percent production, 1900–1925	37	38
Percent production last 15 pre-war years	32	40

This shows that, of the total production since 1800, that of the last 15 pre-war years about equaled that of the preceding quarter of a century and also equaled the entire nineteenth century production of lead and was nearly double that of zinc. Nearly as much lead (48%) and more than as much zinc (55%) have been mined between World War I and World War II as in all preceding history.

The annual production in the 1940's has ranged between 1.6 and 2 million tons of each metal. The chief producing countries in order of rank and their average ratio to world production follows.

LEAD	PERCENT OF WORLD	ZINC	PERCENT OF WORLD
1. United States	27	1. United States	47
2. Australia	13	2. Canada	12
3. Canada	13	3. Russia	6
4. Mexico	12	4. Belgium	6
5. Russia	9	5. Australia	5.5
6. Germany	4	6. Great Britain	5
7. Yugoslavia	4	7. Poland	4
8. Peru	3	8. Mexico	3
9. France	3	9. Germany	2
10. Spain	2.5	10. France	2
11. Italy	2	11. Norway	2
		12. Japan	1.5

The countries with the greatest reserves are the United States, Canada, Australia, and Mexico. About two-thirds of the world's zinc and lead are each derived from four countries.

In the United States, the chief ranking states and their percentage production (in parentheses) for lead are: 1, Missouri (41); 2, Idaho (18); 3, Utah (9); 4, Arizona (7); 5, Colorado (5); 6, Oklahoma (4); 7, California (3); and Montana, Kansas, Nevada, and Virginia between 1 and 3 percent. For zinc: 1, Tri-State (Missouri, Oklahoma, and Kansas) (24); 2, Idaho (12); 3, New Jersey (11); 4, Arizona (8); 5, Colorado and New Mexico (6); 6, New York (5.5); 7, Utah (5); 8, Tennessee and Nevada (4).

Missouri has been the leading state in production of lead since about 1905.

Uses. Lead and zinc rank next to copper as essential nonferrous metals in modern industry. Normally, the chief uses, in percentage of total consumption in the United States, are as shown below.

LEAD	PERCENT	LEAD — Cont.	PERCENT
1. Red lead and litharge	24	10. Bearing metal	2
2. Cable covering	13	11. Other uses	15
3. Storage batteries	10		
4. Tetraethyl lead	9	ZINC	
5. White lead	8	1. Galvanizing	40
6. Solder	6	2. Die castings	26
7. Pipe	5	3. Brass making	19
8. Sheet lead	5	4. Rolled zinc	12
9. Ammunition	3	5. Other uses	3

The lead used in items 2, 3, 7, 8, and 10 largely finds its way back as secondary metal. The use in pigments is being displaced by lacquers and also by lithopone, zinc, titanium, and barium oxide pigments. Antimonial or hard lead is used chiefly for items 1, 2, 6, 7, and 8.

Zinc galvanizing, the process of coating steel with zinc to prevent rusting, is done by dipping sheets into molten zinc. "Sheradizing" is the coating of sheets by zinc vapor. Zinc constitutes 30 to 40 percent of brass and is also a component of certain bronzes and other alloys. Zinc die castings are used for automobile carburetors, pumps, hub caps, and drills, and for many other purposes. Rolled zinc is used for glass jar tops, battery cans, and similar purposes. Zinc also finds application in precipitating gold, and in medicines and chemicals. Little secondary zinc is recovered.

MINERALOGY, TENOR, AND TREATMENT

The mineralogy of lead and zinc ores is simple. Only three lead and six zinc minerals are commercial sources of these metals; two are sulphides and four are oxidation compounds.

Lead Minerals		Percent Pb	Zinc Minerals — *Cont.*		Percent Zn
Galena	PbS	86.6	Hemimorphite	$Zn_4Si_2O_7$	
Cerussite	$PbCO_3$	77.5	(Calamine)	$(OH)_2 \cdot H_2O$	54.2
Anglesite	$PbSO_4$	68.3	Zincite	ZnO	80.3
			Willemite	Zn_2SiO_4	58.5
Zinc Minerals		**Percent Zn**	Franklinite	(Fe, Zn, Mn)	
Zincblende	ZnS	67.0		$(Fe, Mn)_2 O_4$	15–20
Smithsonite	$ZnCO_3$	52.0			

Associations. Galena and zincblende may occur separately but are generally associated, and zincblende rarely occurs without galena. In oxidized ores, however, the two become separated (Chap. 5·8). Galena is rarely free from contained silver; without silver it is called "soft lead." Both minerals may carry gold. Cadmium is a common associate of zinc, and bismuth and antimony of lead. Pyrite and chalcopyrite are common associates, and silver minerals are likely to be present.

Tenor. The tenor of lead ores depends upon the associated metals. In Missouri silver-free ores average 2 to 4 percent lead. Many 2 to 3 percent lead ores containing silver are profitably mined; commonly they average 6 to 8 percent, and rich ores of 20 percent or more are rather common. Zinc ores average 2 to 3 percent zinc in the Tri-State district and range from 4 to 12 percent elsewhere.

Treatment. Simple galena and sphalerite ores of low grade undergo gravity concentration by jig or table, or they may be concentrated by flotation. Complex ores undergo differential flotation, largely eliminating pyrite and yielding lead concentrates low in zinc, zinc concentrates with a little lead, and copper concentrates.

The lead concentrates are smelted, and the molten lead carries down any silver and gold present. The lead is then desilverized and purified. Sphalerite concentrates are roasted, and the zinc is volatilized, condensed to liquid zinc, and cast into slabs. In electrolytic treatment the zinc is dissolved by sulphuric acid from roasted concentrates, copper and cadmium being removed from the solution, and then the zinc is electrolytically deposited.

Kinds of Deposits and Origin

The principal classes of lead and zinc deposits and some examples of each are:

Contact metasomatic: *Hanover* and Magdalena, N. Mex.
  Cavity fillings:
    a. Fissure veins — San Juan, Colo.; *Burma.*
    b. Breccia — *Jefferson City* and *Mascot*, Tenn.
    c. Cave fillings — Mississippi Valley.
    d. Pitches and flats — Upper Mississippi Valley.

Replacement:
  a. Massive — *Leadville,* Colo.; *Bingham* and *Tintic,* Utah; *Sullivan* mine,
       British Columbia; *Flin Flon,* Manitoba; *Cerro de Pasco,* Peru; *Santa
       Eulalia* and Sierra Mojada, Mexico; Trepca, Serbia.
  b. Replacement lodes — *Park City,* Utah; *Coeur d'Alene,* Idaho; *Franklin
       Furnace,* N. J.
  c. Disseminated — *Tri-State-district; southeastern Missouri;* Silesia.
Surficial oxidation — Northern Mexico.

Origin. Most lead and zinc occurs as cavity fillings and replacements formed by low-temperature hydrothermal solutions. Prevail-
ingly they occur in limestones or dolomites. Considerable difference
of opinion has existed as to the origin of many zinc-lead deposits in
limestone, such as those of the Tri-State district. Three views are current. They were formed by (1) descending surface waters, (2) ascending artesian meteoric waters, and (3) hydrothermal solutions of igneous
derivation. Evidence has steadily been accumulating in favor of the
last theory of origin.

In many regions lead and zinc deposits have been oxidized; much
of the Mexican production, for example, comes from oxidized ores.

WORLD DISTRIBUTION OF LEAD AND ZINC

Despite wide distribution, most of the world production comes from
relatively few regions. The belts, however, are less well defined than
those of gold and copper.

North America. The world's greatest concentration of zinc-lead ores
occurs in the Mississippi Valley region, centering around the Tri-State
district of Missouri, Oklahoma, and Kansas and diminishing northward into Wisconsin. A parallel zinc belt lies in the lower Appalachians, embracing the Mascot-Jefferson City region of Tennessee and
Austinville, Va. A great isolated lead district lies between these belts
in southeastern Missouri. Two other small isolated areas occur at
Franklin, N. J., and Edwards, N. Y. A Rocky Mountain province
of lead-zinc ores centers in Colorado and extends into New Mexico and
Utah. The Coeur d'Alene district of Idaho is a rich lead-zinc province
that extends northward into Canada and embraces the greatest lead-
zinc mine of the continent. In Manitoba zinc-copper ores center
around Flin Flon, and similar ores lie around Noranda, Quebec. An
isolated lead-zinc occurrence, not unrelated to those of Gaspé, lies at
Buchans, Newfoundland.

In Mexico, a rich province of lead-silver-zinc ores lies in the center
of the country, expanding in Chihuahua and Coahuila.

About half of the world's lead and zinc comes from the North American Continent.

Other Provinces.   South America contains important local lead-zinc deposits in Peru, Bolivia, and Argentina. Europe is fairly bountifully supplied with lead and zinc. There is a Mediterranean province with deposits in Spain, Roumania, Turkey, and Tunisia; a central European province includes parts of Czechoslovakia, Germany, Poland, and England. A minor northern province has representatives in Sweden, Norway, and Finland, and another one lies in the Urals. Belts of abundant lead-zinc mineralization occur in parts of Australia and in Burma, and minor localized areas occur in Indo-China, Japan, Siberia, Southern Rhodesia, and southwest Africa.

EXAMPLES OF IMPORTANT LEAD AND ZINC DEPOSITS IN THE UNITED STATES

Some of the ranking mines or districts in the United States are listed herewith.

DOMINANTLY LEAD			DOMINANTLY ZINC		
District	*State*	*Percent of Country*	*District*	*State*	*Percent of Country*
1. Southeastern Missouri	Missouri	40	1. Tri-State	Oklahoma Kansas Missouri	24
2. Coeur d'Alene	Idaho	17	2. Coeur d'Alene	Idaho	12
3. Tri-State		7	3. Franklin	New Jersey	11
4. Bingham	Utah	3.5	4. St. Lawrence Co.	New York	5.5
5. Bisbee	Arizona	3.2	5. Central	New Mexico	5
6. Park City	Utah	2.5	6. East Tennessee	Tennessee	4.3
7. Darwin	California	2.5	7. Bisbee	Arizona	4
			8. Upper Mississippi Valley	—	3.2
			9. Red Cliff	Colorado	2.8
			10. Pioche	Nevada	2.7

Southeastern Missouri.   This district, centering around Flat River, also called the "lead belt," is the greatest lead district in the world. Discovered in 1720 and productively mined in 1800, it yields annually up to 180,000 tons of lead with a little silver.

The deposits occur chiefly as disseminated replacements in the flat-lying Bonneterre dolomite of Cambrian age. They lie at shallow depth, within a vertical range of 200 feet, and average 10 to 15 feet in thickness, 200 to 300 feet in width, and 1,000 feet or more in length

(Fig. 13–23). Hundreds of acres are underlain by such flat ore bodies, disturbed here and there by small faults.

Galena, the only important ore mineral, occurs in disseminated grains and crystals up to an inch across. Marcasite is abundant, and minor siegenite (cobalt-nickel sulphide), chalcopyrite, and pyrite are observed, along with a little calcite, glauconite, and dickite.

The deposits are not associated with observable igneous intrusives, and because of this they have been regarded as the products of meteoric waters. However, the mineralogy indicates a hydrothermal

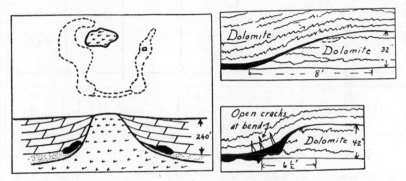

FIG. 13–23. Left: Ores (black) around buried porphyry hill at Missouri Cobalt mine, and shown in plan by dotted lines; Right: Sketches showing how rich ore beds accumulated by removal of 32 to 42 inches of dolomite, No. 16 mine, St. Joseph Lead Co. (*Tarr, Econ. Geol.*)

origin, as maintained by Tarr, from an undisclosed igneous source, the solutions passing through fissures, depositing a little galena in the underlying LaMotte sandstone, and spreading through the Bonneterre dolomite. These ores represent, therefore, low-temperature deposition (probably in late Cretaceous time) and illustrate widespread, lateral, selective penetration of beds by mineralizing solutions uncontrolled by obvious openings, just as in the puzzling cases of the Rhodesian copper deposits and the Rand gold reefs.

Tri-State District. This broad district, centering around Joplin, Missouri, is not only the greatest zinc district of the world, but it yields the lowest-grade zinc ores. Some lead is recovered but no other metals. Its remarkable waste dumps, dotting the flat surface like sand dunes on a desert, attest to the extensiveness of the underground workings (Fig. 13–24). Its area is some 2,000 square miles, and it has yielded over 900 million dollars in metals.

The geology is relatively simple. Nearly flat-lying sediments include the Boone formation, which is separable into 16 beds and is the host rock of the ores. It is unconformably overlain here and there by uneroded patches of Pennsylvanian shale. Beneath the pre-Pennsylvanian erosion surface a karst topography was developed, with sinkholes, underground drainage channels, and caves, in the underlying limestone and particularly along the Grand Falls chert member. Chert and jasper beds are numerous in the Boone. Roy

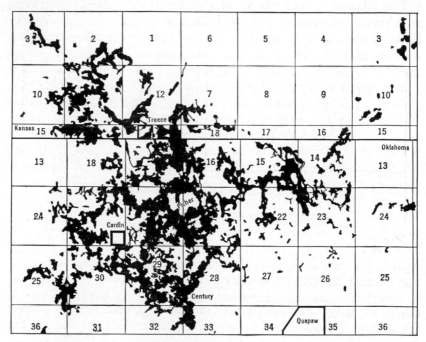

Fig. 13–24. Area of mined ground in Oklahoma-Kansas lead-zinc district.
(*Company map, 1931.*)

concludes that there is an earlier sedimentary chert and later hydro-thermal chert and jasper. Two prominent faults are known, one of which is mineralized. Also, according to Fowler, there are prominent shear zones that brecciated the cherts and localized the ores. These structures appear to be related to major structural features of the Mississippi Valley. Fowler states that the Boone beds are barren of ore except where deformation has created favorable structure for ore. There are no igneous intrusives near the ore.

The deposits consist of (*a*) runs and circles of ore near the surface, and (*b*) sheet or blanket ore along the Grand Falls chert. The first are the most productive. The runs fill and follow shallow solution channels along cherty horizons. They range from 10 to 300 feet in width and are as much as 80 feet thick and 2 miles long. The circles encircle sinkholes and attain dimensions of 800 by 650 feet; the ore is deposited in thicknesses up to 30 feet, around fragments of slumped chert. The sheet ground, which may be 15 feet or more thick, is in the Grand Falls chert where the limestone has been removed from around the chert nodules.

The minerals consist chiefly of zincblende and subordinate galena with minor amounts of wurtzite, marcasite, pyrite, and chalcopyrite. Enargite and millerite have also been identified. The gangue minerals are quartz, chert, carbonates, and a little barite. There is also a large group of oxidation products. The minerals occur in definite sequence.

The origin of these important deposits has evoked extended discussions. For years, a meteoric origin held sway, first by descending surface waters and later by artesian waters that dissolved the metals from basal Cambrian sediments and precipitated them near the old Mississippian surface by loss of dissolved carbon dioxide. Although many supporting data have been advanced, this hypothesis encounters numerous difficulties.

Compelling evidence has been accumulating that the deposits were formed from hydrothermal solutions derived from undisclosed deep-seated igneous centers. Important arguments are (1) the intense silicification; (2) the high-chloride waters of magmatic type found in inclusions in the blende and galena; (3) the temperature of formation of 115° to 135° C, shown by Newhouse from bubble inclusions in the blende; (4) the occurrences of hydrothermal minerals, such as enargite; (5) the sequence of deposition; (6) an indication of mineral zoning; (7) the spectroscopic determination in the ores of numerous elements characteristic of hydrothermal ores elsewhere; and (8) the similarity to many other zinc-lead deposits of undisputed hydrothermal origin. These, and other arguments, may be studied in detail in the extensive literature on the district, notably the paper edited by E. S. Bastin.

Generally similar ores are scattered over the upper Mississippi Valley, particularly in Wisconsin. These ores, now largely exhausted, occur mainly as fillings of caves, pitches and flats, and solution joints in the Galena limestone. They also probably have a hydrothermal origin.

The Appalachian Zinc Provinces. This important belt of straight zinc ores includes Mascot, Jefferson City, and Embree in Tennessee and Austinville, Va., and is the third most important zinc district in the United States. It illustrates deposits formed in breccias.

The Mascot-Jefferson City deposits were formed by the filling and replacement of tectonic and collapse breccias, localized by faults and fault intersections, in dolomitized limestone beds within the gently folded Knox dolomite. Light yellow sphalerite, dolomite, and calcite fill the breccia interstices, and the fragments are partly replaced. Traces of galena, marcasite, and barite are present. The ore bodies follow dipping beds and average 15 to 100 feet in thickness, 125 to 150 feet in width, and have been followed to a depth of 1,000 feet. At Austinville, galena, chalcopyrite, and pyrite are also present, and oxidized ores are abundant.

The origin of these ores, like that of the lead and zinc deposits of the Mississippi Valley region, is controversial. However, the view that they are low-temperature hydrothermal deposits is gaining acceptance.

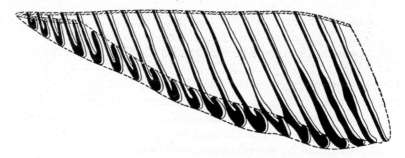

FIG. 13–25.   Stereogram of Mine Hill ore body, Franklin Furnace, N. J.   Black is ore; shaded area is outcrop.   (*After Spencer, U. S. Geol. Survey.*)

Franklin Furnace, N. J.   The unique zinc deposits of Franklin and Sterling Hill have no exact counterpart in the world and make this district the second largest one in the United States. Zinc mining began about 1840. The deposits illustrate high-temperature replacement. The region is made up of Precambrian granitic gneisses and crystalline marble cut by pegmatite dikes.

Both deposits resemble a half-canoe in shape with one side higher than the other and the keel pitching gently underground. In plan and section they are hooklike. They are enclosed within the marble. The thickness of the curved layer ranges from 12 to 100 feet. The inclined keel has been followed over 1,500 feet and to a depth exceeding 1,000 feet (Fig. 13–25). The ore is interrupted by trap and pegmatite dikes.

The mineralogy is unique. The ore minerals are zincite, willemite, and franklinite occurring as abundant grains and bunches in calcite. But the deposit is a mineralogists' paradise, since more than 100 min-

erals have been identified, many occurring nowhere else. Most of
them, such as garnets, amphiboles, pyroxenes, and complex silicates,
resulted from metamorphism induced by the pegmatites. Some sulphides and arsenides occur in post-ore veins. The franklinite is used
for making zinc oxide for paints, and the residual manganese is recovered. The zincite is also used for zinc white, and the willemite for
metallic zinc.

The origin is puzzling. The chief problem is to account for
abundant zinc in the form of oxides and silicates instead of the

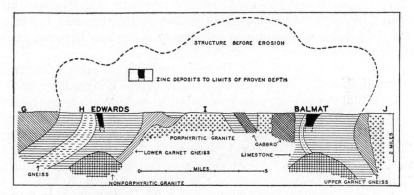

FIG. 13–26.   Longitudinal section, Edwards-Balmat zinc mine, New York.   Black
is ore.   (*Brown, Econ. Geol.*)

customary sulphides found elsewhere. Wolff thought they were
metamorphosed sediments. Spencer thought they resulted by injection of magmatic emanations. Lindgren considered them as pyrometasomatic. Spurr and Lewis thought them to be an ore magma of
sulphides subsequently metamorphosed. Tarr considers them to have
been originally deposited in limestone as sulphides and carbonates,
then weathered to oxides and hydroxides, and later metamorphosed.
Palache suggests that hydrated oxides and silicates replaced limestone
and then were metamorphosed when the limestone was recrystallized.
Ries and Bowen think they were high-temperature replacement deposits. The mineralogy indicates high temperature and pressure, and
the last theory has much to recommend it.

Edwards, N. Y.   At Edwards and Balmat are two unusual zinc deposits. They occur in Precambrian marble, cut by silicic intrusives,
that J. S. Brown believes is part of a large plunging anticline with
Balmat mine on one limb and Edwards on the other (Fig. 13–26).
Both mines are over 2,000 feet deep.

The deposits are hydrothermal replacements along folded limestone bedding resulting in V-shaped lodes. The primary ore consists dominantly of zincblende and pyrite with minor galena and traces of chalcopyrite and pyrrhotite accompanied by carbonate, diopside, talc, quartz, tremolite, hematite, magnetite, phlogopite, chlorite, barite, ilvaite, and garnet, in order of abundance. The Balmat ore averages 16 percent each of sphalerite and pyrite, 1 percent galena, one-third carbonate, and one-third silicates. Some willemite also occurs.

Brown shows that considerable weathering has taken place, resulting in surficial masses of red hematite. He also proposes the bold conclusion, not accepted by all, that the supergene hematite is accompanied by supergene magnetite, specularite, chlorite, and willemite and that about one-fourth of the Balmat zincblende, galena, and chalcopyrite are also supergene sulphides.

Coeur d'Alene, Idaho. This district ranks as the second lead producer, the fourth zinc producer, and the first silver producer in the United States. Mining has progressed since 1885, and the annual production is about 2 million tons ore yielding about 65,000 tons each of lead and zinc, and 7 million ounces silver. The total lead production is over 5 million tons, and the total value of metals exceeds 1000 million dollars. The principal mines are the **Bunker Hill and Sullivan, Morning, and Hecla.**

Folded and faulted Precambrian quartzites and slates are intruded by Cretaceous monzonite. Great faults are barren, but small faults and shear zones in quartzite have been filled and replaced to form replacement lodes or veins. They attain 7,000 feet in length, average 9 feet in width, and have been followed downward 5,300 feet. The ore occurs in definite shoots.

The ore consists of disseminated grains and masses of siderite and galena with minor pyrite, sphalerite, and argentiferous tetrahedrite. Boulangerite and pyrrhotite are present as are also quartz, calcite, dolomite, and barite. Ransome states that sphalerite and pyrite increase with depth. The ores carry 3 to 12% Pb, 3 to 6% Zn, and 2 to 6 ounces Ag per ton. Oxidation is shallow and irregular.

The deposits are closely related to the monzonite intrusions; contact-metasomatic deposits occur close to the monzonite; nearby veins also carry considerable pyrrhotite and magnetite with garnet and tourmaline; but the siderite replacement lodes lie well away from the contact zones. This indicates an intimate gradation between high and intermediate temperatures of formation.

Bingham, Utah. The Bingham district, if it were less overshadowed by the colossal Utah Copper mine, would be better known for its lead-zinc-silver deposits. It ranks fourth in the country in lead production, fourteenth in zinc, and third in silver.   Since 1865 some 650 million dollars has come from these ores in such mines as the Highland Boy, United States, Utah Apex, Yampa, and Jordan.

Around the porphyry copper deposit of the Utah Copper stock are synclinal beds of Pennsylvanian quartzites and intercalated limestones. The limestones near the stock are metamorphosed and contain replacement deposits of pyritic copper ores (13–27).   Zonally distributed outward in the same but unaltered limestone beds are the lead-zinc

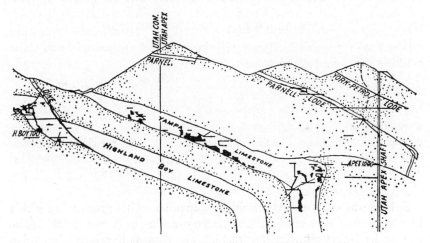

FIG. 13–27. Section through part of Bingham, Utah, showing limestone beds that contain lead-zinc ores.   (*Hunt, 16th Internat. Geol. Cong.*)

ores, and still farther out are siliceous silver ores.   The lead-zinc deposits are fissure veins, replacement lodes, and irregular blanket replacement deposits.   The ore minerals are argentiferous galena and sphalerite with tetrahedrite, pyrite, and chalcopyrite.   Ore body after ore body continues to be found in these beds with remarkable persistency.

The deposits display a close connection with the monzonite intrusives from whose reservoir emanated the waves of hydrothermal solutions that produced the metallization: first, the disseminated copper metallization in the Utah Copper porphyry, then adjacent contact metasomatism, and pyritic copper replacement in altered limestone that displays outward a gradual diminution in alteration coincident with a change from copper, to copper-lead-zinc, to lead-zinc-silver, and then to siliceous silver ores.

Other Districts.   Many other districts contribute appreciable quantities of lead and zinc.   *Park City, Utah,* is an example of prolific lode fissures and bedded replacements in limestone that yield sulphide and oxidized ores. *Leadville, Colo.*, has yielded a half billion dollars from pyritic lead-zinc-silver limestone ores, and illustrates gradations from mesothermal replacements on a grand scale, into epithermal, hypothermal, and even contact-metasomatic deposits. Many veins in the *San Juan* of Colorado contain lead and zinc. Mixed ores are numerous in the *Tintic* and *Cottonwood* districts of Utah and the *Pecos* mine of New Mexico, and contact-metasomatic zinc deposits are found at *Hanover*, N. Mex.

CANADIAN LEAD AND ZINC DEPOSITS

Much of Canada's lead and zinc is derived from mixed ores. The ranking mines and districts are:

DISTRICT OR MINE	LOCATION	CHIEF METALS
1. Consolidated Mining and Smelting Co. (Sullivan mine)	Kimberly, British Columbia	Lead-zinc-silver-copper-gold
2. Flin Flon district	Flin Flon, Manitoba	Copper-zinc-silver-gold
3. Normetal	Noranda, Quebec	Copper-zinc-silver-gold
4. Waite-Amulet	Noranda, Quebec	Copper-zinc-silver-gold
5. Treadwell Yukon	Keno, Yukon	Lead-silver

Sullivan Mine, Kimberly, British Columbia. This great mine yields 98 percent of Canada's lead and three-quarters of her zinc and is the largest lead-zinc mine in the world. The deposit is a lens of massive sulphide some 6,000 feet in length and 270 feet in maximum width, along a zone in tilted Precambrian quartzites and argillites. The body parallels the bedding and has been formed by replacement of the sediments, preserving the original bedding in the ore. The lode contains two ore shoots separated by 700 to 1,100 feet of barren massive pyrite. The ore consists of argentiferous galena and blende, with pyrite dominant in the upper shoot and pyrrhotite dominant in the lower. There are small amounts of garnet, actinolite, tourmaline, and cassiterite. The margins of the body contain less lead and zinc. The ore averages 10% Pb, 4.5% Zn, and is high in silver. The production is 6,500 tons of ore a day. The deposit is illustrative of a high-temperature hydrothermal replacement.

Other North American Deposits. The *Flin Flon* mine, Manitoba, described under copper deposits, is also the second largest producer of zinc in Canada. The pyritic copper-gold deposits of the *Normetal* and *Waite-Amulet* mines, near Rouyn, Quebec, are the third and fourth zinc producers in Canada. Rich silver-lead veins are found at the *Treadwell Yukon* mine at Keno, Yukon. In Newfoundland, the

Buchans mine, discovered by geophysical prospecting, yields complex ore carrying about 8% Pb, 17% Zn, 1.4% Cu, 3 ounces Ag, 0.05 ounce Au, and 30 percent barite, from large mesothermal sulphide replacements in altered tuffs.

In *Mexico* about 80 percent of the lead and zinc production is yielded by four companies — the American Metal, Howe Sound, Fresnillo, and San Francisco. The four most important groups of deposits are the Santa Eulalia, San Francisco, Fresnillo, and Sierra Mojada, all of which are also silver deposits, described in Chapter 12. There are many small deposits scattered through north-central Mexico, which in the aggregate yield much lead, zinc, and silver. Some have been known since the middle 1500's. They are mostly irregular limestone replacements, mantos, and fissure veins, and include such deposits as Ahumada, Santa Barbara, Parral, Mapimi, Mazapil, Taxco, and Angangueo.

AUSTRALIAN LEAD AND ZINC DEPOSITS

Australia ranks second in world lead production and fifth in zinc. The important districts in order of rank, with their percentage of the country's production, are shown below.

DISTRICT	LEAD	ZINC
1. Broken Hill, New South Wales	80	73
2. Mount Isa, Queensland	16	14
3. Rosebery, Tasmania	4	12
4. Lake George, New South Wales		

Broken Hill, New South Wales. The famed Broken Hill lode in the desert of New South Wales has yielded over a billion dollars in lead, zinc, and silver. It yields annually about 200,000 tons of lead and 150,000 tons of zinc from the Broken Hill South, North Broken Hill, the Zinc Corporation, and the New Broken Hill mines.

Precambrian mica and sillimanite schists and quartzites have been forced apart along the bedding planes by pressure lenses of gneiss, amphibolites, and pegmatites. The beds and sills were then sheared and closely folded, forming drag folds that localized the ore and, according to Andrews, were replaced to form large " pressure " lenses, corrugated sheets, and " saddles " of ore.

A central area of intense metamorphism contains the largest lodes composed of lead-zinc sulphides. An outer zone of less alteration lacks zinc but contains platinoid metals, tin, and tungsten. The ore bodies are disconnected lenses and saddles over a length of 3 miles. Two subparallel lodes on the foot wall and hanging wall are known as the Zinc and Silver-Lead Lodes; in places they join. In longitudinal section these are arched, shallow in the center, and deeper (2,200 feet) at the east and west ends. From the foot wall upwards there is a zonal arrangement of (1) rhodonite masses, (2) sulphides and rhodonite, (3) sulphides with silver, and (4) siliceous sulphides with garnet or green feldspar. Rhodonite and garnet abound in the central portion of the arch but decrease toward the ends, where pegmatites are more abundant.

The ore is composed of coarse aggregates of crystals of blende, galena, zinc spinel, rhodonite, green feldspar, manganese garnet, pyroxene, magnetite, gahnite, fluorspar, quartz, and calcite. There are small amounts of iron and copper sulphides and silver minerals. The ore carries 11 to 17% Pb, 10 to 16% Zn, and 3 to 14 ounces Ag per ton.

A heavy gossan, up to 100 feet wide, extends to a depth of 300 feet, below which are oxidized ores of lead and silver followed by a negligible supergene sulphide zone.

The deposits are a high-temperature form of replacement, presumably due to emanations from the underlying magmatic reservoir that supplied the pegmatites.

Mount Isa. This large mine, a newcomer to Australia, yields ores that average about 10% Zn, 9% Pb, and 7 ounces Ag per ton, and illustrates large-scale replacement.

The deposits occur within a 2-mile thickness of crenulated Precambrian shale, lying between quartzites, along a zone of folding and shearing parallel to the bedding, which, according to Blanchard and Hall, resulted from an overthrust. No intrusives occur nearby. The deposits are replacement bodies up to 2,000 feet long and 250 feet wide and are known to depths of 1,500 feet. The ore consists of massive sulphides that exhibit remarkably fine banding inherited from the shale. Zinc, lead, and iron sulphides are the chief metallic minerals, along with minor chalcopyrite, arsenopyrite, tetrahedrite, silver minerals, valleriite, pentlandite, marcasite, and appreciable graphite. The mineralization started with deposition of silica, followed by pyrite and more silica, after which came the other sulphides, ending with carbonate. Deformation continued throughout mineralization. Oxidation extends to a depth of 250 feet, and in this zone are bodies of oxidized lead and zinc ores. Copper bodies underlie the lead-zinc ores. These deposits illustrate noteworthy structural control of ore localization and remarkable replacement preserval of fine shale banding and contortion. Also they constitute another example of large tonnage replacement by solutions of intermediate temperature without other evidence of an igneous source.

At the *Lake George mine,* New South Wales, production of lead-zinc-copper ores started in 1939 from a lode in a shear zone that cuts sheared Paleozoic arkose intruded by quartz-diorite porphyry. The outcrop is 5,000 feet long, and the ore shoot on the 600 level is 1,000 feet long and 20 feet wide. The ore reserves average 7.5% Pb, 13% Zn, 0.7% Cu, 2.25 ounces Ag, and 1.3 dwt Au per ton.

OTHER IMPORTANT LEAD-ZINC DEPOSITS

Europe is fairly rich in lead and zinc deposits, but no major country produces its own requirements of these metals. Asia has only one large representative — Burma.

Poland-Germany, Silesia. The chief zinc-lead district of Europe lies at the former Polish-German boundary. Bedded zinc-lead deposits in slightly warped Triassic dolomites overlie Carboniferous coal seams — a fortunate combination. There are three ore zones, one at 80 to 90 meters deep at the base of the dolomite, another 15 meters above it, with another minor bed above this. The lowest deposit is about 4 meters thick, is regular, extensive, and nearly horizontal; rare pipes and irregular cavities occur. The ore consists of sulphides of iron, zinc, and lead, with rare jordanite and meneghinite along with clay and dolomite, and carries 15 to 23% Zn and 3% Pb. The sulphides exhibit colloform structure. The upper ore beds are largely oxidized. The deposits are mainly replacements with minor cavity filling, formed by low-temperature, hydrothermal solutions.

Italy. The most important deposits of Italy are in Sardinia. Others occur at Raibl and in Trentino and Trieste.

Those of Sardinia were worked by the Phoenicians and Romans. In the Arbus district, according to Wright, granitic intrusives cut Paleozoic schists. Two systems of veins are found, one parallel and another normal to the contact. Of the former, the Main Lode, worked in the Montevecchio and Ingurtoso mines, supplies most of the lead and zinc of Italy. It is 6 miles long and 60 feet wide. In the eastern section lead dominates, and westward in the Brassy section it splits into three lodes and zinc dominates. The ore shoots are from 1,200 to 2,000 feet long, with equal barren intervals. The ore consists of blende and galena, in a gangue of barite, carbonates, and quartz. The ore body is a wide brecciated mass with the ore forming the matrix. It carries 10 to 12 percent of combined zinc and lead. Oxidation extends 20 to 40 feet in depth.

Near *Iglesias, Sardinia,* are large replacement deposits in limestone, and important veins of zinc-lead ores occur in northern Italy at *Raibl* and *Auronzo.*

Spain. Spain was once the leading European producer of lead. The chief district is the Linares-Carolina, in the Sierra Morena, where old Roman workings and dumps are still in evidence. Here, rich silver-lead fissure veins are numerous, the outstanding one being the *Los Guindos* vein, which has been followed underground continuously for 11 kilometers, except for a distance of 1½ kilometers. The vein is enclosed in quartzite and slate and is filled by irregular blocks and fragments of these rocks, surrounded by ore. The inclusions suggest a pre-mineral rubble of a fracture zone around which vuggy ore was deposited,

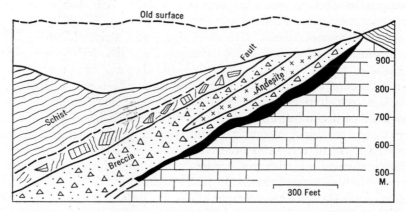

Fig. 13-28. Stantrg mine, Trepca Mines, Ltd., Yugoslavia. Vertical section showing ore along pipelike mass of breccia. (*From Forgan, and private company report.*)

rather than replacement residuals. The width ranges from 5 to 10 feet, in places containing 5 feet of pure galena. Zincblende is absent, and there are traces of pyrite and chalcopyrite in a gangue of quartz, carbonates, and barite. The ore is localized in shoots, one of which is 1,000 feet long and yields 10 to 35 percent lead with 100 to 300 grams silver. Similar veins are found at *La Rosa*. Near Linares, the great *Arrayanes Lode*, over 4 miles long and only 3 feet wide, occurs in granite. Similar ore is in shoots up to 1,000 feet long and 1,500 feet deep. All these veins are part of a silver-lead province of cavity-filling deposits.

Yugoslavia. This country is both an ancient and a recent producer of lead and zinc. Ancient workings around *Trepca* were revived when modern production commenced in 1930 at the *Stantrg* mine after the development of over 3 million tons of ore averaging 8.7% Pb and 8% Zn. In an area explored by 1,000 ancient shafts up to 200 meters deep, an inclined pipelike mass of early Tertiary volcanics has been disclosed between underlying limestone and overlying Paleozoic schist. Hydrothermal solutions followed the contact, and ore replaced the limestone to form an inclined trough-like body (Fig. 13–28), which has been followed on a 40° dip for 1,800 feet vertically. The ore is made up of base metal sulphides, jamesonite, and magnetite with a gangue of quartz, four carbonates, actinolite, garnet, and amphibole and represents a combination of mesothermal and hypothermal metallization.

Burma. The *Bawdwin* mine (Fig. 13–29) in Burma has been one of the world ranking lead mines. Here an ore zone 8,000 feet long and 500 feet wide has a hanging wall of feldspathic grits and a foot wall of rhyolite lavas and tuffs, all of early Paleozoic age. The ore bodies are enormous replacements within a shear zone. Within the ore zone are three lode systems, the principal one of which has been broken by faulting into three segments — the *Shan, Chinaman,* and the *Meingtha.* The ore shoots are up to 1,200 feet long, and in places there are solid sulphides 50 feet wide composed of galena and zincblende, with minor pyrite and chalcopyrite in altered country rock

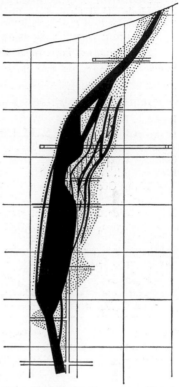

Fig. 13–29. Section across Bawdwin lode, Burma. Black is solid sulphide; stippled, disseminated ore. (*Loveman, A.I.M.E.*)

and quartz. The reserves consist of over 4 million tons averaging 25% Pb, 15% Zn, 0.7% Cu, and 20 ounces silver.

Selected References on Lead and Zinc

Mineral Industry of the British Empire and Foreign Countries: Lead, 2nd ed., 1933, and Zinc, 2nd. ed., 1930. Imperial Inst., London. *Brief summary of occurrences.*

Zinc and Lead Deposits of Canada. F. J. ALCOCK. Can. Geol. Surv. Econ. Geol. Ser., No. 8, Ottawa, 1930. *Résumé of world deposits.*

General References 7, 9, 10, 14, 15.

Lead and Zinc Deposits of the Mississippi Valley Region. Edit. by EDSON BASTIN. G. S. A. Spec. Paper 24, 1939. *An exhaustive compilation of occurrence and origin.*

Southeastern Missouri Lead Deposits. W. A. TARR. Econ. Geol. 31:712–754, 832–866, 1936. *The best article on these deposits; proposes a hydrothermal origin.*

Bingham, Utah; Coeur d'Alene, Idaho; Leadville, Colo.; Park City, Utah; Tintic, Utah. U. S. Geol. Surv. Prof. Papers 38, 62, 77, 107, 148. *Detailed geology.*

XVI International Geological Congress, Washington, 1933. Guidebooks: 17, **Salt Lake Region** (Bingham, Tintic, Park City); A2, **Tennessee;** 8, **Franklin, N. J.** *Descriptions of important lead-zinc deposits.*

Geology and Ore Deposits of Shoshone County, Idaho. J. B. UMPLEBY and E. L. JONES. U. S. Geol. Surv. Bull. 732, 1923.

Flin Flon Mine, Manitoba. G. M. BROWNELL and A. R. KINKEL, JR. Can. Min. and Met. Bull. 279:261–286, 1935. *Large sulphide replacement body.*

Buchans, Newfoundland. W. H. NEWHOUSE. Econ. Geol. 26:399–414, 1931.

Bawdwin Mine, Burma. M. H. LOVEMAN. A. I. M. E. Trans. 56:170–194, 1917.

Trepca, Yugoslavia. H. A. TITCOMB and others. Min. and Met., Sept.–Dec., 1936.

Linares-Carolina, Spain. XIV Int. Geol. Cong., Madrid, 1926. Guidebook A3.

Broken Hill, Australia. E. C. ANDREWS. Econ. Geol. 21:81–89, 1926, and N. S. Wales Geol. Surv. Mem. 8, 1922. *One of the great lead-zinc deposits of the world.*

Sullivan Mine, British Columbia. C. O. SWANSON and H. C. GUNNING. General Reference **15,** pp. 219–230, 1948. *Description of world's greatest lead-zinc mine.*

Mount Isa, Australia, Rock Deformation and Mineralization. R. BLANCHARD and G. HALL. Aust. Inst. Min. Met. 125:1–60, 1942. *Details of geology, occurrence, and origin.*

Structural Control of Ore Bodies, Jefferson City Area, Tennessee. A. L. BROKAW and C. L. JONES. Econ. Geol. 41:160–165, 1946. *Ore control by cross folds.*

Tin

History. Tin is believed to have been one of the first metals employed by man. So far as is known it was first used as a constituent of bronze, probably obtained by accidental smelting of mixed tin and copper ores. It is reported that bronze originated in the Orient, and Umhau states that the bronze industry was well established in China in 1800 B.C. A bronze rod with 9.1 percent tin found in Egypt dates back to 3700 B.C., and many bronze objects with 10 to 14 percent tin have been found in excavations of different ancient civilizations. Tin free from lead and silver was used for an Egyptian mummy wrapping in 600 B.C. The earliest Mediterranean bronzes are presumed to have been made from Asiatic tin, but by 1500 to 1200 B.C. tin was brought from Britain by the Phoenicians; by 500 B.C., 100 tons of tin were exported annually from Britain. Ancient uses for tin were chiefly for making bronze for implements, weapons, and ornaments, but medieval times found greater use for bell casting, bronze armor, and pewter. This supply came mainly from Britain, and later from Saxony and Bohemia, and then from Asia. Malay tin was in high demand from 1400 to 1800, and by that time tin also came from the Netherlands

Indies and China. By the end of the eighteenth century tin production was exceeded only by three metals — iron, lead, and copper. About the middle of the nineteenth century tin production started in Australasia, followed by that in South America and finally in Africa, in 1910. North America is singularly lacking in tin.

Uses. Many of the early uses of tin, such as for pewter, lining of cooking vessels, and tin roofing, have been displaced by the newer uses of bronze bearings and tin-plated food containers. The United States is the largest consumer of tin. Its total annual consumption of 60,000 to 100,000 tons is normally used (in percentage) as follows: tin and terne plate, **33**; bronze and brass, **28**; solder, **21**; babbitt metal, **8.5**; tinning, **2.6**; type metal, **2.5**; bar tin, collapsible tubes, foil, chemicals, alloys, and tubing, **4.2**.

Production. The annual world production of tin ranges normally from 150,000 to 250,000 tons of metal. The trend of geographic change in production may be seen from the table below.

| | | PERCENT OF TOTAL | |
COUNTRY	19*th* Century	1900–1925	1926–1950
Federated Malay States	37.1	38	30
United Kingdom	22.2	3.3	0
Netherlands Indies	19.4	17.7	24
China	6.4	5.6	3.5
Australia	8.5	5.1	1.4
Thailand (Siam)	3.8	5.2	2.0
Bolivia	1.8	18.5	23
Nigeria	0	2.6	5.5
Belgian Congo	0	0	3.5
All others	0.8	4.0	7.1
	100	100	100

Asia, so far, has produced about 68 percent of the total world production. In contrast to production, the normal percentage consumption of tin is: United States 40, Great Britain 13, Russia 12, Germany 7, France 4.7, and Japan 4.5.

MINERALOGY, TENOR, AND TREATMENT

Mineralogy and Associations. Practically all the tin of commerce is obtained from the mineral cassiterite or tinstone; stannite, the tin-copper-iron sulphide, and teallite supply a minor amount in Bolivia. The accompanying gangue minerals are chiefly altered granite, quartz, and white mica. A common associate is tungsten; molybdenum or silver may be present, and gold is generally absent.

Tenor and Treatment. Placer deposits range in tin content from 0.4 pound to 5 pounds per cubic yard, and lode deposits carry from 1 to 8 percent tin.

Pure cassiterite is easily smelted as oxygen is the only constituent to be eliminated. This is done by adding coal, coke, or charcoal. However, iron and silica cause tin to go into the slag, necessitating careful manipulation of fluxes and resmelting of the slag. The molten tin is drawn off and purified by agitation with air or by electrolytic refining.

KINDS OF DEPOSITS AND ORIGIN

The principal classes of tin deposits are placer deposits, stockworks, fissure veins and disseminated replacements.

Origin and Distribution. The lode deposits contain high-temperature minerals and, except for Bolivia, are always in close association with silicic granites, from whose magmatic reservoirs they are generally considered to have been derived by pneumatolytic action. Since Daubree's famed synthesis of cassiterite, which was one of the earliest minerals to be synthesized, it is commonly assumed that the tin was transported from the magma chamber as gaseous tin fluoride or tin chloride, which by reaction with water formed cassiterite, releasing HF or HCl. The granite wall rocks are generally altered to muscovite, quartz, and topaz (greisen), presumably by attack of the acid gases. The associated fluorine-bearing minerals, topaz and fluorite, suggest this mode of origin, which, however, is only a hypothesis. A hydrothermal origin applies to the Bolivian tin deposits and perhaps also to the others.

The greatest tin province of the world is the belt of placers found in a strip of country 1,000 miles long and 120 miles wide along the Malay peninsula, including the Netherlands East Indies to the south, and Burma and Thailand (Siam) to the north. The province also projects into Yunnan, China. Separate provinces occur in Bolivia; Cornwall, England; Erzgebirge, Germany; Nigeria, Belgian Congo; Australia; and South Africa.

EXAMPLES OF TIN DEPOSITS

Malaya. The chief occurrences of tin are in the 519 mines in the Federated Malay States of Perak, Selangor, Negri Sembilan, and Pahang. Perak produces over 60 percent of the output from the rich and accessible deposits of the Kinta Valley. The ore is mostly alluvial, but some lodes are also worked. The bedrock is limestone, granite, and schist, overlain by a thick mantle of tropical residual soil.

The tinstone contained in stockworks and veins resisted weathering and so accumulated in the residual soils. The soils were washed into the valley, leaving the heavy resistant tinstone to accumulate on the pinnacled bedrock, particularly near valley sides. The center of the valley has been only slightly prospected. As it contains valuable agricultural lands, the placer miner is not welcomed because of his absorption of water, destruction of agricultural land, and undesired tailings disposal. Probably much placer ground still remains. In places the rich eluvial deposits in the residual soils of the valley slopes can be recovered by sluicing or open pitting. The reserves of placer tin are large, and little is known as to whether the source lodes may prove productive beneath the soil mantle.

Generally similar conditions exist in other parts of the Federated States except that granite is generally the bedrock. In many places the deeply decomposed granite is directly mined for its tin content.

The Malaya tinstone is treated at smelters located in Penang, Singapore, and Selangor, which smelters also treat the ore from Siam, Burma, and other neighboring provinces.

Netherlands Indies. The three Netherlands islands of Banka, Billiton, and Singkep, lying south of Malaya, have long been producers of tinstone, about two-thirds of it coming from Banka, one-third from Billiton, and small amounts from Singkep and adjacent islands. The Banka deposits are all alluvial, but some lodes are worked on the other islands. The ore goes to smelters at Banka, Singapore, and Arnhem, Holland.

Geologically the tin occurrences are generally similar to those of the Malay States. According to Westerveld, humid, tropical weathering of post-Triassic granite containing stockworks of cassiterite gave rise to residual, hill-slope, eluvial placers (koelits). Part of these in turn were swept into the adjacent valleys, adding tinstone to the residual accumulations already present in the valley bottoms (kaksas). These were covered by alluvial top beds of sands, which in places contain upper " hanging " placer beds. Quaternary submergence drowned the lower parts of the river valleys with their bedrock tinstone to depths of 30 meters below sea level. These have been traced out to sea and recovered on a large scale by sea dredges.

The primary deposits consist of (1) pneumatolytic stockworks in greisen containing cassiterite, along with wolframite, tourmaline, topaz, and minor base-metal sulphides, and (2) contact-metasomatic deposits in the intruded sandstone-shale formation containing cassiterite, magnetite, abundant base-metal sulphides, contact silicates, siderite, and fluorite.

Bolivia. Bolivia, the second-ranking tin country, derives its ores entirely from unique lode deposits, chiefly from *Llallagua-Uncia, Huanuni,* and *Potosi,* but there are ten tin centers. The ores carry about 2–4 percent tin with silver, and normally have been smelted in Europe but are also treated in Texas. Formerly the upper-level ores carried 8 to 12 percent tin.

The *Llallagua-Uncia* group of the Patiño Company produces about 60 percent of Bolivian tin. Turneaure describes the ore as occurring

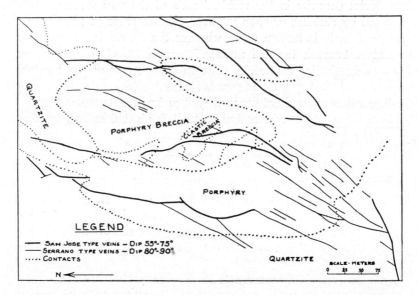

Fig. 13–30. Fissure systems of the Llallagua tin mine, Bolivia. (*Turneaure, Econ. Geol.*)

in a great network of veins within and contiguous to a small quartz porphyry stock of late Tertiary age and near-surface origin, a network which contracts in depth between walls of folded Cretaceous sediments. The veins are of two distinct intersection structural types (Fig. 13–30). One group is composed of strong, continuous, parallel fissures 2 to 6 feet wide, and the other is a maze of discontinuous small cracks. Many do not outcrop. Their close spacing may give rise to stopes 15 to 35 feet wide.

The veins are fracture fillings. The walls are well defined, and the filling is crustified and drusy. In the major veins the ore occurs in distinct shoots, parts of which are phenomenally rich, containing 3 to 4 feet of solid cassiterite. Cassiterite, quartz, pyrite, and marcasite make up 90 percent of the filling. Silver is absent. The episodes of vein formation are: 1, vein fillings of quartz, bismuthinite, cassiterite, and franckeite; 2, replacement of franckeite by pyrrhotite, and 3, replacement of pyrrhotite by wolframite, stannite-sphalerite, marca-

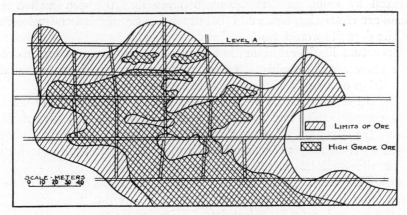

Fig. 13–31. Zonal arrangement of tin ores with high-grade ores in central part, Llallagua, Bolivia. (*Turneaure, Econ. Geol.*)

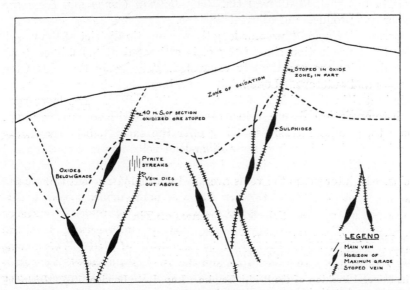

Fig. 13–32. Vertical section of Llallagua tin veins showing position of rich ore shoots. (*Turneaure, Econ. Geol.*)

site, arsenopyrite, pyrite, and siderite; 4, crustification and some replacement by later sphalerite and pyrite, chalcopyrite and gangue; 5, wavellite; and 6, supergene minerals.

The ores are pronouncedly zoned (Fig. 13–31). Vertically, a rich hypogene zone lies under the zone of oxidation but extends up into it (Fig. 13–32), and horizontally there is a central cassiterite zone flanked by sulphides. Turneaure believes that the rich cassiterite zones are clearly hypogene and that supergene tin enrichment is absent.

The deposits were formed under shallow cover by medium- to high-temperature liquid solutions that changed to low-temperature in the late stages, when rapid release of pressure permitted concentrated deposition of a cassiterite core. Telescoping is indicated.

The other tin deposits of Bolivia are mostly Tertiary fissure veins that carry stannite and teallite as well as cassiterite and silver. At *Potosi* the silver-tin veins are richer in tin at depth. At *Oruro*, Chace describes three stages of vein mineralization: 1, quartz-pyrite-cassiterite; 2, silver-sulpho-salt; 3, oxidized ores.

Other Deposits. *China* has long been an important producer of tin from lodes in the Kochiu district, Yunnan, where one mine, operated for 500 years, is over 3,000 feet deep. *Nigerian* tin comes mainly from the Bauchi plateau, where the weathering of veins and stockworks in granite has yielded alluvial tin. The *Belgian Congo* and *Siam* are important producers of alluvial tinstone.

Cornwall, England, was for long the famous tin district of the world. The total production since 500 B.C. is estimated at 3.3 million tons. The deposits are now largely exhausted. The early tin came from placers and outcrops. Lode mining on a large scale began in 1600 and culminated in the middle of the nineteenth century. The lodes were high-temperature fissure veins, carrying cassiterite, occurring in, or close to, granite intrusives. They offer the earliest, best-known examples of mineral zoning; some silver-lead veins at the surface passed downward into copper veins, and copper veins in depth yielded to tin veins. In later years the veins averaged about $1\frac{1}{4}$ percent tin.

Selected References on Tin

Tin Fields of the World. W. R. JONES. London, 1925. *Brief sketches.*

Tin. Pardners Mines Corp., New York, 1935. *A popular story of tin and its extraction.*

The Mineral Position of the British Empire. Imp. Inst., London, 1937. *Distribution and statistics.*

World Reserves and Resources of Tin. C. W. MERRILL. U. S. Bur. Mines Inf. Circ. 6249, 1928.

Geologic Features of Tin Deposits. H. G. FERGUSON and ALAN M. BATEMAN. Econ. Geol. 7:209–262, 1912. *General features of occurrence and origin.*

Geology of Malayan Ore Deposits. J. B. SCRIVENOR, London, 1928.

Tin Ores of the Dutch East Indies. J. Westerveld. Econ. Geol. 32:1019–1041, 1937.

Tin Deposits of Llallagua, Bolivia. F. S. Turneaure. Econ. Geol. 30:14–16; 170–190, 1935.

Tin-Silver Veins of Oruro, Bolivia. F. M. Chace. Econ. Geol. 43:333–383; 435–470, 1948. *Detailed geology of bonanza tin-silver veins of shallow depth of formation.*

Geology of Colquiri Tin Mine, Bolivia. D. F. Campbell. Econ. Geol. 42:1–21, 1947. *High- to low-temperature cassiterite veins.*

Ore Deposits of the Eastern Andes of Bolivia. F. S. Turneaure and K. K. Welker. Econ. Geol. 42:595–625, 1947. *Laterally zoned high-temperature tin-tungsten veins and sulphides.*

Tin Deposits of Carguaicolla, Bolivia. F. S. Turneaure and R. Gibson. Amer. Jour. Sci., Daly Vol. 243–A: 523–541, 1945. *Unusual teallite-franckeite-cassiterite-wurtzite-sphalerite mineralization.*

A Survey of Deeper Tin Zones, Carn Brea Area, Cornwall. B. Llewellyn. Bull. Inst. Min. Met. 477:1–19; Disc. 479:1–28, 1946. *Newer geologic data of the classical tin deposits.*

Tin. C. L. Mantel. 2nd Ed. Mon. 51 Am. Chem. Soc. Reinhold Pub. Corp., New York, 1949. *Occurrence, mining, production, properties, and uses.*

Tin-Bearing Placers near Guadalcázar, State of San Luis Potosí, Mexico. By Carl Fries, Jr., and Eduardo Schmitter. U. S. Geol. Surv. Bull. 960-D, 1949. *Finely divided cassiterite and other metals in huge alluvial fan.*

Aluminum

Aluminum is the latest metal to find general large-scale use in modern industry. Although the most abundant metal in the earth's crust, it long defied commercial extraction in the metallic state. It is an important constituent of such common substances as clays, soils, and rock silicates, but it is obtained commercially almost entirely from only one mineral — bauxite; a little andalusite with 30 to 40 percent Al_2O_3 has been utilized at Boliden, Sweden, and experimental runs have been made on anorthosite, alunite, and clays. Although aluminum was isolated in 1827, it was not until 1886 that Hall and Heroult discovered the cheap process for its extraction by its electrolysis in fused cryolite. It is largely a twentieth-century metal.

Uses. The lightness of aluminum, its high strength compared with its weight, its resistance to atmospheric corrosion, and its electrical conductivity make it a popular modern metal. It enters into almost every conceivable use. It is a desired metal for household products, airplane construction, streamlined trains, and automobile parts, and in the electrical industry it is a strong competitor of copper. It is used in 30 industries and has over 4,000 applications.

The aluminum consumed in the United States is shown by the following percentage distribution for pre- and post-war uses:

Purpose	Pre-War	Post-War	Purpose	Pre-War	Post-War
Transportation	38	9	Building construction	4	29
Electrical conductors	16	8	Foundry and metal		
Cooking utensils	14	13	work	4	19
Machinery and electri-			Chemical	2	2
cal appliances	9	10	Food and beverage	2	2
Metallurgy	8	3	Miscellaneous	3	4

The chief use is in the form of alloys, whose light weight and strength make them particularly desirable for airplanes, motor cars, and trains.

Of the bauxite ore mined about 65 to 80 percent is used for the production of aluminum, about 7 percent for chemicals, and about 10 percent for abrasives, refractories, and alumina cements.

Production. The production of aluminum, unlike that of most other metals, takes place largely in countries distant from the source of the ore. Because of the large amount of electrical power necessary for its reduction the ore moves to available cheap power. Thus, countries such as Germany and Canada are large producers of the metal but not of the ore. Annual production of the metal is around 1 million tons; wartime production exceeded 2 million tons. United States production is around 600,000 tons.

The chief producing countries in order of their output are: United States, Canada, Germany, Japan, Russia, United Kingdom, France, Italy, Norway, and Switzerland.

In contrast, the chief sources of bauxite, the ore of aluminum, are Surinam, British Guiana, United States, France, Hungary, Italy, Yugoslavia, Gold Coast, and Russia.

In the United States the chief production of bauxite is from Arkansas, with minor amounts from Alabama, Georgia, and Mississippi, where large low-grade deposits have been developed.

Mineralogy and Associations. As shown in Chapter 5·7, bauxite is not a mineral species but a colloidal mixture of hydrous oxides of aluminum and iron and water. Commercial bauxite occurs in three forms: pisolite or oölite, sponge ore, and amorphous or clay ore.

Tenor and Treatment. The tenor of bauxite varies with the use to which it is put and the nature and amounts of the impurities present. The figures given below are in percent.

Purpose	Alumina	Allowable SiO_2	Allowable Fe_2O_3	Allowable TiO_2
Aluminum	52	4.5	6.5	
Chemicals	52		3.0	
Abrasives	40–60	3–5	3–5	2.5–4
Refractories	50	3–6	3–5	2.5–4

For aluminum production, bauxite is washed, purified, ground, and digested in steam-jacketed ovens with hot, strong, sodium hydroxide, which dissolves the alumina as sodium aluminate, leaving the impurities. The solution is agitated with aluminum hydroxide, causing precipitation of the metal as hydroxide, which is then filtered, dried, and calcined. This product is converted to metallic aluminum by electrolysis, in a bath of fused cryolite at 950° C, contained in a cell of baked carbon. Four tons of bauxite yield 2 tons of pure alumina and 1 ton of metal.

Kinds of Deposits. All bauxite deposits result from residual weathering and occur as: (1) blankets at or near the surface; (2) interstratified deposits on unconformities; (3) pocket deposits in limestone; and (4) transported deposits. Descriptions of these are given in Chapter 5·7.

World Distribution of Bauxite

Commercial bauxite deposits are widely distributed over the tropical and temperate zones of the continents.

In *North America* the important commercial deposits are confined to Arkansas, with minor ones in Georgia and Alabama, and rather small deposits in Tennessee, Virginia, Mississippi, and New Mexico. *South America* contains an important belt in British and Dutch Guiana (Surinam) and minor deposits in Minas Geraes, Brazil.

The greatest deposits of the world, however, occur in southern *Europe* parallel to the Mediterranean, reaching their maximum development in France and extending through parts of Italy, Yugoslavia, Hungary, Greece, and Roumania. This group formerly supplied about 60 percent of the bauxite of the world. Minor deposits occur in Germany, Spain, and Ireland, and more extensive ones in Russia. *Asia* contains relatively few deposits. Some occur in India, and deposits in the Dutch East Indies, Celebes, and Palau are being worked.

Africa, a tropical continent, contains commercial bauxite only in the Gold Coast. Large deposits are reported from the Atlas Mountains, Morocco, and from the Tichenya Plateau, Nyasaland. Small deposits are worked in Portuguese East Africa, and other deposits are reported in Tanganyika, Sierra Leone, Madagascar, and Ethiopia. *Australia* yields a small tonnage from Victoria and New South Wales.

Examples of Deposits

Arkansas. The bauxite region of Arkansas is one of Tertiary sediments and masses of nepheline syenite that intrude folded Paleozoic beds.

The bauxite deposits, which resulted from the weathering of the
nepheline syenite, lie upon an undulating pre-Tertiary erosion surface
developed across it. The main deposits at Bauxite, as shown by a
mine model, flank the sides and tops of low syenite ridges that project toward a barren, shallow, central valley. Any bauxite formed in
the valley was washed away, and was probably incorporated in nearby
accumulating Eocene sediments. These and the residual ores were

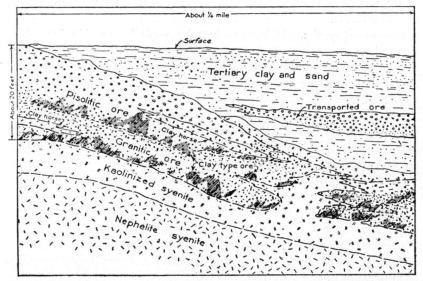

FIG. 13–33. Section of the bauxite deposits of Arkansas showing relation between
syenite, bauxite, and Tertiary beds. (*Branner, 16th Internat. Geol. Cong.*)

slowly covered by advancing Eocene sediments that were subsequently
stripped off in places to disclose small areas of syenite and bauxite.
Some beds of lignite occur in the overlying sediments, and the operators
say that wherever they see lignite they expect good bauxite beneath it
(Fig. 13–33). According to Branner, there are five stages readily discernible to the eye, in the formation of the bauxite: (1) unaltered
nepheline syenite, (2) partly kaolinized syenite — nepheline partly removed, (3) completely kaolinized syenite — nepheline more completely
removed, (4) bauxite high in silica, (5) merchantable bauxite, low in
silica (Fig. 5·7–3). The bauxite is pisolitic, "granitic," or amorphous.
The ore beds reach 35 feet in thickness and average 11.5 feet. The
maximum overburden is 80 feet, and the bauxite is extracted by power
shovel from large open cuts. The annual production has ranged from
a pre-war rate of 400,000 tons to a wartime rate of 6 million tons.
Three grades of ore are produced — aluminum, abrasive, and chemical

ore. The bauxite runs from 56 to 59% Al_2O_3, and some is mined with less than 5% SiO_2 and 2 to 6% Fe. The known reserves of high-grade ore are being rapidly depleted.

Alabama-Georgia. These deposits occur along a 60-mile belt and consist of pockets and irregular masses of residual bauxite lying in and on a mantle of residual clay that marks an Eocene erosion surface on the Knox dolomite. Hayes thought the bauxite was formed by the hydrothermal alteration of the clay; but it seems more probable that it, like other bauxites, resulted from weathering.

France. The French deposits occur mainly in Var and Hérault. They consist of pockets and blankets on karsted erosion surfaces developed on Jurassic and Cretaceous limestones. Consequently, their lower surfaces are highly irregular and discontinuous. They are mostly covered by Upper Cretaceous sediments and are, therefore, of Cretaceous age. Some of the deposits have slumped into collapsed solution cavities; others have been worked over by the transgressing sea (Fig. 5·7–5). They have been formed by the alteration of argillaceous material, some of which may have been residual clay and some shale beds let down by erosion from a higher horizon.

There are three main varieties of bauxite — white, white banded, and dark red. They carry 57 to 60% Al_2O_3, about 20% Fe_2O_3, and 3 to 5% SiO_2, and they are rather low in water. Some of the bauxite is hard and splintery. It is chiefly used for making aluminum and chemicals. Much of it has to be extracted by underground methods of mining.

Reserves have been estimated at 16 million tons with a maximum content of 7 to 8 percent silica.

Hungary, Yugoslavia, Greece, Italy. These countries have more than twice as much bauxite as the United States. The deposits here, as in France, occur in limestones, are really a part of the same belt of deposits, and are of similar origin and occurrence. They lie in pipes, pockets, and lenses (Figs. 13–34, 5·7–5), in residual clay that mantles the old surface and lines solution holes in Jurassic, Cretaceous, and lower Eocene limestones. Many lie in karst depressions. They are unconformably overlain by Upper Cretaceous or Eocene sediments. They were formed under different climatic conditions than exist today and during erosion intervals that preceded the Upper Cretaceous, lower Eocene, or middle Eocene. Those of Dalmatia (Fig. 13–34A) are of Upper Cretaceous and lower Eocene age; those of Gánt, Hungary, are of middle Eocene age. Invariably the deposits lie on unconformities.

These bauxites range from 50 to 65 percent alumina and are low in silica and iron.

British Guiana.   The British Guiana deposits, and also the deposits of Dutch Guiana, according to Harder, were formed on an old base-level surface that truncates schists and gneisses and basic and silicic

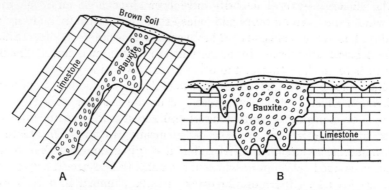

Fig. 13–34.   Sections of typical bauxite.   *A*, bauxite bed in Dalmatia; *B*, bauxite pocket in Istria.   (*After Harder, Aluminum Industry, Vol. 1.*)

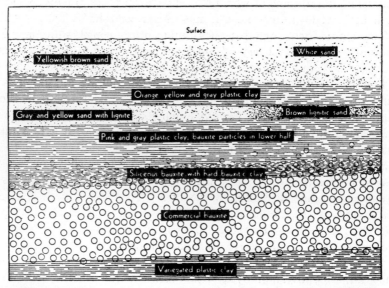

VERTICAL SECTION THROUGH TYPICAL BAUXITE BED, BRITISH GUIANA

Fig. 13–35.   Vertical section through bauxite bed and underlying clay, Trevern mine, British Guiana.   (*Harder, Bull. Can. Min. Inst.*)

igneous rocks.   First, residual clays were formed, which were then changed to bauxite.   A Tertiary coastal depression caused inundation, and sand and clay covered some of the bauxite accumulations; others were washed into the advancing sea. Elevation, still proceeding to-day, permitted erosional cutting through the unconsolidated sediments, exposing the bauxite deposits as beds or flat lenses up to 40 feet thick (Fig. 13–35). Several deposits have been discovered under a cover of white sand. Small pipes and veins extend down into the underlying clay. The top is siliceous. The overburden is up to 60 feet thick. The deposits are worked by power shovel in large open cuts. The ore contains 58 to 63% Al_2O_3, 2 to 5% SiO_2, and 3 to 6% Fe_2O_3.

Gold Coast. The deposits of the Gold Coast, studied by Cooper and others, constitute large resources. They lie on flat erosion remnants of a peneplaned surface developed across folded sediments, volcanics, and shales (Fig. 5·7–6). They mostly rest on a thin band of "lithomarge," a cream-colored colloidal claylike material that represents a residual clay from which the bauxite was derived. The best bauxite occurs on the most dissected peneplain remnants, and Cooper thinks it underwent enrichment during dissection. It is overlain by a few feet of red soil and in places by swampy materials. The bauxite beds are as much as 60 feet thick. One 15-foot bed in Yenahin averages 64% Al_2O_3, 1.9% Fe_2O_3, 2.4% SiO_2, and 4.1% TiO_2. About 60 percent of the Yenahin deposits contain 55 percent alumina. The average thickness is 33 feet. Reserves are estimated in tens of millions of tons.

Other Deposits. *Russia* contains extensive deposits of bauxite in the Leningrad district near Tikhvin, in the Urals, and in the Belovo region, Siberia. The Ural Devonian bauxites lie between marine strata and are thought by Arkhangelskii to be marine sedimentary deposits derived from Paleozoic weathering of lavas and tuffs, a conclusion with which others differ. The Triassic deposits of the Urals and the Carboniferous deposits of the Leningrad district are interpreted to be transported bauxites deposited in lakes. The Ural deposits contain up to 62% Al_2O_3, from $3\frac{1}{2}$ to $5\frac{1}{2}$% SiO_2, and up to 27% Fe_2O_3. The Tikhvin bauxites are much higher in silica. Diaspore and chlorite have developed in the altered bauxite of Ivdel.

The *Netherlands Indies* are large producers of bauxite from the islands of Batam and Bintang. The beds are about 12 feet thick and contain 53% Al_2O_3, $2\frac{1}{2}$% SiO_2, and 13% Fe_2O_3. In *India* bauxites in the Jabalpur and Khaira districts are thought to have been formed from underlying original kaolin beds after the covers had been removed. Fermor states that some of the deposits resulted from the surface weathering of an amygdaloidal trap. In *Germany* bauxite is also reported to be derived from basalt. The bauxite of *Minas Geraes, Brazil*, occurs in lenticular lenses, between itabirite and slate, associated with residual clay. Large deposits in *Jamaica, Haiti,* and the *Dominican Republic* are high in iron (colloidal), which offers treatment difficulties.

Selected References on Bauxite

General references are given under Bauxite, Chapter 5·7.

Arkansas Bauxite Deposits. G. C. BRANNER. XVI Int. Geol. Cong. Guidebook A2:92–102, 1932. *Brief résumé of ores and occurrences.*

Les bauxites de la France méridionale. J. DE LAPPARENT. Minist. Trav. Pub., Mém. Carte géol., Paris, 1930. *French occurrences.*

Bauxite in British Guiana. E. C. HARDER. Can. Inst. Min. and Met. Bull., November, 1936. *Excellent description of occurrence and origin.*

Bauxite Deposits of the Gold Coast. W. G. G. COOPER. Gold Coast Geol. Surv. Bull. 7, 1936. *Occurrence and origin.*

Bauxite and Aluminous Laterite Occurrences of India. C. S. FOX. India Geol. Surv. Mem. 49, 1923.

Aluminum Industry of Europe. J. E. COLLIER. Econ. Geog. 22:75–108, 1946. *Bauxite reserves and production.*

Relations of Bauxite and Kaolin in Arkansas Bauxite Deposits. M. GOLDMAN and J. I. TRACEY, JR. Econ. Geol. 41:567–575, 1946. *Thesis that bauxite is derived from syenite rather than intermediate kaolin.*

Stratigraphy and Origin of Bauxite Deposits. E. C. HARDER. Geol. Soc. Am. Bull. 60:887–908, 1949. *Broad discussion of origin of bauxite; good bibliography.*

General References 7, 11, 16.

CHAPTER 14

IRON AND FERROALLOY METALS

Iron

Although iron is the second most abundant metal in the earth, the character of its natural compounds prevented its use as early as some other metals. It was known by 4000 B.C., and the Egyptian Pharaohs regarded it more highly than gold, but this probably was the rare meteoric iron. Apparently by 1200 B.C. iron was manufactured but was still rare, and its industrial use did not commence before 800 B.C., which dates the start of the Age of Iron. Steel came into use about 800 years later, and the blast furnace in the fourteenth century. During the sixteenth century the forests of Great Britain were denuded to supply charcoal to smelt iron ore, but this waste became unnecessary when the great discovery was made about 1710 that coal could be used to reduce iron ore. This was the beginning of the great industrial age of iron that culminated in the steel age made possible by Bessemer's discovery in 1856. In the nineteenth century Great Britain with her resources of iron and coal became the first and greatest of the modern industrial nations.

In America iron was discovered in Virginia in 1608, and the first large-scale smelting began in Massachusetts in 1664 and in Pennsylvania in 1730. With the use of anthracite for iron making in the early nineteenth century the industry centered around the anthracite region of Pennsylvania, but with the advent of coke as fuel it jumped across the mountains to what is now Pittsburgh. The great Lake Superior iron-ore deposits, discovered in 1844, ushered in the industrial age of the United States. Where coal and iron met, in Pennsylvania and along the Great Lakes, great industrial centers arose, railroads were pushed afar, and a new era of United States development began.

Uses. Iron is the backbone of modern civilization. Few are aware to what extent we have become dependent upon it in homes, farms, cities, machines, automobiles, trains, and ships. Without it we would spin our clothes by hand and travel in wooden carts over dusty roads. When iron, or steel, is not suitable for certain uses, it is alloyed with other substances to make it suitable. To enumerate the various uses

561

of iron would be to compile a history of the innumerable creations of modern civilization and industry. Each of the main types of iron — steel, cast iron, wrought iron, and iron alloys — has its particular sphere of use; steel, of course, exceeds all others.

Production. The world production of iron ore, responding to economic conditions, ranges between 150 and 240 million tons annually. Steel production ranges from 100 to 160 million tons. The chief producing countries and their approximate normal percentage of world production are shown below.

IRON ORE		STEEL	
Country	Percent of World	Country	Percent of World
1. United States	50	1. United States	58
2. France	11	2. Germany	12
3. Russia	10 ?	3. Russia	10
4. Sweden	8	4. Great Britain	10
5. Great Britain	8	5. France	3
6. Germany	5	6. Japan	4
7. Luxembourg	1.5	7. Others	3
8. India	1.5		
9. Spain	1.5		
10. Australia	1.3		
11. Algeria	1.1		

Canada, Newfoundland, Brazil, and Chile each produce under 1% of iron ore, but Canadian and Venezuelan production will increase.

The United States annual production of iron ore ranged during the forties from 60 to 90 million tons, divided as follows:

STATE	PERCENT OF COUNTRY	PERCENT Fe
1. Minnesota	68	51.5
2. Michigan	10	51.8
3. Alabama	9	35.8
4. New York, Pennsylvania	5	62.2
5. Utah	1.5	53.7
6. Wisconsin	1.2	54.1

Mineralogy. The economic iron-ore minerals are shown below.

MINERAL	COMPOSITION	PERCENT Fe	COMMERCIAL CLASSIFICATION
Magnetite	Fe_3O_4	72.4	Magnetic (or black) ores
Hematite	Fe_2O_3	70.0	Red ore
"Limonite"	$Fe_2O_3 \cdot H_2O$	59–63	Brown ore
Siderite	$FeCO_3$	48.2	Spathic, black band, clay-ironstone

Of these, magnetite is the richest, but of minor quantity, hematite is the mainstay of the iron industry, limonite and siderite are of minor importance in America but important in Europe. In an average year the percentages of the different ores produced in the United States are: hematite 94, magnetite 5, and brown ore 1.

Associations. Common impurities in iron ores are silica, calcium carbonate, phosphorus, manganese (especially in hematite), sulphur, alumina, water, and titanium.

Recovery. The making of a usable product of iron involves two steps: first the reduction of the iron ore to pig iron, and secondly the treatment of the pig iron to make cast iron, wrought iron, or steel.

The ore is smelted with coke and limestone. Air, blown in at the bottom, burns coke to carbon monoxide, which removes oxygen from the iron ore, reducing it to the metal. The limestone slags off the silica, alumina, and other impurities. Two tons or less of good iron ore and scrap iron yield 1 ton of pig iron. The pig iron always contains carbon, and the percentage of included constituents determines the use to which the pig is put. Various types of pig iron are: basic open hearth, foundry, bessemer and low phosphorus, malleable, and forge. Specialized pig, high in silicon or manganese, is also produced.

Cast iron is obtained by melting the pig in a cupola furnace. For certain castings the pig iron may be purified or other ingredients added to it. For *wrought iron,* which is rather pure, the pig goes to a puddling furnace, where the impurities are slagged off by stirring, and the red-hot iron is hammered to desired shapes. Wrought iron is malleable and ductile, and it resists corrosion.

Steel is iron with alloyed carbon, which is generally less than 1 percent but may reach 1.6. "Mild" steels with low carbon approach wrought iron. Steel is made by several processes. In the *open-hearth* method, now the most extensively used, molten pig iron, some hematite, and limestone are melted in a tilting furnace; the excess carbon and silicon are slagged off by the oxygen of the hematite; sulphur is vaporized and also unites with manganese to form MnS. Phosphorus is removed by uniting with calcium from the refractory brick lining, to prevent the steel being "cold short" (brittle). Pig high in phosphorus is treated in a basic open hearth or one lined with basic brick. An acid open hearth is used for low-phosphorus pig. In the *bessemer process,* low-sulphur, low-phosphorus pig is rapidly converted into steel in a tilting barrel-shaped furnace through which air is blown to slag off the impurities. Desired amounts of carbon and manganese are added and mixed by air blowers. For the various ferroalloys, proper amounts of alloy metals are added to yield the desired steel.

Kinds of Deposits and Origin

The various types of commercial deposits of iron ore, and some important examples of each are:

1. Magmatic: Magnetite, titaniferous magnetite — *Kiruna* and Taberg, Sweden; Iron Mountain, Wyo.; Adirondacks, New York.
2. Contact-metasomatic: Magnetite, specularite — Fierro, N. Mex.; *Cornwall*, Pa.; Iron Springs, Utah.
3. Replacement: Magnetite, hematite — *Lyon Mountain*, N. Y.; Dover, N. J.; Iron Mountain, Mo.
4. Sedimentary: Hematite, limonite, siderite — *Clinton ores* of New York to Alabama; *Wabana*, Newfoundland; *Minette ores* of central Europe; Jurassic ores of England; Brazil.
5. Residual: Hematite, limonite — *Lake Superior;* Appalachians; Cuba; Bilbao, Spain; *Pao*, Venezuela; Labrador-Quebec.
6. Oxidation: Limonite, hematite — Rio Tinto, Spain.

Their origin is given in Chapter 5. The sedimentary, magmatic, and residual processes give rise to the most important deposits; but some, such as the Lake Superior deposits, have been formed by a combination of two or more processes, e.g., sedimentation, followed by residual enrichment and by metamorphism.

Distribution of Iron Ores

As would be expected of multi-origin deposits, iron ores are widely distributed under various geologic conditions. They occur in basins of sedimentation, with eroded, deep-seated intrusives, and where deep tropical weathering prevails.

In *North America* magnetite deposits occur in the deeply dissected regions of plutonic intrusions, such as the northeastern states and the Cordillera. Hematite deposits outcrop around the margins of the great sedimentary basin from Alabama to New York to Wisconsin and in Newfoundland. They are abundantly and richly concentrated in the Lake Superior region. Residual deposits occur in the eroded Appalachians and in Cuba.

In *Central Europe* great sedimentary deposits underlie parts of Lorraine, Luxembourg, France, Belgium, and Germany. There are rich oxide deposits of igneous and metamorphic origin in Sweden. Farther east and north are the extensive deposits of the Ukraine and European Russia.

In *Africa* good-quality ores lie near the Mediterranean in Morocco and Algeria and large bodies of low-grade magnetite lie to the south in the Transvaal.

In *South America* there are extensive deposits in Brazil, Chile, and Venezuela.   *Asia* has stupendous resources in India and minor ones in China.

EXAMPLES OF DEPOSITS IN THE UNITED STATES

Lake Superior Region.   The iron ranges in Minnesota, Wisconsin, and Michigan (Fig. 14–1), known as the *Cuyuna, Vermilion, Mesabi, Marquette, Menominee,* and *Gogebic,* are the largest and richest hematite

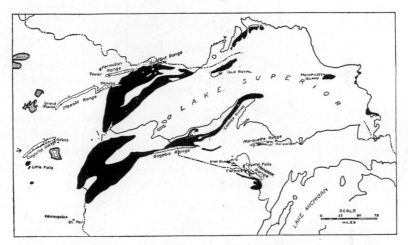

Fig. 14–1.   Lake Superior region showing locations of iron ranges and basic igneous intrusives and extrusives.   (*Gruner, Econ. Geol.*)

deposits of the world, having produced about 2½ billion tons of iron ore.   The Mesabi is the most important.   All the deposits are Precambrian in age, the Vermilion being Archean and the others Middle or Upper Huronian.

The deposits are the weathered portions of former sedimentary iron-ore beds from which silica and other substances have been leached by circulating waters, causing a residual concentration of the iron oxides.   The original beds or iron formations consist of interbanded chert and hematite called jasper, ferruginous chert, or taconite.   The iron was present originally in the form of siderite, greenalite, and hematite and made up about 30 percent of the formation.   The present ores grade downward into these minerals.   Leith considers that the ore was concentrated solely by Precambrian surface waters which leached silica, leaving a residual mass of porous hematite (Fig. 14–2), in a manner similar to the accumulation of residual bauxite and clay.

Gruner, on the other hand, thinks that the leaching was effected by hydrothermal solutions.

The iron formations, up to 1,000 feet thick, are part of a thick, sedimentary series with intercalated volcanic rocks. These have been

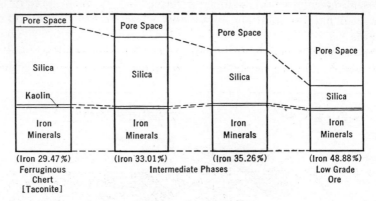

FIG. 14–2.   Alteration of Lake Superior iron formation to iron ore by leaching of silica.   (*After Leith, U. S. Geol. Survey.*)

strongly folded and crumpled in most of the ranges and metamorphosed into slates, quartzites, marbles, schists, cherts, and jaspers. In some places regional and contact metamorphism have converted the iron formations to magnetite-amphibole schists and the hematite to specu-

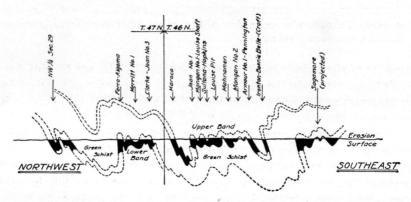

FIG. 14–3.   Geologic section across part of Cuyuna range, Minnesota, showing structure, relation between upper and lower ore formations, and relation to surface.   (*Zapffe, Econ. Geol.*)

larite. Some geologists (Van Hise and Leith) attribute silica in the iron formations to a magmatic source and deposition on the sea floor. Others (Grout and Gruner) believe that they came from the weathering of an adjacent Precambrian land mass and were deposited as regular sedimentary beds. The work of Moore and Maynard demonstrating the natural solution, transportation, and deposition of iron and silica supports this view. Woolnough suggests that such deposits represent epicontinental sediments formed as chemical precipitates from cold natural solutions in isolated closed basins on peneplaned surfaces.

The ore occurs mainly in pitching synclinal troughs (Fig. 14–3) in large bodies that are irregularly shaped and are banded or bedded.

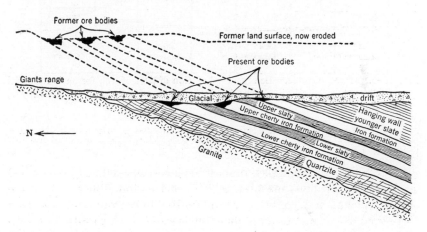

Fig. 14–4. Mesabi range, Minnesota. Diagrammatic cross section showing downward migration of ore bodies with erosion. (*Newhouse, Ore Deposits as Related to Structural Features.*)

They generally bottom on impervious basements that guided and controlled the water circulation. In most of the ranges the enclosing rocks are strongly folded, with high dips, and in such places the ore extends to the surprising depths of 3,000 to 4,000 feet, although the best ore is generally above 1,000 feet. Most of the deep deposits are complex in structure and are mined by underground methods. However, in the Mesabi range, which yields about 80 percent of the Lake Superior ore, the rocks dip gently and the ore has been concentrated over a wide area to depths mostly less than 200 feet (Fig. 14–4), which permits mining on an enormous scale from large open cuts by power shovel; the maximum depth of ore is 900 feet. This range yields 20 to 40 million tons of ore annually, containing around 56 percent iron, and still has large reserves. The other ranges (Figs. 14–5, 14–6) exhibit complexities in rocks and structure that result in diverse details of origin.

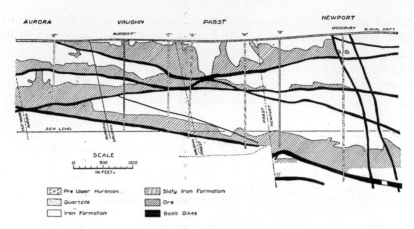

FIG. 14–5. Gogebic range. Longitudinal section of part of iron formation parallel to dip, showing ore relation to basic dikes and iron formation. (*J. W. Gruner, Econ. Geol.*)

The ore is somewhat different in each range. The iron content lies between 50 and 62 percent; sulphur is generally low; phosphorus

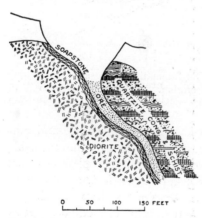

FIG. 14–6. Marquette district, National mine. Replacement deposit in Goodrich quartzite. (*Gruner-Van Hise-Bayley, Econ. Geol.*)

averages about 0.09 percent; silica 7 to 8 percent, alumina about 2 percent, and moisture about 11 percent; other constituents are low. Manganese is important only in the Cuyuna range, where it rarely drops below 5 percent and reaches 17 percent; manganiferous ores of 6 to 8 percent manganese are now produced. The minerals are chiefly hematite with subordinate magnetite and limonite; magnetite has altered to martite (hematite) in the Mesabi. Both bessemer ($P < 0.05\%$) and nonbessemer ($P > 0.05\%$) ores are produced. The Mesabi range yields three grades of ore: (1) blue ore (of magnetite derivation) with 59% Fe, (2) brown ore with 55 to 56% Fe, and (3) low-grade ore with 50% Fe.

Clinton Iron-Ore Beds. The Clinton sedimentary iron-ore beds rank next in importance to the Lake Superior deposits. These beds outcrop across Wisconsin and New York and south to Alabama. They have been mined at numerous places, and at Birmingham, Ala., they have given rise to a large industrial steel center owing to the contiguous coking coal and limestone flux.

The ore occurs as sedimentary beds intercalated with shale, limestone, and sandstone of Clinton (Silurian) age. There may be from one to four beds, which thicken and thin from a few inches to 30 feet.

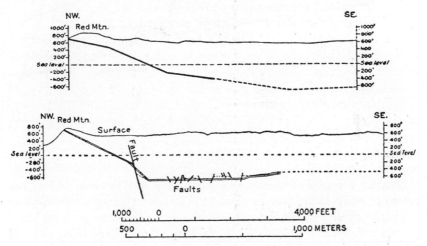

FIG. 14–7. Typical profile of red ore mines, Birmingham, Ala. (*Burchard, 16th Internat. Geol. Cong.*)

Their greatest development is at Red Mountain, Birmingham, where two of four beds are workable, the production being mainly from the Big Seam.

The types of ore are: (1) oölitic ore, composed of round oölites of hematite 1 to 2 mm across, enclosed in a matrix of hematite and calcite; (2) fossil ore, composed of fossil fragments coated and partly replaced by hematite and enclosed in amorphous and oölitic hematite, and high in calcium carbonate; (3) flaxseed ore, composed of flattened concretions of hematite surrounded by hematite mud and replaced fossil fragments. The primary ore is hard and is high in calcium carbonate, which is leached out near the outcrop, producing thereby enriched "soft" ore. The soft ore may carry 50 to 60 percent iron, but the "hard" ore runs only 35 to 38 percent. The high calcium carbonate content makes the ore largely self-fluxing; insoluble matter runs 12 to 14 percent.

At Birmingham, the Big Seam outcrops for a length of 20 miles and dips on an average less than 30° (Fig. 14–7). It is 16 to 30 feet thick, of which 7 to 12 feet is minable; in places it is divided by a shale parting. The ore averages about 36 percent iron and is mined from inclined shafts that extend 4,000 to 6,000 feet down the slopes. The annual production ranges from 5 to 6 million tons of ore, and Burchard estimates the reserves at 1.4 billion tons of good ore and 0.5 billion tons of second-grade ore.

The origin of the Clinton ores has emerged from controversy that had an important bearing upon their development. A former theory of *residual enrichment* retarded their development because it implied near-surface restriction. This theory was discarded when it was realized that the surficial enrichment was merely surface leaching of calcium carbonate, forming richer soft ores. A theory of replacement was also discarded. Today it is generally accepted that they are of straight sedimentary origin (see Chap. 5·5), which implies down-dip continuity and relatively wide distribution, such as would encourage exploitation.

The oölitic sedimentary hematite ores of *Wabana, Newfoundland*, are generally similar to the Clinton ores, except that, according to Hayes, they contain considerable chamosite and siderite. The beds are 10 to 30 feet thick, and one of them has been mined for a distance of nearly 2 miles under the sea.

Cornwall, Pa. At Cornwall, a thick quartz diabase dike and sill intrude Cambrian limestone, resulting in two large contact-metasomatic bodies of magnetite containing subordinate iron and copper sulphides, and some cobalt. The enclosing rock now consists chiefly of diopside, actinolite, phlogopite, chlorite, calcite, and dolomite. Hickok shows that the minerals are arranged zonally outward from the diabase contact, first a diopside and then an actinolite zone within which the magnetite bodies occur. The magnetite occurs in three forms — normal isometric, anisotropic, and maghemite. The exhalations from the diabase, in the order of occurrence, contained potash, silica, iron oxides, base-metal sulphides, and hydrous silicates. Magnetite constitutes 40 to 60 percent of the ore. The western body is 4,500 feet long, 200 to 600 feet wide, dips 25° S., and diminishes in depth. The eastern body is slightly smaller. The deposit, which has been mined since 1742, has yielded over 40 million tons and produces about 1 million tons annually.

Other contact-metasomatic iron deposits occur at *Iron Springs, Utah*, where monzonite porphyry intrudes limestone, forming hematite and magnetite in contact silicates, and at *Hanover, N. Mex.*, where granodiorite porphyry has yielded a large aureole of contact-metasomatic silicates, including bodies of magnetite with hematite and sulphides of iron, copper, and zinc.

Other Magnetite Deposits. Important deposits of magnetite occur in New York, New Jersey, North Carolina, Missouri, and Wyoming. At *Lyon Mountain, N. Y.*, Gallagher describes valuable bodies of non-titaniferous magnetite, that occur as high-temperature pneumatolytic replacements of gneissic granite. The largest body is 5,000 feet long and 20 feet wide and extends 1,600 feet down the dip. The ore carries 60 to 70% Fe, and milling ores run about 30 to 50% Fe. Within the ore are miarolitic cavities, large enough to admit a man, containing huge crystals of orthoclase, quartz, albite, hastingsite, aegirinaugite, magnetite, and titanite and water under high pressure.

Similar ores are mined at *Dover, N. J.*, and at *Mineville, N. Y.*, where magnetite and apatite occur with pegmatitic minerals. A fair-sized replacement body of hematite-magnetite is worked at *Iron Mountain, Mo.* In the *Adirondacks* numerous deposits of titaniferous magnetites are associated with anorthosite and gabbro, and one at Lake Sanford yields concentrates of ilmenite and magnetite. Generally similar ores with anorthosites occur at *Iron Mountain, Wyo.*

Other Hematite Deposits. Replacement bodies of hematite at *Pilot Knob* and *Iron Mountain, Mo.*, constitute important local sources of iron ore. The *Hartville district, Wyoming,* is the chief source of Rocky Mountain iron ores. The *Sunrise deposits* there have yielded over 16 million tons, and reserves are large. Stone has shown that these Precambrian hematites are similar in origin to the Lake Superior deposits, and he found other large bodies hidden beneath Paleozoic limestone.

EUROPEAN IRON ORES

The Lorraine Minette Ores. The " Minette " ores or oölitic limonites of the *Lorraine-Luxembourg* region rank next in importance to those of the Lake Superior region. They also extend into parts of Belgium and Germany.

The ores are sedimentary deposits interbedded with Jurassic shales, limestones, and sandstones and occur in synclinal basins that pitch southwestward. They outcrop in Lorraine along the eastern limb of the Paris Basin and dip beneath France. Formerly the Lorraine outcrops were worked cheaply in open cuts by Germany, but now most of the ore comes from the down-dip portions in France, just west of Lorraine, at depths of 600 feet; but they are known by drilling to a depth of 3,000 feet.

Seven ore beds occur, ranging in thickness from a few inches to 25 feet. They are lenticular along the strike and are not everywhere workable, the Longwy-Briey, Metz-Thionville, and Nancy Basins being the important areas. The ore is soft, earthy, oölitic limonite and hematite with some siderite and iron silicates. The grade is quite low, averaging 30% Fe with 0.5 to 1.8% P, 5 to 12% CaO, and 7 to 20% SiO_2. A little magnetite in the form of oölites is present, and rarely pyrite. The ores are clearly of sedimentary origin, and, although Cayeux thought in 1909 that the iron oölites replaced calcite oölites, they probably are no different from other oölitic iron ores, which are generally regarded as of direct deposition with replacement on the sea bottom.

The deposits cover a wide area. When Lorraine was German there were some 50 mines yielding around 20 million tons of ore annually, which supplied former German needs for iron ore. An equal number of mines yielded similar tonnage in France. Now there are fewer mines, and the French production from these ores has ranged from 32 to 50 million tons. Despite their low grade they are of great economic importance. The reserves are estimated at 5 billion tons of ore.

German Iron Ores. After 1918, Germany fell back on low-grade sedimentary limonites, chiefly in the Siegerland, Salzgitter, Lahn-Dill, and Weser areas and in parts of Bavaria and Württemberg. These ores are mainly sedimentary limonites, but some are siderite or hematite. They are rather thin beds, are distant from coal, and carry about 32 percent iron; higher-grade ores are imported for mixing. The high-phosphorus Salzgitter ores, carrying 30% Fe and 25% SiO_2, are inclined oölitic sedimentary beds which are being mined by surface and underground methods at depths of as much as 2,000 feet. These ores yield about 3 million tons a year, and the reserves are stated to be 1 billion tons. German ores supply 5 to 10 million tons annually.

The deposit of *Erzberg* in *Austria* is a low-grade siderite body which can be calcined up to 50 percent iron and low phosphorus and which yields 2 million tons annually.

Great Britain. Great Britain achieved greatness with her coal and iron, but newer and richer iron deposits elsewhere have provided severe competition to the low-grade English ores. Her production ranges from 12 to 20 million tons annually, nine-tenths of which comes from low-grade Jurassic sedimentary beds, averaging 28 percent iron. This is a phosphorus ore of oölitic siderite and silicates in beds from 6 to 25 feet thick, which are mined by open cuts in the Northampton, Rutland, Frodingham, Cleveland, Leicester, Oxford, and intervening districts. The total reserves are estimated at 3 billion tons of ore.

The nonphosphorus hematite deposits of Cumberland and Lancashire containing 53 percent iron were formerly of great importance but now yield less than a million tons annually.

Spain. The large iron ore deposits of *Bilbao*, northern Spain, have yielded several million tons annually for export to Europe. A 250-foot bed of Cretaceous limestone is cut by minor diabase and trachyte of Tertiary age. The limestone has been hydrothermally replaced by masses of siderite with minor ankerite and pyrite. Weathering has produced surficial residual concentrations of hematite and limonite.

Sweden. Sweden ranks fourth in world iron-ore production and contains the largest deposits of high-grade ore in the world. The annual production is 6 to 14 million tons, most of which comes from the great *Kiruna area* in Lapland.

The Kiruna deposit forms a ridge 2.8 kilometers long and 350 meters high, the backbone of which is the steeply dipping magnetite body, which is from 30 to 150 meters wide (Fig. 5·1–3). The enclosing rocks are Precambrian. The foot wall is syenite porphyry. It is penetrated by dikelets of magnetite which in places form " ore breccia " composed of fragments of syenite in a magnetite matrix. Similar contact relations exist in the hanging wall quartz porphyry.

The ore consists dominantly of magnetite with intergrowths of apatite of simultaneous or later crystallization. Some pyroxene is present. The ore carries from 56 to 71% Fe, 0.03 to 1.8% P, 0.5% S, 0.7% Mn, 1.5% SiO_2, 0.7% Al_2O_3, 3% CaO, and 0.3% TiO_2, and is mostly of bessemer grade. Across the lake from Kiruna is the similar but smaller Luossavaara body which is some 4,000 to 5,000 feet long and 130 feet wide.

The ore is generally considered to be a magmatic differentiation in depth, and the evidence appears conclusive that the ore was injected in a molten condition. It is an extreme differentiation to almost pure iron oxide, with minor titanium, phosphorus, and fluorine. Vogt considered that the differentiation took place by crystal settling and later remelting, a view not in keeping with Bowen's conclusions. Geijer concludes that the " ores represent the last crystallizing parts." Any molten intrusion of magnetite is difficult to account for except by filter pressing. (See Chap. 5·1.)

At *Gellivare*, 50 miles south of Kiruna, are generally similar tabular deposits of steeply dipping lenses 30 to 100 feet thick of magnetite and hematite, which represent a metamorphosed Kiruna type of deposit. These yield about 1.5 million tons annually.

The estimated reserves of the Kiruna deposits are 226 million tons of developed ore or 520 million tons of possible ore, and those of Gellivare are 239 million tons of possible ore. Most of the Swedish ore is normally exported to Germany and England.

Russia. Russia had an annual pre-war output of 20 to 30 million tons of ore, about two-thirds of which came from Krivoi Rog in the Donetz Basin in the Ukraine, and the remainder from the Urals. There are nine important districts.

The *Krivoi Rog* deposits occur in a belt 25 miles long in which are several beds, lenses, and chimneys in metamorphosed Precambrian schists. The ore is martite, hematite, and some magnetite, averaging 57% Fe, 8% SiO_2, 0.25% Mn, and 0.05% P. The deposits resemble the Lake Superior ranges and are mined by underground methods.

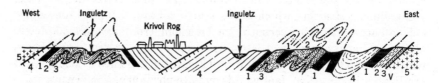

Fig. 14–8. Section across Krivoi Rog. 1, iron-ore beds; 2–4, schists; 5, gneiss. (*Putzer, Zeit. f. ange. Mineral.*)

The *Ural* deposits, chief of which is at Mount Magnitnaya, with an annual output of 6 million tons of ore of 45 percent iron, are mostly magnetite ores of contact-metasomatic origin. Other deposits of decreasing importance are Sverdlovsk, Vysokaya, Bakal, and Blagodat.

The iron-ore reserves of Russia have been claimed to be 10.6 billion tons, exclusive of the enormous tonnages of low-grade quartzitic ores of *Kursk*.

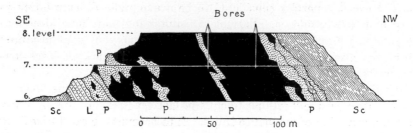

Fig. 14–9. Main ore body, Minas del Rif, Spanish Morocco. Black is ore; *sc*, schist; *L*, sidero-calcite; *p*, altered porphyry dikes. (*Heim, Econ. Geol.*)

OTHER IRON ORE DEPOSITS

North Africa. Algeria yields around 3 million tons of iron ore annually, Tunisia 1 million, and Spanish Morocco 1.5 million tons. The Algerian deposits are mainly of the Bilbao type, i.e., siderite replacement deposits in Cretaceous limestones now residually concentrated to hematite, but sedimentary and contact deposits are also known. The ores average about 55 percent iron and are of bessemer grade. Reserves are about 160 million tons.

The deposits of Spanish Morocco near Melilla (Fig. 14–9) are high-temperature replacement ores of magnetite and hematite carrying 62 percent iron with low phosphorus. Reserves are about 30 million tons.

India. India produces about 3 to 5 million tons. However, vast new reserves described by H. C. Jones and J. A. Dunn indicate greater future production. The chief sources are the Bihar and Orissa (Singhbhum) districts which Jones states contain 4 billion tons of 60 percent ore. The ores occur in the Archean Iron Ore Series (Newer Dharwar) consisting of a sedimentary-volcanic series including a banded hematite-quartzite member with iron-ore bodies. The bedded ores consist of massive, laminated, and powdery hematite with a little magnetite and martite, carrying 60 to 69 percent iron and low phosphorus and manganese. The banded ironstones extend for 30 miles, have a width up to 1,000 feet, and dip 70°.

Jones considers that the banded hematite-quartzites resulted from original deposition, giving rise to a rock containing 28 to 30 percent iron and identical in nature and origin with the original iron formations of the Lake Superior region. Subsequent leaching of parts of this removed silica, leaving residually concentrated iron ores in beds up to 120 feet thick. Dunn believes that the banded ironstones were originally tuffs and sandy beds laid down under subaerial conditions and replaced in part by silica and iron oxides to yield secondary quartzites and iron formation. Part of the silification may have been contemporaneous with deposition. Subsequent removal of silica produced the high-grade hematites. Some low-grade lateritic iron ores also occur.

Chile. Important deposits occur in Coquimbo, near Cruz Grande, where there are 100 million tons of 65 percent iron ore low in silica, phosphorus, and sulphur. Annual production has been 1.5 million tons. The ores consist of hematite and some magnetite, with hematite replacing magnetite. They occur in diorite, and Miller and Singewald believed they are magmatic differentiations of the magma that gave rise to the enclosing diorite.

Brazil and Venezuela. At Itabira, Minas Geraes, Brazil, there is a body of hard hematite, in excess of 200 million tons, containing around 65 percent hematite with low phosphorus. The ore bodies are huge lenses in sedimentary schists of Precambrian age. The purity of the ore makes it ideal for blending with high-phosphorus iron ores of Newfoundland and Europe.

Near Pao, Venezuela, the largest deposit of South America, composed of hard hematite ore, is undergoing development.

Cuba. Residual concentrations of iron ore resulting from the weathering of serpentine occur in the Mayari, Mao, and San Felipe districts. They are surficial mantles on plateaus, which, in Mayari, cover an area 4 by 10 miles to an average depth of 15 feet and contain several hundred million tons of workable ore. The ores consist of hematite, limonite, and some magnetite. Typical ore contains about 54% Fe, 11 to 13% H_2O+, 5–7% SiO_2, 10–12% Al_2O_3, about 1.7% Cr, 0.01 to 0.5% Ni, 0.5% Mn, 0.01% P, and water. The chromium and nickel content is unusual but represents concentrations of original constituents in the serpentine. The deposits are iron-rich laterites. Contact-metasomatic deposits with magnetite, hematite, and high sulphur occur near Santiago.

China. Fairly large reserves of iron ores are found in China. The most important are contact-metasomatic deposits of hematite and magnetite averaging 56 percent iron, with low phosphorus and moderate sulphur. The best of these occur in a belt along the lower Yangtze Valley. Extensive Precambrian sedimentary beds of oölitic hematite in Chihli contain 55 percent iron; low-grade Archean banded ironstones also occur in Chihli. The most extensive reserves occur in Manchuria.

Canada. Extensive deposits of iron ore have been developed at Steep Rock, Ontario, and in Quebec-Labrador where bodies in excess of 300,000,000 tons of high-grade hematite ore have been developed.

Selected References on Iron

Iron Ores, Occurrence, Valuation, and Control. E. C. ECKEL. McGraw-Hill, New York, 1914. *Broad treatment of iron ores of the world.*

Iron Ore Resources of the World. XI Int. Geol. Cong., Stockholm, 1911.

Iron and Steel Industries of Europe. C. W. WRIGHT. U. S. Bur. Mines Econ. Paper 19, 1939. *A brief survey made from personal observations.*

Geology of the Lake Superior Region. C. R. VAN HISE and C. K. LEITH. U. S. Geol. Surv. Mon. 52, 1911. *A classical monographic survey.*

Lake Superior Region. W. O. HOTCHKISS. XVI Int. Geol. Cong. Guidebook 27, Washington, 1933. *A concise summary of the iron ore deposits.*

The Origin of Sedimentary Iron Formation (Mesabi). J. W. GRUNER. Econ. Geol. 17:407–460, 1922.

Hydrothermal Leaching of Iron Ores of the Lake Superior Type. J. W. GRUNER. Econ. Geol. 32:121–130, 1937.

Brown Iron Ores, Tennessee. E. F. BURCHARD. Tenn. Geol. Surv. Bull. 39, 1934. *Residual ores.*

Birmingham District, Alabama. E. F. BURCHARD. XVI Int. Geol. Cong. Guidebook 2, Washington, 1933. *A concise summary of sedimentary Clinton hematite.*

Iron Ores of the Kiruna (Sweden) Type. P. GEIJER. Swedish Geol. Surv. Ser. C, 367, 1931. *Detailed geology of magmatic magnetite deposit.*

Ores and Industry in the Far East. H. FOSTER BAIN. Council of Foreign Relations, 1933. *Iron ores of China and Japan.*

Iron Ore Deposits of Bihar and Orissa. H. C. JONES. India Geol. Surv. Mem. 63, 1934. Also J. A. DUNN, Econ. Geol. 30:643, 1935. *Geology of India iron-ore deposits.*

Solution, Transportation and Precipitation of Iron and Silica. E. S. MOORE and J. E. MAYNARD. Econ. Geol. 24, Nos. 3, 4, 1929. *Experiments on cold water solution and deposition of iron.*

Iron Ore and the Steel Industry. C. M. WHITE, in Seventy-Five Years of Progress in the Mineral Industry. A. I. M. E., New York, 1947. *Survey of world supply and U. S. resources and production.*

Wabana Iron Ores of Newfoundland. A. O. HAYES. Econ. Geol. 26:44–64, 1931. *Description, occurrence, and origin of these sedimentary deposits.*

The Ukraine in U. S. S. R. N. SVITALSKI. XVII Int. Congress Guidebook, pp. 51–76, Moscow, 1937. *Description of important Russian deposits.*

Canadian Iron-Ore Deposits. **General Reference 15.** Pp. 414–429. *General descriptions.*

Precambrian Iron Formations. E. L. BRUCE. Geol. Soc. Amer. 56:589–602, 1945. *Presidential address; review of world Precambrian iron formations and their origin.*

World Iron-Ore Map. H. M. MIKAMI. Econ. Geol. 39:1–24, 1944. *World distribution and world reserves.*

Iron Ore Supply for the Future. W. O. HOTCHKISS. Econ. Geol. 42:205–210, 1947. *Outlook for future iron-ore reserves.*

Iron Ore at Steeprock Lake, Ont., Canada. T. L. TANTON Roy. Soc. Can. Trans. 40:103–111, 1946. *Primary hydrothermal deposit.*

Mineralogy and Geology of the Mesabi Range. J. W. GRUNER. Iron Range Res. Rehab. State Office, St. Paul, 1946. *Concise detailed report on geology, minerals, structure, and types.*

Mineral Resources of China. V. C. JUAN. Econ. Geol. 41:424–433, 1946. *Occurrences, nature, and resources.*

Development of Lake Superior Soft Iron from Metamorphosed Iron Formations. S. A. TYLER. Geol. Soc. Am. Bull. 60:1101–1124, 1949. *Broad discussion of origin and concentration by oxidation and solution.*

See also general references under Iron in Chaps. 5·5 and 5·7.

Ferroalloy Metals

Under this heading are included a group of important metals whose chief use, but not their only use, is for alloying with iron to yield special steels of desired properties. They include *manganese, nickel, chromium, molybdenum, tungsten, vanadium, cobalt,* and *titanium.* Silicon and phosphorus are also used for certain steels. The addition

of a few percent of these metals to steel not only improves its properties but also adds new ones, such as hardness, toughness, strength, durability, lightness, ability to retain temper at high temperatures, and
resistance to corrosion. Despite the small volume in which they are
used they are essential to modern industry, particularly for dependable
high-speed machines. The large industrial nations lack most of them,
and the United States is deficient in all of them except molybdenum
and titanium.

Manganese

Manganese is the most important of the ferroalloy metals. Not
only is it necessary for the making of high-manganese steels but it is
also an absolute essential, for which there is no substitute, in the
making of all carbon steel, some 13 pounds being required for every
ton of steel produced. About 95 percent of the consumption is for
metallurgical purposes, of which a minor amount is used for other
alloys, such as some bronzes. Its chief purpose in steel making is to
remove oxygen and sulphur in order to produce sound, clean metal.
It is added in the form of ferromanganese (80% Mn), for which
ore containing a minimum of 46 percent manganese is desired, but it is
also used in the form of spiegeleisen (20% Mn). For chemical
uses, manganese ore of high purity is required for dry batteries, the
glass industry, paints, pigments, dyes, and fertilizers. Manganese
steel is used where hardness and toughness are desired; it enters into
armor plate, projectiles, car wheels, railway switches, safes, crushers,
cutting and grinding machinery, machine tools, cogwheels, structural
and bridge steel, etc.

Production and Distribution. The world production of manganese
ore is about 4 to 5 million tons, of which Russia produces about one-
half, followed in order by Gold Coast, India, South Africa, Brazil,
Cuba, Morocco, United States, and Japan. Minor production comes
from Austria, China, Czechoslovakia, Netherlands Indies, Hungary,
Italy, Malaya, Philippines, Roumania, Chile, Mexico, North Africa,
and Egypt.

The United States production of commercial ore, which is small,
comes mainly from Montana, New Mexico, Nevada, and Utah.

The annual domestic consumption of manganese ore with $35+$
percent manganese has ranged between 500 and 1,500 thousand tons
and is derived from the following sources: Russia, Gold Coast, India,
South Africa, Cuba, Brazil, and Chile.

Mineralogy and Tenor. The commercial minerals and classes of
manganese ores are:

MINERAL	COMPOSITION	PERCENT Mn	CLASSES OF ORE	PERCENT Mn
Pyrolusite	MnO_2	63	Chemical	82–87
Manganite	$Mn_2O_3 \cdot H_2O$	62.4	Manganese	35+
Psilomelane	$MnO \cdot MnO_2 \cdot 2H_2O$	45–60	Metallurgical	46
Hausmannite	Mn_3O_4	72.5	Ferruginous manganese	10–35
Rhodochrosite	$MnCO_3$	47.6	Manganiferous iron	5–10
Rhodonite	$MnSiO_3$	41.9		
Bementite	$2MnSiO_3 \cdot H_2O$	39.1		

Treatment. Manganese ore is generally hand sorted or washed to bring it to shipping grade. It is then furnaced with iron ore to make ferromanganese. Lower-grade ores are similarly treated to make spiegeleisen. A little metallic manganese is made electrolytically.

KINDS OF DEPOSITS AND ORIGIN

Hydrothermal deposits: *Butte, Philipsburg,* Mont.; Leadville, Colo.; Cuba.
Sedimentary deposits: *Tchiaturi* and *Nikopol,* Russia.
Residual concentrations: *India, Gold Coast, Brazil,* Egypt, Morocco.
Metamorphosed deposits: *Postmasburg,* South Africa; *India;* Olympic Mountains, Washington.

Most of the world's supply is obtained from sedimentary and residual deposits. The latter are surficial deposits that have resulted from the weathering of manganese-bearing minerals of schists, pegmatites, sedimentary beds, fissure veins, and replacement deposits.

EXAMPLES OF DEPOSITS

Russia. The annual production of manganese ore in Russia has ranged between 2 and 3 million tons, about 2½ times that of any other country. Ninety-three percent of it comes from two groups of sedimentary deposits, Tchiaturi in Georgia and Nikopol in the Ukraine; the rest comes from west Siberia, the middle Volga, and south Urals.

The *Nikopol* district consists of two areas of Oligocene beds of sandy clay about 85 square miles in area. The ore bed averages 6.5 feet in thickness and lies at a depth of 50 to 250 feet. It contains oölitic earthy lenses and nodules of pyrolusite, with wad, polianite, manganite, and iron oxides. It is believed that the manganese was precipitated directly from sea water by the activity of algae and bacteria living in the littoral zone of the Oligocene sea. There was some subsequent re-solution, deposition, and replacement of limestone. The deposit is clearly of sedimentary origin. The ore carries 28 to 33 percent manganese and is hand sorted and concentrated to 42 to 52 percent shipping ore. Ore reserves are claimed to be 398 million tons.

The *Tchiaturi* district (Fig. 14–10) has long been the world's largest producer of manganese ore. The deposits occur in a horizontal bed of Eocene marly sand from 6.5 to 10 feet thick in a zone 19 miles long by 5 to 6 miles wide. Canyons permit access by tunnels beneath an overburden of 300 feet or more of sands and shales. The ore bed consists of oölites of psilomelane and pyrolusite imbedded in a cement of wad and calcium carbonate with which small quantities of iron, copper,

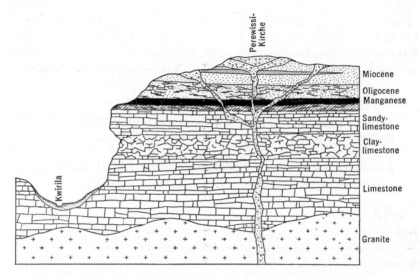

FIG. 14–10.   Section across manganese bed of Tchiaturi, Caucasus, Russia.   (*After de la Sauce, Abh. z. prakt. Geol. u. Berg.*)

nickel, phosphorus, and barium are associated. Some 2 to 3 tons of ore after washing and concentration yield 1 ton of concentrates carrying 50 to 53 percent manganese. Reserves are claimed to be 162 million tons averaging 26 percent manganese.

India.   Extensive and rich manganese deposits occur in the Central Provinces, Madras, Bihar, Bombay, and Mysore, but the Balaghat district of the Central Provinces and the Sandur State in Madras are the most important producers.

The deposits of the *Central Provinces,* particularly, are believed by Fermor to have resulted from an ancient (perhaps Archean) oxidation, chiefly of Archean crystalline " gondite " schists — containing spessartite and rhodonite. Most of the silica and alumina are believed to have been removed during weathering and the resulting manganese minerals are chiefly psilomelane, pyrolusite, and wad. One body is 1,500 feet wide and half a mile long. Fermor believes that originally the manganese was deposited as a sediment that was later metamorphosed to schists with bands of pure manganese oxides and silicates, which in turn underwent surface alteration to form residual enrichment deposits.

The *Sandur* deposits of Madras occur in Dharwar schists that have undergone surface alteration to form principally psilomelane and wad, with minor pyrolusite and manganite, which have in part replaced the country rock. The larger deposits are about 100 by 700 feet, and the district is said to contain 10 million tons of ore. Somewhat similar deposits are near Vizagapatam, where peculiar " Kodurite " rocks containing manganese garnet and manganese pyroxene have been altered and residually enriched to manganese oxides. Many other generally similar deposits have been described by Fermor in a 1,300-page memoir.

South Africa. The *Postmasburg* deposits, northwest of Kimberley, give South Africa an annual output of 600,000 to 700,000 tons of 40 percent ore; shipments average 49 percent. There are at least two belts, an eastern and a western, the latter being 38 miles long. The deposits lie at the contact of an eroded surface of dolomite of the Transvaal system and the unconformably overlying Gamagara shales and quartzites. It is exposed in disconnected troughs and outcrops. The ore forms an irregular bed 5 to 20 feet thick on the dolomite itself or in the overlying cherty ferruginous breccia (Fig. 14–11). The ore is rather hard, carries 56 to 60 percent combined manganese and iron, and occurs in botryoidal and steely crystalline varieties. The minerals are stated by DuToit to be pyrolusite, psilomelane, polianite, braunite, sitaparite, hematite, barite, and diaspore, but under the microscope some of these specific manganese minerals are seen to be intimate mixtures of several oxides, suggesting a former colloidal deposition. The present crystalline character is presumed to be due to metamorphism. The origin is not clear. Nel thinks the deposits are due to residual concentration and subsequent metamorphism, and DuToit advocates a hypogene replacement origin. Reserves may amount to a billion tons.

Gold Coast. Kitson states that the oldest deposits are in sediments, later than these are their metamorphosed representatives, and the youngest are residually concentrated surficial deposits. The manganiferous rocks occur over a distance of 600 miles but only one deposit is worked, namely the Nsuta mine, at Dagwin, 34 miles from the port of Sekondi. This is the largest manganese mine in the world. The ore crops out on a ridge for 2½ miles where it is mined by power shovels. It occurs in ancient altered rocks in massive and concretionary bodies. The capping of the hill, 30 to 35 feet thick, contains

detrital ore of about 40 percent manganese resting upon "black" ore
bodies, which are massive and lenticular and contain 50 to 55 percent manganese.   Kitson states that the deposits have resulted from
residual accumulation through tropical weathering of manganiferous
bedrock that removed silica and alumina and redeposited the man-

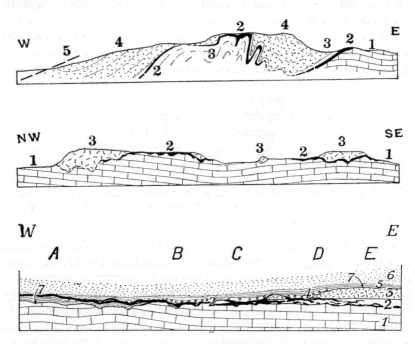

Fig. 14–11.   Sections across manganese deposits of Postmasburg, South Africa.
Upper, Western belt, Martha's Poort, showing relation of ore (2) to dolomite, 1;
shales, 3; quartzite, 4.   *Middle,* section along eastern belt through Klepfontein
Hills with ore between dolomite and breccia.   *Lower,* schematic section showing
unconformable relations of ore (black).   (*DuToit, Econ. Geol.*)

ganese near the surface.   Junner, however, believes that they were
formerly ancient high-grade manganese ores that were only slightly
enriched by meteoric waters.   One theory would indicate a shallow
depth for the ores, the other that they might persist to considerable
depth.   The altered rock is washed from the ores and the resulting
shipping product averages 50 to 53 percent manganese.   The deposit
yields 800,000 or more tons of ore annually and is said to have 10
million tons of ore in sight.

Brazil.   The most important manganese deposits of Brazil are in
Minas Geraes (Fig. 14–12), although large ones occur in Matto Grosso

and Bahia. The chief deposits, at Lafiete, according to Wright, represent residual concentrations from the weathering of ancient crystalline rocks containing lenses of manganese carbonate, tephroite, rhodonite, and spessartite. The ore averages 48 to 51 percent manganese. The Morro da Mina has six ore bodies; the largest (No. 3) is 14,400 feet long and 200 feet wide and is known for 750 feet vertically. Reserves are estimated at 6 million tons of ore.

The undeveloped *Urucum* deposit, near Bolivia, consists of two horizontal beds, one 8 feet thick, containing 15–30 million tons of ore. A larger deposit at *Amapau* is being developed.

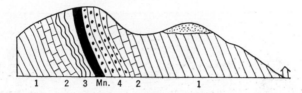

Fig. 14–12. Section of Burnier manganese deposit, Minas Geraes, Brazil. 1, Phyllite; 2, limestone; 3, manganiferous iron ore; 4, itabarite and jacutinga. (*After Scott-Beyschlag, Krusch, Vogt.*)

Egypt. The deposits of *UmBogma*, 7 miles south of Suez, yield about 200,000 tons annually. They occur in Carboniferous limestone over a distance of about 20 kilometers and range from 0.5 foot to 12 feet in thickness. They are horizontal bedlike deposits of irregular shape and size, carrying variable proportions of manganese and iron. The tenor is 28 to 32 percent, but local concentrations carry 60 percent manganese. Fenine considers that the manganese was in part deposited as sedimentary ore but mostly resulted by residual concentration of manganiferous limestones.

Cuba. In the foothills of the Sierra Maestra range, Oriente Province, Cuba, are many small manganese deposits which together have yielded up to 300,000 tons of manganese ore a year. The largest is the Quinto mine near Cristo. These unusual deposits are hydrothermal replacements of either water-laid volcanic tuffs or intercalated limestones of Eocene age. Most of the larger deposits are bedded; others are nonbedded. The ores contain from 15 to 55 percent manganese and according to Park and Cox contain psilomelane and pyrolusite with minor manganite, ranciéite, braunite, orientite, and piedmontite along with calcite, cryptocrystalline silica, and rock remnants. The low-grade ores of the Quinto mine carrying 18 percent manganese are sintered and concentrated to 50 to 52 percent manganese.

The ores are considered to have been deposited from warm waters of the dying phase of volcanism that yielded the tuffs.   In many places fractures localized the sites of replacement.

United States.   The minor United States production of manganese ore is obtained from small deposits that are mostly low in grade and relatively costly to work.   There are, however, extensive emergency reserves of low-grade material.   Minor residual deposits occur through-out the Southern States, but the main production has come from vein deposits in Montana and manganiferous iron ores in Minnesota.

The southern residual deposits are derived from manganiferous sediments, and most of them underlie remnants of dissected erosion

FIG. 14–13.   Brecciated quartzite (*A*) occupied by manganese-limonite (black), Wolfen Gap, Ga.   (*Crickmay, 16th Internat. Geol. Cong.*)

surfaces and consist of nodules in residual clay or replacements of underlying limestones.

In *Arkansas,* the rich residual ores of Batesville (Fig. 5·7–2) occur as "buttons" in the Cason shale and its residual clay and as masses of manganese oxides in residual clays occupying solution depressions in weathered Fernvale and other limestones.   The clays attain 80 feet in thickness, and the ores range from 25 to 50 percent manganese.   The manganese is considered to have been deposited originally in the Cason shale and in the Fernvale limestone and subsequently to have been weathered to form accumulations, residual in the clays.   More recent work suggests that the richer Batesville ores were derived from the weathering of epigenetic rhodochrosite deposits in limestone.

In *Georgia,* the important deposits of Cartersville, according to

Crickmay, consist of (1) boulders of manganese oxides in gravel derived from the weathering of quartzite (Fig. 14–13); (2) hard, concretionary masses; and (3) soft " chemical ore " in veins in clay and gravel. The *Virginia* Blue Ridge deposits, including the Crimora mine, consist of nodular masses of oxides in clays derived from the Shady dolomite and are considered to have grown by replacement of the clays. Other deposits are fillings and replacements of breccia zones in sandstone. The manganese accumulated during or after the early Tertiary peneplaning of this region.

Montana produces about two-thirds of the high-grade ore of the United States from rhodochrosite veins at Philipsburg and the Emma mine at Butte, which carry 35 to 37 percent manganese. This is concentrated and nodularized to a high-grade product suitable for battery ore and for ferromanganese.

In *South Dakota* the extensive low-grade deposits near Chamberlain are estimated to contain in excess of 100 million tons of manganese. Manganiferous iron carbonate nodules, which contain 18% Mn, 11% Fe, 12% CaO, and 11% SiO_2, occur in a 38-foot horizontal shale bed that caps remnants of a dissected plateau. The ore is an isomorphous mixture of the carbonates of Mn, Fe, Ca, and Mg. About 12 to 14 cubic yards of shale yields 1 ton of nodules, and 4.5 tons of nodules are required to make 1 ton of sinter containing 68% Mn. The nodules could be recovered cheaply by power shovel and screening. Other low-grade emergency deposits have been partly developed in Nevada, New Mexico, Arizona, and California.

Selected References on Manganese

Mineral Production and Trade of U. S. S. R. C. W. Wright. Foreign Minerals Quart. 1:No. 2, Pt. 2, June, 1938. U. S. Bur. Mines, Washington. *Brief survey of Russian deposits.*

Mineral Resources, Production and Trade of Brazil. C. W. Wright. Foreign Minerals Quart. 4:No. 1, July, 1941. U. S. Bur. Mines, Washington. *Much information on manganese in Brazil.*

Development of the Low-Grade Manganese Ores of Cuba. F. S. Norcross, Jr. A. I. M. E. Tech. Pub. 1188, May, 1940. *Description of the Quinto deposit.*

Manganese Deposits of Cuba. Charles F. Park, Jr. **Manganese Deposits in Part of the Sierra Maestra, Cuba.** C. F. Park, Jr., and M. W. Cox. **Geology and Manganese Deposits of Guisa-Los Negros Area, Oriente Province, Cuba.** W. P. Woodring and S. N. Daviess. U. S. Geol. Surv. Bulls. 935–B, F, and G, Washington, 1942–1944. *Occurrence and hydrothermal replacement origin of various manganese deposits in Oriente Province, Cuba.*

Some Economic Aspects of Philippine Manganese. W. F. Boericke. Eng. & Min. Jour. 141:54–57, 1940.

Geology of India. D. N. Wadia. Macmillan & Co., London, 1939. *Covers occurrence of Indian manganese deposits.*

Manganese Deposits at Philipsburg, Montana. E. N. GODDARD. U. S. Geol. Surv. Bull. 922-G, 1940. *Geology of one of the more important United States deposits.*

Manganese and Iron Deposits of Morro do Urucum, Brazil. J. V. N. DORR, 2ND. U. S. Geol. Surv. Bull. 946-A, 1946. *A large, undeveloped bedded deposit.*

Manganese. Hearings before Subcommittee on Mines and Mining. 80th Congress, Hearing 38, pp. 497. Feb. 12–27, 1948. *Covers distribution, uses, technology, statistics, and exploration for and distribution of domestic and important foreign deposits.*

The World's Manganese Ore. I. KOSTOV. Min. Mag. 72:265–270, 1945. *Estimates of world's reserves.*

Mineral Development in Soviet Russia. C. S. FOX. Min. Geol. Met. Inst. India Trans. 34:98–210, 1938. *Sketches of Russian occurrences.*

Manganese Deposits of Serra do Navio, Amapá, Brazil. J. V. DORR, II, C. F. PARK, JR., and G. DE PAIVA. U. S. Geol. Surv. Bull. 964-A, 1949. *Brazil's newest deposit.*

Recent Studies of Domestic Manganese Deposits. E. C. HARDER and D. F. HEWETT. A. I. M. E. Trans. 63:3–50, 1920. *A brief survey of domestic deposits.*

Manganese Deposits of India. SIR L. L. FERMOR. India Geol. Surv. Mem. 37, 1909. *A monographic study of Indian manganese deposits.*

Geology of the Nsuta Manganese Deposits. H. SERVICE. Gold Coast Geol. Surv. Mem. 5, 1943. *Geology, origin, and occurrence.*

Manganese Deposits of Postmasburg, South Africa. A. L. DU TOIT. Econ. Geol. 28:95–122, 1933. *Occurrence and origin.*

Manganese. R. RIDGWAY. U. S. Bur. Mines Inf. Circ. 6729, 1934. *Statistical.*

Manganese, 2nd Ed. A. W. GROVES. Imp. Inst. London, 1938. *Résumé of world occurrences and good bibliography.*

Manganese Deposits of Mexico. P. D. TRASK and J. RODRÍGUEZ CABO, JR. U. S. Geol. Surv. Bull. 954-F, 1948. *Broad survey of various types of deposits.*

General Reference 11, Chap. 25.

General references are given under Manganese, Chapter 5·7.

Nickel

Nickel ranks second to manganese among the ferroalloys, but it has also a multitude of other applications. Most of the world reserves are centered at Sudbury, Ontario. New Caledonia, Finland, and Russia are minor producers.

Uses of Nickel. Nickel is pre-eminently an alloy metal, and its chief use is in the nickel steels and nickel cast irons, of which there are innumerable varieties. It is also widely used for many other alloys, such as nickel brasses and bronzes, and alloys with copper (Monel metal), chromium, aluminum, lead, cobalt, manganese, silver, and gold. Nickel steels contain from 0.5 to 7 percent nickel in the " low-nickel steels " and 7 to 35 percent in " high-nickel steels "; nickel-irons from 0.5 to 15 percent; nickel brasses and bronzes from 0.5 to 7.5 percent; copper-nickel alloys from 2.5 to 45 percent; Monel metal 67 percent, and special alloys up to 80 percent nickel. The proportionate uses are as follows:

	PERCENT OF CONSUMPTION
Various nickel steels	42
Nickel cast irons	3
Nonferrous alloys	32
Heat- and electric-resistant alloys	9
Plating	12
Miscellaneous	2

Nickel imparts to its alloys toughness, strength, and lightness and anti-corrosion, electrical, and thermal qualities. Consequently, nickel steels and alloys are preferred for moving and wearing parts of innumerable machines, tools, shafts, bolts, axles, and gears, in automobiles, airplanes, and ships and in railway, power, agricultural, crushing, mining, milling, and pressing equipment. Its use for coinage and plating is widespread.

Production and Distribution. The annual world production of nickel ranges between 100 and 150 thousand tons of metal, of which about 85 percent comes from Canada, and the remainder from New Caledonia, Russia, South Africa, Sweden, and Norway. Finland contains a large deposit, but its yield is unknown. A large reduction plant at Nicaro, Cuba, yielded nickel oxide during World War II but is now closed. The United States is the greatest consumer, followed by the other steel-making countries.

Mineralogy and Tenor. The chief commercial, primary mineral of nickel is pentlandite (Fe,Ni_9S_8), which is always associated with pyrrhotite and chalcopyrite. Hydrous nickel silicate ores, the ore minerals of New Caledonia, carry 1 to 4 percent of nickel. The sulphide ores of Sudbury average about $1\frac{1}{2}$ percent nickel and 2 percent copper.

Treatment. The nickel-copper sulphide ores are first roasted to release sulphur and then smelted to a Ni-Cu-Fe matte, which is bessemerized to a matte of 75 to 80 percent Cu-Ni. Some matte is used directly to make Monel metal, the rest being specially smelted to separate copper and nickel sulphides, the latter being roasted, reduced with carbon, and electrolytically refined to pure nickel.

Kinds of Deposits. There are only two kinds of commercial nickel deposits — residual concentrations of nickel silicates from the weathering of ultrabasic igneous rocks and nickel-copper sulphide deposits formed either by replacement or magmatic injection.

Nickel-Copper Deposits of Sudbury, Ont. The great deposits of Sudbury dominate the world nickel production. The annual production is about 130,000 tons each of nickel and copper, 200,000 ounces platinum, 50,000 ounces gold, 1.5 million ounces silver, 150,000 pounds selenium, and 10,000 pounds tellurium. The ore reserves exceed 240 million tons of ore, with 8.5 million tons of copper-nickel.

Geologic Setting. The region is noted for the norite-micropegmatite intrusion 36 miles long and 16 miles wide (Fig. 14–14) which crops out as a great ellipse and formerly was considered spoon-shaped. Yates

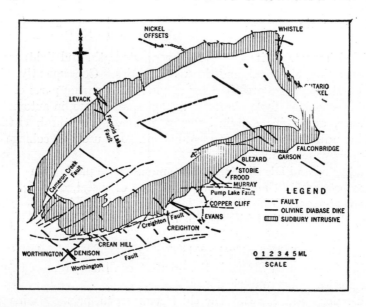

Fig. 14–14. Structural pattern of the Sudbury basin, showing outlines of the norite-micropegmatite and late dikes. (*Yates, Structural Geology of Canadian Ore Deposits.*)

thinks that its position and shape were controlled by the position and shape of a pre-existing syncline of folded and faulted Huronian formations. The rocks present, according to Yates, consist of (1) Keewatin volcanics and quartzites, (2) Sudbury series of volcanics and sediments, (3) Algoman granite, (4) lower Huronian sediments, (5) upper Huronian (?) Whitewater series, (6) norite-micropegmatite intrusive, (7) Murray and Creighton granites, (8) breccia, (9) trap dikes, and (10) olivine diabase dikes. Quartz diorite, the chief host rock of the ore, cuts the dikes. Faulting follows, and, lastly, ore emplacement. The norite intrusive consists of an outer rim of norite, a narrow transition zone, and an inner zone of micropegmatite, each representing products of differentiation.

Ore Deposits. Yates states that there are three characteristic types of deposits, all with the same mineralogy, namely: (1) disseminated bodies largely in quartz diorite, (2) massive sulphide generally along faults or breccia zones, and (3) sulphide stringer zones in shattered and brecciated country rock of all kinds. He emphasizes that the most favorable host rock and the rock in which most of the ore occurs is quartz diorite or quartz diorite breccia, and that very little ore occurs in norite. Massive sulphide ore occurs in granite, quartzite,

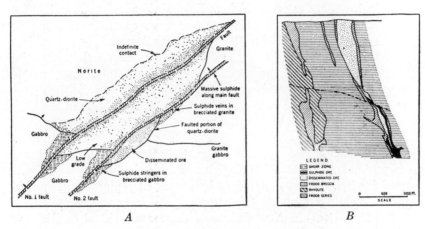

Fig. 14–15. *A*, section through ore bodies, Creighton mine. *B*, section, Frood mine. (*Yates, Structural Geology of Canadian Ore Deposits.*)

andesite, breccia, gabbro, and other rocks where structural conditions permit but never far from quartz diorite. The quartz diorite is apparently related to the norite. Most of the deposits lie about the margins of the norite intrusive (Fig. 14–14). The main deposits such as the Creighton, Frood, Murray, and Falconbridge lie on the south rim; a few, including the Levack, lie on the north rim. The older concept of marginal and " offset " deposits is no longer held. The Levack deposit is enclosed in granite breccia intruded between norite and footwall gneiss; the Creighton bodies, both massive and disseminated (Fig. 14–15), occur along parallel faults in a large slab of quartz diorite and in shattered footwall rocks. The Frood-Stobie (Fig. 14–15*B*) deposit lies in a large dikelike zone of quartz diorite breccia, separated from the norite footwall by a mile of Murray granite. A huge, deep body of massive sulphide in breccia is capped and sheathed by disseminated ore in quartz diorite. The Falconbridge deposit consists of breccia, massive and disseminated ore along a contact shear zone that extends for 9,000 feet in norite, quartz diorite, and greenstone.

The Sudbury ores consist of pyrrhotite, chalcopyrite, pentlandite, minor cubanite, complex arsenides, the nickel-bismuth sulphide parkerite, tellurides of gold and silver, sperrylite, and platinum metals. Considerable selenium is present. Other minor minerals noted are magnetite, polydimite, and in late stage stringers are galena, sphalerite, marcasite, violarite, and silver minerals along with quartz and calcite. The massive ores contain up to 60 percent sulphides. The ratio of nickel to copper in Frood disseminated ore is 1 to 1, but copper in the massive ore is higher; Falconbridge ore ratio is 1 to 0.6.

Origin. The origin of these ores has been controversial, and for detail the reader is referred to the voluminous literature on the subject. The older conception of Barlow and Coleman of magmatic segregation by early settling out of the sulphides into embayments is not tenable. The ore is now known to be definitely later than the solidification of the norite. Between its solidification and the introduction of sulphides there were five stages of igneous intrusion and a period of faulting. Consequently, there could have been no magmatic segregation of sulphides *in situ*, nor does a late magmatic injection of a sulphide melt seem probable. The staff geologists consider that the deposits were formed by hydrothermal solutions by means of replacement, and the evidence certainly supports this view.

Other Deposits. In *New Caledonia* there are residual enrichments of nickel silicates, already described in Chapter 5·7. Similar deposits occur at Nicaro, Cuba, Brazil, Venezuela, and the Celebes.

At *Petsamo, Finland,* a nickel-copper sulphide deposit similar to the Sudbury deposits was developed between 1941 and 1945. It carries 1.6% Ni and 1.3% Cu and contains about 4 million tons of ore.

At *Lynn Lake, Manitoba,* developments have disclosed about 10 million tons of Sudbury-type ore carrying 1.5% Ni and 0.68% Cu.

Near *Hope, British Columbia,* are small sulphide deposits of low-grade nickel ore carrying the Sudbury type of minerals occurring as injections in hornblendite. The ores contain 1.4% Ni and 0.4% Cu. Somewhat similar noncommercial deposits are found on *Chichagof Island* and *Yakobi Island*, Alaska. Another deposit of the Sudbury type of ore has been drilled at *Rankin Inlet* on *Hudson Bay*. It appears to be a segregation in the base of a pyroxenite sill and carries 4.6% Ni, 1.2% Cu, and platinum, but the tonnage is insufficient to justify exploration. Similar small nickel deposits are reputed at *Nittis* and *Kumajie,* and in the *Urals,* in Russia, and a new deposit of high-grade ore has been opened at *Celebes,* Netherlands East Indies. The old *Gap mine, Pennsylvania,* formerly was a producer of nickel from magmatic sulphide ores. *Greece* contains small deposits, and an interesting deposit occurs at Insizwa and in the Bushveld Complex, *South Africa.*

Selected References on Nickel

Report of Royal Ontario Nickel Commission. Toronto, 1917. *General survey.*

Sudbury Nickel Field Restudied. A. G. BURROWS and H. C. RICKABY. Ontario Dept. Mines Rept. 43:Pt. 2, 1934. *General survey of geology and veins; bibliography.*

International Nickel Company Operations. Part II, Geology. Can. Min. Jour. 58: No. 11, 1937.

New Caledonia Nickel, Les Resources Minérales de la France d'Outre Mer. Bur. Etudes Géol. et Min. Colon., Vol. 2, 1934. *Detailed descriptions and mineralogy.*

General Reference 15. Sudbury District. Huron District. H. C. COOKE. **Inter. Nickel Co. Mines.** A. B. YATES. **Falconbridge Mine.** S. DAVIDSON. *The most up-to-date lucid statements of ore genesis and structure.*

La Genèse et L'Évolution des gisements de nickel de la Nouvelle-Calédonie. E. DE CHÉTELAT. Bull. Soc. Géol. France 17:106–160, 1947. *Latest discussion of ores and origin.*

Chromium

Chromium, like manganese, occurs mainly in countries that use little of it, and most of the large consuming countries are deficient in it. It was first mined in Norway in 1820 and then in Maryland in 1827. Formerly, chromite was used chiefly as a refractory mineral, but since the rapid development of stainless steels has become now a prized steel alloy.

Uses. The main uses for chromium are: metallurgical, 50 percent; refractories, 35 percent; and chemical, 15 percent. The metallurgical uses include a great variety of alloys, mainly with iron, nickel, and cobalt. Some of the important ones are:

	PERCENT Cr		PERCENT Cr
Low chrome steels	0.5–5	Super stainless steels	12–30
Low chrome iron	0.2–4	Chrome-nickel ferrous alloys	14–30
Medium chrome steels	3–12	Electrical resistance alloys	8–20
Stainless irons	12–15	Chrome-cobalt alloys	20–35
Stainless steels	12–18		

Chromium imparts to alloys strength, toughness, hardness, and resistance to oxidation, corrosion, abrasion, chemical attack, electricity, and high-temperature breakdown. Consequently it finds manifold uses. The great strength of chrome steels allows reduction in weight of metal in automobiles, airplanes, and trains, for example. The stainless steels, containing 18 percent chromium and 8 percent nickel, have become popular as noncorrosive, tough, hard, strong steels. Special chrome alloys, such as stellite, using also tungsten, molybdenum, and cobalt, yield hard, high-speed tool steel. Chromium plating has become more popular than nickel plating.

As a refractory, the mineral chromite is widely used for furnace linings, and chemical compounds are used for dyeing, tanning, bleaching, pigments, and oxidizing agents.

Production and Distribution. World leadership in chromite production has passed from the United States (1860) to Turkey, to Russia, to New Caledonia, in 1922 to Southern Rhodesia, and again to Russia. The annual world production ranges between 1 and 2 million tons, chiefly from Russia, Southern Rhodesia, Turkey, South Africa, Cuba, Philippines, India, Greece, Japan, and New Caledonia.

Minor production comes from Brazil, Bulgaria, and Canada. The United States production of metallurgical grade ore is almost negligible, but large deposits of lower-grade chemical ore have been developed in Montana; its imports, nearly 1 million tons, normally come chiefly from Rhodesia, Russia, the Philippines, South Africa, Cuba, New Caledonia, and Turkey.

Mineralogy, Tenor, and Treatment. There is only one ore mineral — chromite, which theoretically carries 68% Cr_2O_3 and 32% FeO, but Al_2O_3, Fe_2O_3, MgO, CaO, and SiO_2 may displace some Cr_2O_3, reducing the Cr_2O_3 content to as little as 40 percent. Commercial ores should contain 45% Cr_2O_3. The chrome-iron ratio should be above 2.5 to 1 for metallurgical chrome.

Chrome is marketed as lump chromite after hand sorting or rough concentration. Most chromite ores are not adaptable to concentration processes. The chrome ore is smelted in an electric furnace with fluxes and carbon to ferrochrome, in which form most of it is marketed.

Kinds of Deposits and Origin. Most all chromite deposits are magmatic segregations in ultrabasic igneous rocks, as described in Chapter 5·1. Chromite occurs in the host rock as masses, lenses, and disseminations. In some places disseminated grains have undergone residual concentration.

Examples of Chromite Deposits

Rhodesia. High-grade deposits occur at Gwelo, which includes Selukwe and Victoria, Lomagundi, and Hartley. These yield from 150,000 to 350,000 tons annually. The deposits occur in or near the amazing Great Dyke, which is 330 miles long, averages 4 miles in width, and consists of layers of ultrabasic rocks now largely altered to serpentine. The layers dip toward the center from both sides (Fig. 14–16).

The dike deposits are bands of chromite about 8 inches thick that parallel the pseudostratification. The deposits outside of the dike are lenses in talc schists and serpentine from 150 to 450 feet long. The ore contains from 45 to 56 percent chromic oxide. Keep considers that the deposits are early magmatic segregations formed by crystal settling, but Sampson concludes that they are late magmatic accumulations or even possibly hydrothermal replacement deposits.

FIG. 14–16. Section across Great Dyke, Southern Rhodesia, showing chromite bands C and platinum horizon P. N, norite; E, enstatite rock; S, serpentine. (*DuToit, Geol. of South Africa.*)

South Africa. This country yields from 150,000 to 350,000 tons of chromite annually from stratified bands within the Bushveld Igneous Complex around Rustenburg and Lydenburg (Figs. 5·1–1, 5·1–2,

FIG. 14–17. Occurrence of chromite in norite, Lydenburg, South Africa. (*Schneiderhöhn, Min. Boden i Sud. Afrika.*)

14–17). The bands dip gently and are from a few inches to 3 feet thick. The chromite horizon can be traced tens of miles, and large tonnages are present. Except for a little rich ore, the grade, however, is rather low, averaging only 43% Cr_2O_3. Such ore is useful for chemical purposes rather than for metallurgical use.

Russia. The important deposits of Russia center around the Sverdlovsk region on the east side of the Urals, where they occur as magmatic segregations of lenses, stringers, and disseminations in serpentine and soapstone. The best ores mostly average only 40 to 46% Cr_2O_3; some 56 percent ore occurs. The lower-grade ore is concentrated up to 48 percent. Other deposits occur in South Bashkiriya and the Kazakh Republic. Uncertain estimates of reserves range from 1 to 13 million tons.

Turkey. Turkey contains numerous scattered deposits, the most important ones occurring near Bursa, south of the Sea of Marmora, and along the south Mediterranean coast. The most important deposits are the Guleman mines, near Ergani Maden. All are associated with altered ultrabasic intrusives and are thought to be magmatic segregations. The ores are hard and of high quality; many mines average 50 to 52% Cr_2O_3, others yield 43 to 50 percent ore.

Other Deposits. In *Cuba* important refractory ores carrying 33 to 43% Cr_2O_3 are found in Camaguey, Oriente, and Matanzas. There are also huge tonnages of low-grade chromiferous nickel-iron ores in Mayari and Mao Bay. The *Philippines* have been an important producer of 46 to 50 percent ore. Small deposits of high-quality ore occur in Baluchistan, Mysore, Bihar, and Orissa, *India. Yugoslavia, Greece, Brazil,* and *New Caledonia* are minor producers. In the *United States* small deposits are mined mainly in California, Oregon, Maryland, North Carolina, and Alaska. A deposit is also known in Newfoundland.

Selected References on Chromite

World Production and Resources of Chromite. L. A. SMITH. A. I. M. E. Tech. Pub. 423, 1931. *Statistical.*

Chromite Deposits of the Eastern Part of the Stillwater Complex, Montana. J. W. PEOPLES and A. L. HOWLAND. U. S. Geol. Surv. Bull. 922–N, 1940. *Montana low-grade deposits.*

Chromite. R. RIDGEWAY. U. S. Bur. Mines Inf. Circ. 6886, 1936. *Production and reserves.*

Chromite, Lomagundi and Mayie, Southern Rhodesia. F. E. KEEP. S. Rhod. Geol. Surv. Bull. 16, 1930. *Descriptions of the Great Dyke occurrences.*

Chromite Deposits of the Bushveld Igneous Complex. S. Africa Geol. Surv. Geol. Ser. Bull. 10, 1937. *Low-grade magmatic deposits.*

Chrome Resources of Cuba. T. P. THAYER. U. S. Geol. Surv. Bull. 935–A, 1942. *Geology of refractory chrome deposits.*

Strategic Mineral Supplies. G. A. ROUSH. McGraw-Hill, New York, 1939. *Occurrence, distribution, uses, and substitutes of chromite.*

Varieties of Chromite Deposits. EDWARD SAMPSON. Econ. Geol. 26:833–839, 1931, and 24:632–641, 1929. *A discussion of origin and relation to differentiation.*

Geology and Chromite Resources of Camagüey, Cuba. D. E. FLINT, J. FRANCISCO DE ALBEAR, and P. W. GUILD. U. S. Geol. Surv. Bull. 954–B, 1948. *Geology, occurrence, and origin of refractory deposits.*

Chrome Ores of the Western Bushveld. C. J. F. VAN DER WALT. Geol. Soc. S. Africa 44:79–112, 1942. *Discussion of ores and chemical relations.*

General Reference 11, Chap. 10.

Some Occurrences of Chromite in New Caledonia. JOHN C. MAXWELL. Econ. Geol. 44:525–544, 1949. *Occurrence and origin of residual concentrations.*

Molybdenum

Metallic molybdenum was isolated in 1782 and produced commercially in 1893, but only 10 tons of it had been made before 1900. Large-scale use began in 1913, and it attained importance in 1927.

Uses. Practically all molybdenum produced is used as an alloy in iron and steel, where it is displacing other alloy metals. In steel it acts like tungsten, but more powerfully, requiring only one-half to one-third as much to produce similar results. It is the most potent hardening alloy. It increases the strength of steel and of cast iron and also increases the ductility. It serves best in steels containing nickel, manganese, or chromium. Only small quantities are needed, usually less than 2 percent.

Molybdenum steels are used widely in aircraft, automobiles, oil machinery, shafts, valves, pumps, gears, high-speed tools, pins, bolts, guns, armor plate, dies, wire, and many other materials. It greatly improves cast iron for industrial, railroad, and automotive uses. Minor amounts are used for catalysts, dyes, and lithographic inks.

Production and Distribution. The world production of molybdenite ranges from 30 to 65 million pounds per year, of which the United States produces about 85 percent and the remainder comes from Mexico, Chile, Canada, and Norway. The United States produces from 30 to 60 million pounds, which production is divided about equally between the Climax Molybdenum Company at Climax, Colo., and the Kennecott Copper Corporation from its Utah and New Mexico mines. Minor production comes from Arizona and other New Mexico mines.

Mineralogy, Tenor, and Treatment. Practically all molybdenum is obtained from molybdenite. Wulfenite and ferrimolybdite are minor contributors.

The Climax ore averages 0.6 to 0.7% MoS_2; smaller deposits range from 1 to 3% MoS_2. The Utah Copper copper ores carry about 0.04% MoS_2 as a by-product.

Molybdenite ores are mined by underground methods and are concentrated by differential flotation. The product is generally marketed in the form of molybdenite concentrates. For adding to steel this is roasted with lime to make calcium molybdate (40 to 50% Mo), is converted to ferromolybdenum (50 to 65% Mo), or is made into briquettes of molybdenum oxide.

Kinds of Deposits and Origin.   The types of commercial deposits are:

Pegmatites: Moss mine, Quebec.
Contact metasomatic: Azegour, Morocco; Yetholme, New South Wales; Renfrew, Ontario; Helvetia, Ariz.
Disseminated replacement: *Climax,* Colo.; *Bingham,* Utah; *Santa Rita,* N. Mex.; Cananea, Mex., Braden, Chile.
Fissure veins: Questa, N. Mex.
Pipes: Copper Creek, Ariz.

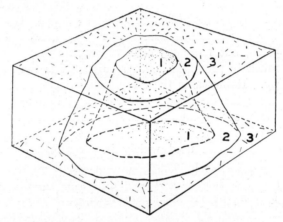

Fig. 14–18.   Idealized representation of Climax molybdenum. 1, Silicified ore; 2, ore zone; 3, slightly silicified outer zone. (*Butler-Vanderwilt, U. S. Geol. Survey.*)   This idealism applies only to upper levels; the lower workings (1941) disclose large ore bodies in parts of the central core.

Examples of Deposits

Climax, Colo.   The greatest molybdenum deposit of the world, at Climax, Colo., illustrates large-scale disseminated replacement.   It lies at an altitude of 11,300 feet and is mined from tunnels in excess of 14,000 tons per day.   It can produce more than 4½ million tons of ore and 30 million pounds of molybdenite annually.   The ore reserves exceed 200 million tons.

The deposit occurs in Precambrian granite that contains large schist inclusions and numerous Tertiary quartz monzonite dikes. Figure 14–18 idealizes the occurrence on the upper level only.   The central core (1) is barren silicified granite 700 feet across at the upper tunnel. The next concentric cone (2) is the ore body, up to 300 feet thick. It consists of altered granite intersected by innumerable, closely spaced, intersecting veinlets mostly less than ½ inch in width. These contain quartz, some orthoclase, and molybdenite borders with some molybdenite and fluorite in the centers. The interveinlet rock is sericitized. Other veinlets carrying a little pyrite, chalcopyrite, zincblende, huebnerite, topaz, quartz, and fluorite cut both the ore body and the other cones. The ore zone grades into an outer zone of moderately altered rock with fewer veinlets and less molybdenite, and this merges into unaltered country rock. The ideal truncated cone does not hold for the lower levels, where much lower-grade ore occurs in the core. Some oxidation of the ore zone has taken place yielding limonite, jarosite, and molybdite.

The deposit was formed by hydrothermal solutions that presumably emanated from the magmatic reservoir that supplied the Tertiary dikes. The deposit must be considered as one variety of the porphyry copper type of mineralization.

Other Deposits. Other producers of molybdenum are the *Chino Copper Company* at *Santa Rita, New Mexico, Colorada mine, Cananea, Mexico,* and *Braden mine, Chile.* In all these, as well as in *Utah Copper,* the molybdenum is a by-product of copper ores, where it is disseminated in small quantities and is recovered during milling by selective flotation. The *Questa Molybdenite* mine, *New Mexico,* is a unique example of filled fissure veins, where molybdenite is the only valuable sulphide, with minor chalcopyrite and sphalerite, and a gangue of quartz, orthoclase, biotite, fluorite, and carbonates. Norway production comes mainly from the *Knaben* mine in *Fjotland,* where deposits consist of quartz veins and disseminations in granite. A deposit carrying 4.8% MoS_2 occurs at *Algruvan, Sweden.* In Australia, contact-metasomatic deposits at *Yetholme,* quartz pipes at *Wonbah, Chillagoe,* in and around Kingsgate, and quartz veins yield small quantities of molybdenite.

Selected References on Molybdenum

Molybdenum. A. V. PETAR. U. S. Bur. Mines Econ. Paper 15, 1932. *Brief survey of distribution, occurrence, and technology.*

Molybdenum Deposits. F. L. HESS. U. S. Geol. Surv. Bull. 761, 1924.

Climax Molybdenum Property of Colorado. B. S. BUTLER and J. W. VANDERWILT. U. S. Geol. Surv. Bull. 846:191–237, 1933. *A great low-grade, disseminated deposit in granite.*

The Occurrence and Production of Molybdenum. J. W. VANDERWILT. Colo. School of Mines Quart. 3:1, 1940.

Questa Molybdenite Deposit, New Mexico. J. W. VANDERWILT. Colo. Sci. Soc. Proc. 13:No. 11, 1938. *A fissure vein of molybdenite.*

Tungsten

Discovered in 1781 and isolated in 1783, tungsten was first used in 1847, but, with the discovery in 1898 that steel containing it could be used to cut other steels at high speed, it came into general use as a steel alloy. Today it is an essential alloy metal.

Uses. The chief use for tungsten is for making high-speed cutting steels which retain their hardness even at red heat. These contain 15 to 20 percent tungsten with some chromium and vanadium. Molybdenum may replace some of the tungsten. Self-hardening steels contain 2 to 12 percent tungsten. The tungsten steels are used for various cutting tools and materials subjected to hard wear. It is widely used in stellite, an alloy of tungsten, chromium, and cobalt, for hard facing materials. A newer use is in making tungsten carbide, which, next to the diamond, is the hardest known cutting agent. Tungsten filaments for electric light bulbs are familiar to everyone. Tungsten is also used for electrical contact parts, electrical apparatus, radio, X-ray, pigments, textiles, armor plate, guns, and projectiles.

Production and Distribution. Production of tungsten, of which it is said " the more tension, the more tungsten," is stimulated by war, and the unsettled conditions in China, the richest tungsten country in the world, has stimulated world-wide small production elsewhere. Some 20 to 60 thousand tons of concentrates are produced annually, of which China produces one-third, and the rest comes from the United States, Burma, Brazil, Bolivia, Japan, Argentina, and Australia. The chief tungsten countries follow the volcanic chain that skirts the Pacific.

The world reserves of tungsten are little known. The United States reserves have been estimated to be 26 million pounds of metal but war stimulus increased this calculation.

The United States production ranges from 5,000 to 10,000 tons of concentrates and comes chiefly from Nevada, California, Idaho, North Carolina, Colorado, and Montana.

Mineralogy, Tenor, and Treatment. Twelve tungsten minerals are known, but the only economic ones are scheelite, wolframite, ferberite, and huebnerite. These are all compounds of calcium, iron, or manganese, with WO_3. The tenor of tungsten ores is mostly less than 1 percent, and the standard grade of concentrates is 60% WO_3.

Tungsten ores are mined by underground methods; the ores are readily concentrated, because of the high specific gravity of tungsten minerals, by means of jigs and tables. The concentrates are treated to yield metal or ferrotungsten. For metal, they are converted chemically to tungstic oxide and reduced to metal by heating in hydrogen or carbon. Ferrotungsten (75 to 80% W) is produced in the electric furnace by heating with carbon and is added to steel in this form.

Types of Deposits and Origin. The types of commercial deposits and examples are:

Pegmatites: Oreana, Nev.; Burma, China.
Contact-metasomatic deposits: *Mill City*, Nev.; *Pine Creek*, Calif.; Mexico; Northern Brazil; *Sandong*, Korea.
Replacement deposits: Silver Dyke, Nev., *Yellow Pine*, Idaho; *Kiangsi*, China; *Malay;* Burma; Australia.
Fissure veins: *Kiangsi*, China; Mawchi and Tavoy, Burma; Silver Dyke, Nev.; Atolia, Calif.; *Bolivia; Hamme*, N. Car.
Placers: Kiangsi, China; Burma; Atolia, Calif.

The fourth, third, second, and fifth types are respectively the most important. Tungsten, like tin, is universally associated with granitic rocks and occurs in or near them. It is formed under conditions of medium to high temperature by gaseous emanations or hydrothermal solutions.

EXAMPLES OF DEPOSITS

China. Tungsten production started in China in 1915 and by 1919 led all other countries. The deposits lie mainly in the Nanling region, including parts of Kiangsi, Hunan, and Kwangtung, and in Kwangsi and Yunan. These have been ably described by K. C. Li. The important Kiangsi area contains 17 districts and 80 mining localities grouped in 5 structural zones localized in major anticlines. The deposits are steeply dipping fissure veins either in granite or in nearby invaded sediments and range up to 3 meters wide, 1000 meters long, and to depths of more than 350 meters. Many pegmatite veins are also present. The ore is mainly wolframite with minor scheelite accompanied by bismuthinite, molybdenite, quartz, muscovite, topaz, fluorite, and common sulphides. Cassiterite is sufficiently abundant in many veins to yield tin concentrates. The wall rocks have been intensely altered, and pneumatolytic minerals have been introduced; granite walls have been greisenized. Tungsten ore reserves in China are about 2 million tons carrying 60% WO_3.

Burma. Burma yields annually from 4,000 to 6,000 tons of concentrates. The producing localities are Mawchi, Merguir, and Tavoy.

The Mawchi mine, Karenni, contains 27 veins ranging from 2 to 6 feet in width and occurring in tourmaline granite and invaded sedimentary rocks. The principal minerals are wolframite, cassiterite, quartz, calcite, and some scheelite. There are also base-metal sulphides, bismuthinite, and arsenopyrite. Other gangue minerals are tourmaline, beryl, fluorite, garnet, and orthoclase. The ores average 2.5 percent tin and tungsten.

At Hermyingyi, Tavoy, are 60 veins from 1 to 5 feet wide and up to 1,100 feet long. They occur in granite but extend upward into overlying sediments. The ore minerals are wolframite, cassiterite, base-metal sulphides, molybdenite, bismuthinite, and native bismuth, with quartz, muscovite, and fluorite.

Dr. J. A. Dunn believes that in both areas the veins are fissure fillings with gradations between the quartz lodes and pegmatites, formed by liquid solutions given off from the granite, and that the metals were carried as colloids.

United States. The United States normally produces about 10 percent of the world supply of tungsten, consumes about 25 percent, and now supplies much of its own needs, mainly from Nevada, Colorado, Idaho, California, and North Carolina. Nevada has yielded nearly one-third of the production, the main districts being Mill City, Mina, and Oreana.

At *Mill City, Nev.*, are four important mines, the Stank, Humboldt, Sutton, and Springer. The deposits are all similar and, according to Kerr, are " veinlike " contact-metasomatic deposits produced in steeply dipping thin limestone members of a siliceous sedimentary series by a post-Jurassic granodiorite intrusive. The ore bodies occur within 2,000 feet of the intrusive and are 5 to 6 feet thick, swelling in places to 20 feet where faulted. The strike length ranges from 500 to 1,000 feet and the present depth is 1,325 feet. The ore beds consist mainly of garnet, epidote, and quartz, in which are scattered small grains of scheelite and a few rich concentrations. Other contact-metasomatic silicates and common sulphides are present. The scheelite is white to pale brown in color and shows up strikingly underground with ultra-violet light. The tenor averages 1 to 1.5% WO_3.

At *Oreana, Nev.*, a unique and important scheelite deposit is described by Kerr as of pegmatitic mineralization. There are two modes of occurrence. In one the tungsten is in vertical pegmatite dikes that terminate at shallow depth. In the other it occurs in a string of pegmatitic lenses (Fig. 14–19) within intrusive metadiorite parallel to its inclined contact with limestone and is fed from underlying channelways. The pegmatites are associated with post-Jurassic granitic intrusives. The pegmatite dikes are irregularly mineralized, mainly with beryl, oligoclase, and masses of fluorite and quartz; scheelite masses are several feet across. The lenses are mainly scheelite, phlogopite, and oligoclase. Other minerals of this unusual ore are orthoclase, albite, rutile, tourmaline, and some pyrite and pyrrhotite. The ore is exceptionally rich. The first shipments ran 30 percent scheelite; the subsequent average is about 5 percent.

At *Silver Dyke, Nev.*, near Mina, are scheelite fissure veins filled by scheelite and comb quartz, with albite and minor sulphides. The veins

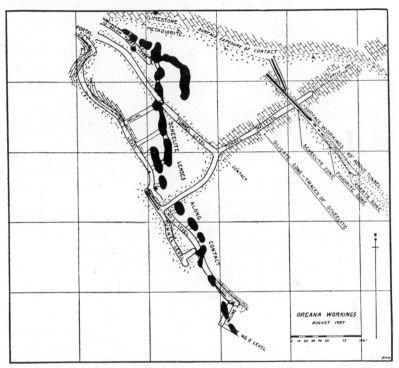

Fig. 14–19. Scheelite ore bodies along limestone-diorite contact disclosed underground, Oreana, Nev. (*Kerr, Econ. Geol.*)

occur along the contact of Triassic volcanics and intrusive diorite that has been hydrothermally altered and kaolinized and cut by albitized dikes. The steeply inclined veins are 2 to 24 inches wide and occur in a zone 25 to 200 feet wide; intervening plates of rock have been replaced by quartz, scheelite, and albite up to widths of 15 feet. Workings extend to 800 feet. The deposits are epithermal to mesothermal fissure fillings.

Korea. The large tungsten production of Korea comes chiefly from the Sangdong mine, which according to Klepper is one of the principal tungsten mines of the world. Six sedimentary lenticular beds have been metasomatically altered into scheelite-bearing tactitelike rocks carrying 1.75% WO_3 and bismuth sufficient to make this one of the world's largest bismuth mines. The main bed is 4.5 meters thick and 525 meters long. The accompanying minerals are bismuthinite, molybdenite, common sulphides, quartz, fluorite, apatite, and biotite. Ore reserves are listed as 3.2 million tons carrying 1.7% WO_3. Quartz-wolframite-molybdenite veins also occur in the vicinity.

Brazil. The discovery of rich scheelite ores in northeastern Brazil in 1942 has made Brazil the second largest tungsten-producing country in South America. Some 60 scheelite localities are known, in most of which the deposits are of contact-metasomatic origin where granitic rocks intrude schists and limestone. Tactites and contact-metasomatic minerals enclose scheelite, molybdenite, and bismuth minerals. Some scheelite also occurs in quartz veins. Cassiterite is rare.

Other Deposits. Economic tungsten deposits occur at *Yellow Pine, Idaho* (now closed), along with gold and antimony in a shear breccia; in highly productive scheelite-quartz veins at *Atolia, Calif.;* in ferberite-wolframite veins at *Boulder, Colo.,* along with chalcedony and wurtzite with structures suggesting gel deposition; in large contact-metasomatic scheelite bodies in tactite at *Pine Creek, Calif.;* in tactites at *Tungsten City, Calif.,* and *Nightingale, Nev.;* in an unusual blanket of tungsten-bearing limonite and manganese of hot spring origin at *Golconda, Nev.;* in quartz veins carrying huebnerite and scheelite in the *Hamme District, North Carolina;* and in the *Paradise Range, Nevada,* where a 4-foot vein carries scheelite and leuchtenbergite.

Bolivia yields considerable tungsten ore from the fissure veins in the tin belt carrying quartz, wolframite, some scheelite, common sulphides, and rarely cassiterite. The occurrence is similar to the tin veins already described. In *Argentina* are many shallow tungsten quartz veins carrying wolframite, with minor scheelite and ferberite, and with bismuthinite, molybdenite, common sulphides, and notable amounts of niobium and tantalum minerals. The *Australian* production comes from contact-metasomatic, pegmatite, and vein deposits in which are wolframite and scheelite along with bismuth and molybdenum. The *Portuguese* veins yield chiefly wolframite, generally accompanied by minor bismuthinite, cassiterite, and columbite-tantalite. In the *Nelson area, British Columbia,* scheelite is won from contact-metasomatic tactites. In *Nigeria* tungsten associated with granite occurs in quartz veins with abundant columbite related to tin mineralization.

Selected References on Tungsten

Tungsten. W. O. VANDENBURG. U. S. Bur. Mines Inf. Circ. 6821, 1935. *General summary.*

Strategic Mineral Supplies. G. A. ROUSH. McGraw-Hill, New York, 1939. Chap. 6. *Occurrence; strategic importance.*

Burma-Mawchi-Tavoy. J. A. DUNN. India Geol. Surv. Rec. 73:209–245, 1938. *Detailed geology of Burmese occurrences.*

Geology of the Tungsten Deposits, Mill City, Nev. P. F. KERR. Univ. of Nev. Bull. 28:2, 1934. **Tungsten Mineralization at Silver Dyke, Nev.** P. F. KERR. Univ. of Nev. Bull. 30:5, 1936. **Tungsten Mineralization at Oreana, Nev.** P. F. KERR. Econ. Geol. 33:390–425, 1938. *These three papers give details of geology and origin.*

Boulder County Tungsten Ores. F. B. LOOMIS, JR. Econ. Geol. 32:952–963, 1937. *Ferberite, huebnerite type.*

Bolivia Tungsten. C. W. WRIGHT. Foreign Minerals Quart. 2:4:30–38, 1939. *Cassiterite-stannite veins, with silver.*

Tungsten Mineralization in the United States. PAUL F. KERR. Geol. Soc. Amer. Mem. 15, New York, 1946. *Occurrences and origin of deposits and summation of all U. S. occurrences; extensive bibliography.*

Tungsten. K. C. LI and C. Y. WANG. 2nd Ed. Amer. Chem. Soc. Mon. 94. Reinhold Pub. Corp., New York, 1947. *Comprehensive treatment of geology, uses, and recovery of tungsten. Covers world occurrences, notably those of China. Good bibliography.*

Sangdong Tungsten Deposit. M. R. KLEPPER. Econ. Geol. 42:465–477, 1947. *Geology and origin of one of the most remarkable tungsten-bismuth deposits of the world.*

Origin of the Tungsten Ores of Boulder County, Colorado. T. S. LOVERING. Econ. Geol. 36:229–279, 1941. *Occurrence, origin, and wall rock alteration of ferberite veins.*

Scheelite in Northeastern Brazil. W. D. JOHNSTON, JR., and F. MOACYR DE VASCONCELLOS. Econ. Geol. 40:34–50, 1945. *Geology of one of the world's important sources of tungsten.*

Argentina Tungsten. C. W. WRIGHT. Foreign Minerals Quart. 3:3:25–30, 1940. *Brief descriptions of important occurrences.* Also, N. B. KNOX, Eng. & Min. Jour. 146:90–93, 1945.

Wolfram in Nigeria, with notes on cassiterite and columbite. H. L. HAAG. Bull. Inst. Min. Met. 458:1–34, London, 1943. *Tin-tungsten-columbite veins.*

Tungsten Deposits of Vance County, North Carolina. G. H. ESPENSHADE. U. S. Geol. Surv. Bull. 948–A, 1947. *Brief descriptions.*

Tungsten Deposits in the Sierra Nevada, Near Bishop, Calif. D. N. LEMMON. U. S. Geol. Surv. Bull. 931–E, 1941. *Large contact-metasomatic deposits.*

Tungsten Deposits in the Republic of Argentina. WARD C. SMITH and E. M. GONZALES. U. S. Geol. Surv. Bull. 954–A, 1947. *Geology of vein deposits.*

Vanadium

Vanadium is a minor alloy that toughens steel desired for axles, pistons, crankshafts, and pins, and has various other uses where strain, shock, and fatigue is involved. Such steels use 1 to 1.25 percent. A little added to any steel helps remove oxygen and nitrogen, and gives uniform grain size. For high-speed steels 4 to 5 percent vanadium is used. It is also added to chromium, molybdenum, and tungsten steels, and to cast iron, brass, and bronze. A newer use is as a catalyst, replacing platinum, and a little is employed in the electrical, chemical, ceramic, paint, dye, and printing industries.

Production and Distribution. From 2 to 8 million pounds of vanadium is produced annually chiefly from Peru and the United States and some from Southwest Africa and Northern Rhodesia. The United States ore is obtained in Colorado, Utah, Idaho, Montana, and Wyoming.

Mineralogy and Treatment. The chief economic vanadium minerals are patronite, roscoelite, vanadinite, and descloizite. Carnotite also

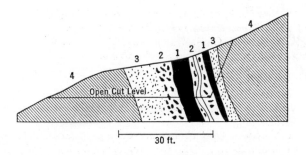

Fig. 14–20. Section through vanadium deposit, Minasragra, Peru. 1, Patronite ore; 2, coke; 3, quisqueite; 4, shales. (*After Hewett, A.I.M.E.*)

carries vanadium. The ores are sorted, concentrated, and largely reduced to ferrovanadium.

Types of Deposits and Origin. Most commercial vanadium comes from deposits that have been weathered, with or without residual concentration. Some may have been formed by concentration through organisms. Hess regards the uranium-vanadium ores of Colorado and Utah as of sedimentary origin, but later work indicates that they are post-sediments.

Examples of Deposits. The greatest deposit known is at *Minasragra, Peru* (Fig. 14–20), where vanadium occurs in a mass 30 feet wide and 350 feet long enclosed in Cretaceous shales and limestone intruded by porphyry dikes. The ore is associated with asphalt and consists of a red calcium vanadate (50% V_2O_5) and patronite filling cracks and fissures in the shale. The richer ore is sorted to a shipping product with 11% V_2O_5 and the low-grade material is calcined and burned to an ash containing 22% V_2O_5. Large reserves of low-grade ore are indicated.

The *United States* production is won from (1) vanadium-bearing carnotite deposits in Colorado and Utah; (2) so-called roscoelite ores of Colorado and Utah; (3) oxidized silver-lead-molybdenum ores in Arizona; and (4) as by-product vanadium from the treatment of sedimentary phosphate rock in Idaho and Wyoming.

The deposits of Colorado and Utah are the most productive. The carnotite deposits of the Morrison formation are discussed on page 177. These deposits occur in two to six beds, one of which was mined for 3,000 feet, yielding ore that averaged 1.2 to 1.5% U_2O_3 and 3.5% V_2O_5 and over 200 grams of radium. Somewhat similar deposits occur in the Shinarump conglomerate.

The so-called roscoelite deposits around Rifle and Placerville, Colo., according to Fischer *et al.*, occur in a wavy layer in the upper 25 feet of the Entrada sandstone in flat lenses 1½ to 5 feet thick, which do not conform exactly with the bedding. The ore carries 1½ to 3% V_2O_5. The principal ore mineral is micaceous; its chemical composition resembles roscoelite, but X-ray studies indicate that it is one of the hydrous mica group of clay minerals; vanoxite and other vanadium minerals are present.

At *Broken Hill, Northern Rhodesia,* vanadium ores, consisting of descloizite and vanadinite, occupy a marginal position to rich oxidized zinc-lead replacement deposits in dolomite.

At *Tsumeb, Southwest Africa,* oxidized replacement deposits of lead-copper-zinc ores in dolomite carry vanadium ores containing descloizite, mottramite, and vanadinite.

Petroleum residues after burning yield vanadium ash, and it is also abstracted from petroleum flue dust.

Selected References on Vanadium

Vanadium in Peru. C. W. WRIGHT. Foreign Minerals Quart. **3:1: 36–38,** January, 1940. *Geologic occurrence and mining.*

Broken Hill, Rhodesia. XV Int. Geol. Cong. Guidebook C22:13–16, Pretoria, S. Africa, 1929. *Vanadium is a by-product of oxidized silver-lead ores.*

Tsumeb, Southwest Africa. XV Int. Geol. Cong. Guidebook C21:38–45, Pretoria, S. Africa, 1929. *An unusual copper deposit with vanadium.*

General Reference 9. Uranium-Vanadium Ores of Colorado, by F. L. HESS. Chap. X, Pt. II. *Sedimentary deposits with vanadium-uranium-radium ores.*

Vanadium Deposits of Colorado and Utah. R. P. FISCHER. U. S. Geol. Surv. Bull. 936-P, 1942. *Tabular bedded deposits in sandstones.*

Vanadium. D. C. McLAREN. Min. Mag. **71:203–212,** 1944. *Occurrence, mineralogy, origin, uses.*

Vanadium Deposits near Placerville, Colo. R. P. FISCHER, J. C. HAFF, and J. F. ROMINGER. Colo. Sci. Soc. Proc., **15:115–134,** 1947. *Flat lenses of uncertain origin in Entrada sandstone.*

Cobalt

Cobalt is another minor element used for special steels and desired particularly for magnet steel (35% of use), stellite steels for metal cutting, and temperature-resisting alloys. It is desired also for high-speed steels, dies and valve steel, welding rods, carbide-type alloys, and corrosion-resisting steels. Its ability to maintain strength at high temperatures makes it sought for jet airplane engines. It is also used for making the blue color in glasses and enamels, as a catalyst, and in paint driers.

The total production amounts to some 12 million pounds annually, of which the Belgian Congo is the largest producer followed by Northern Rhodesia, United States, Morocco, Canada, and Burma. The United States production comes from low-grade ores in Missouri.

Cobalt is recovered as a by-product of other ores, chiefly copper and silver. The commercial minerals are linnaeite, cobaltite, smaltite-chloanthite, and the black oxide. The minerals are roasted, and the residue is treated by a wet chemical process leaving a pure oxide of cobalt, the marketable form.

The *Belgian Congo* supply is derived from the black oxide, which is a residual product of weathering in dolomite, associated with the oxidized copper ores. The *Rhodesian* cobalt comes from the mineral linnaeite, which occurs with the copper sulphide ores of the *N'Kana* mine. The Canadian production, formerly large, came from the cobalt-silver arsenide veins of *Cobalt,* Ontario; that of *Burma* is a by-product of the lead-zinc ores of the Bawdwin mine, and the *Moroccan* source is from smaltite and erythrite in gold ores near *Tarudant.* A small amount comes from Finland, Russia, Chile, and Germany.

Selected References on Cobalt

Cobalt. P. M. Tyler. U. S. Bur. Mines Inf. Circ. 6331, 1930. *Good bibliography.*
Cobalt. R. S. Young. Amer. Chem. Soc. Mon. 108. Pp. 181. Reinhold Pub. Corp., New York, 1948. *Occurrence, uses, properties, and metallurgy.*

Other Minor Ferroalloys

Titanium is also used as a minor alloy for steel, but it is dominantly a pigment mineral and will be described elsewhere. *Ferrosilicon,* with 7 percent silicon, is used for electrical machinery. *Ferrophosphorus, silicomanganese, zirconium,* and *calcium-molybdenum* are other minor alloys.

CHAPTER 15

MINOR METALS AND RELATED NONMETALS

The minor metals are used in relatively small quantities, yet several of them play an indispensable role in modern industry, such, for example, as antimony for type metal or mercury for electrical apparatus. Several are by-products of other ores and are not mined separately.

Antimony

For a minor metal, antimony has many diversified and indispensable uses both for normal industry and for war. Its striking property of expanding, rather than contracting, upon cooling makes it desirable for making type metal alloy, which when cast does not change size. Its main use is to impart stiffness and hardness to various lead alloys, thus permitting lead to be used for purposes not otherwise possible. Antimonial lead or hard lead contains 1 to 25 percent antimony, produced by alloying the two metals or by smelting mixed antimony and lead ores.

The chief uses for antimony-lead alloys are: storage battery plates, sheets and pipe, sheathings for electrical cables, collapsible tubes (toothpaste), foil, bullets, type metals (10 to 15% Sb), solder, and antifriction bearings.

" White metal " alloys include Britannia metal (Pb-Sb-Cu), pewter (Pb-Sn-Sb), Queen's metal (Sn-Sb-Cu-Zn), and Sterline (Cu-Sb-Zn-Fe).

Metallic antimony is little used as such except for ornamental castings and bric-a-brac, but compounds are used for flame-proofing, pigments, enamels, safety matches, glass, vulcanizing, and medicines. Military uses include shrapnel balls and bullet cores, detonating caps, and bursting charges of shrapnel to yield the dense white marker smoke.

Production and Distribution. Antimony production ranges from 25,000 to 50,000 tons, and most of it comes from China, Bolivia, Mexico, United States, South Africa, and Peru. Minor amounts come from Hungary, Austria, Yugoslavia, Italy, Turkey, and Morocco. The United States production comes from Idaho, Alaska, and California. The abode of antimony is with lead ores.

Mineralogy, Tenor, and Treatment. There are many minerals of antimony, but stibnite and lead ores yield most of the commercial metal. Native antimony and oxidation products, such as cervantite, contribute a little. Ores range in tenor from 3 to 8 percent. Ores are hand sorted or concentrated; the sulphide product is enriched by liquidation, whereby the low-temperature melting point of stibnite permits the melt to drain off from unfused enclosing substances. This produces crude antimony. Low-grade ores are roasted to the oxide. Antimony metal is produced by smelting the oxide or the crude or high-grade sulphides in a blast furnace with iron, which unites with the oxygen or sulphur, freeing the antimony. Oxide and electrolytic antimony are now recovered by the "Lee-Muir" process, which separates the components of tetrahedrite.

Types of Deposits and Origin. Most antimony deposits are formed by hydrothermal solutions at low temperatures and shallow depth, giving rise to filled fissures, joints, and rock pores and to irregular replacement deposits. Some primary deposits have been enriched in oxidized products through residual weathering.

EXAMPLES OF DEPOSITS

China. China can yield annually 15,000 to 24,000 tons of antimony metal and for years yielded two-thirds of the world's supply. The principal deposits are in Hunan, near Changsha. Other deposits occur in Yunnan, Kwenchow, Kwangtung, Szechwan, and Kwangsi. The largest field is the Hsi-Kuang-Shan, Hunan, where reserves are in excess of 1.5 million tons of metal. The Panchi is the largest mine. Here, in a belt $3\frac{1}{2}$ by 2 miles, stibnite deposits occur in deformed Silurian sandstone in veins, seams, pockets, and lenses, with the richest deposits, according to Schrader, beneath anticlinal structures. The ore, which is about 6 percent, is hand sorted to 55 or 60 percent.

Mexico. The chief deposits are in San Luis Potosi, Oaxaca, and Queretaro, with minor ones in Sonora and other states. These deposits, which consist of stibnite with some oxide minerals, occur in veins in limestone intruded by porphyry. In Hidalgo, jamesonite replacement deposits occur in Cretaceous limestone cut by monzonite. A unique deposit of livingstonite $HgSb_4S_7$ carrying 1 to 2% Sb and 0.3% Hg is mined at Huitzuco, Guerrero. Considerable antimonial lead is produced at local lead smelters.

Bolivia. In Bolivia large reserves occur in a 150-mile belt from Lake Titicaca to Atocha, and the chief centers are Uncia and Porco. The high-grade deposits are shallow quartz veins in Paleozoic shale containing stibnite and base-metal sulphides.

Algeria. A mineralized belt across Algeria from Morocco to Tunis contains, near Constantine, antimony veins in limestone with primary stibnite and the oxidized minerals cervantite, senarmontite, and valentinite.

United States. Few antimony deposits are worked in the United States except during periods of high prices. Large reserves of low-grade ore occur in Yellow Pine, Idaho, associated with gold and tungsten, and lesser ones in California, Nevada, and Alaska. The main production is a by-product of lead ore smelting.

Other Deposits. Stibnite veins occur at Loznica, Yugoslavia, and very rich, narrow Tertiary veins of stibnite and gudmundite are formed with nickel minerals at Turhal, Turkey. Low-grade ores occur in Slovakia, and minor deposits are in Peru, Japan, Russia, Argentina, Germany, Morocco, and other places.

Selected References on Antimony

Strategic Mineral Supplies. G. A. ROUSH. McGraw-Hill, New York, 1939, Chapter IX. *General occurrence, uses, reserves, production, distribution.*

Antimony Deposits. F. C. SCHRADER. Jour. Wash. Acad. Sci. 20:17:436–438, 1930. *General summary.*

Report on the Wuhsi Antimony Deposit, Yuanling, Hunan, China. C. C. TIEN and H. C. WANG. Hunan Geol. Surv. Bull. 17:31–45, 1934. *Descriptions of Chinese antimony veins.*

Antimony Deposits of Tejocotes, Oaxaca, Mexico. D. E. WHITE and R. GUIZA, JR. U. S. Geol. Surv. Bull. 953–A, Washington, 1947. *Occurrence of rich oxide and stibnite deposits.*

San José Antimony Mines, Wadley, San Luis Potosí, Mexico. D. E. WHITE and JENARO GONZÁLEZ R. U. S. Geol. Surv. Bull. 946–E, Washington, 1946. *Most important antimony district of Mexico.*

Antimony Deposits of Yellow Pine District, Idaho. D. E. WHITE. U. S. Geol. Surv. Bull. 922–I, 1940. *Geology of a large shear-zone deposit with gold and tungsten.*

Arsenic

Arsenic metal is little used except for a few alloys, such as with lead for making shot. However, in the form of calcium and lead arsenate and Paris green it is widely used as an insecticide. Calcium arsenate is employed to fight the cotton boll weevil and other insects and animals. White arsenic (As_2O_3) is a common weed killer, extensively employed by railroads. Arsenic compounds are also used to help fuse glass, as a wood and leather preservative, and in dyes, pigments, medicines, fireworks, and chemicals.

Production and Distribution. Some 60,000 to 70,000 tons of white arsenic are normally produced annually as a by-product of smelter smoke from arsenical ores. The chief producing countries are (in percentage): United States, 37; Mexico, 15; Sweden, 14; France, 11; Belgium, 5.5; Germany, 5; Japan, 4.6; Australia, 3.6. In most places the available material is greater than consumption. Sweden alone could produce enough for the entire world and has difficulty in getting rid of the poisonous material.

Ores and Treatment. The arsenical ores that give off arsenic are mainly gold, silver, copper, lead, and zinc. The chief minerals that yield such arsenic are arsenopyrite, enargite, tennantite, realgar, orpiment, and arsenides of silver and lead. In smelting, the arsenic is volatilized and combines to form As_2O_3, which is condensed. The amount needed for market is refined and shipped as white arsenic.

Selected Reference on Arsenic

Arsenic. P. M. TYLER and A. V. PETAR. U. S. Bur. Mines Econ. Paper 17, 1934.

Beryllium

Beryllium is the new metal of the moment, and its minerals have now passed from the stage of museum collections and gemstones to commercial use. It is a light metal (sp. gr. 1.85) that imparts high strength

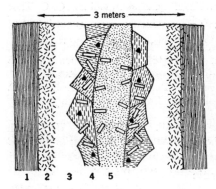

FIG. 15–1. Morada Nova beryl mine, Brazil. 1, schist; 2, muscovite zone; 3, pegmatite; 4, microline with beryl crystals; 5, quartz core. (*After W. D. Johnston, Jr., Geol. Soc. Amer.*)

and fatigue resistance to copper, also to cobalt, nickel, and aluminum. It is so light that an airplane engine constructed of beryllium could be lifted by one man. One and one-half to three percent added to copper increases the tensile strength from 33,000 to 200,000 pounds and gives it fatigue resistance greater than that of spring steel; vibrator springs stressed by 2 billion vibrations show no deterioration. The alloy finds use for various types of instrument springs, control parts, valves, and airplane carburetors and instruments. The alloy is nonsparking, making it desirable for use in explosive and petroleum industries. The oxide is used as a refractory. The metal finds application in X-ray tubes, fluorescent lamps, neon signs, and cyclotrons. Alloys with nickel, cobalt, and aluminum are commanding attention.

The metal is obtained from beryl, which occurs chiefly in pegmatite dikes (Fig. 15-1) and rarely as a gangue mineral of tungsten and tin ores. Chrysoberyl, helvite, and phenacite are possible ore minerals. The output is about 3,000 to 5,000 tons of beryl annually, and the supply is limited. The metal is extracted from beryl with difficulty by electrolysis in a fluoride bath containing sodium and barium.

Present commercial sources of the ore are the Black Hills of South Dakota, Brazil, Argentina, India, and Russia. Small reserves are known in Canada, Mexico, South Africa, and Madagascar.

Selected References on Beryllium

Minerals Yearbook. U. S. Bur. Mines, 1939, pp. 748–751; 1945, pp. 813–817. *Summary of occurrence, new uses, and distribution.*

Beryl-tantalite pegmatites of northeastern Brazil. W. D. JOHNSTON, JR. Geol. Soc. Amer. Bull. 56:1015–1069, 1945. *The largest beryl district of the world; in pegmatites.*

General Reference 5.

Bismuth

Bismuth is a metal consumed in small quantity, but certain peculiar properties create important uses for it. It possesses curative medicinal properties, it has a low melting point, and expands upon solidifying. Its greatest use is for medicinal and cosmetic preparations. Its salts cure wounds and digestive disorders and also form insoluble compounds for internal X-ray examinations. Their smooth unctuous feel makes them desirable for cosmetics. Bismuth salts are also used to give a glaze to porcelain, for enameling, printing fabrics, and making optical glass.

Bismuth forms various alloys with lead, tin, cadmium, and antimony, and, peculiarly, its alloys will melt far below the melting points of the separate metals. For example, bismuth melts at 271° C, but Wood's metal (Bi-Sn-Cd) melts at 60° C and other alloys at even lower temperatures. Such metal, therefore, will melt in hot water or a very hot room. Consequently, the alloys are used for automatic water sprinklers for fire protection. When a room heats up, the plug melts and water is turned on. They are similarly used for electric fuses, boiler safety plugs, and hand grenades. The expansive qualities of bismuth cause it to be used for molds, for making casts of delicate objects, and for electrotypes. Combined with brass and bronze it gives antifriction metals. Bismuth wire is used in electrical apparatus and delicate measuring instruments. There are also innumerable other minor applications.

Production and Distribution. The annual output is only from 1,000 to 1,700 tons of metal. The chief producing countries in order are

United States, Peru, Canada, Mexico, and Bolivia with minor production from Japan, Germany, Spain, and Australia.

Mineralogy, Ores, and Treatment. The commercial ore minerals of bismuth are native bismuth, bismuthinite, and bismuth ochre. Few deposits are mined for bismuth alone. It is obtained (1) from deposits of other metals such as tin, copper, or silver, and (2) as a refinery by-product, chiefly from lead ores.

The extraction of pure bismuth requires a complex process in order to eliminate other elements. Wet and dry methods are used. Ores are reduced with carbon or iron and soda ash in furnaces. The bismuth in lead is dissolved in hydrochloric acid as the oxychloride and smelted. Refinery slimes are fused with caustic soda and soda ash.

Examples of Deposits. Bismuth minerals occur in fissure veins and in replacement deposits formed by hydrothermal solutions.

The *United States* production is obtained from refinery slimes resulting from the refining of lead bullion by the Betts process. The ores containing it come from various deposits in Utah, Nevada, Colorado, and Arizona.

The large resources of *Bolivia* consist of bismuth minerals in the tin deposits (Chap. 14) of Potosi and La Paz, and the Aramayo mines at Tasna, Ukeni, and Chorolgue. The bismuth averages about 1 percent. Richer bismuth ores are obtained from the Tasna mine, where a hand-sorted product contains 25% Bi.

In *Peru,* concentrates from the San Gregorio mine carry 20% Bi and the Cerro de Pasco ore yields bismuth in smelter flue dust. Peru supplies one-fourth of the world's supply. The bismuth of *Mexico* is contained in the lead ores, as in Saxony, *Germany.* In *Australia* and *Japan* bismuth is won from tin, tungsten, and molybdenum ores. In Cordoba, *Spain*, rich bismuth ores are mined in small quantity from fissure veins; in one near Torrecampos the ore carries 25% Bi; those at Azuel carry 1 to 7% Bi along with nickel, cobalt, and silver.

Selected References on Bismuth

Bismuth. P. M. TYLER. U. S. Bur. Mines Inf. Circ. 6466, 1931. *Distribution and production.*

Bismuth in Bolivia and Peru. C. W. WRIGHT. Foreign Minerals Quart. 2:No. 4, 1939, and 3:No. 1, 1940. *Information from first-hand observations.*

General Reference 11.

Cadmium

Cadmium is entirely a by-product metal obtained chiefly from zinc smelting. It has a low melting point, is ductile, is softer than zinc, and emits a peculiar " cry " when bent. It alloys readily with other metals, and this is its chief use.

Alloyed with nickel, or silver and copper, it forms the best high-pressure, anti-friction bearing metal for automobile bearings. It is a constituent of stereotype plates and of many low-melting alloys described under bismuth. Cadmium hardens copper; it makes silver resistant to tarnish; with silver it makes tableware metal; and it is used for making green gold.

An important application is in cadmium plating, particularly on iron, where it forms a thin, rustless surface alloy. Thus automobile bolts, nuts, locks, and other parts are made nonrusting. It also serves as an adhering bond between iron and other plating metals.

Cadmium compounds find use in chemicals, photographic materials, paint, rubber, soaps, fireworks, textile printing, and as pigment for glass and enamels. The sulphide forms an enduring yellow paint.

Production and Distribution. Some 9 to 11 million pounds of cadmium are produced annually, as follows (in percentage): United States, 70; Mexico, 8; Canada and Australia, 5; with lesser amounts from Germany, Belgium, Poland, Norway, and Southwest Africa. It will be seen that cadmium comes from the zinc regions, since its abode is with zinc ores, and minor production comes from other zinc regions.

Minerals, Ores, and Treatment. The ore minerals of cadmium are the sulphides greenockite (= xanthochroite), the carbonate otavite, and the oxide, cadmium oxide. They occur in association with sphalerite ores and their oxidized equivalents. The cadmium is volatilized from them, or is recovered from dust chambers and in electrolytic slimes.

Examples of Deposits. Since cadmium is a by-product, there are no deposits worked for cadmium alone. The zinc ores of the Tri-State district yield most of the United States production. Some also is recovered from the zinc ores of Butte, Mont., and of Utah. Zinc deposits in the countries listed in Chapter 14 also carry cadmium.

Selected Reference on Cadmium
General Reference 11.

Magnesium

Because magnesium is the lightest metal known (sp. gr. 1.74) and is fairly strong it is in demand for the light alloys that are used extensively in airplanes and automobiles and many other materials requiring lightness. Its alloys are noncorrosive in air but not in sea water. The chief alloying element is aluminum, generally with some zinc and manganese. Such alloys have low weight for unit bulk, high rigidity, and greater strength than most aluminum alloys. They are also used for microscope mountings, field glasses, cameras, surveying instruments, artificial limbs, shuttles, bobbins, radio diaphragms, musical instruments, etc. A little magnesium hardens lead for cable sheaths. A remarkably light alloy consists of magnesium and beryllium.

Other important uses for magnesium are for structural shapes and sheets, and for deoxidizing and desulphurizing nickel, Monel metal, brass, and bronze. Because it burns at low temperatures with a strong actinic light it is used for flashlights, photography, fireworks, and signal flares. Incendiary bombs are made of 93% Mg and 7% Al.

Production and Distribution. World production of metallic magnesium has ranged from 12,000 to 238,000 tons, and the United States reached a wartime maximum production of 148,000 tons.

Sources and Treatment. Magnesium is manufactured from (a) natural brines, (b) sea water, (c) magnesite, (d) magnesium chloride (in carnallite deposits), (e) dolomite, and (f) brucite. Consequently supplies are available almost everywhere. Considerable amounts are obtained from brines, and a large sea water plant is operated in Texas.

Brines and potash salts contain $MgCl_2$; this is extracted, and metallic magnesium is obtained from melted $MgCl_2$ by electrolysis. With magnesite, the metal is produced by (1) electrolysis of $MgCl_2$ obtained by chlorination and (2) reducing calcined magnesite with carbon in an electric furnace and chilling the metallic vapor. The impure metal is redistilled in a second furnace and condensed.

Selected References on Magnesium

Magnesium, Magnesite, and Dolomite. J. LUMSDEN. Imp. Inst., Min. Res. Dept. London, 1939. *Most complete information; good bibliography.*
General References 5; 7; 11, Chap. 24.

Mercury

Mercury, or quicksilver (living silver), was known to Aristotle and Theophrastus (315 B.C.). The Chinese apparently knew of it early, since a relief map of China constructed in 210 B.C. depicted the ocean and rivers by liquid quicksilver. Pliny wrote that 10,000 pounds a year were brought to Rome from Almaden, Spain. He also records that later it was used in the recovery of gold. The alchemists delighted in such a unique and mysterious material as liquid metal, and it has found constant use since that time. After the discovery of America quicksilver was used in large quantities for the recovery of gold and silver by amalgamation.

Quicksilver is unique in being the only metal that is liquid under ordinary temperatures. For this reason and for other physical and chemical properties it has become an essential mineral today. Its chief uses in order of consumption are: (1) electrical apparatus, such as switches, clutches, rectifiers, and radio equipment; (2) pharmaceuticals; (3) dry cell batteries; (4) as a catalyst; (5) industrial and control instruments; (6) agricultural insecticides and fungicides; (7) fulminate for munitions and blasting caps; (8) antifouling paint; (9) electrolytic preparations of chlorine and caustic soda; (10) electrical and scientific instruments, thermometers, dental preparations, recovery of gold and silica, vapor lamps, and the Emmett mercury-vapor boiler; and it has many other uses.

Production and Distribution. The annual production of mercury is from 140,000 to 275,000 flasks, of which about 90 percent comes from Spain, Italy, United States, and Mexico. Minor quantities come from Canada, China, Japan, South Africa, Chile, and Peru. The dominance in mercury production belongs to Italy and Spain. In the United States the chief sources are (1) California with 52 mines; (2) Oregon with 13 mines; and (3) Nevada with 17 mines. The total United States output ranges between 15,000 and 25,000 flasks annually, depending on the price, but under war prices this amount was doubled.

Mineralogy, Tenor, and Treatment. The commercial minerals of mercury are cinnabar, metacinnabarite, calomel, and a little native mercury; the ores range in tenor from 0.3 to 8 percent. Mercury is readily extracted by volatilization of its ores, and the vapor is condensed to a liquid in cooling tubes and drained to collecting tanks. It is then bottled for market in steel flasks holding 76 pounds. Because of this simplicity, low-grade ores can be worked.

Types of Deposits and Origin. All mercury deposits are formed from hydrothermal solutions at relatively low temperatures. The chief types of deposits (Fig. 15–2) are: replacement deposits, Almaden, Spain; fissure veins, Monte Amiata, Italy; breccia fillings, Idria, Monte Amiata, Italy; stockworks, New Idria, New Almaden, Calif.; pore-space fillings, Idria, Italy. The mercury deposits may occur in any kind of rock that has been fractured, thus permitting ingress of solutions. They are most typically associated with late Tertiary volcanism.

EXAMPLES OF DEPOSITS

Spain. The Almaden mine, on the slopes of the Sierra Morena, is the greatest and richest mine in the world. Although it has been worked for over 2,000 years it still contains the largest known reserves, capable of supplying the world for the next 100 years.

The region consists of folded and faulted Silurian quartzites and slates, intruded by quartz porphyry and diabase. The ore deposits are parallel, steeply dipping, replacement lodes in quartzite, of which there are three, the San Pedro, San Nicolas, and San Francisco, the last two joining together at one end and all three terminating against a fault. The veins have been worked to the 13th level, a depth of 325 meters. The 11th level at the time of the author's visit had been

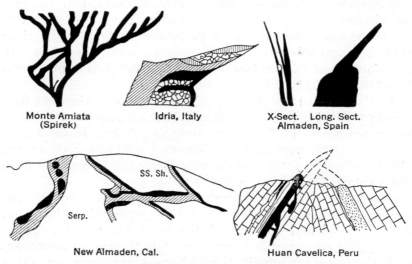

Monte Amiata
(Spirek)

Idria, Italy

X-Sect.   Long. Sect.
Almaden, Spain

SS. Sh.

Serp.

New Almaden, Cal.

Huan Cavelica, Peru

FIG. 15–2.   Cross sections of some quicksilver deposits.
(*Mostly from Schuette, A.I.M.E.*)

worked in a crude way for 44 years. There, the San Pedro vein was 8 to 10 meters wide and in many places averaged 20 percent mercury. The greatest width observed was 25 meters where the two veins joined. The length is 1,000 feet. The three lodes are trending toward each other in depth.

The low-grade ore consists of stringers and blebs of mercury minerals in quartzite. The high-grade ore is made up of massive bands and bunches of ore minerals; a mass of pure cinnabar several feet across was observed in one place. The ore minerals consist predominantly of cinnabar with metacinnabar and some native quicksilver; pyrite is rare and calcite and barite are present. Deposition has been almost entirely by replacement. Ransome showed that the quartzite was first sericitized, then quartz and sericite were replaced by ore minerals. Observations underground of the larger features of occurrence clearly indicate a replacement origin by hydrothermal solutions. The ore averages 5 to 8 percent mercury.

Italy. Italy since 1937 has been the largest producer of mercury. The chief districts are Monte Amiata in Tuscany and Idria in Trieste.

The *Monte Amiata* includes six zones of which the Abbadia San Salvador is the second largest mercury mine of the world. The deposits were worked by the Greeks and Romans. They lie at intersections of minor fractures with a major fracture that cuts Mesozoic and Tertiary sediments capped by a trachyte flow from the Monte Amiata volcano. The deposits, of late Pliocene age, occur in limestone, sandstone, clay-marls, trachyte conglomerate, and trachyte. They are mainly low-temperature, hydrothermal fillings of solution cavities, crushed zones, sandstone pores, and particularly of trachyte conglomerate. Cinnabar is the chief mineral, accompanied by some stibnite. The ores average a little less than 1 percent.

At *Idria,* cinnabar impregnates brecciated blocks of Triassic dolomite overthrust on Carboniferous sediments. Impregnations in sandstones and minor fissure fillings also occur. Some metacinnabarite and native mercury are present. The ore is richest in open-textured sandstones, rich in the dolomite breccia, and lean in the bedding planes and small fractures of the bedded dolomite.

United States. The United States deposits occur in a Pacific Coast belt some 400 miles long in California, projecting into Oregon and Nevada, and in single districts in Texas and Arkansas.

The *California* section has yielded up to 33,000 flasks annually from veins, stockworks, and disseminated deposits in serpentine, and also in sandstone and cherts of the Franciscan group. The chief mines are the New Idria and New Almaden. The New Idria has been in operation since 1850. Here, two fault-fissure veins lying between serpentine and sandstone carry cinnabar and metacinnabarite, much pyrite, quartz, opal, and carbonates, deposited from uprising hydrothermal solutions. The New Almaden in its 120 years of operation has yielded over 75 million dollars worth of quicksilver and reached a depth of 2,450 feet. Its production is now small. The deposits are impregnations in a shatter zone of Franciscan sandstone and shales lying above and about a sloping peridotite intrusive contact. The mineralization is similar to that of New Idria.

Generally similar deposits occur in *Oregon* where the Opalite mine and the rich Brety mine are steady producers. In *Nevada,* also, a few small deposits belong to the same period of metallization.

Near *Terlingua, Texas,* low-grade deposits occur in folded and shattered Cretaceous limestone. The deposits, of which the Chisos mine is an example, are impregnations of cavities and caves which occur along faults or near the crackled crests of anticlines and which are occupied by boulders and clay. The overlying Del Rio clay acted as an impervious barrier to uprising hydrothermal solutions that deposited cinnabar. Surface alterations yielded native mercury, terlinguaite, eglestonite, montroydite, and calomel. The ore averages 0.5 percent mercury but some attains 2 percent.

According to Stein, mercury deposits located in *Pike County, Arkansas*, consist of fault breccia fillings, fault gouge, and shale contact impregnation, vein fillings, and sandstone impregnations. They occur in the shales and sandstones of the Pennsylvanian, Stanley, and Jackfork formations. The introduced minerals are cinnabar, metacinnabar, native mercury, pyrite, stibnite, quartz, opal, siderite, and dickite — the dickite indicating moderate hydrothermal activity. The ores carry from 0.3 to 1.1 percent mercury.

Other Deposits. In *Mexico* are numerous small deposits, of which those around Guadalcázar, San Luis Potosi, Nuevo Mercurio and Sain Alto, Zacatecas, Canoas, and Huitzuco, Guerrero, have been the most productive. At Huancavelica, *Peru*, cinnabar ores were deposited in openings in folded Cretaceous sandstone and limestone intruded by andesite. Nearby hot springs attest to recent igneous activity. Other minor deposits are worked in Japan, China, Czechoslovakia, Algeria, and Russia.

Selected References on Mercury

Occurrence of Quicksilver Ore Bodies. C. N. SCHUETTE. A. I. M. E. Tech. Pub. 335, 1930. *Geology, localization, and origin.*

Strategic Mineral Supplies. G. A. ROUSH. McGraw-Hill, New York, 1939. Chapter 10, Mercury. *Deposits, distribution, uses, strategic features.*

Quicksilver. C. N. SCHUETTE. U. S. Bur. Mines Bull. 335, 1931. *Occurrence, distribution, uses, production.*

Mercury Industry in Italy. EDWIN B. ECKEL. A. I. M. E. Tech. Pub. 2292, 1948. *Geology of the most productive mercury mines in the world.*

Mercury, Mexico. Quicksilver-Antimony Deposits of Huitzuco, Guerrero, Mexico. JAMES F. McALLISTER and DAVID HERNANDEZ ORTIZ. *An unusual deposit with livingstonite.* **Geology of the Cuarenta Mercury District, State of Durango, Mexico.** DAVID GALLAGHER and RAFAEL PEREZ SILICEO. *A productive structurally controlled deposit.* U. S. Geol. Surv. Bulls. 946–B and F, Washington, 1945–46.

Mercury, Western States. Quicksilver Deposits in the Steens and Pueblo Mountains, Southern Oregon. CLYDE P. ROSS. **The Wild Horse Quicksilver District, Lander County, Nevada.** CARLE H. DANE and CLYDE P. ROSS. **Quicksilver Deposits of the Opalite District, Malheur County, Oregon, and Humboldt County, Nevada.** ROBERT G. YATES. **Quicksilver and Antimony Deposits of the Stayton District, California.** EDGAR H. BAILEY and W. BRADLEY MYERS. U. S. Geol. Surv. Bulls. 931–J, K, N, and Q, Washington, 1942. *Gives various modes of occurrence of mercury.*

Radium and Uranium

Radium is a disintegration product of uranium and is one of the extremely rare elements, although uranium is not a rare constituent of the earth's crust, despite the paucity of workable deposits. Radium is prized for curative medicine and is also used for detecting flaws in steels and other alloys, and for luminous paints for clock dials and other objects. Uranium has been desired formerly chiefly to obtain radium, but its salts were used to give yellow to brown colors for glass and glazes and for special alloys of steel, copper, and nickel. Now it is desired for atomic energy.

The main sources of uranium are the Belgian Congo and Canada, and less important sources are the United States and Czechoslovakia. About 60 percent of the pre-war radium demand was supplied by Belgian Congo and about 40 percent by Canada. Canada's production of radium was about 70 to 90 grams annually, United States 8 grams, and Germany 1 to 3 grams.

Minerals. The chief ore mineral of uranium is pitchblende which is found primarily in fissure veins and is the source of major uranium production in the Belgian Congo, Canada, and Czechoslovakia. Uraninite is sometimes associated with pitchblende, but it is most likely to be found in pegmatites. Other hypogene minerals are the uranium-bearing oxides of columbium, tantalum and titanium, such as betafite, euxenite, fergusonite, and samarskite, and their home is in pegmatites where they occur as small nests or pockets. Although these have been mined for uranium on a limited scale, they generally do not occur in large enough concentrations to be significant sources of uranium.

The important supergene uranium minerals are carnotite, tyuyamunite, autunite, meta-torbernite, and torbernite. These, and other brilliant, yellow, brown, and green minerals are oxidation products of pitchblende and other hypogene minerals. More than 100 uranium-bearing minerals are known to exist. Consequently, uranium minerals are not uncommon, but concentrations of them are few.

Deposits. The chief deposits are hydrothermal fissure veins containing pitchblende generally associated with silver, cobalt, nickel, and copper minerals. The deposits next in importance are impregnations of carnotite in sandstones associated with roscoelite and other vanadium minerals. Most of the pegmatite deposits carry such a small amount of uranium that it has not been feasible to mine them for this metal.

Uranium deposits also occur in shales and phosphates over wide areas but the grade is lower than in the carnotite deposits. Deposits in shale occur in Sweden, Russia, and Tennessee, and deposits in phosphates have been found in Idaho, Montana, and Florida. The fourth type deposit, generally of small size, is in river and beach placer sands and gravels.

Uranium minerals decompose readily under weathering. Hence, they are not common in placers, and yellow oxidation products commonly overlie pitchblende.

Belgian Congo. Until the discovery of the Canadian deposits, the valuable deposits of the Shinkolobwe mine, in the copper belt of Katanga, were the main source of the world supply of radium. The ores consist of pitchblende and a large number of its vivid oxidation minerals, along with gold and palladium and the sulphides of nickel, cobalt, and copper. They occur in capricious veins, stockworks, and disseminations in the dolomite schists of the Mine Series (see also Katanga copper deposits, Chap. 14). The mineralization sequence is quartz, uraninite, sulphides, carbonate, oxidation products. The deposits are closely associated with the copper deposits and clearly have a magmatic source, presumably the granitic intrusives with which the copper is related.

Great Bear Lake, Canada. Since the discovery of the deposits on the shore of Great Bear Lake, Northwest Territories, the quantity of available radium has been vastly increased and the price lowered. The deposits, according to Kidd, consist of replacement lodes and stockworks, along fractures and shear zones that traverse Precambrian sedimentary and altered volcanic rocks cut by granodiorite. Several zones are known, the longest being 5,000 feet; three, up to 30 feet wide, are worked at La Bine point. Some 40 minerals have been recognized, but the chief ones are pitchblende, native silver, pyrite, and chalcopyrite, with carbonate and quartz gangue. The others include compounds of iron, cobalt, nickel, copper, lead, zinc, silver, molybdenum, bismuth, and manganese. Kidd and Haycock recognize three stages of mineralization, (1) pyrometasomatic, (2) hydrothermal, and (3) supergene. The hydrothermal sequence is (a) pitchblende-quartz-Co-Ni, (b) quartz-Co-Ni, (c) dolomite-Pb-Zn-Cu, (d) rhodochrosite-Cu-Ag. They believe that the deposits are of magmatic derivation related to granitic intrusives and represent a range from high- to low-temperature conditions. The pitchblende represents colloidal deposition in cavities.

Czechoslovakia. The mines of Joachimsthal, made famous as the source of the pitchblende from which Madame Curie first extracted radium, still produce radium and uranium. The ore occurs in fissure veins along with silver sulphides, quartz, and dolomite and compounds of copper, lead, zinc, bismuth, silver, nickel, cobalt, and iron and is of hydrothermal origin.

United States. The carnotite ores of the Colorado Plateau have been worked, first for vanadium, with uranium and radium as by-products. They once occupied first place in radium production and have yielded 20 grams. Now they are worked chiefly for atomic energy uranium. The uranium ore is carnotite, with associated roscoelite and other vanadium minerals. Mostly it occurs disseminated in the Morrison sandstone where 2 to 6 beds have been mineralized in part; a less amount occurs with the vanadium found in the Entrada and Shinarump sandstones. Fossil logs in the sandstone have been especially rich. They are described in Chap. 5·5.

Other Deposits. Minor radium deposits occur in Gilpin County, Colorado, Germany, Portugal, South Australia, and Madagascar.

Selected References on Radium and Uranium

Concentrations Uranifères du Katanga. J. THOREAU. XVI Int. Geol. Cong., Washington, 1936. *Details of occurrence and mineralogy.*

Great Bear Lake, Canada. D. F. KIDD and M. H. HAYCOCK. Geol. Soc. Am. Bull. 46:879–960, 1935. *Occurrence, mineralogy, origin.*

Prospecting for Uranium. U. S. Atomic Energy Comm. and U. S. Geol. Surv., Washington, 1949. *Minerals, deposits, prospecting, testing, selling, laws.*

General Reference 9. Pp. 455–480. **Colorado-Utah,** by F. L. HESS. Details of *occurrence and origin.*

Minerals Yearbook, 1934. U. S. Bur. Mines, Washington. Pp. 498–502, Belgian Congo.

The Eldorado Enterprise. R. MURPHY. Can. Inst. Min. Met. Trans. 49:423–438, 1946. *Geology and history of the radium-uranium-silver deposits of Great Bear Lake, Canada.*

Uranium-Bearing Sandstone Deposits of the Colorado Plateau. By R. P. FISCHER. **Uranium in Pegmatites.** By L. R. PAGE. **Characteristics of Marine Uranium-Bearing Sedimentary Rocks.** By V. E. McKELVEY and J. M. NELSON. Econ. Geol. 45:1–53, 1950. *These papers give carnotite occurrences, uranium minerals distributed in pegmatites, occurrences in aluminous and carbonaceous sediments.*

Selenium and Tellurium

Selenium, an acidic element related to tellurium, is another minor by-product obtained from the refining of copper ores. It is desired chiefly for making red glass for danger signals but is used also as a decolorizer of green glass, for making red enamels, for vulcanizing rubber, and for alloys. Its property of conducting electricity only when exposed to light permits its use for automatic devices such as electric signs, " electric eyes," street lamps, burglar alarms, and many electronic devices. It is also used in drugs, chemicals, dyes, and in photography, and it forms the only solvent for plastics. The chief North American production is from the copper ores of Sudbury, Noranda, Hudson Bay mines in Canada, and from United States copper refining. Much more could be saved from flue dusts if the demand should justify it.

Tellurium, a semi-metal, closely related to selenium, is also obtained from copper-refining sludges. It is one of the few elements that combine with gold. It also occurs native. It is more abundant than selenium, but only enough is recovered to meet the demand for it. It is used to make tough hard rubber for sheathing hose and cables, as an alloy metal for adding blue and brown colors to glass, to make lead harder and less corrosive, and to improve the creep of tin. Most of the supply comes from the Precambrian copper ores of Manitoba, Ontario, and Quebec and from the United States copper refineries.

Selected References on Selenium and Tellurium

Selenium and Tellurium. R. M. Santmyers. U. S. Bur. Mines Inf. Circ. 6317, 1930.
General Reference 5.

Tantalum and Columbium

Tantalum and columbium are rare metals of restricted usefulness that occur together in pegmatite dikes. They are obtained from the minerals tantalite, samarskite, manganotantalite, and columbite. These are won mainly from the weathered parts of pegmatites or as by-products in extracting other pegmatite minerals. The world production of the minerals is about 2,000 to 3,000 tons, but the war greatly stimulated their production for use in airplane radio sets and for making synthetic rubber. Tantalum comes principally from Brazil, Belgian Congo, western Australia, Uganda, and the Black Hills of South Dakota, and the main sources of columbium are the Tantas-Nigeria mine, Nigeria, Belgian Congo, South Africa, and the Black Hills.

Tantalum and columbium because of their great affinity for gases are widely used in vacuum tubes. Tantalum has high resistance to corrosion and is used for acid-resisting chemical ware (when used cold) and in absorption systems for making hydrochloric acid. It makes extremely hard alloys for abrasive and cutting purposes; tantalum carbide is almost as hard as the diamond, and its melting point is above

3,000° C.   Alloys are used for springs, saws, gun barrels, and surgical and dental instruments.   Tantalum wire is stronger than steel and permits electric current to flow in one direction only.   Columbium-bearing stainless steels are very resistant to high temperature and corrosion.

Selected References

Beryl-Tantalite Pegmatites of Northeastern Brazil.   W. D. JOHNSTON, JR.   Geol. Soc. Amer. 56:1015–1069, 1945. *Highest-grade tantalite deposits of the world — in pegmatites.*
World Survey of Tantalum Ore.   J. S. BAKER.   U. S. Bur. Mines Inf. Circ. 7319, 1945. *Geology, occurrences, uses.*

Titanium

Titanium is a common metal in igneous rocks but is rarely concentrated into workable lode deposits.   It is a common impurity in magnetite, generally rendering it useless as an ore of iron.   It holds to its bound elements with such tenacity that it is difficult to separate.

Titanium has been a latecomer into the field of usefulness, but it holds promise of becoming the outstanding new metal of this century. Demand for it has been increasing rapidly because it makes the whitest of all paints.   Its opacity is twice that of zinc oxide and three times that of lead oxide.   It is used in the form of titanium oxide, called titanium white, and this can be mixed with other pigments without decreasing its opacity.   The titanium paints are pure titanium oxide or are combined with other paint compounds.   They are used for most inside paints and many of the outside white paints.   The pigment is also used in toilet articles, linoleum, artificial silk, white inks, colored glass, pottery glazes, for tinting artificial teeth, and for dyeing leather and cloth.   Metallurgically, ferrocarbon titanium is alloyed with steel for high-speed steels and with chrome steels; the oxide is used for alloys and cemented carbides and also as a paper filler, as a coating for welding steel, and in electrodes for arc lamps.   In the form of chloride, titanium is used for removing colors from cloths, and in the form of tetrachloride for making smoke screens and sky writing.   Metallic titanium, when produced commercially, holds unusual promise for many applications.

Production and Distribution.   The annual production of titanium minerals ranges up to 500,000 tons, most of which formerly came from India and Norway.   In 1942 production of ilmenite with magnetite was initiated at Lake Sanford, N. Y., and government curtailment of Indian exports led to the development in 1948 of a large new deposit at Trail Ridge, Fla. The largest-known deposit, at Allard Lake, Quebec, will be able to supply the entire world's need for titanium. The chief sources of the future will be Canada, United States, India, and Norway.

Mineralogy and Treatment. The ore minerals of titanium are ilmenite, rutile, and subordinate titanite. These minerals are heavy and are resistant to weathering and therefore accumulate as placers. Rutile ores, as marketed, contain 92 to 98% TiO_2; placer ilmenites carry from 51 to 60 percent. Ilmenite ore is first concentrated and is then melted in electric furnaces; the fused product is leached with sulphuric acid to make titanium oxide, which undergoes further purification. Titanium white is generally precipitated along with barium sulphate.

Occurrence and Origin. Titanium minerals occur as accessory minerals in rocks, which upon weathering yield the beach and stream placers that account for most of the titanium used in the past. Ilmenite is also found in metamorphic rocks, in magmatic deposits associated with magnetite, in large massive magmatic deposits associated with minor hematite in anorthosite, and in nelsonite dikes. Rutile and ilmenite also occur as disseminated replacement deposits. Common associations in placer deposits are monazite and zircon.

United States Deposits. Dikes of nelsonite that consist of ilmenite, rutile, and apatite are mined in *Nelson and Amherst Counties, Virginia*. The adjacent syenite, which is impregnated with 3 to 6 percent each of rutile and ilmenite, is also mined. The dikes were earlier considered to be of direct magmatic origin, but C. S. Ross suggests that parts of the dikes have been hydrothermally replaced by the titanium minerals and apatite and that the wall rock disseminations were also formed by replacement. The ores consist of gossan ores, weathered ore, and hard or unweathered ore. The weathered rock contains 18.5% TiO_2. In the unweathered ore are found ilmenite, rutile, apatite, pyrite, pyrrhotite, allanite, and graphite. C. H. Moore considers, however, that the nelsonite dikes are genetically related to nearby hypogene granodiorite and that they represent true dikes rather than hydrothermal alterations. A magmatic origin is likewise held by Davidson, Grout, and Schwartz. Reserves of some 4.4 million tons of TiO_2 are indicated.

Beach deposits formerly yielded rutile south of Jacksonville, Fla., and a large deposit of beach sands at *Trail Ridge, Fla.*, is now the second largest supply of ilmenite in the United States. The Adirondack titaniferous magnetite deposits carry from **7** to **23%** TiO_2. The Tahawus mine at *Lake Sanford* yields about **280,000** tons of ilmenite concentrates annually and is the largest hard-rock ilmenite mine in the United States. The ore is an ilmenite-magnetite mixture in anorthosite and gabbro, and carries vanadium.

India. At Travancore, India, the sands of the south beach, 6,000 feet long, carry 50 to 70 percent ilmenite, under 6 to 8 feet of burden. The ilmenite is removed by magnetic separators from zircon, rutile, and other minerals. The north beach is about 15 miles long and has yielded most of the ore. The occurrence of these deposits has been described in Chapter 5·7, under Beach Placers.

Quebec. Near Allard Lake, Quebec, are several deposits of ilmenite-hematite ore enclosed in anorthosite. The largest body contains over 200 million tons of ore consisting of about 90 percent combined oxides. The ore is dominantly ilmenite with microscopic ex-solution laths of hematite along the ilmenite cleavages. The TiO_2 content exceeds 37 percent. The massive ilmenite bodies are enclosed in coarse granular anorthosite. Around the margins of the deposits coarse-grained dikelets of ilmenite intrude the anorthosite, and coarse fragments of anorthosite are enclosed in coarsely granular ilmenite. The ilmenite is later than the anorthosite and appears to be a late magmatic injection.

These deposits constitute the largest known bodies of ilmenite in the world. The ore is electrically smelted to yield pure iron, and a slag containing 70–75% TiO_2, which will be used in lieu of ilmenite to produce titanium white.

Other Deposits. The *Norwegian* deposit at Blaafjeldite, Sogndal, is a large body of ilmenite-magnetite which, after concentration, yields 42 to 45% TiO_2. The deposit is estimated to contain 30 million tons of ilmenite.

At *St. Urbain, Quebec,* a large deposit consists of dikelike masses of ilmenite-hematite with some rutile, sapphirine, and spinel, enclosed in anorthosite. The ores contain 35 to 40% TiO_2. Mawdsley and Osborne consider them to be magmatic injections, but Gillson concludes that they represent high-temperature replacement deposits. Large ilmenite deposits with 16% TiO_2 are stated to occur in the Ilwen Mts., *Karelia, Russia.*

Beach deposits of the Travancore type are mined on the coasts of *Brazil, New South Wales,* and *Queensland, Australia.* The tin stream placers of *Malaya* also carry ilmenite, which is separated in quantity by magnetic concentration. Beach sands in *Hondu, Japan,* are said to contain 10 billion tons carrying 20–30% Fe, 8–12% TiO_2, and 0.6% V_2O_5.

Selected References on Titanium

General Reference 11. Chap. 49 by J. L. GILLSON. *Mineralogy, geology, distribution, production, uses, controls; good bibliography.*

Vanadium-Bearing Magnetite-Ilmenite Deposits, Lake Sanford, New York. J. R. BALSLEY, JR. U. S. Geol. Surv. Bull. 940–D, 1943. *A magnetite-ilmenite body in anorthosite and gabbro.*

Titaniferous Magnetite Deposits of Lake Sanford, New York. R. C. STEPHENSON. A. I. M. E. Min. Tech. Vol. 9, No. 1, pp. 1–25, 1945. *Petrography, mineralogy, and geology; a late magmatic residual liquid injection origin is advocated.*

Titanium Deposits of Nelson and Amherst Counties, Virginia. C. S. Ross. U. S. Geol. Surv. Prof. Paper 198, 1941. *Nelsonite dikes consisting of ilmenite, rutile, and apatite.*

Notes on the Ilmenite Deposit at Piney River, Virginia. D. M. DAVIDSON, F. F. GROUT, and G. M. SCHWARTZ. Econ. Geol. 41:738–748, 1946. *Conclusions for a magmatic origin.* Reply by C. S. Ross. Econ. Geol. 42:194–198, 1947. *Arguments for a replacement origin.*

Titanium, Its Occurrence, Chemistry, and Technology. J. BARKSDALE. Ronald Press, New York, 1949.

Zirconium

Zirconium is a relatively recent entry into industry. The mineral zircon is worn as a semi-precious stone. The oxide (zirconia) and the metal are used industrially. Zirconia is one of the most refractory oxides known and is employed for crucibles, laboratory wares, chemical-resisting furnace bricks, and high-temperature cements. Its use is rapidly expanding to give opacity to enamelware, paints, and automobile enamels and lacquers. It is also used as an insulator for heat and electricity, as an abrasive, for gas mantles and certain incandescent lamps, as a polisher, for toughening rubber, and for white inks. The metal is used for electronic tubes, flashlight bulbs, electrical condensers, X-ray filters, lamp filaments, spot-welding electrodes, rayon spinnerets, and as an alloy. Zirconium steels make good armor plate, and projectiles and, with nickel, make high-speed and sharp cutting tools. With copper it imparts properties similar to that of beryllium, and with aluminum and vanadium in ferroalloy it serves as a steel refiner in furnaces.

The commercial minerals are zircon ($ZrSiO_4$) and baddeleyite (ZrO_2), the latter being obtained only in Brazil. Zircon is an accessory mineral of igneous rocks and is readily concentrated in placer deposits. Some 9,000 tons are produced annually from Australia, Brazil, and India, where it occurs along with ilmenite, rutile, and monazite in beach sands. A little was formerly obtained from the beach sands of Florida and North Carolina placers. The beach sands at Trail Ridge, Fla., yield zircon as a by-product to ilmenite in quantities sufficient to meet world demand.

Miscellaneous Minor Metals

Barium metal is produced in small quantities and is used as a "getter" inside electronic tubes to promote vacuum.

Boron is used as a hardener for steel and for atomic energy.

Caesium is in demand for photoelectric cells, which are used for talking pictures, television, traffic controls, automatic door opening, and similar purposes. Caesium is also used in radio tubes. It is obtained from the mineral pollucite, which is mined near Custer, S. Dak.

Calcium is considered a coming metal and is already being used as a scavenger in melting steel, copper, nickel, lead, as an alloy with steel and nonferrous metals, as a reducer in the production of rare metals, and in making hard bearing metal.

Cerium is alloyed with 30 percent iron to form *ferrocerium*, which brittle alloy when abraded emits sparks and is used as "flints" or sparkers for cigarette lighters. It is also used in signaling, photography, glassware, and radio tubes, and as a metal scavenger. It occurs in monazite, from which it is extracted as a by-product in obtaining thorium. The monazite is found in beach placers along with ilmenite.

Gallium and **rhenium** are minute by-products of the copper ores of Mansfeld, Germany, and of coal ashes. Some gallium is won from the zinc ores of Joplin, Mo. Gallium is one metal other than mercury that is liquid at low temperatures. It is used in electrical fields, reflecting surfaces, military devices, atomic piles, and thermometry. Rhenium is used for special plating and to impart long life to tungsten filaments in bulbs.

Germanium is won from zinc-refinery slimes and is contained in the minerals germanite and argyrodite. Deposits of these occur at Tsumeb, Southwest Africa, and in Bolivia. Its chief use is medicinal, as a specific for pernicious anemia, and for sleeping sickness. It is used also in radionic devices, optical glass, and alloys with precious metals and as a catalyst.

Indium is also obtained from zinc residues. It is used as a precious-metal alloy, as dental alloy to resist corrosion, as a nontarnishing plating for silverware, for heavy-duty bearing metal, for low-melting-point alloys, to give amber colors to glass, and for atomic work.

Lithium is used as a scavenger in copper-base alloys and for some nonferrous alloys.

Mesothorium is produced from monazite sands in small quantity. It is used in the treatment of cancer and skin diseases and as a luminous paint.

Neodymium and **praseodymium** are recovered from cerium minerals. The former gives glass a delicate violet color, and the latter a fine greenish yellow tint.

Rhenium is produced from molybdenite roaster flue dust. It can be used as an alloy with platinum metals for noncorrosive purposes, electrical parts, and pen-nibs.

Rubidium is used in mercury-vapor lamps.

Thallium is recovered from cadmium-containing flue dusts and is used as a rodenticide and insecticide, for signaling devices, and for certain alloys.

PART III

Nonmetallic Mineral Deposits

NONMETALLIC MINERALS

The materials of the nonmetallic kingdom are more commonplace and familiar than metallic ores. Most of them are rather abundantly distributed throughout the world, and the value of many depends less on the material itself than on the ability to use it near by. Their economic value is determined largely by the cost of transportation. Their specifications vary widely, and, unlike the constancy in a metal, the specifications are in general determined by the uses to which the material is put. Nonmetallic minerals are used essentially in the form in which they are extracted, and few are broken up into elemental parts. The gross value of all nonmetallic products annually exceeds that of metallic ores.

Nonmetallic products defy simple classification. One substance will occupy more than one pigeonhole or be formed by more than one process. The outstanding feature of them, and indeed the one that often determines their value, is the purpose for which they are used. With ores, one of the chief features of interest is the deposit, but with nonmetallics this is often subordinate and interest centers in usefulness. Consequently, in this section they will be grouped according to their chief uses.

For scientific purposes, however, a genetic classification according to process of formation is proposed in Table 17, which shows also the main uses to which the minerals are put. The table also indicates the use groups followed in Part III, and the chapters in which processes and the various nonmetallic materials are considered. No attempt has been made to include innumerable minor nonmetallic minerals.

Many of the nonmetallic materials have already been described in Part I.

TABLE 17

Genetic and Use Classification of Important Nonmetallic Materials

Process of Formation and Important Products	Chap.	16 Mineral Fuels	17 Ceramic Materials	18 Structural-Building Materials	19 Metallurgy, and Refractory Materials	20 Industrial and Manufacturing Materials	21 Chemical Minerals	22 Fertilizer Minerals	23 Abrasives	24 Gemstones, Ornamental	25 Water Supplies
A. Igneous processes:											
I. Rocks:	5.1										
Building				Ⓧ						X	
Soapstone				Ⓧ		X			Ⓧ		
Pumice									Ⓧ		
Corundum					Ⓧ				X	X	
Diamond									Ⓧ	Ⓧ	
II. Pegmatites:	5.1										
Feldspar			Ⓧ		X				X	X	
Quartz (silica)			X		X	X			Ⓧ	X	
Mica				X		Ⓧ					
Cryolite			X		Ⓧ						
Spodumene						Ⓧ					
Gemstones						X				Ⓧ	
III. Magmatic emanations:	5.4										
Pyrite (sulphur)							Ⓧ	X			
Fluorspar					Ⓧ	X	X				
Barite and witherite						Ⓧ	X				
Magnesite					Ⓧ		X				
B. Sedimentary processes:	5.5										
I. Sedimentary rocks:											
Building stones				Ⓧ							
Lime, dolomite, magnesite				Ⓧ	X	X	X	X			
Hydraulic cements				Ⓧ							
Clay, shales			Ⓧ	Ⓧ	X	X					
Phosphates								Ⓧ			
Sand, sandstones				Ⓧ	X				Ⓧ		
Bentonite, fuller's earth, diatomite				X		Ⓧ			X		
II. Chemical precipitates:	5.6										
Rock salt							Ⓧ				
Gypsum				Ⓧ				X			
Potash							X	Ⓧ			
Nitrates							X	Ⓧ			
Borax and borates					X	Ⓧ	X				
Sodium compounds						Ⓧ	Ⓧ				
Miscellaneous chemicals							Ⓧ				
III. Organic deposits:	5.5										
Coal		Ⓧ			Ⓧ						
Oil, gas		Ⓧ					X				
Sulphur							Ⓧ	X			
Bitumens				Ⓧ		Ⓧ					

TABLE 17 (*Continued*)

GENETIC AND USE CLASSIFICATION OF IMPORTANT NONMETALLIC MATERIALS

Process of Formation and Important Products	Chap.	16 Mineral Fuels	17 Ceramic Materials	18 Structural-Building Materials	19 Metallurgy, and Refractory Materials	20 Industrial and Manufacturing Materials	21 Chemical Minerals	22 Fertilizer Minerals	23 Abrasives	24 Gemstones, Ornamental	25 Water Supplies
C. Weathering processes:											
I. Residual concentration:	5.7										
Bauxite			X	X	(X)				X		
Clays			(X)	(X)	X	X					
Mineral pigments						(X)					
Tripoli									(X)		
II. Mechanical concentration:	5.7										
Sands			X		(X)	(X)			X		
Monazite & zircon					X	(X)	(X)				
Ilmenite					X	(X)					
Phosphates							X	(X)			
Gemstones									X	(X)	
D. Metamorphic processes:	5.9										
Asbestos				X		(X)					
Graphite					(X)	X					
Emery, garnet									(X)	X	
Talc			X		X	(X)					
Sillimanite minerals			X		(X)	X					
Gemstones									X	(X)	
Roofing stones				(X)							
E. Ground-water processes:	5.6										
Water supplies							X	X			(X)
Brines (salt)								(X)			
Bromine, iodine						(X)		X			
Nitrates								X	(X)		
Sulphur								(X)	X		
Gemstones										(X)	
Salines (except salt)							X	(X)			

(X) indicates the chief uses.

CHAPTER 16

THE MINERAL FUELS

The sources of utilizable energy in the world are coal, petroleum and natural gas, wood, water power, and solar and atomic energy. Of these, coal outranks all others. Despite the inroads made upon it by petroleum, coal still produces about 70 percent of all energy units. Water power developed to the fullest can never displace coal. Wood is only a local fuel. Petroleum, however, is the preferred fuel for all mobile power units and has encroached greatly on coal. This, in turn, has created greater efficiency in the burning of coal. For example, steam railroad consumption per 1,000 ton-miles hauled has dropped, in the quarter-century 1920–1945, from 170 pounds of coal to 115 pounds; in electric power plants the pounds of coal per kilowatthour have dropped from 3.2 to 1.3 in the same period, and in modern plants it is 0.75 pound. In the United States, of the heat and energy units produced, coal supplies about 47 percent, oil 33 percent, gas 15 percent, and water power 4 percent.

COAL

Coal is widely distributed, and reserves of it are sufficient to last hundreds or thousands of years. It has long been the backbone of industrial life. Those countries endowed with it have risen commercially and politically, those lacking it have mostly become agricultural or handicraft nations.

Coal was known in ancient times and in the ninth century entered household use in England. By the thirteenth century trade in it was active. The invention of the steam engine stimulated active coal mining; and then the industrial age commenced in England, when coal replaced man power, and eventually mechanical power held sway. When iron ore was smelted by charcoal and the forests of England were vanishing, it was discovered that anthracite was a smelting fuel. This was another stimulus to coal mining. Later, when coke made from bituminous coal was found to be a still superior fuel, the coal industry received a great impetus, and huge industrial expansion ensued. A further stimulus to the coal industry occurred when cities

634

began to produce artificial gas from coal for domestic and industrial use. Its high position, which receded under the competition of oil and gas, is again being slowly increased as the efficiency of coal burning is raised.

In North America, the Indians burned coal in 1660; bituminous mining began (in Virginia) in 1787, and anthracite mining in 1805.

Uses of Coal. Coal is a primary source of heat and power. It is estimated that, of the coal produced, 78 percent is used for fuel and 22 percent is used as raw coal in producing pig iron (10 percent), steel (7 percent), and gas (5 percent). The fuel coal used is divided as follows: steam raising, 29 percent; railway transportation, 23 percent; domestic purposes, 17 percent; electric generation, 6 percent; bunker coal, 3 percent.

The coal used to make coke for metallurgical and other purposes yields important by-products. One ton of coal yields about 70 percent coke and by-products as follows: gas $1.48, tar 59 cents, light oils 41 cents, ammonia 33 cents, and other products 25 cents.

Different coals find specific uses. Thus, there are coking coals, gas coals, and steam coals. Coal for locomotives should raise steam quickly and not have too high an ash content or clinker. Domestic coal should not yield excessive smoke.

Production and Resources. The normal world production of coal and lignite amounts to about 1.6 billion tons annually divided as follows:

PERCENT OF WORLD		PERCENT OF WORLD	
1. Germany	25	6. Japan	2.5
2. United States	24	7. Poland	2.2
3. Great Britain	14	8. Czechoslovakia	1.8
4. Russia	6	9. Belgium	1.8
5. France	3	10. India	1.7

The United States annual production ranges from 500 million to 650 million tons, valued at 1.5 billion to 2 billion dollars. The ranking states are, in order: West Virginia, Pennsylvania, Illinois, Kentucky, Ohio, Indiana, Virginia, Alabama, Colorado, and Wyoming. There is a total of 7,000 active mines employing 400,000 miners; 84 percent of the production is yielded by 1,130 mines, each producing over 100,000 tons annually.

The coal resources of the world as computed by the 12th International Geological Congress, Canada, in 1913, are shown in Table 18. New figures assigned to Canada are only 148 billion tons; Ashley estimates those of the United States to be only 1,661 billion tons; and those of Russia have subsequently been greatly increased.

TABLE 18

Coal Resources of the World

Type	Billion tons	Continents	Billion tons	Leading Countries	Billion tons
Anthracite	497	1. Americas	5,105	1. United States	3,839
Bituminous	3,903	2. Asia	1,280	2. Canada	1,234
Other coals	2,998	3. Europe	784	3. China	996
		4. Oceania	170	4. Germany	477
Total	7,398	5. Africa	58	5. Russia	233
				6. Great Britain	190
				7. Australia	160
				8. India	79

Kinds of Coal and Classification

In common usage, coals are divided into four main groups, (1) anthracite, or hard coal, (2) bituminous, or soft coal, (3) lignite, and (4) cannel coal. The last is a special type; the others are divided into ranks. Thus, from the lowest rank upward are lignite, brown coal, sub-bituminous, bituminous (3 ranks), superbituminous, semianthracite, and anthracite. Peat lies below lignite, and graphite above anthracite. The relation of the various ranks of coal is shown in Fig. 16–1.

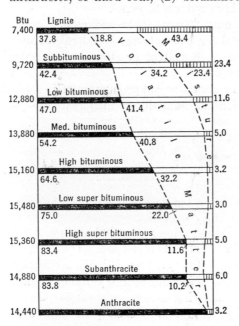

Fig. 16–1. Diagram showing ranks of coals, chemical composition, and heat units in Btu (figures at left). (*After Campbell, U. S. Geol. Survey.*)

Peat. Peat is not coal, even though it is a fuel. It is an accumulation of partly decomposed vegetable matter that represents the first stage in the formation of all coals.

Lignite. Lignite (and brown coal) represents the second stage. It is brownish black and is composed of woody matter embedded in macerated and decomposed vegetable matter. It is banded and jointed and, because of its high moisture content, slacks or disintegrates after drying in the air. It is subject to spontaneous combustion and has low heating value. It is used for local fuels and to make producer gas, and in powdered form for heating and steam raising; it supplies synthetic petroleum in Germany.

Subbituminous. This intermediate coal is often difficult to distinguish from bituminous coal. It is dull, black, and waxy. It shows little woody material, is banded, and splits parallel to the bedding but lacks the columnar cleavage of bituminous coal. Some varieties disintegrate upon exposure. It is a good clean fuel but of relatively low heating value.

Bituminous. Bituminous coal is a dense, dark, brittle, banded coal that is well jointed and breaks into cubical or prismatic blocks and does not disintegrate upon exposure to air. Vegetable matter is not ordinarily visible to the eye. Dull and bright bands and smooth and hackly layers are evident. It ignites readily and burns with a smoky yellow flame. Its moisture is low, volatile matter medium, fixed carbon high, and heating value high. It is the most used and most desired coal in the world and serves for steam, heating, gas, and coking.

Cannel coal is a special variety of bituminous coal that breaks with a splintery or conchoidal fracture, is not banded, does not soil the fingers, and is lusterless. It is made up of windblown spores and pollen. It is clean, burns with a long flame, and is preferred for fireplace coal.

Bituminous coals are subdivided into several varieties, as is shown in Table 19. The higher ranks have the maximum heating power of all coal.

Anthracite. Anthracite is a jet-black, hard coal that has high luster, is brittle, and breaks with a conchoidal fracture. It ignites slowly, is smokeless, burns with a short blue flame, and has high heating value. It is restricted in distribution and is used almost exclusively for domestic heating. It also is subdivided into varieties.

Constitution of Coal

Chemical. Chemically, coals are made up of various proportions of carbon, hydrogen, oxygen, nitrogen, and impurities. As shown in Fig. 16–2, from lignite to anthracite there is a progressive elimination of water, oxygen, and hydrogen, and an increase in carbon. The carbon is present as *fixed carbon* and in *volatile matter*, and the ratio of these to each other is an important characteristic of coal. Thus, $\dfrac{\text{Fixed carbon}}{\text{Volatile matter}}$ = fuel ratio: The fuel ratio is high in anthracite and low in lignite. The fuel ratio is the main feature determining the rank of coal. The volatile matter is that which burns in the form of a gas;

it causes ready ignition, but too much gives a long, smoky flame or poor storing qualities. The fixed carbon is the steady, lasting source of heat. It produces a short, hot, smokeless flame.

Sulphur is an injurious impurity commonly present in most coals in the form of marcasite or pyrite. More than 1.5% S excludes coal for making gas or coke.

Ash is the residual of noncombustible matter in coal that comes from included silt, clay, silica, or other substances. The less present, the better the quality of the coal. Too much ash may put a high-rank coal in a low grade. Too much iron in the ash makes objectionable clinkers.

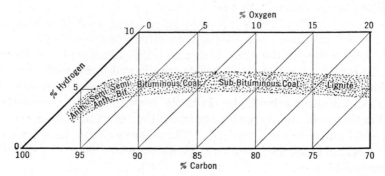

Fig. 16–2. Diagram showing increase of carbon and progressive elimination of oxygen and hydrogen from lignite to anthracite. (*Stutzer, Geology of Coal.*)

There are other minor constituents of coal. In practice, however, only the above main constituents are determined, in what is called a "proximate analysis," along with the heating value (Fig. 16–1), expressed in British thermal units (Btu).

Physical Constitution. Banded coals are made up of partly decomposed and macerated vegetable matter, mainly vascular land plants, of which the following parts have been recognized:

cellulose	(2)	protoplasm	(1)	resin	(10)	pigments	(3)
pentosanes	(2)	starches	(2)	gums	(6)	oils	(5)
lignose	(7)	cutin	(8)	waxes	(9)	fats	(5)

The numbers in parentheses represent, according to David White, the order of assailability to bacterial action. Resin and waxes being the least attacked are present in most coals. The various bands of banded coals are called vitrain, clarain, durain, and fusain.

Fusain, or mineral charcoal, or "mother of coal," is carbonized wood resembling charcoal. It is high in ash. It has been thought to have come from burned wood, but White shows that the ducts full of resin argue against burning.

Vitrain (or *anthraxylon*) constitutes thin bands of bright, glassy-looking, jetlike coal with conchoidal fracture. Woody structure is not visible megascopically. Its brilliance, approaching jet, varies with the rank of coal. Vitrain supplies coking qualities.

Durain is dull coal, lacking luster and having a matte or earthy appearance. It is hard, black to lead-gray in color, and consists of cuticles, spores, etc., formed in water less toxic than for vitrain.

Clarain forms thin bands in coal characterized by bright color and silky luster. It is composed largely of translucent attritus.

Attritus is finely divided plant residue composed of the more resistant plant products.

Classification of Coal

In 1937 a new classification of coal by rank was adopted for American usage by the American Society for Testing Materials and the American Standards Association, based mainly on fixed carbon and Btu content. The higher-rank coals (FC = 69% +) are classified according to fixed carbon on the dry basis, the lower-rank coals according to the Btu on the moist basis. Physical properties of agglomeration and weathering are also used. A simplified form of the classification is given in Table 19.

TABLE 19

CLASSIFICATION OF COAL BY RANK

[FC = fixed carbon; VM = volatile matter; A = agglomerating;
NA = nonagglomerating; W = weathering; NW = nonweathering]

Class	Group	Limits		Physical Properties
		FC	VM	
I Anthracitic	1. Meta anthracite	98+	2–	NA and NW
	2. Anthracite	98–92	2–8	NA and NW
	3. Semi-anthracite	92–86	8–14	NA and NW
II Bituminous	1. Low volatile	86–78	14–22	NA and NW
	2. Med. volatile	78–69	22–31	NA and NW
	3. High volatile A	69–	31+	NA and NW
		Btu		
	4. High volatile B	14,000–13,000		
	5. High volatile C	13,000–11,000		A or NW
III Subbituminous	1. Subbit. A	13,000–11,000		NA and W
	2. Subbit. B	11,000– 9,500		NA and W
	3. Subbit. C	9,500– 8,300		NA and W
IV Lignitic	1. Lignite	8,300–		Consolidated
	2. Brown coal	8,300–		Unconsolidated

For grade of coal, the ash and sulphur contents and temperature of ash softening are added. Thus a designation of 132-A8-F24-S1.6 means a coal with 13,200 Btu, ash of 6 to 8 percent, ash softening at 2,400° to 2,590° C, and sulphur 1.4 to 1.6 percent.

Origin of Coal

That coal is of vegetable origin has been accepted since 1825 but there have been various forms of this theory, namely, accumulation in a sargasso sea, transported vegetable debris, fresh- or salt-water deposition, lacustrian deposition, and growth of vegetation *in situ*. Today it is considered that all ordinary coals are of vegetable origin and that the common banded coals were formed *in situ*. There is some diversity of opinion as to whether the different kinds of coal originated because of different kinds of vegetation or by different degrees of alteration, the latter being the prevailing American opinion. The present concept is that all coal originated in swamps and went through a peat stage.

Source Materials. The microscope reveals that the raw material of banded coals was vascular swamp vegetation not unlike that growing in present-day, peat-forming swamps. Over 3,000 plant species have been identified from Carboniferous coal beds. Roots and stumps found in underclays below coal beds attest that the vegetation grew and accumulated *in situ*. Luxuriant vegetation flourished, and consisted mainly of ferns, lycopods, and flowering plants, with conifers and other varieties. Ferns were treelike, rushes grew 90 feet high, lycopods (small shrubs today) attained 100 feet in height. Bulbous and arched roots attest to trees that lived in water. *Sigillaria* and *Lepidodendron* were among them. Some coals have roots in the roof also. None of the plants found are salt-water species. The same kinds of plant remains are found in coals of all ranks.

Places and Conditions of Accumulation. The extensive distribution of individual coal seams implies swamp accumulation on broad delta and coastal plain areas, on subsiding base-level continental regions, on broad interior basin lowlands that have been nearly base-leveled, and where shallow waters persist throughout the year. Most coals are underlain by carbonaceous shales of lake bottom deposition. Low-lying surrounding lands are necessary, else there would be too much inflow of silt. The fresh-water swamps necessary for the growth of the plant species found in coal may be separated from the sea by only a sand bar, or barrier of vegetation, since marine beds commonly alternate with, or rest unconformably upon, coal. A subsiding shore line would supply the best conditions of accumulation. Coal fields show cycles of deposition. They are not rhythmic.

The rate of accumulation depends upon the vegetation and climate. J. Volney Lewis estimates that it would take 125 to 150 years to accumulate enough material for 1 foot of bituminous coal and 175 to 200 years for 1 foot of anthracite.

Climate. The climatic conditions that favored coal accumulation were mild temperate to subtropical, with moderate to heavy rainfall well distributed throughout the year. Severe frosts were absent, but the climates were not without dry spells, since some of the plants have water containers in their trunks and roots. The lush vegetation and the large growth means, however, a well distributed rainfall also. The Carboniferous growth rings are poor, indicating nonseasonal type of climate. Probably the climate was much like that of the Carolinas or Florida. Coals have been found in places where they would not now be formed, as in Spitzbergen, Greenland, and Antarctica.

Decomposition and Toxicity. The change from plant debris to coal involves biochemical action producing partial decay, preserval of this material from further decay, and later dynamochemical processes. The type of the coal depends upon the environment, kind of plants, and nature and duration of bacterial action. The rank of coal depends upon the degree of metamorphism.

When a tree falls upon dry land it decays. The complex constituents become broken up into CO_2 and H_2O, with which it started, and most of it is returned to the air. No coal accumulates. When, however, vegetation falls into water a similar decay sets in, but it operates more slowly. A necessary step is that the decay be arrested before complete destruction ensues so that the residue can accumulate. This is accomplished by means of decay-promoting bacteria that make the stagnant water toxic, which prevents further decay of the vegetable tissues and permits their preserval and accumulation.

The biochemical changes liberate oxygen and hydrogen and concentrate the carbon. The bacteria are most active near the surface and first attack the most easily decomposable constituents, such as the protoplasm, cellulose, and starches. The resistant substances, such as the waxes, resin, and cutin, along with woody fragments drop to the bottom of the swamps, where the toxicity slows up or prevents further decay and humus accumulates. The kind of bacteria and the duration of bacterial action largely determine what constituents collect in the humus. The bacterial attack may take out certain ingredients of the wood so that it appears to have been chewed up. The parts not destroyed collect in a jellylike mass that becomes more pasty and finally forms the binder of the whole. Under normal conditions of water supply and no agitation of the surface, all but the least resistant of the vegetable matter is preserved. However, heavy rains will dilute the water, lessen the toxicity, and foster further decay. Floods are catastrophic in that toxicity is lessened and material is washed away. In a dry season the water will be lowered and detritus may be lost by exposure to the air, where decay proceeds. The water level thus plays an important part — a stable level gives vitrain; dilution gives thin layers of cutinous material; attritus or wash material gives durain. The accumulation of this detrital material in the swamp gives peat, the initial stage of coal.

Carbonization of Progressive Metamorphism. The progressive change from peat to anthracite involves chemical, physical, and optical changes. After the peat stage bacteria presumably play little part, and most of the changes are chemical, induced by pressure and slight increases of temperature, which result from deposition of overlying sediments. In peat the oxygen has already been reduced about 10 percent. The further progressive elimination is probably caused by combination of O with C to form CO and CO_2, which escape.

The chief changes, according to David White, effected by progressive metamorphism are:

PHYSICAL	CHEMICAL
1. Compaction, drying, induration, lithification.	1. Progressive elimination of water up to anthracite.
2. Jointing, cleavage, schistosity.	2. Progressive loss of oxygen.
3. Reconstruction.	3. Conservation of hydrogen up to graphite stage.
4. Optical changes.	4. Progressive increase of ulmins.
5. Dehydration up to anthracite.	5. Progressive loss of bitumens.
6. Color change — brown to black.	6. Development of heavy hydrocarbons.
7. Increase of density.	7. Large loss of H in anthracite.
8. Change of luster.	8. Increased resistance to solvents.
9. Fracture changes — from bedding to cleavage to conchoidal.	9. Increased resistance to oxidation.
	10. Increased resistance to heat.

The result of these changes is the successively higher-rank coals previously described. The same plant constituents are observed in the higher-rank coals as in peat.

The change in rank is largely a result of pressure and time. The older the coals, the more likely they are to be more deeply buried, which increases the pressure and accelerates the metamorphism. Lateral pressure induced by folding (accompanied by increased temperature) is also important. Folded Tertiary coals of Alaska are high-rank coals; the nonfolded ones of Dakota are lignite. In Pennsylvania the metamorphism of the coal increases with the intensity of the folding; anthracite coals are in closely folded beds; bituminous coals of the same age are in gently folded beds. The more competent the enclosing beds, the greater will be the metamorphism of the included coal. In places where local igneous intrusions occur, the rank of the coal is increased, even to natural coke.

Occurrence and Age

Coal occurs as a sedimentary rock within "coal measures," which consist of alternating beds of sandstone, shale, and clay, mostly of fresh-water origin. The coal beds, or seams, in a large way are flat lenses, although some are remarkably persistent. The Pittsburgh seam, for example, underlays 15,000 square miles. Coal seams are generally underlain and overlain by shale, although clay or fireclay is a common "underlie," and sandstone may constitute the roof. The character of the roof affects mining operations, since a weak shale roof may need much support, and a very strong sandstone roof may not cave uniformly during mining.

The thickness of coal seams ranges from a mere film up to 100 feet. The thickest bed in the United States, at Adaville, Wyo., is 84 feet. The famous Mammoth seam in Pennsylvania is 50 to 60 feet, but most seams range between 2 and 10 feet and are rarely 20 feet; the widespread Pittsburgh seam averages 7 feet. Most coal seams exhibit considerable variations in thickness, as would be expected from a swamp origin. Generally there are several coal seams in a given vertical section, not all of which are workable, however. For example, in Pennsylvania there are 29 seams aggregating 106 feet of coal; in Alabama 55 seams; in Indiana 25 seams totaling 90 feet, of which 9 are minable; in England there is an aggregate thickness of 85 feet, and in Germany 120 feet.

Coal seams exhibit the structural features of sedimentary rocks. They commonly contain "partings," or thin bands of shale or clay, called "bone" or "slate," which interfere with mining of clean coal; thick partings are called splits. The Commentary seam of France on one side of the basin divides into six splits of sandstone, showing inroads of sediment from that side during accumulation. Seams also exhibit "cut outs" caused by erosion along a channel and subsequent filling by later sediments; "rolls" where the floor bulges up into the coal; "horsebacks," where a body of foreign matter lies in the coal; and "clay slips." Folding produces protuberances of roof and floor into the coal and pinches and swells of the soft coal along the limbs and crests of folds; many supposedly thick seams are merely thin seams squeezed into anticlinal bulges. Faulting and slickensiding are not uncommon. In a general way the anthracite coals are in regions of close folding and the bituminous coals and lignite are in regions of gentler folding.

Age. Coal occurs in all post-Devonian periods. The Carboniferous (Mississippian and Pennsylvanian) received its name because of its world-wide inclusion of coal. Permian coals are less widespread but are found in China, Russia, India, South Africa, Australia, and perhaps in Kansas. The Triassic contains coal in Australia, central Europe and eastern Asia, and the Jurassic in Alaska, China, Australasia, and Austria. The Cretaceous ranks next to the Carboniferous in importance, containing extensive beds in western North America and central Europe. The Tertiary yields most of the lignite of the world, although high-rank Tertiary coals occur in Alaska and elsewhere. Miocene coal occurs in Antarctica.

Distribution of Coal

It is rather striking that coal is much more common in the Northern than in the Southern Hemisphere; few countries north of the equator are entirely lacking in it.

In *North America* the world's most bountiful supply is in the United States with northward extensions into eastern and western Canada and a great blank area between the Rockies and the Atlantic north of the Great Lakes. Coal is sparsely distributed in northern Mexico. In *South America* small coal basins occur in Colombia, Peru, and Chile. In *Europe* extensive coal deposits occur in England, west-central Europe, and Russia but are sparse in Scandinavia, the Mediterranean countries, Switzerland, Bulgaria, and Roumania. *Asia* contains vast fields in Russia, China, and India and moderate ones in Japan, Manchuria, Indo-China, and Persia. *Africa* is deficient in coal; moderate-sized deposits occur in South Africa, Congo, Rhodesia, and Nigeria and small deposits in Madagascar, Nyasaland, East Africa, and Ethiopia. *Oceania* holds large deposits in Australia and moderate ones in New Zealand, Netherlands East Indies, North Borneo, and the Philippines.

Coal Fields of the United States

The United States is so bountifully supplied that few states are more than one state away from coal; it is mined in 33 states. The various fields are shown in Fig. 16–3, the Appalachian, Interior, and Rocky Mountain being the most important.

Appalachian Field. This field yields all of the anthracite and 70 percent of the bituminous coal of the United States. It includes the highest-grade coals on the continent and the largest continuous bituminous area in the world. It embraces the coal areas of Pennsylvania, Ohio, West Virginia, Virginia, Maryland, Kentucky, Tennessee, Alabama, and Georgia, or an area of around 70,000 square miles, 75 percent

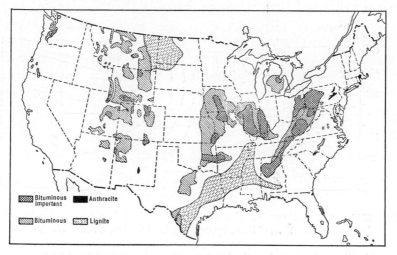

FIG. 16–3. Map of coal fields of United States.

of which contains workable coal. The thick coal measures consist of westward-thinning, intercalated lenses of sandstone, conglomerate, shale, limestone, fireclay, and coal. They are mostly of Pennsylvanian age, but the Mississippian also contains coal. The beds were involved in the Appalachian orogeny and are strongly folded in eastern Pennsylvania, less so in the west, and steeply folded and faulted in the south. Economically, the field may be divided into three units, namely, the anthracite unit, the northern unit, and the southern unit, the southern being included within the Southern States.

The anthracite area occupies 480 square miles in northeastern Pennsylvania, where erosion has left only synclinal residuals in intricately folded coal measures (Fig. 16–4). These residuals represent about 5 percent of the original extent before destruction by erosion. Some 4.5 billion tons have been extracted to date and some 9 billion tons remain. The several coal seams range from a few inches to the 60 feet of the Mammoth seam, which, however, splits in places to form several seams. The close folding has produced much crushed coal or "fines." The steep dips, the hard coal, and many thin beds make anthracite more costly to mine than bituminous coal, but its cleanliness and high fuel ratio make it a desirable domestic fuel despite its high ash content.

The northern bituminous area is the heart of the American coal industry. Here the coal measures form hill cappings or occur in gently dipping basins; northeastward the folding is greater and the bituminous belt is flanked by semi-anthracite coal. Numerous seams are mined,

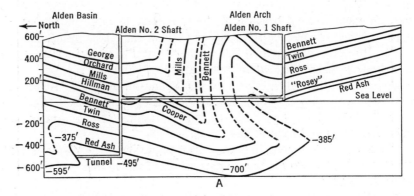

Fig. 16–4. Section through coal seams of Alden mine, Northern Anthracite region, Pennsylvania. Dotted lines, hypothetical. (*Darton, U. S. Geol. Survey.*)

among them the Pittsburgh, in Pennsylvania and Ohio, which averages 7 feet in thickness over an area of 2,500 square miles; the Pocahontas No. 3 bed, underlying 300 square miles in West Virginia, averages 6.5 feet in thickness. The coal seams are persistent, regular, lie at shallow depths, and are gently dipping. Consequently they are cheaply mined. The coals are clean, low in ash, and embrace all ranks. In general, the lower-rank coals lie to the west and the higher-rank to the east, where folding is more accentuated. The coals of western Pennsylvania, Ohio, and Kentucky are mainly noncoking. High-grade coking coals occur in the eastern and central part. Individual districts are widely known for superior coals, such as the Connellsville, Pennsylvania, for coking coals, the Kanawha, West Virginia, for gas coal, and the famed Pocahontas coal of the Virginias for its smokeless steam coals. This field furnishes the great eastern industrial centers with steam and gas coals and much of industrial United States with coke.

The southern bituminous area is one of folding and faulting, with steeply dipping coal seams distributed in four basins. There are many seams, but few exceed 5 feet in thickness. There are good coking and fair steam coals, but numerous partings do not give cheap or clean coal. In proximity with Clinton iron and limestone, however, this coal forms the backbone of the large Birmingham steel industry.

Interior Fields. There are four Interior fields, the Eastern, Western, Northern, and Southwestern.

The Eastern Interior field is an oval-shaped basin in Illinois, Indiana, and Kentucky; mining is restricted to the margins. There are 9 minable seams out of 25, ranging from 2 to 7 feet in thickness. High ash and sulphur make most of the coals undesirable for coking, but they are excellent steam and domestic coals. The production ranks next to the Appalachian field because of the proximity to large industrial regions. Much of this coal is mined by steam shovel stripping.

The Western Interior field lies in the Great Plain states from Iowa to Arkansas, where the coal measures are mostly flat and shallow. The coals are all bituminous, except for some semi-anthracite in Arkansas, but they are of lower rank and grade than eastern coals. They are used for local steam and domestic fuel but suffer strong competition with petroleum and natural gas.

The Southwestern field, in Texas, is an extension of the Western field. There are three workable bituminous seams of noncoking coal of Pennsylvanian age.

The Northern field is a smaller, shallow basin in Michigan, containing irregular Pennsylvanian bituminous seams that yield fuel coal.

Rocky Mountain Field. This extends along the Rockies from Montana to New Mexico. The coals are bituminous and lignite, with some local anthracite where altered by igneous intrusions. They are mostly Cretaceous in age; the Dakota lignite is Eocene. The field consists of several disconnected basins with coals of variable rank, depending upon the amount of folding. Some of the coal is excellent for coking and steam, but much of it is subbituminous; the lignites are mined only locally. The chief bituminous industry is in Colorado, particularly around Trinidad, where good steam and coking coals are mined. Utah also yields coking coal. The reserves are large but the industry is confined to local markets.

Other Fields. The *Pacific Coast* field consists of small Tertiary basins in Washington containing lignite to bituminous coal, some of which is coking. Unimportant basins occur in California and Oregon. In *Alaska*, in the Matanuska and Bering River areas, Tertiary coals range from lignite to high-rank bituminous, depending upon how much folding they have endured. Attempts to mine the bituminous coals failed because of their folded, faulted, and crushed character. Lignite and subbituminous coals are mined for local consumption. The vast resources of good quality coal credited to Alaska are uneconomic. Carboniferous coals occur in the Arctic.

Coal Fields of Other Countries

Canada. There are three coal areas in Canada — Nova Scotia, western provinces, and the Pacific Coast islands.

Nova Scotia contains four areas of fair bituminous coal of Pennsylvanian age, two of which contain coking coals. In two areas the beds are folded and dip steeply. At Sydney, the most important area, there are several seams aggregating 1 to 46 feet of coal. The minable seams dip gently seaward, and a strong roof permits slope mining for 1½ miles out from shore.

The *western provinces*, Alberta and British Columbia, contain Cretaceous coals in the eastern Rockies and the foothills. The measures are folded and faulted; the rank of the coal rises with the folding from low-rank bituminous and even to a little anthracite. Good bituminous coals are mined at Canmore, Frank, Lethbridge, and Bankhead in Alberta, and Crows Nest, British Columbia. There is considerable coking coal, but much is high in ash and low in fuel value. In places there are 15 to 20 seams aggregating 100 feet of coal; the minable seams range from 4 to 7 feet in thickness, although an 80-foot seam is worked at Corbin. The extensive reserves formerly credited to Alberta are now known to be greatly overestimated.

The *Pacific Coast* area, on Vancouver and Queen Charlotte Islands, contains Upper Cretaceous coals around Nanaimo and Comox. The measures are folded and faulted, and individual seams pinch and swell. The rank is subbituminous to bituminous, and some coals are coking. The reserves have been greatly overestimated.

Europe. Extensive coal fields in Europe yield over half of the world production. Carboniferous coals are well distributed, there are a few areas of Mesozoic coal, and Tertiary lignites and brown coal underlie broad areas in central and eastern Europe. Sweden and Italy have only meager amounts of coal, and Norway, Denmark, Greece, and European Turkey are almost entirely lacking in it.

Great Britain, one of the four great coal-producing countries, is second in production of black coal. All her coals are of high rank, from bituminous to anthracite. The coals of England and Wales are Pennsylvanian, and those of Scotland are Mississippian. There are three main coal areas, the southern, central, and northern. In the southern area, in Wales, there are 8 to 40 seams with a maximum of 120 feet of coal, in coal measures 12,000 feet thick. The coal is of high rank, high grade, and low in ash and sulphur, and these qualities have made Wales the premier export field of the world. The central area, in the *Midlands*, has good bituminous coal of lower rank, of which only a minor part is coking. The seams are regular, fairly thick, and are mined to depths of more than 3,000 feet. This coal is the mainstay of the great Midland industrial area of England. The *northern* fields contain thin seams of excellent coking coal, low in ash and sulphur.

Germany has about equal amounts of bituminous and brown coal. The best bituminous coals center around the *Ruhr*, where also are the great industrial centers. Carboniferous coal measures dip northward

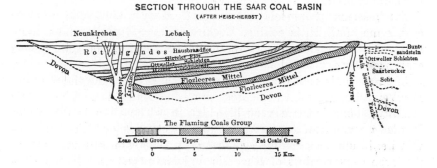

SECTION THROUGH THE SAAR COAL BASIN
(AFTER HEISE-HERBST)

FIG. 16–5. Section through the Saar Basin coal beds, Germany. (*Moore, Coal.*)

beneath younger rocks, and blind outcrops apex below a peneplaned surface. Some 30 meters of coal are known here, divided among many thin but regular seams. The coals are of lower rank than British coal and are gassy. Excellent coking coal makes the Ruhr field more valuable than other German fields. In the *Saar* (Fig. 16–5), Carboniferous strata contain 40 meters of workable bituminous coal, but it is noncoking, high in ash, and of low fuel value. The *Silesian* field underlies the old boundaries of Germany, Poland, and Czechoslovakia and contains numerous thick seams of Carboniferous bituminous coal.

In *France* and *Belgium*, the Valenciennes-Namur basin has long supplied the coal of this industrial region. The coal is a high-volatile, noncoking bituminous variety, occurring in thin beds that are much folded and faulted and deeply buried. Some 69 seams are known but none exceed 2 meters in thickness. The *Aachen-Limburg* field has been extensively worked by the Dutch.

Russia contains enormous quantities of Carboniferous coals in the Moscow Basin, the Donetz, and in the Urals. The Donetz contains excellent low-ash, noncoking anthracite, and bituminous coals. Some

135 workable seams are known in a much disturbed area. Bituminous and cannel coals are deeply buried in the Moscow Basin. Mesozoic bituminous coal and Tertiary lignites are abundant in Siberia and in the Caucasus.

Asia. Extensive bituminous and lignite coals have been actively developed throughout *Siberia*. *China* has vast quantities of good Carboniferous to Tertiary coals ranging in grade from lignite to anthracite. In Hunan one anthracite seam is 50 feet thick, and several average 15 feet. Large resources of high-rank coal have long been mined in *Manchuria*. Some of the Tertiary low-rank bituminous seams are of remarkable thickness; the Chien-Chin-Chai seam ranges from 120 to 200 feet in thickness, with 100 partings aggregating 20 feet. *Japan* has small Mesozoic and larger Tertiary low-rank bituminous coal, some of which is coking. *India* has large deposits of Permian, low-rank, bituminous coal in the Gondwana series, but good coking coals are scarce.

Africa. Coals of Karoo age, in many places overlying the Permian Dwyka tillite, are mined in Rhodesia and South Africa. These coals are bituminous but are very high in ash. The Wankie deposit of *Rhodesia* consists of seams from 1 to 12 feet thick of high-ash, high-volatile, bituminous coking coal that supplies steam coal and coke to the large copper mines of Rhodesia and Katanga. The Karoo bituminous coal of *South Africa* carries the same flora (Permian) as the coals of India and Australia.

Oceania. *Australia* contains the largest coal reserves south of the equator. Valuable Paleozoic coals in *New South Wales* adjoin the coast near Sydney and Newcastle and supply high-grade bituminous coal for steaming, gas, bunkering, and coking purposes. Similar but less extensively developed coals also occur in *Queensland;* one Paleozoic seam at Blair Athol contains 66 feet of clean coal at shallow depth. Most of the Queensland coals now mined are Mesozoic in age. *Victoria* contains the thickest seams of brown coal in the world, at Moreland, where a 1,010-foot bore hole cut three seams 266, 227, and 166 feet thick. This coal is mined by open pit and is utilized for electric generation and briquetting. It contains only 33 percent fixed carbon.

Selected References on Coal

Coal. 2nd Ed. E. S. MOORE. John Wiley & Sons, New York, 1940. *Comprehensive textbook.*

Geology of Coal. OTTO STUTZER. Univ. of Chicago Press, Chicago, 1940. *Excellent for constitution, and for European and brown coals.*

Coal Fields of the United States. M. R. CAMPBELL. U. S. Geol. Surv. Prof. Paper 100, 1929. *A comprehensive and detailed description.*

Some Structural Features of North Anthracite Region, Pennsylvania. M. R. CAMPBELL. U. S. Geol. Surv. Prof. Paper 193–D, 1940.

Evolution of the Mineral Fuels. J. VOLNEY LEWIS. Econ. Geol. 29:1–38; 157–202, 1934. *The origin of coals and transformation into higher ranks.*

Coal: Its Constitution and Uses. BONE and HIMUS. Longmans Green, New York, 1936.

The Origin of Coal. DAVID WHITE and R. THIESSEN. U. S. Bur. Mines Bull. 38, 1913. *A comprehensve discussion of constitution and origin.*

Conditions of Deposition of Coal and Progressive Regional Carbonization of Coal. DAVID WHITE. A. I. M. E. Trans. 71:3–34; 253–281, 1925. *This same volume also contains good papers by R. Thiessen, E. C. Jeffrey, and A. Seyler.*

Bibliography on Coal (general and American). In **Economic Geology.** H. RIES. John Wiley & Sons, New York, 1939. *Very complete.*

Bituminous Coal Fields of Pennsylvania. G. H. ASHLEY. Pa. Geol. Surv. Bull. M6, Pt. I, Harrisburg, 1926.

Coal of Great Britain. 2nd Ed. W. GIBSON. London, 1927.

Classification of Coal. T. A. HENDRICKS. Econ. Geol. 33:136–142, 1938.

Variable Coalification: The Processes Involved in Coal Formation. J. M. SCHOPF. Econ. Geol. 43:207–225, 1948. *A discussion of coalification by incorporation, vitrinization, and fusinization.*

Coal Reserves of Canada, Report of Royal Commission on Coal and Maps. B. R. MACKAY. Pp. 113. Ottawa, 1947. *Canadian coal reserves revised to 10 percent of 1913 estimates.*

The Nature and Origin of Coal and Coal Seams. A. RAISTRICK and C. E. MARSHALL. I–VII, 13–282. English Universities Press, Ltd., London, 1939. *The formation of coals and coal measures with considerable reference to British coals.*

World Coal Resources. C. A. CARLOW. General Reference 14, pp. 634–684. *A new survey of world resources; good bibliography.*

The First Century and a Quarter of the American Coal Industry. H. N. EAVENSON. Pittsburgh, 1942. *Traces the development and progress of the coal industry.*

Nomenclature of the Megascopic Description of Illinois Coals. GILBERT H. CADY. Econ. Geol. 34:487, 1939. *Suggests nomenclature to be used in describing coal.*

Modern Concepts of the Physical Constitution of Coal. GILBERT H. CADY. Ill. State Geol. Surv. Bull. 80, 1942. *Megascopic recognition of four primary type coals — vitrain, clarain, durain, and fusain.*

Coal Geology: An Opportunity for Research and Study. G. H. CADY. Econ. Geol. 44:1–12, 1949. *Problems of the natural history of coal.*

How Much Coal Do We Really Have? A. B. CRICHTEN. A. I. M. E. Tech. Pub. 2428, August, 1948. *Resurvey of resources.*

Appropriate Research in Natural History of Coal Beds. E. C. DAPPLES. Econ. Geol. 44:598–605, 1949. *Problems of coal sedimentation.*

Problems of Coal Geochemistry. H. P. MILLER. Econ. Geol. 44:649–662, 1949. *Relations between chemical and physical properties of coal.*

PETROLEUM AND ASSOCIATED PRODUCTS

Petroleum is the preferred fuel of the twentieth century and has become a necessity of modern civilization. The value of its products exceeds that of any other mineral used by man, and modern industry, transportation, and warfare have become dependent upon it. Being a fluid, petroleum is quickly and cheaply extracted from the earth, and its mobility permits it and its products to be inexpensively transported and handled. The stupendous demand for it has in large part been met by the contribution of geology to oil-finding and extraction.

Petroleum was known in antiquity. Both Neolithic and Paleolithic man used bitumens in building, and the Egyptians used it for mummy preservation and for making boats of woven bullrushes. The Japanese used " rock oil " for light over two millenniums ago, and the Chinese are reported to have drilled for it by 221 B.C. Herodotus in 450 B.C. referred to seeps in Persia and Greece, and Pliny told of how the Romans obtained it for light. The oil of Baku has been exploited for over 300 years.

In America, survivors of the De Soto expedition repaired their ships in 1543 with asphalt from the Sabine river; oil seeps were known in 1627 in southwestern New York; and in 1650 seeps were observed near Cuba. In " Oyl Creek " in Pennsylvania oil was discovered in 1755 and mapped in 1791. In the eighteenth and early nineteenth centuries it was largely a medicinal curiosity, fetching high prices. Up to about 1850 it was quite widely sold for medicinal purposes and thereafter it became used for the extraction of illuminating oil, or kerosene, which gradually displaced the coal oil distilled from coal. The oil for such purposes was obtained from oil seeps and from wells that were drilled for salt. Thus, a salt well yielded some oil in West Virginia in 1806, another salt well in Kentucky in 1829 yielded much oil, and prior to 1850 a 400-foot salt well at Tarentum, Pa., yielded oil.

An historic incident occurred in New Haven, Conn., that really gave birth to the great American oil industry. A sample of seepage oil from Pennsylvania was sent to Benjamin Silliman, Jr., who obtained from it desired kerosene and excellent lubricating stock. Casual discussion of his discovery with three friends led to forming a small company to drill for oil, just as was done for salt. One member supplied a young temporary conductor from his railroad, who was sent to Oil Creek, Pa. To give him dignity in the community they bestowed upon him the title of Colonel by addressing letters in advance of his arrival to Col. E. L. Drake. In 1859, the first well actually drilled for oil in America, the historic Drake well, struck oil at 69 feet with a flow of 25 barrels. An exciting oil boom started, which resulted in many discoveries throughout Pennsylvania, Ohio, West Virginia, Indiana, Colorado, Kansas, Texas, and other states during the 70's and 80's. With the need for gasoline, the industry expanded, and the 2,000-barrel production of 1859 has become the 2-billion-barrel production of today.

Uses of Oil and Gas. The primary use of oil and gas is to produce energy for power or heat, and for lubricants. Formerly petroleum was desired mainly for kerosene; now it is for motor fuel. These uses account for 90 percent of the petroleum consumed.

Oil may be used in the crude state for fuel oil or road oil, but most of it is refined into its component parts. In essence, refining consists of heating crude oils in stills and driving off as vapor first the more volatile constituents, followed by the less volatile. These are condensed to fluids, such as benzene, gasoline, distillate, and kerosene. The residuum is treated for lubricants and other constituents. In practice the procedure is complicated; heavier hydrocarbons are " cracked " under pressure to yield more gasoline.

The products of refining, their average percentages, and the chief uses of refinery products and raw materials are shown below. In addition, hundreds of organic compounds manufactured from petroleum compounds are used for chemicals, medicines, solvents, lacquers, textiles, resins, dyes, explosives, saccharine, antiseptics, rubber, and perfumes.

RAW MATERIALS	USES
Crude oil	Fuel, road oil, lubricants.
Natural gas	Fuel, natural gasoline, carbon black.
Natural gasoline	Blending with gasoline; aviation fuel.
Bitumens	Asphalt, roads, paints.

PRODUCTS	PERCENT OF OIL	USES
Gasoline	40.9	Motor fuel, solvents.
Kerosene	4.7	Range fuel, illuminant, motor fuel.
Gas oil, distillate	14.5	Domestic fuel, Diesel fuel.
Residual fuel oils	27.3	Heavy fuels, Diesel fuel, gas oils.
Still gas	6.0	Refinery fuel.
Lubricants	2.4	Heavy and light oils and greases.
Asphalt	2.3	Paving, roofing, paints, chemicals.
Coke	0.6	Fuel, graphite, carbon products.
Road oil	0.2	Roads.
Wax	0.2	Candles, sealing, waterproofing.
Others	1.1	
Losses	0.4	

Fuel oil averages about 19,000 Btu per pound as compared with about 13,000 for coal.

The change in trend of refinery products is shown in Fig. 16–6.

Production and Resources. The annual world production of petroleum of about 3 billion barrels and that of the United States of about 2 billion barrels are divided as below:

COUNTRY	PERCENTAGE OF WORLD	PROVED RESERVES (billion bbl)	RESERVES (percentage)	POTENTIAL RESERVES (billion bbl)	RESERVES (percentage)
United States	61.5	28.0	31.6	110	18.0
Venezuela	14.4	9.6	14.2	80	13.1
Middle East	10.5	28.8	42.3	155	25.4
Russia	6.5	5.0	7.4	150	24.6
Other South America	2.8	—	—	—	—
Other North America	1.7	1.2	1.7	40	6.6
Europe	1.6	0.6	1.0	13	2.1
Indonesia	0.7	1.1	1.6	30	4.9
Other Asia	0.1	0.1	0.1	24	4.0

STATE	PERCENTAGE OF U.S.	STATE	PERCENTAGE OF U.S.
Texas	44.1	Wyoming	2.1
California	19.1	Arkansas	1.7
Oklahoma	8.1	Mississippi	1.1
Illinois	4.4	Michigan	1.0
Louisiana	7.6	Pennsylvania	0.7
Kansas	5.6	Colorado	1.3
New Mexico	2.2	Others	1.0

Important countries thus far deficient in petroleum are Germany, Italy, Japan, and China; and others lacking it are Sweden, Norway, Belgium, Holland, Spain, Turkey, Australia, South Africa, and New Zealand.

United States Reserves. Accumulated oil reserves discovered up to 1950 amount to about 68 billion barrels; the reserves now stand at about 28 billion, and the annual consumption is about 2 billion. Roughly half of the known reserves are in Texas; the remainder are largely in California, Louisiana, Oklahoma, Wyoming, Kansas, New Mexico, and Illinois.

Of 196 major pools developed between 1900 and 1936, 36 were obvious, 31 were discovered by random drilling, 93 by geological methods, 28 by geophysical methods, and 8 by deeper drilling. From 1937 to 1949, 85 pools in excess of 25 million barrels were discovered. Formerly 10 percent of new reserves came from random drilling, lately

only 2 percent are so found; in the 1920's, 65 percent came by geological methods, in the 30's, 30 percent; in 1930–35, 55 percent resulted

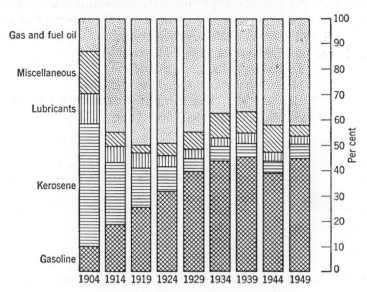

Fig. 16–6.   Change in trend of refinery products of petroleum.   (In part from
Pogue, Econ. of Petrol. Industry.)

from geophysical exploration, but in the 1940's new producing wells
were found by geology as compared to geophysics in the ratio of nearly
2 to 1.

TABLE 20

THE RATE OF DEVELOPMENT OF NEW RESERVES IN THE UNITED STATES

PERIOD	RESERVES A YEAR (*million bbl*)	MAJOR POOLS PER YEAR
Pre-1920	300	2
1920–1925	900	8
1926–1930	2,200	12
1931–1935	700	8
1936–1940	2560	
1941–1945	1900	
1946–1949	4000±	

Constituents, Types, and Properties.   Petroleum is composed of
many compounds of carbon and hydrogen, with minor oxygen, nitrogen, and a little sulphur.   The numerous members of the different CH
series each have different properties; at normal temperatures some

are gases, some are liquids, and some are solid waxes. Since the proportion of these varies in different oils, no two oils are alike in their properties or constituents. A crude oil may contain CH members that give the oil a high gasoline content, or certain ones may be absent and the oil will contain little or no lubricants. Thus, oils may be referred to as paraffin base, or asphaltic and naphthene, or mixed base.

In general, paraffin-base oils are light and yield good lubricants, and asphaltic-base oils are heavy, unsuited for good lubricants, and may be usable only for fuel oil. The chemical composition conveys little practical information, but the yield does (Table 21).

TABLE 21

YIELDS OF CONTRASTING CRUDE OILS IN PERCENTAGE

PRODUCT	PENNSYLVANIA LIGHT	CALIFORNIA HEAVY	MID-CONTINENT CUSHING
Benzenes (gasoline)	23	30	31
Kerosene	42	7	20
Gas oil	0	20	15
Lubricants	31	15	0
Residuals	4	25	27

The terms " light " and " heavy " refer to the gravity, measured in arbitrary units called API (American Petroleum Institute), which replaced Baumé units. A specific gravity of 1.00 equals 10 API; 0.9 equals 25.7 API (a medium oil), and 0.8 equals 45.4 (a light oil).

Geologic and Geographic Distribution of Petroleum

The home of petroleum is in sedimentary rocks; only rarely has it moved out of this environment into adjacent igneous rocks. The containing rocks of commercial pools are sands, sandstones, conglomerates, porous limestones and dolomites, and rarely fissured shale. The enclosing rocks are almost invariably marine strata or associated freshwater beds. Consequently, areas of igneous Precambrian and metamorphic rocks can be eliminated.

Oil is found in rocks of all ages from upper Cambrian to Pliocene, but Tertiary beds the world over are the most prolific, followed by Paleozoic beds outside the United States and Canada. The age of the productive oil horizons of the major petroliferous provinces of the world is shown in Table 22.

The distribution of the oil and gas fields of the United States is shown in Fig. 16–7.

TABLE 22

AGE OF OIL BEDS IN MAJOR PETROLIFEROUS PROVINCES OF WORLD

	Paleozoic							Mesozoic				Cenozoic			
	Upper Cambrian	Ordovician	Silurian	Devonian	Mississippian	Pennsylvanian	Permian	Triassic	Jurassic	Comanchean	Cretaceous	Eocene	Oligocene	Miocene	Pliocene
California											X	X	X	X	X
Rocky Mountains					X	X	X	X	X	X	X	X	X		
Mid-Continent	X	X	X	X	X	X	X			X					
Illinois					X	X									
Gulf Coast											X	X	X	X	X
Appalachian				X	X	X									
Western Canada				X						X					
Mexico										X	X	X	X	X	
Venezuela												X	X	X	
Colombia												X	X	X	
Trinidad													X	X	
Peru												X	X	X	
Argentina									X			X		X	
Galicia											X	X	X	X	
Rumania												X	X	X	X
Russia														X	X
Iran and Iraq													X	X	
Netherlands E. Indies														X	X

Origin of Oil and Gas

It is now universally believed that oil and gas are of organic origin. The inorganic theories of volcanic origin, which involve the reaction of carbides within the earth to form acetylene and ultimately hydrocarbons, have been discredited because of the overwhelming accumulation of evidence favoring an origin from low forms of marine or brackish-water life. There exist, however, differences of opinion as to the details of the processes of conversion into petroleum constituents.

In brief, it is held that organic material buried in marine muds underwent changes to produce natural hydrocarbons, and these subsequently moved into porous reservoir rocks and accumulated to form commercial oil pools.

Source Materials and Source Places. The slow, oxygen-free decomposition of the remains of plant and animal organisms is considered to be the source of petroleum hydrocarbons, a possibility that has been demonstrated by laboratory experiments. The role of plant life in coal formation and the close relationship between oil and diatomaceous shales in California have led to the belief that plants have played a more important part than animals. The higher forms of life that grow and live on the land are unimportant, but the lower planktonic organisms, such as diatoms and algae, that thrive abundantly near the surface of the sea, are now considered the most probable and im-

Fig. 16-7. Distribution of oil and gas fields in the United States. (*Powers, A.I.M.E.*)

portant source materials. Some geologists think that the differences in oils may be due to differences in source material, but this seems quite unimportant, since the complicated processes that follow could yield vastly greater differences.

The organic remains accumulate in the bottom muds of lagoons or in depressions on the floor of shallow seas and there become incorporated in accumulating sediments. Trask has tested thousands of bottom mud samples and has found the organic content to be constant for about 100 miles offshore and to range from 0.3 percent in deep sea oozes to 7 percent off the coast of California. The average organic content is 2.5 percent in recent sediments and only 1.5 percent in ancient sediments. A high organic matter is no longer considered by Trask to be a reliable indication of a source rock.

Conversion to Petroleum. The bacteria that thrive in the upper mud of the sea floor are thought to change the organic matter into the

mother material of oil by removing oxygen and nitrogen and producing other changes. Trask states that the composition of average
plankton is 24 percent protein, 3 percent fat, and 73 percent carbo-
hydrates, and thinks that it is not the fats, cellulose, and simple
proteins that yield the oil but rather the complex proteins and carbo-
hydrates. The next step is more elusive, i.e., whether bacterial processes on the sea floor yield oil droplets, prior to burial, or the mother
of oil is buried and is afterwards changed to oil. With deeper burial,
bacterial action presumably ceases, and then other, as yet unknown,
chemical changes occur to yield oil before consolidation. Further
post-consolidation changes probably take place during migration.
Pressure probably exerts some influence, but temperatures could not
have been high because certain compounds could not exist at 140° to
300° C. Trask states that the later changes to liquid hydrocarbons
may have been aided by polymerization or methylation. The liquid
hydrocarbons that are formed are believed by Brooks and Trask to
be capable of dissolving other organic substances such as pigments,
waxes, and fatty acids. During migration, other organic compounds
might also be dissolved, thus continually changing the composition
of the petroleum, and perhaps giving rise to the differences in oils.
This phase of the subject is far from being completely known.

During the conversion of organic matter into petroleum, natural
gas is also formed. This is dominantly methane, and it is of interest
that methane is the gas formed in peat swamps and in the transformation of low-rank to high-rank coals.

Formation of an Oil Pool

An oil pool is not a subterranean pond but merely an accumulation
of petroleum in rock pores. Dispersed droplets generated in the
source muds do not constitute an oil pool. The conditions necessary
for this are: (1) migration and accumulation, (2) suitable reservoir
and caprocks, (3) suitable traps, and (4) retention.

MIGRATION AND ACCUMULATION OF OIL

Since oil pools are found chiefly in sandstones, it follows that the
dispersed droplets of oil generated within the source muds must have
migrated out of the source rocks into the sands. The forces that
cause migration are (1) compaction of the muds, (2) capillarity,
(3) buoyancy, (4) gravity, and (5) currents.

Compaction. Source muds may contain up to 80 percent water. As
covering sediments are deposited, the accumulating weight gradually
compacts the lower beds and the enclosed fluids are squeezed out into

places of less pressure, such as pore spaces in sands. The fluids may move up, or down, or even laterally. Athy estimates that compaction results in removal of half the water when burial has reached 1,000 feet and 85 percent at 4,000 feet. Compaction is seemingly the most important factor causing migration of oil.

Capillarity. If oil-wet shales are in contact with water-wet sandstones, the water, owing to its higher surface tension, will move from the coarse sandstone pores into the fine, capillary, shale pores, and displace the oil therefrom into adjacent sandstones.

Buoyancy. Oil being lighter than water tends to rise upon a water surface. Free gas similarly rises above the oil. This "secondary migration" takes place within reservoir rocks. It is most effective where the pore spaces are large and where large volumes of fluids are involved; it gives rise to the common stratiform arrangement of water, oil, and perhaps also gas. If the host beds are horizontal, the oil will tend to rise to the tops of permeable beds and no pronounced accumulation may occur; if inclined, the oil tends to migrate up-dip and, as is shown later, accumulation into oil pools may occur.

Gravity. Where water is present, the differences in specific gravity between oil and water give rise to buoyancy, but where water is absent, gravity causes oil to migrate down-dip until arrested by impervious beds; this gives rise to synclinal accumulation.

Currents. Currents of subsurface waters flush oil along with them and accelerate oil migration. Such currents may be caused by compaction or artesian water circulation. The latter may be a very effective means of bringing about large-scale migration over large areas, as in the Rocky Mountain province.

Accumulation. Migration of oil generally leads to accumulation, which is the collection of oil droplets into pools. Oil may migrate without accumulating, or it may accumulate into noncommercial bodies, as at the top of horizontal beds, but concentrated accumulation is essential to produce commercial oil pools, and this in turn is dependent upon requisite reservoir rocks and traps.

RESERVOIR AND CAPROCKS

Reservoir Rocks. Accumulation takes place only in porous and permeable rocks. A rock like clay has high porosity but small permeability and, therefore, is not a suitable reservoir rock. Its numerous minute pores create frictional resistance to flow and cause fluids in them to be held firmly by capillarity. The most suitable reservoir rocks are sands and sandstones. Others are porous or cavernous limestone and dolomite, rarely fissured shale, and rarely jointed igneous rocks. Cavernous limestones may yield sensational and prolific flows of oil, as in Mexico, but the total yield of oil from rocks other than sandstone is relatively small.

Unconsolidated sands of California average about 25 percent porosity. In sandstone, the cementing material decreases the porosity from 12 to 25 percent; an average porosity of many oil sandstones is 17 percent. The capacity of sand with 20 percent porosity is 1,550 bbl per acre-foot, or 155,000 bbl for a sand 100 feet thick. The size, shape, and sorting of the sand grains is important, and the highest porosity results with well-sorted angular grains.

The greater the porosity, the greater the amount of oil that a reservoir rock can contain; and the larger the pore size, the greater the amount of oil it will yield, since with small grains and small pores more oil clings to the rock grains and is not recovered. The percentage of oil recovered from reservoir rocks is surprisingly small. Powers states that the Bradford sand of Pennsylvania yields only 8 percent by pumping and an additional 20 percent by flushing; Oklahoma sands yield 10 to 60 percent, and unconsolidated sands of California, with 25 to 40 percent porosity, yield only 10 to 30 percent of the contained oil. The part of the original pore space that will yield its oil is called the *effective porosity*.

Caprocks. A confining impervious caprock is necessary to retain oil in a reservoir rock. Shale and clay are the most common caprocks, but dense limestone and dolomite and gypsum also serve, and even well-cemented, fine-grained sandstone or shaly sandstones are effective. Water-wet impervious shales, clays, and limestones even retain gas. Good caprocks form effective seals to underlying oil and gas for long periods of geologic time, but poor ones permit slow escape of mobile hydrocarbons.

TRAPS OR "STRUCTURES"

In inclined sedimentary beds the up-dip migration of oil will continue until it escapes at the surface or is arrested in some trap where it accumulates to form an oil pool. If the strata are folded into an anticline, up-dip migration ceases when the oil reaches the top of the arch (Fig. 16–8), where it accumulates to form an anticlinal oil pool. This is the commonest kind of trap.

In the earlier period of oil exploration most of the oil discovered was found in domes and anticlines. Consequently, it was assumed that structures were the controlling loci of oil accumulation and there arose such terms as "anticlinal theory of oil accumulation" (proposed by I. C. White, 1885), now in disuse, and "oil structure"; a

dry hole was said to be " off structure." Now, however, it is realized
that there are also many other kinds of oil and gas traps and these may
be divided into structural and stratigraphic traps, as follows:

STRUCTURAL TRAPS	STRATIGRAPHIC TRAPS
Anticline*	Unconformities*
Dome*	Ancient shore lines*
Monocline*	Sandstone lenses*
Terraces	Shoestring sands*
Synclines	Up-dip wedging of sands*
Faults*	Up-dip porosity diminution
Fissures	Overlaps*
Salt domes*	Reflected buried hills*
Igneous intrusions	Buried coral reefs*

* The more important types of oil reservoirs.

Folds. Up-folds, giving rise to *anticlines* and *domes,* have been the
source of the greater part of the oil so far produced. An elongated
anticline (Fig. 16–9) is an ideal petroleum trap since its extended

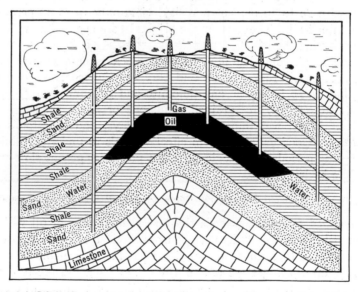

FIG. 16–8. Oil in asymmetric anticlinal structure with most oil on the gentler
dipping limb. (*Leven, Done in Oil.*)

flanks facilitate migration, its arch permits accumulation, and its
crest or " center " generally contains domelike areas in which oil collects from all sides and is retained beneath the caprock (Fig. 16–8).
It is customary to draw structure contours on such a fold, and where

the contours close it is called a "closed fold"; the amount of "closure," expressed in feet, is the height to the crest above the lowest contour that closes. The oil ordinarily accumulates within the closure. Folds with gentle dips offer larger areas for accumulation of oil; generally the dips are less than 30° but they may reach as much

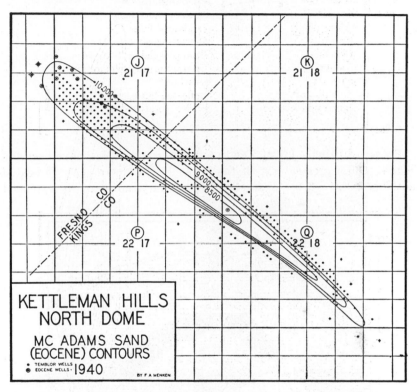

FIG. 16–9. Kettleman Hills, North Dome; McAdams (Eocene) sand contours.
(*Menken, A.A.P.G.*)

as 60°, as at Ventura, Calif. (Fig. 16–10). In asymmetric folds, the gentler-dipping limb often contains the most oil (Fig. 16–8).

Anticlinal pools vary greatly in size. The highly productive Long Beach pool of California has an area of only 1,305 acres; others may attain tens or even hundreds of square miles. Large anticlines are rarely productive throughout.

In ideal arrangement in an anticline, free gas, if present, is at the top, oil lies beneath, and the water below. Commonly, however, in deep wells where pressure is high, the gas is contained within the oil and is liberated only upon drilling. An individual productive sand may range from a few feet to a few hundred, or even to a thousand feet or more in thickness. Also, many productive sands may lie below each other within the same anticline. They may be separated by

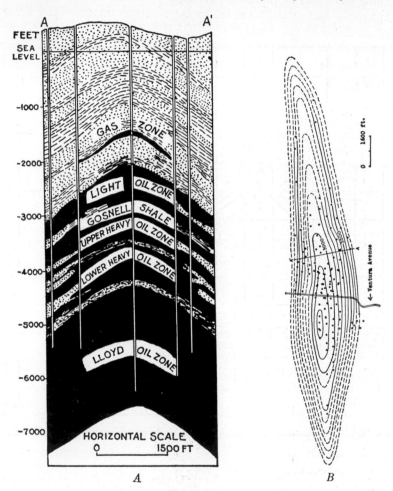

FIG. 16–10. Ventura Avenue oil field, California. *A*, Section illustrating thick "oil zones" below which are deeper zones; *B*, subsurface structure contour map of the field. (*Drawn by Decius; reproduced by Powers, A.I.M.E.*)

large intervening thicknesses of barren beds, or be essentially contiguous, as at Long Beach, Calif. (Fig. 16–24), where drills intersected productive horizons between depths of 2,900 and 6,500 feet.

Anticlinal oil structures have been formed by compressive force, deposition of beds over ancient hills, and by differential compaction where shales have undergone greater compaction than sands. Many details of anticlines, beyond the scope of this chapter, are to be found in numerous books on petroleum geology.

Anticlines and domes form important oil traps in practically all the large oil fields of the world.

Synclines, in a few cases, serve as oil traps where water is absent, and *monoclines* and *terraces* constitute important oil structures in several places.

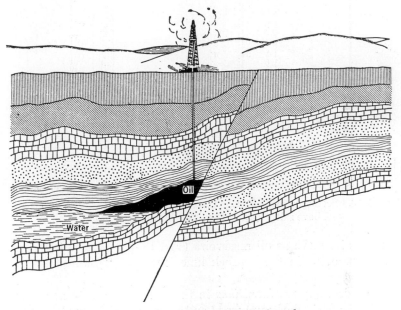

FIG. 16–11. Oil trapped by fault seal.

Faults. Faulting has given rise to many important oil pools as, for example, the Mexia-Powell field, Texas. In inclined beds an impervious shale may be faulted against the up-dip continuation of an oil sand, causing an effective seal and permitting oil accumulation beneath it (Fig. 16–11). Rarely, a thick fault gouge may serve as an effective upward seal to an oil sand. Open faults may permit the escape of oil to upper beds or to the surface.

Salt Domes. A spectacular and prolific type of structure is that associated with salt domes (see Chap. 5·6), particularly in the Gulf Coast region of Texas and Louisiana and in Rumania and Mexico. Over 100 have been found in the Gulf Coast by geophysical methods alone. Salt domes yield most of the oil of that province.

The up-thrust plastic salt plugs (Fig. 5·6–1) that pierce the overlying beds created excellent traps for oil where upturned edges of sandstone beds are sealed against the salt; several horizons may be productive. In some cases arched beds above the salt dome, or porous

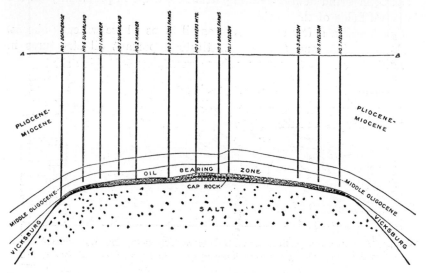

FIG. 16–12. Section across Sugarland salt dome, Texas, showing oil accumulation above salt. (*After Teas, 16th Internat. Geol. Cong.*)

caprock, also serve as oil reservoirs (Fig. 16–12). Economic deposits of sulphur occur in the caprock limestone and gypsum of some salt domes.

In Mexico, *igneous intrusions* in the form of volcanic necks, dikes, and sills have similarly upturned and sealed petroliferous beds to form oil traps.

Stratigraphic Traps. Stratigraphic traps, as distinguished from structural traps, are those formed by conditions of sedimentation in which lateral and vertical variations in thickness, texture, and porosity of beds result. These may result from intercalation of beds, interruptions to sedimentation, and sites of deposition with respect to shore lines. A sandstone bed may wedge out or undergo decrease in porosity and thus make a suitable trap. With the discovery of most of the obvious structural traps, increasing amounts of oil are being found in stratigraphic traps, and these will probably constitute the most important sources of future oil. More study will have to be given to them and new methods devised for locating them.

Unconformities produce an important class of oil reservoirs. They mask tilted strata, overlays, variable porosity, and folding. Underlying tilted sand beds may be sealed at the unconformity by overlying beds, and accumulation can take place. Angular unconformities

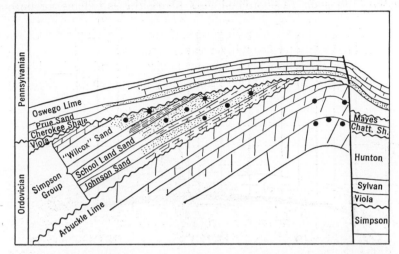

FIG. 16–13. Idealized section across Oklahoma City field, showing relation of oil (black circles) and gas (stars) to unconformity. (*Levorsen, Probs. of Petrol. Geol.*)

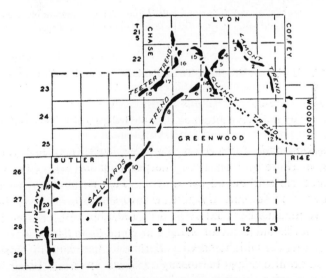

FIG. 16–14. Shoestring sands in the Burbank sand, Kansas.
(*After Bass et al., A.A.P.G.*)

are more effective than parallel unconformities (Figs. 16–13, 18). The underlying beds are commonly deeply weathered and therefore porous, and constitute oil containers. If the erosion surface is developed on limestone, solution caverns may be present. The basal beds above an unconformity are also commonly quite porous and may serve as reservoir rocks. Deeper drilling has disclosed oil associated with two and even three underlying unconformities, giving rise to what Levorsen has aptly termed "layer-cake" geology.

Shoestring sands are long, narrow bodies of sand enclosed within shaly beds (Fig. 16–14). These sands may have been deposited in off-

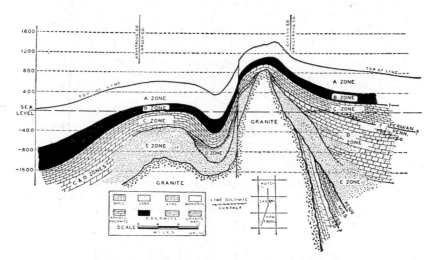

Fig. 16–15. Buried granite ridge, Texas Panhandle, showing sites of oil accumulation in dolomite and granite wash; helium occurs in the gas. (*Rogarty, A.A.P.G.*)

shore bars by tidal currents in lagoons or by meandering streams. Many small pools have been found in them in the Mid-Continent field.

Shore-line sands formed on low-lying submerged coastal plains, tapering in shore and extending deeper offshore may be covered by clays and form suitable oil containers. This is spoken of as "lensing out" of sands. It generally gives rise to small pools, but it accounts for the stupendous East Texas field of 120,000 proven acres.

Buried hills and structures have accounted for several important fields, such as the Oklahoma City, Seminole, and Amarillo pools along the buried granite ridge of Kansas and Oklahoma (Fig. 16–15). Here, projecting granite ridges received a mantle of sediments, thin on top and thicker on the flanks, simulating anticlines, and giving rise to overlap. These stratigraphic features made excellent oil traps.

Other stratigraphic traps caused by porosity variation are *sandstone lenses* (Fig. 16–16) enclosed in shales, shore-line *wedges of sand* that become fine grained and impervious offshore (Burbank pool, Okla.), sandstone overlaps against impervious beds, up-dip seals by asphalt (Fig. 16–17), and buried coral reef rocks (Alberta).

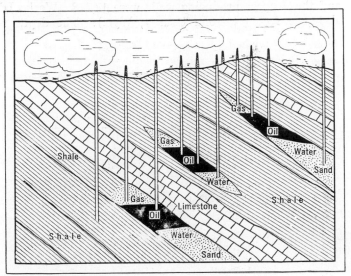

FIG. 16–16. Diagram illustrating three types of stratigraphic traps. Right, pinching sandstone or wedge; center, sandstone lens; left, sandstone grading into shale. (*Leven, Done in Oil.*)

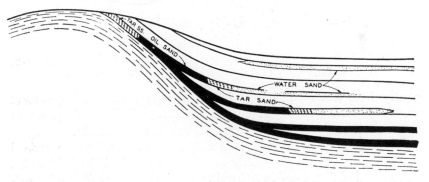

FIG. 16–17. Tar seals causing oil accumulation, also porosity changes. (*Pack, U. S. Geol. Survey.*)

FEATURES OF OCCURRENCE OF OIL AND GAS

Depth. Oil occurs from the surface to the greatest depths yet probed by drilling. As drilling technique has improved, deeper oil has been found. Prior to 1920 a depth of 4,000 feet was unusual; by

1930, 6,000 to 8,000 feet was common; by 1940 wells had reached a depth of 15,004 feet in California and 14,952 feet in the Mid-Continent field; and by 1950 the deepest well was 21,482 feet (Paloma Field, Calif.) and the deepest producing wells were 13,778 (Weeks Island, La.) and 14,309 feet (Natrona County, Wyo.). The depth varies

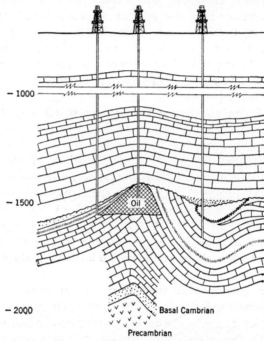

FIG. 16–18. Part of section of Geneseo uplift, Kansas. (*After Clark, Arnett, Royds, Struct. Typ. Am. Oilfields, Vol. 3, A.A.P.G.*)

greatly from field to field and in an individual field. Also, several producing horizons may be cut by a single well. The deepest oils have come in general from regions of thick Tertiary strata.

It has been noted in California and the Gulf Coast where several oil horizons exist that commonly the deeper ones yield the lighter oils.

Pressure. Oil and gas in natural reservoirs are generally under sufficient pressure to cause the oil to flow freely from the well, often with great force, giving rise at times to huge gushers with flows up to 300,000 bbl per day. Shallow wells lacking pressure have to be pumped, and flowing wells ultimately have to be pumped as the gas pressure diminishes. Waste of gas diminishes or dissipates the lifting force. Pressures in the oil sands range from a few pounds up to **8,225** pounds per square inch (Gulf Coast).

The cause of pressure is not simple. As a general rule pressures within a given oil pool increase with depth. Oil-pool pressure is often incorrectly referred to as "rock pressure"; Heroy suggests it be called "reservoir pressure." Oil and gas pressure in many places and at certain depths shows a close relationship to hydrostatic pressure (43.4 lb per 100 ft of depth), but in many other places pressures are abnormally lower or even higher than the hydrostatic head. In addition to the hydrostatic head, other factors affecting pressure are 1, the weight of rock overlying oil-bearing shales; 2, diastrophism, as at Khaur; 3, expansive force of confined gas; 4, pressures due to generation of oil and gas; 5, temperature gradient; and 6, mineralogical alterations in the oil strata.

Mills showed that in a group of wells averaging 2,309 feet, the pressure was almost the hydrostatic head but that in wells below 4,000 feet the average pressure was considerably less. At Seminole, Okla., the pressure at 3,980 feet was 1,520 lb and at 4,217 feet it was only 800 lb. This may mean that the ground water does not penetrate to great depths, a suggestion amply evidenced by deep mines. In the deep Wasco, Calif., well, however, at a depth of 5,000 feet the measured pressure was 2,950 lb or 780 lb greater than the theoretical hydrostatic head, and at 13,175 feet it was only 68 lb greater. Pressures in some 8,000-foot wells at Long Beach, Calif., were only about 12.5 lb per 100 feet of depth.

The pressure is usually greatest at, or shortly after, the opening of a well and tends to diminish fairly rapidly with continued production, particularly if gas is drawn off excessively. High pressures are counteracted by the introduction into the well of a column of drilling mud containing finely ground heavy minerals such as barite or hematite.

Temperature. The deep Wasco well showed a temperature of 268° F at 15,000 feet, but this was probably abnormally low, owing to the cooling influence of the circulating mud; the corrected temperature is estimated to be 297° F. This gives a temperature gradient of 91 feet per degree F. Van Ostrand shows for over 500 wells a general range of temperature gradients in feet per degree F as follows: California, from 40 to 70; Louisiana, 32 to 56; Kansas, 47 to 70; New Mexico, 163 to 180; Oklahoma, 38 to 170; Wyoming, 20 to 75. Some negative gradients have been observed. These temperature gradients are generally higher than those recorded for deep mines.

Metamorphism and Carbon Ratios. Oil, like coal, is sensitive to metamorphism. A suggestion by David White that the nature of the coal in thin seams in the enclosing rocks might be utilized to indicate the degree of metamorphism was formulated into the *carbon ratio* theory. The fixed-carbon content of coal increases with metamorphism; therefore if the coal shows a high fixed-carbon content on ash-free basis, any oil in the associated strata would have been volatilized or destroyed. Thus, in a region where carbon ratios of more than 65 to 70 occur, no commercial oil pools may be expected; between 50 and 65 medium to light oils and gas may be expected; and under 50, fields of heavy oils may be expected. Regions containing anthracite coals would yield no oil, and little or none need be expected with high-rank coals. Broad areas have been mapped with " isocarb " contours. The theory supplies an approximate guide to oil occurrence, but it should be used with caution.

Associated Waters. Salt waters underlie oil in most oil fields. These are considered to be connate waters that have undergone subsequent chemical changes. Some are not altered sea water. The waters of different oil sands may have different compositions and thus may serve as means of correlation.

As oil is withdrawn from a pool the water rises to displace it, and water encroachment into wells marks the decline of a pool. It appears first as " edge water " in the outer and structurally lower holes, which eventually are flooded and have to be abandoned. The rate of rise to the next structurally higher hole gives a rate of decline for the pool. Too rapid oil withdrawal may cause water to surge ahead of the oil with consequent loss. In some old fields, as at Bradford, Pa., water is introduced into selected holes and flushes oil ahead of it to designated discharge wells, thereby greatly increasing the oil recovery.

Fractionation. Oil filtered through bleaching clay loses certain of its heavier ingredients, which suggests that some of the very light oils have been naturally fractionated by passage through clay. Oils recovered from shale are generally light.

Preserval and Destruction. A pool normally accumulated may be destroyed or dissipated by subsequent geological vicissitudes, such as faulting, or long-continued erosion that bares the oil sand at the surface. Diastrophism and resulting metamorphism may gasify or even carbonize the oil contained in the strata.

FEATURES OF OIL FINDING, DRILLING, AND PRODUCTION

In oil finding, if the stratigraphy indicates suitable source, reservoir, and caprocks, search for suitable traps is made by areal and structural contour mapping, subsurface geology, and geophysical methods. Most geological work is now subsurface, utilizing well logs, well samples, heavy minerals, microfossils, sediment analyses, electric logging, and geophysical data. Surface indications, such as oil or gas seeps, paraffin dirt, or soil analyses, rarely give a clue to the presence of oil, and a test or " wildcat " hole must be drilled to test the

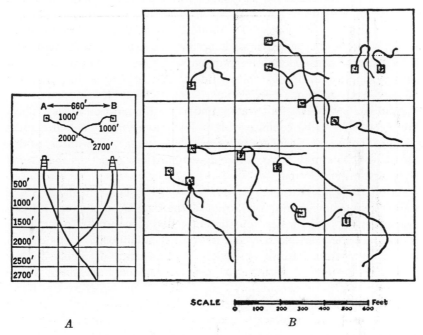

FIG. 16–19. A, Plan and section of two holes that intersect at 2,000 feet at Seminole, Okla.; B, plan of 14 crooked wells drilled to 6,000 feet. (Suman, Elem. of Petrol. Industry.)

" structure." If a discovery is made, a scattered grid of holes is next drilled to the limits of the pool and its productive acreage. The thickness of the oil sand and its porosity and permeability having been determined, an estimate of the oil content can be made.

Wells are commonly spaced with one well to each 10 acres, but present tendency is toward one well to each 20 or even 40 acres. Close spacing, unless carefully controlled, makes for rapid decline and oil loss. The completion of a producing well close to a property line requires the drilling of an " offset " well to prevent oil drainage to the adjoining property. " Town lot " drilling (e.g., Santa Fe Springs) has given rise to the more economic unit pool operation.

Rotary drilling has largely supplanted the older " cable tool " method. Surveying of drill holes has disclosed unexpected deflections from the vertical, up to 30° or 40°. Two holes in California 2,000 feet apart intersected each other at 1,615 feet; another roamed under 9 acres of ground (Fig. 16–19). This discovery prompted directional drilling,

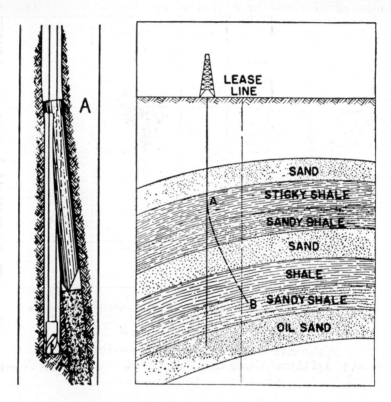

FIG. 16–20.*A.* Application of directional drilling to obtaining oil beyond lease line. (*Suman, Elem. of Petrol. Industry.*)

by means of which a hole may be directed from an onshore location to tap submarine oil, or several holes may be fanned out from a single location (Fig. 16–20).

Production efficiency is increased by conservation of gas, proper well spacing, repressuring, and controlled flow. Gas expansion is fundamental in causing wells to flow, and wells allowed to flow wide open are wasteful of gas and induce rapid decline in pressure. Gas wastage is now generally prohibited, and repressuring practice is common. Proration, invoked to cope with mounting production, tends to conserve reservoir pressure, permits slower and more even water encroachment, and results in larger yields of oil per acre.

Some 40,000 to 45,000 wells are drilled per year, of which about 30 percent are dry holes. Lahee shows that exploratory holes drilled on technical advice are about 4 times as successful as those drilled on

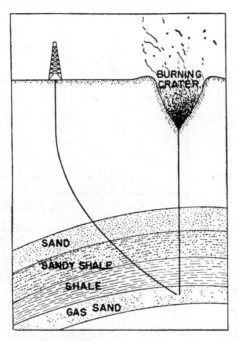

FIG. 16–20B. Application of directional drilling to kill a burning cratered well.
(*Suman, Elem. of Petrol. Industry.*)

nontechnical advice. Pratt shows that the proportion of dry holes to completed holes is increasing, the initial production of successful wells is decreasing, and fewer fields are being found. There are about 428,000 producing oil wells, of which about 7,000 yield one-half the total production; the average of all wells is 13 barrels daily. The average production per acre for 34 of the largest pools has been 13,770 barrels with an average gravity of 30.2. The largest individual gusher was the Cerro Azul No. 4, Tuxpan, Mexico, with 300,000 bbl per day. One well in the Yates pool, Texas, yielded 185,000 barrels per day, and 50,000- to 100,000-barrel wells are not uncommon. Gas wells have yielded over 100 million cubic feet per day.

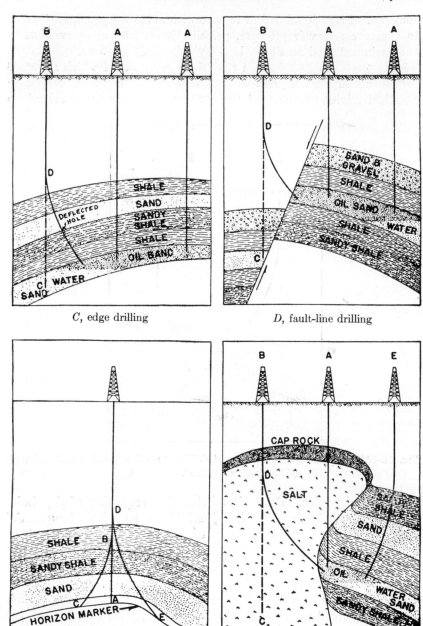

C, edge drilling

D, fault-line drilling

E, exploration drilling

F, salt-dome drilling

FIG. 16–20C–F. Applications of directional drilling. (*Suman, Elem. of Petrol. Industry.*)

Important Oil Fields of the World

UNITED STATES

The United States is underlain by 1,105 million acres of sedimentary rocks, of which about 2¼ million acres are petroliferous. The distribution of the oil and gas fields is shown in Fig. 16–7, the ranking of the individual states is given on pages 653–654, and the producing horizons of the major fields in Table 22. The oil districts of the United States from east to west are:

RANK	DISTRICT	STATES
6	Appalachian	Pennsylvania, New York, eastern Ohio, West Virginia, eastern Kentucky.
4	East Central	Illinois, Michigan, western Ohio, Indiana, central Kentucky, Tennessee.
1	Mid-Continent	Oklahoma, northern Texas, Kansas, Louisiana, New Mexico, Arkansas.
3	Gulf Coast	Texas, Louisiana.
5	Rocky Mountain	Colorado, Wyoming, Montana.
2	California	California.

Within each large district there is a general similarity of controlling geology, oil traps, and nature of oil. In the following pages only a brief outline of the major features of the districts is given; details may be found in the numerous works on oil geology.

Appalachian. The Appalachian, the oldest field of the country, continues to produce a paraffin-base oil rich in lubricants and gasoline. The gravity ranges from 39° to 48°. The field roughly coincides with the Appalachian geosyncline. Some 56 oil horizons are known, but the production comes from 22 sands, chiefly from two, the Berea and Big Injun in the Mississippian. The average thickness is 25 feet, and most of the oil is at shallow depth (1,000 to 2,500 feet); deep test wells have been unproductive.

The oil and gas occur mainly in gently dipping, shallow anticlines and synclines and in the up-dip edges of lensing sands. Bradford is the most important as well as the oldest field in North America. The pools are small but have long life. The field includes some 122,000 wells whose average production is only 0.6 bbl per day, by pumping. The sands yield only 25 percent of their oil, and water flushing has been locally employed with success. The field has produced about 300 million barrels.

East Central. The East Central field embraces small areas of Michigan, Ohio, Indiana, Illinois, and central Kentucky and Tennessee.

Most of the pools are small, but the intensive development in Illinois since 1937 gives this field fourth place in the United States.

The prolific Illinois pools include the Louden, Salem, Clay City, Southeastern, New Harmony, Clay, Richland, Centralia, and others west of the older LaSalle anticline. The oil comes mainly from the McCloskey sand and sandstones of the Chester series of Mississippian age, the Benoist sand, and a Devonian sand at Sandoval; several Pennsylvanian sands also yield oil. The pools are found in domes along large anticlines and in edging sands and yield mostly a good quality paraffin oil.

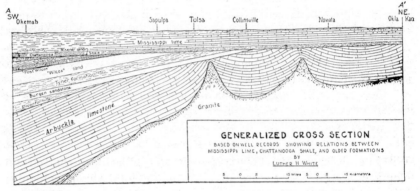

Fig. 16–21. Generalized section across Oklahoma showing unconformities and buried ridges that localize oil pools. (*Wrather, 16th Internat. Geol. Cong.*)

Mid-Continent. This extensive district (Fig. 16–21) is the greatest oil-producing area of the world. Its production nearly equals that of all foreign countries. It yields in large volume and at low cost high-grade paraffin- and mixed-base oils, desirable for lubricants.

The district embraces parts of Oklahoma, Kansas, Texas, Louisiana, Arkansas, and New Mexico and includes scores of pools, some of sensational size. It includes several units such as the Oklahoma-Kansas basin, Wichita-Amarillo, Bend Arch, West Texas-New Mexico, Balcones fault-Ouachita uplift area.

The *Oklahoma-Kansas* unit is bounded by outcrops of early Paleozoic rocks surrounding the Ozark dome in Missouri and the Wichita and Arbuckle Mountains to the south. The area is characterized by buried ridges of granite overlain by Paleozoic sedimentary formations (Fig. 16–15), along the crests and flanks of which occur many oil pools. The most prominent of these are the Nemaha granite ridge and the central Kansas uplift, with which many of the richest oil and gas pools of the district are associated.

In general, the oil is located on anticlinal structures or in stratigraphic traps. There is an important relationship between oil occurrences and unconformities, the most striking being the post-Mississippian and the post-Devonian. Many producing sands are contained in middle and late Paleozoic formations. The anticlines range from small structures with slight closure and gentle dips up to well-developed anticlines with flanking dips of 1° to 5°. The strong anticlines are best developed in the older Paleozoic beds. Stratigraphic traps here include pinching sands, shoestring sands, and unconformities. The most prolific horizons are the Wilcox (Ordovician), Bartlesville and Glenn (Pennsylvanian), and Woodbine (Cretaceous). Some of the well-known pools within this unit are Oklahoma City, Seminole, Cushing, Garber, El Dorado, Glenn, Healdon, Burbank, and Tonkawa.

The *Wichita-Amarillo* unit includes southern Oklahoma, north-central Texas, and the Panhandle. The oil pools of this area are associated with structural features due to Appalachian orogeny. Some of the pools are superimposed upon buried mountains. Others are on the flanks, and some are on small parallel zones of deformed rocks. The Red River uplift lies in the southern part. A series of buried fault blocks have been important features in localizing oil. The chief producing horizons are Pennsylvanian and Ordovician sands. An important producer also is the Big Lime of Permian age and the Gray Lime of Pennsylvanian age. The largest production comes from sands in the Glenn formation. The region is noted for its high gas production. Some of the well-known pools of the area are Healdon, Hewitt, Sayre, Duncan, Burkburnett, and Electra.

The *Bend Arch* region lies in north-central Texas, where pools are aligned close to the highest portion of an arch developed in the earliest Pennsylvanian strata. The chief producing beds are the Strawn sands and the Caddo and Breckenridge lime. Some of the well-known pools are Breckenridge, Ranger, and Desdemona.

The *West Texas-New Mexico* basin lies in southeast New Mexico and adjacent Texas. The pools are located around the margins of a broad, shallow basin that contains thick Permian formations and evaporites. Many of the most prolific pools lie above or around the margin of a buried platform formed during Permian time along the west side of the Permian basin. Faulting produced structural traps, and anticlines and changes in lithology constitute others. The producing horizons are Permian sandy limestones and dolomites and a few intercalated sands, the important formations being the Big, Brown, and White Limes. Important production is also found beneath the Permian. Some of the pools are prolific with yields up to 33,000 bbl per acre. Some of the bonanza pools are the Yates, Hendrick, Big Lake, and Hobbs. The great Scurry field is in reef rock.

The *Balcones-Coastal Plain-Ouachita uplift* regions include many oil accumulations in the Cretaceous Woodbine sand, Comanchean Edwards lime, and Eocene beds. The Balcones-Mexia fault zones, indicating subsurface structural trends, contain many fault-line pools,

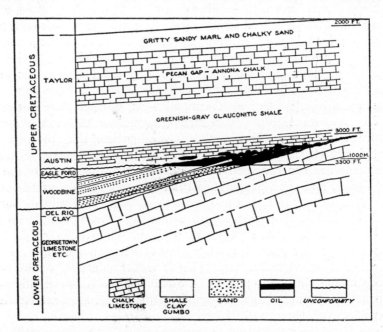

Fig. 16–22. Generalized section across East Texas field showing oil accumulation beneath an unconformity and overlap of the Austin on Woodbine. (*Lahee, 16th Internat. Geol. Cong.*)

of which the Powell, Luling, Mexia, and Salt Flat are prolific. The Sabine uplift area includes the sensational East Texas field, the largest in the world, with its 55 miles of length, 25,000 wells, and estimated yield up to 5 billion bbl oil (Fig. 16–22). Accumulation here is in shore-line sands. Near by is the Van pool, a domed structure probably overlying a buried salt plug. Other important pools are the prolific Smackover and Eldorado of Arkansas, the great Monroe and Richland gas fields, Magnolia and the Homer, Caddo, Bull-Bayou, Zwolle, Haynesville, and others in structural and stratigraphic traps.

Gulf Coast. The Gulf Coast district is a curving belt parallel to the Gulf of Mexico in Texas (246 oil fields) and Louisiana and particularly distinguished because it is the salt-dome region. The region is covered by Tertiary and Quaternary strata that dip and thicken toward the Gulf. Many Miocene and Oligocene and a few Eocene sands are petroliferous. The oil occurs (1) in flank sands that pinch against the salt, (2) in porous calcareous caprocks, and (3) in sands arched above the salt domes. Production is both shallow and very deep. Enormous yields of oil come from many domes and sands. At Spindletop the Lucas gusher yielded 75,000 bbl of oil per day from caprock at 1,139 feet, and lateral sands to depths of 5,100 feet yielded prolific flows amounting to over 300,000 bbl per acre. Other notable domes are Humble, Sour Lake, and Goose Creek. Other oil and gas fields in the Gulf Coast region occur in anticlines and structural traps in Tertiary beds.

Rocky Mountain. The Rocky Mountain oil province embraces parts of Montana, Wyoming, Colorado, and New Mexico and extends northward into Turner Valley, Alberta. Some 100 disconnected pools, all small except for Salt Creek, Wyo., occur throughout the Powder River, Bighorn, Uinta, Julesburg, and San Juan basins. The major production comes from Cretaceous formations, but the underlying Mesozoic Morrison, Sundance, and Embar also yield some oil. A little heavy asphaltic oil is won from Paleozoic formations. The Mesozoic oils are light, high-grade, paraffin-base oils, high in lubricants and gasoline.

Almost all the pools occur on anticlines, but many anticlines include oil sands barren of oil. Powers stated (1931) that 268 anticlines had been drilled and 39 oil fields and 35 gas fields had been found. At Florence, Colo., light oil occurs in fractures in shale in a broad syncline.

The Salt Creek field in Wyoming, well known because of Teapot Dome, is the largest field, covering 21,000 acres. It yields oil from eight sands in the Cretaceous and Jurassic and from the Madison limestone (Mississippian). The Grass Creek, Elk Basin, Oregon Basin, Big Muddy, Lost Soldier, Lance Creek, La Barge, Big Muddy, Rock Creek, and other pools contribute to Wyoming production. The Bighorn basin contains a large reserve of heavy, low-grade, asphaltic oil. In Montana are the Cut Bank, Kevin-Sunburst, and Cat Creek pools. In Colorado are the important Rangley field and the Isles, Moffat, Fort Collins, and the old Florence pools. A carbon dioxide gas well with some 45° gravity oil was brought in at 5,113 feet in northern Colorado. The annual production of the province is about 50 million barrels.

California Province. California, the second ranking oil state, yields annually some 250 million bbl of oil from 60 oil and gas pools in thick Tertiary strata in (1) the San Joaquin Valley, (2) the Los Angeles basin and (3) the Ventura and Santa Maria area. The California oil, except for Kettleman Hills, is mainly heavy asphaltic oil ranging from 14° to 27° gravity, much of which is used directly as fuel oil or is treated to remove some lighter oils. California pools are noted for their prolific production, the great thickness of oil "zones," and the depth of oil within Tertiary beds. Some of the sensational fields are Wilmington, Coalinga, Belridge, Kettleman Hills, Long Beach, Huntington Beach, Santa Fe Springs, Ventura Avenue, Lost Hills, Midway-Sunset, Kern River, Newport, Salt Creek, Paloma, Lompoc, and Santa Maria. The oil occurs in sands associated with organic shales rich in diatoms — the supposed source of the oil.

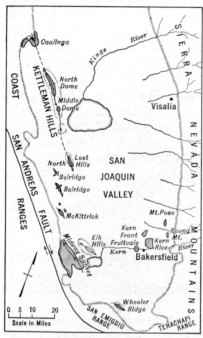

FIG. 16–23. San Joaquin Valley oil fields, California. (*McCollough, Probs. of Petrol. Geol.*)

The *San Joaquin Valley* is a large synclinal basin (Fig. 16–23) composed of thick Tertiary sediments. The oil fields occur in the southern part, mostly in Miocene and Pliocene sands in individual oil "zones" up to 1,500 feet thick. The pools are located mainly in anticlines modified by faulting, but unconformities and overlaps also determine oil loci. The unusual Kettleman Hills discovery well yielded 3,900 bbl daily of crude naphtha (60° gravity) from 6,900 feet. This great pool may prove to be one of the largest in the world, having surpassed the prolific Midway-Sunset field.

The *Los Angeles Basin* is noted for remarkably thick oil zones and large pools, particularly Signal Hill. The basin contains Pliocene and Miocene sediments, and the pools are located around its margin in anticlines and in a monoclinal structure on the downthrown side of a major fault zone. Santa Fe is almost a perfect dome, and 1,500 acres have averaged 220,000 bbl per acre from seven zones in a vertical range of 4,600 feet. At Long Beach (Fig. 16–24) the vertical range of producing Pliocene sands is 3,000 feet, and there are three zones in Miocene sands below this (Fig. 16–25).

The *Ventura district* pools are controlled by closely folded anticlines modified by faults, in Pliocene and Miocene strata, some of which extend out under the ocean. The outstanding Ventura Avenue

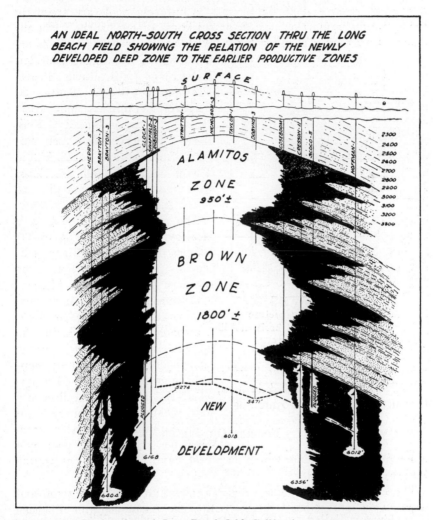

FIG. 16–24. Section through Long Beach field, California. (*Jensen and Robertson, A.A.P.G.*)

pool yields oil from 12 zones throughout 8,000 feet of Lower Pliocene strata (Fig. 16–10). Miocene strata yield prolific flows of oil, and both the Oligocene and Eocene yield oil.

The *Santa Maria* district, which contains two main anticlinal structures modified by faulting, produces from Miocene sands and fractured shale. Some of the pools are partly under the ocean, and wells are drilled from offshore.

The outstanding features of California oil geology are the sharp anticlinal structures, the great thickness of petroliferous sands, the prolific pools, the low-gravity oil, and the great depths of oil occur-

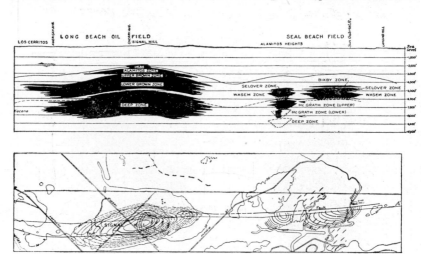

Fig. 16–25. Section and structure contours of Long Beach and Seal Beach oil fields, California. Lowest horizontal line is at 10,000-foot depth. (*After Gale, 16th Internat. Geol. Cong.*)

rence. Many of the deepest wells of the world are in this province. It has also had some noted gushers; the Lakeview gusher uncontrolled flowed 6 million bbl in 5 months.

Foreign Oil Fields

Canada. Canada's oil production comes from three areas: 1, western provinces; 2, Sarnia, Ontario, a northern extension of the Lima-Indiana field; and 3, an extension of the Appalachian field, west of Niagara Falls.

In Turner Valley, noted for its former "white oil" or naphtha (72°), drilling since 1936 has disclosed additional pools. They are located in a drag fold cut off by a fault, and the oil is found in two porous zones down to depths of more than 10,000 feet, some 6,500 feet below the highest well (Fig. 16–26). There is a proven closure of 4,400 feet, of which the upper 1,500 feet is gas cap. The oil-bearing limestone is closely folded and faulted, and the structure is obscured

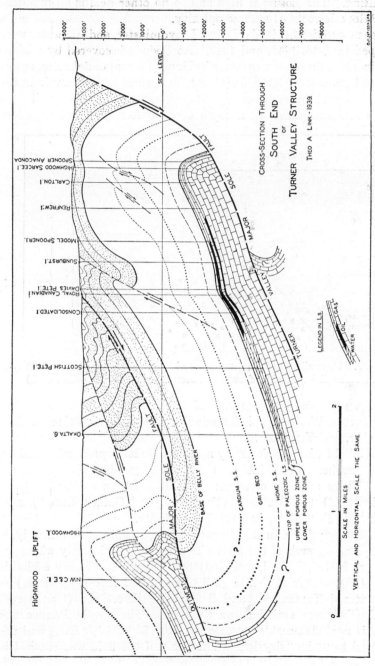

FIG. 16–26. Turner Valley oil field, Alberta, Canada. (*McKenzie, A.A.P.G.*)

by faulting. The closure is high and in no other field is oil produced from Paleozoic rocks of such complexity.

Three new major fields, the Leduc, Lloydminster, and Redwater were developed in 1946, 1947, and 1949. At Leduc, discovered by seismograph in 1947, there are two zones in Upper Devonian dolomite, one at 5,000 and another at 5,150–5,400 feet, in a coral reef reservoir in the form of a broad structural nose. Lloydminster yields heavy oil from Lower Cretaceous sands at a depth of 2000 feet, and Redwater is a

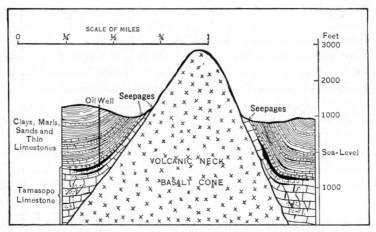

Fig. 16–27. Generalized section through a volcanic neck in oil fields of Vera Cruz and Tamaulipas, Mexico, showing oil occurrences. (*Clapp, Econ. Geol.*)

Devonian reef pool with reserves of 400 million barrels. Other important newer Western pools include Woodbend, South Princess, Conrad, Taber, and Vermilion.

Mexico. Mexico was formerly noted for its large production and for its great gushers, which include three of the greatest oil wells of the world. There are four districts: (1) the Northern or Panuco (near Tampico), (2) the Southern (Tamasopo or "Golden Lane") near Tuxpan, (3) Poza Rica, and (4) Tehuantepec.

The *northern fields* include Panuco, Ebano, Topila, and others lying on the plunging arch of the Sierra Tamaulipas particularly where the flanks are folded, faulted, and shattered. The oil, of heavy asphaltic grade (12°), occurs at depths of 1,400 to 2,500 feet (Fig. 16–27) in the shattered Tamaulipas and San Felipe limestones (Cretaceous) and also in cross anticlines on the arch. The gas in Panuco and Topila is noninflammable, containing from 85 to 97% CO_2, and has been used to make "dry ice." The peak of the field was reached in **1923.**

The *southern fields* form an arc convex westward, some 1,000 meters wide and 82 kilometers long, and include Cerro Azul, Dos Bocas, Alamo, Tepetate, Amatlan, and twelve other fields. These productive fields have produced about 1 billion bbl of low-gravity oil from depths of 1,200 to 2,300 feet from a reef facies of Lower Cretaceous limestone which forms an asymmetrical anticlinal ridge cut by cross faults. Many fossil casts make it a porous container, and fractures and solution enlargements permit long-distance drainage to wells. The ridge is a buried hill covered by east-dipping Tertiary sediments. Three wells alone, the largest in the world, have supplied one-quarter of the entire production; Potrero No. 4 yielded 100 million bbl; Cerro Azul No. 4, 86 million; and Casiano No. 7, 70 million. The pools declined rapidly, as they antedated the days of controlled production. The *Tehuantepec* fields yield small production of lighter oil from sands adjacent to salt domes and salt ridges. The *Furbero* field yields oil from metamorphosed Eocene sediments at the contact with a gabbro intrusive.

Venezuela, Colombia, and Trinidad. Venezuela's production of over 435 million bbl comes largely from a 50-kilometer belt around and under the east shore of Lake Maracaibo. The oils are mostly heavy and asphaltic. The Maracaibo basin is a synclinorium, with 25,000 feet of petroliferous beds from Cretaceous to Pliocene. The pools occur in sands that lens out up dip to the eastward. Tar plugs and unconformities also constitute seals. Most of the production is from Tertiary beds, particularly beneath an Oligocene-Miocene unconformity, and in Pliocene strata. Some gently dipping oil sands have been tested 1,500 feet vertically down dip without encountering edge water. Some of the important Maracaibo pools are Lagunillas, Tia Juana, LaRosa, Mene Grande, Ambrosio, and Benitez. Prolific new pools are being developed in Cretaceous limestones west of Lake Maracaibo.

In the coastal plain to the north of Lake Maracaibo are several small pools of high-gravity oil in Tertiary rocks. A southward extension of the Maracaibo basin into Colombia has yielded the Tarra and Rio de Oro pools in anticlines in Eocene rocks. The nearby El Barco field in Colombia yields a good-quality oil from Tertiary sands.

In *eastern Venezuela*, north of the Orinoco River, several commercial pools have been discovered, such as Anaco, Tabasca, Templador, Salto, El Roble, Tigre, and Oficina. They offer high expectations for the future of this region.

Trinidad contains the famous Pitch Lake, an important source of asphalt. Several pools between Palo Seco and Brighton now yield about 20 million bbl of heavy oil from anticlines and synclines in Miocene sediments.

Colombia production comes mainly from the upper Magdalena Valley, where the Infantas and La Cria pools yield about 25 million bbl annually. Light oil occurs in Eocene beds on domes and anticlines and is piped to the coast. Petroliferous rocks also occur in the lower Magdalena Valley.

Other South American Countries. Argentina produces about 22 million, Peru 13 million, and Ecuador about 2 million bbl of oil annually.

Peru's main field is the Tumbez-Paita, on the north coast. It is about 125 miles long and 30 miles wide, and has produced about 300 million bbl of 38° gravity oil. Under unit ownership it has yielded steady, efficient production. The main area is the Negritos, where a large faulted anticline has three high points. Most of the oil comes from Eocene sands at depths of as much as 3,800 feet. The accumulation is mainly anticlinal. The wells have initial production up to 4,000 bbl, average 300 bbl, and have a long life. Oil discoveries resulted from seeps and asphalt, which were known to the Incas.

Argentina has four main producing fields: (1) Comodoro Rivadavia on the south coast and (2) Salta, (3) Neuquen, and (4) Mendoza, in the eastern Andes front; the first yields about 80 percent of Argentina's oil.

The Comodoro Rivadavia field yields oil from both continental and marine sands of Upper Cretaceous age. In fact, most of the Argentine oil comes from continental and estuarine sands. The oil is not in domes but, according to Fossa-Mancini, is in a " faulted tabular structure " and is obtained from depths of 1,500 to 5,500 feet.

The eastern Andes fields yield oil from faulted anticlines and monoclines in Permian and Mesozoic beds. At Cacheuta (Mendoza) no marine beds are known, and the Salta oil occurs in continental Triassic and Permian (partly glacial) beds. Deeper drilling below known unconformities holds promise.

Small pools yield oil in Bolivia, Chile, and Brazil.

Russia. Russia after 1883 was long the second largest oil producer and has yielded 12 percent of the world's oil. Present production is less than half that of Venezuela. Oil and gas have been known there for 2,500 years and formerly were extracted from pits; the first well was drilled in 1871.

Russia has five main producing fields, the first four in the Caucasian belt, (1) the older Baku district on the Apsheron Peninsula, East Caucasus, (2) the Grozny district, (3) the Maikop-Kuban district, north of the Caucasus near the Black Sea, (4) Dagestan, and (5) the Ural-Volga ("Second Baku ").

Baku has abundant surface indications of oil from productive middle and upper Pliocene strata, which has accumulated in a ring of anticlines around Baku. The productive series (1,015 to 1,463 meters) is divided into an upper division with 34 oil sands, a middle division with water-bearing strata, and a lower division. The upper is largely exhausted and the lower is exploited. Present drilling depth averages 3,200 feet, and the Tertiary strata are known to be 7,000 feet thick. The oils are heavy, black, and asphaltic, averaging about 17° API gravity, but deeper oils rise to 34°. Initial well production has been as much as 100,000 bbl per day. The Caucasus produces about 60 percent of Russian oil.

The *Grozny* field lies north of the Caucasus, 450 kilometers from Baku. The Old Grozny field is a shallow, steep, asymmetric anticline yielding oil from middle Miocene sands (250 to 1,000 meters) ; some oil comes from beneath a thrust fault. The New Grozny field is a simple anticline with 22 producing sands in 1,740 feet. The Grozny oil sands are thinner and much less productive than at Baku, and the wells are shorter lived.

The *Maikop* field production is from anticlinal upper Oligocene and lower Miocene gravels and sands in old erosional valleys. The upper beds are shallow and yield a heavy oil (17°), the lower beds yield medium oils (33°).

The important *Ural-Volga* production is largely from reef limestones of Devonian to Permian age.

Middle East (Iran, Iraq, Arabia, Kuwait). The astonishing oil developments in the Middle East have made this region the greatest potential oil area of the world. Its proven reserves already exceed those of the United States, and its potential reserves are likewise greater. Its mere 225 producing wells average 5,000 bbl a day and yield about ⅕ of the United States production in comparison with 428,000 wells for the United States that average only 13 bbl a day. *Iran* since 1910 has risen to third place in world production, and *Saudi Arabia* is rapidly approaching Iran. The geology of adjacent Iraq and Iran is similar.

Copious oil seepages and bitumens have been known in *Persia* since earliest civilization, but modern production began only in 1910. The main fields of *Iran-Iraq* are: (1) Haft Kel, (2) nearby Masjid-i-Sulaiman (Fig. 16–28), (3) Agha Jari, (4) Gach Saran, (5) Naft Safid, and (6) Lali, all northeast of the head of the Persian Gulf. Naft-i-Shah (7) is 80 miles north of Bagdad. Just across the border in Iraq is (8) Naft Khaneh, and beyond are (9) Kirkuk and (10) Qaiyarah. South Iran oil is piped to the Persian Gulf and Iraq-Iran oil to the Mediterranean; a trans-Arabian pipe line carries Arabian oil to Haifa.

The oil (35°) in the southern fields occurs, according to Lees, in the Asmari limestone (1,000 feet thick) of Oligocene-Miocene age,

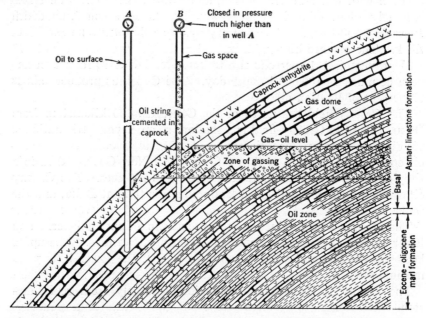

Fig. 16–28. Diagrammatic section through oil reservoir, Masjid-i-Sulaiman, Iran. (*H. S. Gibson, Jour. Inst. Pet., 1948.*)

has migrated up from underlying Eocene or Mesozoic beds, and has been retained by overlying folded Lower Fars anhydrite and salt. This salt formation is not to be confused with the 130 known salt plugs of Cambrian age that pierce the overlying rocks, one even forming a hill 5,000 feet high. The oil accumulation is anticlinal in crackled and fissured Asmari lime. At Naft Khaneh and Naft-i-Shah the accumulation is similar except that the reservoir rock is the equivalent Kalhur limestone, and at Kirkuk it is of upper Eocene-Oligocene age, and to a less extent, lower Miocene. Here, a thrust fault in the folded Foss salt obscures the underlying regular structure.

The Iran fields gave rise to spouting wells at depths of 1,000 to 9,000 feet, one of which flowed at the rate of 12,500 bbl per day. A well developed in 1947 at Lali yielded 28,000 bbl a day. One well at

Kirkuk yielded 35,000 to 95,000 bbl of 34° gravity oil per day from 1,521 feet.

In *Saudi Arabia*, oil was discovered at (1) Dammam, about 90 miles from the Gulf of Persia in 1938, followed by (2) Abqaiq, (3) Qatif, (4) Abu Hydira, and (5) Buqqa. In 1948, 40 wells in the first two fields yielded an average of 7,340 bbl of oil per day. The structure at Dammam is a salt dome; the others are anticlines. The oil horizons are the Cretaceous Bahrein zone and two zones in the Arab oölitic limestone (Jurassic). The Abqaiq structure is 33 miles long and is one of the world's largest fields.

At *Bahrein Island, Arabia*, a huge anticline yields oil from a porous member of a shale-sandstone-limestone series of Cretaceous age at depths of 1,900 and 2,500 feet.

The *Kuwait* region, discovered in 1938, is the largest of the new fields. It yields 30° to 35° API oil from three sands (anticlinal) of middle Cretaceous age at depths of 3,670 to 4,750 feet.

Netherlands Indies. Netherlands Indies obtains about 60 million bbl annual production from fields in Sumatra, Borneo, Java, and Ceram. Except for the unimportant field of Ceram, all the oil is in Tertiary sediments, which are as much as 10,000 feet thick, mainly Miocene and Pliocene. It is entirely confined to anticlines and is at shallow depth. Except for some heavy oil in Borneo, all the oil is light paraffin high in gasoline.

Rumania. The Rumanian fields have yielded oil since 1857. There are three districts, the sub-Carpathian, the Transylvanian basin (gas only), and the Bacau district. The sub-Carpathian area comprises a 20-mile belt in the foothills of the Carpathians and the plains to the south, and includes the Dambovita and Prahova districts. The oil occurs in Miocene and lower Pliocene sands (Meotian, 330 to 580 meters thick) in anticlines, salt domes, and salt anticlines, at depths down to 2,300 meters. Considerable gas is present, and the oil is high in gasoline.

Other European Areas. The other countries of Europe are notably lacking in oil. Small production is obtained in Poland, Germany, Hungary, Yugoslavia, Albania, Austria, and Morocco, and negligible amounts are produced in France, Czechoslovakia, and Italy.

In *Poland* several small declining fields along the Carpathian front yield about 3.8 million bbl annually from broken domes and folds of Cretaceous, Eocene, and Oligocene sands that lie within three complicated nappes. The Boryslav field yields most of the oil.

Germany has three areas that yield some oil, chiefly from the Permian Zechstein (Nienhagen and Wietze) fields, and a little occurs in the Rhine graben and in Bavaria. The annual output of 4.5 million bbl is supplemented by synthetic oil derived from coal. At Nienhagen, the only important field, a heavy oil is obtained from Rhaetic to Lower Cretaceous sands in complex structures associated with a salt intrusion. At Wietze, where much of the oil is mined from shafts, conditions are similar. The geologic conditions of Germany are not favorable for large oil fields.

Hungary since 1937 has obtained a little oil from Budafa-Puszta, near Yugoslavia, and Bukkszek, east of Budapest. The former is on an anticline discovered geophysically and yields a medium-grade oil from Pliocene sands to depths of 1,179 meters. *Albania* yields less than 1 million bbl annually of heavy, black, sulphurous oil from sand lenses in folded Miocene strata. In *Austria,* the Zistersdorf field yields about 5 to 6 million bbl from Miocene sands. *Czechoslovakia* bails about 100,000 bbl of oil annually from two domes in upper Miocene beds at Hodonin and Gbely. *France* has one small oil field at Pechelbronn where about one-half million bbl of light oil is produced annually, nearly half of which is from mines and the remainder from 620 wells in Oligocene strata. *Italy* obtains about 100,000 bbl of light oil from small scattered pools in Tertiary lenses in the Apennines.

Other Areas. Oil is fairly widely distributed throughout Asia, since, in addition to the countries previously described, important production is obtained in Burma, Sarawak, Brunei, Sakhalin, India, and Japan.

Burma and *Assam* together produce some 8.5 million bbl annually. The similar petroleum provinces of the two states lie in two adjacent parallel geosynclines of Tertiary sediments separated by the West ranges. The sediments have been folded into long, narrow, asymmetric folds that contain the oil. The Burma fields follow a single curving structure along the Irrawaddy River, where there are 2,700 wells. The oil beds are Eocene to middle Miocene. The oils range from 21° to 62°. In *Punjab, India,* north of the Salt Range about 2 million bbl is produced from domes in Miocene sands and Eocene limestones. *Sarawak* and *Brunei,* in *Borneo,* yield about 3 million bbl, with much light oil high in gasoline, from anticlines in Miocene beds. *Sakhalin Island* yields about 6 million bbl of heavy oil from anticlinal structures in Pliocene sands, and *Japan* realizes about 2½ million bbl of heavy oil from anticlines in Miocene sands, namely from Echigo and Akita on Honshu, near the Japan Sea. A little light oil is found in *Shensi basin, China.*

The *Egyptian* fields lie 150 miles south of Suez, west of the Gulf, and yield about 11 million bbl from Cretaceous and Miocene beds that have been tilted and squeezed during faulting.

Associated Products

Natural Gas. The origin and occurrence of natural gas have already been described in connection with petroleum. It is a universal associate of petroleum, but it may be limited, abundant, or constitute the only valuable hydrocarbon present. Methane is the most important constituent of natural gas; ethane increases its heating power, and other hydrocarbons may be present, such as butane, pentane, and hexane. Carbon dioxide may also be present; rarely it may be the main constituent. Helium is present in some Texas and Utah gas.

Some 4,000 billion cubic feet are produced annually in the United States. Its chief use is for domestic and commercial fuel, for making carbon black, and for fuel in drilling and refining. It is also liquefied to a limited extent for domestic use. Much is also used for repressuring wells. Pipe lines deliver it thousands of miles to consumer points. About half of the gas produced is processed to yield *natural gasoline*, with an average recovery of about 1 gallon per 1,000 cubic feet of gas. This is used for high-test gas and for blending. The *carbon black* made from burning natural gas is used in tires, printers' ink, and paints.

Natural Asphalt and Bitumens. Asphaltic-base oils at seepages yield natural bitumens, such as asphalt, maltha, rock asphalts, pyrobitumens, and related compounds, and paraffin oils yield natural wax or ozokerite, desired for fine candles. These natural bitumens are also used for road materials, waterproof paints, and many other purposes.

Oil Shale. Oil shale is a shale containing bituminous matter that yields petroleum by destructive distillation. A ton of rich oil shale may yield from 18 to 40 gallons of crude oil. It cannot compete with natural petroleum in cost but may constitute a large future reserve of oil. Large deposits occur in Colorado, Utah, Wyoming, Scotland, and elsewhere.

Selected References on Petroleum

The Science of Petroleum. A. E. Dunstan. Oxford Univ. Press, 1937. *Contains résumés of important oil fields.*

Geology of Petroleum. W. H. Emmons. John Wiley & Sons, New York, 1931. *A general elementary textbook.*

Problems of Petroleum Geology. Edited by W. E. Wrather and F. H. Lahee. A. A. P. G., Tulsa, Okla., 1934. *An excellent monograph covering all phases of theoretical petroleum geology.*

Structure of Typical American Oil Fields. 2 vols. A. A. P. G., Tulsa, Okla., 1929. *A treatise on oil-field structures.* Vol. 3. Pp. 516, 1948. *Unusual oil-field structures.*

Gulf Coast Oil Fields.   A. A. P. G., Tulsa, Okla., 1936.   *Particularly salt-dome oil occurrences.*

Geology of Natural Gas.   Edited by H. A. LAY.   A. A. P. G., Tulsa, Okla., 1929. *Occurrences of natural gas.*

Stratigraphic Type Oil Fields.   Edited by A. I. LEVORSEN.   Pp. 902.   A. A. P. G., Tulsa, Okla., 1941.   *A comprehensive authoritative monograph giving examples and descriptions of stratigraphic trap oil-fields.*

Source Beds of Petroleum.   P. D. TRASK and W. PARNODE.   Pp. 566.   A. A. P. G., Tulsa, Okla., 1942.   *Summation of investigations of carbon content of sediments as possible oil source beds.*

Finding and Producing Oil.   Amer. Pet. Inst., Dallas, Texas, 1939.   *Popular presentation of field procedure.*

Economics of Petroleum Industry.   J. E. POGUE.   Chase Nat. Bank, New York, March, 1939.   *A study of production trends.*

The Birth of the Oil Industry.   P. H. GIDDENS.   Pp. 216.   The Macmillan Co., New York, 1938.   *A story of the history of the development of oil.*

Oil and Gas Field Development in the United States.   National Oil Scouts and Landsmen's Association Yearbooks, Annual volumes, Austin, Texas.   *An annual review of geological and geophysical prospecting, exploration, development; new reserves.*

American Oil Operations Abroad.   L. M. FANNING.   Pp. 270.   McGraw-Hill Book Co., New York, 1947.   *Problems, development, and progress in the international development of the oil industry.*

Our Oil Resources.   Edited by L. M. FANNING.   Pp. 331.   McGraw-Hill Book Co., New York, 1945.   *Eighteen articles by authoritative authors dealing with phases of the development and resources of oil and gas.*

Done in Oil.   D. D. LEVEN.   Pp. 1084.   Ranger Press, New York, 1941.   *History, supply, U. S. and foreign oil fields, finding and producing oil, economics of oil.*

Inferences About the Origin of Oil from the Composition of Organic Constituents of Sediments.   P. D. TRASK.   U. S. Geol. Surv. Prof. Paper 186–H, 1937.   *A discussion of source materials.*

Occurrence of Petroleum in North America.   SIDNEY POWERS.   A. I. M. E. Tech. Pub. 377, 1931.   *An excellent résumé of American occurrences.*

Oklahoma and Texas.   XVI Int. Geol. Cong. Guidebook 6.   Washington, 1932. *Concise descriptions of these oil regions.*

Geologic Formations and Economic Development of the Oil and Gas Fields of California.   OLAF P. JENKINS.   Pp. 773.   Calif. Dept. of Nat. Res. Bull. 118, San Francisco, 1943.   *Comprehensive treatment of the various oil fields of California.*

Petroleum in the United States and its Possessions.   ARNOLD and KEMNITZER. Harpers, New York, 1931.   *American petroleum fields, historical and descriptive; lengthy.*

Hearing before a Special Committee Investigating Petroleum Resources; U. S. Senate.   I, Wartime Petroleum Policy.   Pp. 280;   II, American Petroleum Interests in Foreign Countries.   Pp. 462.   Washington, D. C., 1946.   *Many data regarding reserves, production, foreign occurrences, development, history, and technology.*

Review of the Mineral Industry of India and Burma.   India Geol. Surv. Annual. *Details of Burmese oil occurrences.*

Geology of Venezuela and Trinidad. 2nd Edit. R. A. Liddle. Pp. 890. Paleont. Research Inst., Ithaca, N. Y., 1946. *Detailed descriptions of geology with special reference to oil regions.*

Mexican Oil Fields. Three articles by A. G. Rojas, E. J. Guzmán, and G. P. Salas. A. A. P. G. 33:No. 8, 1336–1409, 1949. *Best information on Mexican oil geology.*

Oil Fields of North America. W. A. VerWiebe. Edwards Bros., Ann Arbor, Mich., 1949. *Detailed descriptions.*

Petroleum Development and Technology. A. I. M. E., New York Annual volumes. *Annual reviews of occurrences.*

American Association of Petroleum Geologists, Tulsa, Okla. Monthly bulletins.

Leduc Oil Field, Alberta. T. A. Link. Bull. Geol. Soc. Amer. 60:381–402, 1949. *Excellent description of unusual field.*

Migration and Accumulation of Petroleum and Natural Gas. F. M. VanTuyl, B. H. Parker and W. W. Skeeters. Quar. Colo. School Mines 40:1–111, 1945. *Survey of the status and problems of migration and accumulation.*

CHAPTER 17

CERAMIC MATERIALS

The ceramic industry utilizes common minerals and rocks and produces lowly brick or tile, porcelains for utility, or beautiful chinaware. Its chief raw material is clay, but other substances are also necessary. In the making of brick, terra cotta, tile, or stoneware, clay alone is needed, but for whiteware or porcelain, feldspar and quartz are necessary. For certain porcelains bauxite is essential, and some even require highly refractory substances, such as andalusite. Bentonite, fuller's earth, and other aluminous substances, and pyrophyllite, zircon, and fluorspar are also drawn upon, but the backbone of the industry is clay.

Clay

Clay is one of the most widespread and earliest mineral substances utilized by man. It carries the records of ancient races inscribed upon tablets, in brick buildings, in monuments, and in pottery. Its products portray the history of man, and by its beautiful wares we trace the development of the delicate artistry of the Chinese, the utility of the Romans, or the humor of the Incas. A wealth of artistic wares culminated in the eighteenth century, but today utility holds sway in the multitudinous utilizations of the varied clay products.

The term " clay " is applied to earthy substances consisting chiefly of hydrous aluminum silicates with colloidal material and specks of rock fragments, which generally become plastic when wet and stonelike under fire. These properties give clays their usefulness, since they can be molded into almost any form, which they retain after firing. Widespread accessibility, ease of extraction, and adaptability to so many uses has resulted in the products of clay entering the wide ramifications of modern industrial civilization. Clay has many other uses than in ceramics, particularly in building and manufacturing, but, for purposes of unity of treatment, all types of clay are considered together in this chapter. Its use as a building material is referred to in Chapter **18**.

Composition and Properties. Clay is not a mineral but an aggregate of minerals and colloidal substances. The constituents are so fine that, until the use of the X-ray for mineral determination, its exact composition was unknown. Some clay minerals can be observed in detail only by the electron microscope at magnifications greater than 5,000 times. Residual clay is often called *kaolin*, but this is now recognized as just one variety of clay. It was formerly thought that kaolin was composed of the mineral kaolinite. It is now known that, although kaolin contains considerable kaolinite, other of the clay minerals are also constituents.

The *clay minerals* are flakelike, lathlike, fiberlike, or hollow-tube-shaped, and they are recognized by the microscope, X-ray, and thermal analyses curves. They have replaceable bases; the one formed is determined by the mode of origin and may change in response to changes in environment. The clay minerals are:

GROUP	COMPOSITION	ORIGIN	OCCURRENCE
A. Kaolinite			
1. Kaolinite	$Al_2Si_2O_5(OH)_4$	W	China clays,
		H	underclays,
			soils, wallrocks
2. Dickite	$Al_2Si_2O_5(OH)_4$	H	Wallrocks. U
3. Nacrite	$Al_2Si_2O_5(OH)_4$	H	Wallrocks. U
4. Anauxite	$Al_2Si_2O_5(OH)_4$	W	Soils. U
5. Halloysite	$Al_2Si_2O_5(OH)_4$	W	Soils
6. Endellite	$Al_2Si_2O_5(OH)_4 \cdot 2H_2O$	W	Soils
B. Montmorillonite			
1. Montmorillonite	$Mg_2Al_{10}Si_{24}O_{60}(OH)_{12}[Na_2,Ca]$	W	Soils, bentonite,
			fuller's earth,
		H	wallrocks
2. Nontronite	$Fe'''Si_{22}Al_2O_{60}(OH)_{12}[Na_2,Ca]$	H	Veins
3. Saponite	$Mg_{18}Si_{22}Al_2O_{60}(OH)_{12}[Na_2]$	H	Veins
4. Beidellite	$Al_{13}Si_{19}Al_5O_{60}(OH)_{12}[Na_2]$	H	Vein gouge
5. Hectorite	$Li_2Mg_{16}Si_{24}O_{60}(OH)_{12}[Na_2]$	W	Clays
C. Hydrous micas (illite)	$K_y(Al \cdot Fe \cdot Mg)(Si_{2-y} \cdot Al_y)O_5(OH)$	W	Soils, marine clays, underclays
D. Miscellaneous			
1. Attapulgite	$Mg_5Si_8O_{20}(OH)_2 \cdot 4H_2O$	W	Fuller's earth
2. Sepiolite-like	$Mg_6Si_8O_{20}(OH)_4 \cdot nH_2O$		
3. Allophane	$Al + SiO_2 + H_2O$	W	Clays, soils

W = weathering; H = hydrothermal; U = uncommon; square brackets surround exchangeable bases.

The kaolin group minerals differ little in composition but have different crystal lattices. Allophane had been regarded as a solid solution of alumina, silica, and water until the electron microscope showed it to be a distinct mineral. It is commonly associated with halloysite.

Clays also contain other substances such as rock fragments, hydrous oxides, and colloidal materials, and their nature often determines the value and use of a particular clay. Obviously a different clay is required to make delicate porcelain from that usable for sewer pipes. The composition of the purest kaolins, from which particles of quartz have been washed out, resembles that of kaolinite. *Quartz* decreases plasticity and shrinkage and helps make clay refractory; if coarse, it must be removed. *Silica* in colloidal form increases the plasticity. *Alumina* makes clay refractory. *Iron oxide,* as well as feldspar, lowers the temperature of fusion and acts as a flux; it is also a strong coloring agent — a little of it makes burned clay buff colored, much of it (over 5 percent) makes a red product; white-burning clays should have 1 percent or less. *Lime, magnesia,* and *alkalies* act as a flux; lime is a bleacher, but it also forms lumps of undesirable quicklime. *Titanium* acts as a flux at high temperatures. *Carbon* and *water* are driven off during burning.

Physical properties of importance are: (1) plasticity, which permits raw clay to be shaped before burning — " fat " or highly plastic clays may be mixed with " lean " clays to produce desired plasticity; (2) transverse strength; (3) shrinkage, both during drying and during burning — if high the clay is useless for firing; (4) fusibility, which starts at 1,000° C with low-grade clays and reaches 1,300° to 1,400° C for refractory clays. Fusible clays may be vitrified below 1,300° C and refractory clays not below 1,600° C. The size of clay particles is of the order of fineness of 0.002 mm diameter.

Uses of Clay. The uses of clay and clay products are too numerous to list completely. In domestic life clay is used in pottery, earthenware, china, cooking ware, vases, ornaments, plumbing fixtures, porcelain stoves, tiles, fire kindlers, oilcloths, linoleum, wallpaper, scouring soaps, and polishing bricks. It even finds a place as an adulterant in foods. In buildings it is used for building bricks, vitrified and enameled brick, building and conduit tile, tiles for floors, walls, and drains, copings, flues, chimney pots, sewer pipes, and foundation blocks. In the electrical industry it is used for conduits, cleats, sockets, insulators, and switches. In refractory ware it is used for fire brick, furnace linings, chemical stonewear, crucibles, retorts, glass-melting equipment, and saggers. Other important uses are for fulling cloth, foundry sands, terra cotta, emery wheels, rubber crucibles, water conduits, paving brick, septic tanks, railroad ballast, portland cement, filtering oils, paper making, and innumerable minor purposes.

Most of the high-grade clays go into paper making to give body to the paper or to furnish a good surface or coating for printing and illustrations. Next in importance is their use in whiteware.

Origin and Occurrence. The origin of all clay deposits is essentially the same, namely chemical disintegration of aluminous rocks. A geological classification by origin, according to Ries, is:

> Residual clays — formed by weathering *in situ:*
>> Kaolins.
>> Red-burning clays.
>
> Colloidal clays — landslide masses.
> Transported clays:
>> Sedimentary:
>>> Marine.
>>> Lacustrine.
>>> Floodplain.
>>> Estuarine.
>>> Delta.
>>
>> Glacial.
>> Wind-formed — some loess.

The origin and occurrence of the sedimentary and residual types have already been discussed in Chapter 5·7. Glacial clays are those resulting from glacial erosion and deposition from meltwater. They are common in glaciated regions. Wind-formed, loess clays are widespread in the Great Plains region of the United States. The various modes of origin yield different types of clays, and since usefulness is the dominant feature of clays, they are generally classified according to types or uses.

Types of Clay. Clays are classified into types as follows:

TYPE	CHIEF USE	CHIEF CHARACTERISTIC
Kaolins:	Whiteware, porcelain fillers, paper making	High-grade, fine-grained, white-burning
China clay		
Paper clay		
Ball clay	Whiteware, mixing	White-burning
Fire clay	Refractories	High alumina
Flint clay		
Diaspore clay		
Stoneware clay	Stoneware	Dense burning
Paving and sewer pipe	Paving bricks, sewer pipes	
Brick and tile clay	Brick and tile	Common clays
Bentonite	Iron and steel works, filtering	
Fuller's earth	Filtering	Absorptive qualities

In addition, certain types of clays carry names designating their use, such as *slip clay,* used for glazing, *pottery clay, retort clay, earthenware clay, terra cotta clay, pipe clay, bleaching clay, bonding clay, foundry clay, sagger clay,* and *rubber clay.*

Kaolins or china clays are the purest, whitest, and most expensive clays. They are residual clays of limited occurrence. The higher-grade washed clays are used for fine coated printing papers, in rubber, refractories, and pottery. They are also used for whiteware, for porcelains of all kinds, and for paper filler. Some so-called sedimentary, nonplastic, Cretaceous kaolins of Georgia and South Carolina are used as paper clays and for some grades of whiteware and are sometimes grouped under " whiteware clays."

Ball clays are good quality, sedimentary, plastic, refractory clays of restricted distribution. They are used chiefly in pottery and stoneware, and they are added to supply plasticity and high bonding qualities.

Fire clays include all refractory clays exclusive of kaolins and ball clays. They are mostly of sedimentary origin. Some are nonplastic. Because they are free from deleterious ingredients and endure high temperature, they enter into all kinds of refractory products, of which fire brick, fire-clay mortar, and foundry and steel works materials are the important ones.

Stoneware clays are semi-refractory, dense-burning clays, closely related to fire clays, and used for stoneware and pottery.

Paving-brick and sewer-pipe clays are fair-quality clays yielding strong products of variable colors.

Brick and tile clays are of the common, low-value type that is used everywhere for these products. They burn to creamy or red colors.

Bentonite and fuller's earth are clayey materials often listed under " Clays." They are described in Chapter 20.

Production. About 90 percent of all clay produced is common brick clay, for which world statistics are not kept. The annual production of clays in the United States ranges from 25 to 35 million tons, valued at 50 to 60 million dollars. Considerable kaolin and ball clay used in the United States is imported.

Clays, exclusive of common clays, produced in the United States amount to about 10 million tons valued at 45 million dollars and are divided approximately as shown below.

Kind of Clay	Percentage of Total Tons	Percentage of Total Value	Chief Use	Percentage of Total
Kaolins	13	30	Refractories	63
Ball clay	2.3	5	Paper	8.4
Fire and stoneware	76	47	Drilling muds	5.3
Bentonite	5.7	9	Pottery and stoneware	5.1
Fuller's earth	3	9	Bleaching clay	4.5
			Rubber, linoleum	2.3
			High-grade tile	1.9
			Chemicals	1.1
			Kiln materials	0.6
			Others	7.8

The annual value of the clay products in the United States exceeds 300 million dollars. Common clays, used chiefly for heavy clay products and cement, constitute about two-thirds of the total clays.

Technology. Common clays are mined in bulk and undergo no processing except for removal of stones. Even ball clays are not purified. Kaolins, however, are washed, screened, settled in water, and filtered. Fire and burning clays are generally processed to remove rock particles. The present tendency is to process all good clays by centrifuging, oil flotation, and bleaching to obtain fine grain, to remove deleterious minerals, and to remove minerals that give undesired colors upon firing.

The clay is then mixed with water in a pug mill, de-aired, and molded into forms. After drying it goes to the kiln, where the temperature is raised gradually to the desired point. Most wares are fired just to the incipient fusion point. Before firing, special surfaces may be given the clay. If salt is added to the fire, the exterior of wares fuse to give " salt-glazed " ware. Glazes and enamels are added after firing, by dipping, spraying, or dusting, and the ware is then refired to fuse and set them. A glaze is transparent, showing the under body color or design; an enamel is opaque, hiding the under color.

For pottery, plastic and nonplastic clays are carefully blended, stored for " curing," then skillfully molded for firing. Decorations and limited colors are added, followed by a second firing; the wares are then glazed. Lusterware is made by adding silver or copper to the glaze, firing at low temperature, and bringing out the metallic iridescent film by smoke from rosemary or gorse wood.

Distribution. Common clays are found everywhere, but high-grade clays are more restricted. Their chief distribution is as follows:

Type of Clay	Foremost Localities	U. S. Distribution
High-grade kaolins	England, Czechoslovakia, Germany, France, China	Minor — North Carolina, Delaware
Paper clays	Same as above	Georgia, South Carolina
Ball clays	England, Germany	Kentucky, Tennessee
Fire clays	Germany, England, Belgium, United States	Pennsylvania, Ohio, Kentucky, Missouri, Indiana, Illinois, Maryland, New Jersey, California
Bentonite	United States	Wyoming, South Dakota

Kaolins are not common. Where they occur, ceramic centers of world renown have sprung up, and fine chinaware made from them has international markets. China is the largest producer, the principal center being in Kiangsi Province, but important deposits occur in other provinces. Fine chinaware has been made there since A.D. 220, and one million people have been employed in the industry. The most important deposits, however, are those of Cornwall and Devon, from which kaolin is shipped all over the world for the making of chinaware and paper. The kaolins of France have made that country famous for its fine Limoges and Sèvres porcelains, but supplies are sufficient only for its own needs. They occur chiefly in Brittany and in the Departments of Allier and Drôme. Large German deposits have given rise to the famed Dresden ware and other high-quality wares and supply paper clay in quantity. Czechoslovakia also has extensive deposits from which its own famous chinawares have been made and which, along with Bavaria, Cornwall, and Georgia and North and South Carolina, supply most of the fine paper clays. In the United States North Carolina is the most important producer of residual kaolin, but deposits occur also in Pennsylvania, Virginia, Maryland, Alabama, Connecticut, and Washington. The chief ranking countries in order of their kaolin production are China, United Kingdom, Bavaria, Japan, United States, and Czechoslovakia.

Refractory or *fire clays* are chiefly of sedimentary origin and in general underlie coal seams. Fire clays occur in Pennsylvania, Maryland, Ohio, Kentucky, Missouri, Indiana, Illinois, Georgia, central Europe, Japan, and many other localities. *Ball clay* of the best quality comes from England and Germany, but a considerable amount is supplied by Kentucky and Tennessee. *Slip clay* is mined in the United States only in New York and Pennsylvania, but it is found in the United Kingdom and central Europe. *Pottery* and *stoneware clays* are also chiefly sedimentary clays associated with fire clays. *Brick* and *tile clays* are the common variety found everywhere.

EXAMPLES OF DEPOSITS OF HIGH-GRADE CLAYS

England. The residual china clay deposits of Cornwall and Devon are the most important of the world because of their purity, standard shipping quality, and access to cheap transportation. They are the products of alteration of a porphyritic granite, in part sericitized. The china clay is absent from hilly granite and is restricted to low-lying areas, where it is abruptly delimited from thin swampy soil covers. Its depth is unknown since the economic limit of open-pit operation is about 300 feet. Lilley states that the raw product averages 20 to 25 percent clay substance, and the remaining "sand" or other granite disintegration products are removed by processing. The "sand" also includes such minerals as cordierite, topaz, andalusite, tourmaline, and, locally, considerable fluorspar. The Stannon Marsh clay belt is more than a mile long and half a mile wide.

The depth, extent, and mineral assemblage have led most field investigators to conclude that clay deposits have been formed by hydrothermal alteration of the granite. Such is also Lilley's conclusion. Others maintain that it was normal weathering of hydrothermally altered granite. Since some deposits can be traced into partly altered granite, and weathering is the common mode of origin and hydrothermal action a rare process of formation, the burden of proof, and stronger proof than has yet been presented, must fall on him who maintains a hydrothermal origin.

Czechoslovakia. The extensive kaolins of Czechoslovakia are in great demand throughout Europe for paper clay. Lilley has described two interesting occurrences, the celebrated deposits of Zettlitz and the large areas of Pilsen.

The *Zettlitz* kaolin deposits lie in a graben south of the Erzgebirge (Fig. 17–1), forming a belt about 12 kilometers long and up to 4 kilometers wide. The kaolin lies on the granite sides and bottom of the graben at depths up to 60 meters beneath Tertiary sediments that contain lignites and basalt flows. The deposit is about 14 meters thick, and the upper part is the best; the lower part grades into granite. It is high-quality kaolin and contains 25 to 40 percent clay substance, the remainder being largely quartz. It has resulted from normal weathering, although the proposal has been made that the thermal Carlsbad spring waters have modified it. The best areas of kaolinization correspond to overlying lignite, so acids derived from the lignite are thought to be the important agencies of kaolinization.

In the *Pilsen* area the kaolin has been derived from the weathering of highly feldspathic Carboniferous arkosic sandstones. The product, which is a very high-grade paper kaolin that is much sought through-out Europe and even in America, consists of 25 to 40 percent clay minerals, the remainder being almost entirely quartz. The impurities are low. The beds are about 20 meters thick but attain 70 meters. Above the sandstones are thin remnants of Tertiary clays and sands with more or less lignite. The kaolinization is thought to have been produced during the Tertiary by solutions in percolating surface waters, derived largely from the decomposition of overlying peat.

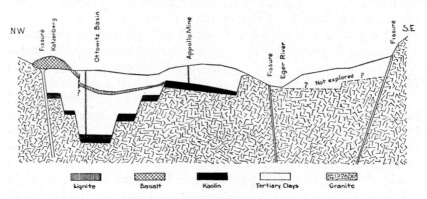

Fig. 17–1. General section through Zettlitz kaolin district, Germany.
(*Lilley, A.I.M.E.*)

Germany. The most important deposits occur in Saxony, north of the Erzgebirge, in the vicinity of the famous ceramic centers of Dres-den, Halle, and Kemmlitz. According to Lilley, in the *Kemmlitz* area the kaolin attains thicknesses of 50 meters in basinlike areas in lower Permian porphyry and is covered by an overburden of glacial drift and alluvium. It is white and contains 30 to 40 percent of kaolin substance. The deposits were formed by normal weathering, but it is thought that former overlying lignite beds contributed acid waters that helped purify the kaolin.

The deposits of *Halle* (Fig. 17–2) are mostly covered by Tertiary sands and clays that now contain or did contain beds of lignite. In places the deposits crop out. The high-quality kaolin has resulted from the alteration chiefly of the older and younger porphyry but also of other rocks. German geologists believe that lignite decomposition has been effective in its formation.

In *Bavaria*, near Arnberg, a group of kaolin deposits, like the Czechoslovakian deposits, was derived from arkosic sandstones.

Here, however, the kaolin is of Triassic age and crops out as a belt 50 to 150 meters wide and 9 kilometers long. The kaolin content ranges from 10 to 25 percent. Permian arkosic sandstones have been less extensively kaolinized. Somewhat similar but relatively unimportant deposits occur in Thuringia. The Bavarian clays are also valuable paper clays.

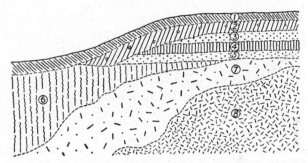

Fig. 17–2. Section of part of Halle kaolin area. 1, Soil; 2, alluvium; 3–5, Tertiary sand and clay; 6, kaolinized lower Permian beds; 7, kaolinized porphyry; 8, granite porphyry. (*Lilley-Behr, A.I.M.E.*)

China. China is the largest producer, but not exporter, of china clay, and is the home of chinaware and the source of the word kaolin (from kaoling — Ries). Several varieties are produced, but the best material is that of the Chimen district, Anhwei, derived from feldspathic rocks.

United States. Residual deposits of low-grade clays are numerous south of the glaciated areas and east of the Rockies, and glacial clays are found in the North. Deposits of high-grade white clays, however, occur mostly in North Carolina, Virginia, and Georgia, where they are derived mostly from pegmatite dikes. Those of North Carolina yield good quality kaolin, but they are generally less desirable than European kaolins. The deposits are as much as 150 feet wide and 100 feet deep. The content of kaolin substance averages about 25 percent but reaches 40 percent. The Virginia deposit is a white, nonplastic clay developed on syenite. An unusual deposit occurs in Litchfield County, Connecticut, where a bed of feldspathic quartzite 1,000 feet long, lying between harder quartzites, has been weathered to kaolin and was preserved from glacial erosion. Other high-grade clays occur in lesser amounts throughout the southeastern states. The white clays or so-called sedimentary kaolins of Cretaceous age in South Carolina and Georgia have been utilized for paper clays.

Selected References on Clays

Clays, Origin, Properties, and Uses. H. RIES. John Wiley & Sons, New York, 1927. *General treatise.*

General Reference 11. Chap. 11, by H. RIES. *Excellent summary and complete bibliography.*

Clays and Other Ceramic Minerals. C. A. PARMELEE. Edwards Bros., Ann Arbor, Mich., 1937. *Mineralogy, formation, origin.*

The Kaolin Minerals. C. S. Ross and P. F. KERR. U. S. Geol. Surv. Prof. Paper 165–E, 1930; also Prof. Paper 185–G, 1934. *Detailed mineralogy.*

Clay. P. M. TYLER. U. S. Bur. Mines Inf. Circ. 6155, 1935. *Occurrence, technology, and production.*

White-burning Clays of Southern Appalachian States. J. H. WATKINS. A. I. M. E. Trans. 51:481–501, 1916. *Some sedimentary clays.*

High Grade Clays of the Eastern United States. H. RIES and W. S. BAYLEY. U. S. Geol. Surv. Bull. 708, 1922. *Detailed geographical descriptions.*

Surface Chemistry of Clays. A. D. GARRISON. A. I. M. E. Trans. 132:191–204, 1939. *Some causes of plasticity and bleaching power.*

Modern Concepts of Clay Minerals. R. E. GRIM. Jour. Geol. 50:225–275, 1942. *Clay concept and descriptions of clay minerals and crystal structures.*

Absorbent Clays. P. G. NUTTING. U. S. Geol. Surv. Bull. 928–C:127–221, 1943. *Summary of knowledge of absorbent and bleaching clays.*

Minerals of the Montmorillonite Group, Their Origin and Relation to Clays. C. S. Ross and S. B. HENDRICKS. U. S. Geol. Surv. Prof. Paper 205–B, 1945. *Comprehensive discussion of the montmorillonite and related clay minerals.*

Amenability of Clay Minerals to Alumina Extraction. R. E. GRIM, J. S. MACHIN, and W. F. BRADLEY. Ill. Geol. Surv. Bull. 69:77, Urbana, Ill., 1945. *Includes discussion of various clay minerals and their properties.*

Diagnostic Criteria for Clay Minerals. W. F. BRADLEY. Amer. Miner. 30:704–713, 1945. *X-ray criteria.*

Relation of Clay Mineralogy to Origin and Recovery of Petroleum. R. E. GRIM. A. A. P. G. 31:1491–1499, 1947. *Crystal structure reasons.*

Kaolins of North Carolina. J. L. STUCKEY. A. I. M. E. Tech. Pub. 2219, 1947. *Distribution, composition, and occurrences.*

General Reference 5.

Glossary of Clay Mineral Names. PAUL F. KERR and P. K. HAMILTON. **Reference Clay Localities — United States.** PAUL F. KERR and J. L. KULP. **Differential Thermal Analyses of Reference Clay Mineral Specimens.** PAUL F. KERR, J. L. KULP, and P. K. HAMILTON. Am. Petrol. Inst. Proj. 49, Clay Mineral Standards Prel. Repts. 1, 2, and 3, Columbia University, New York, 1949. *Clay minerals, history, origin, formulas, properties, occurrences.*

See also references at end of Chap. 5·7.

Feldspar

Feldspar is used in ceramics for the making of pottery, both in the body of the ware and in the glaze. It is also used in enamels for household utensils, tile, porcelain sanitary ware, and other minor ceramic uses. About 44 percent of the feldspar produced in the United States is used for such purposes and 54 percent to supply desired alumina, potash, and soda for glass manufacture. Nepheline syenite, with about 24 percent alumina, however, is now a strong competitor of feldspar in the glass industry. Some feldspar is also an ingredient in scouring soaps, abrasives, roofing materials, and false teeth.

Kinds and Properties of Commercial Feldspars. Both potash and soda feldspars are the commercial varieties, the former being the more important; the plagioclases high in lime are undesirable. The potash feldspars (orthoclase and microcline) always contain some soda from included albite, and the soda feldspar always contains a little lime. Iron, manganese, and sericite are deleterious. Feldspars high in potash or soda, upon cooling from fusion, yield a solid glass, but lime-rich varieties become partially crystallized.

Occurrence. Although feldspars constitute about 50 percent of igneous rocks, commercial varieties are derived chiefly from pegmatite dikes, mainly granitic ones. They occur in areas of granitic and metamorphic rocks and, as is common with pegmatites, generally are irregular in size, continuity, and in the distribution of the feldspar content. Commercial pegmatite dikes are mostly lens-shaped and are as much as a few hundred feet across and a mile or so in length. Most of the feldspar crystals range from a few inches to a few feet in diameter but may attain 20 feet. Quartz, the most abundant associate, and other pegmatitic minerals have to be removed by processing. The cleaned product is finely ground for market.

Production and Distribution. The annual world production is 600 to 700 thousand tons, of which the United States produces about 80 percent. The other chief producing countries in order are: Canada, Sweden, France, Norway, Germany, Australia, and Japan. In the United States the leading states in order are: North Carolina, South Dakota, Colorado, Virginia, Wyoming, Maine, and Connecticut. The leading district in North America is Spruce Pine district in North Carolina, where plagioclase predominates over potash feldspar. This and the crystalline belt extending from South Carolina to Maine yields most of the American production. The large pegmatites of the Black Hills of South Dakota are also important sources of feldspars.

Selected References on Feldspar

General Reference 11. Chap. 15, by B. C. Burgess. *Excellent bibliography*.

Spruce Pine District, North Carolina. C. S. MAURICE. Econ. Geol. 35:49–78; 158–187, 1940. *Detailed descriptions but difficult reading.*

Feldspar. BOWLES and LEE. U. S. Bur. Mines Inf. Circ. 6381, 1930. *Occurrence, distribution, uses, production.*

Feldspar Deposits of the United States. E. S. BASTIN. U. S. Geol. Surv. Bull. 420, 1910. *Descriptions of occurrences.*

Feldspar. H. S. SPENCE. Canada Dept. Mines, Mines Branch Pub. 731, Ottawa, 1932. *A general treatise.*

General Reference 5.

New Jersey Potential Feldspar Resources. J. M. PARKER, III. N. J. Bur. of Min. Res. Bull. 5, Pt. 1, 1948. *Geological occurrence and resources.*

Internal Structure of Granitic Pegmatites. E. N. CAMERON, *et al.* Mon. No. 2. Economic Geology Publishing Co. 1949.

Other Ceramic Materials

Other materials are utilized in minor amounts in the ceramic industry either to supply certain desired ingredients to clay or to make special ceramic products.

Bauxite. Bauxite with its high alumina content is required for certain porcelains to add strength and resistance to heat, corrosion, abrasion, and spalling. Aluminous refractory brick, made of baked bauxite and a binder, are desired for severe furnace conditions. Fused aluminous refractories of the metallic type are cast directly from the melt. (See also Aluminum, Chap. 13.)

Sillimanite Minerals. This group of minerals includes sillimanite, andalusite, kyanite, and dumortierite, which are considered as a group because of similar ceramic uses. These are furnaced at 1,545° C to give silica and mullite. This material, stable up to 1,810° C, makes excellent refractories. Porcelains produced from it are strong, withstand high temperature, have small expansion, are excellent insulators, and resist corrosion. They are used for making spark plugs, electrical and chemical porcelains, enamelware, saggers, and hotel ware.

Refractories containing mullite are used for glass tanks, crucibles, furnace linings, fireboxes, and high-temperature cements. According to Kerr, the Champion Spark Plug Company uses dumortierite from Oreana, Nev., for its spark plugs and markets Champion " sillimanite " made from dumortierite and andalusite. Dumortierite is more favored than the others because of its higher Al_2O_3 content. Kyanite is the least favored because it increases in volume upon changing to mullite. It is, therefore, generally pre-calcined before use. The mineralogy, occurrence, origin, and examples of deposits are given in Chapters 5·9 and 19.

Borax. Borax, although required in minor quantities, is an essential constituent of porcelain enamels used to coat iron and steel household and kitchen utensils, such as cooking utensils, stoves, refrigerators, washing machines, bathtubs, sinks, table tops, tiles, pipes, and many similar products that require an attractive, durable, and sanitary finish and for articles used in hospitals and doctors' offices. Borax also permits addition of pigments to enamels. Since borax lowers expansion and makes durable products it is indispensable for many ceramic products, such as certain pottery, china, colored wares, glazes, and various heat-resisting glasses used in laboratory, kitchen, signal, thermometer, and optical glass. It imparts brilliance to glass as well as strength and durability.

Magnesite. Magnesium oxide or silicate obtained from calcining magnesite (or dolomite) is used in making ceramic bodies for electric stoves. Pure magnesia with 2 percent talc is used to make extruded insulators for radios, and pressed magnesia is made into insulators for high-conductivity copper conductors. Laboratory crucibles made from magnesia are used for refining metals because they are chemically inert and stand high temperature. Fused magnesia is used to make molds for certain glass objects, such as electric light bulbs.

Lithium Minerals. There has been increasing use of lithium compounds in ceramics. They are introduced directly into the batch as lepidolite, spodumene, or as chemically prepared salts, particularly lithium carbonate. Betz states that lithium (1) is a powerful flux along with feldspar, (2) permits the use of much less alkali, (3) has a mineralizing effect on ceramic bodies, (4) increases the fluidity and gloss of enamels and glazes, (5) reduces vaporization of the glaze, (6) permits the manufacture of glass with high electric resistance and the power to transmit ultraviolet light. Lepidolite, with its fluorine and lithium, decreases expansion and increases the strength of ceramic bodies. It also is a good opacifier for opal and white opaque glasses, and is used for nonshattering glass.

Cornish Stone. Cornish stone, or china stone, is a kaolinized granite rich in orthoclase and albite that is used largely by English potters in place of feldspar. It contains from 6 to 15 percent kaolin, 55 to 77 percent feldspar, and 16 to 31 percent quartz. It is obtained in Cornwall, England, near the china clay deposits, and because it occurs in large bodies it can be mined cheaply. It is the principal source of the alumina used in British pottery.

Diaspore. Diaspore (see Refractories, Chap. 19) is used mainly for refractory fire brick, although some of it is added to certain clays to increase the alumina content and to give a hard porcelain.

Bentonite. Bentonite is used mainly for purposes other than ceramic, but it is also added to other clays and ceramic materials as an additional plasticizing and bonding agent to improve the products after burning.

Fluorspar. Fluorspar (Chap. 19) is used in the ceramic industry in the manufacture of enamels, in opalescent, opaque, and colored glass, and in facings for bricks and vitrolite.

Barite. Barite, converted to barium carbonate, is utilized in making optical glass and enamels and " granite ware " for coating metal dishes.

Potash Minerals. Potash minerals are employed for jewelry type enamels to give high brilliancy and luster. The potassium-lead-silicate type use as much as 36% K_2O.

Talc. Talc is used in ceramics for making calcined talc (lava), which is harder than steel and can be tooled and threaded for use in gas tips, refractories, and electrical insulators. For occurrence and origin see Chapter 5·9; for uses, distribution, production, and examples of deposits see Chapter 20.

Pyrophyllite. Pyrophyllite is used as a ceramic body material in wall tile, tablewear, and electrical porcelains (see also Chap. 5·9).

Diatomite. Diatomite is used in the manufacture of glazes and enamels.

Zirconia. Zirconia (see Refractories, Chap. 19) is used for certain porcelains and high refractory materials, such as crucibles for melting platinum and other metals, which withstand chemicals and high temperatures.

Selected Reference

General Reference 5.

CHAPTER 18

STRUCTURAL AND BUILDING MATERIALS

Present-day structural and building operations create large demands for bulk materials of the mineral kingdom. A bridge is built of steel, stone, and concrete, all mineral products; a highway of stone, sand, and cement; and a modern city building from its concrete and steel foundation to its slate or metal roof employs steel for its structure, stone, cement or brick (clay) and mortar for its walls, metals for its plumbing, heating, wiring, elevators, and finishings, plaster (gypsum) for its walls and roofing, cement for floors, roofing stones, or asbestos shingles, mineral wool for insulation, glass (glass sands) for windows, and mineral paints for finishing. Except for a minor amount of wood, it is entirely composed of mineral products. The materials are common substances occurring in large bulk. Mostly they are the stone, clay, and sand used by early man. Some are used directly; some are dressed; and others undergo considerable preparation. The demand for these products has increased so greatly that huge industries have sprung up around them. The group, some materials of which have already been considered, includes:

Building, roofing, and crushed stones.
Hydraulic cements.
Building sand and gravel.
Gypsum.
Lime.
Magnesite.

Mineral pigments.
Heat and sound insulators — mineral wool.
Asphalt and bitumens.
Clay products.
Miscellaneous.

Building, Roofing, and Crushed Stones

The stone industry is widespread, since rock is the most abundant of all material things. Except for a thin mantle of soil and water, it is the earth itself. Making use of it is one of the oldest human activities, and today in the United States there are about 2,800 quarries with an annual production of around 240 million tons, worth about 250 million dollars.

The term " rock " differs from " stone " in that the latter is applied to blocks or pieces broken for use. Not all rock, however, makes commercial stone; certain geological and physical properties are necessary,

711

such as strength, durability, and ease of processing and quarrying. Most stone is used crude and untreated. It may be cut and dressed and is then termed *dimension* stone; it may be broken into aggregates and called *crushed* stone; or it may be treated and made into cement, lime, glass, or refractories. The common rocks used for commercial stone are granite and related igneous rocks, limestone, marble, slate, sandstone, and soapstone, all of which serve more than one purpose.

BUILDING AND STRUCTURAL STONE

For 12,000 years or more man has used stone for building shelters. In the mighty pyramids of Egypt and in the great buildings of the Greeks, Incas, and Mayas he early overcame the problems of transportation of huge blocks, weighing up to 90 tons. Later, beauty followed grandeur, and the magnificent and exquisite cathedrals of Europe reflect the artistic use of building stone.

Today utility follows beauty. Seldom, however, are buildings now constructed of massive masonry, but thin slabs of stone sheathe most steel skyscrapers and more modest buildings. In brick and concrete buildings, sills, trim, facings, and steps are generally of stone. Stone is used for interiors in steps, fireplaces, wainscoting, baseboards, and other ways. If stone were used more in America the fire hazard of buildings would be less.

Most building stone is dimension stone and is used as cut or finished stone, according to drawings supplied; as ashlar or rectangular blocks; as rough building stone; and as rubble, with one flat face. Such stone is sawed or chipped into shape. Dimension stone is also used for other structural purposes, such as bridges, abutments, fences, retaining walls, monuments, flagstones, paving stones, cobblestones, curb stones, switchboards, and blackboards.

Despite the abundance of rock outcrops, few rocks satisfy the requirements for dimension stone. The important ones are ease of quarrying, strength, color, hardness and workability, texture and porosity, and durability. Transportation is generally an economic factor.

Quarrying. The rock exposure must be free from closely spaced joints, cracks, or other lines of weakness or sizeable sound blocks cannot be obtained. Some lines of weakness, such as well-spaced bedding and joint planes, are necessary to assist in quarrying and to permit breaking to one or more flat surfaces, else the quarried rock would have to be dressed on all sides. Deep and irregular weathering is undesirable.

Strength. This is a minor factor because most stones have excess strength for all building purposes. The crushing strength of good building stones ranges from 5,000 to 25,000 lb per sq in. and a 6,000-lb stone would support a structure 1 mile high before failure. For ordinary purposes 5,000 lb is satisfactory; a safety factor of 20 is employed. Transverse strength is important for window caps.

Hardness and Workability. Hardness ranges from that of soft coquina or limestone to that of granite, which exceeds steel. Workability depends in part upon hardness. Limestone is easy and cheap to dress, but granite is expensive. Hardness is generally unimportant except where wear is concerned, such as for steps, flooring, or paving stones.

Color and Fabric. Color is very important. Many architects are inclined to let color predominate over other essential qualities. Color, however, is a matter of taste and even of fashion; the prized brownstone buildings of yesteryear are undesired today. For buildings, reds, browns, buffs, grays, or white are preferred. White is undesirable for smoky cities, and blacks and dark grays are not popular. Most rock colors are permanent and are not affected by weathering, but stones containing minerals, such as pyrite, that oxidize and produce unsightly stains are undesirable.

A pleasing fabric fetches high prices for ornamental purposes, such as columns or wainscoting. Some limestone breccias of varicolored fragments, hard agglomerates, porphyries, or certain banded or schistose stones give pleasing effects.

Porosity and Texture. Porosity affects freezing expansion and solution. Small pores give up water slowly and are, therefore, affected most by frost action. High permeability aids solvent action. Texture affects workability, therefore cost; fine-textured rocks split and dress more readily than coarse ones. In many ornamental and monumental stones, texture is also vital, but architectural trend in building stone is toward uneven texture.

Durability. Durability should be, but is not always, a determining factor in the choice of a building stone subject to weathering, climate, frost, heat, and fire. Cleopatra's Needle survived 3,000 years of Egyptian climate but succumbed to a quarter century of New York climate. Weathering is caused by rain containing CO_2 and SO_2 and by frost action and solar heat. This produces crumbling, spalling, and exfoliation. In some of the English colleges almost every stone has had to be replaced, and the Wren churches of England have suffered demolition of sculptures. Granite and allied rocks resist climate best; silica-cemented sandstone is good; close-textured, low-porosity limestone resists well; but coarse marble and limestone, poorly cemented sandstones, and much arkose succumbs fairly quickly. Fire and heat resistance is high with fine-grained, compact stones, such as limestone and fine sandstone, and less so in coarse crystalline rocks.

The life of building stones in New York ranges from 15 to 50 years for coarse brownstone and 50 to 200 years for granite.

Those rocks that combine several of the qualities listed above have attained widespread eminence as building stone; such are the well-known Indiana or Bedford limestone, Ohio sandstone, Minnesota granite, Vermont marble, Carrara marble (Italy), Mexican onyx marble, Italian travertine, Scottish granite, and Virginia soapstone.

Kinds of Rocks. Some of the rocks particularly desired for building and other dimension stone in North America and their chief uses and important localities are shown in Table 23.

Under " granite " the quarryman includes also syenite, monzonites, diorites and their porphyries, and granitic gneisses. Sometimes rhyolite and andesite are included under granite. Miscellaneous rocks used for building and other structural stones include alabaster, amazonite, greenstone, and mica schist.

Extraction Methods. Building and other dimension stone is obtained by open quarrying, rarely from underground. Explosives are used sparingly so that the blocks will not be harmed. With hard rocks, like granite, wedges are driven into drilled holes to spring out the blocks. This is supplemented by light blasting, advantage being taken of the rift and grain. Smaller blocks may be extracted by wedging or by drilling small holes and using plug and feathers. With limestones, sandstones, and marble narrow, parallel channel cuts are made with a channeling machine, and the blocks are then removed and subdivided by plug and feather or by cutting machines. Wire saws are also used now in many quarries to cut out blocks. For cut stone, saws, planes, and rubbing and polishing machines are used.

CRUSHED AND BROKEN STONE

Owing to increased use of machinery and concrete and accelerated highway construction, the crushed stone industry now vastly exceeds that of dimension stone in tonnage and value.

Raw materials are widespread, the requisite qualities are few, and quarries are widely scattered. Because crushed stone can be produced on a large scale by inexpensive operations, the product is cheap. Since cost of transportation is the chief limiting factor the industry must depend upon local markets.

The rocks used are few, mainly limestone (including marble), with

TABLE 23

NORTH AMERICAN BUILDING STONES

STONE	DESIGNATION	CHIEF USE	IMPORTANT LOCALITIES
Granite	Barre	Monuments	Vermont
	Quincy	Monuments	Massachusetts
	New Hampshire	Building	New Hampshire
	Maine	Paving blocks	Maine
	Westerly	Building and monuments	Rhode Island
	St. Cloud	Ornamental and building	Minnesota
	Wisconsin	Building	Wisconsin
	Winnipeg	Building	Manitoba
	Norwegian Pearl Gray (Laurvikite)	Decorative	Norway
	Rapakivi	Decorative	Finland
Limestone	Indiana (oölitic)	Trim and building	Indiana
	Coquina	Building	Florida
	Missouri	Building	Missouri
	Kasota	Decorative	Minnesota
	Texas	Building	Texas
	Tyndall	Building	Manitoba
Travertine	Salida	Decorative	Colorado
	Montana	Decorative	Montana
	Tivoli	Decorative	Italy
Marble	Vermont	Building, monument	Vermont
	Rutland	Decorative	Vermont
	Georgia	Building, monument	Georgia
	Carrara	Monument, decorative	Italy
	Siena Parian	Monument, decorative	Greece
	Savoie	Decorative	France
	Onyx	Decorative	Mexico
Serpentine	Verde antique	Decorative	
Sandstone	Berea or Ohio	Building, trim	Ohio
	Portland	Building, trim	Connecticut
	Blue stone	Flags	Pennsylvania, Kentucky, New York
Dolerite	Trap	Curbstones	Connecticut, New Jersey, Massachusetts
Soapstone	Soapstone	Switchboards, sinks, trim	Virginia

lesser amounts of trap rock (including dolerite, basalt, and andesite), granite, sandstone, and quartzite.

The particular rocks employed depend largely upon the desired uses. These fall into two main groups, (1) for use in the crude state, and (2) for chemical purposes. The former include highway construction, concrete aggregate, railroad ballast, and riprap or broken stone for retaining walls and sea walls. Minor uses are for sewage filter floors, shingle aggregates, and road ballast. For highway construction, fill, toughness, hardness, and strength are desired, and trap rock fits these requirements best, with granitic rocks next. Sandstone is used for aggregates and riprap.

The chemical purposes for which limestone serves almost exclusively are manufacture of portland cement, and to an important but less extent, for fluxstone, lime, alkali works, agriculture, sugar factories, asphalt filler, refractory stone, glass and paper, calcium carbide, refractories, and rock wool.

The rocks are extracted from huge quarries by churn drilling and large-scale blasting, loaded by power shovel, crushed, and screened for proper sizing.

Roofing Stones (Slate)

Roofing stones are largely confined to slate, but a little sandstone flag is also used. *Slate* is a very durable rock with a pronounced parallel cleavage, differing thereby from phyllites and schists. It is the cleavage, strength, and durability that make it so excellent a roofing stone; it is noninflammable; also it is found in a variety of pleasing colors of gray, blue-black, red, green, purple, and mottled. The colors are mostly permanent. The cleavage of slate rarely coincides with bedding, and thin beds of different compaction can often be discerned as " ribbons " across the cleavage, which makes second-quality stone. Slate is found in metamorphic regions, and most of the commercial product of North America comes from Pennsylvania and Vermont, with lesser amounts from the other Appalachian states.

In addition to roofing, slate is also used as mill stock for baseboards, steps, walks, mantles, shower-bath floors, blackboards, switchboards, school slate, as granules for surfacing shingles and roofing, and as pulverized slate or flour for fillers. The enormous waste attendant upon cleaving shingles and blocks has led to increasing use of granules.

A slate quarry should have little overburden and should be free from siliceous or carbonaceous beds. Extraction involves greater care than with other rocks. Slate is generally sawn by wire saws into blocks that are wedged off, removed, and split by hand into shingles or blocks.

Selected References on Stone

The Stone Industries. OLIVER BOWLES. McGraw-Hill, New York, 1934. *A good general treatise.*

General Reference 11. Chaps. 14 by OLIVER BOWLES on Dimension Stone and 12 by A. T. GOLDBECK on Crushed Stone. *Good summaries of requirements, types, and specifications.*

General Reference 5.

Roofing Granules. G. W. JOSEPHSON. General Reference 16, pp. 460–472. *Types and nature.*

Hydraulic Cements

The vast expansion of highway and building construction has created great demand for cement, and the industry is now the sixth largest in value of output among American mineral industries. Cement is a manufactured material which when mixed with water sets or becomes hard, either in air or under water. Essentially, it is a mixture of about four parts of limestone and one of clay or shale, calcined to near fusion and ground to a powder.

One kind of cement was known to the Carthaginians, and the Romans discovered that quicklime added to the volcanic ash of Puzzuoli gave a cement that set under water. They used it to build aqueducts, baths, the Pantheon, and other structures, and it became known as *Puzzuolan* cement. It is still made in Europe.

In 1756 John Smeaton in England made " lime " by burning argillaceous limestone and discovered that it set under water. This was the start of the manufacture of *natural hydraulic cements* and inaugurated searches for similar limestone in Europe and America. It was first discovered in America during the building of the Erie Canal and supplied the hydraulic cement for making locks. A big natural cement industry sprang up around Rosendale, N. Y., and Lehigh, Pa. Since such limestones are not common, it became the practice to add either limestone or shale to give the correct proportion. Naturally, the composition was variable, which caused natural cement to lose favor.

Portland cement was discovered by Aspdin in England in 1824 and named by him because it resembled the famous Portland stone. It has now almost entirely displaced natural cement.

Ingredients and Manufacture. Portland cement is made by burning to a clinker a finely ground mixture containing about 75% $CaCO_3$, and 25% clayey minerals, the latter consisting of 20% SiO_2, Al_2O_3, and

Fe_2O_3, and 5% magnesia, alkalies, etc.; MgO should not exceed 5 percent of the finished product. Three percent of gypsum is added before final grinding to prevent too rapid setting.

The calcining releases the CO_2, and the remaining constituents combine to form complex silicates, aluminates, and ferrates of calcium, which in turn break down to form other compounds. Tricalcium and dicalcium silicate and tricalcium aluminate are the chief constituents. The addition of water gives rise to a gel of hydrous compounds, which later crystallize and interlock, giving the hard set. Mixed with crushed rock and sand, the cement binds all together into a rocklike mass of great strength, forming *concrete.*

Limestone is the chief rock used to supply the calcium oxide and the nearer it approaches the composition of the cement mixture the better. Pure limestone is neither necessary nor desirable. MgO should not exceed 10 percent, and pyrite and free silica should be absent. CaO is also supplied in part by marl, furnace slag, and oyster shells. SiO_2 and Al_2O_3 are supplied by clay or shale; sandstone or sand, bauxite, or iron ore may be added to supply deficiencies in SiO_2 and Al_2O_3 or Fe_2O_3. The combinations of raw materials used are (1) limestone with clay or shale, (2) cement rock, alone or with high-calcium limestone, (3) blast furnace slag and limestone, (4) marl and clay, (5) oyster shells and clay. In many plants the raw materials are now subjected to flotation to discard objectionable minerals. About 612 pounds of raw materials are used to a barrel of finished cement (376 lb).

There is a tendency to make a variety of specialized cements adapted to particular uses, such as high-early-strength, masonry, low-heat, oil well, and high-alumina cements.

Production and Distribution. The raw materials for cement are so widely distributed that few states or countries lack them. Consequently, cement plants may be located wherever favorable associations of raw materials occur.

The world production normally amounts to 80 to 100 million tons, and the main producing countries in order are United States (30 percent), Germany, United Kingdom, Japan, Russia, and Italy. The United States production of 30 million tons comes largely from Pennsylvania, the midwestern states, California, New York, and Texas, Pennsylvania producing about 20 percent of the total.

Selected References on Cements

The Stone Industries. O. Bowles. McGraw-Hill, New York, 1934. *Comprehensive book on industrial stones, including cement rocks.*

General Reference 11. Chap. 8, by W. M. MYERS. *Good summary and bibliography.*

Cements, Limes, and Plasters. E. C. ECKEL. John Wiley & Sons, New York, 1928. *Good material on industrial limestones used for cement rock.*

General Reference 5.

Sand and Gravel

Humdrum sand and gravel are essential in modern construction, particularly in paving and building. They are widely distributed the world over and are generally used locally because their low value permits only very low cost water transport. Despite their low unit cost, the total sand and gravel produced in the United States amounts to between 250 and 300 million tons, worth from 150 to 200 million dollars, of which about 40 percent is for sand and 60 percent for gravel.

Sand is a broad term used to cover almost any comminuted rock or mineral, but technically it is restricted to quartz sand with minor impurities of feldspar, mica, and iron oxides. There are also black sands (magnetite), coral sands, gypsum sands, and others. Sand grains fall between 0.06 and 2 mm in diameter; coarse sand ranges from 0.6 to 2 mm, fine sand from 0.06 to 0.2 mm, and gravel from 2 to 8 mm. Gravel consists dominantly of quartz pebbles and grains but also includes pebbles of other rocks and minerals. Dry sand ranges from 90 to 110 lb per cu ft, and dry gravel from 90 to 107 lb. Both are of detrital origin.

Occurrence. Sand and gravel occur as sedimentary beds, lenses, and pockets lying at or near the surface or interbedded with other sedimentary beds. They occur as fluvioglacial deposits, stream-channel and flood plain deposits, as present and elevated seashore deposits, as wind blown deposits along and near bodies of water, as desert sand dunes, and as marine and fresh water sedimentary beds. The deposits may be well sorted and may consist of almost pure sand grains or they may consist of irregular sizes and impurities necessitating screening to size them or washing to purify them.

Uses. The chief uses of sand are shown in the table below.

USE	PERCENT OF TOTAL	USE	PERCENT OF TOTAL
Building	51	Engine sand	3
Paving	30	Abrasives	0.9
Railroad ballast	0.7	Molding	7
Glass	5	Others	2.4

Other special sands are glass, furnace, filter, roofing, flooring, quartz powders, and fillers.

For *building and paving*, clean, angular, sharp sand is desired for concrete and mortar; specifications for paving are less rigid. Gravel for concrete is largely replaced by superior crushed stone. For secondary roads, paving gravel is generally used as extracted. About one-half of the sand and gravel produced goes into concrete.

Railroad ballast requires only bulk and packing ability. Crushed stone is now largely replacing gravel for this purpose.

Molding sands are used in foundries for making molds to receive molten metal. Such sands should be refractory to withstand melting; cohesive to take and hold shape, therefore some clay is necessary; porous to allow escape of gases; and low in iron.

Engine sands used for prevention of wheel slipping should be fine and even grained, sharp, and dry.

Abrasive sands should be sharp, clean, sized, and free from clay.

Filter sand for filtering water supplies should be clean, sized, quartzy, and free from clay, lime, and organic matter.

Fire and furnace sand should be refractory, clean, and quartzy. To this a binding clay is added.

Glass sands are treated in Chapter 20.

Selected References on Sand and Gravel

General Reference 11. Chap. 41, by B. NORDBERG. *Comprehensive and good bibliography.*

Non-Metallic Minerals. R. B. LADOO. McGraw-Hill, New York, 1925. *Types, specifications, uses.*

Technology and Uses of Sand and Gravel. W. M. WEIGEL. U. S. Bur. Mines Bull. 266, 1927. *Excellent summary.*

General Reference 5.

Gypsum

Today gypsum is one of the most important nonmetallic minerals. It was known and prized by the Assyrians and Egyptians for making containers and for sculpturing. The Egyptians used it for making plaster for the building of the pyramids. The Greeks and Romans also used it for plaster, and the soft alabaster was desired for sculpturing. In the United States it was first used for " land plaster " in 1808.

Properties and Uses. Gypsum occurs in five varieties; 1, rock gypsum and 2, gypsite, an impure earthy form, both of which are used commercially; 3, alabaster, a massive, fine-grained, translucent variety; 4, satin spar, a fibrous silky form; and 5, selenite, a transparent crystal form. Gypsum is hydrous calcium sulphate containing 20% H_2O. Anhydrite, the other form of calcium sulphate, lacks water and is considered an impurity in commercial gypsum deposits. Commercial gypsum generally contains about 90 percent of hydrous calcium sulphate. Its important industrial property is that, after calcining, the addition of water makes it into plaster of Paris and other quick-setting plasters.

About 90 percent of the gypsum produced is used for building purposes in the form of plasters, cements, and plaster boards. The chief products and uses are:

PRODUCTS (by value)	USES	PERCENTAGE OF PRODUCTION
1. Laths	Building	19
2. Building plasters	Building purposes	23
3. Wallboard	Building	48
4. Tile	Building	2
5. Portland cement retarder	Building and paving	4
6. Industrial plasters		3
7. Agricultural	Fertilizer	1

The various plasters set quickly with smooth finish and are mainly cement plaster, plaster of Paris, stucco, neat plaster, flow plaster, and finishing plaster. The laths, plaster board, and wallboard consist of thin sheets of paper, felt, or thin wood, separated by compressed plaster and are used for walls, ceilings, partitions, roofing, and for fireproofing and sound deadening. The other uses are for embedding plate glass and decorative stone for polishing, for various types of cements and casts, and for fillers, paints, crayons, insecticides, and other minor purposes.

Production and Distribution. The normal world production amounts to 8 to 10 million tons annually, divided in percentage as follows: United States 50, United Kingdom 17, Canada 16, France 5, others 12.

The United States annual production ranges from 3 to 5 million tons (from 52 mines) and is obtained chiefly from Michigan, New York, Texas, Ohio, California, and Iowa, in the order named. About half of the production comes from the eastern states where gypsum occurs mainly in Silurian strata, but the most extensive deposits are in the western states, where it occurs in Pennsylvanian, Permian, Triassic, and Jurassic formations. It is, however, found with rocks of all ages from Silurian to Pleistocene.

Gypsum is widely distributed, but large production is confined to those countries of large local demand or cheap extraction and transportation. The extensive occurrences in the Paris Basin have given rise to the term "plaster of Paris."

Occurrence and Origin. Gypsum occurs as an evaporite in regular beds or lenses, in various states of purity, and in a great range of thickness from a few feet to hundreds of feet. Gypsum may occur singly or with anhydrite as a primary deposit, or as a surficial hydration product of anhydrite. The occurrence and origin are fully dealt with in Chapter 5·6.

Extraction and Preparation. Since gypsum occurs chiefly in flat or gently inclined beds, most of it is mined by underground methods. It is crushed and ground to a powder. That used for portland-cement retarder and agricultural purposes is sold as raw gypsum. The rest is calcined in kettles or kilns at temperatures of 300° to 350° F to remove part, or rarely all, of the water of crystallization. The 2 molecules of water in gypsum are reduced to ½ molecule, or the " half hydrate " as it is called. This gives calcined gypsum or, if pure, plaster of Paris. With addition of water, the plaster sets in 6 to 8 minutes, so a retarder is necessary to delay the set 1 to 2 hours. This consists of glue, fiber, lime, or other materials. The impurities of gypsite commonly serve as effective retarders.

Selected References on Gypsum

Gypsum and Anhydrite. F. T. Moyer. U. S. Bur. Mines Inf. Circ. 7049, 1939. *Occurrence, mining, processing.*

General Reference 11. Chap. 20, by T. R. Lippard. *Succinct treatment.*

Cements, Limes, and Plasters, 3rd Ed. E. C. Eckel. John Wiley & Sons, New York, 1928. *Comprehensive.*

Gypsum Deposits of the United States. R. W. Stone et al. U. S. Geol. Surv. Bull. 697, 1920. *Good bibliography, and descriptions of occurrences.*

General Reference 5.

Gypsum, Texas. In Edwards Limestone. Virgil E. Barnes. **In Hockley Dome.** H. B. Stenzel. Texas Min. Res. Univ. Texas Pub. 4301:35–46; 205–226, 1934. *Evaporite beds and salt-dome occurrences.*

Lime

Since the time of the ancients, lime has been used for mortar. It is simple and cheap to prepare. Limestone or other calcareous rock is heated in kilns to 903° C, when the CO_2 is driven off and quicklime (CaO) remains. This slakes with water, and mixed with sand it makes mortar or plaster. Commonly the lime is prepared as hydrated lime $(Ca(OH)_2)$ by adding the necessary water. One hundred pounds of pure limestone yield 56 pounds of lime. Dolomite may also be used, giving CaO·MgO, which slakes more slowly and gives less heat than CaO lime.

Uses and Production. The utility of lime depends upon the stone from which it is made. With the addition of water it sets in air and finds wide use as mortar and plaster. Much of it, therefore, is consumed in the building trade, but it also has many uses in the chemical industries and is an important fertilizer. The chief uses are shown below.

	Percent		Percent
Building	20.7	Glassworks	3.4
Metallurgy	20.4	Tanning	1.5
Refractory lime (dead-burned dolomite)	17.8	Sugar refining	0.4
Paper mills	11.6	Others	11.3
Agricultural	7.5		—
Water purification	5.4	Total	100.0

Magnesia lime makes a strong, hard, elastic stucco.

Since the raw materials for lime are world-wide in distribution it is produced almost everywhere according to its need. The United States produces 5 to 6 million tons annually valued around 60 million dollars, from 42 states, with Ohio and Pennsylvania leading.

Selected References on Lime

General Reference 11. Chap. 22, by N. C. Rockwood. *Good summary.*
Cements, Limes, and Plasters. E. C. Eckel. John Wiley & Sons, New York, 1928. *Materials, properties, manufacture.*
Lime. O. Bowles and D. M. Banks. U. S. Bur. Mines Inf. Circ. 6884, 1936. *Uses, requirements, technique.*

Magnesite

Magnesite is desired chiefly as a raw material for magnesium compounds and metallic magnesium and is used to some extent in its natural state. Its use expands with activity in the building and metallurgical industries. After calcination it yields material for refractory bricks, cements, and flooring. Dolomite is used in place of magnesite when possible, because it is cheaper.

Properties and Uses. Magnesite occurs as both the crystalline and amorphous (cryptocrystalline) varieties, the latter being generally the purer. It loses its carbon dioxide content upon heating, forming magnesia (MgO), which upon further heating develops into periclase. This resists hydration and carbonation at ordinary temperatures. The magnesite of commerce refers not only to $MgCO_3$ but also to the sintered products, magnesia, and to breunnerite, the carbonate with over 5 percent ferrous carbonate. It is burned to *caustic magnesite,* containing 2 to 7% CO_2, at 700° to 1,200° C, and to *dead-burned magnesite,* with less than 0.5% CO_2, at 1,450° to 1,500° C. Most of the magnesite consumed in the United States goes into these products.

Caustic magnesite is used for oxychloride or Sorel cements, which are employed mainly for flooring and stucco. As a stucco it has many advantages over portland cement, except for weathering qualities. Since this difficulty has been overcome its use for this purpose has increased. Its use for flooring and wall board has also increased greatly. Mixed with magnesium chloride and fillers it makes an inexpensive, hard, nonshrinking, dustless flooring material, which is fireproof, flexible, durable, and takes wax or polish. It is suitable for bathrooms, hospitals, and public buildings. An important use is as a chemical accelerator in rubber. It is molded into pipe coverings for heat insulation, and has some ceramic use.

Dead-burned magnesite is a high-grade refractory in the metallurgical industry (Chap. 19).

Magnesite is extensively used for metallic magnesium. Other minor applications are in the paper, ceramic, and glass industries, as an abrasive, as a source of CO_2, and for magnesium chemicals.

Production and Distribution. The annual world production is about 2 million tons, and the principal producing countries are Russia, Austria, Manchuria, United States, Greece, Czechoslovakia, Yugoslavia, and Korea.

The United States production, amounting to about 325 thousand tons, comes from Washington and California.

Occurrence and Origin. Magnesite has three modes of occurrence:

1. Replacement of dolomite or limestone, e.g., Washington, Austria, Manchuria, Czechoslovakia, Quebec.
2. Veins, e.g., California, Greece, India, Russia, Yugoslavia.
3. Sedimentary beds, e.g., Nevada.

The *replacement* deposits yield the " crystalline " variety and have resulted from progressive replacement (rarely complete) of limestone or dolomite by $MgCO_3$ through hydrothermal solutions. This forms bedded deposits, lenslike or irregular in shape, and of large size. They generally contain some ferrous iron.

The *veins* contain the hard amorphous variety and occupy fractures or crush zones in serpentine or ultrabasic rocks. They result from the breakdown of serpentine by hydrothermal carbonate solutions, accompanied by the release of silica, which forms opal or chalcedony.

Sedimentary beds are uncommon (see Chap. 5·5).

Extraction and Preparation. Magnesite is quarried, hand-sorted, crushed, and then calcined in kilns. For caustic magnesite, the kiln temperature is about 1,200° C, and for dead-burned magnesite, to which iron ore is added for sintering, the temperature is about 1,560° C.

EXAMPLES OF DEPOSITS

United States. The largest deposits are near *Chewelah, Wash.*, where crystalline magnesite occurs as a hydrothermal replacement of Carboniferous dolomite. The deposits are huge, bedded lenses up to 1,000 feet long and 300 feet thick. The material varies in grain size, is white to red in color, and is low in iron. The reserves within 100 feet of the surface are large. The *California* deposits of the Coast Ranges occur as pure amorphous magnesite in serpentine in veins,

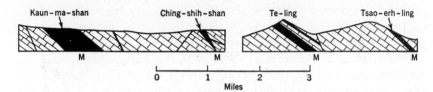

FIG. 18–1. Sections of Manchuria magnesite deposits. (*Niinomy, Econ. Geol.*)

lenses, or masses up to 24 feet wide. One deposit consists of boulders or nodules in serpentine breccia. A deposit in *Kern County* is apparently sedimentary. *Nevada* contains a sedimentary deposit up to 200 feet thick along the Muddy River. Brucite deposits, covered by hydromagnesite and formed as an alteration of dolomite, are worked in Nye County. This is made into "thomasite," a ferrite bonded magnesia refractory.

Russia. Russia contains four large deposits of crystalline magnesite in the Urals; the largest, at Satka, has reserves of 145 million tons. The deposits extend for 5 miles in two parallel series of lenses up to 270 feet wide within a series of sediments.

Austria. Austria was formerly the world's largest producer from deposits in Styria. The magnesite is breunnerite, with 14 percent ferrous carbonate, and occurs as lenticular replacements in metamorphosed Carboniferous sediments. The Veitsch deposit is 1 mile long, 1,500 feet wide, and is a replacement of limestone; partial replacement yielded dolomite. There are 100 years' reserves of workable rock.

Manchuria. The largest deposits of magnesite in the world are worked in Manchuria over a 9-mile belt (Fig. 18–1). They occur as replacements associated with dolomite in a series of metamorphosed Cambrian sediments. Individual deposits, according to Niinomy, attain a length of 1,900 meters and a thickness of 900 meters. The deposits of Korea are generally similar.

Others. The principal deposits of *Greece* are in Euboea, where veins and lenses of magnesite occur in serpentine (Fig. 18–2). One deposit is a series of lenses a mile long, of which individual lenses attain 770 feet in length and 200 feet in width. The largest Mandoudi lens is

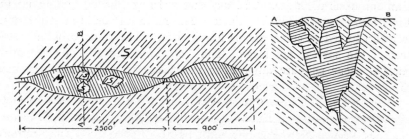

FIG. 18–2.   Sections of magnesite deposits, Mandoudi, Euboea, Greece.
(*Sagui, Econ. Geol.*)

3,200 by 160 feet. Similar but smaller deposits occur in *Yugoslavia*. *Czechoslovakia* has a 75-mile belt containing numerous deposits, apparently of replacement origin in limestones.

Selected References on Magnesite

Magnesium, Magnesite and Dolomite. J. LUMSDEN. Imp. Inst. Min. Res. Dept. London, 1939. *Brief world survey; extensive bibliography.*

General Reference 11. Chap. 24, by R. E. BIRCH and O. M. WICKEN. *Brief technological résumé; good bibliography.*

Types of Magnesite Deposits and their Origin. G. W. BAIN. Econ. Geol. 19: 412–433. 1924. *Good general paper.*

Magnesite. P. M. TYLER. U. S. Bur. Mines Inf. Circ. 6437, 1931. *Occurrence, distribution, uses, preparation.*

Preliminary Report on Magnesite Deposits of Stevens Co., Washington. W. A. G. BENNETT. Wash. Dept. of Cons. and Dev. Repr. Inv. 5, 22 pp., 1941. *Description of deposits.*

Production and Properties of Commercial Magnesias. M. Y. SEATON. A. I. M. E. Tech. Pub. 1496, 21 pp., 1942. *Deposits and description of production processes.*

General Reference 5.

Mineral Pigments

Mineral pigments are utilized directly as paints or to give color, body, or opacity to paints, stucco, plaster, cement, mortar, linoleum, rubber, plastics, or other materials. There are three classes, (1) natural mineral pigments, (2) pigments made by burning or subliming natural minerals, (3) manufactured paints. Combinations of the three are common.

Natural Mineral Pigments. Natural mineral paints contain as their essential color constituents either limonite, hematite (or rarely magnetite), with or without mixtures of clay and manganese oxides. They form ochers, umbers, and siennas. Most of these were used by early man for decoration and drawings and were extensively employed in America in colonial days. Their chief use is for painting steel and ironwork, barns, and freight cars, since they consist of unfading minerals. The various colors and their brilliancy depend upon the proportions of the essential constituents. These natural mixtures are called ochers and are formed mostly by residual weathering.

Mineral red or Indian red is composed of hematite and is formed from the weathering of hematite or iron-bearing minerals. *Vermilion* has long been a New England barn paint. *Persian red,* the original Indian red, contains 65 to 72% Fe_2O_3 and comes from the Persian Gulf. *Spanish red,* containing 82 to 87% Fe_2O_3, is a nearly pure red hematite from Spain. *Venetian red* originally came from Italy.

Mineral yellow-brown is limonite, generally with some manganese oxide.

Ochers are mixtures of hematite, limonite, and clay, with 15 to 80 percent iron oxide, and yield yellow and brown colors. When roasted they yield reddish browns. *Umber* is an ocher with 11 to 25 percent manganese oxide, which gives it the typical brown color. When roasted it takes on a rich brown color of *burnt umber*. *Sienna* is a yellowish-brown ocher with less manganese oxide and more limonite, named from Sienna, Italy. There are several rich yellows, such as mineral yellow, Chinese yellow, and Roman earth. Roasting gives a browner tint, known as *burnt sienna*.

Greens, such as green earth, green ocher, terreverte, and Celadon and Verona greens are all low-tinting pigments derived from ferromagnesian silicates, greenstones, and rocks rich in chlorite.

Whites are obtained from gypsum, barite, talc, white clay, and other substances and are used for plaster, mortars, and cold-water paints.

Ground red and black slate and shale and other colored rocks also serve as natural pigments.

Natural mineral pigments are obtained from several states, led by Pennsylvania, Georgia, and New York, and from Spain, the Caspian Sea, Cyprus, Italy, France, South Africa, and India.

Manufactured Pigments. Iron oxide red and brown pigments are made by roasting ochers, umbers, and siennas and also iron ore and copper ore. Various shades of red, yellow, brown, and black are made from iron salts, such as ferrous sulphate or ferric chloride. Carbon black, for black paints and ink, is made by impinging flaming natural gas upon cold steel plates.

Most chemical paints are made from lead, zinc, titanium, barium, chromium, and carbon.

Lead paints are made chemically and by smelting. The lead compounds used are white lead (basic carbonate), sublimed lead (basic sulphate), red minium oxide, red, orange, and yellow chromates, yellow massicot and orange litharge. About 97 percent of the material used is metallic lead.

Zinc paints are made from zinc ore and from metallic zinc, giving zinc oxide and leaded zinc oxide. Many white paints are made of white lead and zinc oxide.

Barium is used to make *lithopone,* a brilliant white paint composed of 70 percent barium sulphate and 30 percent zinc sulphide and oxide. After a phenomenal rise, lithopone has suffered severe competition from titanium paints.

Titanium yields white paints of such outstanding opacity and covering ability that they are growing competitors to lead and zinc paints. It is used as titanium oxide, or blended with barium sulphate or calcium compounds, or added to lead and zinc paints.

Other *mineral compounds* and the colors obtained follow.

METAL	COLOR	COMPOUND USED
Chromium	Chrome green	Chromic oxide
	Chrome yellow	Lead chromate
	Guignet green	Lead chromate and prussian blue
	Chrome red	Basic lead chromate
	Zinc yellow	Zinc and potassium chromates
Cadmium	Cadmium red	Cadmium sulphide and selenium
	Cadmium yellow	Cadmium sulphide
	Cadmium orange	Cadmium sulphide
Cobalt	Cobalt blue	Cobalt oxide and silicate
Mercury	Vermilion	Mercuric sulphide
Calcium	Venetian red	Iron oxide and calcium sulphate
Carbon	Carbon black	Natural gas soot
	Black	Natural graphite
	Black roofing	Asphalt
	Black rouge	Precipitated or ground magnetite

All pigment minerals must be finely ground and must have opacity or hiding power and ability to absorb oil. The chemical composition is of minor importance. Various combinations of the materials listed may be used to produce intermediate colors.

Occurrence and Distribution. Most natural pigments are formed by residual concentration (see Chap. 5·7). Certain localities have become noted for certain colors. For example, yellow ocher, France and United States; sienna, Italy; umber, Cyprus; red oxides, Spain, Canada, and United States; Persian red, Persian Gulf; and various metallic paints, United States.

Selected References on Mineral Pigments

Non-Metallic Minerals. R. B. LADOO. McGraw-Hill, New York, 1925. *Best article; comprehensive; good bibliography.*
General Reference 11. Chap. 28, by C. L. HARNESS. *Excellent summary.*
Iron Oxide Mineral Pigments. H. WILSON. U. S. Bur. Mines Bull. 370, 1933.
General Reference 5.

Heat and Sound Insulators (Mineral Wool)

There has been a rapidly increasing demand for heat and sound insulators in building construction, particularly in the United States. Heat insulation is now general in buildings to conserve heat in winter and to render them cooler in summer. Various nonmetallic materials are used for this purpose, such as mineral wool, diatomite, exfoliated vermiculite, expanded gypsum, magnesia, mica, pumice, asbestos, and aluminum foil. Except for mineral wool and vermiculite, these materials have other more important uses and are described elsewhere. All the substances mentioned above have the property of holding small pockets of air between nonconducting fibers or plates, and this makes a good heat insulator. Sound insulation, for the improvement of acoustics and reducing noise, depends upon continuous channels connected with many small pores; the channels admit the sound waves and the pores trap them. Also, sound is reflected by means of smooth-surfaced tile, brick, or plaster.

MINERAL WOOL

Mineral wool exhibits the lowest heat conductivity of any material except still air, and ranks below cork. It is also light, weighing only 8 to 12 pounds per cubic foot. It is an artificial product composed of extremely thin silicate fibers, 1 to 10 microns in diameter, and resembles wool. It includes innumerable minute air pockets and hence is a good insulator. The term includes rock wool, slag wool, glass wool, glass silk, and silicate cotton.

Manufacture. According to Lamar and Machin, mineral wool is made by subjecting molten silicate to shearing forces while cooling rapidly enough to prevent crystallization. This is accomplished by 1, blowing steam or air through molten silicate, making droplets that elongate into threads in air; 2, centrifuging on a large rotating disc; 3, allowing droplets to fall from an elevation; 4, spinning a thread upon a rotating drum from extended silicate, or from the molten end of glass rods; and 5, subjecting a flowing stream of molten glass to a steam of air blast so that the glass stream is not broken. The addition of fluorspar gives finer fibers.

Rock Wool. Rock wool is made from (a) "natural woolrocks," and (b) rock mixtures or "composite woolrocks." These are impure carbonate rocks that consist mainly of lime, magnesia, alumina, silica, and carbon dioxide and contain about 40 to 65 percent carbonate. The natural woolrocks may consist of a single rock or a series of interbedded strata (limestone and shale or sand) that together constitute the necessary composition. The groups include impure limestone and dolomite high in silica or silt; cherty carbonate; calcareous or dolomitic sandstone and shale; some glacial clays; some gravels. Suitable mixtures may be compounded from high calcareous rocks and siliceous or argillaceous rocks. Wollastonite rock is used in California. High iron, sulphur, and calcium sulphide are objectionable. The materials are melted in cupola furnaces at as low a temperature as possible ($800°$ C); CO_2 is eliminated, and molten silicate is formed.

Slag Wool. This is made from the slag of iron blast furnaces, which is high in CaO because limestone is used for fluxing iron ore. Copper and lead slags may also be used. A slag with 30 to 50 percent CaO or CaO + MgO would yield mineral wool; others need added ingredients to correct the composition. Much of the mineral wool or "rock wool" of today is made from slag instead of rock.

Glass Wool. Glass wool or glass silk is made from commercial glass or "soda-lime glass," which, so far as is known, is made of the customary raw materials for glass making (see Chap. 20). Glass wools serve the same purpose as other mineral wools. The fibers are strong and flexible and because of this are also used for weaving into glass fabrics that are strong, flexible, lustrous, nonfading, and noninflammable. Curtains, upholstery, and clothing are made of spun glass.

Distribution, Production, and Uses. Mineral wools are made wherever raw materials are available. In the United States much is manufactured in Illinois, Indiana, Wisconsin, New Jersey, Pennsylvania, Maryland, Alabama, and California. Some 800,000 tons per year are produced, worth about 8 million dollars.

Mineral wool is used chiefly for house insulation. Minor uses are for insulating pipes, boilers, tanks, stoves, refrigerators, refrigerator cars, air-conditioned trains, and for insulating boards, blankets, protecting water plants (mulching wool), filters, and sound insulation.

OTHER INSULATORS

Vermiculite. Vermiculite is a mica that exfoliates under strong heat into very thin sheets. It is used for heat insulation in loose form or, with a binder added, is fabricated into forms. It is supposed to represent a hydrothermal alteration of biotite or phlogopite. Deposits occur at Libby, Mont., and in Colorado, Wyoming, North Carolina, Georgia, Newfoundland, and South Africa.

Diatomite. Diatomite or diatomaceous earth, an accumulation of myriad microscopic siliceous shells of diatoms, is very porous and so light that the dry powder weighs only 7 to 16 pounds per cubic foot. Its lightness and its minute cells, filled with immovable air, make it an ideal insulator; it is used in the form of powder, bricks, and blocks. Its chief use, however, is as a filter (see Chap. 20).

Gypsum. Calcined gypsum in an expanded form (Insulex) is also used for heat insulation. Chemicals that produce gas are added to calcined gypsum, and when water is added the gases expand the plaster of Paris to light-weight, cellular powder or flakes. Perforated gypsum board is also used for sound insulation.

Asbestos. Asbestos has other more important uses than for heat insulation, but the short fibers and refuse are utilized in asbestos cement and in covers for boiler and pipe insulation or are made into fireproof shingles, sheet roofing, and wall boards. The fabrics made from the longer fibers are also utilized for fireproof curtains, awnings, warship mats, gloves, aprons, and firemen's clothes. The long fibers, providing air channels between, are suitable for soundproofing.

Basic Magnesium Carbonate. This substance, made from calcined dolomite, is used with 15 percent asbestos to make pipe and boiler coverings for heat insulation.

Pumice. Pumice, a volcanic frothy glass, because of its large cellular space and light weight is an ideal natural insulator. It rarely occurs, however, in a form suitable for insulating use. A massive bed in California yields large blocks that can be sliced directly into sheets suitable for refrigerator insulation. Less massive material and granules are used in stucco and plaster for sound insulation and as a cement admixture. The United States produces about 70,000 tons annually, most of which is used as an abrasive.

Perlite. Perlite, an acid volcanic glass, when heated becomes porous and serves for insulation and light aggregate.

Selected References on Heat and Sound Insulators

General Reference 11. Heat and Sound Insulators (Mineral Wool), by J. E. LAMAR and J. S. MACHIN, Chap. 21. *The best summarization in existence.* Diatomite, by A. B. CUMMINGS and H. MULRYAN, Chap. 13; Asbestos, by G. F. JENKINS, Chap. 2; Pumice, by J. A. BARR, JR., Chap. 36.
General Reference 5.

Asphalt and Other Native Bitumens

Asphalt (Chap. 16) is a semi-solid member of the hydrocarbon family found where oil seepages have evaporated. Asphalt Lake in Trinidad covers 114 acres. A similar lake occurs in Venezuela. Asphalt also occurs as a cement in sandstones, forming bituminous or asphaltic sandstones. Much asphalt is also obtained as a residual or petroleum distillation.

About two-thirds of the asphalt consumed in the United States is for paving. Asphaltic sandstones are also used for this purpose. About one-fourth is used for roofing in the form of asphalt shingles and roll or sheet roofing. The roofing materials are mostly coated with rock granules. Asphalt is also used for waterproofing, blending with rubber, briquetting, pipe coatings, molding compounds, and paints.

Gilsonite, a natural asphaltite of high purity, occurs in veins in Utah and is used in varnishes, japans, stains, inks, roofing, and molded articles. *Wurtzilite* or *elaterite* is an asphaltite that also occurs in veins in Utah and is used for waterproof and insulating paints. *Grahamite,* another asphaltite, occurs in veins in Cuba and is used for similar purposes. *Ozokerite,* a natural wax derived from paraffin oils, is obtained from Galicia and Utah and is used for electrical insulation, waterproofing, and wax candles.

About 8 million tons of asphaltic materials are consumed annually in the United States.

Selected References on Bitumens

General Reference 11. Chap. 31, by A. H. REDFIELD.
General Reference 5.

Miscellaneous Materials

Clays and Clay Products. Clays and clay products are extensively used for building and structural materials. They are also widely used for a great variety of other materials and therefore are treated together as a unit under " Ceramic Materials " (Chap. 17).

Building materials utilize mainly the common, lower-grade, widely distributed types of clays and shales of sedimentary, residual, or glacial origin. The clay products that are employed in building operations are: various types of bricks; tiles in the form of hollow building tile, conduits, roofing, floor, fireproofing, drain, sewer, and mosaic; flue linings; wall coping; chimney pots; toilet fixtures; laundry tubs; enamel stoves; doorknobs; and other porcelain and whiteware. In addition, innumerable products of clay are used about the home. (For full details of clays and clay products see Chap. 17.)

Mica. Since mica is used dominantly for purposes other than structural, it is treated separately and in detail in Chapter 20. A growing minor proportion is utilized for building and structural materials.

In the preparation of industrial sheet mica there is a large amount of scrap and waste. This and ground mica, which is ground up scrap and pulverized mica-rich schists, are utilized in part for backing for roofing rolls and shingles, and in concrete and some cements. The better grades of wet ground mica are used in some paints, and mica is the luster-imparting ingredient in wallpaper. Its other uses are given in Chapter 20.

CHAPTER 19

METALLURGICAL AND REFRACTORY MATERIALS

A group of unrelated minerals finds common use in the large metallurgical industry. Some are used directly for making steel and other metallurgical products; others serve as refractory materials to withstand the high temperatures employed in metallurgical and related furnaces. The group includes fluorspar, cryolite, graphite, and refractory substances, such as magnesite and dolomite, aluminous compounds, refractory clays, diaspore, bauxite, silica, and zirconia.

Fluorspar

Fluorite or fluorspar (CaF_2) is a commercial mineral of critical importance because of its necessity in the steel industry. Since the discovery of this use in 1889 its production has increased rapidly.

Uses. About 50 percent of the fluorite produced is consumed in the basic open-hearth process of steel making. It facilitates fusion and the transfer of objectionable impurities such as sulphur and phosphorus into the slag and gives fluidity to the slag. About 28 percent is used for making hydrofluoric acid, used chiefly in the aluminum industry for making synthetic cryolite. The glass and enamel industry consumes about 15 percent. Other uses are listed below.

METALLURGICAL	CHEMICAL AND MISCELLANEOUS
Iron foundry flux	Extraction of potash from feldspar and portland-cement dust
Electric furnace ferroalloys and alloy steels	Making calcium carbide and cyanamid
Producing nickel, Monel metal, brass	Fluorides and silicofluorides
Iron and manganese fluorides	Making portland cement
Smelting Au, Ag, Pb, and Cu	Bond for emery wheels and electrodes
Refining Cu, Pb, Sb	Insecticides, preservatives, dyestuffs
	Refrigerating fluids
	Ceramic purposes
	Rock-wool manufacture
	Microscope lenses

Production and Distribution. The annual production of fluorite ranges from 900,000 to 1 million tons annually, distributed in the following order: United States, Germany, Russia, France, United Kingdom, Korea, Newfoundland, and Mexico.

The United States output, which ranges between 200 and 350 thousand tons, is produced largely in Illinois, Kentucky, Colorado, and New Mexico. About 30 mines supply 85 percent of the American production.

Occurrence and Origin. Fluorite occurs in fissure veins, and as replacement beds in limestone. It is a common gangue mineral of many cavity fillings and replacement deposits. Fluorite is normally a hydrothermal mineral, and most geologists consider that it is always of igneous derivation. Its occurrences in igneous rocks as well as in limestone indicates that limestone is not essential to its deposition.

Extraction and Preparation. Fluorite is mined in the same manner as underground metalliferous deposits; rarely it is obtained from open pits.

In some deposits fluorite occurs in such purity that a merchantable product is obtained by hand cobbing and sorting. Generally, however, it must be concentrated and cleaned by jigs or tables or by flotation. The last yields a product up to 98 percent pure and enables low-grade deposits to be worked.

The highest-grade product, known as " acid " fluorspar, is used for making hydrofluoric acid and requires 98 percent of calcium fluoride; for glasswork, 97 percent purity is required, and for steel furnaces 85 percent is satisfactory. The products are marketed in gravel, lump, and ground form.

EXAMPLES OF DEPOSITS

Illinois-Kentucky District. This district of 40 square miles spans the Ohio River and is the largest and most productive fluorspar district in the world.

The area is underlain by horizontal Mississippian sediments cut by mica peridotite dikes. It is arched into a broad dome 30 miles across that has collapsed by intricate faulting, some of the faults having displacements up to 1,300 feet.

The deposits are of two types: (*a*) steep replacement veins occupying faults of moderate displacement, and (*b*) horizontal, bedded or blanket replacement deposits. The former are the more important.

The vein deposits are typified by those near Rosiclare, Ill. They are as much as 35 feet wide, and the Rosiclare vein has been mined for 9,000 feet along the strike. The fault in which it occurs is 4½ miles long. The veins contain chiefly fluorite and calcite but also some quartz, barite, and base-metal sulphides. Masses of pure fluorite 10 feet across are not uncommon. Calcite was deposited first, followed by fluorite, which replaced calcite and limestone; in the Daisy mine,

minor calcite was seen to replace fluorite. The veins represent combined replacement and open-space filling; vein matter cements and also
replaces limestone, fault breccia, and wall rock. Vugs lined with
crystals of calcite and fluorite are common. They are typical hydrothermal deposits.

The bedded deposits near Cave In Rock, Ill., occur in the flattish
Fredonia limestone beneath shale. These interesting ore beds are from

FIG. 19–1. Polished surface of high-grade fluorite specimen (90 percent fluorite)
showing dark and light banding. (*Currier, Econ. Geol.*)

a few inches to 12 feet thick, average 4 feet, and contain fluorite of
high purity. The same minerals of the vein deposits, with the addition
of marcasite, are present in minor amounts. Vugs yield handsome
aggregates of fluorite crystals. These deposits exhibit remarkable
parallel banding of fluorite (Figs. 19–1, 19–2), interpreted by Bastin
as rhythmic replacement and by Currier as preservation of bedding and
cross bedding in replaced limestone. Currier believes that the fluorite
resulted from reaction of hydrothermal solutions containing hydrofluoric acid upon limestone, giving calcium fluoride.

The reserves of fluorspar in this field are estimated to be about 15
million tons.

Foreign Deposits.   Important deposits of fluorspar, with ample reserves, occur in *Germany* in the Harz Mountains, Anhalt, Baden, Bavaria, Saxony, and Thuringia, where the spar is an important constituent of base-metal fissure veins.   *French* fluorspar of high purity is obtained from Var, Puy-de-Dôme, Saône-et-Loire, and other places. It occurs in fissure veins associated with chalcedony, barite, and quartz. In *Great Britain* important deposits occur in Derbyshire and Durham in long fissure veins carrying also lead, zinc, barite, and calcite.   Less important deposits occur in Cornwall and Devon in fissure veins carry-

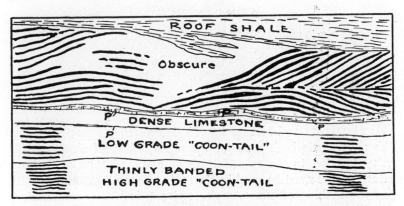

Fig. 19–2.   Sketch of mine face showing banding in fluorite.   Black, impure bands; white, pure bands; *p*, shaly partings.   (*Currier, Econ. Geol.*)

ing ores of lead, copper, tin, and tungsten.   A large production comes from epithermal veins at St. Lawrence, *Newfoundland*.   In *Russia* deposits occur at the Kara Sea and beyond Lake Baikal, and in *China* in Fengtien, Shantung, and Chekiana.   Important fluorite deposits are also found in Bolzano, Italy, Mexico, Spain, Transvaal, New South Wales, British Columbia, and Ontario.

Selected References on Fluorspar

General Reference 11.   Chap. 17, by H. T. Mudd.   *Brief descriptions.*

Fluorspar Industry of the United States.   P. Hatmaker and H. W. Davis.   Ill. Geol. Surv. Bull. 59, Urbana, 1938.   *Occurrence, uses, mining, milling.*   Also General Reference 10, 187–188.

Fluorspar Deposits of Hardin and Pope Counties, Ill.   E. S. Bastin.   Ill. Geol. Surv. Bull. 58, Urbana, 1931.   *Details of occurrence and origin.*

Origin of the Bedding Replacement Deposits of Fluorspar in the Illinois Field.   L. W. Currier.   Econ. Geol. 32:364–386, 1937.   *Reviews theories of origin and proposes replacement.*

Fluorspar in Silicate Melts, with Reference to Rock Wool.   J. S. Machin and J. F. Vanecek.   Ill. Geol. Surv. Bull. 68, Urbana, 1940.

Fluorspar Deposits of the Western United States. J. L. GILLSON. A. I. M. E. Tech. Pub. 1783, 1945. *Survey of western deposits; bibliography.*

Fluorspar Deposits of St. Lawrence, Newfoundland. R. E. VAN ALSTINE. Econ. Geol. 39:109–132, 1944. *Epithermal veins along faults.*

Fluorspar Deposits of the Jamestown Dist., Colo. E. N. GODDARD. Colo. Sci. Proc. 15–1:45–47, 1946. *Fluorite in veins and breccias associated with sulphides.*

Fluorspar Resources of New Mexico. H. E. ROTHROCK, C. H. JOHNSON, and A. D. HAHN. New Mex. Bur. Min. & Min. Resources Bull. 21:245, 1946. *Veins and breccia deposits.*

Bedding Replacement Fluorspar Deposits of Spar Valley, Texas. E. GILLERMAN. Econ. Geol. 43:509–517, 1948. *Replacements in three limestone beds.*

General Reference 5.

General Reference 7, pp. 101–105.

Cryolite

Cryolite is a rare mineral mined in only two places in the world, namely, Miask, Russia, and Greenland. Small amounts of it also occur at Pikes Peak, Colo.

This aluminum and sodium fluoride is desired for flux and solvent in the electrolytic bath used for the manufacture of aluminum. Its scarcity has led to the manufacture of artificial cryolite, but some natural mineral must be used in the bath. A small amount also finds use in the making of enamels, glass, glazes, insulating material, and insecticides.

At Ivigtut, Greenland, on the shore of Arsuk Fiord, cryolite, associated with pegmatite, is quarried from a small mass of porphyritic granite. The deposit is irregular in shape, about 100 by 500 feet in area, and at a depth of 150 feet is increasing in size. The cryolite contains base-metal sulphides, siderite, and fluorite.

Selected Reference on Cryolite

General Reference 11, Chap. 17.

Graphite

Graphite receives its name from *graphein*, " write." Once mistaken for lead, it was called " black lead " (also plumbago), and pencils made from it are still called " lead pencils." Although chemically the same as the diamond and charcoal, it is a crystalline modification of carbon. It ranges in purity from 30 to 98 percent. In commerce it is classified into natural and artificial; the natural varieties are divided into " crystalline " and " amorphous," the only difference between them being that of grain size. Graphite deposits contain from $2\frac{1}{2}$ to 7 percent carbon.

Uses. The high melting temperature of graphite (3,000° C) and its insolubility in acid create many uses for it. The oldest use is in making pencils, which still persists. The softest graphite is finely ground, mixed with clay, and baked, the amount of clay and the time of baking giving the desired hardness. Its chief use is for foundry facings, followed by that for crucibles, lead pencils, lubricants, paints, dynamo brushes, and stove polish. The crucibles are used for melting brass, steel, and other alloys. Graphite mixed with oil makes a good, heavy lubricant. It is also used for electrodes, dry batteries, glazing powder, pipe cement, and other purposes. Material as low as 30 to 35 percent graphite carbon is suitable for paints.

Production and Distribution. The annual production of graphite ranges between 150,000 and 280,000 tons, derived chiefly from Russia, Korea, Bavaria, Austria, Ceylon, Madagascar, and Mexico with minor amounts from Italy, United States, Czechoslovakia, and Norway.

The United States production is derived chiefly from New York, California, Alabama, Georgia, Tennessee, Texas, and Nevada.

Occurrence and Origin. Graphite occurs in metamorphic, igneous, and sedimentary rocks in flakes, lumps, and dust. It originates by:

1. Magmatic concentration — Irkutsk, Siberia.
2. Contact metasomatism — Calabogie, Ontario; Ticonderoga, N. Y.
3. Hydrothermal deposition in veins — Ceylon; San Gabriel, Calif.; Dillon, Mont.
4. Metamorphism — Alabama; Raton, N. Mex.; Bavaria; Madagascar; Sonora, Mexico.

Most graphite originates through metamorphic processes, and these, as well as other processes of origin, are discussed in detail in Chapter 5·9.

Extraction and Preparation. Vein graphite is mined no differently from other vein deposits; the included lump graphite is hand cobbed or screened out. Disseminated flake graphite makes up only a few percent of the enclosing rock and is generally mined from open cuts and concentrated to a product of 80 to 85 percent carbon. The concentrates may be further refined to about 95 percent carbon by finer grinding and screening and rarely by chemical removal of impurities.

EXAMPLES OF DEPOSITS

Russia. The deposits near Irkutsk, Siberia, are among the most important of the world and for long were the chief supply of lump pencil graphite. The graphite occurs in nepheline syenite and adjacent schist.

Other extensive deposits have been developed in the Turukhansk and Yenisei districts, Siberia.

Korea. Korea in 1944 attained the largest production of graphite ever attained by any country. Ninety-six percent of it was amorphous. The deposits occur in schists in layers and lenses, many of which are near the contacts of Cretaceous granites. Several deposits are metamorphosed coal seams; others are metamorphosed graphitic argillites.

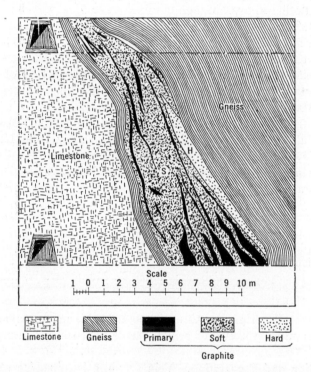

Fig. 19–3. Graphite deposits at Schwarzbach, Czechoslovakia. Black, first quality; *s*, softer; *h*, harder. (*After Stutzer-Pollansch, Lager. Nicht-Erze, V4.*)

Central Europe. The graphite deposits of Bavaria, Austria, and Czechoslovakia, have long been the most productive in the world. The Passau deposits of Bavaria extend into Czechoslovakia (Fig. 7–10). The graphite, which is the flake variety, occurs in large lenses in cordierite gneiss containing bands of schist and limestone, surrounded by granite (Fig. 19–3). It is mainly recovered from the weathered rocks. A large amount of amorphous material surrounds the higher-grade flake. The Austrian deposits are generally similar; those of Styria are thin beds in schist, and those of Mähren are in marble.

Ceylon. The deposits of Ceylon are the most productive of high-grade graphite. They are unusual in that the graphite occurs in fissure veins up to 9 inches wide, filled almost exclusively with large, thick plates and fibrous graphite. Subordinate quartz, pyrite, and calcite are present. The veins cut gneiss and marble and are intruded by granite and pegmatites. Some veins are worked to depths of 1,600 feet. Disseminated graphite also occurs in the gneisses and marble and some occurs in pegmatites.

Madagascar. Since 1917, Madagascar production of flake graphite has exceeded that of Ceylon. The graphite occurs as disseminated flakes or in veins, pockets, beds, and masses in gneisses and schists. The belt extends for 400 miles; there are scores of deposits, and the supply is practically inexhaustible. So far only the weathered portions are worked. The content is 10 to 12 percent.

Mexico. Sonora, Mexico, has been furnishing to the United States a considerable tonnage of graphite, chiefly used for lead pencils, formed by the metamorphism of upper Triassic coal beds by granitic dikes. The beds, according to Hess, are as much as 24 feet thick, and the carbon content averages 86 percent.

Canada. Commercial graphite deposits occur in Ontario and Quebec where flake graphite is disseminated in graphic gneisses, schists, and crystalline limestone. It also occurs in masses enclosed in limestone near igneous intrusions, in narrow veins, and in pegmatite dikes. An unusual deposit is the Black Donald mine, Calabogie, Ontario, where an intrusive cuts Grenville limestones, forming a contact-metasomatic deposit of graphite, calcite, and silicates 15 to 70 feet wide and known to a depth of more than 200 feet. The rock contains up to 60 percent graphite, of which one-fourth is flake and three-fourths is amorphous.

United States. Although graphite has been mined in 27 states, the chief areas of production were Alabama, New York, Pennsylvania, and Texas. Production is now confined to Alabama and New York. The Alabama belt, 60 miles long and 5 miles wide, consists of mica schist and included pegmatites with graphitic bands from 20 to 100 feet wide that average from 2.5 to 3.5 percent flake graphite. Only weathered parts, to depths of 30 to 50 feet, are mined. Prouty thinks that the graphite was derived from former petroleum or petroleum products in the rock.

The New York occurrences lie in the Adirondack region, mostly near Ticonderoga. The important deposits are graphitic quartz-schists carrying 5 to 7 percent graphite. One bed is 8 feet thick, another 3 to 30 feet thick. Workings extend for 2,000 feet along the strike and to a depth of 250 feet. The occurrences are associated with garnet gneiss

and Grenville limestone intruded by granite. Near by, granite pegmatites have intruded limestone and schist producing a contact-metamorphic zone containing contact silicates and graphite.

In Chester County, Pennsylvania, graphite occurs in graphitic schist
and crystalline limestone cut by pegmatites. Similar deposits occur
in the Llano-Burnet region, Texas, where schists contain 10 to 14 percent flake graphite. Ceylon type graphite has been produced from
veins near Dillon, Mont. Near Raton, N. Mex., coal has been metamorphosed to graphite and similarly amorphous graphite has been
formed from coal at Cranston, R. I. Beverly describes deposits in
Los Angeles County, California, in shear zones in schists cut by pegmatites in which the graphite particles replace rock minerals.

Selected References on Graphite

General Reference 11. Chap. 19, by G. R. Gwinn. *Good survey and bibliography.*

Graphite Deposits of Ceylon. E. S. Bastin. Econ. Geol. 7:419–445, 1912. *Excellent geological description.*

Graphite Deposits of Ashland, Ala. J. S. Brown. Econ. Geol. 20:208–248, 1925.
Disseminated type in schists.

Graphite. H. S. Spence. Canada Mines Branch, Bull. 511, Ottawa, 1920. *Excellent general account of graphite.*

Origin of the Graphite of Ceylon. D. N. Wadia. Ceylon Rec. Dept. Miner. Prof.
Pap. 1:15–24, 1943. *Veins, disseminations, and pegmatite occurrences.*

Alabama Flake Graphite in World War II. H. D. Pallister and R. W. Smith.
A. I. M. E. Tech. Pub. 1909, 1945. *Occurrence and developments.*

General Reference 7, pp. 110–111.

Graphite for Manufacture of Crucibles. G. R. Gwinn. Econ. Geol. 40:86, 1945.

General Reference 16. Pp. 583–586. *Sources and uses.*

Refractories

Modern high-temperature metallurgical processes require refractory
materials for furnace linings that will withstand high temperatures; if
the materials show signs of fusion below 1,500° C they are not classed
as refractories. Other requirements are that the refractories will not
react with the materials being melted, that they are strong enough to
withstand the weight and wear of molten metals and slag, that they
resist cracking and spalling under temperature changes, and that they
can be molded into bricks or other forms. Some refractories are required for high-temperature purposes other than furnace linings, such
as for retorts, for electrical purposes, and for ceramics.

Purposes and Uses. Tyler and Heurer estimate the main uses of refractories to be distributed about as shown below.

Industry	Percent	Industry	Percent
Iron and steel	50	Glass making	5
Public utilities	20	Oil refining	4
Nonferrous metals	6	Ceramic	3
Cement and lime	5	Miscellaneous	7

In the iron and steel industries refractories are required for iron ore blast furnaces, steel furnaces, open hearths, soaking pits, reheating furnaces, boilers for generating steam, and coke ovens. Public utilities use them for electrical power and gas plants, locomotives, incinerators, and other purposes. Refractory linings are necessary for the smelting and refining of metals, making alloys, cement kilns, glass furnaces, oil stills, holders or saggers for firing ceramic materials, making spark plugs, and a host of minor purposes in which high temperatures are involved.

The typical products supplied are bricks and other pre-formed standard shapes and numerous special shapes, such as retorts, muffles, tiles, saggers, cupolas, and glass-tank blocks.

Principal Varieties of Refractories

The common refractories are generally grouped under fire clay, silica, high alumina, magnesite, and chrome. The chief materials used in the United States are shown in Table 24.

All the materials, except chrome, are nonmetallic products. Most of them have already been discussed in other sections of this book.

Fire Clays. Fire clays make up the bulk of modern refractories. Most of them are combinations of nonplastic refractory flint clays, moderately refractory plastic clays, and high-alumina clays. Their heat resistance ranges from 2,714° to 2,984° F, and they are utilized for furnaces, boilers, oil stills, and miscellaneous furnace linings. High-grade fire brick made from fire clays and the flint clays of Missouri yield superior brick with anti-spalling properties. Refractory bricks of fire clay are furnished for every purpose needed. Refractory cements are also made for these bricks, some of which contain silica or high-alumina compounds.

The occurrence, origin, and distribution of the clays used are discussed under Ceramics, Chapter 17.

Silica Refractories. Silica in its numerous natural forms is a widely used refractory. The most common forms are quartzite (ganister) and crushed quartz, which, bonded with 2 percent of lime, yield silica brick, an important acid refractory product utilized in many metallurgical processes. Common sand and diatomite are also used, and the Berea sandstone of Ohio is sawn into bricks and blocks. The chief ganister

TABLE 24

CHIEF REFRACTORY MATERIALS USED IN UNITED STATES

Material	Chief Constituents	Upper Temperature Limit, °F	Important Sources
Clays:			England, Germany, Belgium,
Kaolin	Al_2O_3,SiO_2	3,245	Pennsylvania, Georgia, Alabama,
Fire clay	Al_2O_3,SiO_2	3,173	Maryland, Louisiana, Ohio,
			Missouri
Silica:			
Quartz	SiO_2	3,092	Pennsylvania, Ohio, Alabama
Cristobalite	SiO_2	3,092	California
Diatomite	SiO_2	2,939	California, Nevada
High alumina:			
Bauxite	Al_2O_3,H_2O	3,668	Arkansas, Guianas, Gold Coast
Diaspore	Al_2O_3,H_2O	3,632	Missouri
Corundum	Al_2O_3	3,686	Canada, South Africa, artificial
Sillimanite group	Al_2O_3,SiO_2	3,290	California, Nevada, North Carolina, Virginia, India, Kenya
Magnesia:			
Magnesite	MgO,CO_2	5,072	Washington, California, Manchuria, Greece, Austria
Dolomite	MgO,CaO,CO_2	4,505	Numerous places
Periclase	MgO	5,072	Rare, artificial
Spinel	MgO,Al_2O_3	3,812	Artificial
Brucite	MgO,H_2O		Nevada
Chrome	Cr_2O_3,FeO	3,722	Africa, Cuba, Turkey, New Caledonia
Miscellaneous:			
Beryllia	BeO	4,352	Brazil, Argentina
Graphite	C	Infusible	United States, Ceylon, Madagascar, Mexico
Limestone	CaO,CO_2	4,505	Everywhere
Rutile	TiO_2	2,966	Virginia, Brazil, India
Thoria	ThO_2	5,522	Artificial
Yttria	Y_2O_3	4,370	Artificial
Zircon and baddeleyite	ZrO_2,SiO_2	4,178	India, Brazil, Australia

deposits in the United States are the Tuscarora and Baraboo quartzites of Pennsylvania and Wisconsin, respectively.

Silica bricks are highly refractory and do not soften appreciably before their melting point, but they do spall when suddenly heated or cooled.

High-Alumina Refractories.   The *sillimanite group* of minerals (sillimanite, andalusite, kyanite, dumortierite) all convert to mullite with a melting point of 3,290° F.   This yields a high-cost but much desired refractory, especially for highly refractory and tough porcelain bodies used for spark-plug cores, electrical porcelains, saggers, and laboratory porcelains. The geology of these minerals is discussed in Chapter 5·9, and their ceramic uses in Chapter 17.

Bauxite (Chap. 13) with the addition of a clay binder is utilized for making high-alumina refractory brick. Because of impurities present such bricks fuse at a lower temperature than the melting point of alumina. They are used chiefly for linings for open-hearth steel furnaces. Bauxite brick is now largely displaced by diaspore brick.

High-alumina refractory bricks are made from *diaspore* and combinations of plastic and flint clays. "Sixty percent" brick is substituted for fire-clay brick in boilers, and " seventy percent " brick is used for boilers, furnaces, and portland-cement kilns. The diaspore used contains 85 percent alumina and is obtained from Missouri, where it occurs in porous masses or as oölites in clay.

Magnesia Refractories.   (See Chap. 5·5)   Dead-burned *magnesite* is desired chiefly as a basic refractory for basic slags in metallurgical furnaces, for kilns, and for use with corrosive materials. The successful development of the basic converter, basic open-hearth, and electric furnaces is due in no small part to its use. The carbonate is calcined at 2,730° F, the carbon dioxide is driven off, and the magnesia is converted to *periclase*. This is made into magnesite brick or used as granules, which form the hearth beds of basic steel furnaces, but not the roof because of a tendency to expand and spall. Magnesite is also used for " unburned magnesite brick," which resists spalling better than the ordinary magnesite brick. The occurrence of magnesite is given in Chapter 18.

Dolomite, when dead burned, is used for magnesia brick for basic open-hearth furnaces, being preferred to magnesite because it is cheaper.

Spinel is made artificially for refractory purposes by fusing calcined magnesite and alumina in an electric furnace. The product is then mixed with some ball clay and made into refractory brick, which has the formula of spinel. It is used as a special furnace brick and also for refractory cement.

Brucite, in minor quantities from large deposits in Nye County, Nevada, is calcined for a basic refractory for uses similar to those of magnesite brick.

Chrome (Chap. 14). This mineral makes one of the best refractory substances known because of its high melting point and remarkable physical and chemical stability. It is made into high-duty brick that is used for side walls at the slag line in steel, copper, nickel, and chemical furnaces.   Because of its lower cost it is displacing magnesite brick in certain fields.   About 2.5 pounds of chromite is used per ton of steel produced.   It is considered essential in steel furnaces.   Chromemagnesite bricks with about 75 percent chrome are also used.   Chrome ore suitable as a refractory contains from 33 to 48% Cr_2O_3, 12 to 30% Al_2O_3, and about 17% MgO.

Zirconia Refractories.   Zirconia (ZrO_2) is one of the most refractory substances known.   It is strong and hard, resists spalling, and is relatively inert to slags and corrosive substances.   Zirconia crucibles withstand temperatures up to 4,532° F.   They are used for refining precious metals and in electric furnaces.   Zirconia and zircon brick are also made and zircon cement is used to coat other refractories.

The zirconium minerals used are zircon, the silicate, and baddeleyite, the oxide.   They are obtained from beach sands in India, Brazil, Florida, and New South Wales, along with monazite and ilmenite.   For their geologic occurrence see Chapter 5·7.

Miscellaneous Refractory Materials.   *Graphite* and *carbon* are utilized to make refractory carbon brick.   The graphite used is both natural and electric furnace product.   Carbon is obtained from coke and charcoal.   Carbon bricks are used for certain types of furnaces, boilers, and acid vats.   They are, however, consumed in an oxidizing atmosphere at high temperatures, or in contact with fused oxides of metals.

Beryllia, with a melting point of 4,172° F, forms a porcelainlike material suitable for crucibles.   It displaces alumina for higher-temperature work.   *Rutile,* with lime as a binder, is made into brick for use as an acid refractory.   *Thoria* yields a high-temperature refractory, more basic than zirconia, for extra heavy duty.   *Olivine* has been found to be a desirable refractory material to replace chromite in steel furnaces, and for use above the slag line in nonferrous furnaces. Olivine is also mixed with magnesite to form the low-cost " forsterite " brick, which has excellent refractory qualities.   Olivine, the chief mineral of dunite rock, is produced in North Carolina and Washington. *Soapstone* is sawed to desired shapes and used as a refractory for alkali recovery furnaces.   *Pyrophyllite* blocks are also used for refractory purposes.   *Talc* makes a very hard refractory for small shapes, such as gas tips; and *vermiculite* finds some uses.   *Silicon carbide* (trade name, Carborundum) is an artificial product of excellent refractory qualities up to 4,064° F.   It is strong, acid proof, has high heat conductivity, and is used generally for the same purposes as high-alumina brick.   *Fused alumina* (alundum, aloxite) is also widely used as a special artificial refractory for muffles, brick for certain types of furnaces, and for small-scale high-temperature work.

Selected References on Refractories

General Reference 11. Chap. 39, Refractories, by P. M. TYLER and R. P. HEUER. *Best all-around article available.* Chap. 24, Magnesite, by R. E. BIRCH and O. M. WICKEN. Chap. 42, Sillimanite Minerals, by F. H. RIDDLE and W. R. FOSTER.
Clays. H. RIES. John Wiley & Sons, New York, 1927. *Fire clays.*
Refractories. F. H. NORTON. McGraw-Hill, New York, 1931. *Types and uses.*
General Reference 5.

Foundry Sands

Foundry sands are siliceous sands utilized to make forms for casting metals, and molding sands are those used for making molds. Some are used for the cores of castings. Since a bond is necessary to make foundry sands adhere, a certain proportion of clay is essential, but clay or other binder may be added. Core sands are bonded with oil, resin, pitch, or other substances. The specifications for foundry sands are rigid. According to Ries the important properties are fineness, bonding strength, permeability, sintering point, and durability. The grains are partly angular and range in size from 20 microns to 3 mm. The fineness affects permeability, strength, and smoothness of casting. Permeability is necessary to permit escape of gases. If the sintering point is low the sand will " burn " on to the casting.

Most foundry sands are of marine or lacustrine sedimentary origin. Dune sands or sandstones may also be used.

Foundry sands are rather widely distributed. Prominent United States occurrences are the Hudson Valley and Tertiary beds of New Jersey, Ohio, Indiana, Illinois, and California.

The annual United States production ranges from 2 to 3 million tons.

Selected Reference on Foundry Sands

General Reference 11. Chap. 45, by H. RIES.

Limestone and Lime

Limestone and lime are variously employed in the metallurgical industry. Some 25 million tons of limestone is utilized for fluxstone in metallurgical furnaces, particularly those of iron and nonferrous metals. It serves to offset silica, to yield a thin slag, and to produce a basic slag that collects and retains impurities separating from the metal.

Lime is finding increasing metallurgical uses. Of the 6 million tons of lime made annually in the United States, about 20 percent is employed metallurgically for fluxes in furnaces (78%), ore concentration (20%), wire drawing, and foundry uses. In basic open-hearth steel furnaces lime is added (with limestone) to flux off impurities. With increasing use of scrap iron more lime is added. It is also used as a flux in electric steel furnaces and nonferrous smelters. In the flotation process of ore dressing, lime is added to make an alkaline circuit. In cyaniding, lime is added to neutralize soluble acid salts, to coagulate slimes, and to protect against carbon dioxide.

Refractory lime or dead-burned dolomite, which amounts to about 10 percent of all lime made, is used as a furnace refractory.

Other Metallurgical Materials

Furnace Sand. This is used to line bottoms and walls of open-hearth and other furnaces. Such sands are high in silica; either they contain some bonding clay or else some fine clay is added. Mixed grain sizes are desirable so that the voids will be filled. When sintering occurs a hard, resistant lining is formed.

Bauxite. In addition to its many other uses (Chap. 13), bauxite serves as a slag corrective in iron smelting.

Borax. Borax makes a good flux in brazing, welding, and soldering metals and is also utilized in smelting copper, refining gold and silver, and in assaying. Some borides act as deoxidizers to prevent scale and scum in making brass and bronze, and ferroboron is used in small quantities as a deoxidizer for certain steels. Boron steels of remarkable strength are made with as much as 2 percent boron.

Strontium Minerals. Strontium metal is alloyed with copper to increase hardness and decrease blow holes. Strontium and tin added to lead make a harder and more lasting metal for storage batteries. With tin it serves as a scavenger for metals containing copper, lead, zinc, or tin. In Germany strontianite is employed to desulphurize and dephosphorize iron and steel by making a thin slag.

Dolomite is employed as a flux in blast furnaces, particularly in the making of ferrosilicon and ferromanganese because it carries very little of the silica or the manganese into the slag.

Phosphate (Chap. 22) is employed metallurgically as ferrophosphorus, produced by smelting phosphate rock and iron ore, as an addition to special steels. The phosphorus content of pig iron is adjusted by the addition of phosphate rock.

CHAPTER 20

INDUSTRIAL AND MANUFACTURING MATERIALS

A large number of unrelated nonmetallic materials are used for industrial and manufacturing purposes. Wares are made wholly of them, or they form important constituents of manufactured products, or they are employed in the production of other industrial products. This group does not include minerals utilized in the chemical industries, as they are treated in the following chapter. The group includes:

Asbestos	Mineral filters
Mica	Optical minerals
Talc	Lime
Barite and witherite	Meerschaum
Glass sand and materials	Miscellaneous materials
Mineral fillers	

Asbestos

Asbestos is unique among materials of the mineral kingdom in that it consists of delicate, flexible fibers, so soft and silky that they can be spun readily into threads and woven into cloth. Moreover, it resists fire and acid and is a good nonconductor of heat and electricity. These properties make it of considerable value. Although used by the Romans for making lampwick that would not burn out and for cremation cloths that became cleansed by fire, its real utilization did not start until 1878, after the discovery of the large Quebec deposits. During the twentieth century it has become one of the principal industrial minerals, and for many purposes, such as brake linings, it has no adequate substitute.

Uses. The uses of asbestos are so numerous that a listing given by Ross occupies four pages. There are two principal use-classes, spinning and nonspinning fiber. Spinning fiber comprises the longer grades of flexible fibers, and nonspinning comprises the shorter lengths of the same, and other short, matted, or harsh fibers. Spinning fiber is the most valuable and is spun into threads and yarns and woven into textiles. The fibers must possess strength and flexibility; individual filaments may be only 1/25,000 inch in diameter. The largest use of asbestos is for motorcar brake band linings and clutch facings, and an important one is for gaskets for steam fittings. The nonspinning fibers are chiefly valuable for various types of fireproofing and heat insulation materials. Some of the important uses are shown below.

SPINNING FIBER

Asbestos cloth:
  Brake lining, clutch facings
  Theater curtains and scenery
  Gaskets and sheet packings
  Wall linings
  Firemen's clothing
  Fireproof gloves
  Floor rugs and linings
  Blankets and bags
  Welding and acid equipment
  Heat insulators
  Steam and chemical equipment
  Filter pads
  Ovens
Asbestos rope, tape, yarn, cord:
  Fire-department equipment
  Fire mats
  Deck mats for war vessels
  All kinds of pipe and joint packing
  Covering for electrical wires
  Gaskets
  String asbestos cloth
  Laboratory uses
  Fire, acid, electrical equipment
  Wicks

NONSPINNING FIBER

Asbestos millboard:
  Housing, household, and factory wall-
    board
  Gaskets
  Stoves, ovens, boilers
  Kilns, furnaces
  Garages
  Various insulators
Asbestos paper:
  Floor and roofing paper
  Pipe covering
  Numerous heat insulators
  Electrical and steam equipment
Asbestos cements:
  Coverings for furnaces, pipes, etc.
  Floors, walls
  Heat tables, etc.
  Acid and chemical equipment
  Burners
Asbestos shingles
Composition materials:
  Various binders for porcelains, plas-
    ters, hard rubber, phonograph rec-
    ords, electrical insulators

Production and Distribution. The world production of asbestos ranges from 500,000 to 600,000 tons of fiber derived mainly from the following countries in decreasing order: Canada (60%), Russia, southern Rhodesia, South Africa, Swaziland, United States, Finland, Japan, Italy, Korea, Cyprus, and Czechoslovakia.

The United States consumes over half of the world's supply of spinning fiber but produces very little of it. Most of the fine-quality spinning fiber comes from Canada, Russia, Rhodesia, South Africa, and Cyprus. The large Canadian production emanates from a small area in eastern Quebec.

Asbestos Minerals. Asbestos is a commercial term applied to a group of minerals that separate readily into fibers. The minerals differ in chemical composition and in the strength, flexibility, and usefulness of their fibers. Broadly, they fall into two groups — serpentine and amphibole; the former includes the mineral chrysotile and the non-commercial picrolite, and the latter includes anthophyllite, crocidolite, amosite, tremolite, and actinolite. Mountain leather, mountain wood, and mountain cork are varieties of amphibole.

Chrysotile is the most valuable variety. Its fibers are fine, silky, and strong; 32,000 feet of thread can be spun from a pound of it. Some varieties stand temperatures up to 5,000° F.

Crocidolite, or " blue asbestos " from South Africa, is a long, coarse, flexible spinning fiber with low fusibility and high resistance to acids.

Anthophyllite occurs as mass fiber, which is short, brittle, and non-spinning and is used chiefly for insulation. *Amosite* is an iron-rich variety of anthophyllite that occurs in unusually long, splintery, coarse fibers, some of which can be spun but which are used chiefly as a binder for heat insulators.

Tremolite and *actinolite,* except for the Italian tremolite, have little commercial value; some actinolite is ground up for insulating purposes.

The length of fiber is important in determining the utility and value. Long fibers give strong yarns and threads; short fibers cannot be spun and are suitable for lower-cost products, such as boards, shingles, and cements. The fiber is graded on the basis of length, and identical quality ranges in price from $20 to $900 a ton according to length of fibers.

Occurrence and Origin. The mode of occurrence and origin of the different kinds of asbestos are stated in Chapter 5·9.

Extraction and Preparation. Asbestos deposits are mined by large- and small-scale methods, both in open cuts and underground. In Quebec, the fiber rock is extracted from large open cuts and loaded by power shovels. Some of the pits are so deep that underground methods are employed for the deeper rock. The rock as quarried averages 6 percent fiber.

In milling, distinction is made between " crude," " mill fibers," and " shorts." No. 1 Crude is spinning fiber over ¾ inch in length, and No. 2 Crude is from ⅜ to ¾ inch; both are hand cobbed on a picking belt. Mill fibers are freed by crushing and beating the fiber rock, and shorts are the lowest-grade mill products. If fiber ⅜ inch long is worth $100 a ton, ¾-inch length is worth $500 a ton; therefore, in milling, if long fiber is broken its value may be reduced three-quarters. Consequently milling methods aim to separate fibers with minimum breakage. Crude is generally pounded by hand to free attached serpentine. Mill rock is crushed and screened in stages, the fiber being drawn off by suction. The crudes are prepared by mechanical beating and combing to separate and fluff the fiber for threads or yarns.

EXAMPLES OF ASBESTOS DEPOSITS

Examples of chrysotile in serpentinized ultrabasic rocks are: Thetford-Black Lake, Quebec; Shabani, Rhodesia; Barberton, Transvaal; Bajenova, Saja and Uss, Russia; and Cyprus.   Examples in serpentinized limestone are: Arizona; Laiyuau, China; Carolina, Transvaal; Madras, India; and Minusinsk, Russia.   Crocidolite and amosite are confined to South Africa and Australia, and anthophyllite to the United States, Newfoundland, and Finland.   Some tremolite is mined in Italy.

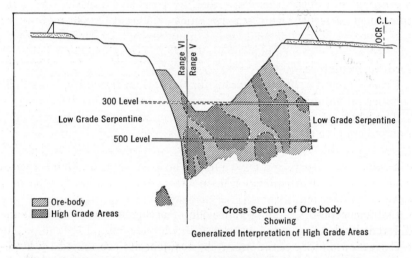

FIG. 20–1.   Section of King asbestos mine, Quebec, showing ore body and high-grade areas (cross hatched).   (*Ross, Bull. Can. Min. Met.*)

Quebec.   The Canadian deposits occur around Thetford-Black Lake and East Broughton in eastern Quebec, where sills of serpentinized dunite cut metamorphosed Cambrian sediments.   This forms a belt 70 miles long by 6 miles wide, which projects into Vermont.   In places the dunite has been completely serpentinized; all of it is partly altered, and chrysotile occurs only in serpentine.   According to Cooke all the deposits (Fig. 20–1) occur close to the northeast borders of the three structural types: (1) those associated with a single fault or fault system; (2) those in blocks of ground bounded by downward converging faults; (3) those in blocks bounded by upward converging thrust faults. Individual deposits are as much as 800 feet in length.

The chrysotile is mostly soft cross fiber, but slip and harsh fiber occur (Fig. 20–2).   The bulk of the production is obtained from veinlets ¼ to ½ inch wide; a little is found up to 5 inches long.

The yield of fiber is about 5 percent of the rock mined and 7.7 percent of the rock milled. Crude amounts to 1½ percent of the total fiber, and 99 percent of the fiber is less than ⅜ inch long. Pits extend 500 feet, drilling has revealed commercial fiber to a depth of 1,700 feet. The mills of the district can handle about 1,000 tons of fiber rock per hour.

Russia. The asbestos of Russia occurs chiefly in the Bajenova district, Urals, with minor amounts in Minusinsk, Yenisei River, and Maikop and Laba River in the Caucasus. In the Bajenova district, according to Keyser, there are some 20 pits developed in ellipsoidal masses of a serpentine ultrabasic rock, 3,500 by 1,000 feet, which

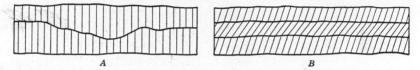

FIG. 20–2. Types of Quebec asbestos veins. *A*, Cross fiber, irregular medial; *B*, bent fiber.

disclose cross-fiber chrysotile veins. The fiber contains sufficient ferrous iron to cause discoloration upon weathering. The quality is lower than Canadian, but, although the percentage of total fiber is about the same, the proportion of spinning fiber is higher. The reserves are large. Similar but smaller deposits have been found in the Sayan Range, west of the Altai Mountains, and in the Aktovrak and Uss districts.

Southern Rhodesia. This country contains much good spinning chrysotile in three districts, the Bulawayo (including the Shabani mine), Victoria, and Lomagundi.

The Shabani deposit occurs in the central part of a serpentine mass 10.5 by 1 to 3 miles, which has been altered from dunite, perhaps as the result of a granite intrusion. The fiber zone is up to 3 miles long and 600 feet wide, but the best fiber occurs in the lower 20 to 200 feet of the inclined serpentine body. Veins of chrysotile are up to 6 inches wide, but because of partings they rarely yield more than 3-inch fiber. The fiber seams parallel the contact. About 25 to 30 percent of the fiber produced is of spinning grade, and the shorts are not utilized. The depth is unknown but reserves are stated to be 7 million tons. Similar deposits are mined at Victoria and Lomagundi along the Great Dyke.

South Africa. This country first produced the unique crocidolite and amosite. It also yields high-grade chrysotile from two types of occurrence: (1) cross-fiber ribbon rock in serpentinized dunite, in the Barberton district, and (2) cross fiber in a bed of serpentinized dolomite, in the Carolina district.

The *Barberton* occurrence exhibits remarkable and puzzling geological features. Near Kaapsche Hoop is a gently dipping (10° to 20°) serpentinized peridotite sill that is fiber bearing for 3 miles in length and which, in the New Amianthus mine, has been followed down-dip for over 2,000 feet. There is an unusual zone or " ribbon line " or

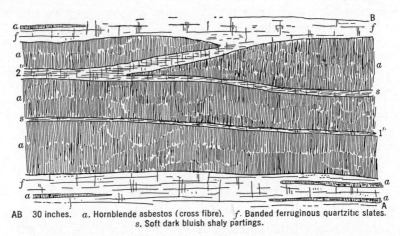

AB   30 inches.   *a*. Hornblende asbestos (cross fibre).   *f*. Banded ferruginous quartzitic slates.
s. Soft dark bluish shaly partings.

Fig. 20–3. Amosite asbestos occurrence, Penge mine, Transvaal, South Africa. Section, 30 *m*; *a*, hornblende asbestos; *f*, ferruginous quartzitic slates; *s*, shaly partings.

rhythmic banding, from 5 to 20 feet high, with cross-fiber veins parallel to the sill contact. The bottom 6 to 8 feet contains widely spaced fiber seams from 2 to 9 inches thick that make up about 10 percent of the entire rock. Above these are thinner (¼ to ¾ inch), more closely spaced seams, and then a zone of closely spaced ribbon rock with glistening seams of silky chrysotile, averaging 15 to 30 seams to the foot, that aggregate 30 percent of the rock. Above this is a dolerite sill. The fiber recovery is about 20 percent of the rock mined. About 8 percent of the output is of spinning grade and 5 percent is over 1 inch in length. The remarkable parallelism and rhythmic alteration of the ribbon rock is baffling.

The *Carolina* district contains long, spinning chrysotile, within a 5-foot zone of dolomite, altered to serpentine, where it overlies a dolerite sill. Serpentine and fiber are of replacement origin.

The amosite deposits (Fig. 20–3) occur near Penge, Transvaal, as cross-fiber veins interbedded with banded siliceous-ferruginous slates.

The fiber attains the remarkable length of 11 inches and averages 6 inches, but it is harsh and splintery. Its origin is an enigma.

The blue crocidolite is found near the amosite, under similar conditions, in a belt 250 miles long. The fiber length is up to 4 inches, about 12 to 20 percent of the total fiber being of spinning grade; the percentage of fiber is high. The reserves are large.

Cyprus. Chrysotile occurs at Amiandos, where a dunite plug is peripherally altered to serpentine, which contains irregular veinlets of " shingle-stock " fiber too short for spinning. The asbestos content runs only 1 to 2 percent.

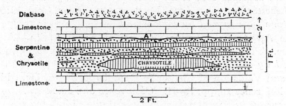

FIG. 20–4. Asbestos bands in serpentine bands in dolomite, Sierra Ancha, Ariz. (*Bateman, Econ. Geol.*)

United States. *Vermont* contains a few small deposits of the Quebec type, but with negligible spinning fiber. Most of it is mill fiber that is used for molded brake linings, shingles, paper, and pipe covering.

In the Sierra Ancha of *Arizona* are small deposits of spinning chrysotile, free from iron. The deposits contain one or more horizontal seams of cross fiber within dolomite limestone beds of which certain bands have been altered to serpentine (Fig. 20–4). The fibers, which are as much as 6 inches long, occur only where the limestone has been cut by or overlies dolerite sills. Discontinuous seams occur en echelon. A replacement origin seems indicated. Somewhat similar occurrences are found in the Grand Canyon region.

In *California*, small quantities of mill fiber with a little spinning fiber occur in serpentinized dunite.

Anthophyllite is produced commercially at Sall Mountain and Hollywood, *Georgia*, where 90 to 95 percent of the rock quarried is short fiber that is used for fireproofing, insulating, and in paint.

Selected References on Asbestos

Asbestos. OLIVER BOWLES. U. S. Bur. Mines Bull. 403, 1937. *Summation of geology, occurrences, and industry.*

General Reference 11. Chap. 2, by G. F. JENKINS. *Concise geology and technology; good bibliography.*

Chrysotile Asbestos in Canada. J. G. Ross. Canadian Mines Branch Mem. 707, 1931. *Detailed occurrence.*

Asbestos Deposits of Thetford Area, Quebec. H. C. Cooke. Can. Geol. Surv. Sum. Rept. 1931:1–24; 1932:41–59; Roy. Soc. Can. Trans. 29:7–19, 1935. *Geology, origin.* Also in Econ. Geol. 31:355–376, 1936.

Geology of the Shabani Mineral Belt, Southern Rhodesia. F. E. Keep. S. Rhodesia Geol. Surv. Bull. 12, 1929. *Description of Rhodesian occurrences.*

Asbestos in South Africa. A. L. Hall. S. Africa Geol. Surv. Mem. 12, 1918. *Occurrences of crocidolite, amosite, and chrysotile.*

Geology of Quebec, Vol. 2. J. A. Dresser and T. C. Denis. Geol. Rept. 20: 413–442. Dept. of Mines, Quebec. 1944. *Occurrence and origin.*

Asbestos in the Barberton District, South Africa. A. L. Hall. South Africa Geol. Soc. Trans. 1921:168–181; 1923:31–49. *Unusual chrysolite occurrence.*

An Arizona Asbestos Deposit. Alan M. Bateman. Econ. Geol. 18:663–683, 1923. *A limestone type formed by replacement.*

Origin of the Amphibole Asbestos Deposits of South Africa. A. L. DuToit. Geol. Soc. S. Africa Trans. 48, pp. 161–206, 1946. *Crocidolite and amosite developed by alteration of sediments.*

Examination and Valuation of Chrysotile Deposits in Serpentine. M. J. Messel. **General Reference 16,** pp. 79–84. *Occurrence and valuation.*

General Reference 6.

Mica

The amount of mica consumed annually is relatively small, but it is essential to the electrical industries because of its form and physical properties. No other mineral combines in itself so many desirable properties for the qualifications imposed by the electrical industry.

Mica Minerals. The chief mica minerals are listed below.

Mineral Name	Commercial Name	Chief Constituents in Addition to Al and SiO_2
Muscovite	White mica	Potash
Phlogopite	Amber mica	Magnesia, potash
Biotite	Black mica	Magnesia, iron, potash
Vermiculite	Jeffersite	Magnesia, iron, potash
Lepidolite	Lithia mica	Lithium, fluorine, potash
Zinnwaldite	Lithia-iron mica	Lithium, iron, potash, fluorine
Roscoelite	Vanadium mica	Vanadium, magnesium, iron
Fuchsite	Chrome mica	Chrome

The first three are the only micas employed commercially as mica, and the first two are the only principal ones; the lithium and vanadium micas are used as sources of those elements, and vermiculite is used chiefly for sound and heat insulation.

Properties and Uses. An important property of mica is its perfect basal cleavage, which permits it to be split into sheets or films. Muscovite and phlogopite are split into sheets 1/1,000 of an inch in thickness; biotite does not split so well. This property, combined with low heat conductivity, high dielectric strength, flexibility, resilience, and toughness, make the light-colored micas outstanding electrical insulators. Muscovite withstands temperatures up to 550° C and phlogopite up to 1,000° C. At higher temperatures they lose their water content. Crystallographic imperfections of twinning, intergrowth, or distortion, and foreign mineral inclusions spoil their splitting qualities, and much of the mica mined can be used only for scrap and grinding mica, to make mica powder.

Most of the sheet mica produced is used for electrical insulation, chiefly in condensers, tubes, radio, radar, and television. Large sheets go into electrical heater elements, such as toasters, and smaller sizes into segments, discs, and washers for electric-light sockets, and fuses. Films are bonded with shellac into plates that are cut to desired shapes. Airplane spark plugs are built of bonded phlogopite. Clear sheets are used for heat windows and gas masks.

Wet ground mica gives luster to wallpaper and some paints; it is used as a filler in rubber and some plastics. Dry ground mica is used for roofing materials, stucco finish, lubricants, rubber dusting powders, and molded electric insulation. Biotite, obtained from biotite schists, is little used in sheets, and most of it goes into ground mica.

Most mica is won by quarrying with carefully controlled blasting to prevent damage to the crystals. It is hand-cobbed, sorted, freed of adhering rock, and broken into sections. Skilled trimmers hand-split and trim the sheets into marketable sizes. The larger the sheet the more valuable the mica. Much waste and scrap result from the production of sheet mica. Mica is marketed as bloc, sheet, splittings, punch, and wet or dry ground.

Production and Distribution. The annual production of mica ranges between 50,000 and 65,000 tons. The United States, whose large production is mainly scrap and ground mica, produces only a small part of its needs of sheet mica. The deposits of Bihar and Madras in India supply most of the world output of sheet mica for electrical purposes, chiefly for the United States. Brazil, Madagascar, United States, and Canada supply most of the remainder; and small amounts come from Russia, Argentina, South Africa, Korea, Ceylon, Scandinavia, East Africa, Rhodesia, and Bolivia. India supplies muscovite, and Madagascar and Canada phlogopite only.

Occurrence and Origin. Commercial muscovite occurs only in silicic pegmatites, mostly in association with granitic intrusives. Commercial phlogopite comes from quartz-free, basic pegmatites in Canada and Madagascar. Biotite is obtained from metamorphic biotite schists. Lithia mica is confined to granitic pegmatites of limited areas. Vermiculite is regarded as hydrothermally altered biotite, which at Libby, Mont., represented originally rich concentrations in a basic intrusive. Roscoelite occurs in sandstones in Colorado, Utah, and Arizona. Commercial mica occurs as zonally distributed " books " in the pegmatites. A single crystal weighing over 2 tons was found at Spruce Pine, N. C.; a crystal 12 feet across was uncovered in South Africa, and a crystal 6.5 by 9.5 feet weighing 7 tons was found at Eau Claire, Ontario. Mica rock carries about 10 percent of mica, of which only 1 percent is of sheet size.

The mica of pegmatites has been formed from magmatic solutions. For long it was considered to be an igneo-aqueous deposit, later it was generally held to be formed by hydrothermal replacement of earlier minerals, but the trend of thought is swinging back toward original deposition.

Selected References on Mica

General Reference 11. Chap. 26. E. H. DAWSON. *Comprehensive, good annotated bibliography.*

Mica. H. S. SPENCE. Can. Dept. Mines, Mines Branch, No. 701, Ottawa, 1929. *Canadian and worldwide occurrences.*

Mica in India. C. S. Fox. India, Geol. Surv. Rec., Vol. 70, 1935. *Occurrence and production.*

Mica. H. W. HARTON. U. S. Bur. Mines Inf. Circ. 6822, 1935. *Review of American industry.*

Mica. J. A. DUNN. Rec. Geol. Surv. India, Vol. 76, Bull. 10:80, 1942. *Indian mica; uses, occurrences, mining, preparation, and statistics.*

Muscovite Mica in Brazil. D. D. SMYTHE. A. I. M. E. Tech. Pub. 1972:1–24, 1946. *Occurrence, types, quality, and evaluation.*

The New England Mica Industry. H. M. BANNERMAN and E. N. CAMERON. A. I. M. E. Tech. Pub. 2024:1–8, 1946. *Occurrence, production, yield, and future outlook.*

Mica in War. R. G. WAYLAND. Min. Tech. Tech. Pub. 1749, 1944. *Qualities, uses, and supplies.*

Muscovite in the Spruce Pine District, North Carolina. T. L. KESLER and J. C. OLSON. U. S. Geol. Surv. Bull. 936-A, 1942. *Geology of most important American occurrences.*

Internal Structure of Granitic Pegmatites. E. N. CAMERON, *et al.* Mon. 2. Econ. Geol. Publishing Co., 1949.

General Reference 7, pp. 139–142.

General Reference 5.

Talc

Talc, the softest of all minerals, is known to everyone because of talcum powder. The pure, soft mineral in trade is called talc, but the term is also used to include *steatite,* a massive compact variety, and *soapstone,* which consists essentially of talc but includes other minerals. Pyrophyllite is oftentimes improperly included with talc because it is used for some of the same purposes.

Properties and Uses. The softness, flakiness, and stability of talc make it a desirable substance for many purposes. In industry, it is roughly classified into hard and soft, and fibrous and flaky, and is marketed as crude, ground, and sawed; about 90 percent of the United States talc is marketed as ground talc. The fibrous variety contains tremolite. In the United States talc is utilized in various industries as follows (Minerals Yearbook):

	PERCENT		PERCENT		PERCENT
Paint	23	Roofing	12	Toilet articles	4
Ceramics	14	Rubber	14	Foundry facings	2
Insecticides	14	Paper	7	Other uses	10

In paint, talc ("asbestine") serves as an extender and is the white pigment of cold-water paints. In ceramics, ground talc is used for wall tile, electrical porcelain, and dinnerware. Calcined talc or "lava," formerly used for gas tips, is harder than steel and is made into blocks that can be intricately tooled and threaded for use in electrical insulators, radio tubes, and refractories. Only the finest grades of talc are used in toilet powders, lotions, and face creams.

Soapstone is also utilized for ground talc but goes chiefly into the smelting furnaces of Kraft paper mills. It is also used for electrical switchboards, acidproof flooring, table tops, sinks, tanks, mantles, fireless cookers, and crayons. Pyrophyllite is used for similar ceramic purposes.

Production. World production of talc and soapstone amounts annually to about 800,000 tons, of which the United States supplies over half, the remainder coming chiefly from Korea, France, Italy, Austria, Canada, Spain, China, India, Norway, and Sweden.

The United States production of about 450,000 tons, worth 6 million dollars, comes mainly from New York, North Carolina, California, Vermont, Virginia, and Georgia. The New York talc goes to the paint, ceramic, and paper industries; North Carolina to the rubber industry; and Vermont to roofing materials. The soapstone comes from Virginia, and pyrophyllite from North Carolina. Imports come mostly from Italy, France, and Canada. Steatite also comes from India and Sardinia.

Occurrence and Origin. Talc and soapstone occur as lenses in metamorphosed dolomites, schists, or gneisses, or as large bodies in altered ultrabasic intrusives. The occurrences and origin are treated in detail in Chapter 5·9.

Extraction and Preparation. Talc is mined by underground and quarrying methods, and soapstone is quarried in large blocks ready for sawing. The talc is dry ground and air separated and is carried through a fine grinding stage. Cosmetic talc is hand sorted, screened, ground very fine, and bolted through silk cloth.

Distribution and Examples of Deposits

Commercial talc deposits are confined to regions of metamorphism and occur along the Appalachian axis from Canada to the Carolinas; in the Canadian Shield; in the folded rocks of the western Americas; in the metamorphosed masses of the Alps and Pyrenees; in the Italian Piedmont; in the shield areas of Russia, Scandinavia, South Africa, and India; and in Manchuria, China, and Japan.

United States. The deposits of St. Lawrence County, *New York,* occur in contact-metamorphosed Grenville limestone cut by granite, in a zone of tremolite and enstatite; the tremolite has been altered to talc. Some twenty lenticular bodies are known, ranging in width from 20 to 146 feet, one of which, at Talcville, has reached a depth of 600 feet.

The *Vermont* deposits occur in serpentine or in adjacent schists or gneisses. Associated minerals are dolomite, actinolite, chlorite, biotite, apatite, magnetite, and pyrrhotite. Gillson believes the chlorite replaced the host rocks and was later altered to talc by hydrothermal solutions.

Virginia supports the largest soapstone industry. The deposits occur in a belt 30 miles long, centering at Schuyler. There the soapstone bodies occur in altered ultrabasic intrusions in the Loudon schists, are sheetlike in form, and are 180 to 300 feet thick and 1,500 to 2,000 feet long. The soapstone consists of talc with minor quantities of chlorite, carbonate, amphibole, pyroxene, and magnetite. The original rock was altered progressively to hornblende, actinolite, chlorite, and then to talc, by hydrothermal solutions. The product is sawed into slabs and blocks.

In *California,* irregular bodies of white talc and associated tremolite and serpentine occur at a diorite-limestone contact, and in altered dolomites or magnesites in *Washington.* In *North Carolina* and *Georgia* a belt of Cambrian marble, in which are large tremolite crystals, contains minable lenses of talc, and the pyrophyllite deposits of *North Carolina* replace a volcanic-sedimentary belt.

Canada. In the Modoc district, Ontario, massive, white talc occurs in lenses in contact-metamorphosed Grenville dolomite near a granite intrusion. The deposits are lenslike masses 25 to 40 feet wide and up to 500 feet long.

France and Italy. The finest grades of toilet talc come from the Pyrenees and the Alps. The deposits occur in two beds intercalated with schists and slates, and Gillson suggests they may represent replacements of included dolomite or serpentine. The beds are 50 to 200 feet thick, have been traced several miles, and contain blocks of dolomite and granite. Some Italian talc occurs in serpentine.

Selected References on Talc and Soapstone

General Reference 11. Chap. 48, by A. E. J. ENGEL. *Comprehensive; four pages of bibliography.*
Origin of Talc and Soapstone. H. H. HESS. Econ. Geol. 28:634–657, 1933.
Origin of Talc and Soapstone Deposits of Virginia. J. D. BURFOOT. Econ. Geol. 25:805–826, 1930. *Problems of origin.*
Talc Deposits of Canada. M. E. WILSON. Can. Geol. Surv., Econ. Geol. Ser. No. 2, 1926. *Detailed.*
Talc, Steatite, and Soapstone; Pyrophyllite. H. S. SPENCE. Can. Dept. Mines, Mines Branch, No. 803, Ottawa, 1940. *Comprehensive treatise of all phases.*
Block Talc. J. E. EAGLE. Am. Ceramic Soc. 26:272–274, 1947. *Use and occurrences of block talc.*
General Reference 5.

Barite and Witherite

Barite ($BaSO_4$) and to a minor extent witherite ($BaCO_3$), also known as barytes and heavy spar, are heavy, inert, and stable. These properties make them valuable. Barite formerly was used chiefly as a filler and adulterant, but it is now important in the paint and oil-well drilling industries.

Properties and Uses. The three chief products of the barium minerals are lithopone paint, 10 percent; ground barite, 79 percent; and barium chemicals, 11 percent.

Barite is the chief constituent of lithopone paint (70% $BaSO_4$, 30% ZnS), an interior white or light-colored paint. Ground barite is used mostly as an inert volume and weight filler in well-drilling mud, rubber, glass, linoleum, oilcloth, paper, plastics and resins, and as a paint extender. It is used in tanning, making highly surfaced papers, such as playing cards, for buttons, poker chips, printer's ink, and face powder. It is also employed in making glass, glazes, and enamels.

Barite yields most barium chemicals, of which there are many, although witherite is also used. The chief one is blanc fixe or precipitated barium sulphate, used for paints. Mixed with TiO_2 it makes a white paint of outstanding opacity. Precipitated barium carbonate is used in ceramics, in making optical glass, " granite ware " enamel, and for flat velvety paints.   Barium peroxide is used to make hydrogen peroxide and other chemicals.

Production and Distribution.   The annual production of barite is about 1 million tons.   The chief sources in order of production are: United States, Germany, United Kingdom, Canada, France, India, Italy, Greece, Korea, Cuba, Spain, and Eire.   Important Russian production is unreported.   In the United States over three-quarters of the production comes from Missouri and Arkansas; the remaining quarter comes from Georgia, Tennessee, Nevada, Arizona, California, North and South Carolina, Texas, and Virginia.

Occurrence and Origin.   Commercial barite deposits are formed as (1) fissure fillings — Appalachians, California, Germany, and Great Britain; (2) breccia fillings — Nova Scotia and Virginia; (3) replacement deposits — Virginia, Arkansas, and Germany; and (4) residual deposits — Missouri, Georgia, and Virginia.   The residual deposits have been derived from the other three types.

Barite is also a gangue mineral of many metalliferous veins but they yield only a little by-product mineral.   Veins and replacement deposits are the important sources in Germany and Great Britain, but residual deposits yield most of the barite of the United States. These are accumulations of barite in residual clays that resulted from the weathering of bedrock with included barite deposits (Chap. 5·7).

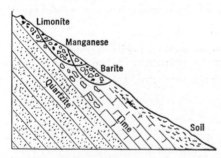

FIG. 20–5.   Residual barite in Georgia. (*After Hull, Geol. Survey Ga.*)

EXAMPLES OF DEPOSITS

In *Missouri*, according to Tarr, the primary bedrock deposits are hydrothermal fillings of fissures or solution cavities and disseminated replacement deposits.   The important deposits of Washington County occur in residual clay derived from Cambrian dolomite, in which the primary deposits occur.   Resistant barite nodules and fragments occur in the clay to depths of 20 feet.   Some deposits occur in leads or runs 10 to 20 feet wide and several hundred feet long.   Others are in sinkholes.   The residual deposits can be traced mineralogically and positionally to underlying primary deposits.

The deposits in *Georgia* (Fig. 20–5), *Virginia*, and *Tennessee* are similar.   *Magnet Cove, Ark.*, is the most productive area of the United States.   Here barite occurs in shale on both limbs of a syncline.   The ore bed is 30 to 40 feet thick and carries 60 to 70% $BaSO_4$. In *Nevada,* near *Battle Mountain,* barite occurs as large replacement deposits in limestone; the *California* barite is also a replacement of limestone lenses. The deposits of the *Rocky Mountain states* are mainly veins that yield high-grade barite.

In *Germany,* large barite vein deposits occur particularly in Thuringenwald and the Harz Mountains, where witherite is also mined. In *Great Britain* important barite vein deposits are mined in Ayrshire, Scotland, and in Shropshire, Devonshire, and Derbyshire, and witherite in quantity is obtained from the Settlingstones mines in Northumberland. Vein deposits also occur in Italy, Greece, France, Spain, Canada, and other countries.

Selected References on Barite

Barite Deposits of Missouri. W. A. TARR. Econ. Geol. 14:46, 1919. *Description of occurrences and origin.*

Barite Deposits of Virginia. R. S. EDMUNDSON. A. I. M. E. Tech. Pub. 725, 1937. *Detailed descriptions and sources.*

Barite and Barium Products. R. M. SANTMYERS. U. S. Bur. Mines Inf. Circ. 6221, 1930. *Distribution, production, uses.*

General Reference 11. Chap. 3, by F. J. WILLIAMS. *Good, general discussion of barite.*

Barite Production in the United States. A. C. HARDING. A. I. M. E. Tech. Pub. 2412, 1948. *Brief descriptions of United States deposits.*

Barite Deposits of Central Missouri. W. B. MATHER. A. I. M. E. Tech. Pub. 2246:1–15, 1947. *Fissure, breccia, circle, solution channel, replacement, and residual deposits.*

General Reference 5.

Glass Sands

The chief constituent of glass is silica sand and minor constituents are soda and lime. Sands have been used for glass making for over 5,000 years.

Composition. Glass sands have a silica content of 95 to 99.8 percent, but the impurities are important; the kind of glass produced depends upon them. *Alumina* is not harmful for common glass, but it should be less than 4 percent, and for optical glass less than 0.1 percent; a little aids in preventing devitrification. *Iron oxide* imparts green and yellow tints; only for amber glass can it reach 1 percent; small amounts can be neutralized by decolorizing agents. Manganese, after long exposure to the sun, imparts delicate amethystine or lavender colors, much prized by glass collectors. *Lime, magnesia,* and *alkalies* are desired; they are constituents of all common glasses.

The grains of glass sands should be even-sized of less than 20 and more than 100 mesh. Coarse sand produces batch scum.

Occurrence and Distribution. Glass sands are obtained chiefly from cemented sandstones but also from unconsolidated deposits. The sandstone must be friable and break readily around the grains. Glass sands range in age from Devonian to Recent.

Glass sands are found in the United States in the Devonian Oriskany sandstone of West Virginia and Pennsylvania and in the Ordovician St. Peter sandstone of Illinois and Missouri, which furnish three-fourths of the supply. They range in thickness from 60 to 200 feet and in places are very pure and so friable that the sand can be hydraulicked and pumped out. In other places they are quarried. New Jersey Tertiary sands are less pure and are utilized mainly for bottle and window glass. Some 5 million tons of glass sands are produced annually in the United States.

Preparation and Glass Manufacture. The prepared sand and the other proper ingredients to make the type of glass desired are mixed into a " batch." Common window glass consists of about 72% SiO_2, 14% CaO + MgO, 12% Na_2O, and 1 to 2% $Al_2O_3 + Fe_2O_3 + SO_3$. This is fused to a melt; CO_2 escapes, and the soda and lime unite with silica to form silicates. The batch is cooled and poured, pressed, rolled, or blown into desired shapes. Salt cake and soda ash (Chap. 21) are used for bottle, sheet, and plate glass, and magnesite is used as an opacifier in glazes.

Pyrex laboratory and cooking glass contains 80% SiO_2, 12% B_2O_3, and 8% Na_2O and Al_2O_3. *Flint glass* contains lead oxide, which gives it brilliance for cut glass. *Quartz glass* is desired for crucibles, tubes, dishes, quartz lights, and ultraviolet windows. *Optical glass* is the highest-quality glass made; in it the potash content is twice that of soda. Best-quality *crystal glass*, with high brilliance and good musical tone, is a potash, lead-silica glass with 7 to 8% K_2O. Potash also imparts chemical durability and resistance and is used for *bulb glass*. In X-ray *lead glass* and *church-window glass* practically all the alkali is potash. *Colored glasses* contain coloring agents; metallic gold gives gold ruby glass; selenium gives red glass; chromium and copper give greens; cobalt gives blue; cadmium and uranium give yellows; manganese gives violet; iron oxide gives browns; and calcium fluoride and tin oxide give opal glass.

Selected References on Glass Sands

American Glass Sands. C. R. Fettke. A. I. M. E. Trans. 73:398–423, 1927. *Types, occurrences.*

Introductory Economic Geology. 2nd Ed. W. A. TARR. McGraw-Hill, New York, 1938. Pp. 491–496.

General Reference 11. Chap. 45, by H. RIES. *General summary.*

Glass Sands and Glass Making Materials in Georgia. Georgia Div. Mines, Mining, and Geol., Atlanta, 1940.

Some Glass Sands from New Jersey. A. S. WILKERSON and J. E. COMEFORO. Econ. Geol. 44:63–67, 1949. *Occurrences and analyses.*

Mineral Fillers

Mineral fillers are finely ground and generally inert cheap mineral substances that are added to such manufactured products as paint, paper, rubber, linoleum, and other materials to give body, weight, opacity, wear, toughness, or other useful properties. They generally change the properties of the product but are themselves unchanged. They are not necessarily adulterants, but some are.

Many fillers are waste products; others are used solely or partly as fillers. All are produced cheaply because generally they need only to be ground and sized; a few are prepared. Their number is large, and their applications are even larger. In addition to natural fillers there are several manufactured fillers. A list of the more important fillers, modified from Ladoo, is given below.

NATURAL ROCKS

anhydrite	flint	rottenstone
bentonite	fuller's earth	sand
chalk	gypsum	serpentine
clay	limestone	shale
coal	magnesite	slate
diatomaceous earth	marble dust	soapstone
dolomite	pumice	talcstone

NATURAL MINERALS

apatite	graphite	sulphur
asbestos	iron oxides	talc
barite	mica	tripolite
calcite	ocher	umber
celestite	pyrophyllite	vermiculite
feldspar	quartz	witherite

MANUFACTURED MINERAL PRODUCTS

blanc fixe	magnesia	sublimed white lead (sulphate)
calcium sulphate	pearl filler	
carbon black	pearl hardening	whiting
gypsum (artificial)	portland cement	zinc oxide
lime (hydrated)	satin white	
lithopone	white lead (carbonate)	

The natural fillers are described in other parts of this book.

Products and processes in which fillers are used (after Ladoo) are listed below.

Artificial stone	Insulators	Plastics
Asphalt surfacing*	Leather	Polishes
Cements*	Linoleum and oilcloth	Putty
Ceramics	Lubricants	Roofing materials*
Cordage	Paint*	Rubber*
Dry batteries	Paper*	Soap
Dynamite*	Phonograph records	Textiles
Fertilizers*	Pipe coverings	Tooth powder and paste
Insecticides	Plasters*	Wood finishing

* Indicates more important consuming industries.

Some of the mineral fillers most used in the more important products follow:

SHEET ROOFING	FERTILIZERS	PAINTS	PAPER
Mica	Sand	Diatomite	Clay
Talc	Limestone	Calcium carbonate	Talc
Slate dust	Dolomite	Whiting	Asbestine
Silica dust	Phosphate rock	Ground limestone	Gypsum
Dolomite		Mica	Barite
Cement		Barite	Blanc fixe
Diatomite		Clay	Calcite
		Amorphous silica	Magnesite

PLASTICS	RUBBER	COSMETICS
China clay	Carbon black	Talc
Gypsum	Clay	Lithopone
Asbestos	Barite	Barite
Mica	Chalk	Chalk
Diatomite	Diatomite	Kaolin
Barite	Magnesite	Fuller's earth
	Whiting	Diatomite

Selected References on Mineral Fillers

General Reference 11. Chap. 27, by A. B. CUMMINS. *General treatment.*
Nonmetallic Minerals. R. B. LADOO. McGraw-Hill, New York, 1925. Pp. 364–367. *Gives details of fillers.*
General References 5, 16.

Mineral Filters and Purifiers

Certain mineral substances possess properties of base exchange and selective adsorption that make them highly efficient filters and purifiers of mineral, animal, and vegetable oils, sugar, fats, water, chemicals, and other materials. The oil and sugar industries in particular consume large quantities of them. The ones used in considerable volume are: fuller's earth or bleaching clays, diatomite, bentonite, and filter sand; those less commonly used are bauxite, tripoli, and alunite. Some of these substances have been described in other chapters of this book.

Purposes Used. Mineral filters are used for the following purposes:

1. Bleaching and filtering the products of petroleum refining, i.e., gasoline, kerosene, furnace oils, waxes, lubricants.
2. Deodorizing animal and vegetable oils and food products.
3. Removing undesirable sediments and sludges.
4. Decolorizing oils, chemicals, medicines, varnishes, and food products.
5. Desulphurizing mineral oils.
6. Clarifying wine, beer and liquor, chemicals, and medicines.
7. Clarifying and refining sugar.
8. Extracting certain chemicals, medicines, and vitamins.
9. Clarifying turbid waters and sewage.
10. Filtering water supplies.
11. Bacterial filters.

DIATOMITE

Diatomite consists of microscopic siliceous tests of diatoms and rarely of other silica-secreting organisms. It resembles chalk or clay but contains chiefly silica and 3 to 10 percent water with a little alumina, iron oxides, and alkalies. It is also known as diatomaceous earth, diatomaceous silica, kieselguhr, and (incorrectly) infusorial earth. It has gone under 22 different names.

Properties and Uses. Diatomite is friable, porous, and so light that when dry it will float on water. Its specific gravity is about 2.1, and the dried powder is only 0.45. Its absorptive power is such that it may carry 25 to 45 percent of water. It is insoluble in acids but soluble in alkalies. Its important commercial properties are its porosity, fineness of pores, absorptive power, light weight, and low heat conductivity.

It is used as a filter and filler, for heat and sound insulation, abrasives, in building materials, ceramics, and for miscellaneous chemical purposes.

Its chief use, according to Cummins and Mulryan, is as filter of the materials listed.

Various sugars and glucose	Water and sewage
Mineral oils and products	Gas purification
Molasses	Varnishes and lacquers
Fruit juices and beverages	Starches and pastes
Beer, wine, and liquor	Trade waste and effluents
Vegetable and animal oils and fats	Vinegar and flavoring extracts
Malt products and extracts	Dyestuffs
Liquid soap	Medicines
Gelatin glue and adhesives	Cosmetics and perfumes
Metallurgical slimes and solutions	Vitamin extracts

Occurrence and Distribution.   Diatomite occurs in marine and fresh-water sedimentary beds and in existing lake and swamp bottoms.   The beds are mostly a few inches to a few feet thick.   The largest deposits are in the marine Monterey series near Lompoc, Calif., where beds up to 1,400 feet thick are worked on a large scale.   Other thick deposits occur in Tertiary beds of the California Coast Ranges.   They range in age from Cretaceous to Recent but are most abundant in the Tertiary. The diatom remains must have accumulated under conditions where the deposition of other sediments was temporarily inhibited.   The silica is thought to have been extracted by the microscopic plants from silicates, such as clay, suspended in the water.   The United States production, the largest in the world (200,000 tons), comes mainly from California, Washington, and Oregon, with some from Idaho, Nevada, Florida, New York, and Massachusetts.   It is fairly common in many Pleistocene and Recent lake beds.   It is also produced in 26 other countries, the most important of which are Denmark, Japan, Algeria, and France.

Extraction and Preparation.   Diatomite is dredged or quarried and removed by mechanical shovels or scrapers.   After drying, it is milled and prepared into powder, aggregates, or brick, depending upon its use.   For filtering, powder is desired, and for this the important specifications are microscopic structure, particle size, chemical purity and inertness, low density, and filter performance.

Selected References on Diatomite

General Reference 11.   Chap. 13, by A. B. CUMMINS and H. MULRYAN.   *Good, brief survey.*

Diatomite.   P. HATMAKER.   U. S. Bur. Mines Inf. Circ. 6391, 1931.   *Distribution, uses, and statistics.*

Diatomaceous Earth.   R. CALVERT.   Am. Chem. Soc. Mon. 52, New York, 1930.

Diatomaceous Earth of California.   R. ARNOLD and R. ANDERSON.   U. S. Geol. Surv. Bull. 315, 1907.   *Detailed local geology.*

Geology, Mining and Processing of Diatomite, Lompoc, Calif.   H. MULRYAN. A. I. M. E. Tech. Pub. 687, 1936.   *Good.*

Diatomite, its Occurrence, Preparation and Uses.   V. L. EARDLEY-WILMOT.   Can. Mines Branch Bull. 691, 1928.   *Comprehensive.*

Diatomites of the Pacific Northwest as Filter-Aids.   K. C. SKINNER *et al.*   U. S. Bur. Mines Bull. 460, 1944.   *Occurences, properties, and tests.*

FULLER'S EARTH OR BLEACHING CLAY

Fuller's earth is a variety of clay, so named because it has been used by fullers to full, or remove grease from cloth.   Its marked ab-sorptive powers, however, have caused it to be more widely used to filter and decolor oils, fats, and greases. The term has come to be applied to certain natural bleaching clays which possess high absorptive capacity for oil without activation and which are mostly found in the southeastern United States. It is one of the varieties of bleaching clay. Fuller's earth is chemically little different from other clays, and its composition is no indication of absorptive power. The composition of clays is given in Chapter 17.

Properties and Uses. Bleaching clays fall into two broad groups — those naturally active and those that become active after artificial activation by acid treatment. Fuller's earth belongs to the former class, and does not respond satisfactorily to activation. Bleaching clays are rarely plastic, and most of them are nonslaking and generally disintegrate in water. Used fuller's earth can generally be treated by heat and reused about 20 times. It is characterized by large water content, foliated structure, and tendency to adhere strongly to the tongue when dry. Not much is known of the seat of decolorizing power; a test is necessary. Nutting thinks that the replacement of a base in the surface of a mineral by the hydrogen of water (or of acid), followed by removal of the hydrogen to leave an open bond, appears to be essential to produce an active absorbing surface. Charged particles unite, positive with negative, and this union of acidoid particles with the basic colored ions in oil is thought to be the basis of the bleaching action in active clays. If an active clay is heated higher than the temperature necessary to drive off the combined water the absorbing power is lost.

The dominant use (about 70 percent) of fuller's earth is in petroleum refining for filtering and clarifying petroleum products, mainly lubricants, and about 10 percent is utilized in the refining of vegetable oils and animal fats. Its use is growing in domestic water purification. It also removes putrescence, odors, and even coliform bacteria from oily waste waters.

In oil refining, colored oils passed through fuller's earth come out colorless. In addition, it removes naphtha gum, and improves the sludge and carbon content, acidity, and viscosity of lubricating oils. It removes color, odor, and taste from vegetable and animal oils. Fuller's earth cannot be reused after filtering vegetable oils.

Minor uses include printing, abrasives, detection of coloring agents in food products, filler, and cosmetics. In oil refining fuller's earth is being displaced by activated bentonite.

Production and Distribution. The United States is the leading producer of fuller's earth (some 300 thousand tons annually), of which one-half is from Florida-Georgia, and a third from Texas. Considerable quantities are produced in England, Germany, Russia, France, and Japan.

Occurrence and Origin. Fuller's earth, according to Ries, is mainly of sedimentary origin. Exceptions are the residual earth in Arkansas and glacial silt in Massachusetts. It occurs interstratified with geologically late sands and clays. The large deposits of Georgia and Florida occur in Miocene strata, at shallow depth, and were formed in coastal embayments; some deposits occur in lake beds.

It is thought that most fuller's earth originated from volcanic ash. Montmorillonite and attapulgite are the principal constituents, and montmorillonite is characteristic of bentonite. Nutting believes that it was derived from bentonite by natural leaching in surface water, assisted perhaps by plant acids and bacteria. This is further suggested by the occurrence of bentonite at the base of thick beds in Georgia and Florida. "Ash deposited *in situ* becomes bentonite; that which is gradually washed in, with leaching of grain surfaces, became fuller's earth." Extended weathering of fuller's earth may yield kaolin.

Extraction and Preparation. Fuller's earth is mined in open cuts by power shovel or scraper. Tunneling may be used for deep clays. It is dried to about 15 percent moisture, crushed, ground, and screened. If the fuller's earth is used for percolation, absorption granules are used; if by contacting method, powder is agitated with the oil and later filtered off.

BENTONITE

Bentonite is a clay composed dominantly of montmorillonite and minor beidellite, with small amounts of igneous rock minerals. It contains about 5 to 10 percent of alkaline earths and alkalies and 3 percent ferric iron.

According to Bechtner there are two classes of bentonites: (1) those that absorb about 8 times their own volume of water and swell enormously in water, remain in suspension in thin water dispersions, and are nonbleaching; (2) those that do not swell, that settle in water, and that have absorptive properties. Type 2 is analogous to fuller's earth, is used as a bleaching clay after activation, and is displacing fuller's earth as an oil clarifier. Activated bentonitic clays are not excelled in activity by any other type of bleaching clay.

Properties and Uses. The activable group of bleaching clays are mostly of bentonitic origin. Those that respond to artificial activation with sulphuric acid are several times more efficient absorbents than natural active clays. Apparently the chemical composition does not determine if a clay will respond to activation.

The bentonitic bleaching clays are characterized by a waxy appearance and by rapid slaking in water without swelling. They are heavier and denser than fuller's earth. Some of these clays, according to Nutting, are naturally active to start with and also show increased activity with acid treatment. Activation consists simply in the removal of ions loosely held at the surface, so that in filtration other chosen ions may be absorbed and thus removed from suspension.

The uses of activated bentonite are the same as for fuller's earth; smaller quantities are required, and it has a lower oil retention. Its use has more than trebled in the last ten years in contrast with that of other bleaching clays.

Type 1, or the swelling type of bentonite, is used for reconditioning and revivifying molding sands (27 percent of United States total bentonite production); for rotary oil drilling mud; for medicinal, cosmetic, and pharmaceutical preparations; for leakage prevention; in insecticides and concrete admixture; as emulsifying agent in asphalt; gelatinizing agent; for dewatering wood pulp; and as a wine clarifier and a paste-forming agent.

Production and Distribution. Figures on world production of bentonite are not available, but that of the United States is around 600,000 tons, of which about one-quarter is used as a bleaching clay. The main production of bleaching clay comes from Texas, Oklahoma, California, Arkansas, Mississippi, Kentucky, Tennessee, and New Mexico. The swelling nonactivable type of bentonite comes mainly from Wyoming and South Dakota. Bleaching bentonites also occur in Germany, Russia, France, and Japan, and nonbleaching bentonites occur in Canada, Mexico, Italy, and China.

Occurrence and Origin. Bentonite is deposited mechanically in lakes or ocean waters, but it is also air borne and air deposited. It occurs intercalated with lake and marine shales and sandstones in beds from an inch to 4 feet thick, which are noted for their remarkable uniformity of thickness and extent. Even the thinner beds can be traced for many miles, and some beds in Pennsylvania less than a foot thick extend for over 100 miles.

Bentonite, as shown by D. F. Hewett, is an altered volcanic ash which can be recognized under the microscope by residual volcanic glass shards. In some bentonites, angular pieces of accessory feldspar, quartz, biotite, pyroxene, zircon, and other minerals of volcanic igneous rocks have been recognized. The main constituents, however, are the clay minerals montmorillonite and beidellite.

Extraction and Preparation. The following steps are involved for bleaching clay (Bell and Funsten): (1) preparation of raw clay for treaters; (2) activating; (3) washing out impurities; (4) mechanical dewatering; (5) drying; (6) grinding, and (7) packaging. For activation, water is added to make a slurry, then sulphuric acid (rarely hydrochloric) is added. The mixture is kept agitated at a temperature of 212° to 220° F for 2 to 12 hours. The material is then dried and ground to a powder.

OTHER MINERAL FILTERS

Filter Sand. Sands are widely used for the filtering of municipal and industrial water supplies to remove sediment and certain bacteria. Filter sands should be even-grained, high in silica, with little soluble material. A certain ratio of fine to coarse grains is desirable. Such sands are obtained from the same deposits that yield glass sands (Chap. 18).

Bauxite. Bauxite (Chap. 13) since 1937 has been growing in use as a mineral filter. It is effective in decolorizing, deodorizing, and desulphurizing mineral oils and other liquids, and is particularly desired for filtering kerosene, lubricants, and paraffinic oils. It is largely used in Pennsylvania and the Mid-continent oil regions. It has to be activated, and, despite its high cost, it competes with cheaper fuller's earth because it can be revivified indefinitely.

Alunite from Marysville, Utah, is used to a minor extent for decolorizing and deodorizing oils.

Activated magnesia is a new compound used as a decolorizing, neutralizing, or absorbing agent in industrial processes. It has an absorptive power many times that of bentonite.

Selected References on Bleaching Clays

General Reference 11. Chap. 6, Bleaching Clay, by A. D. RICH. *Good for technology and bibliography.* Chap. 5, Bentonite, by P. BECHTNER. *Occurrences and uses.*

Geologic Features of Some Bleaching Clays. G. A. SCHROTER and IAN CAMPBELL. A. I. M. E. Min. Tech., Jan. 1940:1–31. *Excellent articles; geology, properties, mineralogy, and extensive bibliography.*

Bentonite, Its Properties, Mining, Preparation, Uses. U. S. Bur. Mines Bull. 438, 1928. Revised by J. E. CONLEY as Tech. Paper 609, 1940. *This is the most comprehensive article on properties, occurrences, and uses.*

Bleaching Clays. P. G. NUTTING. U. S. Geol. Surv. Circ. 3, 1933. *Properties and investigations of absorption properties.*

Fuller's Earth. J. H. TAYLOR. Sands, Clays and Minerals, 1:29–32, 1932. *Occurrences and uses.*

Bleaching Clays of Georgia. H. X. BAY and A. C. MUNYAN. Georgia Geol. Dept. Circ. 6, 1935. *Local geology.*

Absorbent Clays. P. G. NUTTING. U. S. Geol. Surv. Bull. 928–C:127–221, 1943. *Physical and chemical properties, tests, and distribution.*

Geology of Bentonite Deposits, Caspar, Wyo. G. DENGO. Geol. Surv. Wyo. Bull. 37, 1946. *Occurrence, origin, and properties.*

General Reference 5.

Optical Crystals

Various mineral crystals are in demand because of certain optical properties they possess. These are used mainly in microscopes, spectroscopes, and other scientific instruments and in radios and radar. Although demand for them is small, they are essential for specific purposes.

Iceland Spar. This is calcite of uncommon transparency, purity, and perfection of crystallization. It possesses the property of making a dot viewed through it appear as two dots; i.e., it is doubly refracting.

This property is utilized in making Nicol prisms, which constitute the analyzer and polarizer of polarizing microscopes used for the examination of rock slices and polished ore surfaces; Nicol prisms are also employed in scientific instruments for measuring the sugar content of a solution, comparative colors, intensity of light, dichroism of crystals, in polariscopes and polarimeters for studying polarized light, determining index of refraction, examining spectra, and determining crystal structure by X-ray analysis.

Iceland spar for such optical purposes must be free from flaws, bubble inclusions, and twinning, and only a very small percentage of the spar recovered is suitable. Most of it is obtained from clay pockets in cavities in basic igneous rocks. Iceland was long a source of the best optical material. The Cape Province of South Africa has also been supplying optical grade spar; there it occurs embedded in soft clay filling cavities in weathered diabase. A small quantity comes from the United States and Mexico. Polaroid is replacing the Nicol prism for some uses.

Fluorite. Fluorite has a low index of refraction, disperses light faintly, and normally displays no double refraction. These properties make clear fluorite crystals desired for high-power microscope lenses to correct spherical and chromatic errors. Obviously, optical fluorite must be free from flaws and inclusions and preferably should be water-clear. Fluorite is also used in the spectrograph and other instruments where transparency of both ends of the spectrum is desired.

Quartz. Quartz crystals of fine quality and true optical character are indispensable in radio, radar, and telephone and telegraphic operations. For radio transmission, very thin plates of absolute uniformity control the wave frequency. For microscopic examination of minerals and rocks in transmitted, polarized light, quartz wedges, with the flat sides cut parallel to the optic axis and the long edge inclined 45° to it, are essential. Crystals must be flawless and transparent.

Tourmaline. Tourmaline is also a double refracting mineral, and plates of it cut parallel to the principal axis, and set normal to each other obliterate all light. It is used for study of polarized light, and also for " crystals " in radio transmitters to give a definite frequency of sending waves.

Mica. Basal plates of clear, flawless mica are made for use with the polarizing microscope.

Selenite. Cleavage plates of clear, flawless selenite, known as the sensitive plate, are accessories of polarizing microscopes to reveal the slightest double refraction in mineral sections.

Polaroid. This is a recent synthetic material that polarizes light in one plane, and, as with the Nicol prism, when two plates are crossed no light is transmitted. It is beginning to replace the Nicol prism for some purposes in the optical field, but its greatest use so far is for polaroid sun glasses, and on automobile headlights and windshields to cut down automobile light glare.

Selected References on Optical Crystals

General Reference 11. Chaps. 29, by P. M. TYLER, and 38, by R. B. McCORMICK. *Good summaries.*

Iceland Spar. OLIVER BOWLES. U. S. Bur. Mines Rept. Inv. 2238, 1920.

Iceland Spar and Optical Fluorite. H. H. HUGHES. U. S. Bur. Mines Inf. Circ. 6468, 1931.

Deposits of Quartz Crystal, Brazil. F. L. KNOUSE. General Reference 16, 173–184. *Occurrences in vugs, veins, replacement bodies, and pegmatites.*

Quartz Crystals in Brazil. W. D. JOHNSTON, JR., and R. D. BUTLER. Geol. Soc. Amer. Bull. 57:601–650, 1946. *Occurrence and origin of main source.*

Miscellaneous

Meerschaum. Meerschaum (sea foam), also called sepiolite, is so light that when dry it will float on water. It is a hydrous magnesium silicate that occurs as an alteration product of magnesite or serpentine. Its chief use is for making pipe bowls and cigar holders, an industry begun in Budapest in 1723. Also some of it is carved, and it is an ingredient of some porcelains. Most of the world's supply comes from Turkey, where it has been obtained for centuries. About 1,000 tons has come from New Mexico. The Turkish meerschaum occurs as scattered nodules, from the size of an egg to that of a football, in a cemented valley fill that presumably came from the adjacent mountain sides.

Lime. (See also Chap. 18) Nearly one-fourth of the 4 million tons of quick and hydrated lime manufactured annually in the United States is consumed industrially; this, together with chemical and metallurgical uses, accounts for over 52 percent of the lime made. Paper mills and glass works take most of it, but it finds innumerable miscellaneous applications.

In paper mills the sulphite process utilizes milk of lime along with sulphur dioxide for its calcium and magnesium bisulphites to obtain the acid-digesting liquor. It is also used in the soda and sulphate process of paper manufacture.

In glass works, it is the alkaline-earth constituent of the batch for union with silica. In the tanning industry lime is a dehairer and retards putrefaction. Among the many other industrial uses of lime, some of the most important are: sugar refining, brick, gas purification, insecticides, fungicides and disinfectants, magnesia, paints, petroleum refining, gelatin, glue, baking powder, grease, bleaching liquids and powder, soap and fat, salt refining, rubber, and polishers and buffers.

Magnesite (Chap. 18) in the caustic form is used extensively as an accelerator in rubber; it also increases resilience and tensile strength.

Aluminous cement is a rapid hardening cement, resistant to chemicals and heat, that is made from bauxite (Chap. 13).

Carbon dioxide finds considerable industrial use as Dry Ice for a refrigerant for ice cream and perishable goods, for candy and varnish making, for golf-ball centers, and freezing water pipes for repairs. The gas is used for soda and various carbonated waters and beverages and in refrigerating machines. The liquid is used as a " safe " explosive. Most of the gas used for making both liquid and solid carbon dioxide is obtained from natural gas wells and springs, but some is obtained as a by-product from coke and from chemical and metallurgical works. Dry Ice is obtained directly by expansion of CO_2 emitted from CO_2 wells in Mexico and Colorado.

Nitrates (Chap. 22) are used for refrigerants, explosives, fireworks, and flares. Many other items listed under chemical uses (Chap. 21) also have industrial application.

Phosphates (Chap. 22) find a wide application in minor industrial products.

CHAPTER 21

CHEMICAL MINERALS

There are a number of more or less unrelated nonmetallic substances that are used chemically. Some are utilized in the raw state, chemically or medically, others must be prepared, and others supply desired ingredients. Of course, many minerals are sought for their chemical composition rather than for their physical properties, which are utilized dominantly or entirely for purposes other than chemical, such as potash and fluorite. Some of the chemical minerals are also used in other industries, and many other minerals find minor application in the chemical industry.

Those treated in this chapter as chemical minerals are:

Salt and brines.	Sulphur.
Borax and borates.	Strontium minerals.
Sodium carbonate.	Lithium minerals.
Sodium sulphate.	Bromine compounds.
Calcium and magnesium chloride.	Iodine.
Potash.	Miscellaneous minor substances.

Salt and Brines

Salt is the most familiar of all minerals. Its production started with the beginning of man, and each person consumes about 12 pounds of it a year. Because it is a primary human need its acquisition entailed exploration, warfare, conquest, and barter. It became a precious substance in those countries that lacked it. For a long time it has been a basis for government taxation and monopoly, with ensuing power and inevitable political unrest and upheaval. Now, however, its value as a food and food preservative is overshadowed by an industrial demand that makes it one of the most essential raw materials in modern chemical industry.

Properties and Uses. Natural salt and salt brines generally contain impurities. For food salt, even small quantities of calcium sulphate, and calcium and magnesium chlorides, are objectionable; they also make the salt deliquesce. The chemical control of its preparation is, therefore, important.

Salt is so widespread in its uses that the materials made with it or by it are continuously encountered in everyday routine. The chief fields of most of the consumption are:

In Industry:
   Raw products for chemicals and acids — soda, sodium bicarbonate, caustic soda, caustic ash, sal soda, acids.
   As a chemical in manufacturing:
      Metallurgical industries — treating, smelting, and refining of ores and metals.
      Chemical industries — packing-house uses, soaps, dyes, emulsions, tanning, food and wood preservative, cements, blasting, water purification, lacquers, cotton and paper bleaching, road surfacing.
      Ceramics — glazes, shrinkage prevention, dental, vitrifiers.
      Refrigeration agent — ice, ice cream, chemical works, oil refining, cold storage, household refrigerators.
In Agriculture: cattle food, fertilizers, hay preservative, weed eradicators, insecticides, soil amenders, dairying.
In Medicine: drugs, medicines, cleansers, both internal and external.
In the Home: food preservative, refrigerant, cleansers, whiteners, stain removers, miscellaneous.

Production and Distribution. About 35 million tons of salt is produced in the world annually from 77 countries. Those countries with an annual production in excess of 1 million tons are: United States, Great Britain, China, Germany, France, and India, the output of the United States being over $2\frac{1}{2}$ times greater than that of any other country. The United States production of 15 million tons from 75 plants comes mainly from Michigan, New York, Ohio, Louisiana, Kansas, and California.

The amount of salt is almost limitless, and the present producing localities more than meet the demand for it. Barton estimated that the salt domes alone would supply world demand for 30,000 years, and they constitute only 5 percent of the world reserves.

Geologically salt is distributed from the Silurian to Recent. The most widespread deposits of the United States belong to the Permian, but the Silurian deposits of New York, Michigan, and Ohio are the most productive.

OCCURRENCE AND ORIGIN

Commercial salt is obtained from five sources: (1) sedimentary bedded deposits, (2) brines, (3) sea water, (4) surface playa deposits, and (5) salt domes. Salt is generally referred to as rock salt or brine.

The first includes solid salt deposits. Brines include ocean water, salt lake waters, and subsurface brines, both natural and artificial. The brines are of all degrees of saturation and include bitterns with other chlorides, bromides, iodides, and sulphates. Of the United States production, about 56 percent is sold or used as brine, 22 percent is obtained from evaporation, and 22 percent from rock salt.

Bedded Deposits. The sedimentary beds occur intercalated with common strata, and with gypsum, anhydrite, or potash minerals. The beds range from a few inches to several hundred feet in thickness. Generally several beds lie above each other. One well in Kansas, for example, penetrated 32 salt beds; New York state has 7 beds aggregating over 400 feet, and Phalen states that at one place in Michigan there is a thickness of 900 feet. The Southwestern Permian basin, in Texas, according to Sellards and Baker, contains 1,000 feet of salt, and near Carlsbad, N. Mex., one drill hole disclosed over 1,200 feet of salt and included potash minerals.

Most bedded rock salt is associated with gypsum or anhydrite and in places contains potash minerals.

Brines. Ocean and salt lake brines are less used in America than elsewhere. Subsurface brines in sandstones and other porous rocks are regarded as connate or buried sea water. Some brines form locally by solution of rock salt beds. The most important subsurface brines of the United States are those in Mississippian and Pennsylvanian beds in Michigan, Ohio, Pennsylvania, New York, and West Virginia. They supply a considerable part of the salt produced and also yield bromine and calcium chloride. Artificial brines are made by introducing water into salt beds and recovering the resulting brine.

Playas. Salt, along with borax, potash, and other chemicals, occurs as surficial deposits from desiccated salt lakes, such as the extensive deposit of the Great Salt Lake Desert, Utah. It also occurs in the sands and clays of playa marshes.

Salt Domes. Salt domes contain great thicknesses of salt associated with gypsum and anhydrite. Drills penetrated 5,000 feet of salt at Humble, Tex., and 7,700 feet at Calcasieu, La. Phalen states that the base of some salt plugs has been shown by torsion balance surveys to lie at depths of 17,000 to 20,000 feet. For the occurrence and origin of salt domes see Chapter 5·6.

Origin of Salt Deposits. Salt beds originate by evaporation of saline waters. The process is simple; reduction in volume of water results in saturation, with early deposition of gypsum or anhydrite, next of salt, and later of potash and other minerals, as described in detail in Chapter 5·6.

Extraction and Preparation. Salt is marketed as brine, evaporated salt, or rock salt; pressed blocks of salt may be made from the last two. Most of the salt consumed in the chemical industry is used as brine, either natural or artificial. It contains impurities of hydrogen sulphide, iron compounds, and calcium and magnesium salts that have to be removed.

Evaporated salt is obtained from brines by artificial and solar evaporation. For pure salt, the impurities of the brines are removed by aeration and by carefully controlled chemical treatment. Solar salt is produced by progressive evaporation in ponds. Rock and evaporated salt are compressed under great pressure into blocks of 50 to 60 pounds for salting cattle.

Examples of Deposits. Examples of salt deposits occur in practically every country in the world; some have rock salt deposits, some playa deposits, and others brines or sea water. The rock deposits, except for salt domes, are generally similar the world over. The beds display variable attitudes, and because of the solubility of salt, they rarely outcrop.

In the *United States*, the bedded deposits of Michigan occur in the synclinal Michigan basin between Lakes Huron and Michigan, and near Detroit in the Silurian Saline formation. Natural brines occur in salt beds, higher beds, and even in glacial till. The salt beds of New York and Ohio, of the same age, parallel the Great Lakes and dip southward beneath Pennsylvania. Mississippian salt beds occur in Michigan, Pennsylvania, and Virginia; and Pennsylvanian beds in southern Ohio and West Virginia. The salt domes occur in Texas and Louisiana. Playa deposits are important in Utah and California. Extensive flat-lying beds of Permian salt occur throughout the Permian basin. No commercial salt occurs in New England nor in the southeastern or northwestern states.

In *Canada*, salt beds similar to those of Michigan and New York occur in Ontario, Nova Scotia, New Brunswick, and the Prairie Provinces. In *Europe* the chief deposits are Triassic salt beds in England and the Permian deposits of Central Europe. Oligocene beds yield salt from the Rhine trench in Alsace. The considerable salt industry of *India* is a government monopoly. The salt comes from the salt lake brines of Rajputana and rock salt of the Punjab. Salt is also a monopoly industry in *China*, which is one of the large producers of the world.

Selected References on Salt

General Reference 11.   Chap. 40, by W. C. PHALEN.   *Good general account, particularly on industrial phases.*

Salt Resources of the United States.   W. C. PHALEN.   U. S. Geol. Surv. Bull. 669, 1919.   *Broadly descriptive.*

Geology of Texas.   Vol. 2.   E. H. SELLARS and C. L. BAKER.   Tex. Bur. Econ. Geol. Bull. 3401, Austin, 1934.   *Good description of Permian basin.*

Recent Developments in the Use of Salt.   C. D. TOOKER.   A. I. M. E. Tech. Pub. 723, 1936.   *Brief.*

Marketing of Salt.   F. E. HARRIS.   U. S. Bur. Mines Inf. Circ. 7062, 1939.

General Reference 5.

Borax and Borates

Borax, one of the most important chemical minerals. and boric acid are the two principal commercial boron compounds. Borax is obtained in part from the natural borax and in part from borates, which also are the source of boric acid. Schaller points out that the borate industry has been featured by the successive discarding of one mineral for another more usable. First borax from lake muds was used, then ulexite and borax were obtained from playas and borax marshes. These impure products were displaced by bedded deposits of purer colemanite and ulexite from Death Valley — made known by the "Twenty Mule Team Borax." In 1925 a new borate, kernite, was discovered in high-grade deposits along with borax, and these, along with boron compounds extracted from lakes, are the present source of refined borax.

Boron Minerals. Of the 60 known boron minerals, only 7 are used commercially, namely, the water-soluble sodium borates, borax and kernite; the water-insoluble calcium borates, colemanite, ulexite, and priceite; the magnesium borate, boracite; and sassolite (boric acid). The first four, along with brine, have been the supply minerals of the United States; priceite supplies the Turkish, sassolite the Italian, and boracite, the German production. The solubility of borax and kernite makes them preferred. The noteworthy feature about kernite is that it dissolves readily in water and upon evaporation yields normal borax. These two minerals today supply most of the borax of the world.

Uses. Chemically refined borax and the many compounds made from it are used in many articles of everyday life. Borax is a well-known household commodity but is of greater importance in industry. According to Schaller "the cleanser, the pharmacist, the paper and textile maker, the metallurgist, the brazier, and the jeweler all use borax. Hardly any other substance enters into so many diversified lines of manufacture."

It is a good cleanser, either directly or in soaps. It is an ingredient in baking powder, food preservatives, flavoring extracts, syrups and pickles, and insecticides. Medicinally it is a mild antiseptic. It prevents rancidity in cosmetics, pastes, and glues; mold in fruits, leather, textiles, paper, and lumber; and disease in sugar beets and celery. It aids tanning and gives a smooth soft finish to leather. It fireproofs wood and textiles. It gives glaze to glazed papers and is a flux for glass. It is used in manufacturing paper, candles, implements, carpets, drugs, shoes, hats, dyes, ink, jewelry, oil, paints, polishes, tobacco, and tools. Other boron compounds are used in oils, varnishes, inks, linoleum, and electric equipment.

Production and Distribution. The world production of borax amounts to about 400,000 tons, of which the United States produces about 93 percent and the remainder comes from Argentina, Italy, Turkey, with a little from Chile, Bolivia, Germany, and Tibet. The United States source materials are obtained from California, Nevada, and Oregon. The famed borax of Death Valley is now of the past.

Occurrence and Origin. Boron compounds are obtained commercially from: (1) bedded deposits beneath old playas; (2) brines of saline lakes and marshes; (3) encrustations around playas and lakes; (4) hot springs and fumaroles.

The bedded deposits consist of ulexite, colemanite, borax, or kernite, along with other minor boron minerals, in Tertiary and later clays. The kernite deposits consist of about 25 percent clay and 75 percent of equal amount of kernite and borax. Details of occurrence are given in the individual examples. These have been and are the main sources of the United States boron. Lake brines are next in importance. They represent concentrations of former larger lakes and contain borates along with soluble chlorides, carbonates, and sulphates, such as occur in Searles Lake, and described in Chapter 5·6.

Encrustations around saline lakes and playas form impure surficial deposits a foot or so thick. These formerly were important in Death Valley and around Searles and Borax Lakes and still are in Argentina, Chile, Bolivia, and China. Hot springs and fumaroles carrying boric acid yield boron compounds in Italy.

The origin of the various types of borate deposits involves simple concentration and evaporation followed by many chemical mineralogical transformations before final fixation in the borate minerals now found.

The original source of the boron of the United States deposits was presumably fumaroles and hot springs associated with Tertiary volcanism. Boric acid was probably emitted, which reacting with lime and soda, formed ulexite in clay. Colemanite was formed from ulexite by leaching out the soluble sodium borate, which accumulated in lakes where evaporation either yielded borax brines or encrustations. Some borax became buried under later sediments and part of this was transformed from borax with $10H_2O$ to kernite with $4H_2O$ by partial dehydration, perhaps brought about by burial. Some kernite was later hydrated to tincalconite with $5H_2O$ and back again to borax. Kernite exposed during mining was hydrated to borax.

Extraction and Preparation. Bedded borate deposits are extracted by underground mining methods, and kernite and borax are crushed together, roasted to remove the water, separated from the clay, and refined to borax.

Brines are pumped out and the various constituents are separated by complicated chemical treatment, which is essentially evaporation followed by fractional crystallization with careful control of temperature and concentration. During evaporation, the sodium carbonate, sulphide, and chloride are precipitated; then, when saturation with potassium chloride occurs, rapid cooling causes it to be precipitated, and further cooling gives borax and other salts, which are then refined to pure borax.

Examples of Deposits. The *United States* has now only two borate-producing localities — bedded deposits near Kramer and Searles Lake, both in California. The Kramer area in the Mojave Desert contains borax and kernite in a basin 1 mile wide by 4 miles long. The borates occur in clay and shale in a basin of folded Tertiary sediments and volcanic tuffs, at depths between 325 to 1,000 feet and in bodies from 85 to 114 feet thick. Borates constitute about 75 percent of the deposits, and in places there are large masses of pure kernite or pure borax or mixtures of the two. Some tincalconite and probertite are present, and ulexite and colemanite occur in the wall rocks. Ten-foot layers of borax lie above and below kernite. The kernite occurs only in the disturbed portions of the deposit; where borax occurs without kernite the bedding is more regular. Veinlets of borax cut kernite, but Schaller and Hoyt Gale think that the kernite is a transformation from borax.

Searles Lake, in the Mojave Desert, yielding borax and also sodium carbonate, sodium sulphate, and potassium chloride, is described in Chapter 5·6.

Impure borate encrustations, colemanite, and ulexite, occur in several localities in California and Nevada and sparsely in Oregon.

Argentina, second in world production, contains layers and nodules of ulexite in playas. Similar deposits occur in the Atacama Desert of

Chile, at Las Salinas, *Peru*, and in *Bolivia*. In *Turkey* borax is obtained from deposits of priceite (pandermite) in buried playa beds in Anatolia. The unique *Italian* occurrences, in the province of Pisa, are volcanic steam vents, carrying boric acid, emitted from fissures that cut Cretaceous and Eocene sediments. The superheated steam is condensed, yielding a weak solution of boric acid, which is evaporated by natural steam and fumarolic vapors to yield borax.

Selected References on Borates

General Reference 11. Chap. 7, by G. A. CONNELL. *Most complete brief article, and good bibliography*.
Mineral Resources of the Region Around Boulder Dam. Heavy Chemical Minerals. U. S. Geol. Surv. Bull. 871, 92–113, 1936. *Good brief sections on California and Nevada borates*.
Borates. Imp. Inst. London, 1932. *Brief world review and bibliography*.
Boron. R. M. SANTMYERS. U. S. Bur. Mines Inf. Circ. 6499, 1931. *Statistical survey*.
Borate Minerals of Kramer, Mojave Desert, Calif. W. T. SCHALLER. U. S. Geol. Surv. Prof. Paper 158–I, 1930. *Descriptive mineralogy of borates*.
General Reference 5.

Sodium Carbonate and Sodium Sulphate

The carbonate and sulphate of soda are two mainstays of the chemical industry, and the production of their compounds is referred to as the "alkali" industry. The bicarbonate ($NaHCO_3$) is known to all households as soda or baking soda. In industry the anhydrous carbonate (Na_2CO_3) is known as *soda ash*, the sulphate as *salt cake*, and *soda* refers to either the carbonate or oxide of sodium. The purified carbonated salts are marketed as soda ash, bicarbonate, natron ($Na_2CO_3\cdot10H_2O$), and trona ($Na_2CO_3\cdot NaHCO_3\cdot2H_2O$), and the sulphate as salt cake, Glauber's salt ($Na_2SO_4\cdot10H_2O$), and niter cake. The natural compounds occur abundantly in arid regions, but these are now in large part being displaced by soda ash and salt cake manufactured from salt.

Properties and Uses. Sodium carbonate is one of the most important chemical minerals. Soda ash is used for making soap, glass, sodium compounds, and aluminum; caustic soda is used for soaps, dyes, cleansing agents, oil refining, synthetic rubbers, resins, insecticides, and chemicals. The hydrated sodas are used for washing soda and detergents, and the bicarbonate for baking soda, "soda water," in fire extinguishers, and as a chemical.

Sodium sulphate is desired chiefly in making Kraft paper and paper board, in rayon and textiles, and heavy chemicals. It is also utilized for glass, dyes, soap, paint, explosives, and fertilizer. The hydrous salt,

Glauber salt, is used for dyes and as a medicine. Glauber salt is dehydrated for shipment as " crude salt cake," which dissolved in water and recrystallized below 30° C gives Glauber salt again, or if recrystallized above 35° C gives soda ash.

Production and Distribution. World production of sodium compounds is not known, but the United States produces annually some 225,000 tons of natural carbonate and, from all sources, some 5 to 6 million tons of soda ash, 225,000 tons of bicarbonate, and 850,000 tons of sulphate. The other chief producing countries are Germany, Canada, Chile, Peru, Russia, Persia, India, Egypt, Kenya, and South Africa. The United States production of carbonate is from Searles, Mono, and Owens Lakes in California and of sulphate from California, Texas, Wyoming, North Dakota, Arizona, Utah, and Nevada. Canadian supplies come from Saskatchewan, Alberta, and British Columbia lakes which are estimated to contain 115 million tons of hydrous salts, mainly sodium sulphate, and are the largest known reserves.

Occurrence and Origin. Sodium is released during the weathering of igneous rocks and in arid regions reaches local basins, where it is precipitated as the carbonate or sulphate. Sulphur is contributed by the oxidation of pyrite in igneous rocks, by volcanic sources, hot springs and, locally, from gypsum beds. If a solution of gypsum is mixed with a solution of sodium carbonate, evaporation will give calcium carbonate and sodium sulphate.

The natural sodium compounds are obtained from the alkali and bitter lakes described in Chapter 5·6. In Searles Lake an area of some 12 square miles in the midst of a larger playa is underlain by spongy salt. It was formerly 640 feet deep but is now only a marsh and at times is dry. Brine from the upper salt crust is withdrawn by wells 10 to 20 feet deep. Beneath this to depths of 75 to 200 feet is a relatively solid mass of intermingled saline minerals. At greater depths bicarbonate increases and there is a thin bed rich in borax. Drill holes show hard bottom reefs of almost pure bicarbonate with some burkeite. For many years it yielded large quantities of borax and now yields sodium sulphate, potash, borax, and soda.

Owens Lake, 100 square miles in area, has a salinity of over 200 parts per thousand and yields sodium carbonate, of which it is estimated to contain 8 million tons. Mono Lake contains 300 million tons of salts. Soda Lake and Columbus marsh in Nevada were former sources of sodium carbonate and borax. A drought in North Dakota disclosed eight dried-up lake bottoms that aggregate 25 million tons of sodium sulphate. One playa in Kenya, Africa, contains 200 million tons of soda.

The conditions of deposition, occurrence, and mineral associations are described in Chapter 5·6.

Extraction and Preparation. Searles Lake brine contains the following percentages of important constituents, NaCl 16.3, Na_2SO_4 6 to 9, KCl 4.7, Na_2CO_3 3.4, $NaHCO_3$ 0.77, $Na_2B_2O_4$ 1.39, H_2O 65.6. The sodium, potassium, and boron compounds are recovered. The brine is pumped to multiple evaporators in which NaCl, Na_2SO_4, and Na_2CO_3 separate. The remaining solution is quickly cooled to deposit KCl, and the residue yields borax.

Solid salts are harvested on the surface, mined, or later leached. The crude salts are generally refined.

Manufactured salt cake is a by-product in making hydrochloric acid. NaCl treated with H_2SO_4 yields HCl and Na_2SO_4 (salt cake). By another process NaCl is treated with SO_2, O, and H_2O, yielding Na_2SO_4 and HCl. Manufactured Glauber salt is made from salt cake, as previously described. Niter cake ($NaHSO_4$) is a residual product in the manufacture of nitric acid from saltpeter (sodium nitrate) and sulphuric acid.

Examples of alkali lakes and deposits are given in Chapter 5·6.

Selected References on Sodium Sulphate and Carbonate

General Reference 11. Chap 44, by O. C. RALSTON. *Brief résumé and bibliography.*

Sodium Sulphate. P. M. TYLER. U. S. Bur. Mines Inf. Circ. 6833, 1935. *Occurrences, distribution, technology, statistics.*

Sodium Carbonate. C. L. HARNESS and A. T. COONS. U. S. Bur. Mines Inf. Circ. 7212, 1942. *Natural and synthetic salts, uses, and production.*

General Reference 5.

See also references at end of Chap. 5·6.

Calcium and Magnesium Chloride

Calcium and magnesium chloride are obtained from natural brines during the extraction of salt, and calcium chloride is also a by-product in the manufacture of sodium carbonate. Bromine is extracted with the chlorides.

Calcium chloride obtained from brines generally contains magnesium chloride and other salts and in this form is used directly for many purposes. That obtained from sodium carbonate manufacture receives its chlorine from salt and its calcium from added limestone. Some is also recovered in Germany from carnallite.

Calcium chloride is a mineral — hydrophyllite — that occurs at Vesuvius, and hydrous calcium chloride is artificial. The anhydrous form has great avidity for water, which property determines its chief use.

Calcium chloride is used chiefly for laying dust on highways, stabilizing roads, and for ice control on highways and railway tracks. It forms a refrigerating brine and an antifreeze in cooling condensers in oil fields. It dustproofs coal and coke, cures concrete, and preserves wood. It is a dehydrant for laboratory uses, pipe lines, gas, and air conditioning. It has some use in curing and canning vegetables, as a deodorizer, and in many chemicals.

Magnesium chloride is a source of metallic magnesium and it also allays dust.

The chief sources of these chlorides are brines of Michigan, Ohio, West Virginia, California, and the oceans.

Selected References on Chlorides

Calcium Chloride. P. M. TYLER. U. S. Bur. Mines Inf. Circ. 6781, 1934. **General References 5, 11.**

Bromine and Iodine

Bromine is a brownish corrosive liquid that occurs as bromide, along with salt and magnesium chlorides. Formerly it was obtained chiefly from the potash deposits of Germany and France but now it is extracted chiefly from brines in Michigan, Ohio, Pennsylvania, West Virginia, the ocean off North Carolina, the Dead Sea, and in Russia, Japan, and Tunisia. Of the world production of 60 million pounds, the United States produces 95 percent, which is mainly extracted from sea water off North Carolina and Texas. It is marketed chiefly as *ethylene dibromide,* but also as potassium and sodium bromide and other chemicals.

The inexhaustible oceanic supply contains 1 pound of bromine per 2,000 gallons, from which it is liberated by acidification and oxidation with chlorine and blown out by compressed air. It is recovered in absorption towers using soda ash and is made into ethylene dibromide for market.

Bromine is used chiefly to make ethyl or anti-knock gasoline. Bromides are used in the dye, photographic, and motion picture industries, and for many medicinal purposes. Bromine compounds are also used for chemical, metallurgical, ore dressing, tear gas, and hand-grenade purposes.

Iodine is the heaviest nonmetallic element, and its vapor is one of the heaviest known gases. It is widely distributed in minute quantities in the mineral, vegetable, and animal kingdoms and is obtained as a by-product of Chilean nitrates, from natural brines, and from kelp (seaweed). About 2 million pounds is recovered annually, 90 percent of which comes from Chile. The United States production, which is about one-third of its needs, comes from oil-well brines in California and Louisiana.

Iodine is used for biologic and industrial purposes. Its lack in diet is a cause of goiter, and a preventive is the addition of potassium iodide to table salt and water supplies. It is also added to animal feedstuffs and fertilizers. Medicinally, it is an antiseptic, and is also used internally. It finds use as a laboratory reagent, in many chemicals, for sensitizing solutions for photographic plates, films, and papers, and in the dye, tanning, and other industries. Iodine can be used only once.

The Chilean nitrate deposits contain the iodine minerals dietzite and lautarite, and the iodine content averages about 1 pound per ton of nitrate; rarely it reaches 1 percent. The California brines contain 30 to 70 parts of iodine per million, and Java salt wells contain 0.1 gram per liter. Dry seaweed contains less than 0.3 percent, and 25 tons of French drift weed dried to 5 tons yield 1 ton of ash, from which 15 to 35 pounds of iodine is obtained.

Selected References on Iodine

Strategic Mineral Supplies. G. A. Roush. Chap. 13, Iodine. McGraw-Hill, 1939.
General References 5; 11, Chap. 29.

Potash

Potash (see Chap. 22) is an important chemical compound, but the chemical industry accounts for only about 10 percent of the potash produced; the remainder is used in agriculture.

Potassium nitrate is the substance most used to yield the many potassium compounds that emerge from chemical works. This is either obtained from the raw material in Chilean nitrate or is manufactured from crude potash salts. The crude nitrate is used for fertilizer, but the refined nitrate is employed for curing meats, for black powder, fireworks, and tobacco products, and for minor uses in ceramics and drugs.

Potassium carbonate and bicarbonate, formerly obtained from wood ashes, are now manufactured from potash salts, coal, and limestone. The carbonate is employed to make hydroxide and other potassium compounds. The *hydroxide* or *caustic potash* is one of the strongest bases known and is employed for making soap, dyes, laundry and pharmaceutical preparations, disinfectants, laboratory reagents, and for other chemicals, such as potassium chromate and oxalic acid.

Potassium iodide is obtained from the hydroxide or carbonate, and is employed in medicine, photography, and reagents. *Potassium ferricyanide* is employed to make blueprint paper and for blue pigments. *Potassium chlorate,* a strong oxidizing agent, is made from the chloride and is used in making explosives, caps, fireworks, toothpastes, medicines, and dyes. *Perchlorate,* a stronger oxidizing agent, is used mainly for railroad fuses. *Potassium permanganate,* another strong oxidizer, is also made from the hydroxide, and is used as an antiseptic, in gas mask canisters, and has other industrial uses. *Potassium bitartrate,* another derivative, is used for baking powder and cream of tartar.

About 75,000 tons of potash salts are used annually for chemical purposes.

Sulphur

Sulphur is perhaps the most important chemical mineral and one of the most widespread elements. It occurs as native sulphur and in sulphides and sulphates. Native sulphur is the chief commercial source, and pyrite is the only other mineral mined for its sulphur. Substantial amounts of sulphur gases are also recovered from smelters and industrial plants.

Properties and Uses. Cold raw sulphur is relatively inert, but at 478° F it burns to SO_2 or SO_3, and these gases may be used directly or converted into sulphuric acid — the chief use of sulphur. Raw sulphur and sulphuric acid find multitudinous uses and, according to *Chemical and Metallurgical Engineering,* enter into the industries shown below.

RAW SULPHUR	PERCENT USED	SULPHURIC ACID	PERCENT USED
1. Heavy chemicals	50	1. Fertilizer	35
2. Fertilizer and insecti-		2. Oil refining	11
cides	21	3. Chemicals	21
3. Pulp and paper	10	4. Rayon and film	6.4
4. Paint and varnish	3.6	5. Paints and pigments	6.1
5. Explosives	3	6. Coal products	6
6. Dyes and coal tar		7. Iron and steel	5.5
products	2.7	8. Other metallurgical	3.2
7. Rubber	2.2	9. Explosives	2.0
8. Food products	0.2	10. Textiles	0.8
9. Miscellaneous	7.3	11. Miscellaneous	4.0

The raw sulphur utilized for heavy chemicals goes mainly into sulphuric acid, whose chief use in turn is in making fertilizers from phosphate rock. The fertilizer uses of sulphur are given in Chapter 22. In the paper industry sulphur is burned to SO_2, which is absorbed by milk of lime solution to make calcium bisulphite and sulphurous acid for digesting wood pulp. Hard rubber for storage batteries contains 30 percent sulphur, and tires about 1.5 percent.

Sulphuric acid is also employed in petroleum refining to remove gums, tars, and corrosives, to scrub coking gases, and to purify benzol and tuluol. Numerous chemicals are made from it. Metallurgically it is used for pickling iron and steel before galvanizing, and in making electrolytic zinc and copper. In the paint industry it is used for making titanium dioxide. In explosive manufacture, sulphur is used for black powder, and nitrocotton is made in a bath of sulphuric and nitric acids.

Production and Distribution. The world production of sulphur, both crude and from pyrite and smelters, is about 4 million tons annually, of which the United States produces 92 percent, the remainder coming from Italy, Japan, Chile, France, Mexico, Spain, and Netherlands Indies. Russia produces a considerable but unknown amount.

The United States produces from 3 to 4 million tons annually, mostly from Texas, and some from Louisiana. Additional by-product sulphuric acid from zinc and copper smelters amounts to about 1,000,000 tons annually, equivalent to about 230,000 tons of sulphur. The annual acid consumption in the United States is 8 to 9 million tons, most of which is made from native sulphur.

The pyrite production of the world is about 7 million tons, which comes from United States, Spain, Norway, Japan, Canada, Italy, Cyprus, Portugal, France, Australia, and Greece. About half of the world supply of sulphur outside of the United States is obtained from pyrite.

Occurrence and Origin. Native sulphur occurs (1) as depositions around volcanoes — Japan, Mexico, Chile; (2) in salt-dome caprocks — Gulf Coast, United States; and (3) as sedimentary beds — Russia, Sicily.

Pyrite (sulphur-ore type) occurs as massive pyritic replacement bodies with associated pyrrhotite, chalcopyrite, and some sphalerite. Examples are Rio Tinto in Spain, Portugal, Cyprus, Norway, Urals, Noranda, Quebec, and Ducktown, Tenn. Some flotation by-product pyrite is obtained from other types of lode deposits.

Native sulphur originates by several different means. Deposits associated with volcanism may be formed by (a) condensation of sulphur vapors; (b) reaction between H_2S and SO_2; (c) oxidation of H_2S to H_2O and S. Sulphur may be deposited by oxidation and by sulphur bacteria from thermal springs containing H_2S. One or other of these processes has probably given rise to most of the sulphur of volcanic regions and also contributed the sulphur of sedimentary beds, such as those of Sicily, Russia, and some Japanese lakes (Chap. 5·5). Sulphur may also be formed from H_2S eliminated during decay of organic matter and changed to S by sulphur bacteria. Such sulphur would accumulate along with sediments.

The sulphur of salt domes is associated with limestone, gypsum or anhydrite, carbonaceous matter, and sulphur water (as are also the Sicily deposits). This type of occurrence is thought by some investigators to have originated through the reduction of gypsum or anhydrite by bacteria that derived energy from carbonaceous matter and oxygen from sulphates. Calcium sulphate would be reduced to calcium sulphide, which with water and CO_2 would form $CaCO_3$ and H_2S; the H_2S would be oxidized to S. This, however, requires oxidizing conditions, and has been advanced to account for the Sicily deposits. Another suggestion is that anaërobic bacteria reduced the sulphates to sulphur.

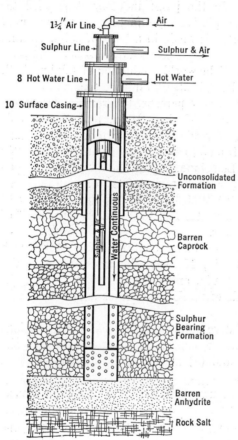

FIG. 21–1. Frasch process of extracting sulphur. (*Lundy, Indust. Min. and Rocks.*)

Extraction and Preparation. The Gulf Coast salt-dome sulphur is extracted ingeniously and simply by the Frasch process (Fig. 21–1). A well with 10-inch casing is driven through the overlying formations to the sulphur deposit. Inside of it is an 8-inch hot water pipe, and inside of this is a return sulphur pipe containing an air pipe. Water heated to 300° F is discharged into the sulphur bed under pressure, melts the sulphur at 283° F and the thin molten sulphur collecting at the bottom is forced by hot compressed air up the discharge pipe. The red liquid sulphur is piped to a stock pile, where it solidifies to pure yellow sulphur.

Sicilian sulphur is mined by underground methods to a depth of 800 feet. It is placed in a battery of connected furnaces where hot gases volatilize the sulphur out of the ore, which is then collected and condensed to marketable sulphur. The extraction is 80 to 85 percent.

Pyrite is generally roasted in furnaces to sulphur dioxide, which is recovered directly for sulphuric acid manufacture, or for sulphite paper pulp, or is compressed into a liquid for shipment. Lump ore or concentrates containing from 40 to 50 percent sulphur are used. The Westcott process recovers elemental sulphur from pyrite concentrates by contacting pyrite with chloridizing gases, forming ferrous chloride

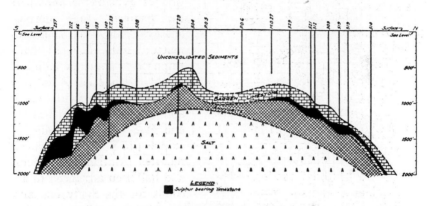

FIG. 21–2.   N–S section across Hoskins Mound salt dome with sulphur cap (black) mostly in limestone above anhydrite.   (*Marx, A.A.P.G.*)

and gaseous sulphur that is condensed to liquid sulphur. The ferrous chloride is oxidized to Fe_2O_3, and the chlorine is recovered. Two tons of pyrite yield about 1 ton of sulphur and 1.5 tons of iron oxide.

Examples of Deposits. In the *United States* the sulphur deposits of Texas and Louisiana are the most important of the world. Their occurrence is as unique as it is puzzling, for they lie in the caprock of salt domes (Chap. 5·6) that have been intruded upward from thousands of feet in depth.

The core of salt containing some anhydrite, is overlain directly by anhydrite or gypsum, above which is cavernous calcite rock from 50 to 1,000 feet thick that possibly is an alteration product of anhydrite (Fig. 21–2). The anhydrite is now generally believed to be the residual accumulation of anhydrite originally present in the upper part of the soluble salt plug; it has been altered in part to gypsum.

The sulphur occurs in the lower part of the cavernous limestone or calcite rock and also in the upper part of the gypsum, but only that in the calcite rock is recovered. It occurs in bunches, seams, specks, and crystals and, according to Wolf, constitutes 20 to 40 percent of the volume. Some barite, celestite, and strontianite, and rarely galena,

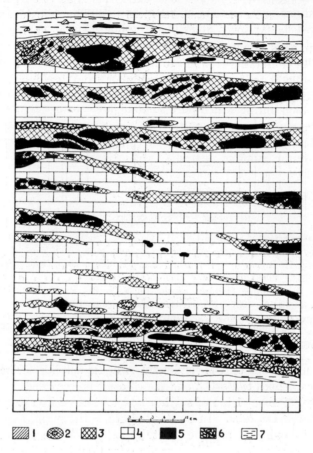

Fig. 21-3. Sedimentary sulphur deposit at Kuibyshev, Russia. 1, Lutecite; 2, calcite vugs; 3, calcite; 4, bituminous limestone; 5, pure sulphur; 6, bituminous sulphur. (*Murzaiev, Econ. Geol.*)

sphalerite, and manganese sulphides occur in the caprock series. The sulphur zone ranges from 25 to 300 feet in thickness; a thickness of 100 feet is common. The mineralized caprock is overlain by 5 to 200 feet of barren caprock, and this by unconsolidated sediments up to the surface, which may lie 1,500 feet or more above the caprock. Caprock areas cover as much as 2,000 acres. Sulphur Dome, La., has only 75 acres and has yielded 10 million tons of sulphur; Boling Dome, Tex., with 1,200 acres, is estimated to contain 40 million tons (Wolf).

Of some 300 known salt domes in the Texas-Louisiana region, only 12 so far are known to contain proven commercial deposits of sulphur.

The sedimentary deposits of Sicily and Russia (Fig. 21–3) are described in Chapter 5·5. Most of the other deposits of the world are associated with fumaroles or hot springs connected with volcanism, where sulphur fills fissures, vugs, and rock pores. Such deposits are found in Japan, Java, Mexico, and the Andes. Watanabe records remarkable eruptions of molten sulphur from Siretoko-Iôsan volcano in Japan in 1936 during which pure sulphur flows, with a temperature of 120° C, cascaded down valley to form a deposit 1,500 meters long, 20 to 25 meters wide, and 5 meters thick.

The pyritic sulphur deposits are described under Copper in Chapter 13.

Selected References on Sulphur

General Reference 11. Chap. 47, by W. T. LUNDY. *Good résumé and exhaustive bibliography from which adequate references may be selected.*

Sulphur. R. H. RIDGEWAY. U. S. Bur. Mines Inf. Circ. 6329, 1930. **Pyrites.** Inf. Circ. 6523, 1930. *Occurrence, technology, statistics.*

Genesis of the Sulphur Deposits of the U. S. S. R. P. M. MURZAIEV. Econ. Geol. 32:69–103, 1937. *Sedimentary and volcanic types.*

The Sulphur Mines of Sicily. C. SAGUI. Econ. Geol. 18:278–287, 1923.

The Boling Dome, Texas. A. G. WOLF. XVI Int. Geol. Cong. Guidebook 6:86–91, 1933. *A sulphur-bearing salt dome.*

Eruptions of Molten Sulphur from Siretoko-Iôsan Volcano, Japan. T. WATANABE. Jap. Jour. Geol. and Geog. 17:289–310, 1940. *An amazing cascading flow of molten volcanic sulphur.*

General Reference 5.

Nitrates and Nitrogen

Nitrogen compounds are dominantly fertilizers and as such are treated in detail in Chapter 22. However, the 15 percent of the supply that is used chemically plays an important role in the chemical industry.

The sources of the nitrogen compounds used chemically are the natural sodium and potassium nitrate from Chile, by-product ammonia and ammonium sulphate from coking ovens, synthetic ammonia and nitric acid from the air, and synthetic cyanamide and cyanogen from the air.

The important nitrogen compounds made or used chemically and their chief uses are:

Compound	Chief Chemical Uses	Important Minor Uses
Sodium nitrate	Explosives, nitric acid, chemicals	Glass, minor chemicals
Potassium nitrate	Explosives, glass, pickling meat	Acids, chemicals, medicine
Nitric acid	Nitration of hydrocarbons; aniline dyes, explosives, plastics, rayon, leather solvent, textiles, sulphuric acid	Photography, lacquers, enamels, varnishes, many chemicals, textiles
Various nitrates	Explosives, hydrogen peroxide, inks, dyes, chemicals	Chemicals, lamp filaments, photography, mirrors, medicines
Ammonia	Refrigerant, household, chemical reagent, silks, dyes, steel hardening	Hydrogen, welding, water purification, neutralizer
Ammonium compounds	Medicines, tanning, dyeing, printing, galvanizing iron, chemicals, oxidizer, celluloid, hydrogen peroxide, sodium carbonate	Batteries, soldering, photography, textiles, baking, scouring, smelling salts, " laughing gas "
Cyanamide	Chemicals, dyes, explosives	
Cyanogen	Various cyanides, gold extraction, indigo, photography, electroplating	Dyeing, painting, printing, blueprint paper, fumigating

Lithium Minerals

Lithium, the lightest metal known and with a melting point of 180° C, is finding increasing chemical and industrial uses. It is a constituent of many minerals but only three, spodumene, lepidolite, and amblygonite, are of commercial importance.

Properties and Uses. Lithium minerals are the source materials for numerous lithium compounds, many of which are used in pharmaceutics, particularly in mineral waters and lithia tablets. Lithium is employed for making Edison storage batteries, and it supplies the red color in flares, fireworks, and signal rockets. Lithium compounds are used for photography, welding aluminum, dental cement, curing meat, making ammonia, making rayon, reducing agents, and purifying helium and other gases. The hydroxide is one of the most important commercial compounds of the metal and is the one used in heavy-duty storage batteries. The chloride is finding growing use in dehumidifying air for air conditioning and industrial dyeing.

The metal finds use as an alloy of aluminum, magnesium, and zinc, for light airplane metals. Bearing-metal alloys, copper electrodes, and lead sheathings are benefited by a little lithium. It serves as a deoxidizer and purifying agent for refining nickel, iron, and copper metals and alloys.

Occurrence and Distribution. The commercial lithium minerals occur in pegmatite dikes but considerable amounts of lithium compounds have also been obtained lately as a by-product from the brines of Searles Lake, Calif.

The chief source of lithium is Searles Lake, and pegmatites of the Black Hills of South Dakota, notably the Etta mine, where spodumene crystals up to 47 feet long and 5 feet in diameter occur; one yielded 90 tons of spodumene, and crystals 5 to 10 feet long are common. Amblygonite and some lepidolite occur in the same deposits. Much lepidolite and some amblygonite is won from pegmatites in California and from the Harding mine in New Mexico. Important spodumene deposits occur in Knox County, Maine, and scattered localities in North Carolina. Hess describes some spodumene dikes in North Carolina up to 1,000 feet long and 100 feet wide with spodumene that carries 6½ percent lithia. The only important occurrence outside of the United States is in southern Manitoba where a pegmatite dike 75 feet wide is said to contain 350,000 tons of spodumene, lepidolite, and amblygonite. Some of the material contains substantial quantities of caesium and rubidium. Other foreign occurrences are Southwest Africa and France, with minor amounts in Czechoslovakia, Sweden, Germany, Spain, and Western Australia. The supply of lithium minerals seems more than adequate to take care of the demand.

Selected References on Lithium

Lithium. Imp. Inst. London, 1932. *Good summary and bibliography.*

General Reference 11. Chap. 23, by R. W. MUMFORD. *Good review of minerals and occurrences.*

Lithium in New England. F. L. HESS and O. C. RALSTON. Eng. and Min. Jour. 139:48–49, 1938.

Spodumene Pegmatites of North Carolina. F. L. HESS. Econ. Geol. 35:942–966, 1940. *Good detailed description of occurrence.*

Exploration of Harding Tantalum-Lithium Deposits, Taos, N. Mex. JOHN H. SOULÉ. U. S. Bur. Mines Rept. Inv. 3986, 1946. *Spodumene in large pegmatite.*

General Reference 5.

Strontium Minerals

The two commercial strontium minerals are celestite (sulphate) and strontianite (carbonate), the former being the common ore and the latter the more valuable because of its lower manufacturing cost. Strontianite is named from Strontian, Scotland, its type locality.

Properties and Uses. Strontium is desired for its salts, used chiefly
to desaccharize beet-sugar molasses. The nitrate is in constant demand for pyrotechnics — the red color of fireworks, rockets, and railroad signals. It is used for electric batteries, paints, rubber, waxes,
purifying caustic soda, for making glasses, glazes, and enamels, for
refrigerant solutions, and for medical preparations. Celestite is a
substitute for barite as a rubber filler, and the carbonate is used in
steel furnaces.

Occurrence and Distribution. Celestite occurs as a gangue mineral
in vein deposits with galena, barite, and calcite; as sedimentary beds
associated with gypsum; as disseminations in limestone; and in solution cavities in limestone. Strontianite occurs as a secondary alteration product of celestite and as replacements of limestone. Strontium
is given off in solfataras and is a constituent of sea water. Most
deposits of strontium minerals are enclosed within sedimentary rocks;
a few occur in igneous and metamorphic rocks.

The most important deposits are the celestite deposits of *Gloucester*
and *Somerset, England,* where celestite occurs as irregular masses and
lenticular bodies in Triassic marl associated with gypsum. It also
occupies flow openings under the marl. In *Westphalia, Germany,* are
bedded deposits in Zechstein dolomite, underlying which are irregular
masses in the Kulm. Near Münster are small strontianite deposits in
veins in Cretaceous marls. In *Sicily* fairly large deposits of celestite
are associated with sulphur and gypsum that occur in fissures underlying upper Miocene beds. Several celestite vein deposits occur in
limestone in *Ontario.* Other deposits occur in *Mexico* and *Spain.* In
the *United States* a number of deposits are known, many of which are
large, but they cannot compete with the imported product from England. Strontianite deposits are few and small. The total world
production seldom exceeds 10,000 tons a year, all of which normally
comes from England and Germany.

Selected References on Strontium

Strontium. R. M. SANTMYERS. U. S. Bur. Mines Econ. Paper 41, 1929. *Brief
       survey;* and Inf. Circ. 7200, by C. L. HARNESS, 1942.
Strontium. L. SANDERSON. Can. Min. Jour. 61:726–728, Nov. 1940. *Brief sum-
       mary.*
General References 5, 11, Chap. 46, by C. L. HARNESS. *Adequate treatment.*

Miscellaneous Chemical Minerals

Alum Minerals. The alums are a series of double sulphate isomorphs
with potash alum. Some are natural minerals; others are artificial

chemicals made from natural raw materials. Some of the commercially used natural alums are common alum or kalinite (K,Al), soda alum or mendozite (Na,Al), ammonium alum or tschermigite (NH$_4$,Al), alum rock or alunite (Al,K), alunogen (Al). All contain water in addition to SO$_4$. They are formed by the evaporation of alum waters.

Manufactured alums are made by treating bauxite with acid or from clay, cryolite, pyritic rocks, lignite, alunite, and leucite. About 400 to 500 thousand tons of aluminum salts are produced annually in the United States.

Potash alum is used for treating furs, and for medicinal and special uses. Ammonium alum is used for stucco, cast stone, statuettes, and for sundry purposes. Sodium-aluminum sulphate is used for baking powder. Aluminum sulphate (not true alum) is used for dyeing, pigments, tanning, water purification, paper, fire-fighting compounds, and for decolorizers and deodorizers.

Barium (Chap. 20). About 22 percent of the barite mined is used for barium chemicals. Barite is furnace reduced with carbon to " black ash " or barium sulphide, or to barium chloride. The principal substance made from this is precipitated barium sulphate or blanc fixe (Chap. 20). Precipitated barium carbonate is used in ceramics or for barium peroxide. Other important barium chemicals are chloride, chlorate, chromate, dioxide, hydrate, nitrate, and peroxide.

Bauxite (Chap. 13). About 175,000 to 200,000 tons of bauxite is used annually in the United States for many chemicals. The most important of these are aluminum sulphate, aluminum hydrate, aluminum chloride, and sodium aluminate. These substances enter into many chemical uses.

Carbon Dioxide (Chap. 20) is obtained as a natural gas and as a by-product. In addition to its industrial uses, it has many chemical uses.

Dolomite is utilized in the sulphite process of paper making to supply the acid liquor, which is a solution of magnesium and calcium bisulphites prepared from dolomite or dolomitic limestone. Roasted dolomite is used to neutralize acid water, and it is also used in the preparation of certain magnesium salts.

Epsomite (MgSO$_4$·7H$_2$O), or epsom salts, is obtained from spring waters and brines to make commercial epsom salts.

Kieserite (MgSO$_4$·H$_2$O), found with potash deposits, is also used for the same purpose, as are also *dolomite* and *magnesite*.

Fluorite (Chap. 19) that is utilized for making hydrofluoric acid must contain not less than 98 percent calcium fluoride. It has many other chemical applications. It is utilized in the extraction of potassium from feldspar and portland-cement flue dust; in making calcium carbide and cyanamide; inorganic and organic fluorides and silico-fluorides, including insecticides, preservatives, and dyestuffs; and in making "Freon," an important noninflammable, nonexplosive, and nontoxic refrigerator fluid.

Greensand (marl) is extensively used chemically as a water softener. Because of its high rate of base exchange, water is softened as rapidly as it can be forced through the greensand, which regenerates equally speedily.

Helium, an absolutely inert gas, is extracted from natural gas at Amarillo, Tex., by liquefying all the components of the gas except helium. Two plants also operate in Kansas and Colorado. Helium has been used mainly for lighter-than-air craft, having a lifting force of 66 lb per 1,000 cu ft compared with 71 for hydrogen. Its main uses now are to prevent deep-sea divers' "bends," for certain metallurgical operations, in radio tubes and electric signs, for filling toy balloons, and for mixing with oxygen for treating pneumonia.

Magnesite is one of the principal raw products from which various magnesium salts used in chemistry and pharmacy are produced. Magnesium citrate is used for saline beverages and by the dispenser. The oxide and carbonate are used for cosmetics, toothpaste, antiacid tablets, the hydroxide in water gives milk of magnesia. Other pharmaceutical salts include epsom salts, salicylate, lactate, phosphate, chloride, and iodide.

Monazite in small quantities is used to obtain thorium, which converted into nitrate is used to manufacture incandescent mantles for gas, gasoline, or kerosene lamps, for arc light carbons, and for rare-earth oxides. The accompanying mesothorium is a substitute for radium. The residual salts of cerium and lanthanum are made into an alloy of iron to make cigarette sparkers.

Lime is widely employed in the chemical industry, which uses about one-eighth of all that produced in the United States. The largest consumption is in water purification. When lime is added to "hard" water, it combines with the excess CO_2, forming $CaCO_3$, which, with the bicarbonate already present, is precipitated and removed by filtration. Lime and soda ash are added to remove "permanent" hardness caused by magnesium and calcium sulphates. Lime also serves as a coagulating agent and as a disinfectant. Another growing use is for sewage and trade-waste purification.

The next most important use is for making calcium carbide and cyanamide. Among some of the minor chemical uses are: for making ammonium, potassium, and sodium chemicals, calcium carbonate, baking powder, salt refining, pigments, and varnish constituents.

Phosphate rock is utilized chemically for numerous products, of which the most important are phosphorus and phosphoric acids. These are used to make many acids, phosphides, and phosphates that have a wide use in metallurgy, photography, medicine, sugar refining, soft drinks, preserved foods, food products, ceramics, textiles, and matches. Elemental phosphorus is used in safety matches, medicines, shells, tracer bullets, grenades, smoke screens, and fireworks. (See also Chap. 22.)

CHAPTER 22

FERTILIZER MINERALS

Another group of unrelated nonmetallic minerals have one thing in common, namely, they are utilized dominantly as fertilizer materials. Also with this group are included other minerals that are utilized in part as fertilizers but whose dominant use lies in other fields. The important ones are:

CHIEF FERTILIZER MINERALS
Potash
Nitrates
Phosphates
Gypsum
Lime
Sulphur

MINOR FERTILIZER MINERALS
Greensand
Magnesite and dolomite
Borax
Epsomite

The fertilizer minerals may be added directly to the soil in the crude state or they may undergo chemical preparation to make them more suitable. In addition, organic substances and nitrogen compounds recovered as by-products or manufactured from the atmosphere are utilized.

Functions of Fertilizers. Fertilizers perform several functions. They provide direct food for plant life, or special foods for specific plants, and replace soil depletions. Some help transform natural soil substances into more soluble form for plant food. Others serve to neutralize unhealthy acidity or alkalinity of soils or to add desired acid and alkaline materials. Some supply substances originally deficient in certain soils, such as the sulphur supplied in " land plaster " (gypsum). Still others are added to neutralize some objectionable ingredient of soil or to eliminate soil pests.

Calcium, potassium, phosphorus, and nitrogen are essential to all plant life, and sulphur to most. The calcium may be supplied by lime, limestone, marl, or gypsum; the potassium by potash, glauconite, or feldspar; the phosphorus by mineral phosphates, slags, and organic substances; nitrogen by nitrates or other nitrogen compounds; and sulphur by native sulphur or gypsum.

The soils of young countries with short agricultural life generally contain sufficient original plant food to support crops, but even such soils give a higher yield with the addition of fertilizers. A ton of wheat takes out of the soil, on an average, about 17 pounds of nitrogen, 18 pounds of phosphoric acid, and 12 pounds of potash. These substances must be replenished else the soil will become barren; it is necessary to add about 200 pounds of mixed fertilizer per acre to older European soils.

Potash

Potash is named from ash made in pots, when iron pots were used to evaporate the leachings of wood ashes for soap making. The product was mostly potassium carbonate, but the term later became applied to caustic potash obtained from treating pot ashes with lime. Commercially, potash is a general term for potassium compounds, but true potash is K_2O, a compound that is neither a natural mineral nor a manufactured product but is used to denote, for comparative purposes, the potassium content of different compounds.

Since the discovery of potash minerals in Germany in 1861, they have become one of the most important fertilizers, and over 90 percent of all potash mined is used for agricultural purposes.

The natural potash minerals result from evaporation of sea water or other brines, and potash brines result from concentration by evaporation of surface waters in arid regions. The various potash minerals and their origin and occurrence are discussed in Chapter 5·6.

Properties and Uses. Wood ashes were the principal source of potash in the United States until 1872, when potash minerals were introduced for fertilizers, but until 1890 their chief use was for dyeing and tanning and making soap, glass, pottery, matches, and explosives.

Weathering releases available potash to the soil for moderate plant growth. Some of this is returned to the soil from decaying vegetation; but as leaching, erosion, fixation, and crop removal deplete it, more must be added. Most plant life requires potash, and this is particularly true of such crops as potatoes, cotton, tobacco, and citrus fruits.

The chief purpose of potash is to form starches and sugars in plants. But it also promotes plant growth and improves crop quality. It strengthens the cell walls of grain straw and plant stalks, rendering them more vigorous and resistant to disease and pests. It improves the shape, flavor, color, and shipping and lasting qualities of vegetables, berries, and fruits. Potash is best when combined with other fertilizers.

Other minor uses of potash are for the chemical purposes given in Chapter 21, ceramics, glass, dyeing, tanning, soap, meat curing, matches, and photography.

Production and Distribution.   The world production of potash ranges between 3 and 4 million tons of K_2O, which is equivalent to about seven times as much crude salts, distributed among Germany, France, United States, Russia, Poland, and Palestine.   Germany supplied most of the world with potash until discoveries were made in other countries after about 1925.

In the United States, the discovery and operation of potash deposits in New Mexico has made this country independent of foreign supplies. The annual production is about 1 million tons of K_2O equivalent, of which 98 percent comes from the underground mines of New Mexico and the brines of Searles Lake, California.   Small amounts are produced in Utah, Wyoming, and Nebraska.   The New Mexico deposits are capable of considerable expansion.

Occurrence and Origin.   Potash deposits have three modes of occurrence: (1) as marine evaporites, (2) playa deposits, and (3) potash brines of saline lakes.

The evaporation beds, described in Chapter 5·6, represent the residual products of evaporation of sea waters concentrated in the deeper parts of extensive basins that once were filled with salt water.   The playa deposits were formed in similar manner (Chap. 5·6) but on a much smaller scale; they also contain numerous other salts (Chap. 21).   The brines of saline lakes represent the later stages of playa deposition not yet carried to completion.   The origin and occurrence of such potash deposits are discussed in detail in Chapter 5·6.

Preparation and Extraction.   Potash deposits are mined by underground methods, reaching to depths of 2,000 feet in Germany and 1,000 feet in New Mexico.   The crude salts range in K_2O content from 9 to 26 percent, and these must be concentrated into higher-grade salts for shipment.

In New Mexico, the products produced are: (1) a high sylvite screen product containing 25 to 30% K_2O that is sold directly as " manure salts "; (2) muriate, containing more than 99% KCl, formed by the removal of halite by solution and fractional crystallization; (3) a sylvite concentrate containing from 96 to 99% KCl, resulting from the removal of halite by flotation.   The Stassfurt methods are entirely chemical.   For agricultural purposes, as long as the K_2O is in soluble form the composition of the salt is immaterial.

EXAMPLES OF DEPOSITS

Germany.   The great German deposits occur in the Zechstein salt series of Permian age, which underlies an area of 77,000 square miles, of which 24,000 square miles is considered potentially productive.   This basin encircles the Harz, Flechtigan, and Thüringer Wald uplifts and extends under the Hanoverian lowlands and into Thuringia. The deposits are worked in six districts — Stassfurt, Hanover, South Harz,

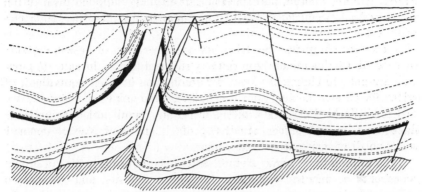

FIG. 22–1. Section across Magdenburg-Stassfurt Basin. Black, potash beds. (*Rice-Everding, U. S. Bur. Mines.*)

Werra-Fulda, Wesel, and Baden. Stassfurt is the best-known one, although now it is little mined.

The beds are intensely folded, and the plastic anhydrite and rock salt have undergone local thickening and thinning (Figs. 22–1, 2). Consequently, there is considerable variation in thickness and depth of the productive beds within the basin. The beds have been rendered accessible to mining, mostly at depths of 1,000 to 2,500 feet, because of upwarped areas or salt domes; in the intervening areas the beds lie at depths of many thousands of feet.

The stratigraphic section of salts at Stassfurt and the minerals present are given on pages 191 to 192. The section shows that there are three potash-bearing zones, which are called the polyhalite, kieserite, and carnallite zones. Halite is present throughout all the zones. The

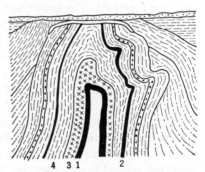

FIG. 22–2. Section through a typical salt horse of the Hanover region, Germany. 1 and 2 are potash beds; 3, main anhydrite; 4, lower rock salt. (*Lilley, adapted from Fulda, Eng. and Min. Jour.*)

carnallite zone averages 9.27% K_2O, with 55 percent carnallite, 17 percent kieserite, 25 percent rock-salt, and sylvite and kainite. The kieserite zone averages only 2.17% K_2O as a whole, and contains 65 percent rock salt, 17 percent kieserite, 13 percent carnallite, and 3 percent bischofite. The polyhalite zone contains about 7 percent polyhalite. The first two zones are the ones mined. Formerly, carnallite contributed most of the potash, but now it yields only about 11 percent of the production and the remainder is obtained from the more easily treated sylvite and kainite salts. They yield about 13% K_2O, whereas only about 9% K_2O is obtained from the carnallite. About 15 percent of the mineral product is shipped as crude salt; the remainder goes to chemical plants for conversion into higher potassium salts.

Before World War I most of the world supply of potash came from these deposits, but since then many of the mines have been closed owing to discoveries elsewhere. Estimates of reserves range from 20 to 3,500 million tons; about 15 million tons were formerly produced per year from about 40 square miles.

France. Somewhat similar deposits occur in Oligocene sediments within the Rhine graben in Alsace, at depths of 1,600 to 2,850 feet. The basin is estimated to contain 1,470 to 1,800 million tons of crude salts containing 17% K_2O. A lower bed 6 to 18 feet thick contains 15 to 21% K_2O; an upper bed 2 to 6 feet thick contains 20 to 25% K_2O. The salts consist largely of sylvite in halite; carnallite is subordinate. The structure is simple, the beds are nearly horizontal, and mining is confined largely to the lower bed. About 40 percent of the production is used in the crude form and 60 percent goes to refining plants.

Spain. Potash deposits are exploited at Suria and Cardona, in Catalonia, where Tertiary sediments of the Pyrenees foothills have been folded, exposing Oligocene salt beds. In most places the depths are beyond workable limits and only the anticlines are accessible. The basin is about 75 miles long and 25 miles wide, and the reserves amount to about 275 million tons of K_2O. Of 2,000 million tons reserves of salts estimated for Suria, 1,600 is carnallite and 400 is sylvite-bearing salts. Two potash zones occur, above and below a 1,000-foot bed of rock salt; the lower is 7 to 26 feet thick, and the upper one, 200 feet thick, contains only thin beds of potassium salts. The sylvite salts contain 18 to 34% K_2O; the carnallite salts, 12% K_2O; the salts as mined average 15% K_2O.

Poland. Two areas of potassium salts are known in Poland, one a little-known extension of the German Zechstein beds containing carnallite; the other, near the Carpathians, yields kainite and sylvite from folded Miocene sediments. Small quantities of carnallite and polyhalite are also present. The beds are 4 to 60 feet thick and average 16% K_2O; the sylvite salts yield 20 to 30% K_2O, and the kainite salts

16% K_2O. Estimates of reserves range from 100 to 250 million tons of crude salts.

Russia. Explorations at Solikamsk, north of Perm, have disclosed a Permian basin of 386 square miles, estimated to contain 15,000 million tons of potassium salts in beds that lie at depths as great as 1,000 feet. Above 800 to 1,300 feet of rock salt are two sylvite zones separated by a zone of carnallite salts; the upper zone is 65 feet thick, the

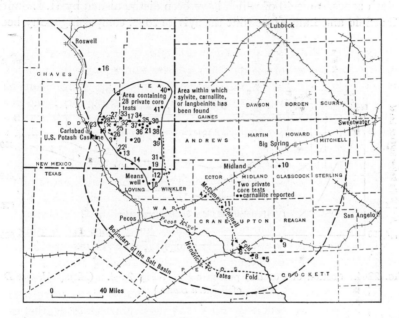

FIG. 22–3. Outline of Texas-New Mexico Permian salt basin containing potash.
(*Mansfield, Ore Dep. of Western States.*)

middle is 200 to 203 feet thick, of which it is claimed 115 feet contains workable carnallite salts that average 20% KCl, and the lower zone, 100 feet thick, contains 25 to 50 feet of workable beds that average 24% KCl.

United States. Scarcity of potash during World War I stimulated United States exploration in the southwestern Permian basin, where studies of oil-well cuttings disclosed widespread polyhalite, and later, sylvite, first identified by R. K. Bailey in 1925. Twenty-two government core-drilling tests revealed sylvite, carnallite, and langbeinite, and eventually resulted in the outline of the basin shown in Fig. 22–3. Private industry continued the work and in all, 95 core tests were drilled through potash beds. The first shipment of crude salts, with

26.8% K_2O, was made in 1931.   Production is around 900,000 tons of K_2O equivalent.

A prospective potash basin of 40,000 square miles is indicated; some 3,000 square miles is known to contain potash salts, and 33 square miles near Carlsbad, N. Mex., is proved to contain commercial deposits of sylvite.   Present mining is restricted to this area.

Nearly horizontal salt beds consist of anhydrite and halite with potash zones, some 40 of which have been distinguished by H. I. Smith. Mansfield and Lang state that the sylvite zone, comprising many beds,

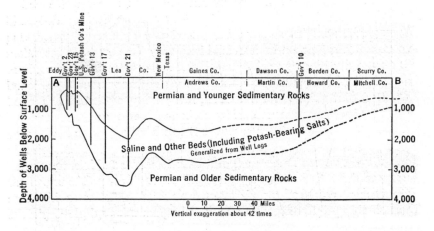

FIG. 22–4.   Section of Permian potash basin of Fig. 22–3.   (*Mansfield, Ore Dep. of Western States.*)

" contains one section 82 feet thick and another 59 feet thick that average about 9% K_2O.   The bottom 10 feet range from 25 to 30% K_2O."   Below the sylvite zone is a polyhalite zone with several potash beds, of which one, at a depth of 1,250 feet, is 9 feet thick and averages 13.5% K_2O.   A typical section, according to Schaller and Henderson, is, from bottom to top: clay, anhydrite, polyhalite, halite with some polyhalite, and halite enclosing sylvite.   Additional potash minerals recognized are: langbeinite, carnallite, and minor leonite, kainite, and glaserite.   Polyhalite is the most abundant and most widespread.   The same authors recognize definite evidence of replacement: polyhalite replaces anhydrite and halite; and sylvite and carnallite have probably replaced halite.

Mining is chiefly on the bottom sylvite zone, at Carlsbad, at a depth of about 1,000 feet, where a sylvite bed (Fig. 22–5) from 5 to 12 feet thick consists of large cubic crystals of halite and sylvite.   Only

3 to 4 percent of impurities are present, consisting of iron oxide, a little clay, and traces of calcium and magnesium.   Halite constitutes a little

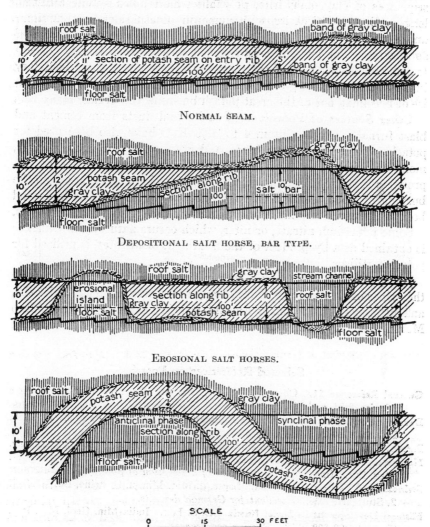

FIG. 22-5. Potash bed, Carlsbad, New Mexico. Upper, normal seam; lower, erosional salt horses. (*Ageton, A.I.M.E.*)

more than half of the mixture, which averages 26 or 27% K_2O.   This single area is estimated to contain 100 million tons of crude salts that average higher in K_2O than those of any other commercial deposits.

The thickness of the beds in the Permian basin ranges from 4,000 to 11,000 feet; that of the salt series is from 1,000 to 2,600 feet. Repeated sequences of clay, anhydrite, polyhalite, halite, and sylvite attest to long-continued desiccation with numerous fluctuations, interruptions, and nonuniform deposition. The richer portions occur in marginal sub-basins. The bitterns are presumed to have been largely concentrated in a deeper part of the main basin near its southern boundary in New Mexico. The greater basin, as yet sparsely tested, gives promise of becoming one of the great potash basins of the world.

Other Sources of Potassium. Leucite and dusts from cement and blast furnaces which contain 4 to 20% K_2O have also been used for potash. Sugar cane, sugar beets, and grapes used for making sugar, molasses, alcohol, and wine yield potassium-bearing wastes or by-products that are utilized as fertilizers. Also kelp and ashes from the burning of wood, bamboo canes, tobacco stalks, cottonseed hulls and bolls, and sunflower stalks yield potassium for refining into fertilizers.

Some potassium nitrate, or niter, which occurs naturally in Chile, or is obtained as a by-product of the treatment of nitrates, is utilized for potash fertilizer.

An appreciable amount of potash in the form of muriate is won from the brines of Searles Lake, Calif., along with borax, soda ash, salt cake, and lithium salts. A minor amount is being extracted from Solduro Marsh, Utah, by solar evaporation, and refined to high-grade muriate.

Selected References on Potash

General Reference 11. Chap. 34, by H. I. SMITH. *Good for technology, New Mexico field, and bibliography.*

Developments Affecting the Potash Industry. H. I. SMITH. A. I. M. E. Tech. Pub. 722, 1936. *Technology and developments.*

Potash. B. L. JOHNSON. U. S. Bur. Mines Econ. Paper 16, 1933. *Brief survey.*

Nonmetallic Resources of New Mexico. S. B. TALMAGE and T. P. WOOTTON. N. Mex. Bur. Mines Bull. 12, 1937. *Brief description of potash deposits.*

Steinsalz und Kalisalz. E. FULDA. Lager. d. Met. Min. u. Gesteine, Bd. III, Teil 2, Stuttgart, 1938. *Excellent for German deposits.*

Mineral Development in Soviet Russia. C. S. FOX. India Min. Geol. Met. Inst. Trans. 34:98–200, 1938. *Russian occurrences.*

Potash Reserves of the United States. S. H. DOLBEAR. Amer. Potash Inst., Washington, D. C., 1946. *Review of reserves.*

Potash Mining in Germany. J. H. EAST, JR. U. S. Bur. Mines Inf. Circ. 7405, 1947. *Description of fields and production.*

Potash in North America. J. W. TURRENTINE. Amer. Chem. Soc. Mon. Series 91:186, 1943. *Occurrence, origin, and treatment.*

General References 5; 7, Chap. 39.

Other general references will be found at the end of Chap. 5·6.

Nitrates and Nitrogen

Nitrogen compounds are necessary for fertilizers. Organic matter, largely through the action of nitrifying or nitrogen-fixing bacteria, absorbs nitrogen from the air and adds it to the soil. Certain legumes, such as alfalfa and clover, thus enrich thin soil with nitrogen. Crop rotation makes up nitrogen deficiency, but, where that is not practicable, nitrogen must be added in the form of fertilizer. About 85 percent of all nitrogen is so used. Supplies formerly were obtained solely from the nitrate deposits of Chile. Later they were obtained in part from coal distillation. The demand for nitrates for explosives during World War I and the cutting off of Chilean supplies forced upon Germany the development of synthetic processes for extracting nitrogen from the atmosphere. Subsequently, other nations also made themselves independent of a single distant source by using fixation processes. The United States developed huge reserve plants, hence Muscle Shoals. Chilean exports dropped accordingly, but more refined methods of handling and treating the natural nitrates have brought them again into competition with the synthetic products. The United States now draws upon Chilean nitrates for only a minor part of its nitrogen needs as against a former 100 percent, although many countries still depend entirely upon Chilean salts.

Sources and Uses. There are five sources of commercial nitrogen compounds:

1. Natural sodium nitrate with minor potassium nitrate, from Chile.
2. The ammonium sulphate by-product in making coke and gas from coal.
3. Synthetic combination of nitrogen and hydrogen to form ammonia, which may be used as such or converted to any desired ammonia salt.
4. Synthetic production of calcium cyanamide ($CaCN_2$) from atmospheric nitrogen, which may be used directly as a fertilizer, or treated with steam to form ammonia, or converted into other cyanide compounds.
5. Organic refuse and bird guano, a negligible amount.

The first two are referred to as " chemical " or " mineral " nitrogen and the next two as " synthetic " nitrogen. Item 3 is the most important and together with 4 now yields about three-quarters of all the nitrogen consumed. The ultimate source of most nitrogen is probably the atmosphere, although it may be of igneous origin.

The uses of nitrogen are shown by statistics of consumption. The normal world consumption of nitrogen, according to Roush, is divided as follows:

	PERCENT
Ammonium sulphate and ammonium for fertilizers	42
Other synthetic nitrogen fertilizers	26
Nitrogen for industrial purposes (except Chile nitrate)	12
Calcium cyanamide	10
Chile nitrate	9

The world peace-time agricultural consumption is 85 percent of all nitrogen. Roush has shown that in the United States over a period of 13 years from 1914 to 1935 the consumption was divided as follows:

FERTILIZER USES	PERCENT	INDUSTRIAL USES	PERCENT
As sodium nitrate	27.8	Explosives	9.0
Ammonium sulphate	31.0	Nitric acid	5.8
Calcium cyanamide	5.6	Chemical salts	5.1
Other forms	8.3	Refrigeration	3.3
	—	Miscellaneous	4.1
Total	72.7		—
		Total	27.3

In addition to the three main fertilizer forms shown above, the others include combined ammonium sulphate and nitrate, calcium nitrate, potassium nitrate, ammonium phosphate, and urea. Of the industrial nitrogen used, 44 percent is sodium nitrate, 33 percent ammonium nitrate, and 23 percent other forms.

Production and Distribution. The normal world production of pure nitrogen amounts to several million tons, divided among all the industrial countries of the world. The source materials of the production are:

	PERCENT		PERCENT
From synthetic	64	From coal	17
From cyanamide	11	From Chile nitrate	8
	—		—
Total from air	75	Total from minerals	25

Atmospheric nitrogen of course can be produced wherever power is available. That from coal is yielded by large makers of industrial coke, and the natural salts come from Chile, where reserves are in excess of 100 years' supply. Outside of Chile, small amounts of potassium nitrate come from India, and negligible amounts from Egypt, China, and Spain. Natural nitrates are known in Argentina, Bolivia, and at Quatsap, South Africa. A negligible amount of " cave niter " occurs in the southwestern United States, and a deposit of sodium nitrate is reported near Safford, Ariz.

Occurrence. In the rainless, treeless, fuelless deserts of northern Chile, in a narrow strip 450 miles long and 10 to 50 miles wide, are numerous sodium nitrate deposits, locally called *caliche*. The deposits occur at elevations of 1,000 to 3,000 feet on gentle valley slopes in layers from a few inches to several feet thick, and beneath a few inches to 40 feet of overburden. The nitrate beds are rudely stratified sandy gravels cemented by soluble salts. The sodium nitrate content averages about 25 percent, and is accompanied by 2 to 3 percent potassium nitrate, common salt, sulphates of sodium, calcium, and magnesium, and minor amounts of iodates, borates, bromides, and phosphates. Some of these vital substances, along with lithium and strontium, make Chilean natural nitrate an ideal fertilizer.

The occurrence and divergent views of the origin of the Chile deposits are discussed under Ground Water evaporation in Chapter 5·6.

Extraction and Preparation. The salt bed is mined by hand or power shovel, and in the " Shanks " process the nitrate is recovered by boiling water from which it crystallizes upon cooling. The product contains 96 percent sodium nitrate and valuable fertilizer impurities. The Guggenheim process is more refined, utilizes power equipment and waste heat, operates with lower-grade crude salts, and produces a lower-cost product with valuable by-products of iodine and sodium sulphate.

By-product ammonium sulphate is recovered from coking ovens by washing the gases, distilling the wash water, and absorbing the liberated ammonia in dilute sulphuric acid, from which ammonium sulphate is crystallized.

Atmospheric nitrogen is obtained by, (1) using a catalyst at high temperature and pressure to combine N with H to form NH_3, which is then converted to a soluble salt or oxidized to nitric acid; (2) by high-tension discharge of electricity combining N and O to form NO, which oxidizes to NO_2 and absorbed in water makes nitric acid.

Calcium cyanamide is formed by passing N over CaC_2 at 1,000° C. This may be steamed to form ammonia or may be converted into other cyanide compounds.

The various products appear on the market as sodium nitrate, potash nitrate, ammonium sulphate, ammonia liquor, cyanamide or cyanides, nitric acid, calcium nitrate, or sodium nitride, some of which are used in chemical industries (Chap. 21).

Selected References on Nitrates

Strategic Mineral Supplies. G. A. ROUSH. McGraw-Hill, New York, 1939. Chap. 14. *Good for supplies, uses, technology, production, and consumption.*

General Reference 11. Chap. 32, by H. R. GRAHAM. *Restricted to technology of Chilean nitrates.*

Nitrogen and its Compounds. B. L. JOHNSON. U. S. Bur. Mines Inf. Circ. 6835, 1931. *Technology, occurrence, statistics.*

Chile Mineral Resources. C. W. Wright. Foreign Minerals Quart. 3: No. 2, 1940. *First-hand data.*
General References 5; 7, Chap. 35 (*U. S. resources*).
See also references under Chap. 5-6.

Phosphates

Phosphate, an important plant food, is produced in great quantity, and almost all of it goes into fertilizer. Since the discovery about a century ago that phosphate of lime was the beneficial material contributed to soils by greensand the use of phosphate as a fertilizer has steadily increased.

Phosphorus has a varied geologic cycle. It is supplied from an igneous rock source to the soil, from which it is absorbed by plants, thence by animals, which return it in excreta and bones, from which it is dissolved, carried to the sea, and again deposited in beds or taken into sea life to start the cycle anew.

Materials. The materials of commercial phosphate deposits are:

1. Phosphate rock:
   Nodular " land " rock.
   Hard rock deposits.
   Soft rock deposits.
   Land pebble deposits.
   River pebble deposits.
2. Phosphatic marls.
3. Phosphatic limestones.
4. Marine phosphate beds.
5. Apatite.
6. Guano.
7. Basic blast-furnace slag.

The important substance of phosphatic deposits is tribasic calcium phosphate. The phosphate minerals of sedimentary deposits are amorphous to cryptocrystalline and by X-ray are identified as fluorapatite, hydroxyapatite, staffelite, and dahllite. The fluorine content may be replaced by chlorine, carbonate, or hydroxide. Department of Agriculture investigators state that the phosphatic material is mostly fluorapatite or hydroxy-fluorapatite.

Phosphatic acid is known in sea water, phosphate nodules occur on the sea bottom, and calcium phosphate is a common constituent of many sedimentary rocks. From 3 to 4 percent of fluorine is generally present in phosphate rock.

The phosphatic content is referred to commercially in percentage of tricalcium phosphate, or " bone phosphate of lime," abbreviated BPL. Commercial rock does not ordinarily fall below 66 percent of BPL; most of it contains 70 to 80 percent.

Properties and Uses. About 90 percent of all phosphate produced is utilized for fertilizer and is chemically changed into more usable forms. Treatment with sulphuric acid gives " acid phosphates," known as superphosphate, triple superphosphate, and dicalcium phosphate, in which forms the phosphorus is readily available for plant food and some sulphur is supplied. A small amount is dried and used in the raw state.

Phosphate is also used in phosphorus and phosphoric acids for making phosphorus compounds, the iron and steel industry, rustproofing, safety matches, vermin exterminators, fireworks, shells, grenades, tracer bullets, smoke screens, distress signals, medicines, soft drinks, cements, baking powder, self-rising flour, yeast, table salt, photography, and ceramics.

Production, Distribution, and Reserves. The world production of phosphates ranges around 12 million tons, of which the United States produces over one-half and North Africa one-third, the remainder coming from Russia, Egypt, France, Pacific Islands, Spain, and Sweden.

Some 26 countries produce in excess of 10,000 tons annually. The largest consumers are United States, Germany, Japan, Italy, France, Great Britain, Holland, and Spain, and the largest reserves occur in United States, Tunisia, Morocco, and Algeria.

The United States production, amounting to about 7 million tons annually, and the reserves as estimated by Mansfield are:

PRODUCTION	Percent of Country	RESERVES	Known (million tons)	Total Possible (million tons)
Florida	70	Florida	2,059	5,082
Tennessee	20	Tennessee	195	195
Western States	7	Western States	6,570	6,570
Others	3	South Carolina, Kentucky and Arkansas		30

Occurrence and Origin. Phosphate rock occurs as:

1. *Marine sedimentary phosphate beds,* such as those of the Western States, which are marine chemical depositions in large enclosed basins. This type also yields the important North African production.

2. *Phosphatic marls and limestones,* which are merely sedimentary beds with high phosphate content. They are mostly too low grade for treatment, but the crude material is used locally as fertilizer. They are, however, important source materials for secondary concentrations.

3. *Calcareous beds replaced by phosphates.*

4. *Land pebble,* which represents marine reworking of underlying phosphatic limestones, and deposition of hard, resistant pebbles of phosphate in a gravel bed, subsequently exposed by erosion. Those of Florida supply most of the phosphate of the United States.

5. *River pebble* occurs as bars and banks in stream channels and consists of hard pieces of phosphate gathered by the streams where they have crossed phosphatic beds.

6. *Residual concentrations.* These are derived from the erosion of underlying phosphatic beds and are accumulations of heterogeneous pieces of phosphate that have suffered some solution and redeposition. This is the origin of the "hard rock" deposits of Florida and the "brown rock" of Tennessee.

7. *Apatite deposits,* which are concentrations of apatite in pegmatites, veins, and magmatic segregations of igneous rocks. They are little worked today except in Russia and Sweden.

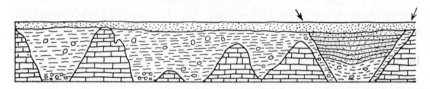

Fig. 22–6. Pebble phosphate in pockets in Peninsular limestone underlying phosphatic sandstone and covered by Pleistocene sand. (*Matson, U. S. Geol. Survey.*)

EXAMPLES OF DEPOSITS

United States. *Florida* yields nearly half of the world production of phosphate, and 95 percent of this production is land pebble.

The *land pebble* of Polk and Hillsborough Counties is in the Pliocene Bone Valley formation, which is a gravel composed of 10 to 50 percent of phosphate pebbles in a matrix of sand, clay, and soft phosphate, along with fossil teeth and bones of land and marine animals. The beds are almost horizontal, 3 to 18 feet thick, occupy an area of 30 by 30 miles, and are overlain by sand and clay up to 40 feet thick. The pebbles were derived by the disintegrating and sorting effect of ocean waves acting on underlying coastal phosphatic beds, mainly of the Miocene Hawthorn formation. (Fig. 22–6.)

The *hard rock* deposits are in the Pliocene Alachua formation. They are residual accumulations, with successive deposition and redeposition of phosphate dissolved from overlying phosphatic beds, and deposited in underlying calcareous beds by replacement and cavity filling. They are associated with sand, clay, chert, and soft phosphate. They reach a thickness of 100 feet and are overlain by 50 feet or so of overburden.

The pebble contains 66 to 73 percent BPL and the hard rock from 77 to 83 percent.

The *Tennessee* deposits are known as " brown " rock, " blue " rock, and " white " rock. The blue rock is an unaltered phosphatic shale or sandstone of late Devonian or early Mississippian age that averages 65 percent BPL, and is from a few inches to 6 feet thick. The brown rock deposits, the most important, are residual products of weathering of arched Ordovician phosphatic limestones. On hill slopes the residually enriched outcrops form " rim " deposits which extend 60 feet down dip to where unaltered limestone is encountered. Accumulations on low ground give blanket deposits up to 50 feet thick, and the cleaned material contains 70 to 78 percent BPL. White rock deposits are small secondary replacements and cavity fillings of limestones by material leached from overlying formations.

The *Western* sedimentary deposits underlie parts of Idaho, Montana, Wyoming, and Utah and extend into Canada (Fig. 22–7). The thickest and richest are in Idaho. The surveyed phosphate area in the western United States amounts to 300,000 acres and is estimated to contain 6,000 million tons of phosphate, of which about five-sixths is in Idaho. Near Banff is another known million tons. Another 2,000,000 acres remain to be surveyed. The area is thus a large one. The commercial beds are in the Phosphoria formation of Permian age. The phosphate zone is 100 to 150 feet thick; the main bed is around 5 feet thick and contains about 32 percent of phosphoric acid, equivalent to about 70 percent of tricalcium phosphate. Other low-grade beds attain a thickness of 75 feet and carry 30 to 50 percent of tricalcium phosphate. Mansfield describes the main phosphate bed as oölitic, dark colored owing to included hydrocarbons, often giving a petroliferous odor and yielding a bluish-white bloom upon weathering. The beds are folded and are devoid of mud cracks, ripple and current marks, or cross bedding. The phosphate is amorphous; collophanite is probably present, and included phosphatized bones and shells probably contain dahllite and staffelite.

Canada. The Phosphoria formation extends into the Canadian Rocky Mountain area but the beds there are thin, of low grade, and are little used. Some apatite deposits of Ontario were formerly mined.

Morocco. Some 500 kilometers south of Gibraltar and 200 kilometers from the Atlantic are the world's second largest phosphate deposits. They occur in horizontal eroded Eocene limestones, marls, and clays, in 3, and in places 4, beds up to 2.5 meters thick. The phosphate, which ranges between 70 and 78 percent BPL, is mixed with limestone and flint and carries about 10 percent water. The deposits are considered to be extensive sedimentary marine beds that were chemically precipitated.

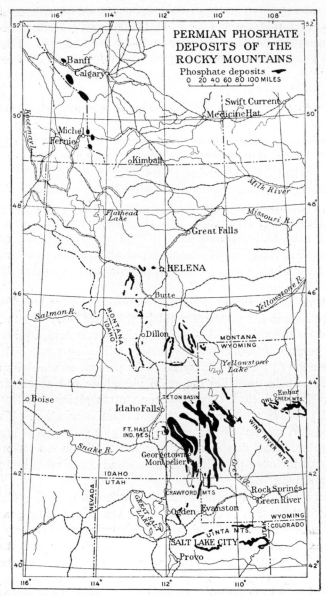

FIG. 22-7. Map of Permian phosphate deposits of the Rocky Mountains.
(*Mansfield, Econ. Geol.*)

Algeria-Tunisia. Near the common boundary of these two countries there is an extensive series of folded Eocene limestones, marls, and conglomerates that carry 1 to 3 oölitic phosphate beds of economic importance (Fig. 22–8). They are up to 4.5 meters thick. The oölites carry 70 percent BPL and the cement about 25 percent, and they are,

Algeria

Fig. 22–8. General cross section of phosphate beds (black) of southern Tunisia, toward Algeria, contained in folded Eocene sediments. (*Cayeux, Res. Min. France d'Outre-Mer, IV.*)

therefore, separated from the cement for shipment, and yield from 58 to 70 percent BPL. The beds are of sedimentary origin. There are also some minor deposits of guano in caves and phosphate masses deposited in limestone. The production amounts to about 2 million tons, of which three-quarters comes from Tunisia. The known reserves

Fig. 22–9. Pinnacles left on Nauru after removal of phosphate. (*Power, Econ. Geol.*)

of this low-grade phosphate exceed 1,350 million tons; probable reserves are given as 10,000 million tons.

Egypt. Cretaceous marls, limestones, and cherts contain thin beds of sedimentary phosphate carrying 63 to 70 percent BPL near the Red Sea. They are the fifth ranking deposits of the world.

Pacific Islands. Nauru, Ocean, Angaur, and Makatea Islands are large producers of high-grade phosphate carrying 80 to 90 percent BPL. Their histories are generally similar. On Ocean and Nauru (Fig. 22–9) thick deposits of guano from innumerable birds supplied phosphatic solutions that reacted with underlying coral rock to form tricalcic phosphate. The bedrock is sculptured into a series of pinnacles and chasms characteristic of limestone under a residual soil cover, and the phosphate fills the chasms to depths of 40 feet on Nauru and 65 feet on Ocean Island. The upper surface forms a mat of rich guano phosphate. The deposits of Angaur and Makatea are similar and are about 6 feet thick; Christmas Island, in the Indian Ocean, has about 2 feet of phosphatized limestone.

Russia. In the Kola Peninsula are important deposits of apatite, which is concentrated to a high-grade phosphate product. Low-grade phosphate rocks occur in central Russia.

Selected References on Phosphates

Phosphate Deposits of the United States. G. R. MANSFIELD. Econ. Geol. 35:405–429, 1940. *Concise résumé of geology and reserves.*

Recent Studies of Reserves of Domestic Phosphates. G. R. MANSFIELD. A. I. M. E. Tech. Paper 1208, Feb. 1940.

General Reference 9. Western Phosphate Field, by G. R. MANSFIELD, pp. 491–496. *Phosphoria sedimentary beds.*

Phosphate. Les Ressources Minérales de la France d'Outre-Mer, IV. Bur. d'Études Géol. et Min. Col. Paris, 1935. In Morocco, by A. BRANGÉ, pp. 2–48; in Algeria-Tunisia, by P. REUFFLET, pp. 71–111. *Detailed geology.*

Phosphate Deposits of the Pacific. F. D. POWER. Econ. Geol. 20:266–281, 1925. *Origin.*

General Reference 11. Chap. 33, by C. A. FULTON. *Sketchy résumé, good bibliography.*

Phosphate Rock. B. L. JOHNSON. U. S. Bur. Mines Inf. Circ. 6256, 1930. *Occurrence, technology, statistics.*

Les phosphates de chaux sédimentaire de France et d'outre-mer. L. CAYEUX. Vol. I:349, Serv. Carte Géol. France, Paris, 1939. *Description of phosphate deposits of world, particularly French.*

Phosphate Resources of Florida. G. R. MANSFIELD. U. S. Geol. Surv. Bull. 934:82, 1943. *Discussion of types, occurrences, and reserves.*

General References 5; 7, Chap. 37.

Agricultural Limestone and Lime

Agricultural liming materials include ground limestone, lime, marl, and oyster shells, all of which are soil conditioners. Of these, pulverized limestone, because of its low cost and availability, is the most used, although its effectiveness is less and its control not so simple as for lime.

Since soils result mainly from weathering, and calcium is soluble in surface waters, most of the calcium salts formed in soils are leached and removed. Consequently most soils are acidic, and even soils derived from limestone may be deficient in lime. In untilled regions, leaching is retarded and soil acidification may be small, but soil cultivation promotes leaching and crops absorb lime which, therefore, must be added.

Uses. The chief purposes of agricultural lime are to correct soil acidity, to granulate heavy clay soil, and to provide plant food. Also, it promotes digestion of other fertilizers, decay of vegetable matter to form nitrates, and counteracts some soil poisons. Few crops favor a strongly acid soil; most of them prefer a slightly or moderately acid soil, and some, such as clover, alfalfa, and sugar beets, thrive best with a neutral soil. A test of the soil acidity indicates the amount of corrective lime to be added. In Wisconsin this amounts to about 65 pounds of lime per acre but in Kansas 0.3 pound.

Materials and Production. In the United States, of a total of about 5 million tons of lime made annually, approximately 7.5 percent is used for agriculture. However, some 9 million tons of total liming materials are used annually, consisting of 94 percent crushed limestone, 4.0 percent lime, 1.1 percent oyster shells, and 0.3 percent marl.

Limestone for fertilizer has few restrictions on purity as long as it has high calcium carbonate equivalent or neutralizing value. Some magnesia is desirable. The lime used may be either hydrated or quicklime (see Building Limes, Chap. 18) because it keeps better. Oyster shells are locally ground for direct soil application or are converted into lime. Marl is available only locally and generally contains desirable phosphate.

The effective lime content of the different liming products is quicklime 84 percent, hydrated lime 70 percent, crushed limestone and oyster shells 43 percent, and marl 45 percent.

The materials available for agricultural limestone and lime are so widely distributed that they are available almost everywhere.

Selected References on Agricultural Lime

General Reference 11. Chap. 22, by N. C. ROCKWOOD. *General résumé.*
Lime. OLIVER BOWLES and D. BANKS. U. S. Bur. Mines Inf. Circ. 6884, 1936.

Sulphur

Although sulphur is dominantly a chemical mineral, agriculture is the next largest consumer. Of the raw sulphur produced annually in the United States, about one-fifth is used directly for fertilizer and insecticides. Further, one-third of the acid produced is utilized for converting phosphates into superphosphates, requiring about 1 ton of sulphuric acid for each ton of phosphate rock. Thus, sulphur is closely bound up with the fertilizer industry, accounting altogether for about one-half of the total sulphur produced.

The uses, production, distribution, and occurrence of sulphur are discussed in detail in Chapter 21.

Uses as Fertilizer Mineral. Raw sulphur dust and sulphuric acid are added to the soil to neutralize alkalinity and to correct sulphur deficiency, since sulphur is a plant food. According to Lundby, sulphur in the soil also prevents the absorption of selenium by plant life. Certain plants selectively absorb selenium when sulphur is absent and poison cattle forage. The insecticidal and fungicidal uses of sulphur are expanding. Sulphur and lime-sulphur sprays are of value in controlling fruit-tree and crop pests. Sulphur dioxide has application in mushroom cultivation and in greenhouses. Carbon bisulphide is a fumigant for stored grains and is a pest and weed killer. Ten percent sulphuric acid is a weed killer.

Sulphur is of greatest value to the fertilizer industry, however, in the form of sulphuric acid for making superphosphates from phosphate rock. Indirectly sulphuric acid is of value for scrubbing ammonia gases obtained from the coking of coal to convert them into ammonium sulphate, another fertilizer.

Miscellaneous Mineral Fertilizers

Greensand occurs in Cretaceous and Tertiary formations of the Atlantic Coastal Plain and is produced for local fertilizer in New Jersey and Virginia. It occurs as marine beds consisting in large part of glauconite, the hydrated silicate of iron and potash. It also contains a little phosphoric acid. The New Jersey greensand, averaging about 6 percent potash, is added to the soil in the raw state, but some of the Virginia material is dried and ground as a commercial fertilizer.

Magnesite (Chap. 18) and magnesium compounds serve as direct and indirect fertilizers. It has been shown that, although the magnesium content of most plants is small, a satisfactory crop yield can be obtained only if the magnesium content does not fall below a minimum value. The carbonate helps to neutralize soil acidity, and both carbonate and sulphate overcome magnesium deficiency.

Borax has been found to stimulate plant growth and prevent certain plant diseases. Small quantities of borax are included in mixed fertilizers for this purpose.

Epsomite serves as a fertilizer for tobacco in Virginia and the Carolinas.

CHAPTER 23

ABRASIVES AND ABRASIVE MINERALS

Many minerals and rocks of diverse composition but with one thing in common — hardness — are employed as natural abrasives. These are used in the natural state, except for processing and bonding. Natural abrasives, however, are being supplanted by artificial abrasives, which in turn are made from mineral products. Most of the abrasive materials are dealt with incidentally in different parts of this book.

Abrasive Materials. Natural abrasives are divided into three groups: (1) high-grade natural abrasives, which include in order of hardness, diamond, corundum, emery, and garnet; (2) siliceous abrasives, consisting of various forms of silica; and (3) miscellaneous abrasives, including buffing and polishing powders. The materials utilized are:

HIGH GRADE	SILICEOUS	MISCELLANEOUS	
diamond	sandstone	bauxite	siliceous carbide
corundum	quartzite	magnesite	(Carborundum,
emery	novaculite	magnesium oxide	etc.)
garnet	flint	ground feldspar	fused alumina
	chert	chalk	(Alundum, aloxite,
	silicified limestone	lime	etc.)
	quartz	china clay	boron carbide
	quartz mica schist	talc	metallic oxides
	sand	oxides of tin, iron,	lampblack
	tripoli	chromium, man-	carbon black
	pumice	ganese	
	diatomite		

Use and Preparation. Natural abrasives may be used (1) in the natural form, e.g., sand and pumice; (2) after shaping, e.g., millstones; and (3) after being ground into grains or powders and made up into wheels or papers.

Abrasives have wide application in all phases of industry, and their use rises and falls with industrial trends. The automobile industry is the largest consumer and the employ of abrasives parallels the fluctuations of automobile and steel output. Industries like the airplane industry increase the call for certain types of abrasives, such as industrial diamonds.

The chief uses of some of the natural abrasives are given in Table 25 (after Eardley-Wilmot).

Abrasives used in the soap industry (including scouring soap) include pumice, pumicite, feldspar, diatomite, bentonite, talc, silica, chalk, and clays. Metal polishes use pumice, emery, diatomite, silica, tripoli, chalk, fuller's earth, clay, bauxite, magnesite, and oxides of metals. Many of the materials that serve as abrasives have greater use for other purposes. Manufactured abrasives have replaced many of the natural abrasives for certain applications, particularly in industrial metalwork.

Industrial Diamonds

One thinks of the diamond primarily as a gemstone, and it is so treated in Chapter 24. However, some 5 to 12 million carats of industrial diamonds are imported into the United States annually. There are two types of industrial stone: the carbonado or carbon, the hard black diamond; and the bort, which includes small stones, fragments, and badly colored or flawed stones.

Carbonados. These are hard, tough, black diamonds that come from Bahia, Brazil, and are utilized for dies in wire drawing, for diamond-drill bits used in exploring for ores and oil, and for tools employed for dressing and truing abrasive wheels. Diamond-tipped tools are used for boring metals and other hard substances, such as Bakelite, in the auto and electrical industries where precision work is necessary.

Bort. Bort has come to replace carbonados for many industrial purposes because of its lower cost. It is extensively employed in the precision manufacture of airplane and motor car engines, both for boring and for abrasion of surfaces. War activity greatly stimulates the consumption of industrial diamonds.

Diamond rock-drill bits are now largely mechanically set with many small borts, which makes them cheaper than carbon bits and which greatly lowers the cost of diamond drilling. Borts are even employed extensively for drilling blasting holes. A typical drilling bit of this type contains an average of 178 small stones weighing 7 carats.

Diamond dust made by grinding bort is used as an abrasive for the cutting of gemstones, minerals, and carbides. Crushed diamonds are now impregnated in a molten metal bond for fast cutting wheels, and similar cutting wheels are made with diamond fragments in a powder-metal bond. Bort comes mainly from the Belgian Congo, Gold Coast, Angola, Sierra Leone, and South Africa.

For the occurrence, origin, distribution, and references on diamonds, see Chapter 24.

TABLE 25

CHIEF NATURAL ABRASIVES AND THEIR USES

MATERIAL	ABRASIVE	CHIEF USES
High-Grade Natural Abrasives		
Diamond	Crystal	Cutting, boring, wheel truing
	Dust	Airplane engines, gem and rock cutting
Corundum	Wheels	Metal cutting, lens polishing
	Papers and cloths	Metals and hardwood; optical
Emery	Wheels	Snagging metals
	Papers and cloths	Metals and hardwood
	Loose grains	Finishing and polishing metals; glass grinding
Garnet	Papers and cloths	Hardwoods; paint and varnished surfaces
	Loose grains	Glass grinding
Siliceous Abrasives		
Sandstone	Grindstone	Grinding saws, knives, metals, etc.
	Pulpstones	Grinding wood for paper pulp
	Sharpening stones	Hand sharpening
	Oilstones	Fine sharpening of steels
Quartz, flint	Burrstones	Grinding flour, pigments, etc.
	Pebbles	Grinding ores in mills
	Slabs	Fine hand sharpening
	Crushed	Soft wood (sandpapers)
Sand	Grains	Sand blasting, glass grinding
	Papers and cloths	Woods, metals
Pumice	Blocks	Rubbing paint and varnish
	Grains	Glass, scouring powders
Diatomite	Powder	Metal polish, dental powder
Tripoli	Powder	Metal buffing
Volcanic dust	Grains	Scouring powders, cleansers
Rottenstone	Powder	Scouring powders
Soft Abrasives		
Feldspar	Powder	Scouring and cleansing
Clays	Powder	Buffing of metals
Dolomite	Calcined	Buffing of metals
Lime	Powder	Buffing of metals
Bauxite	Powder	Buffing of metals
Chalk	Powder	Buffing of silver and metals
Lampblack	Powder	Buffing silverware
Black rouge	Powder	Buffing metals and mineral surfaces
Red rouge	Powder	Buffing metals, optical glass, and mineral surfaces
Green rouge	Powder	Buffing hard metals and mineral surfaces
Tin oxide	Powder	Buffing metals and mineral surfaces

Corundum

Corundum (Al_2O_3) has a hardness of 9 as compared with 10 for the diamond. There are three varieties: gems (ruby, sapphire), corundum, and emery, the last being a mixture of corundum and magnetite. Its hardness makes it a natural abrasive, but it is being supplanted by Carborundum. It is used as loose grain in optical grinding, on paper and cloth, and in the form of abrasive wheels.

Occurrence and Origin. Corundum is formed as (1) magmatic segregations in quartz-free igneous rocks, such as nepheline syenite (Chap. 5·1), (2) as a reaction casing between pegmatite dikes and intruded basic igneous rocks by a process of desilication of the dike.

The desilication type is the most common occurrence and is the origin ascribed by Hall to the South African deposits, and by others to the peridotite occurrences.

Distribution. Commercial deposits of corundum are known in Canada, South Africa, United States, India, Madagascar, and Russia.

In eastern Ontario, *Canada,* are three intrusions of nepheline syenite that contain corundum. The important one is in Renfrew County, near Craigmont, where nepheline syenite pegmatite carries crystals of corundum in commercial quantities. In other places, disseminated grains of corundum occur in the syenite and larger grains flank intrusive pegmatite dikes.

South Africa is the chief corundum area of the world. Brown and ruby corundum occur in deposits flanking pegmatite dikes that intrude basic igneous rocks. The border zone is a highly altered mass, called plumasite, with much dark mica and corundum. Corundum also occurs as a placer mineral. In *India,* the home of the ruby and sapphire, isolated crystals of corundum are picked up over weathered syenite and in schists intruded by pegmatites. In *Madagascar* the mineral is found in the weathered portions of a metamorphic rock containing graphite, mica, and corundum. In *Russia* corundum is obtained from an anorthositic rock called kyschtynnite and from dikes of corundum syenite. These occurrences are probably of magmatic origin. In the *United States* corundum occurs in the southern Appalachian belt at the contact of dunite intrusions with gneiss in North Carolina, and in Georgia where feldspar dikes intersect peridotite masses. This type is probably formed by desilication.

Selected References on Corundum

Corundum, Emery, and Diamond. V. L. EARDLEY-WILMOT. Part II, Abrasives, Can. Dept. Mines, Mines Branch, No. 675, Ottawa, 1927. *General survey.*

Corundum in the Union of South Africa. W. Kupferbinger. Geol. Surv. S. Africa, Geol. Ser., Bull. 6, 1935. *Occurrence and origin of desilication type.*

Corundum in South Africa. R. W. Metcalf. U. S. Bur. Mines Min. Trade Notes, Vol. 24, No. 2, 1947. *General survey.*

Corundum in North Carolina. W. A. White. N. Car. Div. Mines and Res. Rep. Inv., 1943. *Associated with basic igneous rocks.*

General Reference 11, Chap. 1, by R. B. Ladoo.

Emery

Emery is a natural mixture of corundum, magnetite, and some hematite and spinel, named from Cape Emeri, Greece. Spinel emery contains considerable spinel, and corundum may be lacking (American). Feldspathic emery contains much plagioclase. Three commercial grades of emery are recognized: Greek, Turkish, and American.

The hardness and cutting quality of emery depend upon the amount of corundum present. It is coarse or fine grained and tough, and it withstands intense heat. It is used chiefly as grinding wheels, as emery cloth, and as grains and flour for glass polishing. The Grecian is generally the hardest, and the Turkish next, but the American varieties are soft and are used mainly in pastes and composition.

Most of the commercial emery comes from Naxos in Greece, Turkey, and United States.

Occurrence and Origin. Emery is formed mainly by contact metasomatism and occurs in irregularly shaped bodies in crystalline limestone, altered basic igneous rocks, and chlorite and hornblende schists. In New York some also occurs in veins.

Distribution. On the island of *Naxos, Greece,* are lenticular masses of emery 300 feet long and up to 150 feet wide, enclosed in crystalline limestone and formed by contact metasomatism. In *Aidin, Turkey,* are irregular masses of emery, 200 by 300 feet, enclosed in crystalline limestone interfabricated with schists and gneisses. The main supply comes from boulders in residual clay. Similar deposits are worked in the *Urals, Russia.* The *United States* contains deposits of emery in New York, Massachusetts, and Virginia. At Peekskill, N. Y., spinel emery occurs in the Cortlandt basic igneous complex near mica schist inclusions, in the form of sharply defined veins contained in sillimanite-cordierite-garnet-quartz rocks. One vein, according to Zodac, is 50 feet wide and a mile long. At Chester, Mass., emery occurs in pockets in a band of sericite schist. In Virginia, spinel emery occurs in lenticular bodies in (a) schist and quartzite, and (b) in granite cut by pegmatites. Watson considers that these were formed by high-temperature replacement akin to contact metasomatism.

Selected References on Emery

Corundum, Emery, and Diamond. V. L. Eardley-Wilmot. Part II, Abrasives, Can. Dept. Mines, Mines Branch, No. 675, Ottawa, 1927. *Comprehensive.*

Peekskill, N. Y., Emery Deposits. J. L. Gillson and J. E. A. Kania. Econ. Geol. 25:506–527, 1930.

Mineralogy, Composition, and Origin of the Emery of Turkey. J. De Lapparent. Jour. Amer. Cer. Soc., Vol. 29, No. 11, 200, 1946. *Abstract of occurrences.*

General Reference 11, Chap. 1, by R. B. Ladoo. *Adequate; good bibliography.*

Garnet

Garnet is a group name applied to seven different species with generally similar characteristics but of different composition, and with metals that are replaceable with each other. The varieties are listed below.

Varieties	Composition	Geologic Environment
Almandite	Fe-Al-garnet	Mica schists and gneisses — metamorphic origin
Andradite	Ca-Fe	Contact-metasomatic limestone
Grossularite	Ca-Al	Contact-metasomatic impure limestone
Pyrope	Mg-Al	Eclogite — deep-seated igneous origin
Spessartite	Mn-Al	Granitic rocks and pegmatites
Uvarovite	Ca-Cr	Serpentine
Rhodolite	Fe-Mg-Al	

Almandite, or common garnet, is the one generally used as an abrasive, but andradite and rhodolite are also utilized. Production of garnet is relatively small.

Qualifications and Uses. For abrasive purposes, the important properties are hardness, toughness, and fracture. Sharp angular fractures with brittle edges that form new cutting edges, are necessary. Rounded fragments are not desired. The grains should be at least the size of a pea, and the garnet content of the rock should be not less than 10 percent.

Abrasive garnet is used as garnet cloth or paper, and as loose grains, and is employed particularly for hard woods. Its cutting power is 2 to 6 times that of sandpaper. Garnet paper is also employed for finishing hard rubber, celluloid, leather, felt and silk hats and for rubbing down varnished and painted surfaces, automomobile bodies, and copper and brass. Garnet grains are used for surfacing plate glass and ornamental stones and for gang-sawing marble and slate.

Occurrence and Distribution. Garnets are constituents of igneous rocks and are formed by metamorphism in schists and by contact metasomatism in calcareous rocks. They are widely distributed, but commercial deposits are few. Most of the production comes from the United States, the largest mine in the world being in New York, with smaller ones in New Hampshire, North Carolina, and Massachusetts. Negligible amounts come from Spain, India, Canada, and Madagascar.

Examples of Deposits. *New York* includes garnet deposits in Warren, Essex, and Hamilton Counties; the important deposits are those of the Barton mines and the North River Garnet Company. The garnet is almandite and occurs in garnet gneiss in grains and crystals mostly between bean size and an inch or more in diameter. Crystals up to a foot are common, and a few are 3 feet in diameter. The garnet content of the Barton quarry averages 13 percent. At the North River quarry on Thirteenth Lake similar gneiss carries 4 to 8 percent garnet in crystals up to 3 inches in diameter. In *New Hampshire,* near North Wilmot, almandite garnet constitutes 40 to 60 percent of the enclosing garnet schist. The deposit is 100 by 180 feet and 25 feet deep. The garnets are thought to have been formed by granitic emanations. In *North Carolina,* at Sugar Loaf Mountain, is a deposit of rhodolite garnet, which averages 20 to 25 percent of the enclosing schist. In *Idaho,* garnet production comes from Fernwood.

In *Canada* many small deposits are known but the only production has come from Lennox and Addington Counties, Ontario, where almandite occurs in gneiss. In *Spain* pink garnets have been concentrated into alluvial deposits in Almeira. Some alluvial garnet is known in Madras, Deccan, and Mysore, India, and in Madagascar, Ceylon, and Czechoslovakia.

Gem garnet is treated under Gemstones, Chapter 24.

Selected References on Garnet

General Reference 11. Chap. I, by R. B. LADOO.

Abrasives, Part III, by V. L. EARDLEY-WILMOT, Can. Dept. Mines, Mines Branch, No. 677, Ottawa, 1927. *The best article on commercial garnet.*

Classification and Occurrence of Garnets. W. I. WRIGHT. Amer. Min. 23:436–445, 1938. *Largely mineralogical.*

Natural Silica Stone Abrasives

Siliceous rocks are quarried and shaped into forms suitable for use, such as grindstones, pulpstones, millstones, and hand stones of various types. The rock used is dominantly sandstone, but quartzite, quartz-mica schist, silicified limestone, flint, chert, and novaculite are also used.

Grindstones. Grindstones are made mostly from sandstones. The sandstone must be of uniform hardness, possess sharp and even grain, and be sufficiently cemented to insure tenacity but at the same time permit crumbling away to prevent glaze. The larger the grain the faster the cutting power, but with corresponding roughness. Sound blocks must be available in sizes from 2 to 5 feet across. Relatively few sandstones have all of these qualifications.

The main sources are Carboniferous sandstones of the United States, Great Britain, Canada, and Germany. In the *United States* most of the grindstones come from the Dunkard and Berea sandstones of Ohio and to a less extent of West Virginia and Michigan. In *England* the celebrated Newcastle stones come from Coal Measures sandstones near Newcastle, where excellent fine-grained stones suitable for edge tools are quarried. Other similar stones are known as Derbyshire, or Peak, and Yorkshire stones, which are obtained from the Millstone Grits. In *Canada* an old industry obtained grindstones in Nova Scotia and New Brunswick from Carboniferous and Permo-Carboniferous sandstone beds.

Grindstones are used mainly for sharpening saws, machine knives, scythes, shears, and harvesting and die-making machinery. Production in the United States is about 10,000 tons but is waning because of artificial abrasives.

Pulpstones. Pulpstones are used for the grinding of pulp logs to make paper pulp. They are made of sandstone, of much the same type as grindstones, but the beds must be capable of yielding blocks 5 feet in diameter, and with a face from 3 to 10 feet wide. The qualifications are generally similar to those of grindstones but the cementing should be weak enough to permit the bond to wear and the harder quartz grains to protrude. Coarse grain yields too coarse a wood fiber. After quarrying, the stones are seasoned for 1 to 2 years.

Pulpstones may come from the same beds as grindstones. The Millstone Grits of England yield the best, and West Virginia and Ohio yield the next best grade; Canada produces good stones in British Columbia and New Brunswick.

Millstones. Millstones are large circular stones run horizontally or on edge and include burrstone and chaser stone. They are made from any hard suitable sandstone, quartzite, quartz conglomerate, granite, or basalt. True burrstone is chalcedonic silica, originally employed for grinding grain but now used for grinding paints, fertilizers, and graphite. Chaser stones are large heavy stones desired mainly for grinding feldspar, quartz, barite, and other minerals. Most of these stones are produced in Italy, Great Britain, United States, Canada, and Germany, and yearly production has amounted to over 400,000 tons.

Grinding Pebbles and Liners. Pebbles of selected flint and quartzite are used in large rotating ball mills for the grinding of metallic ores, paints, cement, gypsum, clays, ceramic minerals, fillers, and powder abrasives. Steel balls have supplanted stone for many uses where steel cuttings do no harm, but stone must be employed for ceramic and certain other materials. The best-known flint pebbles come from deposits in Denmark, others from the shores of Belgium, France, and England; the United States and Canada yield small amounts.

Sharpening Stones. This group includes the familiar scythestones, whetstones, waterstones, honestones, oilstones, razor hones, and holystones. They are made from fine sandstones, siliceous argillite, and schist. Natural Belgian honestone is desired for hones and fine sandstone and novaculite for oilstones. Hindostan stone from Indiana and Queen Creek stone from Ohio serve for waterstones. Holystones or rubbing stones, made of coarse sandstone, are employed for rubbing auto bodies, furniture, and concrete.

Selected References on Silica Abrasive Stones

General Reference 11. Chap. I, by R. B. LADOO.
Abrasives, Part I, by V. L. EARDLEY-WILMOT. Can. Dept. Mines, Mines Branch, No. 673, Ottawa, 1927.

Natural Silica Abrasives

Natural silica abrasives are silica in various natural forms. Some of them are used as mined, but mostly they are processed and ground to grains or powder. Most of them serve purposes other than abrasives and have been considered elsewhere in this book. This group has a wider industrial application than natural silica stones; it includes diatomite, tripoli, abrasive sand, ground sand and sandstone, ground quartz, pumice, and pumicite.

Diatomite (Chap. 20). Diatomaceous earth, dominantly a filler, serves as an abrasive principally in metal (silver) polishes, powders and paste, automobile polishes, dental powders and paste, and as an abrading agent in match heads and box sides.

Tripoli (Chap. 5·7). Tripoli, a porous, earthy substance that is nearly pure silica, and rottenstone, a siliceous argillaceous limestone, are classed in trade as " soft silicas." About 30,000 tons is produced annually, of which 30 to 50 percent is used as abrasives. It is used chiefly for buffing blocks and powders, for the buffing of metals and plated wares, but also for rubbing down auto bodies and painted surfaces and in scouring and cleaning soaps and powders. Tripoli results from residual weathering of chert and cherty and siliceous rocks (see Chap. 5·7). The largest deposits are in Missouri-Oklahoma, Tennessee Valley, and Illinois. Near Seneca, Mo., the deposits occur in beds 2 to 20 feet thick under thin overburden and were derived from cherty Boone limestone. Those in the western Tennessee Valley are described by E. L. Spain as a well-compacted 25-foot bed of incoherent tripoli that crops out in bluffs under 100 feet of cover. The Butler deposit contains 98 percent or more silica and was derived from the weathering of Mississippian cherty limestones from which the calcareous material has been leached. The southern Illinois tripoli is in compact beds, contains much chert, and has to be ground.

Abrasive Sand. Abrasive sands are employed for sandblasting, sandpaper, grinding or surfacing plate glass, and for scouring stone. United States production comes mainly from Cape May, N. J., Illinois, Ohio, and the Virginias. These sands must be sharp, clean, and fairly uniform in grain. Coarse varieties are used for heavy cast iron and steel work.

Ground Sand and Sandstone. Friable sandstone and sand are ground to finer sizes for plate glass grinding, sandblasting, sandpaper, and other finer abrasive products. About 3 tons of sand are required to surface 1 ton of plate glass. The Ottawa sandstone of Illinois furnishes much of the raw product, and Illinois, New Jersey, Ohio, and Pennsylvania are the chief United States sources. Most other countries supply all their needs. The ceramic industry consumes about 40 percent of the ½ million tons or so of ground sand and sandstone produced. Abrasive use also includes cleansing and cleaning powders and pastes for household and industrial use.

Ground Quartz. Clean quartz, crushed and graded, is used for " flint " sandpaper, harsher metal polishes, in metallurgical works, and various scouring compounds. About 75,000 tons a year are produced in the United States. The quartz is derived from veins, pegmatite dikes, and pure quartzite beds. California, Virginia, North Carolina, Maine, Maryland, and New York are important producers.

Amorphous Silica. Natural amorphous silica from southwestern Illinois, Tennessee, and Georgia is sometimes listed with tripoli. It is used in buffing and polishing compounds. Chemically precipitated amorphous silica serves the same purposes.

Pumice and Pumicite. These substances are silicates and not silica but are generally included under silica or siliceous abrasives. Pumice is the natural volcanic product or frothy glass of steam-expanded siliceous lava, and pumicite is volcanic ash. About 16 percent of an annual production of about 300,000 tons in the United States is devoted to abrasive purposes.

Lump pumice is used for dressing furniture and musical instruments, preparing metal surfaces for silver plating, cleaning lithographic stones, and rubbing and polishing fine tools and instruments. Ground pumice and pumicite make excellent cleansers because of the thin sharp glass shards contained in them. Much of this material is made up into scouring soaps. It is also an ingredient in various polishing compounds for celluloid, hard rubber, and bone and is included in rubber erasers.

The United States supply of pumice comes from California, Kansas, Nebraska, New Mexico, Oklahoma, and Oregon; pumicite comes from Kansas, Nebraska, Nevada, and Oklahoma. Canada, Italy, Japan, New Zealand, and other countries are also producers.

Selected References on Siliceous Abrasives

Abrasives, Part I, Siliceous Abrasives, by V. L. EARDLEY-WILMOT. Can. Dept. Mines, Mines Branch, No. 673, 1927. *Complete survey.*
Tripoli. R. W. METCALF. U. S. Bur. Mines Inf. Circ. 7371, 1946. *General survey.*
General References 5; 11, Chaps. 1, 36, 50.

Miscellaneous "Soft" Abrasives

Calcite ground to a fine powder is used to a small extent as a non-scratch metal polish.

Chalk or whiting is a widely used soft polisher for silverware, gold, nickel, and chromium plate, brass, buttons, etc.

China clay and pipe clay are used as polishing powders for soft metals. Pipe clay was once the standard polish for military and naval uniform buttons and belts.

Dolomite calcined to unhydrated calcium and magnesium oxides, called *Vienna lime,* is a common buffer for various metals, pearl, and celluloid. It is of particular value for nickel plate, to give the deep "under-surface" blue color in highly polished nickel articles.

Sheffield lime is a similar product.

Feldspar ground to powder makes a desirable scouring powder and soap for cleansing porcelain, glass, and enameled surfaces. It is fairly hard, but, being softer than these surfaces, it does not scratch them. It is said to be the chief abrasive constituent of " Bon Ami " cleanser. It is also a constituent of bonds for cementing vitrified abrasive wheels.

Fuller's earth is a soft abrasive for grease removal and high-grade polish for silver and chromium wares.

Lampblack, made from the burning of oil, is a fine, soft buffer and polish for burnishing silverware, black celluloid, and buttons.

Lime is a constituent of Vienna lime.

Magnesia is used as a soft polisher for metal surfaces and also a final polisher for polished surfaces of metals and minerals prepared for microscopic study.

Magnesite serves as a bonding agent and abrasive for making emery wheels.

Metallic oxides are widely employed as final buffing and polishing agents for metals, mineral surfaces, and optical glass.   *Crocus* is a hard, purple-red hydrated iron oxide made up as a paste for buffing tin, steel, cutlery, and other metal surfaces requiring a high finish.   *Rouge* is a hydrated oxide of iron used as a soft metal polisher and for polishing lenses of eyeglasses, microscope lenses, and other optical products.

Green rouge is an oxide of chromium and is a very fine, fast buffer and polisher for platinum and gives the high polish on stainless-steel wares.   It is also a polisher of hard minerals in polished ore sections.

Black rouge, or precipitated magnetite, is used as a buffer on cloth for the final finish on plate and cut glass.   It is also an excellent polisher to remove final scratches from soft minerals in polished ore surfaces.

Tin oxide is employed to give a high finish to certain metals, polished ore surfaces, and stone glazing.

Manganese dioxide is a good polishing medium but it is dirty.

Talc is used as a polisher for rice grains, soft metals, and leather.

Manufactured Abrasives

Artificial abrasives are supplanting many natural abrasives.   Some of them are extremely hard and sharp and are of uniform quality.

The chief mineral products involved in their manufacture are: bauxite, silica, carbon, lime, magnesia, clay, salt, boron and tungsten minerals, iron and steel, rubber, natural gas and oil, and some natural abrasives.   The manufactured products are divided into silicon carbide, fused alumina, boron carbide, and metallic abrasives.   In the new Mohs' scale of hardness, where now quartz is **8**, garnet 10, corundum 12, and diamond 15, the carbides rank between 10 and 14, boron carbide being 14, the next hardest substance to the diamond.   Silicon carbide and fused alumina are the most used.

Some 300,000 tons of artificial abrasives, worth up to 20 million dollars, are made annually in the United States.

Silicon Carbide (SiC).   This is known under such trade names as Carborundum, Crystolon, and Carbolon and is made by the fusion of crushed petroleum coke and silica sand in an electric furnace for 36 hours at 2,200° C.   It is used mainly in cutting wheels and papers and cloths.   Its hardness is 13, Mohs' scale.

Fused Alumina.   This is sold under the trade names of Alundum, Aloxite, etc., and is made of bauxite fused in an electric furnace with coke and iron for 24 hours at 3,500° F.   Corundum may be used in place of bauxite. The fused product is used in the form of vitrified wheels and powders.

Boron Carbide. This product (Norbide), the next hardest substance to the diamond, is replacing the diamond and diamond dust for many abrasive purposes. It is made into molded products for extrusion dies, thread guides in textile machinery, and sandblast nozzles. The powder is suitable for many grinding and lapping operations where extreme hardness is important. It is an electric furnace product made from coke and dehydrated boric acid.

Metallic Abrasives. These include crushed steel, steel shot, angular steel, and steel wool, all of which are made from special irons and steels. They are used for metal blasting, for cleaning castings, forgings, and metals, core drilling, and stone sawing. These abrasives can be reclaimed and used 200 or 300 times. Steel wool is a nonclogging abrasive for woodwork, paint and varnish, and aluminum ware.

General References on Abrasives

Abrasives. V. L. EARDLEY-WILMOT. Parts I–IV. Can. Dept. Mines, Mines Branch, Nos. 673, 675, 677, 699, Ottawa, 1927–29. *Exhaustive treatment and complete bibliography to those dates.*

Boron Carbide, the New Abrasive. N. BARKERIS. Jour. Am. Ceram. Soc. 30, No. 1, 1947.

General Reference 11. Chap. I, by R. B. LADOO. *Broad coverage of all abrasives; good bibliography.*

CHAPTER 24

GEMSTONES

Gemstones are minerals with special physical properties that make them desirable for personal adornment and decorative purposes. They are among the most valuable of all substances and have been prized since early civilization. Ball states that some 15 varieties were known more than 9,000 years ago. The earlier stones belonged to the quartz family (100,000 to 75,000 B.C.); then much later came the emerald (2,000 B.C.), sapphire and ruby (600 to 500 B.C.), and diamond (480 B.C.).

Uncut stones are known as gemstones, and the term "gem" is restricted to cut stones. Gemstones are divided into precious stones, or noble gems — diamond, emerald, sapphire, ruby, and precious opal (and pearl by courtesy) — and semi-precious stones. About 100 minerals have been classed as gemstones. To be prized they must possess beauty, purity, durability, and rarity. Fashion is also a factor.

The beauty may lie in a stone's color or play of color and its brilliancy or "fire." Brilliancy is dependent upon the refraction of light through a transparent stone, and "fire" is dependent upon high dispersing power. The diamond possesses both of these in marked degree. Other stones exhibit pleochroism or a change in color when viewed in different directions, as tourmaline. Color is the chief attraction in the semi-precious stones. The play of color in precious opal is due to interference of light from internal films. Proper cutting, of course, accentuates these properties, but cutting alone cannot make a gem unless the qualities are present within it. No skill can make a gem out of a black carbonado diamond.

Rarity is essential for a precious stone, for obviously no common thing can be precious. Durability and resistance to abrasion are important. Nothing except a diamond can deeply scratch a diamond, but the pearl and opal must be treated with care.

Varieties of Gemstones. A list of the more important gemstones is given in Table 26.

The semi-precious stones have been much in vogue. Aquamarine (beryl), moonstone (feldspar), and topaz are popular. The ones sold in greatest volume are the various members of the quartz family.

Table 26 gives some idea of the distribution of various gemstones.

Uses of Gemstones. In addition to personal adornment gemstones are also used for vases, statuettes, and *objets d'art*, particularly in India, China, and the Urals.

TABLE 26

IMPORTANT GEMSTONES

Stone	Chief Constituents	Common Color	Hardness	Chief Source
Precious Stones				
Diamond	C	Colorless to straw	10	South-central Africa, Brazil
Emerald	Be, Al, Si, O	Green	7.5–8.5	Colombia, Egypt
Ruby	Al, O	Red	9	Burma, Ceylon
Sapphire	Al, O	Blue	9	Ceylon, Burma, Siam
Precious opal	Si, O	Variegated	5.5–6.5	Australia, Hungary, Mexico
Semi-Precious Stones				
Amethyst	Si, O	Purple	7	India, Persia, Brazil
Beryl	Be, Al, Si, O	Various	7.5–8.5	U.S., Africa, Brazil
Benitoite	Ba, Ti, Si	Deep blue	6.5	
Chrysoberyl	Be, Al, O	Green, yellow	8.5	
Feldspar	K, Na, Ca, Al, Si	Various	6	
Garnet	Al, Fe, Mg, Si	Red, green	6.5–7.5	Arizona, South Africa, Urals
Jade — nephrite	Ca, Mg, Fe, Si	Green to white	5.5	Burma
Jade — jadeite	Na, Al, Si	Green	6.5	Burma, China
Kunzite	Li, Al, Si	Lilac	6.5–7	California, Madagascar
Lapis lazuli	Na, Al, S, Si	Dark blue	5–5.5	India, Greece, California, Siberia
Peridot	Mg, Fe, Si	Green	6.5–7	Levant, Egypt, Burma
Quartz	Si, O	Various	7	Worldwide
Spinel	Mg, Al	Reddish	8	Ceylon, Burma, Siam
Topaz	Al, Fe, Si	Yellow	8	Brazil, Ceylon, Montana
Tourmaline	Bo, Si, +	Green, pink	7–7.5	Urals, Madagascar, California, Maine
Turquois	Al, P, O, H	Blue	5–6	New Mexico, Persia, Turkestan, Egypt
Zircon	Zr, Si	Red, orange	7.5	Ceylon, Siam

Ball states that about 15 percent by value and 75 percent by weight of gemstones are normally taken by industry, the diamond being the most important industrial stone (see Chap. 23).

Rubies and sapphires formerly used for watch jewels are now replaced by artificial corundum. Likewise, lapis lazuli as a base for ultramarine is now replaced by artificial colors. Quartz and tourmaline are used in optical equipment. Agate is used for mortars, pestles, and balance bearings. Beryl is a source of beryllium, and zircon is an ore of zirconium.

Precious stones, because their price has long been upward, are treated in the Orient as investments; they serve as concentrated wealth, and are easily transported or hidden during national upheavals.

Occurrence and Origin. The variety of gemstones implies a variety of occurrences and origin. Dominantly, however, they are obtained from igneous rocks, pegmatite dikes, and alluvials. Veins supply quartz; metamorphism yields garnet, lapis lazuli, ruby, and nephrite; hydrothermal solutions yield opal and agate; and supergene processes yield turquois. Alluvial stones are obtained from stream gravels, beach gravels, and desert gravels. Chemical weathering plays an important role in releasing unharmed gemstones, and making it profitable to work many otherwise unprofitable pegmatites.

Most gemstones are found in alluvial gravels. The diamond is obtained in part from diamond pipes; and the emerald, opal, turquois, malachite, and fluorspar are obtained almost exclusively from solid rocks. Tourmaline, beryl, quartz, spodumene, topaz, zircon, and feldspar gemstones are obtained mainly from pegmatites.

Extraction and Preparation. Alluvial diamonds are recovered by well-organized companies by means of native labor and mechanical or hand methods. The gravels are washed, screened, concentrated in " diamond pans," jigged, and then hand sorted or passed over grease tables. The extraction from pegmatite deposits is mostly from weathered portions, the dirt being treated like alluvial gravels. Unweathered pegmatites and rocks are blasted, coarse crushed, hand sorted or screened, concentrated, and hand sorted again. Diamond pipes are mined by open cut or by underground methods from shafts sunk in the country rock, the rock being blasted, crushed in stages with corrugated rolls, and concentrated in large horizontally revolving diamond pans in which the heavy materials in a thin mud medium are thrown by centrifugal force to the outside and the lighter materials are drawn off in the center. The concentrates are then jigged, and the jig concentrates are passed over sloping bronze tables covered with grease. The diamonds adhere to the grease; the tailings do not. The grease with the adhering diamonds is scraped off, boiled in caustic soda, and the freed diamonds are graded into parcels.

The stones are cut round or faceted. Faceting is done on rotating laps fed by diamond dust, the stones being temporarily embedded in solder.

Artificial Stones and Imitations. These are (1) synthetic stones, (2) a stone substituted for another, perhaps the color being altered, (3) imitations in glass or other artificial products, and (4) composite stones.

The diamond cannot be made synthetically in commercial sizes, but the ruby and sapphire can, and the synthetic stones can be distinguished only by experts. They are made by fusing pure alumina, mixed with a metallic oxide for color, in a jet of oxy-coal gas flame. Substitutions rarely are successful with precious stones but are with the semi-precious gemstones. Glass imitations known as " paste " or " strass " are common. The diamond is transparent to X-ray, by which means it may be distinguished from paste.

The Diamond

The diamond is unique among stones in that it is the hardest substance in the world, it is composed of a single element, it is durable, its high index of refraction and dispersion give it unexcelled brilliance and " fire," it comes from the deepest rocks known in the earth's crust, it has been consistently the most desired of stones for 1,900 years, and it is so valuable that $10,000,000 worth of stones could be secreted about one's person. In contrast, the same value of gold would weigh nearly 12 tons.

Apparently the first stone was found in India about 800 B.C. Diamonds formed the eyes of a little Greek goddess dating about 480 B.C. Although valued by the Romans, the diamond did not come into its own until the art of diamond cutting was developed in the Middle Ages. Since then about 100 tons of diamonds have been produced.

	Percent of World			Percent of World
Belgian Congo	60		Southwest Africa	1.5
South Africa	12		Tanganyika	1.2
Angola	8		French West Africa	0.9
Sierra Leone	7		French Equatorial Africa	0.9
Brazil	3		British Guiana	0.2

Production and Distribution. The early production of diamonds came mainly from India, and Ball states that, even by the middle nineteenth century, production amounted to only 4 million dollars. With the discovery of South African mines, production in the 1890's amounted to 20 million, and by 1910 to 48 million dollars. Later it reached 85 million dollars. The annual world production is normally about 10 to 15 million carats (1 carat = 0.2 gram) or about 2½ metric tons worth 40 to 80 million dollars, and divided approximately according to the table.

During several years the Gold Coast has exceeded South Africa in quantity output. Alluvial mines yield 96 percent by weight and 91 percent by value. The British Empire normally produces 28 percent by weight and 70 percent by value of total production; one-fifth by weight are gemstones. Most of the stones of central Africa are small and are used as industrial diamonds, which fetch only $1 to $3 per carat.

The weight of stones produced annually might give the impression that gem diamonds are common. However, of the annual production about one-half is bort; one-fourth is off color and flawed but can be cut into "flashy" jewels; and one-fourth is gem material, much of which is small. Only about 5 percent consists of 2-carat rough stones capable of yielding 1-carat cut stones, or about 100,000 fine large diamonds a year.

Occurrence and Origin. The primary home of the diamond is in kimberlite pipes or other forms of ultrabasic intrusives, and a secondary home is in stream and beach gravels.

The diamond pipes are vertical, carrot-shaped bodies of ultrabasic rock with diameters up to 2,300 feet and depths of 3,500 feet (Fig. 24–1). The kimberlite contains many inclusions of overlying and underlying rocks. The diamonds occur as crystals sparsely disseminated in the kimberlite; a few occur in eclogite inclusions. Many pipes are known; some contain a few diamonds but those with commercial content are few. The pipes decrease in size with depth and the diamond content diminishes.

The astonishing rock inclusions packed in the South African pipes are of particular interest. Some of them containing fossils are the only evidence of a former cover of Karoo beds. According to Williams, some inclusions in the pipes now lie 2,500 feet below their source and these include fragile fossiliferous shales that could not have sunk in such a heavy magma. Masses of igneous rocks have been lifted 2,700 feet; one 500-foot block of granite has been raised 244 feet. These inclusions record a history of origin of the pipes. They lead Williams to conclude that a first stage was gas explosion following which masses and fragments fell back into the cavity. Then came a slow upwelling of kimberlite magma that floated up blocks from below and slowly closed around the dropped-in fragments but was not hot enough to affect them.

Three views have been held for the origin of the diamonds: (1) they crystallized *in situ;* (2) they were originally contained in underlying layers of eclogite, which was fused by upwelling kimberlite that carried along the released diamonds; (3) the diamonds crystallized in the original magma chamber and were carried up as crystals in kimberlite magma to the present site of cooling. Williams argues for the last view with much conviction. One astonishing bit of evidence supports this view — two broken parts of a rodlike diamond crystal were found at different levels of the De Beers mine and when placed together they

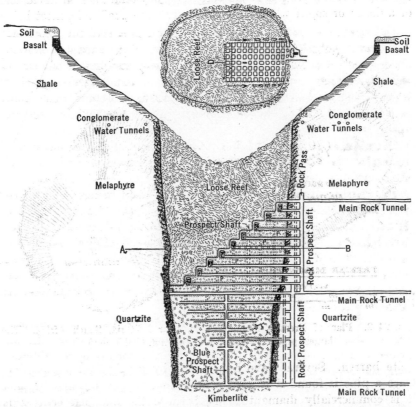

Fig. 24–1. Plan and section of Kimberley diamond pipe (3,600 feet deep), Kimberley, South Africa. (*After DuToit, 15th Internat. Geol. Cong.*)

fitted exactly. This diamond obviously had crystallized before reaching its present site.

The origin and occurrence of alluvial stream and beach diamonds and the occurrence of the remarkable Lichtenburg placers are described in Chapter 5·7.

EXAMPLES OF DEPOSITS

South Africa. Diamonds were discovered in South Africa in 1867 and the " Star of South Africa," a magnificent 83½-carat diamond, was found in 1869. Since then the country has produced about 240 million carats, worth over 1,800 million dollars. The annual production ranges between 1 and 1¼ million carats, of which 85 percent is won from pipe mines (Fig. 24–2).

Scores of pipes, from 50 feet to half a mile across, are known from the Cape Province to the Congo border. They tend to occur in clusters of a dozen or so, of which only one or two are profitable, most being

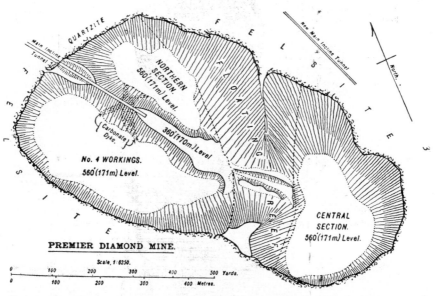

FIG. 24–2. Plan of Premier diamond pipe, near Pretoria, South Africa. The largest pipe. (*Wagner, 15th Internat. Geol. Cong.*)

quite barren. Several have been found by geophysical prospecting. Once a pipe is found it is an expensive matter to determine whether it is commercially diamantiferous, because the diamond content is spotty and low. The profitable pipes carry only about 7 to 20 carats per 100 loads (1,600 lb each). Consequently a large mill test has to be made. Ten profitable pipes cluster around Kimberley, and another group, including the Premier, occurs near Pretoria.

The pipes are funnel-shaped bodies, oval on the surface but becoming elongated or fissurelike at depth (Fig. 24–3). The famous Kimberley mine is 1,600 feet across and 3,500 feet deep; Dutoitspan is 2,550 by 1,200 feet; Bultfontein is 2,400 by 1,900; Wesselton is 1,500 feet across. The giant Premier is 2,800 feet across and at 600 feet has diminished little in diameter.

The pipes intrude a typical section as follows:

Upper Dwyka shales
Dolerite sills
Melaphyre and quartzite ⎫
Quartz porphyry   ⎪
Quartzite and conglomerate ⎬
Old granite, schists, amphibolites ⎭

Mesozoic
Upper Carboniferous or Karoo
Ventersdorp system
   (Precambrian?)

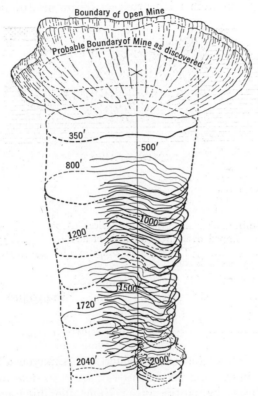

Fig. 24–3. Isometric projection of De Beers mine, South Africa, showing contours of the pipe to a depth below 2,000 feet. (*Williams, Genesis of the Diamond.*)

The pipes are occupied by kimberlite, a greenish-blue peridotite rock with much olivine, brown mica, chrome garnet, and ilmenite. The upper part is weathered to "yellow ground." The kimberlite is variable in composition. Much of it is a breccia, and the intrusion was composite. Inclusions of two kinds are numerous, one type is eclogite pebbles and boulders carried up from depth, and the others, already referred to, are fragmental specks and masses dropped down from above or floated up from below.

The diamonds are so sparsely disseminated that a stone in place is rarely seen by a miner. The richest mine, the Bultfontein, averages 28 carats per 100 loads, the Wesselton 23, the Dutoitspan 13, and the West End 7. The Premier averaged 16 to 19 carats per 100 loads (0.0000052 percent) of rock crushed, and has produced 5½ tons from about 100 million tons of rock excavated, or about 0.2 carat per ton. The percentage of fine stones in one pipe may be large and in another small. The Jagersfontein mine produces 50 percent of stones over 1

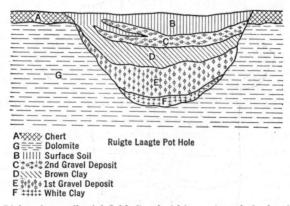

A ✕✕✕✕ Chert
G ≡≡≡ Dolomite   Ruigte Laagte Pot Hole
B |||||| Surface Soil
C ⋄⋄⋄⋄ 2nd Gravel Deposit
D ＼＼＼ Brown Clay
E ⋄⋄⋄⋄ 1st Gravel Deposit
F ⋮⋮⋮⋮ White Clay

Fig. 24–4. Lichtenburg alluvial field, South Africa. A pothole developed in dolomite into which slumped overlying two beds of alluvial stream gravel deposits, shown in circles. (*Williams, Genesis of the Diamond.*)

carat, and the Premier 17 percent, compared with 5 percent for the Belgian Congo detrital stones. The Jagersfontein produced 6 stones of over 500 carats, and the famous Cullinan diamond of 3,106 carats (1.3 lb) was found in the Premier mine in 1905.

Alluvial stones are recovered from Vaal and Orange Rivers and from the remarkable Lichtenburg ridge and sinkhole alluvial (Fig. 24–4) described in Chapter 5·7.

A notable discovery is the great Namaqualand beach deposits, where quantities of good-sized, high-quality stones are obtained. Prospecting operations there yielded 12,240 carats worth $740,000, one of the finest assemblages of stones obtained, and containing one stone of 71 carats worth $35,000, the finest detrital stone found in South Africa. Details of this occurrence are given in Chapter 5·7.

A few diamonds have been found in the gold deposits of the Rand. These are detrital stones of pre-Witwatersrand age that represent an earlier origin than those of the post-Karoo pipes.

Belgian Congo-Angola. This highly productive area yields alluvial stones from the basin of the Kasai River which, with its numerous tributaries, extends from the Congo southward into Angola and occupies an area of 150,000 square miles. Some diamonds also occur in the Lulua, Zambesi, and other basins. This combined area produces over 65 percent by weight of the world output of diamonds, and the Congo yields three-quarters of the world output of bort.

The diamond gravels occur in stream beds, in river flats, elevated terraces, and in ancient stream channels that cross the present drainage. The gravels range from 3 to 7 feet in thickness, and the overburden may be a little thicker. There are two types of deposits, those in narrow valleys 20 to 100 meters wide and about 2 kilometers long, and those in flats about 500 meters across and 1 kilometer long. The streams lie on an old plateau underlain by Archean and Jura-Triassic rocks, the latter consisting of horizontal sandstone and conglomerate, with diamonds in some of the beds.

The gravel averages about 0.6 carat per cubic yard of gravel washed. Only about 5 percent of the diamonds are gemstones, the remainder being industrial stones. The Congo stones average 8 to 17 to the carat, the largest being 44 carats, and the Angola stones 8 to 9. The Angola gravels carry also staurolite, kyanite, ilmenite, tourmaline, chrysoberyl, and garnet. The Congo stones have similar associated minerals and also chalcedony, agate, jasper, and flint artifacts.

The source of the stones is obscure. Some pay gravel apparently has come from lenses in the Mesozoic sediments and these stones are thus older than the diamond pipes of South Africa. It was recognized in 1948 that one kimberlite pipe was responsible for the Bakwanga placers of the Kasai district, Belgian Congo.

The region has been carefully developed since the first discoveries in 1910 to 1912. It yields about 10 million carats annually and still has extensive reserves.

Gold Coast. The Gold Coast produces about ¾ million carats annually and ranks fourth in annual production. Diamonds were discovered in 1919 in the Birrim River, which led to other discoveries in the Bonsa basin. The rocks of the region are Precambrian altered sediments and lavas intruded by granite, amphibolite, and dolerite dikes. The heavy minerals associated with the diamonds are angular, suggesting a nearby source for the diamonds.

The diamond gravels which are in stream valleys, average 2.5 feet thick. The stones are of good quality but are small, averaging about 20 to the carat, and are mainly industrial stones. Only one 4.5-carat stone has been found. About 30 percent are gemstones, 42 percent bort, and the remainder diamond sand. The diamonds are thought to have been derived from nearby metamorphosed basic igneous rocks containing inclusions of carbon-bearing phyllite.

Sierra Leone. From a discovery made in 1930 by the Sierra Leone Geological Survey, the diamond production of Sierra Leone has jumped to between 600,000 and 900,000 carats annually. The stones are larger and of better quality than the Gold Coast stones; one stone of 177 carats was found. Chrome garnet occurs in altered ultrabasic rock near the fields, and small fragments of ruby and zircon crystals occur in the gravels.

Southwest Africa. Although 40 kimberlite pipes and dikes are known, diamond production is from marine-terrace sandstones and gravels extending for a distance of 70 miles near Lüderitz Bay. The gravels are from a few inches to 30 feet thick and average only 0.2 carat per cubic meter, but rich pockets have been formed by wind action in underlying limestone. In one year 7 million tons of gravel yielded 7.5 carats per 100 loads. The stones are brilliant and average over a carat. Later discoveries north of the Orange River are similar to the diamond beach terraces of Namaqualand, yielding stones in quantity that average over 1 carat, with one stone of 246 carats. These stones, like those of Namaqualand (p. 243), probably came down the Orange River. Production ranges from 100,000 to 160,000 carats a year.

French Africa. French West Africa produces some 90,000 carats from alluvial gravels. There are a few good gemstones, but they are mostly bort. In French Equatorial Africa some 90,000 carats are won from the basins of the Marnberé and Loyabe Rivers. The stones are presumed to have been derived from a conglomerate.

Tanganyika. This territory contains some 40 occurrences of kimberlite and represents a northward extension of the South African pipes. They are mostly barren, but the productive ones yield stones of good quality averaging about ½ carat. The workable pipes are Makubi, Shinyanga, and Mzumbi south of Lake Victoria. The Williamson mine at Mwadvi yields gravels overlying a large pipe and has caused Tanganyika production to rise to over 120,000 carats annually.

Brazil. Brazil is the only large diamond region outside of Africa. It has an annual production of about 150,000 carats, about one-fifth of what it used to be when Brazil led the world in production. Since the original discovery in 1720, about 20 million carats have been produced. The chief occurrences are in Bahia, Minas Geraes, Matto Grosso, and Piavi.

The stones are found as placers in present stream channels and ancient terraces, in old conglomerates, and in igneous breccias. The

Minas Geraes alluvial deposits occur in a region of Precambrian metamorphic rocks and later Precambrian folded sediments, overlain by metamorphosed fluvioglacial deposits of Precambrian or early Cambrian age known as the Lavras beds, which are diamantiferous. These sediments are thought to have yielded many of the alluvial deposits.

The Anga Sija deposit in Minas Geraes is in a breccia of angular blocks of sediment enclosed in a soft, claylike matrix presumed to have been altered from an ultrabasic igneous rock like kimberlite. Most of the output, however, is from detrital deposits.

The Matto Grosso occurrences resemble the Minas Geraes.

The Bahia deposits are unusual in that most of the stones are carbonados which, because of special industrial uses, fetch a higher average price than any except Namaqualand stones. One carbonado weighed 3,078 carats. These stones are alluvial and are obtained mostly from the gravels of the Paraguassú River basin. The stones occur only where the streams cut the Lavras beds, which are thought to have supplied both carbonados and gemstones.

The Brazilian stones are of high quality but vary considerably in the numerous occurrences. It is rather striking that in nearly 225 years of mining, during which time less than ten large stones have been found, the largest stone, the " Presidente Vargas " of 726 carats, was not discovered, according to Ball, until September 1938 and the " Darcy Vargas " of 455 carats the following year.

British Guiana. This country yields about 20,000 to 30,000 carats a year from gravels in the Mazaruni and Potaro River basins. The stones occur in a difficultly accessible belt across the country, some 10 to 30 miles wide and 100 miles inland. The gravels yield less than 1 carat per cubic yard, but the stones are of fair size and quality, averaging about 6 to the carat. The larger stones are found nearest the Pakaraima Mountains, and some are found in the talus slopes of sandstone and conglomerate beds, suggesting that they may have been derived from the sediments. No ultrabasic rocks are known. The fields are apparently fairly extensive, but they are little known geologically and are worked by sporadic and primitive methods.

India. India was the home of the diamond until the discovery of the Brazilian fields. Until the eighteenth century any diamond not from India was regarded with suspicion, and the first Brazilian stones could find no market until the prejudice was overcome by the subterfuge of shipping them to India and selling them from there. India, which formerly produced several hundred thousand carats, now produces less than two thousand carats a year. The stones are all alluvial, from the Panna district, and occur in Recent gravels of Precambrian conglomerates, the original source being unknown. One kimberlite pipe was explored at Panna. India has been a great producer of large stones, such as the Koh-i-noor and others.

Other Deposits. *Borneo* was the other source of diamonds in the Middle Ages, but production has dwindled to less than that of India. The stones are alluvial, of fine luster and fair size, with many colored ones and a black diamond.

In *Russia* geological expeditions found 52 gemstones in the western Urals in 1938, and 90 in six occurrences in 1939. The deposits are claimed to be commercial.

Venezuela yields a few thousand carats each year from the Caroni and Cayuni Rivers. A few stones are found in Uganda, Rhodesia, Liberia, and New South Wales. The few *Uganda* stones are alluvial presumably derived from the Kager-Amklean sedimentary rocks at Buhweza and Kibale. A $2\frac{1}{4}$ carat stone was found in 1938. In *Southern Rhodesia* six kimberlite bodies are known, two of which contain low-grade rock, and a small production of alluvial stones is won from Upper Karoo gravels. *New South Wales* yields a few hundred carats a year from gravels at Copeton. Of scientific interest was the finding of two stones in solid dolerite or hornblende-diabase.

In the *United States* several thousand small stones were obtained from a kimberlite intrusion near Murfreesboro, Ark. Occasional stones have been found in California and Oregon gold placers, and in the southern Appalachian states. A few chance stones have been picked up in glacial gravels in Wisconsin, Illinois, Indiana, Michigan, and Ohio. They are good stones of fair size and are presumed to have come from an unknown source in Canada.

Selected References on Diamonds

Gemstones. Imp. Inst., London, 1933. *A brief geographical survey of gemstones in all countries — good bibliography.*

Genesis of the Diamond. A. F. WILLIAMS. 2 Vols. E. Benn, Ltd., London, 1932. *Detailed descriptions of South African pipes and occurrences; magnificent illustrations.*

General Reference 11. Chap. 35, by S. H. BALL. *Concise summary and interesting history; good bibliography.*

The Diamond Industry. S. H. BALL. Annual "Jewelers' Circular." *A concise up-to-date summary.*

Geologic and Geographic Occurrence of Precious Stones. S. H. BALL. Eron. Geol. 17:575–601, 1922; also 30:630–642, 1935.

General Reference 5.

Ruby and Sapphire

These two beautiful stones are the red and blue transparent varieties of corundum (Al_2O_3). "Pigeon's blood" red is the preferred color, for which the term ruby is reserved, and sapphire includes colors other than blue, but the deep royal blue is most prized. It is the colors that make them attractive because, since corundum has low dispersion, they lack fire. Large rubies are not common but large sapphires are. Both stones are dichroic and exhibit different shades of color in different directions. Both change color upon exposure to radium radiation. Both stones possess a faint fibrous structure by which the expert identifies their source. Extreme fibrous structure gives *star ruby* and *star sapphire,* which display the optical effect of six rays of light emanating from a center. Colorless corundum is called " white sapphire"; yellow is called " oriental topaz"; green and purple are called " oriental emerald " and " oriental amethyst." The red of ruby is thought to be due to traces of chromium, and the blue of sapphire to titanium.

The finest rubies come from Mogok, *Burma,* where ruby and spinel occur in a contact-metamorphosed Archean limestone intruded by silicic and basic igneous rocks. The limestone is too lean to work, and the stones are obtained from alluvium and detritus from it and from solution cavities in the limestone. The floor of the valley was formerly a lake, and stones are obtained from a pebbly clay lake deposit beneath 15 feet or so of overburden. The Burmese kings worked these deposits for centuries, and larger operations commenced in 1889. The concentrates also contain sapphires, spinel, tourmaline, gem sillimanite, and many other semi-precious stones. Since the decline of the ruby the mines are now worked by natives, and about 200,000 carats of rubies are produced annually.

The sapphire is more widely distributed. Some are found with the Burmese rubies, but not in the limestone. *Ceylon* has yielded sapphires, as well as rubies and most other gemstones, for 2,500 years. Practically all the stones are alluvial, occurring in gravels that have resulted from the denudation of Archean schists, gneisses, and crystalline limestones. Pelmadulla is the best-known locality. The gravels occur in beds, patches, and pockets down to a depth of 120 feet, the best ground being on bedrock. The sapphires and rubies were derived from the limestones and the other stones from the gneisses and schists.

Fine sapphires occur in a pegmatite dike in the Zanskar Range, *Kashmir, India,* and in the corundum pits of *Madras* and *Mysore.*

Rubies and sapphires also occur in crystalline limestone in *Afghanistan*. *Siam* and *Cambodia* yield both stones from gravels. Old elevated gravel deposits in *Anakie, Queensland,* yield dark blue sapphires, and a 412-carat stone of greenish blue was found at *Rubyvale, Queensland,* in 1938.

Pale sapphires are obtained from gravels near Helena, *Mont.,* and from a monchiquite-camptonite dike in Yogo Gulch, about one-fourth of which are gemstones.

Emerald and Beryls

Emerald, the most valuable of all precious stones, is a transparent variety of beryl with the well-known emerald-green color. Other gemstone varieties of beryl are *aquamarine,* the sea-green and bluish stones; *yellow beryl* or *heliodor,* the yellow variety; *aquamarine chrysolite,* the yellowish-green variety; and *morganite,* the pinkish variety. All owe their beauty to their color and transparency; their dispersion is low, and they lack " fire."

Fine, unflawed emeralds are rare, and large unflawed stones are unknown. One of 1,384 carats, but badly flawed, has been found. Large flawless aquamarines are common.

All the fine emeralds come from *Muzo* or *Cosquez* in *Colombia,* where the mines were worked by the Spaniards. The Muzo deposits occur in the precipitous slopes of a valley in a high jungle, enclosed in high-temperature calcite veins that traverse Lower Cretaceous black shales. They are accompanied by pyrite, parisite, dolomite, and quartz with some apatite, fluorspar, and barite. The Chivor deposits are generally similar, but the accompanying minerals are pyrite, albite, and quartz in veins up to 8 inches wide and 200 feet long. The Cosquez deposits are little known. About 12,000 to 14,000 carats are produced annually.

The *Murchison Range, Transvaal,* yielded about 670,000 carats of emeralds from 1930 to 1940. They occur in shoots and pockets in crystalline schists near pegmatites, which contain beryl but no emerald. The yield is about 3 carats per cubic yard.

The *Ural Mountains,* near Takovaya, yield Russia's most important gemstone; good emeralds occur in schists accompanied by beryl, phenacite, and chrysoberyl. Emeralds occur in altered marble along with quartz and calcite in *Bahia, Brazil,* and in mica schist intruded by pegmatites along with much beryl, tourmaline, quartz, feldspar, apatite, biotite, and molybdenite at *Leydsdorp, Transvaal.* In *Egypt,* near the Red Sea, emerald mines were worked 2,000 years ago. Hundreds of shafts were sunk in mica schist, some to 800 feet. The stones are unsuitable for gems. A few stones are obtained from mica schist in *Austria*.

The other gem varieties of beryl are found in pegmatites or their disintegration products in *Minas Geraes, Brazil,* in pegmatites in *India;* in many localities in *Madagascar* (particularly morganite); in numerous pegmatites rich in many accompanying minerals in Southwest Africa, where fine varieties of aquamarine, heliodor, and the uncommon aquamarine chrysolite occur; and the Urals and Transbaikal, Russia.

Opal

Opal may be precious or semi-precious. It differs from other precious stones in that it is not transparent and its beauty depends upon its lovely play of colors. Since it is amorphous and translucent it must be viewed in reflected light. It is probably of gel origin, and drying caused myriads of cracks, filled by later opal, and these reflect and refract the light to give the great range of colors. It is variable in color; the *black opal* that yields effective flashes of color is highly esteemed. *Fire opal* is a more transparent variety with a particularly fine play of colors. All types of precious opal display vivid color, and the opal exhibits greater diversity than any other stone. Some superstitious people regard it as unlucky, which affects its popularity.

Opal is a near-surface deposit in cracks and cavities formed by hydrothermal waters of silicic volcanic origin.

The chief locality for opal is *Australia*, where it is that country's most important gemstone. It occurs in New South Wales as cavity fillings, fragments in the soil, and in narrow seams in sandstone to depths of 100 feet. The occurrences of Queensland and South Australia are similar. In the latter, at the Coober Pedy field, pieces of opal are found over a distance of 40 miles. In *Czechoslovakia*, near Eperies (old Hungary), precious opal occurs near hot springs in crevices in andesite, accompanied by common opal, barite, yellow sulphides, and stibnite. *Querétaro, Mexico,* contains much common opal and a little fire opal, in layers and cavities in trachyte. A little precious opal is found in Virgin Valley, Nevada, and Latah County, Idaho.

Chrysoberyl, Alexandrite, and Cat's-eye

Chrysoberyl, an aluminate of beryllium, is a greenish-yellow stone that used to be included under chrysolite. The most beautiful and remarkable variety is the delightful *alexandrite*. It has exceptional dichroism, being bluish green to dark green by daylight and raspberry red by artificial light. It is the most intriguing semi-precious stone.

It occurs with emerald in the Urals, in the gem gravels of Ceylon, and in Madagascar. Cat's-eye (true), or cymophane, is a peculiar fibrous variety which, cut in a certain manner, exhibits chatoyance, with a whitish pupil and a groundmass of honey yellow. This stone is not to be confused with quartz cat's-eye.

These stones are of pegmatitic origin and occur in the pegmatites or intruded gneiss or schist.

Jade

Jade is a generic term that includes two somewhat similar minerals, nephrite, or true jade, and jadeite, and sometimes incorrectly bowenite, saussurite, and green garnet. Nephrite is a calcium magnesium amphibole, and jadeite a sodium aluminum pyroxene.

Jade is the most mystical of the semi-precious stones. Its toughness, color, and supposed magical properties have for long made it a treasured stone. Ancient beliefs have descended through the centuries to modern Chinese, who revere the stone because they believe that it wards off accident and ill fortune from the owner and that it is a symbol of purity in private and official life. The traditions of jade are now passing to the western lands, enhancing the popularity of the stone there.

Nephrite, or the Chinese " yu," was first used in China in the reign of Huang-Ti (2697-2598 B.C.), where it was conferred upon officials to show dignity, rank, and authority. It was even at that time brought into China from the West (Turkestan). Through the time of the " Five Emperors " (2647-2206 B.C.) and " Three Dynasties " (2205-247 B.C.) jade was required for official appointment, diplomatic intercourse, funerals, worship, and war. Among old qualities mentioned of jade and not known today was the distinctive sound, which was " slightly vibrating " and " far reaching."

Jade ranges in color from white through all shades of green, and even to red, the green of the peacock's feathers being the most precious.

Most of the precious jade comes from the Myitkyima district, Burma, and is sold in Canton. According to J. Coggin-Brown, it is found in bedrock and as boulders in alluvium. There are two types of bedrock occurrences, one in igneous rock and the other as waterworn boulders in the Uru boulder conglomerate. The denudation of the latter has given rise to accumulation of boulders in the Uru River, which are obtained by native divers.

The chief occurrence is in the Tawmaw dike, which is an albite-jadeite dike intruded into serpentinized ultrabasic igneous rock. The jadeite is in lenses 5 to 7 feet thick in the albite of the footwall side.

Here the dike is 21 feet wide, and the hanging wall side is a breccia composed of albite, jadeite, and amphibolite in a calcareous matrix. Dr. Bleeck believes that the jadeite was formed by metamorphism of an albite-nepheline rock, and Dr. Chibber that it is an original magmatic segregation product. There are similar dikes in serpentine, and the Uru conglomerate boulders and alluvium have come from them.

The Quartz Gemstones

The ubiquitous quartz occurs in numerous forms and beautiful colors and many varieties, which, although fairly common and inexpensive, are used as semi-precious stones. There are two main groups — one composed of single and visible crystals, such as rock crystal, and another, of cryptocrystalline varieties, such as agate. The former are transparent and the latter translucent or opaque. The gemstone varieties include:

MACROCRYSTALLINE		CRYPTOCRYSTALLINE		
Amethyst	Rose quartz	Agate	Chrysoprase	Onyx
Aventurine	Rutilated quartz	Agatized	Egyptian	Plasma
Cat's-eye	Sapphire quartz	wood	jasper	
Citrine	Smoky quartz	Bloodstone	Eye agate	Prase
Rock crystal	Tiger's-eye	Carnelian	Jasper	Sardonyx
		Chalcedony	Moss agate	

The quartz gemstones are prized chiefly for their colors, but in part for structures or inclusions. The prized colored groups include the clear limpid *rock crystal*, the purple or violet *amethyst*, the transparent or yellow *citrine* that is often confused with topaz, the clove-brown or smoky gray *smoky quartz* or *morion*, and the pink to rose-red *rose quartz*.

The varieties of special interest, because of fibrous texture or inclusions, are *aventurine*, the speckled brown variety flecked with iron oxide inclusions; and the *green aventurine*, flecked with chrome mica; *cat's-eye*, the milky green or translucent variety showing chatoyance; *tiger's-eye*, the golden-brown, fibrous, chatoyant mineral formed by the silicification of crocidolite asbestos, and the gray variety *hawk's-eye; prase*, the translucent leek-green variety with minute acicular inclusions of actinolite and other varieties containing inclusions of rutile and other minerals.

The popular monocrystalline chalcedonic varieties include *chalcedony*, the pale colored variety, and *carnelian* and *plasma*, the red and green varieties; *sard* is an orange brown carnelian, and *bloodstone* or *heliotrope* has red spots or patches; *agate*, with its narrow, concentric bands of different color is ornamental, and *onyx* is a variety of agate with black and white bands and *sardonyx* with red and white bands; *chrysoprase* is an esteemed translucent apple-green variety with splintery fracture; and *jasper* has many colors, the most typical being red or brown.

The quartz stones lack fire, but they are more brilliant and durable than " paste." They are used in rings and necklaces and particularly as ornamental stones for cameos, seals, boxes, mosaics, and other objects. Rock crystal is used in abundance for radio and radar.

Other Semi-Precious Stones

Another group of gemstones desired particularly for brilliance, color, and structures are used for jewelry and ornamental stones.

Turquois. This opaque mineral, attractive for its robin's-egg-blue color, was early mined by the Egyptians in the Sinai Peninsula and by American Indians in New Mexico and Colorado. It is now obtained in Nevada, Colorado, and Arizona. Much comes from Persia, and a little from Mexico. Varieties are *variscite* and *odontolite* or *bone turquois*, the fossil teeth of animals. Turquois is a supergene mineral.

Spinel. Gemstone spinel is in two main forms, the rose-red *balas ruby*, and the ruby-colored *spinel ruby*. Less desired colors are the orange variety *rubicelle*, the purplish *alamandine-spinel*, the blue *sapphire spinel*, and the black *pleonaste*. One fine spinel ruby, considered by some as one of the most beautiful gems in the world, is set in the English crown and is said to have been given to the Black Prince in 1367. Spinels occur in contact-metamorphic limestones or their alluvium and come mainly from Ceylon, Afghanistan, Burma, and Siam.

Topaz. The esteemed gemstone variety is a wine-yellow topaz, but the blue, pale green, and pink varieties are also gemstones. A rose pink variety is a yellow stone changed by heating. These varieties exhibit dichroism. Topaz is a durable stone with some fire and considerable beauty.

Gemstone topaz occurs in pegmatites or their alluvium and is found mainly in Brazil, Siberia, and Japan, but also in California, Arizona, New Hampshire, and in other countries. Three large stones came from Minas Geraes, Brazil, in 1939, the largest weighing 596 pounds.

In the jewelry trade many stones have been called topaz, such as "oriental topaz" (yellow corundum). "Bohemian topaz" or "Spanish topaz" are citrine; "false topaz" is fluorspar, and the true stone is called "Brazilian topaz."

Spodumenes. *Kunzite* is a lilac- or rose-colored variety from California and Madagascar, and *hiddenite* is a rare yellowish-green variety from North Carolina; common yellow spodumene comes from Brazil. The two gem varieties are dichroic and have become popular American gemstones. They are of pegmatitic or alluvial origin.

Zircon. Zircon is a beautiful gemstone that is becoming more popular. Its high refraction and dispersion give it a " fire " exceeded only by the diamond. It occurs colorless ("matura diamond "), but the yellowish-red variety *jacinth* or *hyacinth,* the straw-yellow *jargoon,* along with green shades are more common. Its color changes upon heating to lighter hues; some brown Siamese stones become greenish blue. Spain, Ceylon, and Australia are the principal sources.

Tourmaline. This is a beautiful gemstone in many delicate tints. The alkali varieties, greenish-yellow, honey-yellow, dark blue, red, and particularly the deep green and rose pink, are the preferred stones. Tourmaline is commonly zoned and is cut to display the different colors. The various colors go by many different names in the jewelry trade, such as *achroite* (colorless), *rubellite* (pink, red), *indicolite* (indigo blue), *Brazilian emerald* (green), and other names.

Tourmalines are pegmatite minerals and are widespread in distribution, the principal commercial sources being Maine, California, Russia, Burma, Madagascar, and Brazil.

Garnet. Of the garnet family the principal gemstones are: pyrope, the blood-red variety that often passes for ruby and is known as *Bohemian garnet, Cape ruby,* or *Anyma ruby;* almandite, a deep crimson variety called *carbuncle; rhodolite,* the attractive raspberry-red species from North Carolina; grossularite, an orange or yellow member known as *hessonite, cinnamon stone;* and *grossular,* which is passed off as jade; spessartite, the orange-red member known as *spessartine;* andradite, in the uncommon beautiful green form called *dematoid,* which has good fire and is also often incorrectly called " Arabian emerald " and " olivine."

Garnets are widespread and occur in pegmatites, ultrabasic, and other igneous rocks, and in many kinds of metamorphic rocks and gravels resulting from them.

Peridot. Peridot is the transparent olive-green variety of olivine, and *chrysolite* is the yellowish-green variety. Peridot is sometimes called the *evening emerald.* It does not stand wear very well. It is a constituent of basic igneous rocks, and the best comes from Zebirget Island (Red Sea), Madagascar, and Burma.

Lapis Lazuli. This mineral, known also as *lazurite,* has been valued as an ornamental stone since ancient times, particularly for beads and mosaics. Its intriguing quality is its prussian-blue color. It is also used to make the pigment, "ultramarine." Lazulite has a different composition and its color is azure blue. Jasper stained blue passes off as "German" or "Swiss" lapis. Afghanistan is the chief source.

Feldspar stones. Some varieties of feldspar used for minor gemstones are *moonstone*, a milky opalescent variety of orthoclase that comes from Madagascar; *amazon stone*, an apple-green variety of microcline; *labradorite*, with its iridescent blue greens and grays; and *sunstone*, a reddish oligoclase.

Pearl is often listed under precious stones but is not called a gemstone.

Amber is a fossil resin from extinct pine trees and has been used since early times for beads. Its chief source is Oligocene glauconitic sediments on the coast of East Prussia.

Jet is a dense black variety of lignite coal that takes a high polish.

Other minerals occasionally used for minor gemstones include *benitoite*, a bluish silicate; *euclase*, a beryllium-aluminum silicate resembling aquamarine; *phenacite*, a yellowish or rose-red beryllium mineral, and *beryllonite*, which is pale yellow. *Malachite* and *azurite* are used more for ornamental stones; and *rhodonite*, when it is a fine red color; *andalusite* is reddish to amber in one direction and light green in another; *pyrite*, and *hematite*, are both known as *marcasite*.

Selected References on Gemstones Other Than Diamonds

Most of the references given under Diamond.

Gems and Gem Materials. E. H. Kraus and E. F. Holden. McGraw-Hill, New York, 1931. *General reference.*

Mineral Resources of Burma. H. L. Chibber. Macmillan, London, 1934. *For rubies and emeralds.*

A visit to the Gem Districts of Burma and Ceylon. F. D. Adams. Can. Inst. Min. Met. Bull., Feb. 1926, pp. 213–246. *A popular description of interesting gem localities.*

U. S. Bur. Mines Inf. Circs., by I. Aitkens: Rubies and Sapphires, 6471, 1931; Emeralds, 6459, 1931; Quartz Gemstones, 6561, 1932; Turquois, 6491, 1931; Tourmaline, 6539, 1931; Topaz, 6502, 1931; Garnets, 6518, 1931; Feldspar Gems, 6533, 1931. *General information.*

Jadeite. J. Coffin Brown. India Geol. Surv. Rec. No. 64, 1930.

Popular Gemology. R. M. Pearl. Pp. 316. John Wiley & Sons, New York, 1948. *Popular but accurate knowledge of gems and gemstones.*

The Muzo Emerald Zone, Colombia. V. Oppenheim. Econ. Geol. 43:31–38, 1948. *Mineralized shale beds.*

General Reference 5.

CHAPTER 25

GROUND-WATER SUPPLIES

From the viewpoint of economic geology, ground water or subsurface water must rank as a mineral resource of prime importance. From earliest time in many parts of the world the development of civilization has been dependent upon it. Despite our familiarity with water, the principles of occurrence and laws of ground water are inadequately known to many, and with few other subjects has so much superstition and misinformation been handed down through the centuries. Meinzer remarks that one of the truly great achievements of civilized man has been his ability to provide adequate and reliable supplies of good water in those parts of the earth where surface waters are lacking or unusable. Those fortunately situated in regions of abundant supplies of surface water little realize how extensive are the areas entirely dependent for life and agriculture upon underground supplies of this essential mineral. Too often the supplies are meager — a meagerness due as much to a lack of knowledge of ground water as to a lack of supply. It has been estimated by Cox that 20 million acres in India alone is irrigated entirely by subsurface water, an acreage comparable to the total irrigated area of the United States. In arid regions the world over pumping subsurface water is one of the major industries.

The demand for water supplies is mounting steadily with advancing civilization. Meinzer states that the average per capita consumption of water in the cities and towns of the United States is 100 gallons a day. Growing cities are reaching farther and farther back into rural areas for available surface waters, with overlapping from adjacent cities. Rural towns face lessened supplies for their own growth. More and more subsurface waters will need to be tapped, even in regions of surface supplies. Growing demands are made upon water supplies by industrial plants, and in many coastal areas the limit of withdrawal of subsurface water has already been reached or exceeded. Ground water is also important to mining, in another way. Huge volumes of water have to be pumped to free underground workings; in many coal mines 10 tons of water has to be removed for each ton of coal extracted.

The ocean is the great residual reservoir of water, and most of this external water has come from within the earth throughout geologic ages. The land water — that in lakes and ponds, in streams and in soils — is ever returning to the ocean. The sun evaporates water from the oceans into the atmosphere; it falls upon the lands; much of it runs over the surface into the oceans again; part of it sinks into the ground and forms a temporary reservoir of ground water, which in turn seeks an ocean outlet. Thus, the ever-recurring hydrologic cycle continues, and under normal conditions ground water should be present.

It is beyond the scope of this chapter to consider surface waters, and even the subject of ground water, with its voluminous literature, can be treated only briefly. The reader, therefore, is referred to several good books and papers on the subject. This chapter assumes an elementary knowledge of the subject and will recall only briefly a few broad features of the occurrence and will deal with some of the laws and the utilization of ground water.

Occurrence of Ground Water

Ground water or subsurface water is the water that fills all pores and openings within the zone of saturation; it is not, as popularly supposed, an underground stream or pool. Its top is called the *water table*. Between this and the surface is the *zone of aeration*, where descending waters sink to meet the water table and the interstices are alternately occupied by air and water. In dry seasons the water table sinks; in wet seasons it rises. If it reaches the surface, the ground is swampy. If a depression extends beneath it, ground water occupies the depression; if the depression is an open-ended valley, a flowing stream is fed, and if it is an enclosed depression a lake results. An ordinary well is thus merely a miniature lake. If the water table sinks in a dry season a shallow lake or well may go dry, unless the well extends beyond the lowest seasonal oscillation. If water is withdrawn from the well, the water table is temporarily lowered at that point until the water table is reestablished by the ground water seeking its own level. Underground water, like surface water, merely follows the little appreciated elementary principle that the laws of gravity determine its movements.

Relation to Topography. In a flattish area the water table parallels the surface, and in a rolling, humid region the water table roughly parallels the hills and vales, although it is less accentuated (Fig. 25–1). Water, therefore, will always lie closer to the surface under valleys than under ridges. The motion of the water from hill toward valley is continual and slow, and it would seek a level, were it not that surface recharge occurs. In a humid mountainous region the water table is still less accentuated than the topography; and steep valleys in places cut beneath it, giving valley-side seeps or springs. Also, because mountainous regions are commonly much fissured, a more rapid underground movement occurs from high to low places.

Relation to Climate.   In humid regions the water table generally lies at shallow depths beneath flat areas, at less shallow depths beneath hilly areas, and at or near the surface in the valleys.   In arid regions it commonly, if present, lies at considerable depth, beneath valleys as well as hills.   In two-season climates there is a wide fluctuation in depth between the dry and the rainy seasons.   A change from humid

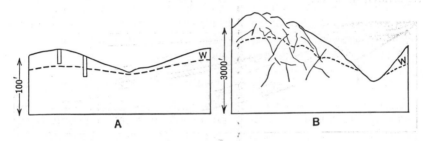

Fig. 25–1.   Relation of water table W to surface in, A, undulating topography and, B, mountainous topography where large fissures carry water that discharges to lower regions.

to dry climate causes a slow sinking or even disappearance of the water table, and the reverse change causes a rising.

Relation to Openings.   The amount of ground water can be no greater than the openings in the rocks and soils.   Generally the pore spaces hold the greatest part of the volume of water, and fractures or other openings the least; in fractured rocks of low porosity and in cavernous limestones the reverse is true.   A well sunk in such rocks obtains a fair volume of water only if interconnected fractures or similar openings are encountered.   A porous rock or soil may be intercalated between impervious rocks or soils, giving separate or " perched " bodies of ground water.   (For discussion of openings and flow of fluids see Chaps. 5·4 and 16.)

Movement of Ground Water.   The movement of ground water is from the places of high pressure to those of the least, for a given area.   Thus, it continually moves downslope from the intake at a rate determined by the head and by the permeability of the rock or soil.   In a mountainous region where the head may be great, the movement may be rapid; with low head it may be slow.   Each particle of water moves from its intake to its discharge point, whether a natural seep or spring, or plant evaporation, or an artificial one.   Its path may be short or scores of miles in length and its time may be short or many centuries.

According to Meinzer, field tests have shown the rate of movement to be as much as 420 feet a day, and the lowest rate determined in the laboratory was 1 foot in 10 years. The natural range may be even greater. However, in water-bearing strata or aquifers, the natural rate is generally not greater than 5 feet a day or less than 5 feet a year.

The hydraulic permeability of a porous medium is its property of transmitting water through its interstices. The *coefficient of trans-*

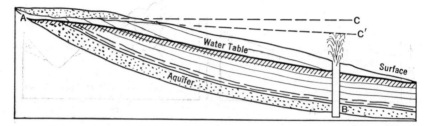

Fig. 25–2. Artesian water conditions. *AC* is level of intake; *AC′* is pressure gradient; *BC′* is the head under which water is forced from aquifer *AB* at *B*.

missibility (Theis) is the permeability (not porosity) with temperature correction, multiplied by the thickness of the saturated aquifer; water strata range between 10 and 5,000.

Artesian Conditions. It is a common popular mistake to call any drilled water well an artesian well. Artesian conditions obtain where subsurface water is confined within inclined porous strata between impervious beds, as in the case of water within the mains of a water system. It is subject to a pressure or head, equivalent to the difference in height of its intake over a point of discharge, less a small correction because of frictional resistance due to flow. If, in Fig. 25–2, *A* is the intake and *B* a point of discharge, the head will be *BC*, and such water would rise theoretically to *C*, but because of friction and leakage it will only rise to *C′*. If a fissure intersects the aquifer at *B*, an artesian spring will result; if a cased well penetrates it at *B*, an artesian well will result; and water will rise in the well to *C′*. If the collar of the well is considerably below the intake elevation it will be a flowing well; if the reverse, the water will rise part way up into the well.

The line *AC′* to which the water will rise is called the *pressure gradient* (or piezometric gradient) of the confined water and indicates the *static level* of all wells that may lie along this section. The *pres-*

sure surface or *piezometric surface* or *pressure indicating surface* is an imaginary surface to which the water from a given aquifer would rise in tightly cased wells that penetrate it. This is generally indicated for given areas by a *piezometric contour map* (Fig. 25–11), which shows for that area the approximate elevation to which artesian water from a given aquifer will rise.

Structure contour maps of artesian aquifers are plotted, in which contours depicting the elevation of the aquifer are superimposed upon surface contours. Thus, if at a given point the surface contour shows an elevation of 2,000 feet and the structure contour for an aquifer at

FIG. 25–3. Section of the Dakota artesian basin. (*Simpson-Upham, U. S. Geol. Survey.*)

the same point is 1,200 feet, it means that a well drilled to a depth of 800 feet will there cut the aquifer. Further if the pressure or piezometric contour for the same point is 2,100 feet, it means that water in a tightly cased well would rise to a height of 100 feet above the surface — or be a strongly flowing well. Such maps are made of different areas by the United States Geological Survey and permit a land owner to determine the depth, head, and flow of artesian water that he might expect on his land.

Notable artesian areas in the United States are the area underlain by the Dakota sandstone aquifer emanating outward from the Black Hills (Fig. 25–3), and the parts of Florida and adjacent states underlain by the Ocala limestone aquifer (Fig. 25–9).

Coastal Areas. In coastal areas, particularly low-lying shores or islands such as Holland, the New Jersey coast, and Long Island, a knowledge of the occurrence and laws of ground water is particularly pertinent. In such places fresh water rides on top of salt water, and lack of care in withdrawal of the fresh water may bring encroachment of salt water and ruination of ground-water supplies.

To those unfamiliar with ground water it is generally thought that it would be impossible to obtain a supply of fresh underground water from sands near the ocean. This is not the case. Fresh water that falls upon the land floats on top of and depresses underlying salt water by an amount proportional to the height of the water table above sea level. This, according to the Herzberg formula (Fig. 25–4) is $h = t/(g - 1)$, h being the depth of fresh water beneath sea level, t being the height of the water table above sea level, g the specific

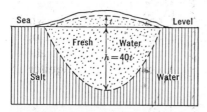

FIG. 25–4. Fresh water below sea level beneath an island, according to Herzberg formula. Fresh water floats on salt water; t is height of water table above sea level; h is height of column of fresh water below sea level ($h = 40t$).

gravity of salt water, and 1 the specific gravity of fresh water. If the density of sea water is 1.025, then $h = 40t$. This means that on Long Island, for example, if the water table is 20 feet above sea level there is a depth of 800 feet of fresh water below sea level, but there is also the disturbing reverse that if the water table should be lowered 5 feet by pumping there is a corresponding rise of 200 feet of salt water, hence, care is needed that pumping does not exceed recharge.

Arid Conditions. In arid regions, because of small intake by rainfall, the water table is apt to have a low stand except in piedmont valleys, and its surface is much less accentuated than the topography. Intake is small in interstream areas because the proportion of rainfall runoff is relatively large due to lack of vegetation, and small run-in; the water of brief arid-region showers does not penetrate deeply into air-filled pores of desert soils. Most of the intake is from the streams.

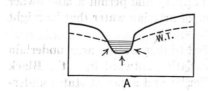 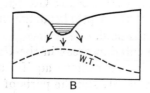

FIG. 25–5. *A*, Effluent stream; *B*, influent stream. (*After Meinzer, U. S. Geol. Survey.*)

Consequently, in general, the water table is higher under arid-region streams and lower away from them, the reverse of water-table profiles under humid-region stream valleys. The former are termed *influent streams* (contributory to ground water) and the latter *effluent streams* (Fig. 25–5). Most arid-region streams are (1) *intermittent*, or those that flow only part of the year, receiving water from melted snows on hills, (2) *ephemeral*, or those that flow only in response to a rainfall, and (3) *through-flowing*, or those that receive water from year around sources beyond the arid area. Effluent and artesian flow into the Tennessee River are shown in Fig. 25-6.

Geologic Changes. Reference has already been made (Chap. 5·8) to changes of level of the water table brought about by climatic changes. There are also other factors that bring about changes of

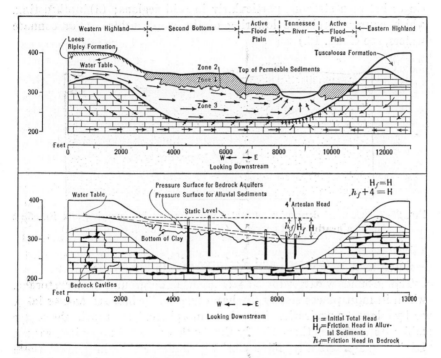

Fig. 25-6. Geologic sections at Kentucky dam site, Tennessee River, showing flow of ground water toward river. (*Rhoades, Econ. Geol.*)

water table during geologic time. Such changes are a part of the geologic history of a region and are bound up with its physiographic development. They are of prime importance in understanding oxidation and sulphide enrichment of ore deposits (see Chap. 5·8), and they even affect the sustenance and migration of peoples.

A rise in the water table may be brought about by: (1) valley filling, causing a slow rise in the water table above the former valley floor; (2) downfaulting of a block of the earth's crust into the water table, causing a relative rising of the water table; (3) a change from arid to humid climate; (4) damming of a stream by landslide, faulting, or uplift athwart its course, causing impounding, lake-bed deposition and rise of water table; (5) rupture of a confining bed of an artesian aquifer with ensuing rise of the water table.

Sinking of the water table may be brought about by: (1) change from humid to arid climate, such as has taken place in Egypt and North Africa; (2) valley cutting, which lowers the water table of the region, and if it is rapid the water table may be lowered deeply under ridges between streams, particularly in arid regions; (3) upthrusting, which causes a relative lowering of the water table. (For connate waters, see Chap. 16.)

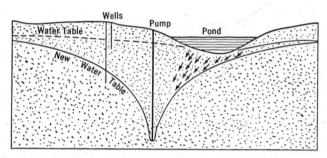

FIG. 25–7. Cone of depression of water table formed by pumping and its effect upon a nearby pond or well. (*Veatch, et al., U. S. Geol. Survey.*)

Withdrawal and Recharge of Ground Water

The ground water serves as a great underground water storage, which in most places slowly discharges naturally into streams or lakes or by plant transpiration into the atmosphere. Its top is the water table; its bottom is theoretically the depth where rock openings cannot persist, but practically, mine workings indicate that in general there is little or no visible unconfined water at depths below 2,000 and 3,000 feet. It is this great reservoir that is drawn upon for useful subsurface water supplies.

Withdrawal. In a flattish region with a corresponding water-table surface, a well beneath the water table will contain standing water at the level of the water table. If the well bottom does not penetrate beneath the zone of fluctuation, the well will go dry during droughts, but if it is beneath the bottom of this zone and has access to the great underground reservoir it will not be affected by drought.

When a well is pumped, the water table is lowered around the well and a hydraulic gradient from all directions is established toward the well. The water table in homogeneous rocks assumes the form of an inverted cone called the *cone of depression*. When pumping ceases, inflow occurs from the unwatered area close to the well (Fig. 25–7).

If pumping is continuous the cone of depression deepens and spreads out laterally, and inflow, traveling longer distances, becomes slower. Continued draw-down gradually lowers the water table at a decreasing rate and nearby shallow wells may go dry. The cone of depression may extend laterally to the limits of the water formation but recharge may halt the extension of the cone of depression. Cessation of pumping causes slow percolation and refill of the unwatered part. If pumping is in excess of recharge, the storage is drawn upon until exhaustion. A *safe yield* is indicated if the recovery by inflow, during periods of pumping, is complete. Consequently, it is important that the rate of pumping be controlled unless the storage is so great that it is little affected over the life of the well. Meinzer points out that the Carrizo sand in a winter garden area in Texas transmits 24 million gallons a day at a rate of 50 feet a year; accordingly a mile of the outcrop contains a storage of 24 million gallons a day for a century.

Under artesian conditions generally similar conditions prevail. The first unwatering, however, is accompanied by compaction of the aquifer, and the squeezing out of water by compaction delays the development of the cone of depression, after which it develops more rapidly.

Recharge. Natural recharge takes place in humid regions by run-in from rainfall and snow. This is aided greatly by forests and grass lands, the vegetation retarding the runoff and facilitating water seepage into the soil. In deforested areas, on the other hand, the proportion of runoff is higher and recharge is less. Plowed land also retards in-seep of rainfall by puddling the topsoil layer and silting up the intake openings.

It is often erroneously said that in humid regions the streams and lakes cause recharge; on the contrary, they are fed by ground water.

In semi-arid regions the paucity of vegetation makes for less recharge and greater runoff than in humid regions. Also the air-filled voids of the soil tend to retard run-in. Most of the recharge occurs from influent streams and in mountain heights where vegetation and also snow may occur.

Recharge is never uniform but is concentrated in favorable localities with the result that the water-table surface is characterized by mounds, giving the opposite effect from withdrawal. Such mounds do not flatten out because of the slow movement of ground water, and of fresh supply.

Artificial recharge is now being practiced in many regions dependent upon ground-water supplies. On the whole such recharge must remain small as compared with natural recharge, but it is of impelling im-

portance in specific areas. It is accomplished by impounding of surface water over spreading grounds to permit in-seep; by the regulation
and diversion of stream flow, particularly in arid regions; and by discharge into wells. In the Los Angeles water system, in one spreading
basin of 47 acres, 106 cubic feet of water per second percolated underground at the rate of 3 to 10 feet per day and eventually reached the
North Hollywood pumping station. Over-pumping near Brooklyn,
Long Island, developed a serious situation with respect to saline contamination by uprising salt water, and recharge into wells was followed.

Ground-Water Supplies

In the United States alone there are some 62,000 large irrigation
wells, and countless other large wells are utilized for municipal and
private industrial purposes. About 6,500 public waterworks are supplied from wells, and in California some 250 million dollars is invested
in irrigation wells. Ground water furnishes domestic supply for more
than half the population of the United States; supplies are, therefore,
of great economic importance.

Location of Water Supplies. The search for underground water is
akin to that for other minerals, but water being a special kind of mineral substance requires special conditions for search. Since water is
widespread it is not difficult to find; in fact it is more difficult to miss
it. The finding of abundant supplies, however, requires a careful study
and proper interpretation of the geology, particularly the nature, distribution, porosity, and structure of water-bearing formations.
Straight geology is, therefore, the best approach. Much of the geological information for this purpose is now made available, through
governmental surveys over many large areas, and by maps that depict
water strata. In the case of artesian basins, such maps indicate the
depth to aquifers and the height to which water will rise in wells that
tap them, the expected flow, and the quality of the water. By such
maps alone, one can locate water supplies with a high degree of success
(see Figs. 25–9 to 25–11).

Where maps are lacking, unconfined water is generally looked for
in surficial gravels and sands where opportunity for natural charging
is available. For example, an alluvial cone consists of gravels and
sands of high porosity and permeability and the streams that enter or
terminate in it provide recharge water. Similar reasoning applies to
valley alluvials. Catchment basins with known precipitation, where
the total infall is not discharged in streams, may be expected to have
underground supplies. Records of nearby wells yield information.

In crystalline rocks the problem is more difficult and involves greater risk of obtaining inadequate supplies. Since the permeability of such rocks is low, an abundant flow must be sought from fractures.

Geophysical methods of water finding are successful in places by locating favorable stratigraphic and structural features. Favorable structural features have been determined by gravitational, magnetic, seismic, and electrical methods. Stratigraphic features are located by conductivity and resistivity methods.

Water divining, although dismissed by Agricola in *De Re Metallica*, about 1546, is still believed in by many people. It is true that many water diviners, using the forked twig, do find water, but it is also true that they would probably have greater difficulty in missing it. Common sense and some knowledge of water occurrence, whether transmitted consciously or unconsciously to the twig, is probably the greatest cause for most of the water diviner's successes. As stated by Rossiter Raymond in 1883, " a whole library of learned rubbish about it which remains to us furnishes jargon for charlatans, marvelous tales for fools, and amusement for antiquarians."

Various nongeological features may serve to indicate the presence of hidden water. Aligned or concentrated tree growth in an area of sparse trees indicates at least moisture for tree growth. Ground mists under certain conditions may be due to nearby subsurface water, and seepages, of course, reveal ground water. " Rock pools " or " rock tanks " indicate places where surface pools of water accumulated to form them and suggest water beneath. Certain types of plants whose roots thrive only in ground or fringe water may serve as water indicators in semi-arid and arid regions.

Recovery of Ground Water. Most ground water is recovered through wells, which may be dug, driven, bored, or drilled. With dug wells, which usually are shallow, care must be used to prevent pollution from top-layer ground water. Their position should be such that they are not on the down-gradient side of sources of refuse disposal.

Bored wells are excavated by hand or power-driven augers from 6 to 30 inches in diameter. Casing is run in the hole, the lower part being perforated for water inflow.

Driven wells are obtained by driving pointed steel pipe with perforations in it to the desired depth; this can be done only in unconsolidated materials.

Drilled wells are made by cable tool or rotary methods, which are necessary for all deep wells or those through rocks. Such wells are generally cased with perforated casing at the bottom. The water is raised by pumping, or by artesian pressure.

In many large wells adits or collecting galleries are driven from the bottom. In suitable places adits are driven into hillslopes beneath the ground water and serve as collecting and discharge galleries, as in Hawaii. Water supplies are extracted, in places, from arroyo beds by sinking a shaft to bedrock on either side and connecting the two with a tunnel, which serves as a collecting gallery.

Quality of Water. The quality of water is of vital importance whether it is for domestic or industrial purposes; but an abundant supply of impure water is more important than a sparce supply of pure water, since the impurities may often be removed.

Impurities of water consist of pollution and contamination. In general, bacterial pollution is absent from ground water because its percolation through strata is an effective means of natural filtering and purification. The contaminations fall mainly into five broad groups: salt, carbonates, sulphates, iron and manganese, and gases.

Salinity assumes importance in some coastal and arid-region ground water. As soon as the salt content passes a certain limit water becomes unfit for human use. The salinity is generally expressed in parts per million (ppm), sea water being 35,000. Some of the tolerances are:

Ppm	Nature
400	No taste of salt.
500	Slightly brackish taste.
1,000–2,500	Strong taste, but bearable.
3,300	Usable domestically.
3,500–5,000	Almost unbearable.
5,000 +	Unfit for human use.
6,250	Horses live on it in good condition.
7,800	Horses can live on it.
9,375	Cattle can live on it.
15,625	Sheep can live on it.
16,000	Beyond tolerance limit for most grasses.

Carbonates of calcium and magnesium give water *temporary hardness,* a characteristic made familiar by the difficulty of using soap. This feature is troublesome in the home, in commercial laundries, and particularly in steam boilers because of the deposits of " scale " that choke up the tubes. It is called temporary because it can be removed by simple chemical processes, called water softening, which is done with lime and soda or exchange silicates. Ground water is ordinarily of higher hardness than surface water. The average hardness of ground-water supplies for large cities is 225 ppm of $CaCO_3$, as compared with 99 for all supplies. Soft waters have as much as 55 ppm;

slightly hard, 56 to 100; moderately hard, 101 to 200; hard waters, over 200 ppm.

Sulphates (alkalies) give *permanent hardness,* which can be removed only by complicated chemical processes. This is removed from municipal supplies by means of base exchange silicates, such as zeolites and greensands.

Iron and manganese are present in many ground waters up to several parts per million and must be removed to give satisfactory domestic and industrial water. Iron is objectionable because of its appearance after air exposure, because of stains it leaves on white goods, clogging of water pipes, and interference with industrial uses, such as dyeing. Manganese has somewhat similar objections. These substances are removed by aeration and filtration, or manganese by filtration and the use of pyrolusite as a catalyst.

Dissolved gases, such as H_2S, methane, and CO_2, prove objectionable but are easily removed by aeration.

Estimation of Supplies. It is pointed out by Meinzer that the most urgent problem in ground water is the determination of the rate at which water strata will supply water to wells. To do this it is necessary to determine additions to the supply and discharges from it, or the *ground-water inventory.* Some water strata are merely reservoirs, others are conduits. If water is withdrawn from storage without being replenished, the supply will eventually fail and it is necessary to determine the *safe yield* or the practicable rate of withdrawal without depletion.

The factors affecting increment to ground water are: (1) rainfall penetration to the water table; (2) natural influent seepage from surface waters; (3) artificial influent seepage; (4) inflow of subsurface water outside the given area. Factors affecting decrement are: (1) effluent seepage and spring flow of free ground water; (2) effluent seepage and spring discharge of artesian water; (3) leakage from aquifers; (4) evaporation, transpiration, and artificial removal; (5) discharge by pumping, and (6) subsurface discharges. Special series of measurements are set up to supply these data.

A method of determining the supply is the quantity of water that can be pumped without a continued lowering of the water table. This is a measure of the water available, since a stationary water table in a pumping confined well indicates a balance between withdrawal and supply.

Important hydrologic features bearing on water supply are permeability and *specific yield.* The latter is the volume of water that a saturated stratum will yield by gravity compared to the volume of the stratum, and is expressed in percentage. Obviously the specific yield of a clay with its small pores will be greatly different from a coarse-textured porous gravel.

For details of the numerous methods employed for estimation of subsurface water supplies and the mathematical treatment thereof, the reader is referred to the extensive literature on this subject, notably the writings of O. E. Meinzer.

Some Outstanding Examples of Ground-Water Supplies

Some features of the occurrence and behavior of ground water may best be illustrated by a brief consideration of a few of the outstanding ground-water areas.

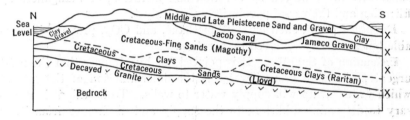

FIG. 25–8. Generalized section across Long Island, near Jamaica, showing water-bearing strata, marked *x*. (*Adapted from Thompson-Wells-Blank, Econ. Geol.*)

Long Island, N. Y. Long Island, approximately 115 miles long and up to 20 miles wide, is a low-lying outlier of the Coastal Plain of New York and Connecticut. It is populated by over 4 million people who depend largely upon ground water for water supply. In addition, there is a heavy industrial demand upon the water supply in the vicinity of Brooklyn.

The island is underlain by a crystalline complex that crops out on the western end and plunges eastward and southward, so that in places it is 2,000 feet beneath the surface. Above this are Cretaceous sands, clays, and gravels containing three aquifers that dip beneath the island toward the ocean on the south (Fig. 25–8). The northern half is covered by unconsolidated glacial moraines that merge into fluvioglacial outwash gravels and sands toward the south. The maximum elevation is 420 feet. These surficial deposits contain a great reservoir of ground water supplied only by rainfall since the strata do not extend across to the mainland. The rainfall is about 42 inches a year, of which one-third to one-half reaches the water table.

A thick body of fresh water floats on salt water, and over 200 million gallons a day has been pumped from it, with concentrated pumping toward the western end.   A delicate balance is thus maintained between the fresh and salt water.   Over-pumping brought about some saline contamination.   According to Legget the water table under Brooklyn dropped to an elevation of 40 feet and much of the water there is now unfit for drinking purposes.   Natural recharge near Brooklyn has been restricted because of paved streets and sewer discharge of rainfall.   In places compulsory artificial recharge of industrial water into wells is practiced.   This small area is an excellent example of the concentrated utilization of a valuable water supply.

A similar occurrence pointed out by Legget is a floating fresh-water supply in the coral limestones and sands of the Bahama Islands.

Florida Artesian Province.   Florida is largely underlain by a vast storage of fresh and nearly fresh water.   Within its stratigraphic column of Mesozoic and Tertiary sediments are several aquifers, notably the Eocene Ocala limestone and the higher Miocene Hawthorn formation.   The Ocala is very pervious owing to large and small cavities and is the important aquifer.   The Ocala is upwarped to a gentle elliptical dome with its crest near Ocala, from where it dips seaward to the east, south, and west.   It constitutes a great artesian aquifer that extends at depth under much of the Florida coast out beneath the sea (Fig. 25–9).

The United States Geological Survey has made excellent maps that give structural contours on the Ocala, the area underlain by artesian water, and a piezometric map showing contours of the ground-water head in the Ocala (Fig. 25–10).   There are three piezometric highs from which water flows out radially, the main one in central Florida, another one to the north, and a subsidiary one north of Tampa.   A similar one extends north of Savannah, Ga. (Fig. 25–11).   This great aquifer must discharge into the sea.   A back-up of discharge causes great inland springs.   Where cut by wells it yields copious artesian flow of fresh water, but in the south where the Ocala is about 1,000 feet deep, the water is highly mineralized and is unfit for domestic use.   This may be due to infiltration of sea water or to incomplete flushing of salt connate water.   Miami, therefore, has to obtain its water supply from wells in Quaternary surface gravels that tap open ground water supplied by rainfall.

The great Florida water supply is another example of both artesian water and a " floating " supply of fresh water that extends to about 2,000 feet beneath sea level.

The Dakota Sandstone Artesian Province.   The domally uplifted strata of the Black Hills of South Dakota include a great economic aquifer — the 100-foot Dakota sandstone, which dips outward beneath the surrounding states (Fig. 25–3). Its impervious cover permits intake water from the Black Hills to be transmitted hundreds of miles, and thousands of flowing wells tap this artesian supply for domestic, rural, and urban use in the Dakotas, Nebraska, Iowa, and Wyoming. Owing to large discharge, the head has dropped appreciably in many

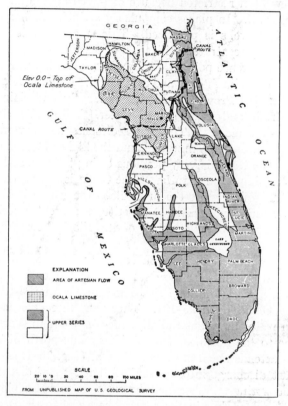

Fig. 25–9.   Area of artesian flow in Florida.   (*Paige-Stringfield, Econ. Geol.*)

places and much water is being drawn from storage. A study of this great aquifer led Meinzer to conclude that compaction of the Dakota sandstone takes place upon withdrawal of water and much of the early discharge is water squeezed out by such compaction.

Other Examples.   Other examples of notable subsurface water supplies are to be found throughout the world. The extensive glacial deposits of northeastern United States are a source of large supply. The sands of New Jersey supply thickly populated areas. The prodigious supplies of the Great Valley of California, as well as those of other western United States provinces, and the unusual ones of the

Hawaiian Islands, have been adequately summarized by C. F. Tolman. The dune-sand, subsurface supplies of low-lying coastal Hol-

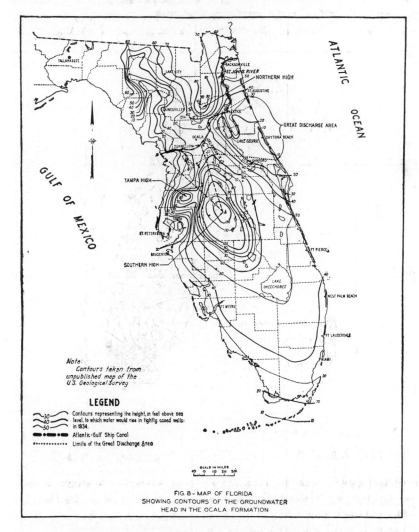

FIG. 8 – MAP OF FLORIDA
SHOWING CONTOURS OF THE GROUNDWATER
HEAD IN THE OCALA FORMATION

FIG. 25–10. Florida artesian basin showing piezometric surface of artesian water, or height above sea level to which water will rise. (*Paige-Stringfield, Econ. Geol.*)

land, in particular, present many interesting problems of "floating" fresh water. Arid region supplies of Australia, North America, northern Africa, and southern Asia offer many problems to tax the skill of the hydrologist.

In the development of concepts of occurrence, movement, estimation of supplies, and economic utilization of ground water, the world owes

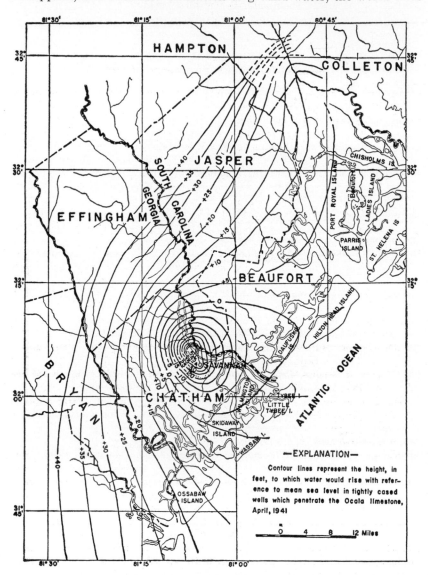

FIG. 25–11. Piezometric surface of artesian water around Savannah, Ga., in 1941. (*Stringfield-Warren-Cooper, Econ. Geol.*)

much to the indefatigable labors of O. E. Meinzer, formerly in charge of the Ground Water Division of the United States Geological Survey.

How can we know, we have not seen,
But we believe the sands will yield
Their water to the thirsty field,
And all the desert turn to green.

Selected References on Ground Water

Occurrence of Ground Water in the United States, with a Discussion of Principles. O. E. Meinzer. U. S. Geol. Surv. W.S.P. 489, 1923. *General résumé.*

Outline of Ground-Water Hydrology. O. E. Meinzer. U. S. Geol. Surv. W.S.P. 494, 1923. *Comprehensive outline and definitions.*

Outline of Methods for Estimating Ground-Water Supplies. O. E. Meinzer. U. S. Geol. Surv. W.S.P. 638–C, 1932. *Full treatment of methods of estimation.*

Present Status of Our Knowledge Regarding the Hydraulics of Ground Water. O. E. Meinzer and L. K. Wenzel. Econ. Geol. 35:915–941, 1940. *A theoretical treatment.*

Ground Water. C. F. Tolman. McGraw-Hill, New York, 1937. *First comprehensive textbook on this subject; principles, occurrence, and western examples.*

Practical Handbook of Water Supply. Frank Dixey. Murby, London, 1931. *Occurrence, principles, recovery, with application entirely to African conditions.*

Geology and Engineering. McGraw-Hill, New York, 1939. Chap. 17, by R. F. Legget. *Brief, rambling treatment with interesting examples.*

Artesian Water in the Florida Peninsula. V. T. Stringfield. U. S. Geol. Surv. W.S.P. 773–C, 1936. *A fine example of an artesian basin.*

Significance and Nature of the Cone of Depression in Ground Water Bodies. C. V. Theis. Econ. Geol. 33:889–902, 1938.

The Theory of Ground Water Motion. M. K. Hubbert. Jour. Geol. 48:785–944. Nov.–Dec. 1940 (Jan. 1941). *Highly mathematical treatment.*

Recent Geologic studies on Long Island with respect to Ground Water Supplies. D. G. Thompson, F. G. Wells, and H. R. Blank. Econ. Geol. 32:451–470, 1937.

Water Supply Papers. Series of U. S. Geological Survey.

Hydrology in Relation to Economic Geology. O. E. Meinzer. Econ. Geol. 41:1–12, 1946. *Scope of hydrology and relation to engineering.*

General Principles of Artificial Ground-Water Recharge. O. E. Meinzer. Econ. Geol. 41:191–201, 1946. *Hydraulics of recharge; indirect methods and recharge from surface and wells.*

Artificial Recharge of Ground Water on Long Island, N. Y. M. L. Brashears, Jr. Econ. Geol. 41:503–516, 1946. *Recharge by wells and pits to prevent overdevelopment.*

Problems of the Perennial Yield of Artesian Aquifers. O. E. Meinzer. Econ. Geol. 40:159–163, 1945. *Procedures of investigation.*

Quantitative Studies of Some Artesian Aquifers in Texas. W. F. Guyton and N. A. Rose. Econ. Geol. 40:193–226, 1945. *Methods of making predictions of draw-down rate through pumping.*

Ground-Water Investigations in the United States. A. N. Sayre. Econ. Geol. 43:547–552, 1948. *Problems of water supply, control, and conservation.*

Geology and Ground-Water Resources of Hawaii. H. T. Stearns and G. A. Macdonald. Pp. 363. Hawaii Div. Hydrol. Bull. 9, 1946. *Outstanding example of ground-water study.*

Recent Groundwater Investigations in the Netherlands. W. F. J. M. Krul and F. A. Liefrink. Elsevier Pub. Co., New York, 1946. *Ground-water problems in a country near sea level.*

Elements of Applied Hydrology. Don Johnstone and William P. Cross. Ronald Press Co., New York, 1949. *Elementary practical data for ground-water engineering.*

Hydrology. Edited by Oscar E. Meinzer under auspices of National Research Council. Dover Publications, Inc., New York, 1949. *A complete reference on hydrology by 24 experts.*

Hydrology. C. O. Wisler and E. F. Brater. John Wiley & Sons, New York, 1949. *Practical procedures for minimum flow and average yield of drainage basins.*

Applied Hydrology. Ray K. Linsley, Jr., Max A. Kohler, and Joseph L. H. Paulhus. McGraw-Hill, New York, 1949. *A text-reference book for students and engineers on general data, theory, and applications.*

GENERAL REFERENCES ON ECONOMIC GEOLOGY

The list below is a group of general references with which the student of economic geology should have familiarity. It is not intended to be a comprehensive bibliography, since specific references for each subject are given at the end of each section or chapter; rather, it covers those general references that embrace many subjects, to save their repetition in the Selected References given throughout the volume.

1. **Journal of Economic Geology.** Economic Geology Publishing Co. New Haven, Conn. Eight numbers per year. Vol. 1, 1905, to present. *The outstanding journal of the world relating to economic geology and containing original articles on all phases of economic geology.*

2. **Annotated Bibliography of Economic Geology.** Economic Geology Publishing Co., Urbana, Ill. Two numbers per year. From Vol. 1, 1933, to present. *Gives abstracts of all important papers in the world relating to economic geology.*

3. **Bibliography and Index of North American Geology.** U. S. Geological Survey, Washington, D. C. Bi-yearly, from 1785 to present; consolidated each 10 years. *Gives authors and titles of all articles on geology in North America.*

4. **Bibliography and Index of Geology Exclusive of North America.** Geol. Soc. America, New York. Yearly, from 1933 to present. *Covers foreign geology, by authors and titles, with brief annotations.*

5. **Minerals Yearbook.** Annual volumes. U. S. Bureau of Mines, Washington, D. C. *Annual volumes giving statistics and reviews for each metal and nonmetal, covering production, consumption, trade, prices, and reviews by states and countries.*

6. **Transactions, Amer. Inst. Min. Met. Engrs.** New York. Yearly since 1871. *Contains many papers related to economic geology.*

7. **Mineral Resources of the United States.** By staffs of the U. S. Bureau of Mines and U. S. Geological Survey. Pp. 212. Public Affairs Press, Washington, D. C., 1948. *Summary of domestic reserves and sufficiency and separate treatment of 39 minerals.*

8. **Mineral Deposits.** WALDEMAR LINDGREN. 4th Ed. Pp. 930. McGraw-Hill Book Co., New York, 1933. *The leading text and reference book on advanced economic geology; covers metallic and several nonmetallic deposits.*

9. **Ore Deposits of the Western United States — Lindgren Volume.** Pp. 797. Amer. Inst. Min. Met. Engrs., New York, 1933. *Contributions by 44 authors relating to the various groups of mineral deposits in the western United States.*

10. **Ore Deposits as Related to Structural Features.** Edited by W. H. NEWHOUSE. Pp. 280. Princeton Univ. Press, Princeton, N. J., 1942. *Brief résumés of most important mineral deposits that exhibit localization of ore by structural features.*

11. **Industrial Minerals and Rocks.** 2nd Ed. Edited by S. H. DOLBEAR and OLIVER BOWLES. Pp. 1156. Amer. Inst. Min. Met. Engrs., New York, 1949. *Articles by a group of authorities covering each of the industrial nonmetallic minerals and rocks.*

12. **Mineral Raw Materials.** U. S. Bureau of Mines. Pp. 342. McGraw-Hill Book Co., New York, 1937. *Treatment of 32 minerals giving use, technology, and distribution; survey of mineral position of important industrial countries.*

13. **Minerals in World Affairs.** T. S. LOVERING. Pp. 394. Prentice-Hall, New York, 1943. *The uses, technology, distribution, geology, country, occurrences, and political features of 16 important metals and minerals.*

14. **Seventy-five Years of Progress in the Mineral Industry — Anniversary Volume.** Edited by A. B. PARSONS. Amer Inst. Min. Met. Engrs., New York, 1947. *Authoritative contributions by group of authors on various phases of the mineral industry.*

15. **Structural Geology of Canadian Ore Deposits — A Symposium.** Pp. 948. Geology Division Can. Inst. Min. & Met., Montreal, 1948. *A comprehensive assemblage by belts or areas of the geology and structure of all Canadian mining districts or mines with emphasis on their structural controls.*

16. **Industrial Minerals** (Nonmetallics). Pp. 650. A. I. M. E. Industrial Minerals Division, Vol. 173, 1947. *Includes papers presented before the Institute between 1942 and 1947.*

INDEX

(Bold face denotes chief reference; asterisk denotes illustration.)